Mechanisms and Machine Science

Volume 79

This book series establishes a well-defined forum for monographs, edited Books, and proceedings on mechanical engineering with particular emphasis on MMS (Mechanism and Machine Science). The final goal is the publication of research that shows the development of mechanical engineering and particularly MMS in all technical aspects, even in very recent assessments. Published works share an approach by which technical details and formulation are discussed, and discuss modern formalisms with the aim to circulate research and technical achievements for use in professional, research, academic, and teaching activities.

This technical approach is an essential characteristic of the series. By discussing technical details and formulations in terms of modern formalisms, the possibility is created not only to show technical developments but also to explain achievements for technical teaching and research activity today and for the future.

The book series is intended to collect technical views on developments of the broad field of MMS in a unique frame that can be seen in its totality as an Encyclopaedia of MMS but with the additional purpose of archiving and teaching MMS achievements. Therefore, the book series will be of use not only for researchers and teachers in Mechanical Engineering but also for professionals and students for their formation and future work.

The series is promoted under the auspices of International Federation for the Promotion of Mechanism and Machine Science (IFToMM).

Prospective authors and editors can contact Mr. Pierpaolo Riva (publishing editor, Springer) at: pierpaolo.riva@springer.com

Indexed by SCOPUS and Google Scholar.

More information about this series at http://www.springer.com/series/8779

Delun Wang · Victor Petuya ·
Yan Chen · Shudong Yu
Editors

Recent Advances in Mechanisms, Transmissions and Applications

Proceedings of the Fifth MeTrApp Conference 2019

IFToMM

N S F C

CMES

Springer

Editors
Delun Wang
School of Mechanical Engineering
Dalian University of Technology
Dalian, China

Victor Petuya
Mechanical Engineering
University of the Basque Country
Bilbao, Spain

Yan Chen
Tianjin University
Tianjin, China

Shudong Yu
Department of Mechanical
and Industrial Engineering
Ryerson University
Toronto, ON, Canada

ISSN 2211-0984 ISSN 2211-0992 (electronic)
Mechanisms and Machine Science
ISBN 978-981-15-0141-8 ISBN 978-981-15-0142-5 (eBook)
https://doi.org/10.1007/978-981-15-0142-5

This Springer imprint is published by the registered company Springer Nature Singapore Pte Ltd.
The registered company address is: 152 Beach Road, #21-01/04 Gateway East, Singapore 189721, Singapore

Preface

Following the successful editions in Timisoara (2011), Bilbao (2013), Aachen (2015) and Trabzon (2017), the 5th IFToMM Conference on Mechanisms, Transmissions and Applications (MeTrApp) is organized by Dalian University of Technology in Dalian, China, October 9–11, 2019. MeTrApp conference series is organized under the patronage of International Federation for Promotion of Mechanism and Machine Science (IFToMM), and it is joint event of two technical committees of IFToMM namely "Linkages and Mechanical Controls" and "Gearing and Transmissions."

This book gathers the technical papers that are presented in the 5th Conference on Mechanisms, Transmissions and Applications organized by Dalian University of Technology, China. MeTrApp 2019 aims to offer a stage for original research presentations for researchers, scientists, industry experts and students in the fields of mechanisms and transmissions with special emphasis on industrial applications in order to stimulate the exchange of the new and innovative ideas. Hence, one of the expected outcomes of this conference is to promote the exchange of ideas and collaboration. Another one is this edited proceedings collection within the Springer Mechanism and Machine Science series, which provides a sharing platform to all researchers, scientists, industry experts and students that are interested in the fields of mechanisms and transmissions and their applications.

This book collection contains 63 contributions from the initial 69 submissions, which is a result of a peer-review process intended to select the newest and the most original achievements in the field of mechanisms and transmissions and their applications. The content is presented in nine parts for the readers to readily navigate the certain subject areas, namely, linkage synthesis, mechanical transmissions, engineering applications, machinery dynamics, parallel manipulators, control applications, compliant mechanism, robots and mechatronics, and education on theory of mechanisms and machines.

First and foremost, we thank all the authors for their strong will to contribute to this conference with their enthusiasm and strong commitment. We also sincerely thank all the reviewers, i.e., Martin Bilek, Jinkui Chu, Bo Dong, Li Dong, Yuan Gao, Youcheng Han, Jun He, Alfonso Hernandez, Zharilkassin Iskakov, Gökhan

Kiper, Haiyang Li, Mingxiang Ling, Jian Liu, Wenrui Liu, Erwin Lovasz, Yi Luo, Yali Ma, Koji Makino, Heikki Martikka, Domenico Mundo, Victor Petuya, Haibo Qu, Giuseppe Quaglia, Aitor Riera, Rūta RimaŠauskienĖ, Huiping Shen, Yukio Takeda, Hidetsugu Terada, Monica Urizar, Jia Wang, Wei Wang, Guanglei Wu, Xiaoyong Wu, Kun Xu, Yundou Xu, Yutao Yan, Yan-An Yao, Bin Ye, Xiao-Yan Yu, Chu Zhang, Haiqiang Zhang, Yanbin Zhang and Fuqun Zhao, for putting their precious time and effort in the review process timely with their strong expertise and scientific background. We are grateful to Springer, especially to Li Shen, Arun Kumar and Beate Siek, who have made this publication possible. Above all, we are grateful to our younger colleagues and postgraduates from Institute of Digital Design, School of Mechanical Engineering (SME), DUT, who performed a magnificent job in preparing the book for publication and the organization of the conference. Without them, things would not have taken place as efficiently and rapidly. We finally gratefully acknowledge the generous support provided by the sponsors of MeTrApp 2019: NSFC, DUT press, SME of DUT.

We hope that this new MMS book will again reach the shelves of scholars, researchers and students who are attracted to the field of mechanisms and transmissions and their applications around the world, and participation to the next MeTrApp will be much higher.

July 2019

Delun Wang
Victor Petuya
Yan Chen
Shudong Yu

Organization

Conference Chair

Delun Wang	Dalian University of Technology, China

Conference Co-chairs

Victor Petuya	University of the Basque Country, Spain
Yan Chen	Tianjin University, China
Shudong Yu	Ryerson University, Canada

International Scientific Committee

Martin Bilek	Technical University of Liberec, Czech Republic
Mehmet Ismet Can Dede	Izmir Institute of Technology, Turkey
Marco Ceccarelli	University of Rome Tor Vergata, Italy
Yan Chen	Tianjin University, China
Burkhard Corves	RWTH Aachen University, Aachen, Germany
Jian S. Dai	King's College London, UK
Xilun Ding	Beihang University, China
Feng Gao	Shanghai Jiao Tong University, China
Ashitava Ghosal	Indian Institute of Science, India
Weizhong Guo	Shanghai Jiao Tong University, China
Uwe Hanke	Technische Universität Dresden, Germany
Tian Huang	Tianjin University, China
Gökhan Kiper	Izmir Institute of Technology, Turkey
Chin-Hsing Kuo	University of Wollongong, Australia
Wei Lin	Singapore Institute of Manufacturing Technology, Singapore
Erwin-Christian Lovasz	Politehnica University of Timisoara, Romania
J. Michael McCarthy	University of California (Irvine), USA

Domenico Mundo	University of Calabria, Italy
Takaaki Oiwa	Shizuoka University, Japan
Victor Petuya	University of the Basque Country, Spain
Giuseppe Quaglia	Politecnico di Torino, Italy
Jaroslaw Szrek	Wroclaw University of Technology, Poland
Yukio Takeda	Tokyo Institute of Technology, Japan
Hidetsugu Terada	University of Yamanashi, Japan
Monica Urízar	University of the Basque Country, Spain
Miroslav Václavík	Research Institute of Textile Machines, Czech Republic
Shudong Yu	Ryerson University, Canada
Xianmin Zhang	South China University of Technology, China

Local Organizing Committee

Guanglei Wu	Dalian University of Technology, China
Zhi Wang	Dalian University of Technology, China

Contents

Linkage Synthesis

Optimized Design of an RSSR Mechanism for an Unpowered Assistive Aid 3
Juewei He, Guangshun Liang, Zhi Lan, and Zhixiu Hao

Dimensional Synthesis of Non-periodic Function Generation of a Spatial RSSR Mechanism Using a Numerical Atlas Method 12
Wenrui Liu, Jianwei Sun, and Jinkui Chu

Kinematic Acquisition of Quadratic Curve Constraints for Finite Position Generation of Planar Mechanisms 21
Hao Lv, Ke Shi, and Xiangyun Li

Path Synthesis of Crank-Rocker Mechanism Using Fourier Descriptors Based Neural Network 32
Xiaojuan Mo, Wenjie Ge, Donglai Zhao, and Yaqing Zhang

Function Generation Synthesis with Planar 4-Bar Linkage as a Mixed Problem of Correlation of Crank Angles and Dead-Center Design 42
Gökhan Kiper and Melih Erez

Mechanical Transmission

Contact Analysis for Spiral Bevel Gear Based on Machine Parameters . . . 51
Yongqiang Zhao, Xiangyang Liu, and Ming Liu

Dimensional Synthesis of the Drawing-In Mechanism in an Automatic Warp Tying Machine 62
Guanglei Wu, Shimin Chen, and Ji Li

Transmission Design and Experimental Research of a Non-contact Nutation Reducer 70
Jiaxin Ding, Ligang Yao, Yongwu Cai, Xiaqi Shan, and Jun Zhang

Releasable Locking Mechanism Using Friction and a Cuneus for Holding a Posture 81
Koji Makino, Tatsuya Kaneko, Kenji Matsumura, and Hidetsugu Terada

Optimization Design of Flange Connection Bolt Set in Large-Scale Rolling Bearings Under Complicated Working Conditions 91
Huimin Dong, Jing Wang, and Delun Wang

Modal Analysis of Nutation Drive with Double Circular Arc Spiral Bevel Gear 99
Simin Ouyang, Ligang Yao, Yongwu Cai, Jun Zhang, and Zhiyu Xie

Engineering Applications

Simulation Analysis on Key Mechanisms of the Automatic Tobacco Harvester Based on ADAMS 113
Jie Yu, Jinkui Chu, and Yang Li

Design and Analysis of a Novel Primary and Secondary Embedded Deployable Offshore Cage 123
Zi-Yu Liu, Lu-Bin Hang, Le-Le Bai, Xiao-Bo Huang, Peng-Cheng Zhang, Su-Long Li, Shi-Sen Lin, Jun-Rong Chen, and Ming-Yuan Wang

An Active/Passive Joint for Reconfiguration Applications 133
Mykhailo Riabtsev, Victor Petuya, Aitor Riera, and Mónica Urízar

Kinematics Analysis and Dynamics Simulation Optimization of Lower Limb Rehabilitation Robot for the Stroke 141
Li Dong, Lu Li, and Shouqi Chen

A Configuration Synthesis Method of Deployable Mechanism Cells Used for Large Supporting Structures in Space 152
Chenhan Guang, Zilu Chen, and Yang Yang

Parameter Analysis and Optimal Design of Energy Storage Device for Non-powered Walking Aid 161
Fanyu Meng, Wenxiang Zhao, Rui Wang, and Zhixiu Hao

The Design and Analysis of Autonomous Underwater Pipeline Inspection Robots 172
Fujie Yu and Yuan Chen

Novel Solution for the Mechanism Used in the Actuation of Intelligent Roof Windows 182
Carmen Sticlaru, Petrisor Laurentiu Tuca, Erwin-Christian Lovasz, Corina-Mihaela Gruescu, Inocenţiu Maniu, and Cristian Emil Moldovan

Design and Kinematic Analysis of a Novel Wire-Driven Spherical Scissors Mechanism 192
Ruiqin Li, Ziwei Liu, Wenbo Mi, Shaoping Bai, and Jianwei Zhang

Machinery Dynamics

Numerical Simulations of the Dynamic Pressure Oil-Gas Separator ... 203
Yutao Yan, Yao Yu, Zhili Sun, and Xulei Su

A New Dynamic Model to Predict Vibration of Cylindrical Roller Bearings with Clearance and Off-Sized Roller Defects ... 213
Huimin Dong, Kui Zhang, and Chu Zhang

The Analysis for Stress Intensity Factor of Three-Dimensional Crack Under Combined Action of Friction Force and Tension ... 223
Yutao Yan, Yibo Zhang, Xin Wu, and Wendong Li

A Directly Coupled Finite Element – Meshless Method with Varying Dilation Parameter ... 233
Yaqing Zhang, Wenjie Ge, Ziang Zhang, Xiaojuan Mo, and Yonghong Zhang

Parallel Manipulators

Kinematics and Dynamics Simulation Analysis of a 3-DOF Parallel Mechanism for Application in Hybrid Machine ... 247
Haiqiang Zhang, Hairong Fang, Dan Zhang, Xueling Luo, and Fuqun Zhao

Kinematics and Singularity Analysis of a Novel Uncoupled 2-DOF Rotational Parallel Mechanism ... 259
Keming Wang, Yanbin Zhang, and Xianling Jing

Performance Analysis and Comparison of Three Planar Parallel Manipulators ... 270
Xiaoyong Wu, Ran Yan, Zhaowei Xiang, Fangyan Zheng, and Rulong Tan

Singularity Loci of a New Four-Limb Parallel Schönflies-Motion Generator ... 280
Guanglei Wu, Huiping Shen, Wenkang Zhao, and Zirong Lin

Kinematics and Multi-mode Application of a Novel Self-assembling Robot with Parallel Structure ... 290
Fuqun Zhao, Sheng Guo, Xiangyang Wang, and Majun Song

Design and Analysis of a New 5-DOF Hybrid Serial-Parallel Manipulator ... 301
Yundou Xu, Zhenghe Xu, Fan Yang, Youen Mei, Yi Yue, Yulin Zhou, Jiantao Yao, and Yongsheng Zhao

Kinematics Research of Chewing Robot Based on Six-Axis Parallel Mechanism ... 312
Xingyu Zhou and Jinghu Yu

Dynamic Model and Simulation Analysis of a 3-DOF Parallel Mechanism 322
Jiaxun Li, Jian Liu, Hongjuan Hou, and Liyong Huang

U-Joint Induced Torsional Instabilities of a Family of 3-DOF Partially-Decoupled Spherical Parallel Manipulators 332
Guanglei Wu and Stéphane Caro

Conceptual Design of a 2-DoF Planar High-Speed Industrial Parallel Manipulator 344
Emre Uzunoğlu, Merve Özkahya, Erkan Paksoy, Barış Taner, M. İ. Can Dede, and Gökhan Kiper

Compliant Mechanism

Design of a Controlled-Stiffness Flexure Mechanism Based on the FACT Method 357
Kaiyu Wu, Fan Zhang, Minhua Zheng, Hecheng Li, and Jing Sun

Compliant Mechanism Based Backbones for Continuous Manipulators and Robots 365
Haiyang Li, Qiaoling Meng, Guangbo Hao, Shujun Li, Yunpeng Gong, and Yong Zhao

Modelling and Prediction of Lifetime of Metal Rubber Isolator Under Random Vibration 376
Hongrui Ao, Yong Ma, Xianbiao Wang, and Hongyuan Jiang

A Recursive Integral-Based Dynamic Stiffness Matrix for Notch Flexure Hinges Used in Compliant Mechanisms 385
Mingxiang Ling

Control Application

Passivity-Based Force/Position Active Disturbance Rejection Control of Dual-Arm Space Robot Clamping Capture Spacecraft 397
Ai Haiping, Chen Li, and Yu Xiaoyan

Motion Planning of a 5-DOF Anthropomorphic Robotic Arm Under ROS Environment 409
Sida Zhang, Zirong Lin, and Guanglei Wu

T-S Fuzzy Compensation Based Sliding Mode Control for a Free-Floating Space Manipulator 419
Zihao Chen, Xiaoyan Yu, and Li Chen

Controlling Strategy of Rail Grinding Mechanism for the Constant Working Power 430
Jin Xie, Zhongqing Yi, and Zhaohui Liu

Terminal Sliding Mode Adaptive Fuzzy Controller of a Free-Floating Space Manipulator Based on Time Delay Estimation 440
Jianyu Zhang, Xiaoyan Yu, and Li Chen

Robots and Mechatronics

Dimensional Synthesis for the Chang'e-Type Legged Mobile Lander Based on Performance Atlas 453
Youcheng Han and Weizhong Guo

Kinematic Modeling and Simulation of a Leg-Wheel Robot for Unexplored Rough Terrain Environment 464
Xu Cai, Jun He, and Feng Gao

Mechanism Design and Leg Kinematics Analysis of Four-Legged Robot 474
G. M. Wang, G. Y. Ma, Y. L. Yao, and R. C. Liu

Optimization Design and Analysis of an Adaptive Variable Magnetic Adsorption Climbing Robot 484
Denghui Guo and Yuan Chen

The Design of a Kind of Two-Joint Mechanical Fish 494
Hui-hui You, Chang-tang You, Xiao-yan Yu, Chang-ling Zhang, and Kai-long Wang

Research on Knowledge Template Technology for Body Design of Transfer Robot 503
Wei Wang, Wenlin Gao, and Hongmin Yu

Kinematics Analysis and Reconfiguration Planning of the Reconfigurable Robot 512
Bin Ye, Feng Gao, and Weixing Chen

Development of Robotic Wiping System for Piezo Components 524
Lulu Chu, Yi Luo, and Xiaodong Wang

Position and Pose Optimization of Work Piece Based on Image Processing Data 533
Feng Qian, Sichen Cai, and Huimin Dong

Development of Conductive Hydrogel Driven Dielectric Elastomer 542
Xiaoru Niu, Jianzhong Shang, Zirong Luo, and Tao Jiang

Kangaroo-Inspired Jumping Robot with Controllable Takeoff 556
Yonghong Zhang, Sizhe Liu, Lei Zheng, Xiaojuan Mo, and Wenjie Ge

Hydrodynamic Coefficients Calculation of Complex-Shaped AUV Based on Lattice Boltzmann Method 566
Yinpo Yan and Yuan Chen

Study on the Development of Toner Xerography Digital Printing Technology . . . 576
Gaosheng He and Yan Li

Education on Theory of Mechanisms and Machines

Project-Oriented Learning/Teaching Applied to Theory of Mechanisms and Machines . . . 589
Guanglei Wu, Delun Wang, Huimin Dong, Yuan Gao, and Feng Qian

APP Based Teaching/Learning Mode and Quality Evaluation of MMT Course . . . 598
Huimin Dong, Delun Wang, Yuan Gao, Feng Qian, and Guanglei Wu

Experimental Study on Mechanical System Dynamics Performance . . . 607
Yuan Gao, Hong Zhang, Delun Wang, Huimin Dong, and Cong Chen

Construction of Golden Course Group for the Innovative Talent Cultivation in New Engineering of Machinery . . . 617
Kedong Bi, Yajing Kan, Ruiming Qian, Guodong Yin, and Zhonghua Ni

Exploration and Practice of Project-Based Teaching in the Course Theory of Machine Design and Mechanism . . . 627
Yuanxiong Cheng, Jiajun Yang, Jianfeng Xu, and Kuanmin Mao

Project-Based Teaching Skills for University Course of Mechanism and Machine Science . . . 636
Weizhong Guo

An Approach to Foster Student Creative Thinking Ability in the Lectures of Mechanical Principles . . . 641
Yunpeng Gong, Shujun Li, Zhang Yu, Haiyang Li, Cuiling Li, and Yu Guo

Author Index . . . 647

Linkage Synthesis

Optimized Design of an RSSR Mechanism for an Unpowered Assistive Aid

Juewei He[1], Guangshun Liang[1], Zhi Lan[2], and Zhixiu Hao[1(✉)]

[1] Department of Mechanical Engineering,
Tsinghua University, Beijing 100084, China
haozx@tsinghua.edu.cn
[2] National Research Center for Rehabilitation Technical Aids,
Beijing 100176, China

Abstract. This paper proposes the use of RSSR space four-bar mechanism to realize the motion and power transmission of a non-powered assistive aid. A three-dimensional modeling, mathematical model analysis and solution function image combination method was proposed to design and analyze the RSSR space four-bar transmission mechanism in this accessory, and the key parameters such as length and rotation angle was optimized. The RSSR mechanism prototype was produced and demonstrated. The results show that the designed RSSR mechanism can effectively transmit the change of the angle of motion to the follower and achieve a better torque transmission effect.

Keywords: Unpowered assistive aids · RSSR mechanism · Optimized design

1 Introduction

Supported by the exercise aids, rehabilitation training for patients with impaired motor function or partial paralysis is an important means of rehabilitation for medical aids. We use sports aids to help or replace the limbs for rehabilitation training, so that patients can restore a certain of the original motor function on the basis of physiological "structural rehabilitation" to achieve a "functional rehabilitation" state. The non-powered lower limb assistive aid is a wearable assistive device without external power source. The patient can rely on his own physical energy to walk with it.

The RSSR mechanism is a typical space four-bar mechanism. This mechanism has the advantages of small number of links, compact structure, flexible motion, high transmission precision. However, due to the diversity and complexity of the motion pairs, the motion characteristics of this space four-bar mechanism are more complex than those of the planar four-bar mechanism, making design and analysis calculations more difficult. Compared with the plane mechanism, the space mechanism can also realize some forms of motion that the plane mechanism cannot present. For this reason, the space linkage mechanism is widely used in light industry (sewing, knitting, shoemaking, tanning, etc.) machinery, agricultural machinery, transportation, chemical machinery, instrumentation and various control devices. Especially in robots and manipulators, space linkages occupy a dominant position. However, at the same time, the description of the motion of space-moving components is far more difficult than

D. Wang et al. (Eds.): MeTrApp 2019, MMS 79, pp. 3–11, 2020.
https://doi.org/10.1007/978-981-15-0142-5_1

that of planar motion. Therefore, the motion analysis of spatial mechanisms is much more difficult than that of planar mechanisms, which affects the development and promotion of spatial mechanisms [1]. The motion analysis of spatial mechanisms mainly refers to the displacement (including the trajectory of points), velocity and acceleration analysis of the mechanisms, and the displacement analysis is the most basic and complex.

In order to take advantage of the space linkage mechanism, researchers have studied the motion and dynamic characteristics of the space linkage mechanism through various analysis and solution methods, and continuously improved the research methods in the process. They put forward some theories and methods, laying a theoretical foundation for the promotion and application of space linkages.

Tao et al. [2] analyzed the spatial input and output equation of the corresponding angular displacement of the driving member and follower of RSSR mechanism, and then they deduced the mathematical model for solving the corresponding angular displacement of the driving member (six groups) and follower (eight groups) of RSSR mechanism. The program was written to solve the model. De-Juan et al. [3] provided a valuable tool for obtaining the best solution to meet the requirements of input/output (I/O) and optimized the four-bar spherical mechanism with RSSR linkage, based on the dimension synthesis technology of local optimization. Miao and Qiao [4] established the motion analysis equation of RSSR mechanism by analytic method based on visualization, and obtained the angular displacement, angular velocity and angular acceleration map of the follower by Matlab. Secondly, based on SolidWorks three-dimensional modeling, motion simulation was carried out by using COSMOS Motion, and the corresponding motion analysis results were obtained. Tanık et al. [5] introduced a new type of flexible spatial four-bar (RSSR) mechanism and used pseudo-rigid body model (PRBM) to analyze and design it. Zhou et al. [6] established the motion analysis model of RSSR mechanism by analytic method, solved the motion curve between input and output of mechanism by Matlab software, and carried out parametric modeling and optimization design of mechanism by ADAMS software. The sensitivity was calculated. As a result, the optimal design variables were determined. Wei et al. [7] established a theoretical analysis model for spatial linkage RSSR mechanism, and obtained the input and output displacement equation of the mechanism. Combining with the kinematic analysis, the restraint counter-force in the kinematic pairs was solved by establishing the static equilibrium equation. The three-dimensional model of the mechanism was established, and the angular relationship and the force of the motion pair were analyzed in the simulation software ADAMS. Therefore, the angular relationship curve and the counter-force curve of the motion pair were obtained. This paper combines three-dimensional modeling, mathematical analysis and function image solving to optimize the RSSR transmission mechanism without power aids so as to achieve better motion and power transmission effect.

2 Design and Optimization Methods

2.1 Model Building

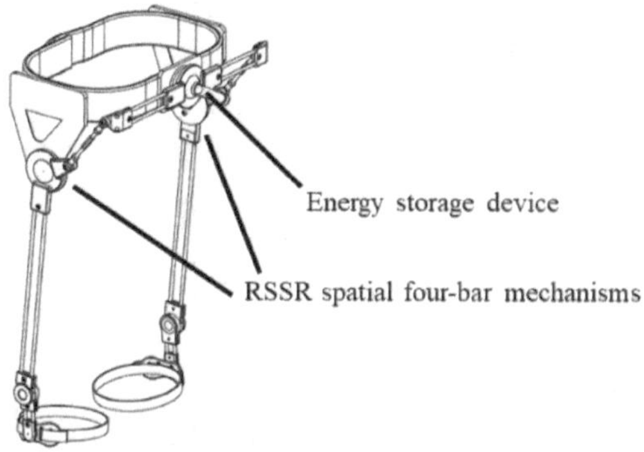

Fig. 1. A sketch of unpowered walking aids

As is shown in Fig. 1, it is an unpowered walking aid designed for hemiplegic patients. It includes two sets of RSSR spatial four-bar mechanisms, which are connected by an intermediate energy storage device and are symmetrically distributed on both sides. According to the original design parameters and working principle of the unpowered walking aids, the three-dimensional model of the RSSR mechanism is built, and the one-side three-dimensional model of the unpowered walking aids is obtained as shown in Fig. 2. R_1 and R_2 are rotating pairs, S_1 and S_2 are spherical pairs, bar L_3 rotates around R_1 in the $y_1o_1z_1$ plane, bar L_1 rotates around R_2 in the $x_2o_2z_2$ plane, L_1 and L_2 are connected by S_2, and L_3 and L_2 are connected by S_1.

According to the working characteristics of unpowered walking aids, the rotation of L_3 needs to reach a certain angle to ensure that the auxiliary aids can assist the user during the walking process. Therefore, the angular displacement and torque of L_1 need to be effectively transmitted to L_3. In this paper, we will refine the design requirement. The optimization goal is that when L_1 rotates, the angular displacement and torque of L_3 should be close to those of L_1 to ensure that the RSSR mechanism can achieve the auxiliary effect on patients with hemiplegia.

In order to effectively achieve the optimization goal of angular displacement transmission and torque transmission, it is necessary to analyze RSSR mechanism in detail, find out the key parameters affecting the RSSR spatial four-bar mechanism, and optimize the key parameters to find out the reasonable design parameters.

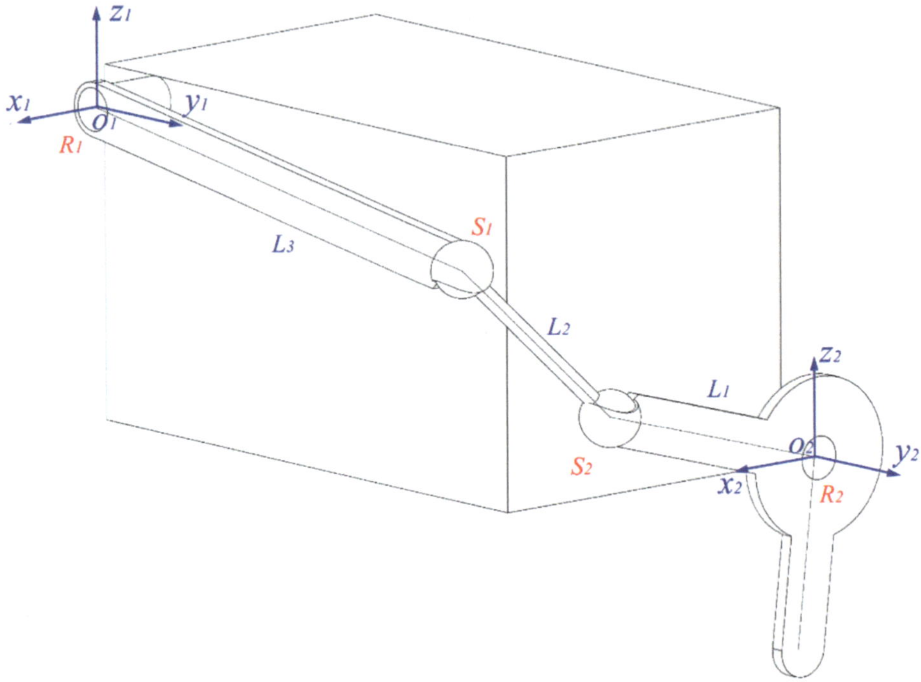

Fig. 2. Three-dimensional model of RSSR mechanism

2.2 Calculate the Freedom of the Transmission Mechanism of the Accessory

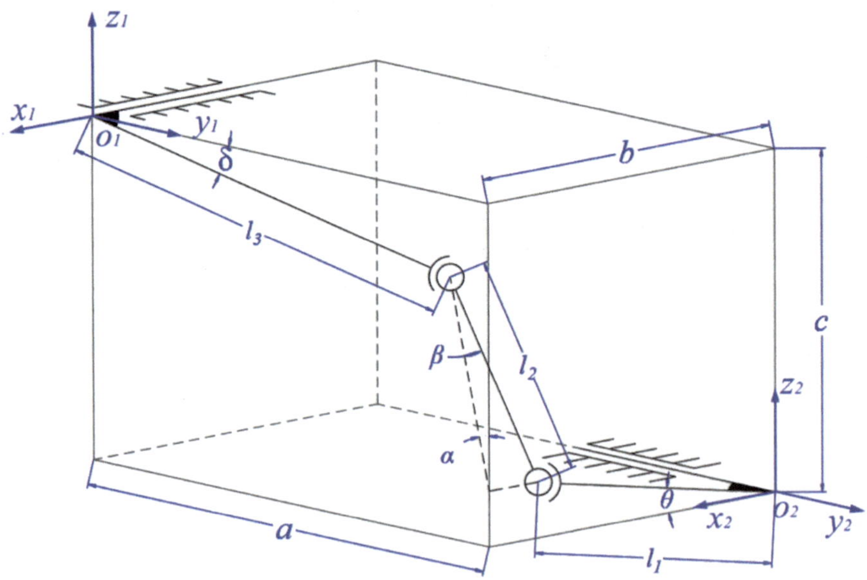

Fig. 3. A schematic diagram of RSSR mechanism

The RSSR spatial four-bar mechanism is further abstracted and simplified to obtain the mechanism sketch, as is shown in Fig. 3. The mechanism has two rotating pairs and two spherical pairs, and the degree of freedom of RSSR spatial four-bar mechanism is:

$$f = 6 \times 3 - 5 \times 2 - 3 \times 2 = 2 \tag{1}$$

That is to say, the spatial four-bar mechanism has two degrees of freedom, one of which is the degree of freedom of the bar L_2 to rotate along its own axis. The degree of freedom has no influence on the transmission of the mechanism, which can be neglected, and the other degree of freedom is left to meet the design requirements.

2.3 Mathematical Modeling of Spatial Four-Bar Linkages

The lengths of the three movable bars L_1, L_2, and L_3 of the RSSR spatial four-bar mechanism are l_1, l_2, l_3 respectively; the length of the frame in the y_1 direction is a, the length of the frame in the x_1 direction is b, and the length of the frame in the z_1 direction is c. The bar L_1 rotates only in the $x_2o_2z_2$ plane, and has a degree of freedom. The angle between the L_1 and the x_2-axis is θ. The bar L_3 rotates only in the $y_1o_1z_1$ plane. The angle between the L_3 and the y_1-axis is δ, the angle between the bar L_2 and the $y_1o_1z_1$ plane is β, and the angle between the bar L_2 and the $x_1o_1z_1$ plane is α. According to the basic parameters of human body, the design parameters a,b,c would be set as design constraints.

In the direction along the three coordinate axes, the geometric relations are as follows:

$$\begin{cases} l_2 cos\beta sin\alpha + l_3 cos\delta = a \\ l_2 sin\beta + l_1 cos\theta = b \\ l_2 cos\beta cos\alpha + l_1 sin\theta + l_3 sin\delta = c \end{cases} \tag{2}$$

Combine the three equations in (2) and eliminate the irrelevant variables α and β:

$$\begin{aligned} &\tfrac{1}{2}\left(a^2 + b^2 + c^2 + 697 + l_1^2 + l_3^2\right) \\ &= cl_1 sin\theta + bl_1 cos\theta - l_1 l_3 sin\delta sin\theta + al_3 cos\delta + cl_3 sin\delta \end{aligned} \tag{3}$$

2.4 Analysis and Optimization of Mathematical Model

The optimization goal is that when $\Delta\theta = \pm 20°$, $\Delta\delta$ can be closer to $\pm 20°$, so the Eq. (3) can be processed first to obtain the formula of the variation of δ with θ, that is, $\frac{\partial\delta}{\partial\theta}$ is calculated, and the result:

$$\frac{\partial\delta}{\partial\theta} = \frac{cl_1 cos\theta - bl_1 sin\theta - l_1 l_3 sin\theta sin\delta}{l_1 l_3 sin\theta cos\delta + al_3 sin\delta + cl_3 cos\delta} \tag{4}$$

It can be seen from Eq. (4) that l_2 does not affect the value of $\frac{\partial\delta}{\partial\theta}$, that is, it does not affect the angle variation of l_1 and l_3, so the length variation of l_1 and l_3 is mainly considered in the optimization process to achieve the ultimate optimization goal.

In the formula (3), the three parameters *a*, *b*, and *c* have been determined, and the size of *c* can be adjusted within a certain range. The value of the parameter *c* is not changed in the optimization design. Set $a = 220\,\text{mm}, b = 150\,\text{mm}, c = 115\,\text{m}$. To facilitate analysis and calculations, the unit is changed to centimeters, then $a = 22\,\text{cm}, b = 15\,\text{cm}, and\, c = 11.5\,\text{cm}$. Since l_2 does not affect the value of $\frac{\partial\delta}{\partial\theta}$, we set the parameter $l_2 = 12\,\text{cm}$ unchangedly and directly substitute the existing parameters into the Eq. (3). The mathematical model of the RSSR studied in this paper is:

$$\begin{aligned} &\tfrac{1}{2}\left(697 + l_1^2 + l_3^2\right) \\ &= 11.5l_1 sin\theta + 15l_1 cos\theta - l_1 l_3 sin\delta sin\theta + 22l_3 cos\delta + 11.5l_3 sin\delta \end{aligned} \tag{5}$$

The mathematical model (5) is an implicit function, and the parameters cannot be directly optimized. Therefore, the angle change under different parameters is obtained by computer drawing. The data is acquired by the image to optimize the parameters according to the data.

The design and optimization methods is shown in Fig. 4.

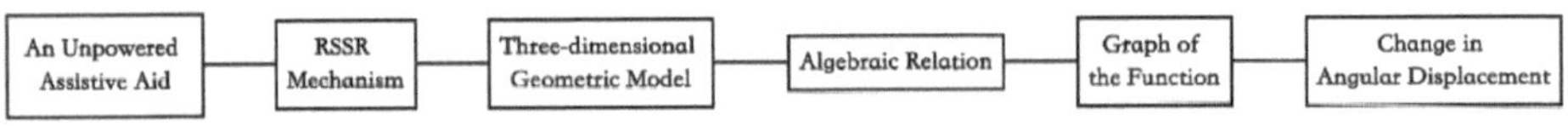

Fig. 4. Flow diagram of the optimized design

3 Results and Discussion

Changing the length l_1 and l_3, then bringing it into Eq. (5), and drawing the graph of function by Matlab, we get the curve showing the relationship between θ_0 (the value of θ when $\delta = 0$, that is, when the patient is standing, the angle between L_1 and the z_1-axis) and $\Delta\delta$ under different length of l_1 and l_3. The variation curve of angle when $l_3 = 20\,\text{cm}, l_1 = 20\,\text{cm}$ is shown in Fig. 5 as an example. In the figure, the abscissa indicates the angle of θ, and the ordinate indicates the angle of δ. The figure indicates the value of θ when $\delta = 0$, that is, the angle between L_1 and the x_1-axis when the patient is standing (Figs. 6 and 7).

Similarly, the function images under different combinations of bar lengths can be drawn. The range of angle variation of δ is calculated. We summary the data in Table 1.

Group according to different l_3, take l_1 as the abscissa, and plot the curve of $\Delta\delta$ as:

According to the obtained data and images, under the combination of the bar lengths that $l_1 = 20$ and $l_3 = 20$, the range of δ reaches 40°, and the variation is positive and negative symmetry. Therefore, this set of bar lengths meets the optimization goal of angle change. The torque transmission effect of this set of bar lengths is tested.

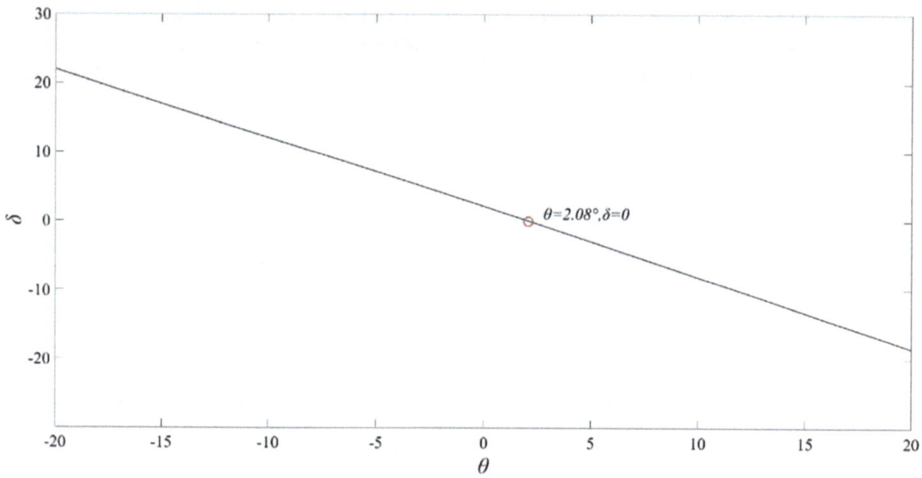

Fig. 5. The variation curve of angle when $l_3 = 20$, $l_1 = 20$

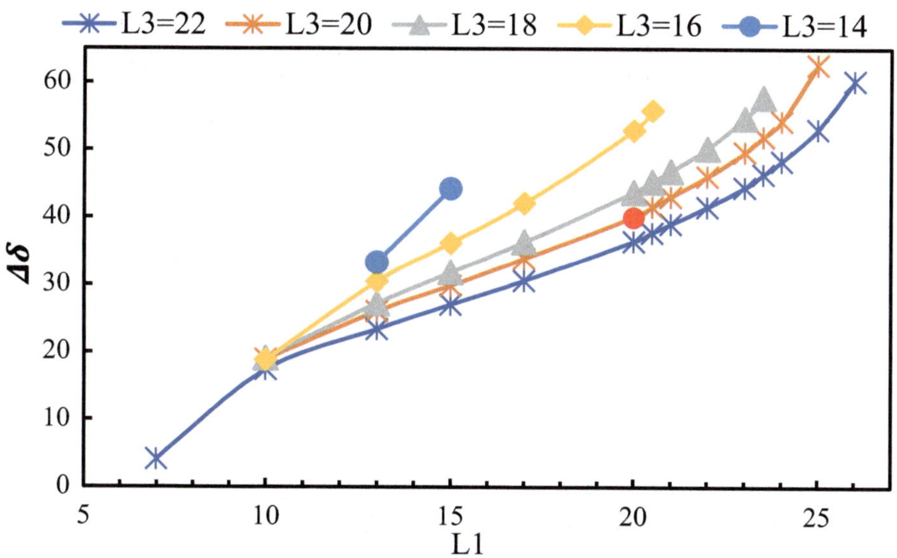

Fig. 6. Relationship between $\Delta\delta$ and l_1 under different l_3

Bringing this set of bar lengths into the original SolidWorks 3D model, the dynamics simulation can be obtained. The ratio of the torque M_1 at the rotating pair R_1 and the torque M_2 at the rotating pair R_2 ($M_1 = kM_2$) is shown in Table 2.

According to the torque ratio relationship, the torque transmitted by the four-bar mechanism of the RSSR space is basically unchanged under the combination of the bar lengths that $l_1 = 20$ and $l_3 = 20$. Therefore, the torque optimization under this set of bar lengths is in line with the target.

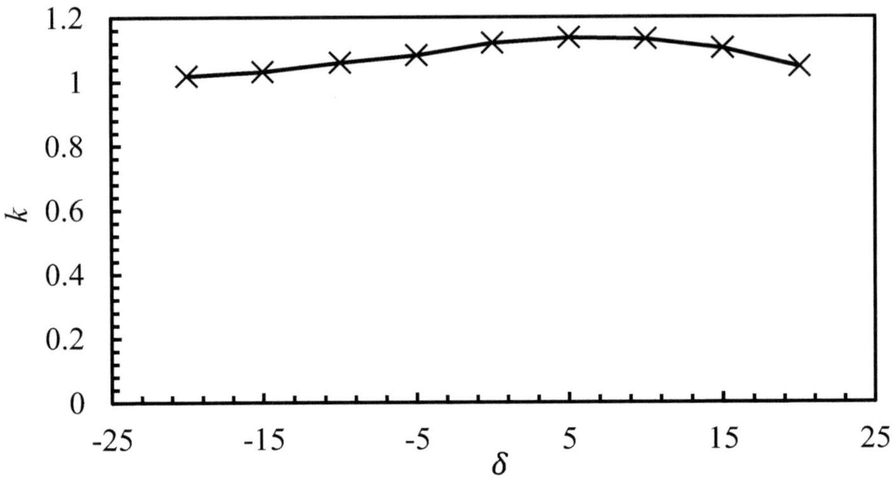

Fig. 7. Torque relationships at different gait angles

Table 1. $\Delta\delta$ under different length combinations

l_1	l_3	θ_0	δ_{min}	δ_{max}	$\Delta\delta$	l_1	l_3	θ_0	δ_{min}	δ_{max}	$\Delta\delta$
7	22	30.0	−0.4	3.7	4.1	24	20	8.5	−27.8	26.4	54.2
10	22	3.5	−7.9	9.5	17.4	25	20	10.0	−29.6	33.0	62.6
13	22	−1.0	−11.5	11.9	23.4	10	18	8.0	−7.9	11.2	19.1
15	22	−2.0	−13.4	13.6	27.0	13	18	2.0	−12.6	14.6	27.2
17	22	−1.0	−15.7	14.9	30.6	15	18	0.7	−14.9	16.9	31.8
20	22	2.0	−19.5	16.9	36.4	17	18	1.2	−17.4	18.9	36.3
21	22	3.0	−21.0	18.0	39.0	20	18	3.7	−21.2	22.5	43.7
22	22	4.5	−22.4	19.1	41.5	21	18	5.0	−22.8	24.0	46.8
23	22	6.0	−24.0	20.4	44.4	22	18	6.5	−24.4	25.8	50.2
24	22	8.0	−26.5	21.8	48.3	23	18	8.2	−26.3	28.4	54.7
25	22	10.0	−29.0	24.0	53.0	23.5	18	9.0	−27.0	30.8	57.8
10	20	4.5	−8.2	10.7	18.9	10	16	15.0	−6.8	12.0	18.8
13	20	−0.7	−12.4	13.6	26.0	13	16	5.7	−12.9	17.5	30.4
15	20	−1.3	−14.4	15.4	29.8	15	16	4.2	−15.8	20.3	36.1
17	20	−0.5	−16.8	17.0	33.8	17	16	4.4	−18.7	23.3	42.0
20	20	2.0	−20.0	20.0	40.0	20	16	6.7	−23.0	29.8	52.8
21	20	3.5	−22.0	21.0	43.0	20.5	16	7.3	−23.8	32.0	55.8
22	20	5.0	−23.8	22.2	46.0	13	14	13.0	−12.3	21.0	33.3
23	20	6.7	−25.8	23.8	49.6	15	14	9.8	−16.2	28.0	44.2

In summary, under the combination of the lengths that $l_1 = 20$ and $l_3 = 20$, the transmission of angular displacement and the transmission of torque have reached the goal of optimizing design.

Table 2. Torque relationships at different gait angles

δ	−20°	−15°	−10°	−5°	0°	5°	10°	15°	20°
k	1.017	1.031	1.058	1.081	1.119	1.135	1.131	1.102	1.044

4 Summary

In this paper, a kind of unpowered walking aids is modeled and analyzed by means of combining three-dimensional modeling, mathematical model analysis and function image analysis. According to the analysis results, the key parameters of RSSR spatial four-bar mechanism used in the design are optimized to meet the design requirements. Under the design constraints, by choosing a suitable combination of bar length, better angle and torque transmission can be achieved between the active and driven parts of RSSR mechanism.

The highlight of this paper is to collect data from function image to solve the problem that implicit function cannot be solved directly. As the parameters change, the angle function images under different combination of parameters can be obtained, and the data needed can be measured from the image. Then, the angle data under the discrete combinations of parameters are fitted to obtain the spectrum between the angle range and the combination of parameters.

The limitation of this paper is that it takes a lot of work to get the function image by changing the parameters one by one, and then measure the data from the image. It is feasible and efficient to automatically draw function images and acquire key data by program.

Acknowledgement. Thank Fanyu Meng and Yang Jiang, who are members of our group, for their help in the research work.

References

1. Xie, K., Mao, J.: Research and application of spatial linkages (in Chinese). New Technol. New Process **7**, 16–18 (2003)
2. Tao, Y., Wen, Z., Ding, Y.: Analytical synthesis of spatial RSSR mechanism (in Chinese). J. Agric. Mech. Res. **3**, 33–35 (2001)
3. Sancibrian, R., De-Juan, A., Garcia, P., Fernandez, A., Viadero, F.: Optimal synthesis of function generating spherical and RSSR mechanisms. In 12th IFToMM World Congress, pp. 18–21 (2007)
4. Miao, H.-B., Qiao, F.-L.: Kinematics characteristics analysis research of spatial RSSR mechanisms (in Chinese). Mach. Des. Manuf. **2**, 11–13 (2008)
5. Tanık, E., Parlaktaş, V.: A new type of compliant spatial four-bar (RSSR) mechanism. Mech. Mach. Theor. **46**(5), 593–606 (2011)
6. Zhou, X., Song, M., Wang, X., Yao, W.: Motion Analysis and optimization design of RSSR spatial mechanism (in Chinese). Mach. Tool Hydraulics **43**(9), 1–4 (2015)
7. Chen, W., Song, M., Wang, X., Yao, W., Xi, Y.: Analysis and simulation of spatial RSSR mechanism (in Chinese). Machine Building & Automation, 2 (2018)

Dimensional Synthesis of Non-periodic Function Generation of a Spatial RSSR Mechanism Using a Numerical Atlas Method

Wenrui Liu[1], Jianwei Sun[1](✉), and Jinkui Chu[2]

[1] Changchun University of Technology, Changchun 130012, People's Republic of China
avensun@tom.com
[2] Dalian University of Technology, Dalian 116023, People's Republic of China

Abstract. In this paper, a wavelet feature parameters method is proposed for non-periodic function generation of a spatial Revolute-Spherical-Spherical-Revolute (RSSR) mechanism. The mathematical model for the output function of an RSSR mechanism is established. The wavelet coefficients, including wavelet approximations and wavelet details, are extracted from the output function curve using a Daubechies wavelet. Based on the geometric significance of the wavelet coefficients, the wavelet details are used to describe the output function of an RSSR mechanism to eliminate the influence of translations of output function curve. The level of wavelet feature parameters is determined according to the relationship between the original output function curves and the reconstruction curves generated by wavelet coefficients. A numerical atlas database is established, which contains more than ten million groups of basic dimensional types. Based on a fuzzy reorganization method, the desired RSSR mechanism can be obtained. The non-periodic function synthesis of the RSSR mechanism is realized. An example is presented to demonstrate the efficacy and practicality of the proposed theory.

Keywords: RSSR mechanism · Non-periodic function generation · Wavelet feature parameters

1 Introduction

Generally speaking, dimensional synthesis contains function generation, path generation and motion generation. The task of function generation is to design a linkage mechanism that efficiently and precisely satisfies the desired function curve. Several methods for the function generation have been developed by scholars. An important input-output formula of planar four bar mechanism is established by Freudenstein [1]. Based on the classical Freudenstein equation, Bai and Angeles [2] proposed a unified input–output formula for the spherical four-bar mechanism. For the same type of mechanism, Yang et al. [3] presented a method for selecting precision points for function generation. Then the design results can be obtained using a fractal algorithm method. Wang et al. [4] established a mathematical model of saddle-point programming for spherical four-bar mechanism function generation. Based on closed form

D. Wang et al. (Eds.): MeTrApp 2019, MMS 79, pp. 12–20, 2020.
https://doi.org/10.1007/978-981-15-0142-5_2

method, Rao et al. [5] proposed a function synthesis method for spatial RCCC and RSSR mechanism. The method can be used to solve eight precision points function generation problem. Dhall and Kramer [6] established a computer aided design program for function generation of spatial HCCC, RCCC and PCCC mechanism. Based on the Fourier series, Sun and Chu [7] proposed a numerical atlas method for function generation of an RCCC mechanism.

As an important design method, the numerical atlas method has been widely studied over the last thirty years. The basic idea of the numerical atlas method is to store the feature parameters of output function or coupler curves generated by linkage mechanism in a database through some mathematical tools, for example, Fourier series [8], B-splines [9] and wavelet coefficients [10]. Then, the geometric parameters of the desired linkage mechanism can be obtained by comparing the feature parameters between prescribed curve and each group stored in the database. The numerical atlas methods can be used to solve dimensional synthesis problems with multiple positions [11]. Since each group of geometric parameters stored in the database can generate linkage mechanisms without limit in the design interval and desired mechanisms are obtained by fuzzy recognition process, the numerical atlas method can provide multiple solutions without branch, order and circuit defects [12]. The accuracy of design results obtained through the numerical atlas method mainly depend on the capacity of geometric parameters stored in the database and the feature parameters extraction method. However, with the increase in the amount of geometric parameters, efficiency of design process lower. Mullineux [13] introduced a normalization method by rotating the central axis of a spherical curve, which is generated by spherical four-bar mechanism, into a coordinate axis. The coupler curve is described by the projections on a coordinate axis and a plane, the influence of frame rotation variations has been eliminated. The dimension of geometric parameters stored in the database is lower, and the burdens of computation are reduced without sacrificing the accuracy of the design results. Sun et al. [14] established a variable step-size numerical atlas for path generation of a spherical crank slider mechanism. The Fourier series method may be the most widely used tool for the linkage mechanism dimensional synthesis problem with the numerical atlas [15–19]. However, in many applications our interest lies in the open curves as opposed to the closed curves. Given the limitations of the Fourier series, it is not a good tool for problems with non-periodic design requirements.

The primary idea of wavelet transform is extracting feature parameters from prescribed curve in the various different frequency bands using scale and location parameters. For this reason, we have used wavelet feature parameters to describe non-periodic output function curves of planar four-bar mechanism, spherical four-bar mechanism and spatial RCCC mechanism. The corresponding numerical atlas databases have been established. In this paper, we extend the wavelet feature parameters method to spatial RSSR mechanism. A numerical atlas database is established, which contains more than ten million groups of basic dimensional types. Based on a fuzzy reorganization method, the desired RSSR mechanism can be obtained. The non-periodic function synthesis of the RSSR mechanism is realized. In Sect. 4, an example is presented to demonstrate the efficacy of the proposed theory.

2 Wavelet Coefficients of the Output Function of an RSSR Mechanism

Figure 1 shows a spatial Revolute-Spherical-Spherical-Revolute (RSSR) mechanism, which is constructed from four links (frame, R-S link, S-S link and S-R link) connected by two revolute and two spherical joints. The length of each link of RSSR mechanism (*AB*, *BC*, *CD*, *EF*, *FA* and *DE*) is noted as a_1, a_2, a_3, a_4, S_1 and S_4. α_1 is the angle between z_4–axis and z_1–axis; θ_1 is the input angle; θ_4 is the output angle. Based on the relationship of geometric parameters of RSSR mechanism, the input-output equation can be expressed as follows:

$$\theta_4 = 2\arctan\left[\left(k_1 + \sqrt{k_1^2 + k_2^2 - k_3^2}\right)/(k_2 - k_3)\right] \tag{1}$$

where

$$k_1 = S_1 \sin\alpha_1/a_1 - \cos\alpha_1 \sin\theta_1 \tag{2}$$

$$k_2 = a_4/a_1 + \cos\theta_1 \tag{3}$$

$$k_3 = \left(a_1^2 - a_2^2 + a_3^2 + a_4^2 + S_1^2 + S_4^2 + 2S_1S_4\cos\alpha_1\right)/(2a_1a_3) + (a_4\cos\theta_1 + S_4\sin\alpha_1\sin\theta_1)/a_3 \tag{4}$$

Based on Eq. (1), it is obvious that the output function of an RSSR mechanism is determined by input angle and seven geometric parameters: a_1, a_2, a_3, a_4, S_1 S_4, and α_1. In this paper, above seven geometric parameters are defined as basic dimensional types, briefly called BDTs.

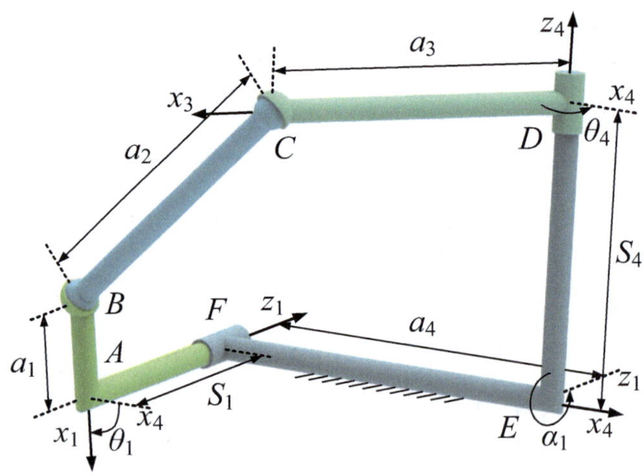

Fig. 1. An RSSR mechanism.

Based on reference [20], the output function of an RSSR mechanism can be described by wavelet coefficients. We assume that the number of sample points extracted from output function curve is 2^j ($j \in \mathbf{N}^+$), then, the j level wavelet expansion can be expressed as follows:

$$\theta_4(\theta_1) = a_{(j,l)}\phi_{(j,l)}(\theta_1) + \sum\nolimits_{J=1}^{j} \sum\nolimits_{l \in Z} d_{(J,l)}\psi_{(J,l)}(\theta_1) \tag{5}$$

where $a_{(j,\ l)}$ is the wavelet approximation; $d_{(J,\ l)}$ is wavelet details; $\phi_{(j,\ l)}(\theta_1)$ is a scaling function; $\psi_{(J,\ l)}(\theta_1)$ is a wavelet function. All wavelet approximations and wavelet details are defined as wavelet coefficients. The scaling function and wavelet function of Daubechies wavelet can be expressed as follows:

$$\phi(\theta_1) = \sum\nolimits_{l=0}^{2N-1} p_l \phi(2\theta_1 - l) \tag{6}$$

$$\psi(\theta_1) = \sum\nolimits_{l=2-2N}^{1} (-1)^l p_{1-l} \phi(2\theta_1 - l) \tag{7}$$

where N is the number of vanishing moments. p is scale coefficients.

Based on Eq. (5), for an arbitrary RSSR mechanism, the wavelet coefficients can be used to describe the feature information of the output function curves. It provides a condition for solving the function generation problem of an RSSR mechanism using the numerical atlas method.

3 Wavelet Feature Parameters of the Output Function of an RSSR Mechanism

In a previous study [7], we have studied the function synthesis problems of the periodic design requirement of a spatial four-bar mechanism using Fourier series method. Moreover, we found that 64 sample points can describe the characteristics of a periodic function curve generated by a spatial four-bar mechanism, accurately. Therefore, in this paper, we choose $j = 6$. 64 sample points are extracted from output function curve and six level wavelet decompositions are used to obtain wavelet coefficients. A relatively complicated output function curve, which is generated by a spatial RSSR mechanism, is shown in Fig. 2(a). The BDTs of the prescribed RSSR mechanism are $a_1 = 21$, $a_2 = 16$, $a_3 = 21$, $a_4 = 11$, $S_1 = 16$, $S_4 = 15$ and $\alpha_1 = 166°$. Then, 64 sample points are obtained and six level wavelet decompositions are used to extract wavelet coefficients from the original function curve using a Daubechies wavelet ($N = 2$). The function curves of the wavelet approximations and wavelet details are obtained and shown in Fig. 2(b) to (h). Figure 3(a) shows the reconstructed function curve by sixth level wavelet approximations and details (the reconstruction curve is denoted by red dot). In Fig. 3(b), the fifth level wavelet details are added in, and so on, with Fig. 3(f) giving the result of all the wavelet coefficients and including the first level wavelet details. Table 1 lists the error between the original and reconstructed curves where average absolute values of two level and three level reconstruction curves are 0.2047° and 0.6531°, respectively. Since the values of the first three levels wavelet details (i.e. first-level to third-level) are very small (see Fig. 2(f) to (h)), the periodic function curve can be reconstructed to approximately match the original curve using wavelet approximations and the last three levels wavelet details. This does not significantly affect the accuracy of the results, but the burdens of the storage and computation can be reduced (the first three levels wavelet details contains nearly ninety percent of wavelet coefficients).

Table 1. Errors between prescribed output function curve and reconstruction curve.

Reconstruction	Average error/°	Maximum error/°
Six level reconstruction	13.4810	35.8367
Five level reconstruction	5.3494	19.7218
Four level reconstruction	1.8370	8.9968
Three level reconstruction	0.6531	3.8133
Two level reconstruction	0.2047	2.1109
One level reconstruction	0	0

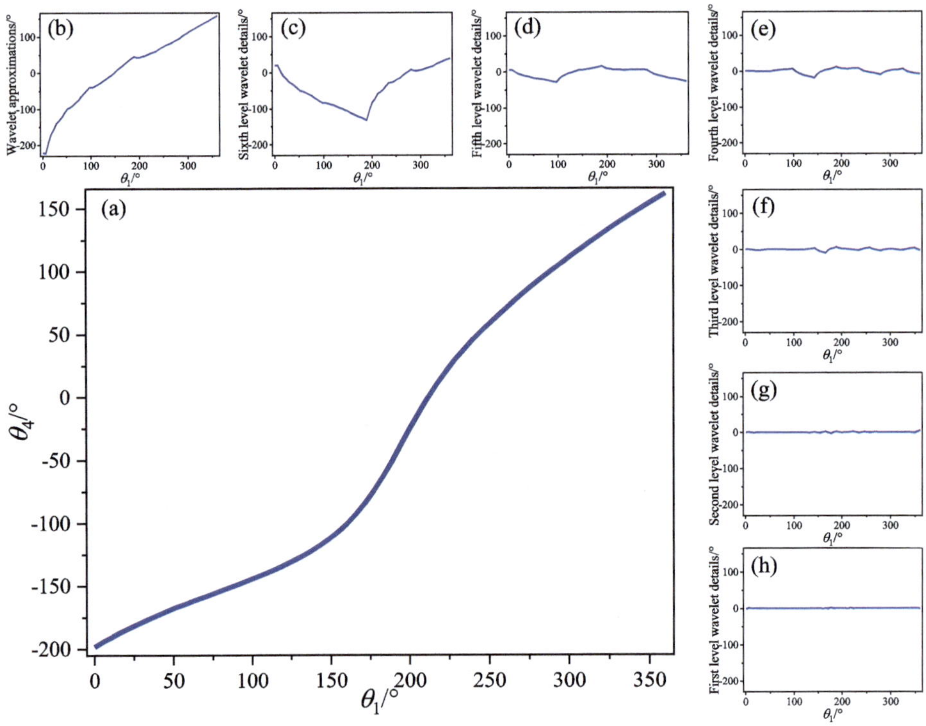

Fig. 2. Prescribed output function curve and the wavelet coefficients curves.

Based on the geometric significance of wavelet coefficients, the wavelet approximations can be used to describe the general trend of the output function curve in the corresponding scale, and the wavelet details can be used to describe the changes of the output function curve in the corresponding scale. Therefore, the translations of output function curve along θ_4-axis have no influence on wavelet details [20]. Based on the relationship of wavelet approximation between the prescribed function curve and the output function curve generated by BDTs, the translation of output function curve along θ_4-axis can be calculated by following equation:

$$\Delta\theta = a_{(j,l)} - a'_{(j,l)} \tag{8}$$

where $a_{(j,\ l)}$ is the wavelet approximation of prescribed function curve; $a'_{(j,\ l)}$ is the wavelet approximation of output function curve generated by BDTs. In this paper, the last three levels wavelet details are defined as wavelet feature parameters (WFPs) to describe the characteristics of the output function curve. Based on above analysis, a numerical atlas database is established, which contains more than ten million groups of basic dimensional types and the corresponding wavelet feature parameters. Then, the desired mechanism can be obtained by comparing the WFPs between the prescribed function curve and each group stored in database, rather than through all the wavelet coefficients. The function synthesis of the RSSR mechanism is realised. The error of WFPs between the prescribed function curve and the output function curve generated by BDTs can be expressed as follows:

$$\delta = \sum\nolimits_{J=4}^{6} \sum\nolimits_{l \in Z} \sqrt{\left[b_{(J,l)} - b'_{(J,l)}\right]^2} \tag{9}$$

where $b_{(J,\ l)}$ is the WFP of prescribed function curve; $b'_{(J,\ l)}$ is the WFP of output function curve generated by BDTs.

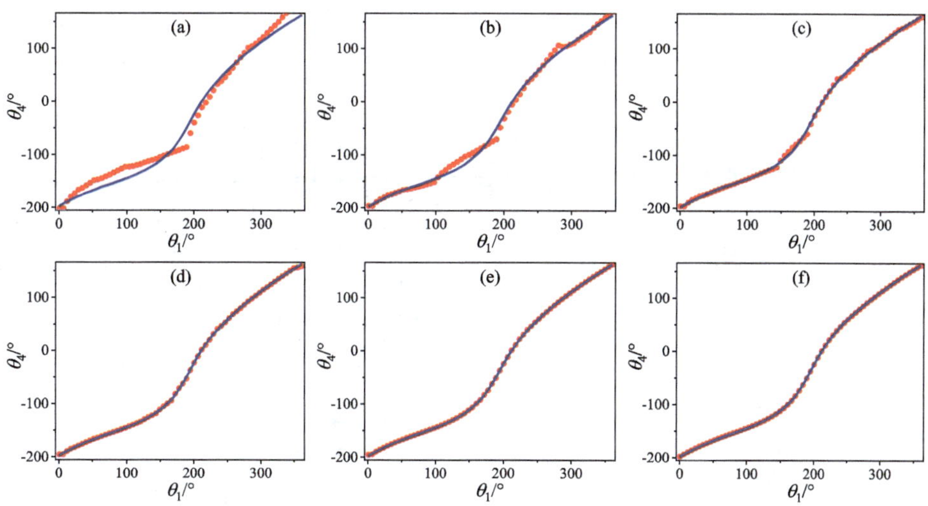

Fig. 3. Comparison between original function and reconstruction curves.

4 Examples

In this example, the prescribed function curve is generated by following expressions:

$$y = \cos x \cos(1.2x + 22.5^\circ) \times 180^\circ/\pi$$

where $x \in [0, 80^\circ]$. Based on the methods proposed in this paper, a numerical database is established. Three groups of design results are obtained and listed in Table 2, where θ_1^1 is the initial angle of input link. The comparison and error graphs for the second group of design results are shown in Fig. 4(a) and (b), respectively. The maximum error and average error of the second group design results are −0.7896° and −0.4872°, respectively. The running time of the synthesis process, including database establishment and matching recognition, is 253.067576 s, which is implemented in MATLAB 2014a on a computer running Window 7 with an Intel processor (Xeon E5-1630 v4) with velocity of 3.7 GHz and 16 GB of memory. The numerical atlas method is also effective for the other type output function curve for function synthesis.

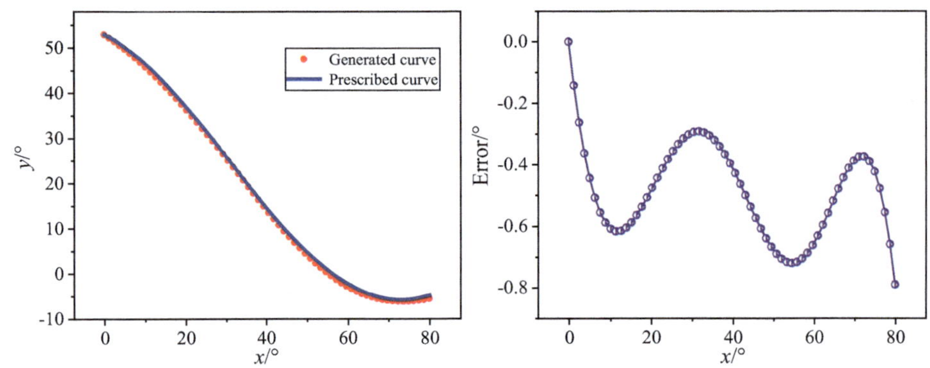

(a) The comparison graph (b) The error graph

Fig. 4. The comparison and error graphs for the second group design results of example one

Table 2. Design results of example one.

Input angle and BDTs	First group	Second group	Third group
a_1/cm	11	11	6
a_2/cm	26	31	26
a_3/cm	16	26	21
a_4/cm	11	16	11
S_1/cm	21	1	21
S_4/cm	15	15	15
α_1/°	130	85	166
θ_1^1/°	179	137	155
$\Delta\theta$/°	−109.7308	−83.4835	−72.2077
δ/°	0.0068	0.0094	0.0097

5 Conclusion

In this study, a new method for the non-periodic function generation of a spatial RSSR mechanism is proposed using a Daubechies wavelet. The mathematical model for the output function of an RSSR mechanism is established. The output function curve is described by wavelet coefficients. Based on the relationship between original curve and reconstruction curve generated by WFPs, a numerical atlas database is established. Then, the non-periodic function generation of the RSSR mechanism is realized.

The WFP method proposed in this paper is applicable for non-periodic function generation of the spatial RSSR mechanism with multiple positions. Moreover, the method overcomes the Fourier series method's inability to describe non-periodic function curve accurately. Since the problem of nonlinear equations solving and initialization selection are avoided, the proposed method can provide design results without branch, order and circuit defects, efficiently. In future work, we plan to extent the WFP method to dimensional synthesis of other spatial linkage mechanism.

References

1. Freudenstein, F.: Approximate synthesis of four-bar linkages. Trans. ASME **77**, 853–861 (1955)
2. Bai, S., Angeles, J.: A unified input–output analysis of four-bar linkages. Mech. Mach. Theor. **43**(2), 240–251 (2008)
3. Yang, S., Yu, H., Tian, G.Y.: Optimal selection of precision points for function synthesis of spherical 4R linkage. Proc. Inst. Mech. Eng., Part C: J. Mech. Eng. Sci. **223**(9), 2183–2189 (2009)
4. Wang, D.L., Wang, S.F., Zhang, B.Y.: New approach to function synthesis of spherical mechansims. Chinese J. Mech. Eng. **40**(2), 45–49 (2004)
5. Rao, A.V.M., Sandor, G.N., Kohli, D., Soni, A.H.: Closed form synthesis of spatial function generating mechanism for the maximum number of precision points. J. Eng. Ind. **95**(3), 725–736 (1973)
6. Dhall, S., Kramer, S.N.: Design and analysis of the HCCC, RCCC, and PCCC spatial mechanisms for function generation. J. Mech. Des. **112**(1), 74–78 (1990)
7. Sun, J.W., Chu, J.K.: Fourier method to function synthesis of an RCCC mechanism. IMechE Part C: J. Mech. Eng. Sci. **223**(2), 503–513 (2009)
8. Wu, J., Ge, Q.J., Gao, F., Guo, W.Z.: On the extension of a Fourier descriptor based method for planar four-bar linkage synthesis for generation of open and closed paths. J. Mech. Robot. **3**(3), 031002 (2011)
9. Unruh, V., Krishnaswami, P.: A computer-aided design technique for semi-automated infinite point coupler curve synthesis of four-bar linkages. J. Mech. Des. **117**(1), 143–149 (1995)
10. Galán-Marín, G., Alonso, F.J., Castillo, J.M.D.: Shape optimization for path synthesis of crank-rocker mechanisms using a wavelet-based neural network. Mech. Mach. Theor. **44**(6), 1132–1143 (2009)
11. Yu, H.Y., Tang, D.W., Wang, Z.X.: Study on a new computer path synthesis method of a four-bar linkage. Mech. Mach. Theor. **42**(4), 383–392 (2007)
12. Lin, S., Wang, H.C., Liu, J., Zhang, Y.: Geometric method of spatial linkages synthesis for function generation with three finite positions. J. Mech. Des. **140**(8), 082303 (2018)

13. Mullineux, G.: Atlas of spherical four-bar mechanisms. Mech. Mach. Theor. **46**(11), 1811–1823 (2011)
14. Sun, J.W., Lu, H., Chu, J.K.: Variable step-size numerical atlas method for path generation of spherical four-bar crank-slider mechanism. Inverse Prob. Sci. Eng. **23**(2), 256–276 (2015)
15. Li, X., Wu, J., Ge, Q.J.: A Fourier descriptor-based approach to design space decomposition for planar motion approximation. J. Mech. Robot. **8**(6), 699–708 (2016)
16. McGarva, J.R., Mulineux, G.: Harmonic representation of closed curves. Appl. Math. Model. **17**(4), 213–218 (1993)
17. McGarva, J.: Rapid search and selection of path generating mechanisms from a library. Mech. Mach. Theor. **29**(2), 223–235 (1994)
18. Ullah, I., Kota, S.: Optimal synthesis of mechanisms for path generation using fourier descriptors and global search method. J. Mech. Des. **119**(4), 504–510 (1997)
19. Nie, X.C., Krovl, V.: Fourier methods for kinematic synthesis of coupled serial chain mechanisms. J. Mech. Des. **127**(2), 232–241 (2005)
20. Sun, J.W., Liu, W.R., Chu, J.K.: Dimensional synthesis of open path generator of four-bar linkage using the Haar wavelet. J. Mech. Des. **137**(8), 082303 (2015)

Kinematic Acquisition of Quadratic Curve Constraints for Finite Position Generation of Planar Mechanisms

Hao Lv, Ke Shi, and Xiangyun Li(✉)

Southwest Jiaotong University, Chengdu 610031, China
xiangyunli.app@gmail.com

Abstract. This paper presents a quadratic constraint acquisition method to address the problem of finite position generation of planar mechanisms, with the benefit of simultaneous determination of type and dimensions. The key to this approach is the development of general mathematical formulations of quadratic curve constraints that are not directly dependent on the complete choice of a planar mechanism type. The Homotopy algorithm is applied to extract the geometric constraints and then the type and dimensions of their corresponding 1-DOF mechanisms can be obtained. Two examples are provided at the end of the paper to demonstrate the validity of the proposed method.

Keywords: Quadratic curve constraint · Position generation · Planar mechanism

1 Introduction

Ever since the development of modern kinematics, mechanism synthesis for motion generation has always been widely studied by many researchers [1–3]. The existing and long-standing approach is to decompose the synthesis problem into type synthesis–the selection of mechanism type for a given task requirement and dimensional synthesis–the determination of dimensions of a selected mechanism type from the numerical aspect of the task requirement.

The current state of art in kinematics and mechanism design is predicated on creating artificial boundaries between planar mechanisms that apparently facilitate selection of the type of linkage mechanisms during the synthesis process. However, this convenience comes at a great price of picking the wrong type at an early stage of the design process. Before carrying out the dimensional synthesis, one has to reach at a decision regarding the type of mechanism to be employed for a specified motion requirement. While dimensional synthesis has been a subject amenable to mathematical treatments, identifying a mechanism type that matches with a given motion task is often driven by the designer's past kinematic experience. Therefore, in order to bridge the gap between type and dimensional synthesis, researchers have been working towards combining

D. Wang et al. (Eds.): MeTrApp 2019, MMS 79, pp. 21–31, 2020.
https://doi.org/10.1007/978-981-15-0142-5_3

the two processes that is driven by the analysis of a given task. Hayes [4] earlier adopted kinematic-mapping theory to carry out simultaneous type and dimensional synthesis of four-bar linkages for motion approximation problem. Lately, Ge and Zhao [5–7] combined the kinematic-mapping theory with algebraic fitting to provide a new integrated synthesis method for the motion synthesis problem of planar four-bar and six-bar linkages.

However, the aforementioned works mainly focus on four-bar or six-bar linkages with acquisition of simple geometric constraints of line and circle. In this paper, the acquisition of quadratic constraints can be used for the type and dimensional determination of planar mechanisms that go well beyond the limitation of four-bar or six-bar linkages. In addition, we present a unified framework for generating mechanism motion for a combination of positions and engineering constraint (installation position).

2 Basic Geometric Constraint

According to the kinematic characteristics of planar mechanisms, the constraint elements can be divided into points, straight-lines, circles, circular-arcs, ellipses, hyperbolas, parabolas, etc. Moreover, the mathematical expressions of these constraints can be obtained easily from plane geometry. Figure 1 shows that one point can be constrained by straight-line, circle, circular-arc, ellipse, hyperbola, and parabola, and also enumerates some 1-DOF mechanisms, and they are classified based on the type of geometric constraints they generated. For example, a circle or a circular-arc can be easily realized with a revolute joint. Ref. [8] has given out a comprehensive discussion on the generation of planar algebraic curves utilizing planar mechanisms.

3 Unified Mathematical Model for Mixed Poses and Practical Constraints

As shown in Fig. 2, the planar motion of a rigid body can be represented as $d(d_x, d_y)$ and θ, where d denotes the Cartesian coordinates of a point on the moving rigid body, and θ represents the rotational angle of the rigid body. Then for given planar poses represented as a group of Cartesian space parameters (d_x, d_y, θ), the planar displacement of the rigid body can be expressed with homogeneous coordinates as

$$[X_1\ X_2\ X_3]^T = \mathbf{H}[x_1\ x_2\ x_3]^T = \begin{bmatrix} cos\theta & -sin\theta & d_x \\ sin\theta & cos\theta & d_y \\ 0 & 0 & 1 \end{bmatrix} [x_1\ x_2\ x_3]^T \tag{1}$$

Also the algebraic equation of a general quadratic curve is

$$a_0x^2 + a_1y^2 + a_2xy + a_3x + a_4y + a_5 = 0 \tag{2}$$

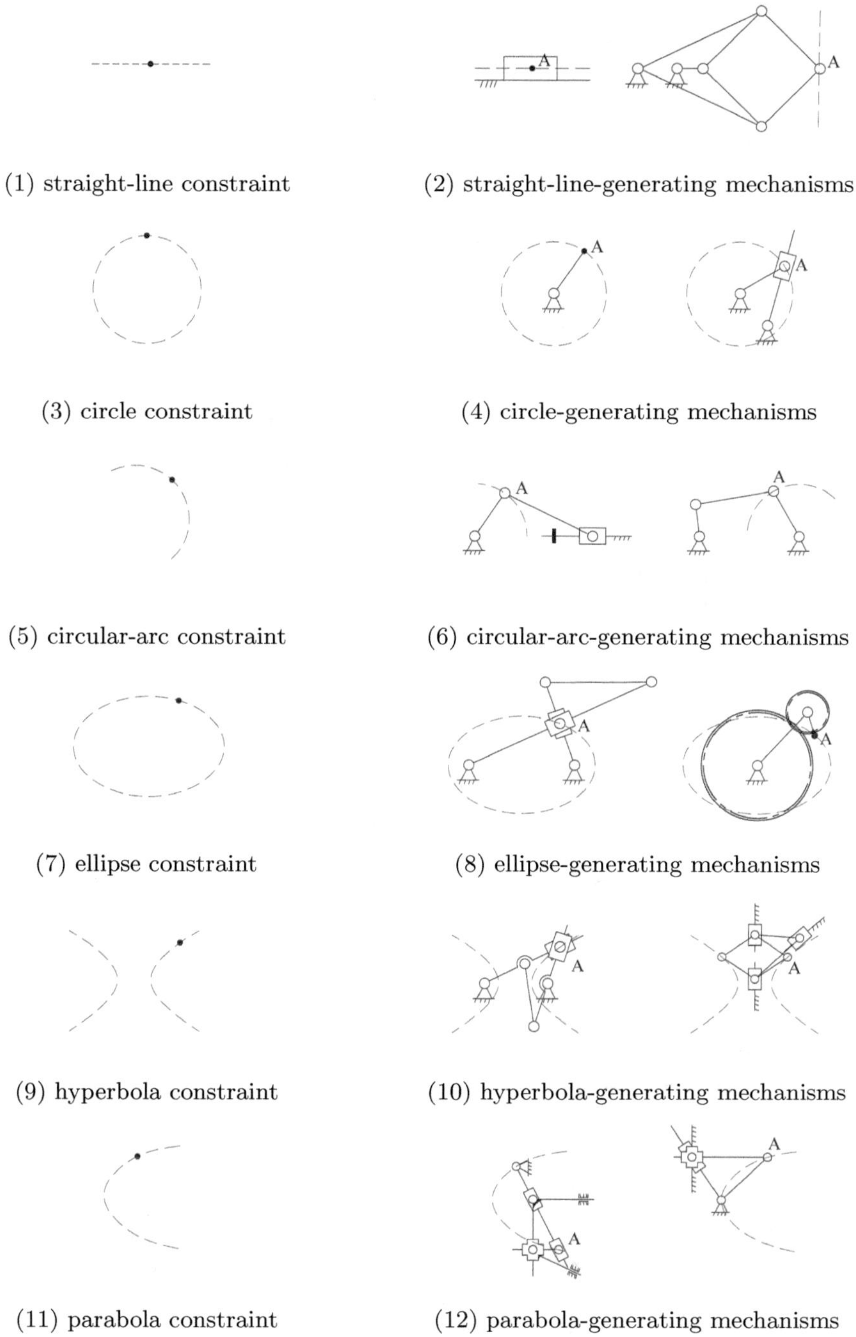

Fig. 1. Geometric constraints and their generating mechanisms

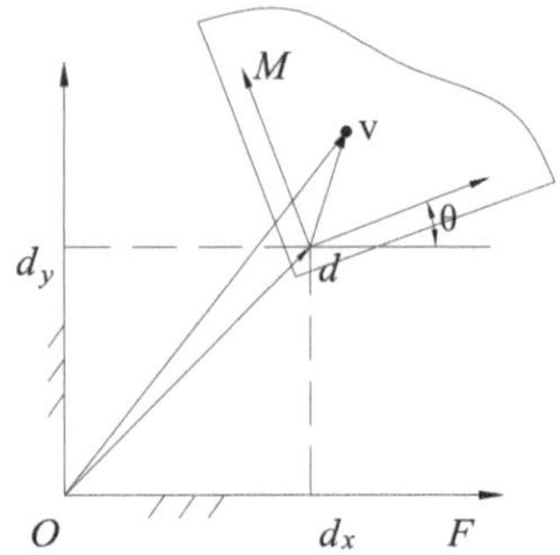

Fig. 2. A planar displacement

The corresponding expression using homogeneous coordinate is

$$a_0X_1^2 + a_1X_2^2 + a_2X_1X_2 + a_3X_1X_3 + a_4X_2X_3 + a_5X_3^2 = 0 \tag{3}$$

It's clear that Eq. (3) contains 9 unknowns, including 6 unknown coefficients $a_0 \sim a_5$ and 3 position coordinates $X_1 \sim X_3$. For planar mechanisms, let $X_3 = x_3 = 1$; for quadratic curves, since the constant term does not change the character of the curve, but only affects the position. For the convenience of calculation, let $a_5 = 1$. So Eq. (3) can be simplified as

$$a_0X_1^2 + a_1X_2^2 + a_2X_1X_2 + a_3X_1 + a_4X_2 + 1 = 0 \tag{4}$$

According to Eq. (1) and the above discussion

$$\begin{cases} X_1 = \cos\theta x_1 - \sin\theta x_2 + d_x \\ X_2 = \sin\theta x_2 + \cos\theta x_1 + d_y \\ X_3 = 1 \end{cases} \tag{5}$$

Substitute X_1, X_2, X_3 into Eq. (3) to get the constraint equation for the prescribed pose as follows

$$\begin{aligned} &a_0d_x^2 + a_2d_xd_y + a_3d_x + a_1d_y^2 + a_4d_y + (2a_0x_1 + a_2x_2)\, d_x\cos\theta \\ &+ (a_2x_1 - 2a_0x_2)\, d_x\sin\theta + (2a_1x_2x + a_2x_1)\, d_y\cos\theta \\ &+ (2a_1x_1 - a_2x_2)\, d_y\sin\theta + \left(a_1x_2^2 + a_2x_1x_2 + a_0x_1^2 + 1\right)\cos^2\theta \\ &+ \left(a_2x_1^2 - a_2x_2^2 - a_0x_1x_2 + a_1x_1x_2\right)\cos\theta\sin\theta \\ &+ (a_3x_1 + a_4x_2)\cos\theta + \left(a_1x_1^2 - a_2x_1x_2 + a_0x_2^2 + 1\right)\sin^2\theta \\ &+ (a_4x_1 - a_3x_2)\sin\theta = 0 \end{aligned} \tag{6}$$

Since Homotopy algorithm is adopted for the solution, according to the characteristics of the algorithm, the number of equations and the number of unknowns in the system must be equal, so the number of prescribed poses should not exceed seven. Therefore, this paper mainly studies the situations that the number of prescribed poses is five or six.

When the number of prescribed poses is six or five, the number of constraint equations obtained from the prescribed poses is also six or five(less than seven), and it is necessary to increase constraint equation to solve the unknown parameters. It can be seen from the Sect. 2 that the positions of fixed pivots of the generating mechanism are often determined by the positions of the focus, vertex and geometric center of the curve. Hence, additional constraints of mechanism installing positions could be imposed on the planar motion to construct the equation system.

Six Poses. For the six poses, the corresponding homogeneous coordinates are

$$\begin{cases} X_{1i} = \cos\theta_i x_1 - \sin\theta_i x_2 + d_{xi} \\ X_{2i} = \sin\theta_i x_2 + \cos\theta_i x_1 + d_{yi} \\ X_{3i} = 1 \end{cases} \tag{7}$$

Therefore, the system to be solved is

$$\begin{cases} a_0 X_{11}^2 + a_1 X_{21}^2 + a_2 X_{11} X_{21} + a_3 X_{11} + a_4 X_{21} + 1 = 0 \\ a_0 X_{12}^2 + a_1 X_{22}^2 + a_2 X_{12} X_{22} + a_3 X_{12} + a_4 X_{22} + 1 = 0 \\ \vdots \\ a_0 X_{1i}^2 + a_1 X_{2i}^2 + a_2 X_{1i} X_{2i} + a_3 X_{1i} + a_4 X_{2i} + 1 = 0 \\ f(a_0, \ldots, a_4) = 0 \end{cases} \tag{8}$$

where $i = 1, 2, \ldots, 6$; $f(a_0, \ldots, a_4) = 0$ is the constraint equation for the mechanism installing position.

Five Poses. For the five poses, the corresponding homogeneous coordinates are the same as Eq. (7). And the system to be solved is

$$\begin{cases} a_0 X_{11}^2 + a_1 X_{21}^2 + a_2 X_{11} X_{21} + a_3 X_{11} + a_4 X_{21} + 1 = 0 \\ a_0 X_{12}^2 + a_1 X_{22}^2 + a_2 X_{12} X_{22} + a_3 X_{12} + a_4 X_{22} + 1 = 0 \\ \vdots \\ a_0 X_{1i}^2 + a_1 X_{2i}^2 + a_2 X_{1i} X_{2i} + a_3 X_{1i} + a_4 X_{2i} + 1 = 0 \\ f_1(a_0, \ldots, a_4) = 0 \\ f_2(a_0, \ldots, a_4) = 0 \end{cases} \tag{9}$$

where $i = 1, 2, \ldots, 5$; $f_1(a_0, \ldots, a_4) = 0$ and $f_2(a_0, \ldots, a_4) = 0$ are the constraint equations for mechanism installing positions.

In this paper, HOM4PS2 [9], which is a polyhedral homotopy solver, is used in the actual solving process. In addition to HOM4PS2, there are other free homotopy packages, including PHCpack [10], POLSYSGLP [11], Bertini [12], etc.

4 Constraint Equation for Mechanism Installation Position

According to Sect. 3, when the number of prescribed poses is seven, a finite number of solutions satisfying the design conditions can be obtained from the constraint equations directly, and then the corresponding generating-mechanism can be obtained. When the number of prescribed poses is less than seven, the number of equations is less than the number of unknown parameters, and there are infinite solutions. In this case, additional constraints should be imposed on the coordinates of special points(geometric center, focus, vertex, ...) for different geometric constraints, thus limiting the mechanism installing positions. Therefore, in the design of planar mechanisms, not only the prescribed poses are realized, but also the problem that the constraint of the mechanism installing position is considered. The motion synthesis problem of five and six poses is also transformed into the motion synthesis problem for mixed-task which includes five poses or six poses. Now we give the coordinates of the special points of each quadratic curve.

Circle: From Eq.(2), the coordinates of the circle center can be expressed as

$$\begin{cases} x_o = -\frac{a_3}{2a_0} \\ y_o = -\frac{a_4}{2a_0} \end{cases} \tag{10}$$

Ellipse: For ellipse, the fixed pivots are generally located at the two foci or the geometric center. Therefore, the coordinates of the two foci and geometric center of an ellipse are

$$\begin{cases} x_F = \dfrac{a_2a_4 - 2a_1a_3 \pm \sqrt{2T\left(a_1 - a_0 + \sqrt{(a_1-a_0)^2 + a_2^2}\right) - a_2^2}}{4a_0a_1 - a_2^2} \\ y_F = \dfrac{a_2a_3 - 2a_0a_4 \mp \frac{|a_2|}{a_2}\sqrt{2T\left(a_0 - a_1 + \sqrt{(a_1-a_0)^2 + a_2^2}\right) - a_2^2}}{4a_0a_1 - a_2^2} \end{cases} \tag{11}$$

$$\begin{cases} x_o = \frac{a_2a_4 - 2a_1a_3}{4a_0a_1 - a_2^2} \\ y_o = \frac{a_2a_3 - 2a_0a_4}{4a_0a_1 - a_2^2} \end{cases} \tag{12}$$

where $T = a_0a_4^2 + a_1a_3^2 - a_2a_3a_4 + a_5(a_2^2 - 4a_0a_1)$.

Hyperbola: For hyperbola, the fixed pivots also install on the two foci or geometric center. So coordinates of the foci and geometric center of a hyperbola are

$$\begin{cases} x_F = \dfrac{a_2a_4 - 2a_1a_3 \pm \sqrt{2T\left(a_1 - a_0 + \sqrt{(a_1-a_0)^2 + a_2^2}\right)}}{4a_0a_1 - a_2^2} \\ y_F = \dfrac{a_2a_3 - 2a_0a_4 \mp \frac{|a_2|}{a_2}\sqrt{2T\left(a_0 - a_1 + \sqrt{(a_1-a_0)^2 + a_2^2}\right)}}{4a_0a_1 - a_2^2} \end{cases} \tag{13}$$

$$\begin{cases} x_o = \frac{a_2a_4 - 2a_1a_3}{4a_0a_1 - a_2^2} \\ y_o = \frac{a_2a_3 - 2a_0a_4}{4a_0a_1 - a_2^2} \end{cases} \tag{14}$$

Parabola: For generating a parabola trajectory, the fixed pivots usually are located at the focus or vertex. Hereby, the coordinates of the focus and vertex of a parabola are

$$\begin{cases} x_F = \dfrac{\sqrt{a_1}\left[a_3^2 + a_4^2 - 4a_3(a_0+a_1)\right]}{4\left(a_3\sqrt{a_1} - \frac{|a_2|}{a_2}a_4\sqrt{a_0}\right)(a_0+a_1)} - \dfrac{a_3}{2(a_0+a_1)} \\ y_F = \dfrac{\sqrt{a_0}\left[a_3^2 + a_4^2 - 4a_5(a_0+a_1)\right]}{4\left(a_4\sqrt{a_0} - \frac{|a_2|}{a_2}a_3\sqrt{a_1}\right)(a_0+a_1)} - \dfrac{a_4}{2(a_0+a_1)} \end{cases} \tag{15}$$

$$\begin{cases} x_v = \frac{-2PQ\sqrt{a_0}+\frac{|a_2|}{a_2}(P^2-4a_5)\sqrt{a_1}}{4Q(a_0+a_1)} \\ y_v = \frac{-2\frac{|a_2|}{a_2}PQ\sqrt{a_1}-(P^2-4a_5)\sqrt{a_0}}{4Q(a_0+a_1)} \end{cases} \tag{16}$$

where $P = \frac{a_3\sqrt{a_0}+\frac{|a_2|}{a_2}a_4\sqrt{a_1}}{a_0+a_1}$, $Q = \frac{\frac{|a_2|}{a_2}a_3\sqrt{a_1}-a_4\sqrt{a_0}}{a_0+a_1}$.

Constraints of installing position. When these special points are on a straight line, their coordinates satisfy the linear equation

$$L_1x + L_2y + L_3 = 0 \tag{17}$$

By substituting the coordinates of the above special points, i.e., setting $x = x_o, x_F, x_v$ and $y = y_o, y_F, y_v$ respectively, the constraint equation for mechanism installing position can be acquired, that is $f(a_0, \dots, a_4) = 0$.

5 Examples

5.1 Six-Pose Synthesis

In this section, we give six prescribed poses and apply the above design method for mechanism synthesis. Table 1 lists the six prescribed poses.

Table 1. Six prescribed poses

i	d_{xi}	d_{yi}	θ_i
1	24.5693	4.9954	91.23°
2	23.2742	3.5226	81.11°
3	22.3892	−0.5111	73.04°
4	21.0779	−5.3117	72.37°
5	20.1899	−9.7120	77.77°
6	19.9957	−13.1093	87.50°

Since there are only six prescribed poses, we choose to constrain the center of the circle and the geometric center of the ellipse to make the fixed pivots of the generating mechanisms always be on the line $y = -4x + 75$. Substituting the coordinates of the geometric center of the ellipse(Eq. (12)) into the linear equation, the constraint equation for the installing position of the ellipse-generating mechanism can be obtained as

$$f(a_0, \dots, a_4) = 4a_2a_4 - 8a_1a_3 + a_2a_3 - 2a_0a_4 - 300a_0a_1 + 75a_2^2 = 0 \tag{18}$$

The coefficient equation system for this six poses example can be obtained by substituting Eq. (18) into Eq. (8). And there are 19 solutions to the coefficient equation system, among which there are two solutions that meet the requirements, as shown in Table 2.

The equations of the two geometric constraints can be obtained as

$$\begin{cases} 0.00087x^2 + 0.00087y^2 - 0.0435x + 0.0435y + 1 = 0 \\ 0.0045x^2 + 0.0062y^2 - 0.0031xy - 0.0875x - 0.0410y + 1 = 0 \end{cases} \tag{19}$$

According to these two geometric constraint equations, a planar mechanism can be designed as shown in Fig. 3, where the dotted lines are the geometric constraint curves. It can be seen that in order to meet the prescribed pose requirement, the planar motion of the mechanism is constrained by both a circle and an ellipse.

Table 2. Two valid solutions for six-pose synthesis

order	a_0	a_1	a_2	a_3	a_4	a_5
1	0.00087	0.00087	0	−0.0435	0.0435	1
2	0.0045	0.0062	−0.0031	−0.0875	−0.0410	1

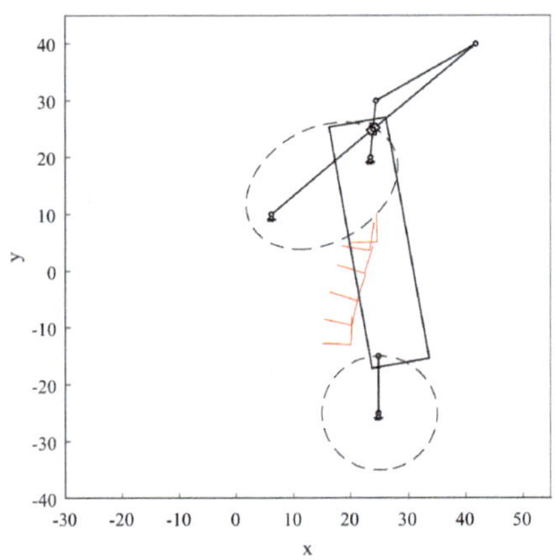

Fig. 3. The synthesized mechanism for six poses

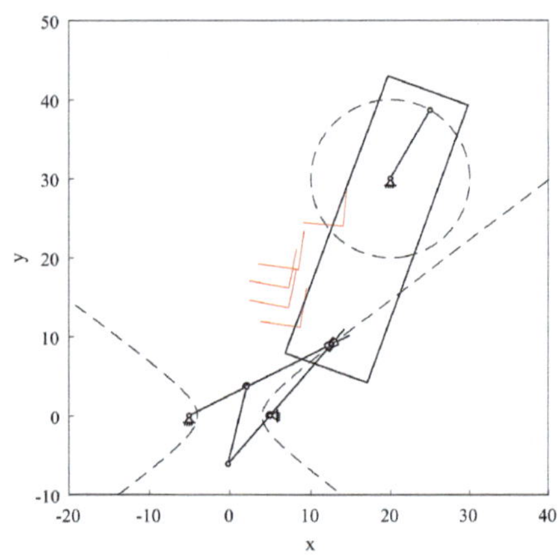

Fig. 4. The synthesized planar mechanism for five poses

5.2 Five-Pose Synthesis

In this section, we will apply the above design method for five-pose synthesis. Table 3 lists the five prescribed poses.

Table 3. Five prescribed poses

i	d_{xi}	d_{yi}	θ_i
1	13.9998	23.9425	84.77°
2	8.4020	18.4442	81.74°
3	7.2165	16.1040	79.17°
4	7.1967	13.6239	78.68°
5	8.6238	11.1009	81.41°

Because there are only five prescribed poses, we choose to constrain the center of the circle and the geometric center of the hyperbola to make the fixed pivots of the generating mechanisms always be on the line $y = 3x/2$. Substituting the coordinates of circle center (Eq. (10)) into the linear equation $y = 3x/2$, the constraint equation for the installing position of the circle-generating mechanism can be obtained as

$$f_1(a_0, a_1, \ldots, a_4) = 3a_3 - 2a_4 = 0 \tag{20}$$

the constraint equation for installing positions of the hyperbola-generating mechanism can be obtained by substituting Eq. (14) into the linear equation as:

$$f_2(a_0, \ldots, a_4) = 2a_2a_3 - 4a_0a_4 - 3a_2a_4 + 6a_1a_3 = 0 \tag{21}$$

The coefficient equation system for this example can be obtained by substituting Eqs. (20) and (21) into Eq. (9). And there are 15 solutions to the system, among which there are two solutions that meet the requirements, as shown in Table 4.

Table 4. Two valid solutions for five-pose synthesis

order	a_0	a_1	a_2	a_3	a_4	a_5
1	0.00083	0.00083	0	−0.0333	−0.0500	1
2	−0.0625	0.1111	0	0	0	1

The equations of the two geometric constraints can be obtained as

$$\begin{cases} 0.00083x^2 + 0.00083y^2 - 0.0333x - 0.0500y + 1 = 0 \\ -0.0625x^2 + 0.1111y^2 + 1 = 0 \end{cases} \tag{22}$$

According to these two geometric constraint equations, a planar mechanism can be designed as shown in Fig. 4. It can be seen that in order to meet the prescribed pose requirement, the planar motion of the mechanism is constrained by both a hyperbola and a circle.

6 Conclusion

In this paper, we mainly focus on the finite position generation of planar mechanism based on quadratic constraints acquisition. With the use of quadratic curves representing type and dimensions of a variety of planar mechanism, we manage to build a unified mathematical framework for realizing a combination of task poses and engineering constraints. Through Homotopy algorithm, the feasible quadratic constraints can be efficiently identified and then matched with their corresponding mechanisms, thus enabling the simultaneous type and dimensional synthesis. Finally, the feasibility of the method is verified by two synthesis examples of six poses and five poses.

Acknowledgment. The work has been financially supported by National Natural Science Foundation of China(Grant No. 51805449), Sichuan Science and Technology Program(Grant No. 2018HH0144), and the Fundamental Research Funds for Central Universities of China(Grant No. 2682017CX037). All findings and results presented in this paper are those of the authors and do not represent those of funding agencies.

References

1. Reuleaux, F., Kennedy, A.B.: The Kinematics of Machinery: Outlines of a Theory of Machines. Kessinger Publishing, Whitefish (1876)
2. Bottema, O., Roth, B.: Theoretical Kinematics, vol. 24. Courier Corporation, North Chelmsford (1990)
3. McCarthy, J.M., McCarthy, J.M.: An Introduction to Theoretical Kinematics, vol. 2442. MIT Press, Cambridge (1990)
4. Hayes, M., Luu, T., Chang, X.W.: Kinematic Mapping Application to Approximate Type and Demension Synthesis of Planar Mechanisms (2004).https://doi.org/10.1007/978-1-4020-2249-4_5
5. Ge, Q., Purwar, A., Zhao, P., Deshpande, S.: A task driven approach to unified synthesis of planar four-bar linkages using algebraic fitting of a pencil of g-manifolds. Journal of Computing and Information Science in Engineering (January 2015) (2013). https://doi.org/10.1115/1.4035528
6. Zhao, P., Li, X., Zhu, L., Zi, B., Ge, Q.: A novel motion synthesis approach with expandable solution space for planar linkages based on kinematic-mapping. Mech. Mach. Theor. **105**, 164–175 (2016). https://doi.org/10.1016/j.mechmachtheory.2016.06.021
7. Zhao, P., Li, X., Purwar, A., Ge, Q.J.: A task-driven unified synthesis of planar four-bar and six-bar linkages with r- and p-joints for five-position realization. J. Mech. Robot. **8**(6), 003–061 (2016). https://doi.org/10.1115/1.4033434
8. Artobolevskii, I.I.: Mechanisms for the Generation of Plane Curves. Elsevier, Amsterdam (2013)
9. Lee, T.L., Li, T.Y., Tsai, C.H.: Hom4ps-2.0: a software package for solving polynomial systems by the polyhedral homotopy continuation method. Computing **83**(2–3), 109 (2008)
10. Verschelde, J.: Algorithm 795: Phcpack: a general-purpose solver for polynomial systems by homotopy continuation. ACM Trans. Math. Softw. (TOMS) **25**(2), 251–276 (1999)

11. Su, H.J., McCarthy, J.M., Sosonkina, M., Watson, L.T.: Algorithm 857: Polsys_glp-a parallel general linear product homotopy code for solving polynomial systems of equations. ACM Trans. Math. Softw. (TOMS) **32**(4), 561–579 (2006)
12. Bates, D.J., Hauenstein, J.D., Sommese, A.J., Wampler, C.W.: Bertini: Software for numerical algebraic geometry (2006)

Path Synthesis of Crank-Rocker Mechanism Using Fourier Descriptors Based Neural Network

Xiaojuan Mo, Wenjie Ge(✉), Donglai Zhao, and Yaqing Zhang

School of Mechanical Engineering,
Northwestern Polytechnical University, Xi'an 710072, China
gwj@nwpu.edu.cn

Abstract. In this paper, the first five harmonics components of Fourier descriptors are used to describe the desired coupler curve with a high precision. Neural network is trained with the normalized Fourier descriptors of the coupler angle functions of relative length four-bar mechanisms and corresponding linkages lengths in the established database. Trained neural network is used to predict the relative linkage lengths of desired crank-rocker mechanism with the normalized Fourier descriptors of the desired coupler curve. With the inner connection between harmonics components of the desired coupler curve and the coupler angle function of predicted relative lengths of four-bar mechanism, the linkage lengths, coupler and configuration parameters of the desired mechanism are calculated. A group of desired coupler trajectories with known mechanism configurations are generated randomly and are used to test the proposed method. The results show that with generated neural network, a high precision four-bar mechanism can be obtained without searching the whole database.

Keywords: Fourier descriptors · Neural network · Path synthesis · Crank-rocker mechanism

1 Introduction

Dimensional synthesis consists of function generation problem, path generation problem and motion generation problem [1]. Over last fifty years, various methods have been developed for solving dimension synthesis problems [2]. The path synthesis problem is to generate a specific mechanism of which the coupler can trace a series of desired points with prescribed timing or not [1]. Numerous path synthesis methods developed for path synthesis might be separated into two different categories, one-phase synthesis and two-phase synthesis. One-phase synthesis attempts to optimize the shape, size, location and orientation information of desired trajectory simultaneously, while two-phase synthesis firstly focuses on the shape of desired trajectory and secondly realizes the scaling, translation and rotation of desired trajectory [3, 4].

One-phase synthesis normally uses the summation the squared position errors between a set of target points that should be passed by the coupler of the desired four-bar mechanism and a set of positions of the coupler of the designed four-bar mechanism as optimization goal. The shape optimization is firstly realized in two-phase path

D. Wang et al. (Eds.): MeTrApp 2019, MMS 79, pp. 32–41, 2020.
https://doi.org/10.1007/978-981-15-0142-5_4

synthesis of four-bar mechanism. B-spline, Fourier descriptors, wavelet descriptors et al. are used to describe the shape of a curve [5–9]. McGarva and Mullineux firstly used normalized Fourier descriptors to describe the shape of coupler curve of four-bar mechanism [6], established a library consists of different mechanisms. The established library can realize rapid search and selection of path generating mechanisms by comparing the normalized Fourier descriptors of desired coupler curve and all stored mechanism inside library. The shape of coupler curve is optimized by a global search method with an effective objective function based on Fourier descriptors and the size, orientation and location are obtained lately by determining analogous points on the desired and candidate curves [10]. Wavelet analysis is firstly applied to path synthesis problem of coupler curve of four-bar mechanism [11]. The lower frequency wavelet coefficients are applied to represent the shape of coupler curve and an electronic atlas is established. By comparing the closeness of wavelet coefficients of the desired coupler curve and that of the established atlas, a series of possible mechanism can be predicted efficiently. Smaili and Diab proposed a method combined cyclic angular deviation vector and ant-search algorithm to optimize the shape of desired curve firstly and realize the scaling, rotation, and translation of obtained mechanism with a simple mathematical approach [12]. Wu et al. proposed finite Fourier descriptors and extended the path synthesis based on Fourier descriptors to open curves [13].

The artificial neural network (ANN) is firstly applied to solve the inverse problem of path synthesis problem which can provide an "fine-tuned" initial value for the following gradient-based optimization [14]. The power spectrum of a curvature plot of the curve is used to provide a scale- and orientation independent input representation and is used as the input of the proposed network. This research reveals the ability of ANN in solving inverse modeling of multi-dimensional, highly nonlinear systems [15]. Fourier descriptors [16] and wavelet analysis [17] are used to solve shape optimization for path synthesis of crank-rocker mechanisms using neural network and their performance are compared.

This paper presents the path synthesis with Fourier descriptors. Neural network is trained with the normalized Fourier descriptors of the coupler curve functions of relative length four-bar mechanisms and corresponding linkages lengths in the established database. Trained neural network is used to predict the relative linkage lengths of desired crank-rocker mechanism with the normalized Fourier descriptors of the desired coupler curve. The results show that with generated neural network, a high precision four-bar mechanism can be obtained using the trained neural network.

2 Fourier Descriptors for Path Synthesis

2.1 Fourier Descriptors of Closed Coupler Curve

Due to the full-rotation property of crank-rocker mechanism, the coupler curve of crank-rocker mechanism is a periodic closed curve. The trajectory function can be described as

$$F(t) = x(t) + \boldsymbol{i}y(t) \tag{1}$$

where $\mathbf{i} = \sqrt{-1}$, $x(t)$ is the function of x coordinates of coupler curve, $y(t)$ is the function of y coordinates of coupler curve, F(t) is the specified complex coordinates function of the prescribed coupler curve.

With Fourier transform, the trajectory the coupler of four-bar mechanism can be descripted with the first few harmonics including amplitude and phase parameters.

$$F(t) = \sum\nolimits_{n=-\infty}^{+\infty} D_n(\cos(\tau + nwt) + \boldsymbol{i}sin(\tau + nwt)) = \sum\nolimits_{n=-\infty}^{+\infty} D_n e^{i\tau} e^{inwt} \tag{2}$$

where w is fundamental frequency in Fourier series, D_n is the amplitude parameter of nth harmonic and τ is the corresponding initial phase angle of nth harmonic.

To illustrate the efficiency of Fourier, transform in describing the coupler curve, we picked out 64 points from each coupler curve and applied the fast Fourier transform (FFT) to those points. Using the first five harmonics, those discrete points coordinates are calculated with inverse Fourier transform. The results show that with first five harmonics, the 64 discrete points on one coupler curve can be described with 22 harmonics parameters with a high precision.

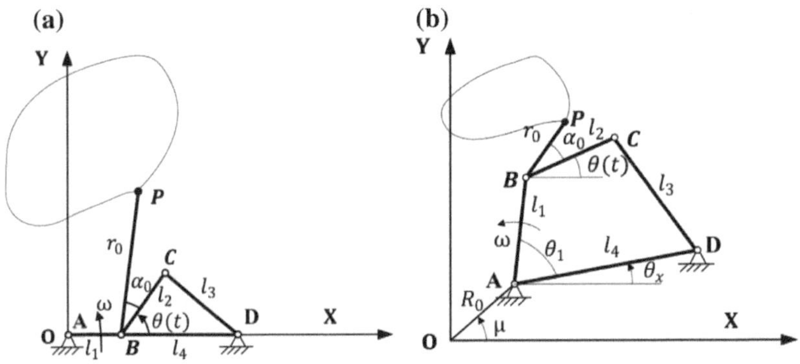

Fig. 1. (a) The four-bar mechanism under standard configuration and corresponding coupler trajectory; (b) The four-bar mechanism under general configuration and corresponding coupler trajectory.

2.2 Fourier Descriptors of Closed Coupler Curve Under Standard and Normal Configurations

As illustrated in Fig. 1(a), the lengths of four linkages in four-bar mechanism is (l_1, l_2, l_3, l_4). The position of joint B is described by $r_{\mathbf{B}}(t)$. In coordinate system that the located origin in joint B and the x axis coincides with coupler BC, the coordinates of P is $r_0 e^{i\alpha_0}$. When the crank AB rotate, the coordinates of P in coordinate system XOY can be described as

$$\boldsymbol{r_P}(t) = \boldsymbol{r_B}(t) + r_0 e^{\mathrm{i}[\alpha_0 + \theta(t)]} = \boldsymbol{r_B}(t) + r_0 e^{\mathrm{i}\alpha_0} \cdot e^{\mathrm{i}\theta(t)} \tag{3}$$

where $e^{\mathrm{i}\theta(t)}$ is a periodic function as the angle function $\theta(t)$ of BC with independent variable t, as both are decided by the lengths of four linkages in four-bar mechanism. If we normalized the linkages length of four-bar mechanism, then the normalized linkages length show one-to-one correspondent with angle function $\theta(t)$ and $e^{\mathrm{i}\theta(t)}$.

Figure 1(a) shows the standard configuration of four-bar mechanism, where (1) joint A locates in origin O; (2) frame linkage AD coincides with x axis; (3) the initial angle of crank AB corresponding to AD is zero. The Fourier transform results of $\boldsymbol{r_B}(t)$ only contains zero and first harmonics due to joint B rotates relative to joint A. Under standard configuration, due to joint A coincides with O and AB coincides with AD, then Fourier descriptor of $\boldsymbol{r_B}(t)$ only contains first harmonic and the initial phase of first harmonic is zero. Fourier descriptor of $e^{\mathrm{i}\theta(t)}$ is

$$\begin{aligned} e^{\mathrm{i}\theta(t)} &= \sum\nolimits_{n=-\infty}^{+\infty} c_n e^{\mathrm{i}\varphi_n} e^{\mathrm{i}nwt} = c_0 e^{\mathrm{i}\varphi_0} + c_1 e^{\mathrm{i}\varphi_1} e^{wt} + c_{-1} e^{\mathrm{i}\varphi_{-1}} e^{-wt} \\ &\quad + c_2 e^{\mathrm{i}\varphi_2} e^{2wt} + c_{-2} e^{\mathrm{i}\varphi_{-2}} e^{-2wt} + \ldots + c_n e^{\mathrm{i}\varphi_n} e^{nwt} + \ldots \end{aligned} \tag{4}$$

Under standard configuration, Fourier descriptor of point P is

$$\boldsymbol{r_P}(t) = \boldsymbol{r_{AB}}(t) + \boldsymbol{r_{BP}}(t) = l_1 e^{\mathrm{i}wt} + r_0 e^{\mathrm{i}\alpha_0} e^{\mathrm{i}\theta(t)} \tag{5}$$

If crank AB rotates with an initial angle θ_{10}, then Fourier descriptor of point P is

$$\begin{aligned} \boldsymbol{r_P}\left(t + t'\right) &= l_1 e^{\mathrm{i}\theta_1} e^{\mathrm{i}wt} + r_0 e^{\mathrm{i}\alpha_0} (c_0 e^{\mathrm{i}\varphi_0} + c_1 e^{\mathrm{i}(\varphi_1 + \theta_{10})} e^{\mathrm{i}wt} \\ &\quad + c_{-1} e^{\mathrm{i}(\varphi_{-1} - \theta_{10})} e^{-\mathrm{i}wt} + c_2 e^{\mathrm{i}(\varphi_2 + 2\theta_{10})} e^{\mathrm{i}2wt} + c_{-2} e^{\mathrm{i}(\varphi_{-2} - 2\theta_{10})} e^{-\mathrm{i}2wt} \\ &\quad + \ldots + c_n e^{\mathrm{i}(\varphi_n + n\theta_{10})} e^{\mathrm{i}nwt} + \ldots) \end{aligned} \tag{6}$$

It is noted that the phase of nth harmonics has an initial angle of $n\theta_{10}$, while the trajectory of coupler point P keeps consistent with the four-bar mechanism under standard configuration.

Under general configuration, the whole four-bar mechanism rotate θ_x corresponding to x axis, position of joint A is $R_0 e^{\mathrm{i}u}$, and the initial angle of crank AB is θ_{10}, combined the zero harmonic and +1 harmonic components, we can obtain that

$$\begin{aligned} \boldsymbol{r_P}\left(t + t'\right) &= [R_0 e^{\mathrm{i}u} + r_0 c_0 e^{\mathrm{i}(\theta_x + \alpha_0 + \varphi_0)}] \\ &\quad + \left[l_1 e^{\mathrm{i}\theta_{10}} e^{\mathrm{i}\theta_x} + r_0 c_1 e^{\mathrm{i}(\varphi_1 + \theta_{10} + \theta_x + \alpha_0)}\right] e^{\mathrm{i}wt} \\ &\quad + r_0 e^{\mathrm{i}(\alpha_0 + \theta_x)} [c_{-1} e^{\mathrm{i}(\varphi_{-1} - \theta_{10})} e^{-\mathrm{i}wt} \\ &\quad + c_2 e^{\mathrm{i}(\varphi_2 + 2\theta_{10})} e^{\mathrm{i}2wt} + c_{-2} e^{\mathrm{i}(\varphi_{-2} - 2\theta_{10})} e^{-\mathrm{i}2wt} \\ &\quad + \ldots + c_n e^{\mathrm{i}(\varphi_n + n\theta_{10})} e^{\mathrm{i}nwt} + \ldots] \end{aligned} \tag{7}$$

Comparing each component of Eqs. (4) and (7), except the zero harmonic and +1 harmonics components, the nth harmonic component in Eq. (8) has a amplitude r_0 times of that in Eq. (4), and the initial angle increases $\theta_x + \alpha_0 + n\theta_{10}$. Based on this, even with the same linkage lengths (l_1, l_2, l_3, l_4), if the coupler parameter (r_0, α_0) differs, it will cause the shape and size of coupler become different; if $R_0 e^{iu}$ and θ_x differ, it will cause the location and orientation of coupler curve become different.

If the effect of (r_0, α_0) and $(R_0 e^{iu}, \theta_x, \theta_{10})$ on the coupler curve trajectory is removed, the Fourier descriptor of coupler point P is found to be connected with the Fourier descriptor of $e^{i\theta(t)}$ of the four-bar mechanism under standard configuration. Based on this, the Fourier descriptor of $e^{i\theta(t)}$ of the four-bar mechanism under standard configuration can be used to link one specific four-bar linkage type with a large amount of coupler trajectories of different shapes, sizes, locations and orientations. This finding helps to reduce the size of four-bar mechanism database.

In Eqs. (4) and (7), using −1 harmonic component to normalize other harmonic components, and multiply the amplitude with 100, Eqs. (4) and (7) can be illustrated as following equations. Comparing Eqs. (8) and (9), it is found that except zero and +1 harmonic components, the amplitude of nth components in Eq. (8) is the same as that in Eq. (9) and the phase angle of nth components in Eq. (9) increases $(n+1)\theta_{10}$ comparing to Eq. (8).

$$\sum\nolimits_{n=-\infty}^{+\infty} 100 \frac{c_n}{c_{-1}} e^{i(\varphi_n - \varphi_{-1})} e^{inwt} \tag{8}$$

$$\begin{aligned} &\sum\nolimits_{n \neq 0,1} 100 \frac{c_n}{c_{-1}} e^{i[(\varphi_n - \varphi_{-1}) + (n+1)\theta_{10}]} e^{inwt} \\ &+ \frac{100}{r_0 c_{-1}} \left[R_0 e^{i(u - \theta_x - \alpha_0)} + r_0 c_0 e^{i\varphi_0} \right] \\ &+ \frac{100}{r_0 c_{-1}} \left[l_1 e^{i(\theta_{10} - \alpha_0)} + r_0 c_1 e^{i(\varphi_1 + \theta_{10})} \right] e^{i(\theta_{10} - \varphi_{-1})} e^{iwt} \end{aligned} \tag{9}$$

Based on the derivation above, each known trajectory can be linked with the $e^{i\theta(t)}$ of the four-bar mechanism with desired relative linkage lengths. Then a database consists of a series of crank-rocker mechanisms and corresponding Fourier descriptors and normalized Fourier descriptors of $e^{i\theta(t)}$ is established. The Fourier descriptor of desired trajectory and normalized the Fourier descriptor are calculated based on Eq. (10), then the normalized Fourier descriptor of goal trajectory is compared with the normalized Fourier descriptors of $e^{i\theta(t)}$ of the mechanisms in the established database. When the normalized Fourier descriptor of desired trajectory and standard mechanism in the database are very similar, the four-bar mechanism selected from the database can be used to calculate the configurations of desired mechanism that can obtained a very close coupler trajectory as desired trajectory. In addition, we can use neural network to train the weight matrix and the threshold matrix used to establish a relation between normalized Fourier descriptors of $e^{i\theta(t)}$ of the four-bar mechanisms with relative linkage lengths in the database. With the established network, the normalized Fourier descriptor of goal trajectory can be used to predict the linkage lengths of standard four-bar mechanism that can be used to calculate the size and configuration parameters of desired four-bar mechanism.

3 Reduction of Design Space and Atlas Database

The four-bar mechanisms with linkages length of (l_1, l_2, l_3, l_4) and $k(l_1, l_2, l_3, l_4), k > 0$ have the same couple angle function $\theta(t)$ and the same Fourier descriptors of coupler angle function $e^{i\theta(t)}$. In order to reduce the database space, the linkage lengths are unified with following method. If the real linkages length of four-bar mechanism is (l_1, l_2, l_3, l_4) and the average of four linkages length is $L = (l_1 + l_2 + l_3 + l_4)/400$, then the relative lengths of four-bar mechanism are $a = l_1/L, b = l_2/L, c = l_3/L, d = l_4/L$ and the sum of the relative linkage lengths $a+b+c+d = 400$. To be one part of closed quadrilateral, each part of relative linkage lengths cannot be bigger than the sum of other three, then

$$0 < \boldsymbol{a}, \boldsymbol{b}, \boldsymbol{c}, \boldsymbol{d} < 200 \tag{10}$$

To obtain a crank-rocker mechanism with crank AB, the relative linkage lengths have to confirm to the Grashof condition listed below.

$$\begin{cases} \mathbf{a} + \mathbf{b} < \mathbf{c} + \mathbf{d} \\ \mathbf{a} + \mathbf{c} < \mathbf{b} + \mathbf{d} \\ \mathbf{a} + \mathbf{d} < \mathbf{b} + \mathbf{c} \end{cases} \tag{11}$$

Based on the theory of Roberts-Chebyshev, the coupler trajectory of the mechanism with relative linkage lengths of (a, b, c, d) can be obtained by two cognate mechanisms with relative linkages lengths of (a, c, b, d) and (b, a, c, d). To avoid repetition and reduce the size of the database, only one group of cognate mechanisms is kept $(a \le c \le b \le d)$ and those two corresponding cognate mechanisms can be obtained with the obtained linkages lengths.

With a step size of 1, 109428 groups of four-bar mechanisms with different linkages lengths combination and corresponding Fourier descriptors and normalized Fourier descriptors of corresponding coupler angle functions $e^{i\theta(t)}$ are obtained with FFT and are used as the training dataset of proposed neural network.

4 Neural Networks for Synthesizing Path Generating Mechanisms

4.1 Training of Desired Neural Network

The structure of the proposed neural network is the most commonly used a multilayer feed forward network with back propagation algorithm. The inputs of neural network are the amplitudes of the normalized harmonics components (except $0, \pm 1$ harmonics) of desired coupler trajectory, and the outputs of the neural network are the relative linkages lengths of the standard four-bar mechanism. In the established neural network, 2 hidden layers of 15 neurons each are adopt, then an 8-15-15-4 neural network is established. A linear transfer function is applied to the input and output neurons to avoid a scaling procedure. For the two hidden layers, the log-sigmoid function is used. The main difference of proposed neural network is that only the relative linkage lengths of the desired coupler trajectory are used as output and only the amplitudes of the normalized harmonics components (except $0, \pm 1$ harmonics) of desired coupler trajectory are used as inputs. Based on the derivation in Sect. 2.2, the strong relation between normalized harmonics amplitude of desired trajectory and the relative linkage lengths of standard mechanism are established, which makes the established neural network work in a simplified and efficient way.

For the learning process, the inputs of the neural network are the amplitudes of the normalized harmonics components (except $0, \pm 1$ harmonics) of the coupler angle function $e^{i\theta(t)}$ and the output is the corresponding linkages lengths of four-bar mechanism in the established database. The leaning and utilization phase of the established neural network is illustrated in Fig. 2(a) and (b). The routines of the Matlab neural network toolbox are then applied to train the network. With the established neural network, a set of relative linkage length of standard linkages lengths are predicted and the real size, orientation parameters of desired mechanism could be obtained with the Fourier descriptors of both the coupler angle function $e^{i\theta(t)}$ of the predicted four-bar mechanism and desired coupler curve.

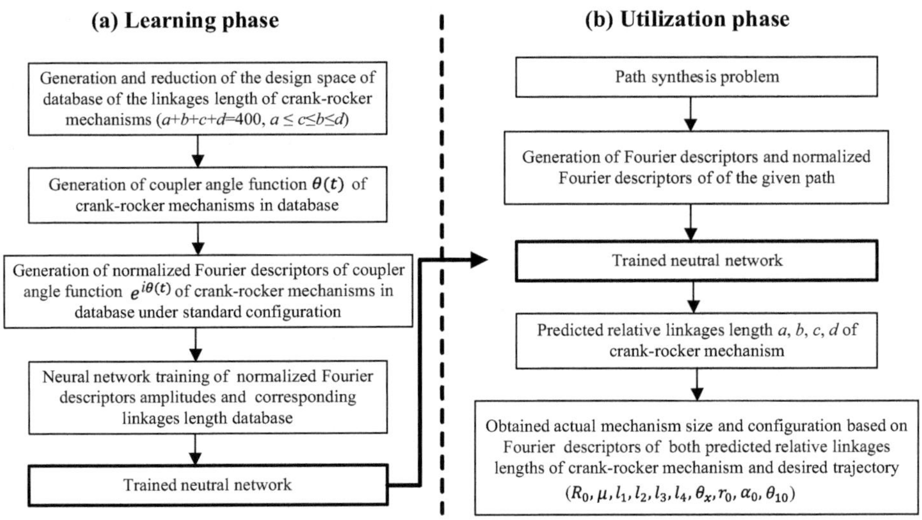

Fig. 2. (a) Scheme of the generation the training data set and learning process of the neutral network; (b) Scheme of the utilization phase of the trained neutral network and calculating the actual mechanism size and installation location.

4.2 Determination of Size, Coupler and Configuration Parameters of Desired Four-Bar Mechanism

The Fourier descriptors of desired coupler curve is consisted of amplitude D_n and phase angle τ_n. With the established neural network, the relative length of linkages (a, b, c, d) are obtained and corresponding Fourier descriptors of the coupler angle function $e^{i\theta(t)}$, consists of amplitude c_n and phase φ_n are obtained. Comparing the harmonics components in Eqs. (4) and (7), the linkage length (l_1, l_2, l_3, l_4), the coupler parameters (r_0, α_0) and configuration parameters $(R_0, u, \theta_x, \theta_{10})$ of the desired mechanism are obtained. The calculations procedures of above parameters are listed in Table 1. The proposed neural network combined the calculation can provide a nearly optimal crank-rocker mechanism solutions for path synthesis.

Table 1. The calculations procedures of size, coupler and configuration parameters of desired mechanism

Step	Basis	Equations
(1)	$\tau_{-1} = \alpha_0 + \theta_x + \varphi_{-1} - \theta_{10}$ $\tau_n = \alpha_0 + \theta_x + \varphi_n + n\theta_{10}, n \neq 0, 1, -1$	$\theta_{10} = \frac{1}{n+1}(\tau_n - \tau_{-1} - \varphi_n + \varphi_{-1})\, n \neq 0, 1, -1$
(2)	$r_0 c_{-1} e^{i(\alpha_0 + \theta_x + \varphi_{-1} - \theta_{10})} = D_{-1} e^{i\tau_{-1}}$	$r_0 = D_{-1}/c_{-1}$ $\alpha_0 + \theta_x = \tau_{-1} - \varphi_{-1} - \theta_{10}$
(3)	$R_0 e^{iu} + r_0 c_0 e^{i(\theta_x + \alpha_0 + \varphi_0)} = D_0 e^{i\tau_0}$	$R_0 = real(D_0 e^{i\tau_0} - r_0 c_0 e^{i(\theta_x + \alpha_0 + \varphi_0)})$ $u = imag(D_0 e^{i\tau_0} - r_0 c_0 e^{i(\theta_x + \alpha_0 + \varphi_0)})$
(4)	$l_1 e^{i\theta_{10}} e^{i\theta_x} + r_0 c_1 e^{i(\varphi_1 + \theta_{10} + \theta_x + \alpha_0)} = D_1 e^{i\tau_1}$	$l_1 = real(D_1 e^{i\tau_1} - r_0 c_1 e^{i(\varphi_1 + \theta_{10} + \theta_x + \alpha_0)})$ $\theta_x = imag(D_1 e^{i\tau_1} - r_0 c_1 e^{i(\varphi_1 + \theta_{10} + \theta_x + \alpha_0)}) - \theta_{10}$ $k = l_1/a, l_2 = kb, l_3 = kc, l_4 = kd$ $\alpha_0 = \alpha_0 + \theta_x - \theta_x$

4.3 Test of the Established Neural Network

In order to test the efficiency and accuracy of proposed method, 10000 coupler trajectories with known mechanism parameters are generated. The linkage lengths of mechanism are obtained by the same method adopted to generate the four-bar mechanism of relative linkage lengths with the step size changed and multiplied with a random positive number. The coupler and configuration parameters are randomly generated. The test trajectories group and corresponding Fourier descriptors and normalized Fourier descriptors are calculated and the obtained amplitudes first five normalized harmonics components (except 0, ±1 harmonics) are used as the input of trained neural network and the size, coupler and configuration parameters are obtained with the proposed method. By comparing the predicted mechanism parameters and known mechanism parameters, the relative prediction errors of each mechanism parameter are calculated and listed in Table 2. For example, if the desired crank length is l_{1d}, and predicted crank length is l_1, then the relative prediction error of crank is defined as

$$error_{l_1} = |l_1 - l_{1d}|/l_{1d} \tag{12}$$

5 Results

Artificial neural network offers a new approach to solve the approximate inverse prediction of crank-rocker mechanism parameters based on the known trajectories. Comparing to the Atlas method, the trained neutral network is consisted of some weight and threshold matrices and can offer a very precise prediction of mechanism parameters. While to obtain the same prediction precision, the atlas method have to store a large amount of dataset and compare all stored dataset with the desired coupler curve each time which makes it time consuming and hard to transfer to other platform or develop related software.

Table 2. Relative prediction errors of each mechanism parameter in the test dataset

Parameters	l_1	l_2	l_3	l_4	r_0
Relative error	(0.93 ± 1.23)%	(0.54 ± 0.84)%	(0.97 ± 1.27)%	(1.01 ± 1.29)%	(0.67 ± 1.03)%
Parameters	α_0	R_0	u	θ_x	θ_{10}
Relative error	(0.65 ± 1.12)%	(0.28 ± 0.64)%	(0.33 ± 0.85)%	(0.88 ± 1.39)%	(0.74 ± 1.2)%

6 Conclusions

The optimization can be separated into two steps: first one is to predict the relative linkages lengths of desired four-bar mechanism using only amplitudes of eight harmonics of normalized Fourier descriptors of desired coupler trajectories as the input of the trained neural network. The second step is to calculate the actual linkages length, coupler and configuration parameters of desired four-bar mechanism using the Fourier descriptors of desired coupler curve and coupler angle function of predicted relative linkages lengths. Based on the derivation in Sect. 2.2, Fourier descriptors of closed coupler curve under normal configurations show a strong relation with the Fourier descriptors of coupler angle function of relative linkage lengths. The normalized Fourier descriptors remove the effect of size, coupler and configuration on the coupler trajectories and linked with the relative linkage lengths of desired mechanism. Then the dataset used to train the neural network in this paper only contains amplitude of eight harmonics components of normalized Fourier descriptors of coupler angle function and the relative linkages lengths, which makes the neural network has a simplified structure of 8-15-15-4, comparing to the network structures proposed by Vasiliu and Yannou and are Galán-Marín et al. are 17-22-22-5 both. Using 10000 randomly generated coupler trajectories with known mechanism parameters, the prediction error of the proposed method are calculated and listed in Table 2. The proposed method shows a very high accuracy and efficiency comparing to the previous atlas dataset method, which makes it can be an alternative of one-phase optimization which is normally time consuming.

Acknowledgement. This work was supported in part by the National Key Research and Development Program of China under Grant 2017YFB1300102. Funders had no role in the study design, data collection and analysis, decision to publish or preparation of the manuscript.

References

1. Norton, R.L., Wang, S.S.-L.: Design of machinery: an introduction to the synthesis and analysis of mechanisms and machines. McGraw-Hill Higher Education, New York (2004)
2. Laribi, M.A., Mlika, A., Romdhane, L., Zeghloul, S.: A combined genetic algorithm–fuzzy logic method (GA–FL) in mechanisms synthesis. Mech. Mach. Theor. **39**, 717–735 (2004)
3. Lin, W.-Y.: Optimum path synthesis of a geared five-bar mechanism. Adv. Mech. Eng. **5**, 757935 (2013)
4. Lin, W.-Y.: Optimum synthesis of planar mechanisms for path generation based on a combined discrete Fourier descriptor. J. Mech. Robot. **7**, 41023 (2015)
5. Bulatović, R.R., Miodragović, G., Bošković, M.S.: Modified Krill Herd (MKH) algorithm and its application in dimensional synthesis of a four-bar linkage. Mech. Mach. Theory **95**, 1–21 (2016)
6. Chuang, G.-H., Kuo, C.-C.: Wavelet descriptor of planar curves: theory and applications. IEEE Trans. Image Process. **5**, 56–70 (1996)
7. McGarva, J., Mullineux, G.: Harmonic representation of closed curves. Appl. Math. Model. **17**, 213–218 (1993)
8. Osowski, S.: Fourier and wavelet descriptors for shape recognition using neural networks—a comparative study. Pattern Recognit. **35**, 1949–1957 (2002)
9. Skoglund, K.: Fourier Series, the DFT and Shape Modelling, Month, 16 (2004)
10. Ullah, I., Kota, S.: Optimal synthesis of mechanisms for path generation using fourier descriptors and global search methods. J. Mech. Des. **119**, 504–510 (1997)
11. Yuebo, W.C.J., Daoheng, S.: Research on applying wavelet analysis to path synthesis for coupler curves of planar four-bar linkages. Chin. J. Mech. Eng. **8**, 6 (2004)
12. Smaili, A., Diab, N.: A new approach to shape optimization for closed path synthesis of planar mechanisms. J. Mech. Des. **129**, 941–948 (2007)
13. Wu, J., Ge, Q.J., Gao, F., Guo, W.Z.: On the extension of a Fourier descriptor based method for planar four-bar linkage synthesis for generation of open and closed paths. J. Mech. Robot. **3**, 31002 (2011)
14. Hoskins, J.C., Kramer, G.A.: Synthesis of mechanical linkages using artificial neural networks and optimization. In: 1993, IEEE International Conference Neural Networks, IEEE, p. 822 (1993)
15. Hoskins, J.C., Kramer, G.A.: System and method of global optimization using artificial neural networks (1994)
16. Vasiliu, A., Yannou, B.: Dimensional synthesis of planar mechanisms using neural networks: application to path generator linkages. Mech. Mach. Theory **36**, 299–310 (2001)
17. Galán-Marín, G., Alonso, F.J., Del Castillo, J.M.: Shape optimization for path synthesis of crank-rocker mechanisms using a wavelet-based neural network. Mech. Mach. Theor. **44**, 1132–1143 (2009)

Function Generation Synthesis with Planar 4-Bar Linkage as a Mixed Problem of Correlation of Crank Angles and Dead-Center Design

Gökhan Kiper[1(✉)] and Melih Erez[2]

[1] İzmir Institute of Technology, 35430 İzmir, Turkey
gokhankiper@iyte.edu.tr
[2] Mecalac İş Makineleri Sanayi Ve Ticaret Ltd Şti, 35410 İzmir, Turkey

Abstract. The function generation synthesis of planar four-bar linkage is formulated as a mixed problem of correlation or crank angles and dead-center design. The synthesis is based on three desired configurations of the linkage, where the input and output joint angles are specified for two of the configurations and the coupler link is folded upon the output link for the third configuration. The formulation results in a set of nonlinear equations which can be solved analytically. The formulations are tested with numerical examples and similar mixed function generation problems are briefly discussed.

Keywords: Function generation synthesis · Correlation of crank angles · Dead-center design

1 Introduction

In many applications linkage design problem can be formulated as function generation. In function generation synthesis, the link lengths of a linkage are determined according to a desired relationship between the input and the output joint variables. Formulation of the function generation synthesis problem depends on the whether it is desired that the input and output joint variables should satisfy certain values (precision points) or it is desired that the output joint variable should vary in a given interval while the value of the input joint variable can be continuously varied. For the planar four-bar linkage Söylemez [1] names the first type of formulation as the "correlation of crank angles" whereas the second type is known as "dead-center design". Many classical text books such as [2–4] do not cover the dead-center design problem. Hall [5] and Norton [6] treats the dead-center design and correlation of crank angles problems separately. Mallik et al. [7] presents the dead-center problems as fourth type of synthesis problem in addition to motion, path and function generation problems. Pennestrì and Valentini [8] summarized the analytical function generation synthesis methods for the four-bar and the slider-crank mechanisms for two and three prescribed positions including the dead-center design problems. According to the author's knowledge, there is publication

D. Wang et al. (Eds.): MeTrApp 2019, MMS 79, pp. 42–47, 2020.
https://doi.org/10.1007/978-981-15-0142-5_5

that treats the correlation or crank angles and dead-center design problems as a mixed synthesis problem.

In certain applications, a planar four-bar linkage needs to satisfy certain input/output joint values, but also the coupler link becomes collinear with either of the cranks in a limiting position. As an example, for the loader mechanism of a construction machine, the bucket is required to have a certain maximum angle in the lowest and highest positions of the boom. In these two positions, when the boom and actuation cylinders are already designed, the four-bar attached to the bucket need to satisfy two sets of given input and output angles. However, the minimum angle of the bucket at the highest position of the boom is obtained for the dead-center (or near dead-center) position of the four-bar-loop attached to the bucket. This paper presents an analytical solution to such a mixed problem.

2 Problem Definition and Formulation

For four-bar function generators since the input and output joint variables are angles, the linkage can be scaled without effecting the input/output relationship, so without loss of generality let fixed link length be 1 unit. For the problem at hand, the input and output joint angles ϕ_1, ψ_1, ϕ_2, ψ_2 are specified for position #1 and position #2 and only the output angle ψ_3 is specified for position #3, where the coupler link is folded upon the output link (Fig. 1). The conventional dead-center design problem is posed in terms of the desired swing-angle between the folded and extended dead-center positions and corresponding crank rotation (or time ratio), but in this case only the folded dead-center position is specified with a given desired output link angle.

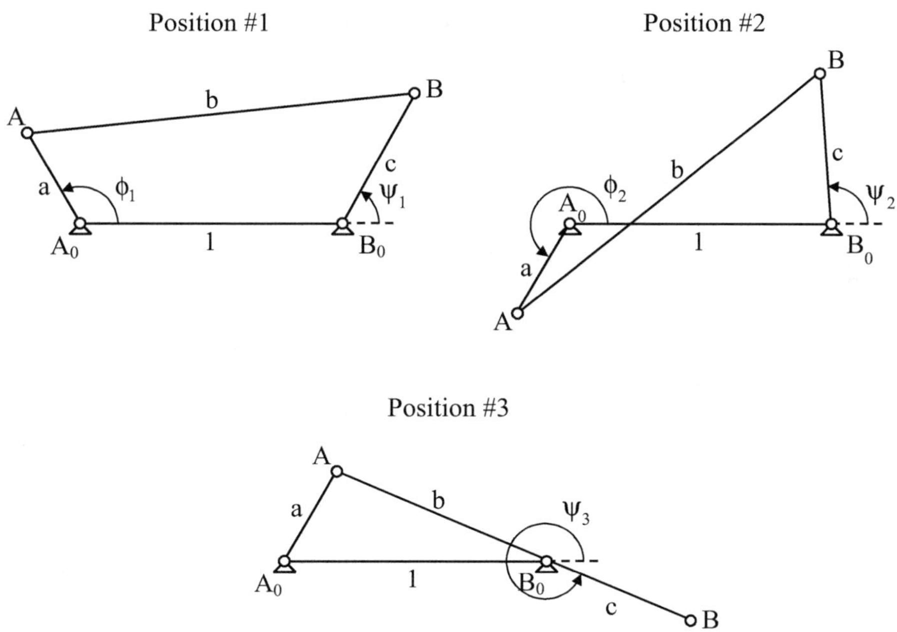

Fig. 1. The three desired positions of a four-bar linkage

The input/output equation for the four-bar linkage can be derived as follows:

$$\left|\overrightarrow{AB}\right| = \left|\overrightarrow{A_0B_0} + \overrightarrow{B_0B} - \overrightarrow{A_0A}\right| \Rightarrow b^2 = (1 + c\cos\psi - a\cos\phi)^2 + (c\sin\psi - a\sin\phi)^2 \tag{1}$$

Expanding Eq. (1) and writing for positions #1 and #2:

$$1 + a^2 - b^2 + c^2 - 2a\cos\phi_1 + 2c\cos\psi_1 = 2ac\cos(\phi_1 - \psi_1) \tag{2}$$

$$1 + a^2 - b^2 + c^2 - 2a\cos\phi_2 + 2c\cos\psi_2 = 2ac\cos(\phi_2 - \psi_2) \tag{3}$$

Writing cosine theorem in triangle A_0AB_0 for position #3 (ϕ_3 is unknown):

$$1 + (b - c)^2 - 2(b - c)\cos\psi_3 = a^2 \Rightarrow 1 - a^2 + (b - c)^2 = 2(b - c)\cos\psi_3 \tag{4}$$

For given angle values Eqs. (2)–(4) constitute 3 equation in 3 unknowns (a, b, c). For a correlation of crank angles problem with three positions all equations are alike and the problem can be transformed into a set of linear equations (Freudenstein equations [9]) in newly introduced coefficients. However in this case, Eq. (4) is different from Eqs. (2) and (3). A simple way to solve these equations is to solve for two of the link lengths from Eqs. (2) and (3) and substitute the solution into Eq. (4) to obtain a single equation in single unknown. Let $\lambda = \frac{b^2 - 1 - a^2 - c^2}{2ac}$. Then

$$b^2 = 1 + a^2 + c^2 + 2ac\lambda \tag{5}$$

Dividing Eqs. (2) and (3) by 2ac and rearranging:

$$\cos\psi_1\frac{1}{a} - \cos\phi_1\frac{1}{c} = \lambda + 2ac\cos(\phi_1 - \psi_1) \tag{6}$$

$$\cos\psi_2\frac{1}{a} - \cos\phi_2\frac{1}{c} = \lambda + 2ac\cos(\phi_2 - \psi_2) \tag{7}$$

Equations (6) and (7) are linear in 1/a and 1/c. a and c solved from Eqs. (6) and (7) as

$$\begin{aligned} a &= \frac{f_1}{f_2\lambda + f_3} = \frac{\cos\phi_1\cos\psi_2 - \cos\psi_1\cos\phi_2}{(\cos\phi_1 - \cos\phi_2)\lambda + \cos\phi_1\cos(\phi_2 - \psi_2) - \cos(\phi_1 - \psi_1)\cos\phi_2} \\ c &= \frac{f_1}{f_4\lambda + f_5} = \frac{\cos\phi_1\cos\psi_2 - \cos\psi_1\cos\phi_2}{(\cos\psi_1 - \cos\psi_2)\lambda + \cos\psi_1\cos(\phi_2 - \psi_2) - \cos(\phi_1 - \psi_1)\cos\psi_2} \end{aligned} \tag{8}$$

Substituting b^2 from Eq. (5) into Eq. (4):

$$\begin{aligned} b^2 = 1 + a^2 + c^2 + 2ac\lambda = a^2 - 1 - c^2 + 2bc + 2(b - c)\cos\psi_3 \\ \Rightarrow 1 + c^2 + ac\lambda = bc + (b - c)\cos\psi_3 \end{aligned} \tag{9}$$

b is eliminated by leaving b alone in Eq. (9), taking square and using Eq. (5) again:

$$\begin{gathered} (c + \cos\psi_3)^2 b^2 = (c + \cos\psi_3)^2\left(1 + a^2 + c^2 + 2ac\lambda\right) = \left(1 + c^2 + ac\lambda + c\cos\psi_3\right)^2 \\ \Rightarrow \sin^2\psi_3 + a^2c^2\left(\lambda^2 - 1\right) + c^2 + 2\sin^2\psi_3 ac\lambda - \cos^2\psi_3 a^2 - 2\cos\psi_3 ac(a + c\lambda) = 0 \end{gathered} \tag{10}$$

Substituting a and c from Eq. (8) into Eq. (10) and rearranging:

$$\begin{gathered} (f_2f_4\sin\psi_3)^2\lambda^4 + 2f_2f_4\left(f_1^2 + f_3f_4 + f_2f_5\right)\sin^2\psi_3\lambda^3 \\ + \left\{ \begin{array}{c} \left[(f_3f_4 + f_2f_5)^2 + 2f_2f_3f_4f_5\right]\sin^2\psi_3 \\ + f_1^2\left[f_1^2 + f_2^2 - 2f_1f_2\cos\psi_3 + 2(f_3f_4 + f_2f_5)\sin^2\psi_3 - (f_4\cos\psi_3)^2\right] \end{array} \right\}\lambda^2 \\ + 2\left\{f_3f_5(f_3f_4 + f_2f_5)\sin^2\psi_3 + f_1^2[(f_2 + f_5)f_3 - (f_3 + f_4)(f_1 + f_5\cos\psi_3)\cos\psi_3]\right\}\lambda \\ + (f_3f_5\sin\psi_3)^2 - f_1^2\left[(f_1 + f_5\cos\psi_3)^2 - f_3^2\right] = 0 \end{gathered} \tag{11}$$

Equation (11) is a fourth order polynomial in λ and has analytical solution with four or two or zero real solutions. For a real λ solution the link length values are solved as

$$a = \frac{f_1}{f_2\lambda + f_3} \tag{12}$$

$$c = \frac{f_1}{f_4\lambda + f_5} \tag{13}$$

$$b = c + \frac{1 + ac\lambda}{c + \cos\psi_3} \tag{14}$$

3 Numerical Simulation

The formulations in Sect. 3 are implemented in Microsoft Excel (Fig. 2). Note that all angle values can be negated without effecting the result because the cosine function is an even function and sine function is always squared in the formulations. The negated angle values correspond to the horizontal mirror image of the linkage.

For the specific example in Fig. 2 the prescribed positions are defined by the angles $\phi_1 = 90°$, $\psi_1 = 40°$, $\phi_2 = 140°$, $\psi_2 = 80°$ and $\psi_3 = -20°$. There are four real solutions for λ in this case. For several worked out examples it is seen that usually there are two real and two complex solutions. For two or four real solutions it is always observed that at most one of the solutions satisfies the desired 3 positions in the same assembly mode of the linkage. The other real solution(s) for λ would correspond to solutions with negative link lengths, which are not acceptable.

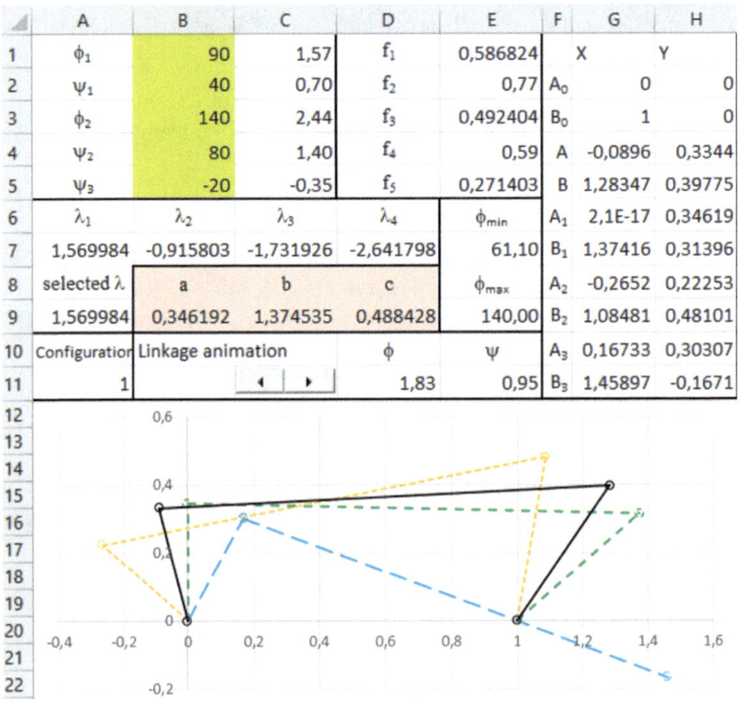

	A	B	C	D	E	F	G	H
1	ϕ_1	90	1,57	f_1	0,586824		X	Y
2	ψ_1	40	0,70	f_2	0,77	A_0	0	0
3	ϕ_2	140	2,44	f_3	0,492404	B_0	1	0
4	ψ_2	80	1,40	f_4	0,59	A	-0,0896	0,3344
5	ψ_3	-20	-0,35	f_5	0,271403	B	1,28347	0,39775
6	λ_1	λ_2	λ_3	λ_4	ϕ_{min}	A_1	2,1E-17	0,34619
7	1,569984	-0,915803	-1,731926	-2,641798	61,10	B_1	1,37416	0,31396
8	selected λ	a	b	c	ϕ_{max}	A_2	-0,2652	0,22253
9	1,569984	0,346192	1,374535	0,488428	140,00	B_2	1,08481	0,48101
10	Configuration	Linkage animation		ϕ	ψ	A_3	0,16733	0,30307
11	1			1,83	0,95	B_3	1,45897	-0,1671

Fig. 2. Numerical implementation in excel

4 Discussion

This paper presents formulations for the synthesis of a planar four-bar linkage for given two sets of input and output joint variables and a limiting position where the coupler link is folded upon the output link with a given output angle. Such a problem differs from the conventional correlation of crank angles type function generation problem in the sense that not all the equations have the same form. The solution of such a mixed function generation problem can easily be found using a CAD software, but when the four-bar design is merely a part of the design of a multi-loop mechanism, it is necessary to implement solution of a parametric model and an analytical solution is required for immediate computation. This paper presents such an analytical formulation.

Similar formulations can be easily derived for extended case instead of the folded case (just use b + c instead of b – c in Eq. (4)) and also for the cases where the coupler is in-line with the input link (just interchange ϕ and ψ) instead of the output link. Similarly, one can address a mixed function generation synthesis problem with the positions where two of the positions are the folded and extended dead-center position and the remaining position is given as a correlation of crank angles. The given two dead-center positions might be specified as one for the input link and one for the output link as well. Or else, one or two of the positions might be specified for a position where the input or the output link is in-line with the fixed link (this condition is usually evaluated for maximum deviation of the transmission angle from 90°). Furthermore, for the cases where the exact angle of the input or the output angles is not important, but a desired angular displacement of the links is specified (as in a typical function generation problem), it is possible to introduce initial input and or output angle values as additional design parameters and design for four or five positions (See for ex. [2]). But in this case the nonlinearity of the problem increases and there may be no analytical solution. For instance, for given extended/folded dead center output angles and correlation of crank angles for a general configurations results in a sixth order univariate polynomial, which has no analytical solution but can be rapidly solved using a simple root finding algorithm. For more than 5 positions analytical optimization methods such as least square or Chebyshev approximation methods or numerical optimization methods might be utilized. A mixed function generation synthesis problem may be formulated for all these cases, but these formulations are kept out-off scope of this study and left for further studies for potential applications.

Acknowledgment. The authors thank to Prof. Eres Söylemez for his comments and suggestions.

References

1. Söylemez, E.: Mechanisms, 5th edn. METU Press, Ankara (2018)
2. Hartenberg, R.S., Denavit, J.: Kinematic Synthesis of Linkages. McGraw-Hill, New York (1964)
3. Sandor, G.N., Erdman, A.G.: Advanced Mechanism Design: Analysis and Synthesis, vol. II. Prentice Hall, New Jersey (1984)
4. McCarthy, J.M., Soh, G.S.: Geometric Design of Linkages, 2nd edn. Springer, New York (2011)
5. Hall, A.S.: Kinematics and Linkage Design. Prentice-Hall, Englewood Cliffs (1961)
6. Norton, R.L.: Design of Machinery, 6th edn. McGraw-Hill, Boston (2019)
7. Mallik, A.K., Ghosh, A., Dittrich, G.: Kinematic Analysis and Synthesis of Mechanisms. CRC Press, Boca Raton (1994)
8. Pennestrì, E., Valentini, P.P.: A review of simple analytical methods for the kinematic synthesis of four-bar and slider-crank function generators for two and three prescribed finite positions. In: Buletin Stiintific Seria Mecanica Aplicata, pp. 128–143. University of Pitesti, Pitești, (2009)
9. Freudenstein, F.: An analytical approach to the design of four-link mechanisms. Trans. ASME **76**, 483–492 (1954)

Mechanical Transmission

Contact Analysis for Spiral Bevel Gear Based on Machine Parameters

Yongqiang Zhao(✉), Xiangyang Liu, and Ming Liu

School of Mechatronics Engineering, Harbin Institute of Technology, Harbin 150001, China
zhyqhit@hit.edu.cn

Abstract. In traditional machining process, the calculation of machine tool adjustment parameters only focuses on calculating the middle point of tooth surface, so it will lead to contact defects such as root contact or diagonal contact of spiral bevel gears. It is necessary to carry out multiple hobbing checks on gear pairs and adjust the setting parameters of machine tools to meet the applying requirements. Based on the above problems, the mathematical model of the machine and the cutter is proposed by studying the adjustment parameters and the processing principle of the spiral bevel gear machine. According to the kinematic relationship between the parts of machine tools and the meshing principle of gears, the tooth surface equation is established. The meshing method is used to discretize the tooth surface. The calculation method of initial value of cutting depth and rotation angle of cutter spindle is studied, and the tooth surface equation is analyzed and solved. Based on the tooth surface equation and gears assembly model, the contact equation is established to calculate the contact path and transmission error. The influence of machine tool adjustment parameters on contact path and transmission error is investigated.

Keywords: Spiral bevel gear · Tooth contact analysis · Parameters of machine tool · Contact adjustment

1 Introduction

The mathematical model of spiral bevel gears reflects the processing method of spiral bevel gears and determines the setting of machine tool parameters. It is the basis of tooth surface modification, tooth contact analysis (TCA) and loaded tooth contact analysis (LTCA). Litvin proposed a mathematical model of spiral bevel gears processed by machine tools with cutter tilting mechanism [1]. Shih proposed a unified mathematical model of spiral bevel gears processed by continuous machining of face hob [2]. Fan established a spiral bevel gear mathematical model processed by the face hob continuous machining method. In this model, the machine tool adjustment parameter is the sixth-order polynomial of cradle angle [3].

Spiral bevel gears can be analyzed by tooth contact to obtain contact point path, contact ellipse and instantaneous transmission error. Litvin proposed a local synthesis method and established the tooth surface contact equation of the spiral bevel gear according to the contact conditions. The tooth surface contact equation was solved by

D. Wang et al. (Eds.): MeTrApp 2019, MMS 79, pp. 51–61, 2020.
https://doi.org/10.1007/978-981-15-0142-5_6

the Newton-Raphson method. The tooth surface contact path and the transmission error curve were obtained [4]. Fan improved the existing TCA algorithm by replacing vector equality with geometric relationship between the tooth vectors, and proposed a new TCA algorithm [5]. Fan applied topological surface to tooth surface contact analysis, and proposed an algorithm for tooth surface contact analysis, which was applied to spiral bevel gears manufactured by Face-hobbing (FH) and Face-milling (FM) machining methods [6]. Astoul used finite element analysis software to analyze the transmission error of spiral bevel gears, and proposed a new method to improve transmission error and topology analysis [7]. Simon investigated the production of spiral bevel gears by FH machining method, and the influence of various machine tool adjustment parameters on gear transmission error and tooth surface bearing capacity [8].

Many scholars have conducted in-depth research on the modification of gear tooth surface, contact adjustment and the influence of machine tool adjustment parameters on the shape and contact of gear tooth surface. Shih et al. proposed a tooth surface correction and modification method for spiral bevel gears and hypoid gears [9]. Based on the complete conjugate mismatched topological tooth surface theory, NIE studied the influence of the topological tooth surface parameters on the unloaded contact and dynamic performance of spiral bevel gears [10]. Using the principle of high order transmission error and complete conjugation, Mu modified the tooth surface of spiral bevel gear by using the method of modifying the hobby ratio [11]. According to the topological tooth surface theory, Shih studied the influence of different machine tool adjustment parameters on the tooth surface deformation [12].

The above research involves the modeling, contact analysis and tooth surface modification. However, when solving the tooth surface equation and meshing equation of spiral bevel gears, the initial value is not given. For the research of contact area adjustment, the proportional correction method is limited to the modification of single machine tool adjustment parameters, and the corrected contact condition cannot be predicted.

In view of the above problems, this paper will be organized as follows.

In the Sect. 2, the mathematical model of the spiral bevel gear machine tool and cutter is established. Based on the adjustment parameters of the machine tool, the tooth surface equation is formulated. The tooth surface is discretized by meshing method, and the solution method of initial value for rotation angle of cutter spindle and cutting depth is proposed according to the machine tool settings. In the Sect. 3, the contact equation is established and solved, and the TCA analysis of the spiral bevel gear is carried out to study the influence of different factors on the tooth surface contact.

2 Mathematical Model of Spiral Bevel Gear

2.1 Tooth Surface Equation

At present, most of the machine tools for processing spiral bevel gears are Cradle-Type generators. According to the machine tool structure, the machine tool coordinate system is established (Fig. 1). As shown in Fig. 1, the coordinate system S_t is rigidly connected to the cutter head, and the z_t is the cutter head rotation axis. The coordinate system S_1 is rigidly connected to the gear blank, and the z_1 is the rotation axis of the gear blank.

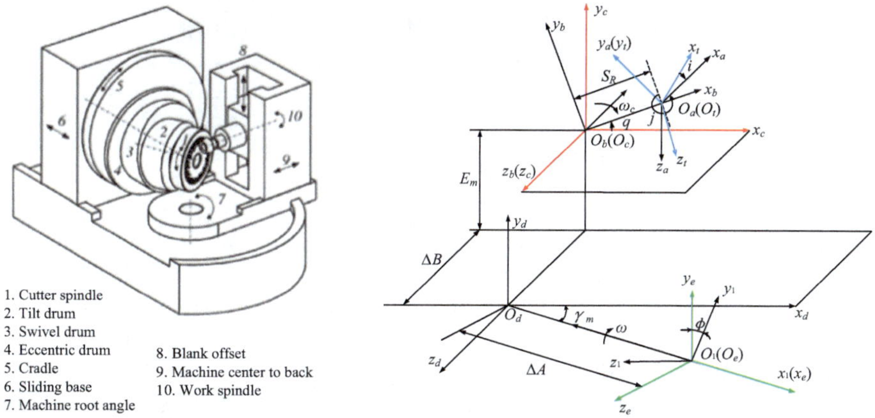

Fig. 1. Coordinate systems of Cradle-Type machine tool

The position vector of the cutter in the coordinate system S_t is represented as

$$\mathbf{r}_t(s,\beta) = \begin{bmatrix} (R \pm s\sin\alpha)\cos\beta \\ (R \pm s\sin\alpha)\sin\beta \\ -s\cos\alpha \\ 1 \end{bmatrix} \tag{1}$$

where s is the cutting depth, β is the rotation angle of cutter spindle, α is the cutter blade angle, R is the cutter point radius. The "±" sign should be regarded as "+" sign for the outter cutter blade, and "-" sign for the inner cutter blade.

The transformation matrices S_t to S_1 yield the following surface locus for the cutting tool in coordinate system S_1:

$$\mathbf{r}_1(s,\beta,\phi) = \mathbf{M}_{1e}(\phi)\mathbf{M}_{ed}(\gamma_m,\Delta A)\mathbf{M}_{dc}(E_m,\Delta B)\mathbf{M}_{cb}(q)\mathbf{M}_{ba}(j,S_R)\mathbf{M}_{at}(i)\mathbf{r}_t(s,\beta) \tag{2}$$

where ϕ is the angle of blank work spindle, γ_m is the machine root angle, ΔA is the increment of machine center to back, E_m is the vertical offset, ΔB is the Sliding base feed, q is the sum of cradle rotation angle, j is the swivel angle of cutter tilt, S_R is the radial setting of cradle, i is the tilt angle of cutter spindle with respect to cradle rotation axis. The transformation matrices expressions in Eq. (2) and later equations are referred to in reference [2].

The transformation matrices S_t to S_1 yield the surface locus $\mathbf{r}_1(s,\beta,\phi)$ for the cutter represented in coordinate system S_1, and the tooth surface of the machined gear can be regarded as the envelope surface of the family of curved surfaces. From the envelope theory, the tooth surface equation can be obtained by connecting the $\mathbf{r}_1(s,\beta,\phi)$ with the meshing equation.

$$\begin{cases} \mathbf{r}_1(s,\beta,\phi) = \mathbf{M}_{1t}(\phi)\mathbf{r}_t(s,\beta) \\ f(s,\beta,\phi) = (\frac{\partial \mathbf{r}_1}{\partial s} \times \frac{\partial \mathbf{r}_1}{\partial \beta}) \cdot \frac{\partial \mathbf{r}_1}{\partial \phi} = 0 \end{cases} \quad (3)$$

2.2 Solution of Scatter Point

According to the Eq. (3), the vector of tooth surface can be represented by $\mathbf{r}_1(s,\beta)$. Since the tooth surface of the spiral bevel gear is relatively complicated, it is necessary to discretize the tooth surface, and the method of mesh division is selected here. As shown in Fig. 2, the mesh is divided along the direction of tooth height and tooth length on the axis section of the gear. According to the calculation accuracy, the number of meshes in the direction of tooth height is 5, and in the direction of tooth length is 9. Thus, the whole tooth surface is discretized into 45 discrete points [13]. R_a is the pitch cone distance, b is the face width, δ is the pitch cone angle, δ_f is the root angle, δ_a is the face angle.

The point M is the midpoint of the tooth surface, and its position vector can be written as $\mathbf{r}_M(s,\beta) = \{x_M, y_M, z_M\}$. The position vector is rotated and projected onto the axial section of gear blank, and the coordinates of point M are obtained.

$$\begin{cases} R_M = x_M(s,\beta) \\ L_M = \sqrt{y_M(s,\beta)^2 + z_M(s,\beta)^2} \end{cases} \quad (4)$$

Solve function in MATLAB is used to solve the equation of tooth surface, and s_I and β_I for the initial values of s and β are given before the calculation. When the cutting tool cuts the midpoint of the gear tooth surface, the axial section of the cutter and the machined tooth is shown in Fig. 3.

The initial value of cutting depth can be obtained by

$$s_I = \frac{h}{2\cos\alpha} \quad (5)$$

where h is the total height of tooth, α is the blade angle.

Figure 4 shows the initial values of β for left-hand and right-hand cutter head respectively. According to the relationship between the angles and the helix angle of the machined gear, the initial values of β for left-hand and right-hand cutter head can be represented by

$$\begin{cases} \beta_{Il} = q - \beta_a \\ \beta_{Ir} = \pi - q + \beta_a \end{cases} \quad (6)$$

where β_{Il} is the initial value for left-hand, β_{Ir} is the initial value for right-hand, β_a is the helix angle of machined gear, q is the sum of cradle rotation angle, $q = \theta_c \mp \phi_c$, θ_c is the initial cradle angle setting, $\phi_c = \phi/Ratio$, *Ratio* is the Roll ratio of cradle and gear blank.

According to scatter solution plan of the tooth surface in Fig. 5, all scatter points of the tooth surface are sequentially solved.

2.3 Contact Equation

In order to describe the contact of spiral bevel gears, the coordinate systems of the assembly gears are shown in Fig. 6. The coordinate systems $S_1(x_1, y_1, z_1)$ and $S_2(x_2, y_2, z_2)$ are rigidly connected to the gear and pinion respectively. The coordinate systems $S_{f1}(x_{f1}, y_{f1}, z_{f1})$ and $S_{f2}(x_{f2}, y_{f2}, z_{f2})$ are fixed to the gear and pinion respectively.

The contact equations in coordinate system $S_2(x_2, y_2, z_2)$ are represented by

$$\left\{\begin{array}{c} \mathbf{r}_2^g(s_1, \beta_1, \phi_p, \phi_g) = \mathbf{r}_2^p(s_2, \beta_2) \\ (\mathbf{n}_2^g(\beta_1, \phi_p, \phi_g) \times \boldsymbol{\tau}_2^g(s_1, \beta_1)) \cdot \mathbf{n}_2^p(\beta_2) = 0 \\ \boldsymbol{\tau}_2^g(s_1, \beta_1) \cdot \mathbf{n}_2^p(\beta_2) = 0 \end{array}\right. \tag{7}$$

where ϕ_g and ϕ_p are the rotation angles respectively, the superscripts "g" and "p" indicate gear and pinion.

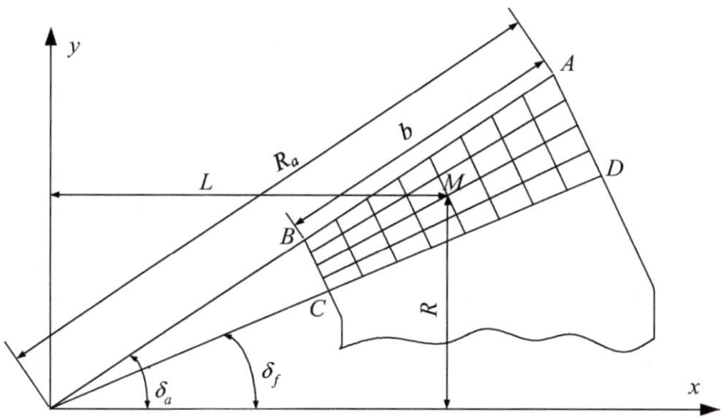

Fig. 2. Meshing of tooth surface

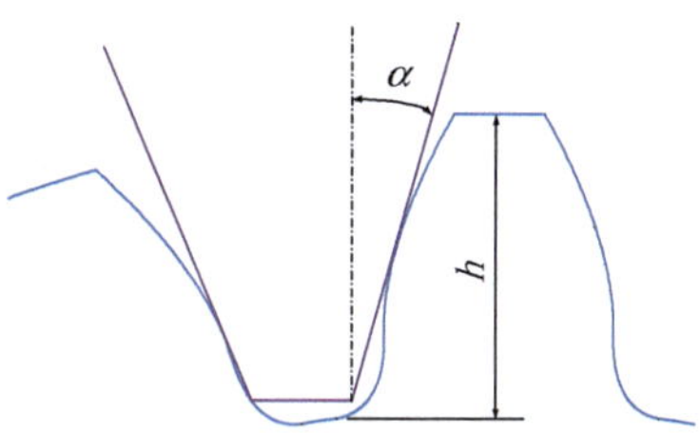

Fig. 3. The axial section of cutter and machined tooth

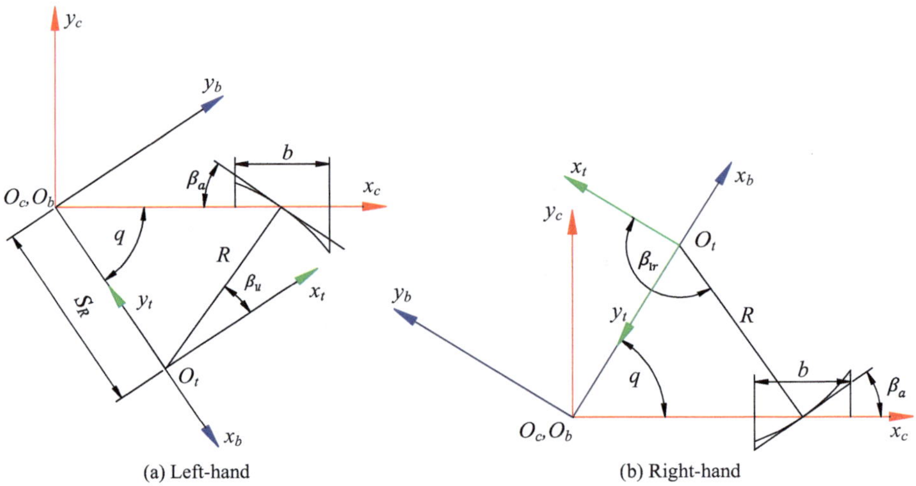

Fig. 4. Initial setting angle of the cutter head

3 Contact Analysis of Spiral Bevel Gear

Taking a pair of spiral bevel gears with a ratio of 21:41 as an example, the parameters of spiral bevel gears (Table 1) and machine tool adjustment parameters (Table 2) are given for contact calculation.

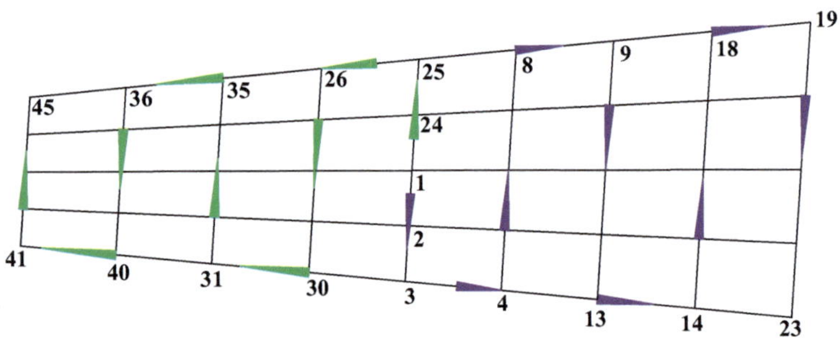

Fig. 5. Planning of scatter point on tooth surface

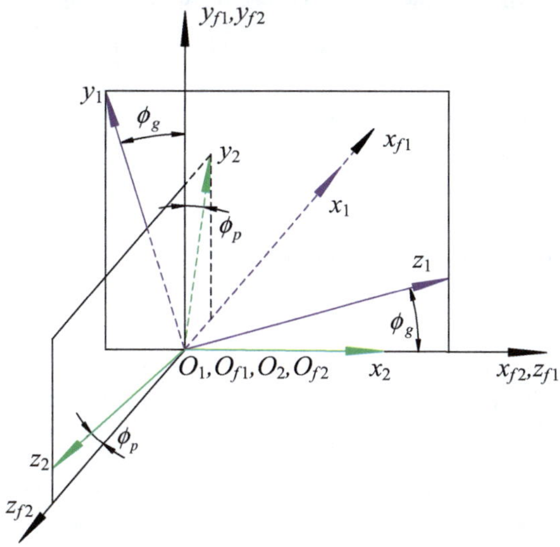

Fig. 6. Coordinate systems of assembly gears

3.1 Contact Calculation

Contact path. According to the method of solving the initial value for the tooth surface equation in Sect. 2.2, the initial contact point is determined by assuming the midpoint contact of the tooth surface for gear and pinion. From the initial contact point, the solver function in MATLAB is used to iteratively solve the position of the contact point when the pinion rotates through different angles, and the contact point is projected onto the tooth surface mesh of gear or pinion, so that the contact paths of the gear and pinion can be obtained.

Transmission error. When the rotation angle of the pinion is ϕ_{p0}, the position of the pinion is the initial position ($\phi_{p0} = 0$), and the gear rotates the angle of ϕ_{g0} to engage with the pinion, which is the initial contact point. Assuming that the pinion rotates the angle of ϕ_p, and the two tooth surfaces are re-engaged after the gear rotates the angle of ϕ_g. Therefore, the rotation angles of the gear and pinion from the initial contact position to the re-contact position are shown in the Eq. (8).

$$\begin{cases} \Delta\varepsilon_p = \phi_p - \phi_{p0} \\ \Delta\varepsilon_g = \phi_g - \phi_{g0} \end{cases} \tag{8}$$

When the transmission ratio is the theoretical value and the pinion rotates the angle of $\Delta\varepsilon_p$, the rotation angle of the gear should be $(Z_2/Z_1)\Delta\varepsilon_p$, so the transmission error should be the actual rotation angle minus the theoretical rotation angle.

$$\Delta\varepsilon = \Delta\varepsilon_g - (Z_2/Z_1)\Delta\varepsilon_p \tag{9}$$

Table 1. Basic parameters of the gear transmission

Items	Pinion (Concave)	Ring gear (Convex)
Hand	LH	RH
Shaft angle Σ	90 deg	
Tooth number Z	21	41
Pressure angle α/°	20	20
Spiral angle β_d/°	35	35
Pitch cone distance R_a/mm	138.196	138.196
Face width b/mm	42	42
Outside diameter d_a/mm	138.15	249.08
Pitch cone angle δ/°	27.117	62.883
Face angle δ_a/°	30.417	64.75
Root angle δ_f/°	25.25	59.583

Table 2. Universal cradle-type machine settings

Items	Pinion (Concave)	Ring gear (Convex)
Cradle test angle Q/°	313.617	79.133
Eccentric angle β_e/°	34.167	36.017
Increment of machine center to back ΔA/mm	−2.17	−1.73
Sliding base feed ΔB/mm	1.02	1.49
Vertical offset E_m/mm	−5.94	1.59
Machine root angle γ_m/°	25.25	59.583
Roll ratio	2.1258	1.1217
Cutter point radius R/mm	224.63	225.81
Blade angle α/°	18	22

3.2 Sensitivity of Machine Tool Adjustment Parameters

By adjusting the machining parameters of machine tool for pinion, the ideal contact form of gears can be achieved. Taking the left-hand pinion and the right-hand gear as an example, the actual working tooth surface is the contact between pinion concave surface and gear convex surface. The influence of different machine adjustment parameters on contact is shown in Fig. 7.

The left side of the figure shows the contact path, and the right side shows the transmission error. The nomenclature of adjustment parameters is shown in Table 2. As shown in (a) and (c) of Fig. 7, β_e and ΔB have a large influence on the inclination of the tooth surface along the tooth length direction. When the β_e and ΔB increase, the contact path moves toward the toe of the tooth surface, and when decrease, the contact path moves to the heel of tooth surface. The ΔA has a great influence on the inclination of the tooth surface along the direction of tooth height. As shown in Fig. 7(b), the contact path moves to the root of the tooth when the ΔA increases, and vice versa, to the top of the tooth. As shown in (d) and (e) of Fig. 7, when the E_m and γ_m increase, the contact

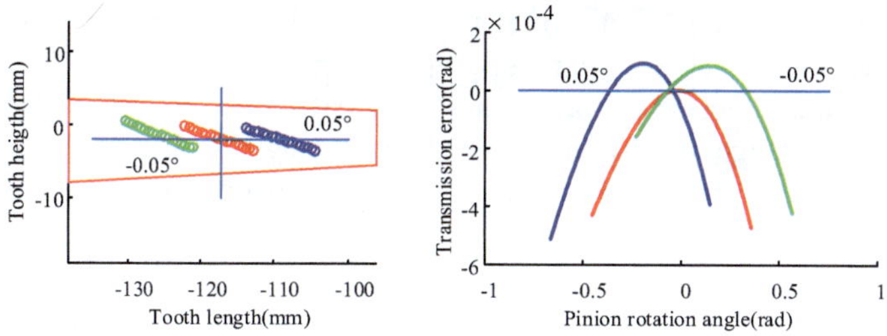

(a) Variation of β_e in range of ±0.05°

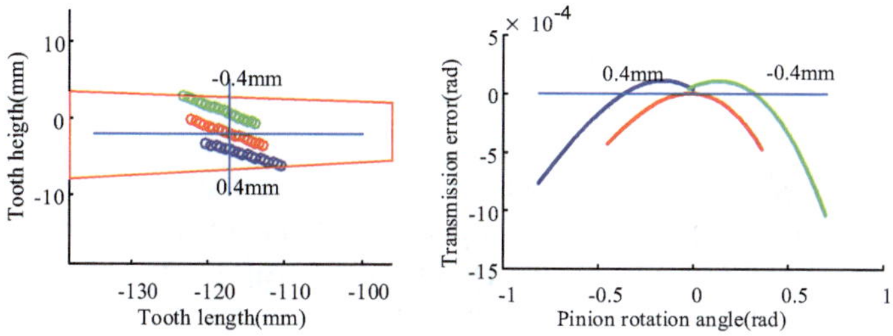

(b) Variation of ΔA in range of ±0.4mm

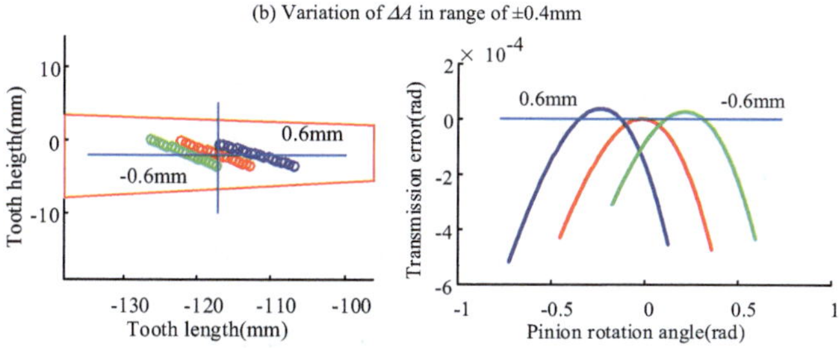

(c) Variation of ΔB in range of ±0.6mm

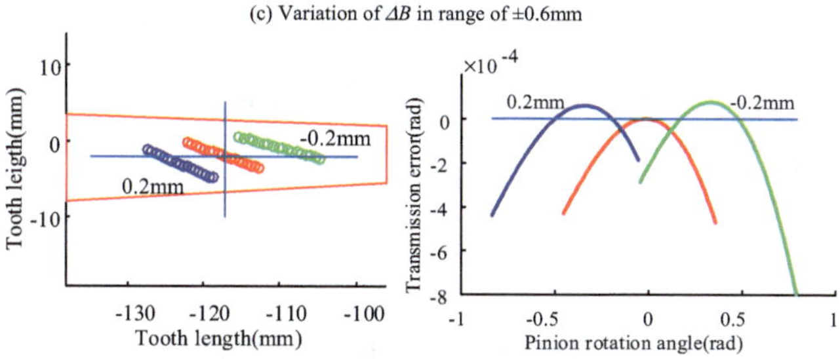

(d) Variation of E_m in range of ±0.2mm

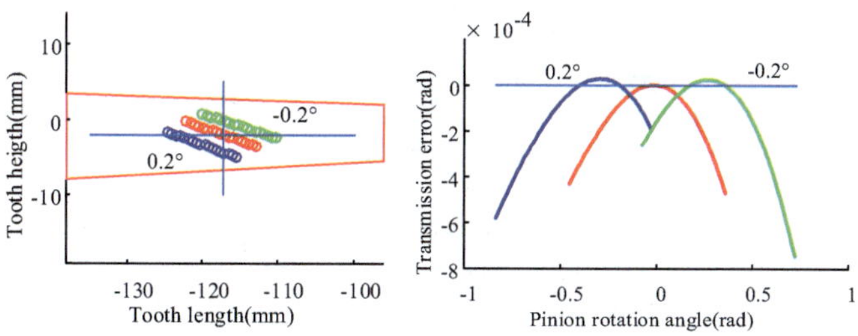

(e) Variation of γ_m in range of ±0.2°

Fig. 7. Contact influences of machine tool setting parameters path moves toward the heel and root of the tooth surface, and when they are reduced, the contact path moves to the toe and top of the tooth surface. For the E_m, the change of contact path in the direction of tooth length is greater than that of tooth height, while the contact path changes in the direction of tooth height is greater than that of tooth length for the γ_m. It can be seen from (a)–(e) of Fig. 7 that different parameters will lead to the asymmetry of transmission error curve.

4 Conclusions

In this paper, the influence of adjusting parameters of spiral bevel gear machine tool on the contact is studied based on the tooth surface contact analysis. The specific research results and conclusions are as follows:

(1) Mathematical models of machine tools and cutters are established according to machine tool processing parameters and cutting tool parameters. Tooth surface equations are established based on the kinematic relationship between cutter and wheel blank. Through meshing, the tooth surface is discretized, and according to the machine tool settings, a solution method for solving the initial value of rotation angle of cutter spindle and cutting depth is proposed. The tooth surface scatter coordinates are obtained by solving the tooth surface equation.

(2) Based on the actual installation of spiral bevel gears, a virtual assembly model of gears is established in the assembly coordinate system. According to the geometric relationship among the tooth vector, normal vector and tangent vector of the gear and pinion, the contact equation is formulated, and the contact path and transmission error curve are obtained by solving the contact equation.

(3) The influence of adjusting parameters of machine tool on tooth contact is investigated. It is found that sliding base feed and eccentric angle have great influence on the position of tooth contact along the tooth length direction, and the distance from center to back of machine tool has great influence on the position of tooth contact along the tooth height direction.

References

1. Litvin, F.L., Fuentes, A., Hayasaka, K.: Design, manufacture, stress analysis, and experimental tests of low-noise high endurance spiral bevel gears. Mech. Mach. Theor. **41**(1), 83–118 (2006)
2. Shih, Y.P., Fong, Z.H., Grandle, C.Y.: LIN: mathematical model for a universal face hobbing hypoid gear generator. Trans. ASME **129**(1), 38–47 (2007)
3. Fan, Q.: Computerized modeling and simulation of spiral bevel and hypoid gears manufactured by gleason face hobbing process. J. Mech. Des. **128**(6), 1315–1327 (2005)
4. Litvin, F.L., Donno, M.D., Peng, A., et al.: Integrated computer program for simulation of meshing and contact of gear drives. Comput. Meth. Appl. Mech. Eng. **181**(1), 71–85 (2000)
5. Fan, Q., Wilcox, L.: New developments in tooth contact analysis (TCA) and loaded TCA for spiral bevel and hypoid gear drives. Gear Technol. **5**, 26–35 (2007)
6. Fan, Q.: Ease-Off and application in tooth contact analysis for face-milled and face-hobbed spiral bevel and hypoid gears. Theor. Pract. Gear. Trans. **34**, 321–339 (2016)

7. Astoul, J., Mermoz, E., Sartor, M., et al.: New methodology to reduce the transmission error of the spiral bevel gears. CIRP Ann.-Manuf. Technol. **63**(1), 165–168 (2014)
8. Simon, Vilmos: Design of face-hobbed spiral bevel gears with reduced maximum tooth contact pressure and transmission errors. Chin. J. Aeronaut. **26**(03), 777–790 (2013)
9. Shih, Y.P.: A novel ease-off flank modification methodology for spiral bevel and hypoid gears. Mech. Mach. Theor. **45**(8), 1108–1124 (2010)
10. Nie, S., Deng, J., Deng, X., et al.: A flank modification method for spiral bevel gears based on mismatch topography adjustment. J. Adv. Mech. Des. Syst. Manuf. **12**(2), JAMDSM0057 (2018)
11. Mu, Y., Li, W., Fang, Z., et al.: A novel tooth surface modification method for spiral bevel gears with higher-order transmission error. Mech. Mach. Theor. **126**, 49–60 (2018)
12. Shih, Y.P., Fong, Z.H.: Flank modification methodology for face-hobbing hypoid gears based on Ease-Off topography. J. Mech. Des. **129**(12), 1294–1302 (2007)
13. Cao, Xuemei, Deng, Xiaozhong, Wei, Bingyang: A novel method for gear tooth contact analysis and experimental validation. Mech. Mach. Theor. **126**, 1–13 (2018)

Dimensional Synthesis of the Drawing-In Mechanism in an Automatic Warp Tying Machine

Guanglei Wu[1,2(✉)], Shimin Chen[1], and Ji Li[1]

[1] Alida (Changzhou) Intelligent Technology Co., Ltd, Changzhou, China
gwu@dlut.edu.cn
[2] School of Mechanical Engineering,
Dalian University of Technology, Dalian 116024, China

Abstract. This paper deals with the dimensional synthesis of a cam-follower based drawing-in mechanism for the automatic warp tying machine. The drawing-in mechanism is modeled and introduced. The corresponding drawing-in processes in the textile products is depicted in 4-5-6-7 polynomial equation. The approach of the kinematic geometry is adopted to design the cam profile and a comparison among three design cases is discussed. The comparison results show a suitable design to allow the smooth motion of the drawing-in mechanism.

Keywords: Dimensional synthesis · Drawing-in mechanism · Warp tying machine · Textile industry · Cam-follower

1 Introduction

The textile industry develops rapidly with the increasing populations as well as the social and economic developments [3]. Figure 1 shows the complete production processes of the textile products, for which a critical process is the drawing-in procedure. In this procedure, the yarns will be assembled as textile spindles in the right sequences subject to the color and patterns of the shell fabric. Therefore, the quality and efficiency of the shell fabric and clothes production heavily depends on the drawing-in procedure. China has about 40,000 textile manufacturers, which is the largest market of the textile industry worldwide. However, 99% of the Chinese manufacturers adopt the manual operation to accomplish the drawing-in procedure. Potential demand for the automatic warp tying machine shown in Fig. 1, is very large, to upgrade the textile industry [2].

An automatic warp tying machine is a complex mechanical system, made of plenty of modular mechanisms. The drawing-in procedure is accomplished by means of its corresponding drawing-in mechanism that needs to be redesigned to differ from the current machines. As displayed in Fig. 2, the new drawing-in mechanism consists of the cam mechanism, section gear, rack, etc. The drawing-in motion is realized through the section gear and rack driven by the cam repeatedly. The dimensional synthesis of the drawing-in mechanism, particularly, the

D. Wang et al. (Eds.): MeTrApp 2019, MMS 79, pp. 62–69, 2020.
https://doi.org/10.1007/978-981-15-0142-5_7

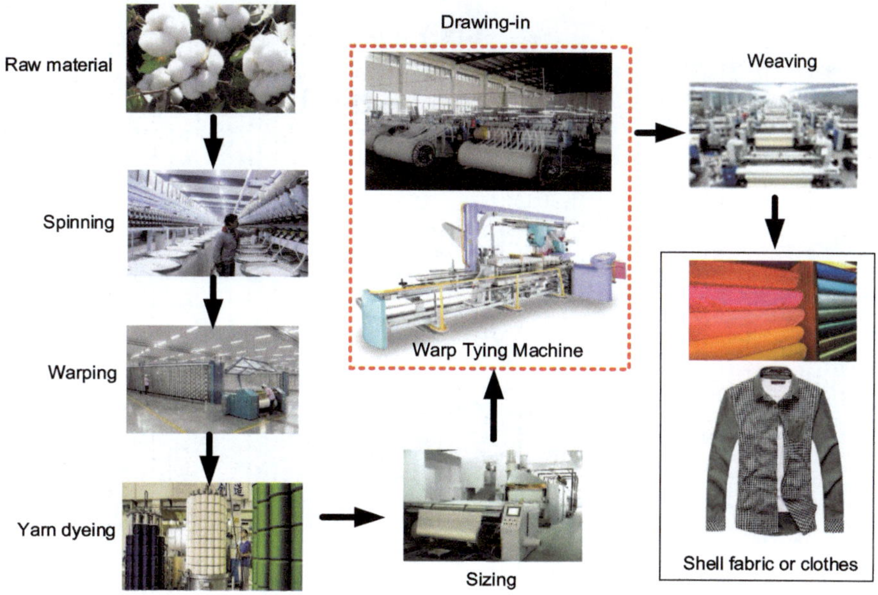

Fig. 1. The drawing-in procedure in the production processes of the textile products.

cam profile, will be the focus of this work. Design of the cam mechanism is a classical topic [1,4–6]. Regarding the motion of follower, the cam mechanism with a planar motion follower in this work refers to the follower moving in combination of translation and oscillation [9]. The design of the cam mechanisms have been extensively carried out with various approaches [7,8,10]. The dimensional synthesis of the cam mechanism in this work will be based on the approach of the kinematic geometry [11].

In this paper, the dimensional synthesis of a drawing-in mechanism, based on the cam-follower mechanism, is introduced for the automatic warp tying machine. The drawing-in process is depicted in 4-5-6-7 polynomial equation. The approach of the kinematic geometry is adopted to design the cam profile and a comparison among three design cases is discussed.

2 Drawing-In Process

Figure 3 depicts the drawing-in process with respect to the rotation of the cam mechanism. The drawing-in motion starts from the initial angle 0 and ends at 85°. In the rang of (85°, 90°], the guided rack keeps static, and then the machine exits the drawing-in motion till 195°. Afterwards, the guided rack will be static until the next working cycle repeats.

According to the drawing-in actions in Fig. 3(a), the movements of the guided rack can be expressed in the 4-5-6-7 polynomial equation, due to its highly frequent working cycle:

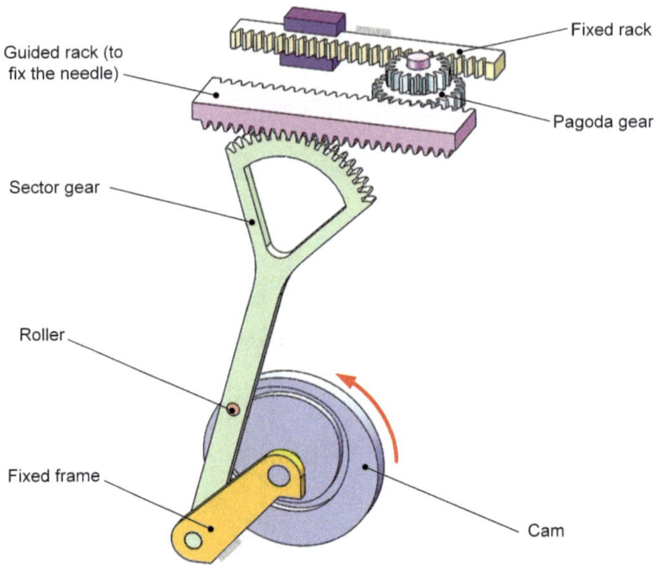

Fig. 2. Simplified CAD model of the drawing-in mechanism.

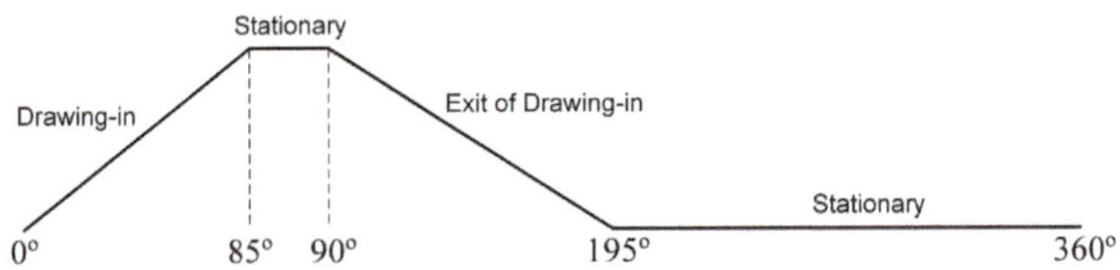

(a)

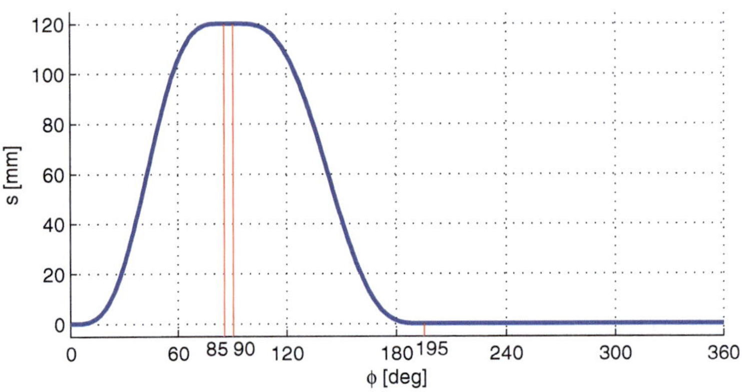

(b)

Fig. 3. The drawing-in procedure with respect to the cam mechanism: (a) drawing-in actions; (b) movement of the guided rack (needles).

- $0 \leq \phi \leq \phi_1$

$$s = h(-20t^7 + 70t^6 - 84t^5 + 35t^4),\ t = \frac{\phi}{\phi_1} \tag{1}$$

- $\phi_1 < \phi \leq \phi_2$

$$s = h \tag{2}$$

- $\phi_2 < \phi \leq \phi_3$

$$s = h(-20t^7 + 70t^6 - 84t^5 + 35t^4),\ t = \frac{\phi_3 - \phi}{\phi_3 - \phi_2} \tag{3}$$

- $\phi_3 < \phi \leq \phi_4$

$$s = 0 \tag{4}$$

where ϕ is the rotation angle of the cam from its initial orientation. Moreover, the motion parameters are listed in Table 1.

Table 1. Motion parameters of the drawing-in mechanism.

h [mm]	ϕ_1	ϕ_2	ϕ_3	ϕ_4
120	85°	90°	195°	360°

3 Dimensional Synthesis of the Drawing-In Mechanism

The parameterization of the drawing-in mechanism is depicted in Fig. 4. The reference coordinate frame $\mathcal{F}_b$ is built on the fixed frame, with the origin O located at the center of rotation of the cam. The center of rotation of the section gear in the frame $\mathcal{F}_b$ is located at position A, with the radius of the pitch circle equal to R, and the horizontal movement of the guided rack is equal to s. Moreover, the length of the follower and the radius of the roller are denoted by $r = |AB|$ and r_0, respectively. In order to precisely describe the shape profile of the cam mechanism, the cam profile is represented by finite discrete positions in its inertial frame, namely, the Cartesian coordinates (x_1, y_1), (x_2, y_2), ..., (x_n, y_n) in the frame $\mathcal{F}_p$, for the interpolation in the CNC machining. Here and after, $\hat{\mathbf{x}}$ and $\hat{\mathbf{y}}$ represent the unit vectors of x- and y-axis, respectively.

With the coordinate system in Fig. 4, the Cartesian coordinates of the discrete points on the cam profile in the reference frame $\mathcal{F}_b$ is derived as

$$\mathbf{p}_i = \mathbf{Q}\mathbf{p}'_i; \quad \mathbf{p}'_i = \begin{bmatrix} x_i \\ y_i \end{bmatrix},\ i = 1, 2, ..., n \tag{5}$$

with the rotation matrix of the cam $\mathbf{Q}$ under an arbitrary orientation angle ϕ given by

$$\mathbf{Q} = \begin{bmatrix} \cos\phi & -\sin\phi \\ \sin\phi & \cos\phi \end{bmatrix} \tag{6}$$

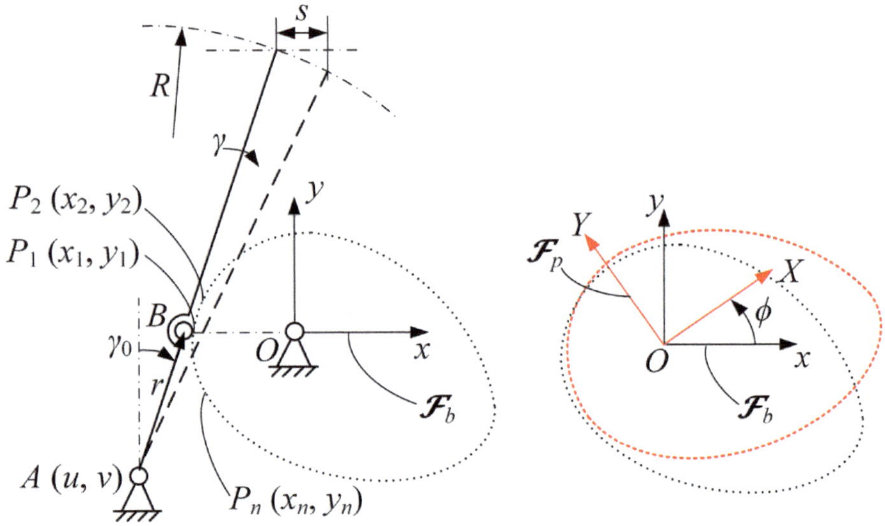

Fig. 4. Parameterization of the drawing-in mechanism.

The Cartesian coordinates of point B in the frame $\mathcal{F}_b$ is expressed as

$$\mathbf{b} = \mathbf{a} + r\mathbf{R}\hat{\mathbf{y}} \tag{7}$$

with

$$\mathbf{a} = \begin{bmatrix} u \\ v \end{bmatrix}, \quad \mathbf{R} = \begin{bmatrix} \cos(\gamma_0 + \gamma) & sin(\gamma_0 + \gamma) \\ -\sin(\gamma_0 + \gamma) & \cos(\gamma_0 + \gamma) \end{bmatrix} \tag{8}$$

where γ_0 and γ stand for initial and rotation angles of the follower, respectively, and γ can be solved from the following geometric constraint equation:

$$\cos\left(\gamma_0 + \frac{\gamma}{2}\right) - \frac{s}{L} = 0; \quad L = 2R\sin\frac{\gamma}{2} \tag{9}$$

Taking the radius of the roller r_0 into account, the geometric constraint equation of the drawing-in mechanism can be deduced by making use of Eqs. (5) and (7) below:

$$\|\mathbf{p}_i - \mathbf{b}\|_2 = r_0,\ i = 1,\ 2,\ ...,\ n \tag{10}$$

To this end, the equation to depict the geometric profile of the cam mechanism is expressed as

$$x_p = (u + r_0)\cos\phi + v\sin\phi + r\sin(\phi + \gamma_0 + \gamma) \tag{11a}$$

$$y_p = -(u + r_0)\sin\phi + v\cos\phi + r\cos(\phi + \gamma_0 + \gamma) \tag{11b}$$

3.1 Design Case 1 of $\gamma_0 = 0$

According to the layout of the drawing-in mechanism in the entire machine system, the radius of the pitch circle of the section gear, the length of the follower and the radius of the roller are set to $R = 500$ mm, $r = 200$ mm and $r_0 = 10$ mm, respectively. If $\gamma_0 = 0$, i.e., the section gear link is vertical at its initial position, based upon Eq. (2) when $n = 360$, the Cartesian coordinates of the discrete points of the cam profile in the reference frame $\mathcal{F}_b$ are computed, as depicted in Fig. 5. Some critical positions are listed in Table 2. It is noteworthy that the profile has a cusp at 85°, namely, at the discrete position $n = 86$.

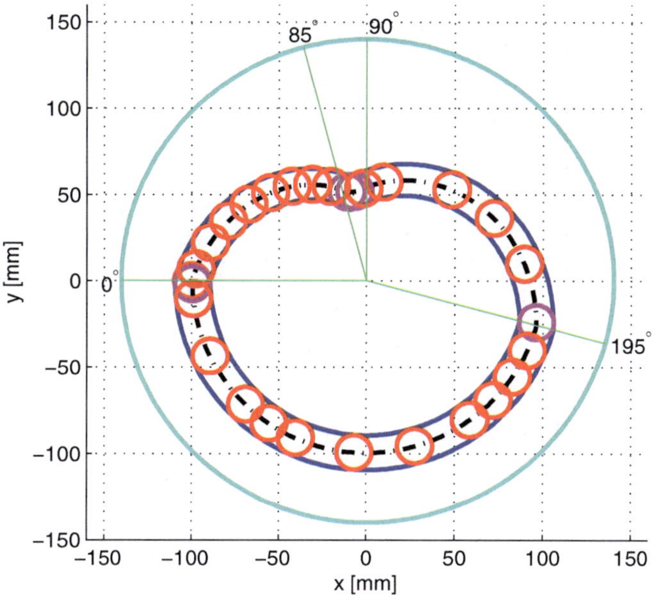

Fig. 5. Cam profile of the drawing-in mechanism with initial angle $\gamma_0 = 0$.

Table 2. Critical discrete positions of the cam profile.

position n	$[x_i, y_i]$ [mm]	ϕ [°]
1	[−90, 0]	0
86	[−9.4837, 41.3307]	85
91	[−5.8454, 42.0000]	90
196	[86.4875, −23.1737]	195
360	[−89.5384, −0.0005]	359

3.2 Design Case 2 of $\gamma_0 = -15°$

With the identical geometric parameters, the cam profile can be formed when $\gamma_0 = -15°$, as displayed in Fig. 6. In this design, the cusp disappears and the profile becomes more smooth, although the overall dimension increases significantly. This means that this design can reduce the mechanical shock and vibration when the machine works.

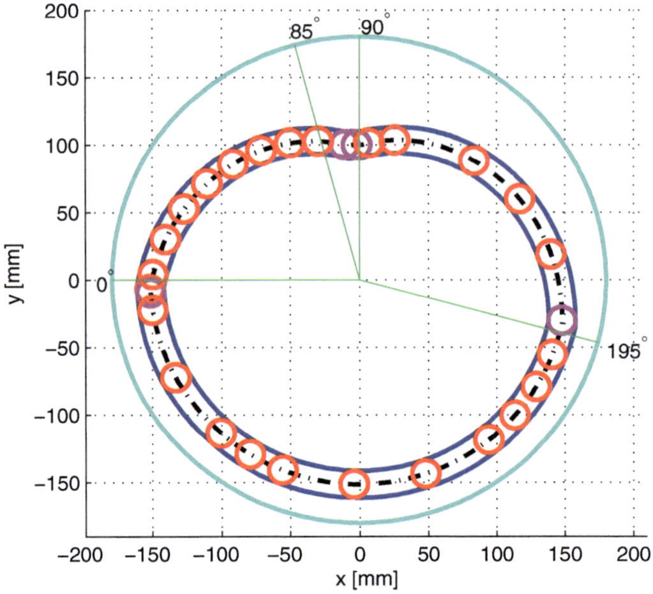

Fig. 6. Cam profile of the drawing-in mechanism with initial angle $\gamma_0 = -15°$.

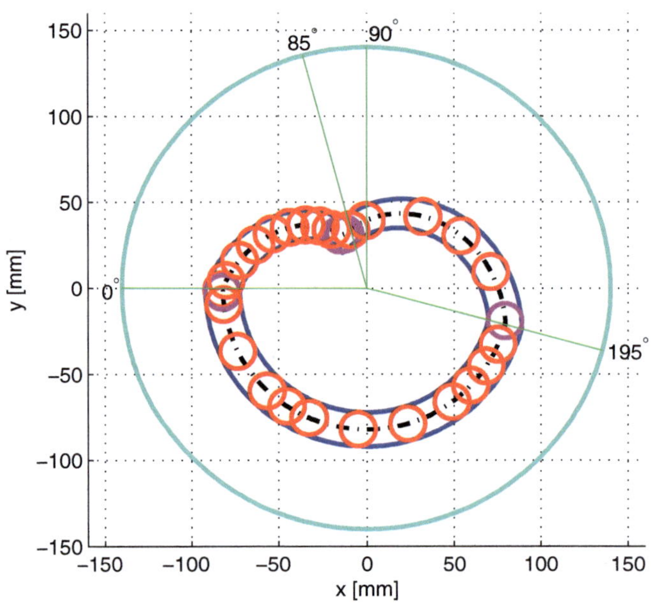

Fig. 7. Cam profile of the drawing-in mechanism with initial angle $\gamma_0 = 5°$.

3.3 Design Case 3 of $\gamma_0 = 5°$

Similarly, the cam profile with $\gamma_0 = 5°$ is depicted as shown in Fig. 7. Compared to the design in Fig. 5, the dimensions of the cam reduce. It should be noted that the cusp occurs and becomes sharp.

From the comparison of the previous three designs, the case 2 could be a better design candidate, as this design lead to smooth cam profile for machine running. The three designs imply that the negative angle γ_0 depicted in Fig. 4 is suitable for the design of the cam profile.

4 Conclusions

This paper deals with the dimensional synthesis of a cam-follower based drawing-in mechanism for the automatic warp tying machine. The drawing-in mechanism is modeled and introduced. The corresponding drawing-in processes in the textile products is depicted in 4-5-6-7 polynomial equation. The approach of the kinematic geometry is adopted to design the cam profile. A comparison among three design cases with respect to the initial angle of the section gear, namely, $\gamma_0 = -15°$, $\gamma_0 = 0$ and $\gamma_0 = 5°$, is discussed. When the angle γ_0 is greater than zero, cusp on the cam profile occurs, which may induce shock and vibration of the mechanism. The comparison results show the design $\gamma_0 > 0$ suitable for the drawing-in mechanism to ensure its smooth motion.

Acknowledgement. The supports from the Applied Basic Research Programs of Changzhou (No. CJ20180017) and the Longcheng Talent Program of Changzhou city (No. PT-CXY-CZ2019-WGL-01) are gratefully acknowledged.

References

1. Beggs, J.S.: Mechanism. Mcgraw-Hill, New York (1955)
2. China Business Data Network, Beijing: Development Trend Forecast and Investment Consultation Report of China's Warp-piercing Rack and Reed-piercing Machine Industry in 2018–2023 (2017)
3. China Economic Vision, Beijing: Analysis and Forecast Report on Market Prospect of China Textile Industry (2019)
4. Faires, V.M.: Kinematics. McGraw-Hill, New York (1959)
5. Hain, K.: Optimization of a cam mechanism-to give good transmissibility, maximal output angle of swing and minimal acceleration. J. Mech. **6**(4), 419–434 (1971)
6. Hinkle, R.T.: Kinematics of Machines. Prentice-Hall, New Jersey (1953)
7. Norton, R.L.: Effect of manufacturing method on dynamic performance of camsan experimental study Part–I eccentric cams. Mech. Mach. Theor. **23**(3), 191–199 (1988)
8. Norton, R.L.: Design of Machinery. McGraw-Hill, New York (2004)
9. Qian, Z.: Research on constant-diameter cam mechanism with a planar motion follower. Mech. Mach. Theor. **42**(8), 1017–1028 (2007)
10. Tsay, D.M., Huey, C.O.: Application of rational B-splines to the synthesis of cam-follower motion programs. J. Mech. Des. **115**(3), 621–626 (1993)
11. Wang, D., Gao, Y.: Theory of Mechanisms and Machine (in Chinese). China Machine Press, Beijing (2011)

Transmission Design and Experimental Research of a Non-contact Nutation Reducer

Jiaxin Ding, Ligang Yao(✉), Yongwu Cai, Xiaqi Shan, and Jun Zhang

School of Mechanical Engineering and Automation, Fuzhou University, Fuzhou, Fujian, China
ylgyao@fzu.edu.cn

Abstract. Robot joint reducer is the most basic and important part of industrial automation. At present, the commonly used joint reducers are harmonic reducers, RV reducers and so on, which exist the disadvantages of large weight, large shape size, poor bearing capacity and limited service life. In view of these shortcomings, a non-contact nutation reducer is proposed based on the principle of nutation and the principle of magnetic drive. The reducer has the advantages of large transmission ratio, smooth transmission and low noise, and is more suitable for use in robot joint reducer. In this paper, the working principle, structure design, manufacture, assembly and experiment of the reducer are studied, and the experimental results under different input speeds and different output loads are obtained and discussed. The experiment verifies the advantages of the transmission ratio of the nutation reducer, and the transmission is very stable under high speed and load conditions.

Keywords: Nutation reducer · Magnetic drive · Experiment · Non-contact transmission

1 Introduction

Robot joint reducer as one of the most basic and important transmission parts in industrial automation production, plays a more and more significant role with the continuous progress of industrial technology. There are inevitably defects such as lubrication, wear and overload damage in the traditional mechanical contact gear transmission [1]. With the increasing complexity of the exploration environment, people hope to develop a reducer with small size, light weight, small wear, large transmission ratio and high drive accuracy in order to adapt to the increasingly harsh industrial production requirements [2, 3].

Nutation transmission is a new form of coaxial transmission, which can be used in industrial automation inspection, aerospace, automobile, navigation and other fields. Its advantages of large transmission ratio, small volume, compact structure and high bearing capacity are very suitable for using in robot reducer with small working space [4]. Currently, for the research work of the nutation transmission, experts and scholars at home and abroad mainly focus on the innovative design of the nutation mechanism,

D. Wang et al. (Eds.): MeTrApp 2019, MMS 79, pp. 70–80, 2020.
https://doi.org/10.1007/978-981-15-0142-5_8

such as the nutation traction movement, the nutation stepper motor, the nutation hydraulic motor and so on [5–8].

Magnetic drive is a new type of transmission to realize the force and torque transmission using the principle of magnetic traction [9]. Because of the non-contact characteristic between the magnetic parts, the transmission needs not to consider the problems of friction, wear and lubrication. At the same time the magnetic drive also has the advantages of smooth transmission and overload protection [10–13].

Combining with the advantages of the above two transmission modes, this paper puts forward a new non-contact nutation reducer utilizing permanent magnet coupling instead of gear meshing to realize the transmission. The reducer not only has the advantages of large transmission ratio, smooth transmission, simple structure, easy installation, but also the preponderance including low noise and no wear, resulting in achieving overload protection.

2 Mathematical Modelling of the Transmission Ratio

Nutation motion is a new type of deceleration transmission originated from the principle of gyroscope motion, which has the characteristics of large transmission ratio, compact structure and high transmission efficiency. The process of nutation transmission can often be demonstrated by the rotation of a coin on the desktop shown in Fig. 1. The initial coin axis and the desktop normal angle is 90°. Due to the actions of the desktop resistance and the gravity of the earth and other external forces at the same time, the coin in the rotation process will produce a deflection. The angle between the axis and the desktop normal gradually reduced with the movement, and finally stopped on the desktop. The deflection movement in this process is called nutation motion. Throughout the movement, the edge of the coin has only one point in contact with the desktop, and the angle between the coin axis and the desktop normal is what we call the nutation angle.

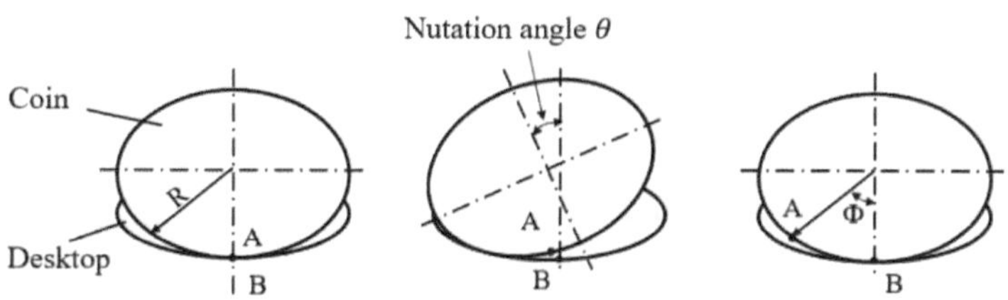

Fig. 1. Principle of nutation motion.

As shown in Fig. 1, coin axis and desktop normal is sharp angle, assuming that the coin and desktop contact points are A and B respectively. In the process of nutation, point A away from point B, there is a certain dislocation angle between the coin and the desktop track circle. When other points on the coin come into contact with point B again, there is an arc difference between point A and point B. Assuming that the radius of the coin is R, the nutation angle is θ, that is the radius of the trajectory in contact with the edge of the coin on the table is less than that of the coin.

According to the hypothesis, the circumference of the coin and the desktop track are 2πR and 2πRcosθ, respectively, so the arc difference S between the fixed point A and B can be calculated:

$$S = 2\pi R(1 - \cos\theta) \quad (1)$$

The stepping angle Φ can be expressed as:

$$\Phi = \frac{S}{R} = 2\pi(1 - \cos\theta) \quad (2)$$

Due to the existence of stepping angle and stepping displacement, there is a certain dislocation angle between the track circles of the coin and the desktop. The stepping angle will be increased with the increase of the nutation angle, according to the design experience of the nutation reducer, the design range of the nutation angle is generally in 0° to 20° [3].

Combining the principle of nutation movement and push-pull principle of permanent magnet, the coin and the desktop are designed as magnetic nutation gears. As illustrated in Fig. 2, the permanent magnets are arranged in alternating order of N pole and S pole. And there is a constant nutation angle θ between the axes of the magnetic gear and the output shaft. The angle can be realized by using the oblique axis or the straight shaft with the eccentric nutation sleeve. There is an unequal spacing air gap between the two magnetic gears, which relies on the non-contact magnetism to realize the transmission of power and motion, rather than relying on the mechanical contact of gears to realize the transmission of power and motion.

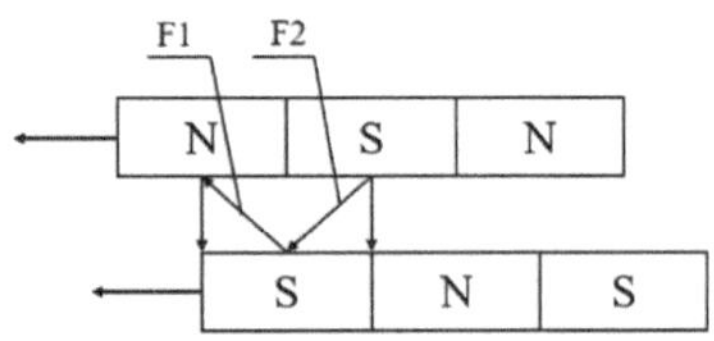

Fig. 2. Push-pull principle of permanent magnet.

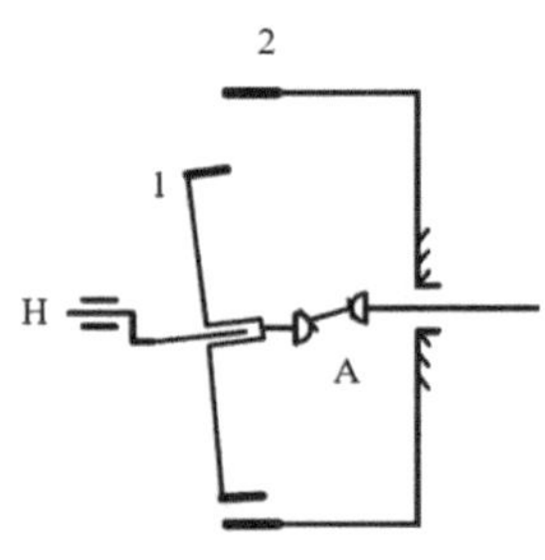

Fig. 3. Single-stage non-contact nutation drive mechanism.

According to Fig. 1, the coin is replaced by a rotating gear 1 with p_1 numbers of permanent magnets around the nutation disk. And the desktop is replaced by a fixed gear 2 containing p_2 numbers of permanent magnets around. There is a constant nutation angle θ between the axis of the nutation disk and the axis of the gear 2. Then a single-stage non-contact nutation transmission mechanism is formed, as presented in Fig. 3. Thus, the transmission ratio of a single-stage non-contact nutation mechanism can be calculated as:

$$\mathrm{i} = \frac{2\pi}{\Phi} = \frac{1}{(1-\cos\theta)} = \frac{p1}{p1p2} \tag{3}$$

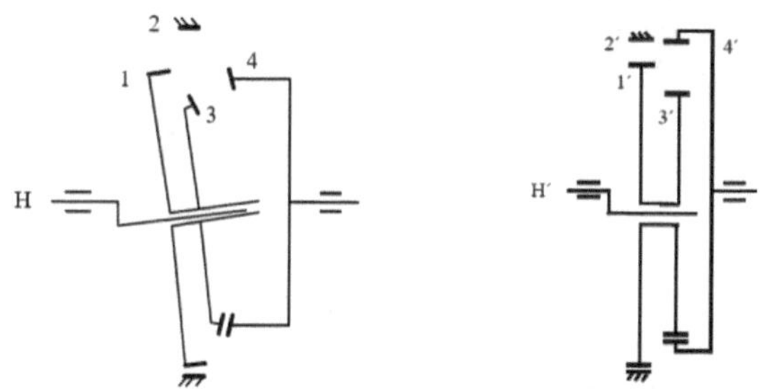

(a) The structure of two-stage reducer. (b) The planetary gear transmission.

Fig. 4. Two-stage non-contact deceleration mechanism.

Similarly, the brief structure of two-stage reducer is shown in Fig. 4(a), when the nutation angle θ is 0°, its structure is transformed into the planetary magnetic gear transmission shown in Fig. 4(b). The number of magnetic poles set on the magnetic nutation gear 1, magnetic gear 2, magnetic nutation gear 3 and magnetic gear 4 is p_1, p_2, p_3 and p_4, respectively. The corresponding angular velocities are: ω_1, ω_2, ω_3 and ω_4, and the input axis angular velocity is ω_H. When making the magnetic gear 4 as the output component, the magnetic gear 2 is fixed, that is $\omega_2 = 0$. According to Fig. 4(b), the planetary magnetic gear transmission can be calculated at:

$$\mathrm{i}_{H4}^{2} = \frac{\omega_2 - \omega_H}{\omega_4 - \omega_H} = \frac{-\omega_H}{\omega_4 - \omega_H} = (-1)^0 \frac{p_1 p_4}{p_2 p_3} = \frac{p_1 p_4}{p_2 p_3} \tag{4}$$

After finishing, the total transmission ratio i is:

$$\mathrm{i} = \frac{\omega_H}{\omega_4} = \frac{p_1 p_4}{p_1 p_4 - p_2 p_3} \tag{5}$$

It can be seen from formula (4) that as the absolute value of p_1p_4 - p_2p_3 decreases, the gear ratio will gradually increase. When p_1p_4 - $p_2p_3 = \pm 1$, the transmission ratio reaches the maximum. Generally, the transmission ratio of the two-stage nutation transmission can reach 10:1-3000:1.

3 Structural Design and 3D Model

The new non-contact nutation reducer is shown in Fig. 5, including magnetic nutation gear 1, magnetic gear 2, magnetic nutation gear 3, magnetic gear 4, nutation sleeve, input shaft and other parts. The structure of the input shaft and the nutation sleeve is shown in Fig. 6. The magnetic nutation gear 1 and the magnetic gear 2 constitute the first stage nutation transmission. The magnetic gear 3 and magnetic gear 4 constitute the second stage nutation transmission. When the external power drives the input shaft to rotate, the power is passed through the key to the nutation sleeve to make it rotate. Because the magnetic gear 2 is a fixed link, the magnetic nutation gear 1 does the nutation movement, which constitutes the first stage of deceleration, as shown in Fig. 7. Similarly, the magnetic nutation gear 3 and the magnetic gear 4 are slowed down for the second stage of deceleration, and the power output is finally carried out by the magnetic gear 4, as shown in Fig. 8.

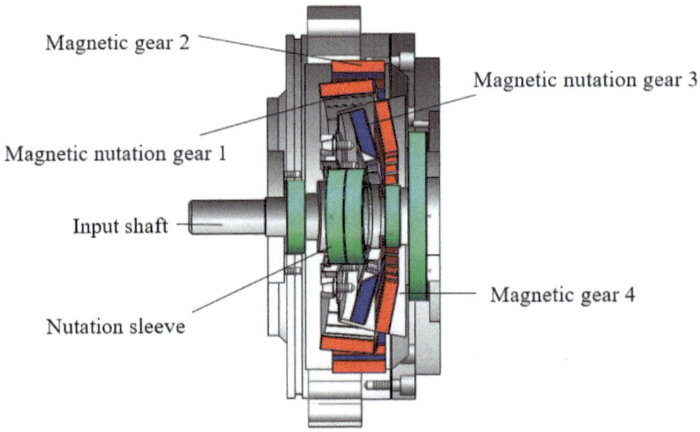

Fig. 5. 3D sectional view of non-contact nutation reducer.

Fig. 6. 3D model of input shaft and nutation sleeve.

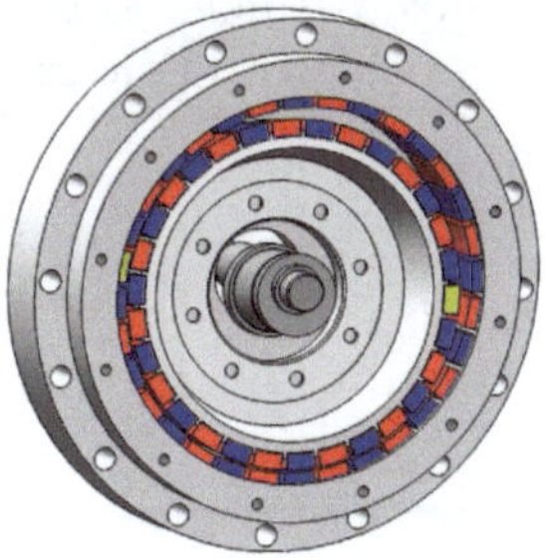

Fig. 7. 3D model of first-stage deceleration.

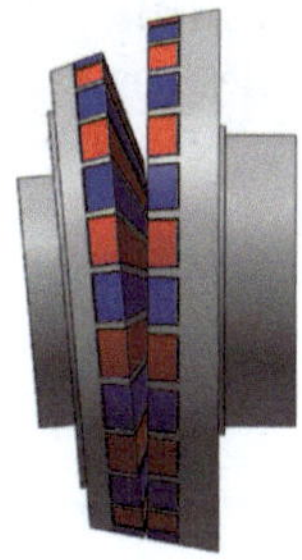

Fig. 8. 3D model of second-stage deceleration.

The main parameters of the non-contact nutation reducer include: nutation angle θ, the magnetic pole number p_i of each magnetic gear, the air gap magnetic field spacing l and so on. The nutation angle affects the stepping angle of the non-contact nutation reducer, and the magnetic pole number affects the transmission ratio of the reducer. According to the formula (4), the total transmission ratio of the non-contact reducer is calculated to 105. Table 1 lists the main structural parameters of the non-contact nutation reducer.

Table 1. Main structural parameters of non-contact nutation reducer.

Parameter	Value
Nutation angle (θ)	5°
Maximum diameter of box (D)	162 mm
Length of box (L)	65 mm
Diameter of input shaft (d)	12 mm
The magnetic pole number of the magnetic nutation gear 1 (p_1)	30
The magnetic pole number of the magnetic gear 2 (p_2)	32
The magnetic pole number of the magnetic nutation gear 3 (p_3)	26
The magnetic pole number of the magnetic gear 4 (p_4)	28
The air gap magnetic field spacing (l)	1 mm
The total transmission ratio (i)	105

This non-contact reducer has two types of poles, a tile and a fan, as shown in Fig. 9. The structure of the tile-shaped magnetic pole is mainly determined by two radii, arc and length, and the thickness of the magnetic pole is also determined after the two radii sizes are determined. The structure of the sector pole is mainly determined by two radii, curvature and thickness, and the difference between the two radii determines the length of the pole. The tile magnet is magnetized radially and the fan-shaped magnet is magnetized axially. The sizes of two magnetic poles and two magnetization forms are shown in Fig. 9. In this paper, permanent magnets are made of N40UH and the magnetic pole sizes are shown in Table 2. It is worth mentioning that, due to the small angle of the magnetic pole, the processed magnetic pole is similar to the rectangular parallelepiped, that is, the magnetic pole thickness can be used to indicate the difference in size between the two radii.

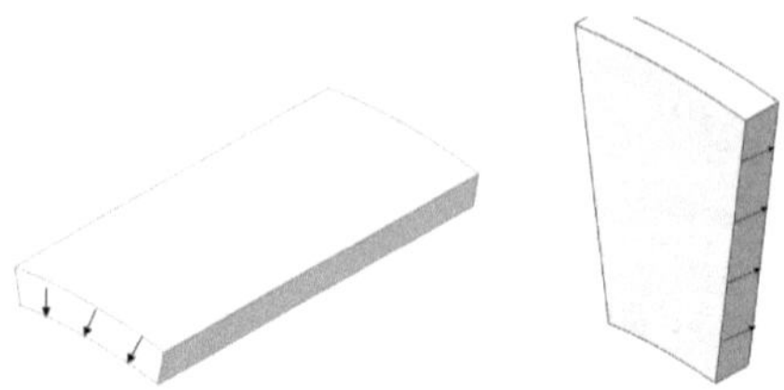

(a) Tile-shaped magnet. (b) Fan-shaped magnet.

Fig. 9. Magnetic pole shape and magnetization form.

Table 2. The sizes of the magnetic pole.

Parameter	Value
The length of the magnetic nutation gear 1 (l_1)	20 mm
The thickness of the magnetic nutation gear 1 (d_1)	5 mm
The angle of the magnetic nutation gear 1 (φ_1)	12°
The length of the magnetic gear 2 (l_2)	20 mm
The thickness of the magnetic gear 2 (d_2)	5 mm
The angle of the magnetic nutation gear 2 (φ_2)	11.25°
The length of the magnetic nutation gear 3 (l_3)	20 mm
The thickness of the magnetic gear 3 (d_3)	5 mm
The angle of the magnetic nutation gear 3 (φ_3)	13.8°
The length of the magnetic nutation gear 4 (l_4)	20 mm
The thickness of the magnetic gear 4 (d_4)	5 mm
The angle of the magnetic gear 4 (φ_4)	12.8°

4 Transmission Performance Experiment

4.1 Prototype Processing and Assembly

Based on the above structural design, the parameters are processed in Tables 1 and 2, and the prototype is shown in Fig. 10

Fig. 10. Prototype of the non-contact nutation reducer.

4.2 Construction of Transmission Performance Experimental Rig

The transmission performance experimental rig is shown in Fig. 11, containing loading device, speed torque sensor, non-contact nutation reducer prototype, speed torque sensor, input motor. The input shaft of the prototype is connected to the input motor with the coupling, and the output shaft is connected to the loading device with the coupling disk. During the experiment, the two speed torque sensors measure the speeds and torques of the input shaft and output shaft in real time, respectively.

Fig. 11. Transmission performance experimental rig.

4.3 Analysis of Results

In the process of the experiment, the output speeds are measured under the different input speeds such as 500 rpm, 725 rpm, 1000 rpm and 1250 rpm, respectively. Moreover, the actual transmission ratio is calculated by $i_e = \omega_{in}/\omega_{out}$, which is compared with the theoretical transmission ratio, and the contrast curve is illustrated in Fig. 13.

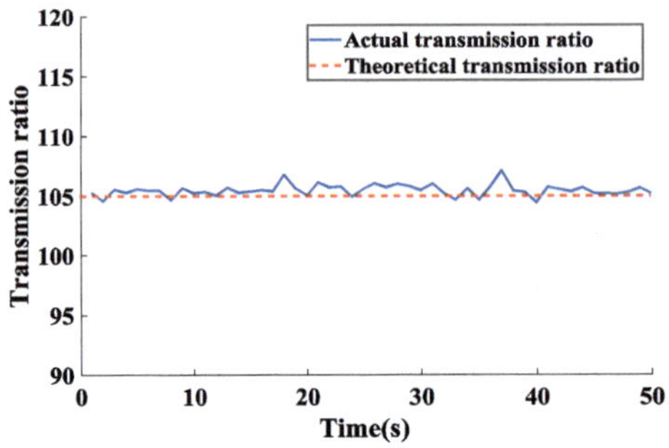

(a) Actual gear ratio of 500 rpm.

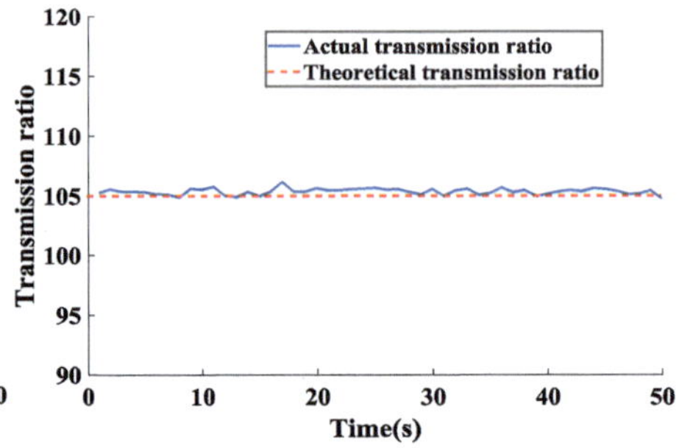

(b) Actual gear ratio of 725 rpm.

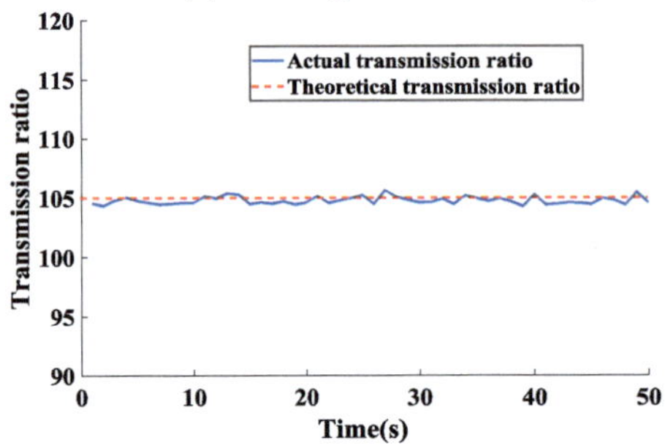

(c) Actual gear ratio of 1000 rpm.

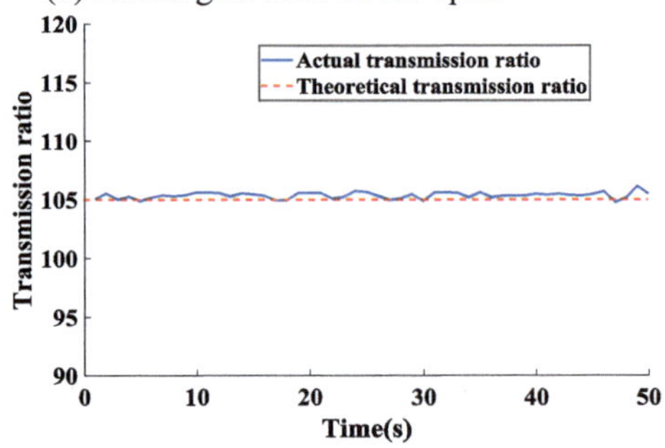

(d) Actual gear ratio of 1250 rpm.

Fig. 12. Transmission ratio comparison of theoretical gear ratio and actual gear ratio.

Figure 12(a)–(d) are easy to compare, with the increase of input speed, the actual transmission ratio is constantly approaching the theoretical transmission ratio of 105. When the input speed is low, such as 500 rpm, the output speed fluctuation is large, this aspect is due to the large transmission ratio, the permanent magnet between magnetic gears has the movement of attracting magnets or hysteresis magnets in advance, which leads to the fluctuation of rotational speed in low speed state, and on the other hand, it is caused by the lack of precision of magnetic pole installation and the absence of special tooling fixture.

Figure 13 shows the transmission ratio curve of different input speeds in the case of no-load. In the no-load state, the actual transmission ratio of the non-contact nutation reducer fluctuates near the theoretical transmission ratio 105, which meets the design requirements.

Then, we conducted a transmission ratio experiment under different loads at an input speed 1000 rpm. The actual gear ratio of the reducer was tested under different load conditions of 0 N•m, 5 N•m, 10 N•m, and the experimental results are shown in Fig. 13. The gear ratio still has a small range of fluctuations, in line with the transmission design.

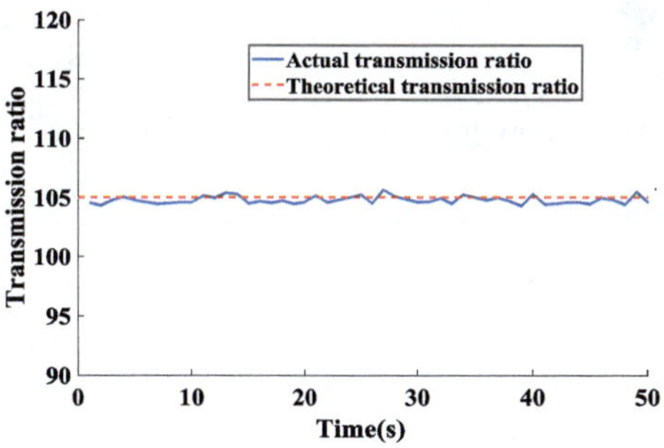

(a) Actual gear ratio of 0 N·m.

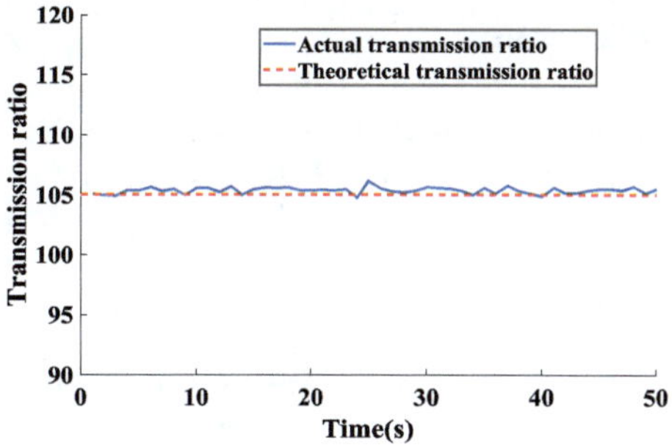

(b) Actual gear ratio of 5 N·m.

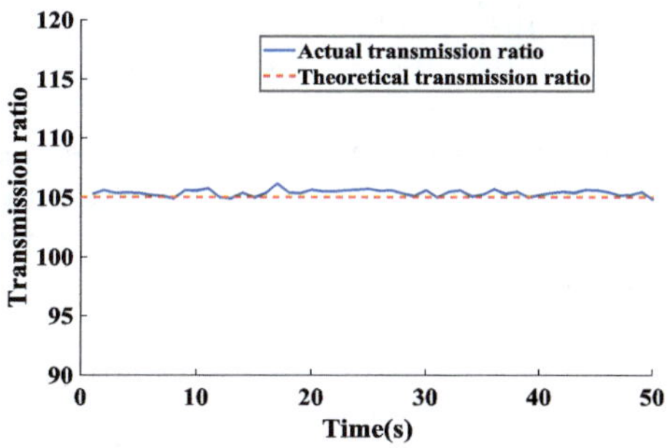

(c) Actual gear ratio of 10 N·m.

Fig. 13. Actual transmission ratio under different loads.

5 Conclusion

Based on the principle of nutation transmission and the principle of magnetic drive, a new type of non-contact nutation reducer is proposed in this paper, which has the advantages of large transmission ratio, smooth transmission. At the same time, the transmission ratio formula of the reducer is deduced, and the transmission ratio is verified by experiments and the curve of transmission ratio in the transmission process is measured. The experimental results show that the transmission ratio of non-contact nutation reducer fluctuates greatly under the low speed and no-load state, and with the increase of rotational speed, the transmission ratio is stable near the ideal transmission ratio, but there are occasional large fluctuations due to errors caused by machining and assembly. Under the load state, the transmission ratio is smoother and stabilizes with the increase of the load.

Acknowledgments. The authors acknowledge the support of the National Nature Science Foundation of China (Grant Nos. 51775114 and 51875105).

References

1. Ge, T., Gong, Z.: Research status of noise and vibration of gears and gearboxes. Noise Vibr. Control **1**, 26–30 (1989)
2. Sensinger, J., Lipsey, J.: Cycloid vs. harmonic drives for use in high ratio, single stage robotic transmissions. In: 2012 IEEE International Conference on Robotics and Automation. IEEE, Saint Paul, MN, USA (2012)
3. Cai, Y., Yao, L., Zhang, J., et al.: Feasibility analysis of using a two-stage nutation drive as joint reducer for industrial robots. J. Mech. Sci. Technol. **33**(4), 1799–1807 (2019)
4. Hong, J., Yao, L., Ji, W., et al.: Kinematic modeling for the nutation drive based on screw theory. Procedia CIRP **36**, 123–128 (2015)
5. Saribay, Z.B., Bill, R.C.: Design analysis of pericyclic mechanical transmission system. Mech. Mach. Theor. **61**, 102–122 (2013)
6. Shi, J.: Simulation of variable speed curve of nutation traction CVT. Sci. Technol. Innov. Herald **15**, 95–97 (2017)
7. Fang, T., Ke, T., Fang, J.: Design and test of nutation reducer used in screw pump. J. Mech. Transm. **29**(1), 50–52 (2005)
8. Yao, L., Gu, B., Wei, G.: Analysis and modeling of double circular spiral bevel gear nutation drive. In: Proceedings of the 2005 Annual Meeting of the Chinese Society of Mechanical Engineering. CMES, Chongqing, CHN (2005)
9. Cao, J., Yao, C., Wang, J., et al.: Modeling and controller design of a magnetic-geared double rotor motor. J. Mech. Transm. **42**(2), 151–162 (2018)
10. Li, G., Jia, R., Ji, Q.: Magnetic drive and magnetic pump. J. Chin. Soc. Rare Earths. **S1**, 534–537 (1994)
11. Mattia, F., Piergiorgio, A., Gregor, G., et al.: Magnetic transmission gear finite element simulation with iron pole hysteresis. Open Phys. **16**(1), 105–110 (2018)
12. Lee, Y.H., Park, G.S.: Analysis of the resonance characteristics by a variation of coil distance in magnetic resonant wireless power transmission. IEEE Trans. Magn. **54**(11), 1–4 (2018)
13. Eiki, M., Noboru, N., Katsuhiro, H.: Magnetic-geared motor with a continuously variable transmission gear ratio and its control method. Electron. Commun. Jpn **138**(2), 164–172 (2018)

Releasable Locking Mechanism Using Friction and a Cuneus for Holding a Posture

Koji Makino(✉), Tatsuya Kaneko, Kenji Matsumura, and Hidetsugu Terada

University of Yamanashi, Takeda 4-3-11, Kofu, Japan
kojim@yamanashi.ac.jp

Abstract. Maintaining the posture of the upper limbs for a long time without any support is hard work. Physical strain can be reduced by holding the joint of a part of the body; however fixing it rigidly can be dangerous. This study proposes a releasable locking mechanism to hold a posture using friction and a cuneus. Moreover, we verify that the maximum torque applied to the instrument is controllable, both by theoretical calculations and experiments.

Keywords: Locking mechanism · Friction · Cuneus · Wearable robot

1 Introduction

The development of a machine or a robot to reduce physical strain during heavy work is of high importance. An automatic robot, e.g., a humanoid robot or a robot arm coupled to an intelligent mobility, is an effective way to this end. However, a flexible decision of a human is necessity in various cases such as nursing work, plant work and so on. Therefore, it is desirable to develop an assist robot of a wearable type, since this gadget can easily reflect the intention of the user. The wearable assist robot is classified to three types. One of those is a power–assist robot. Numerous power–assist robots were developed and studied [1,2]. The second type is a timing–assist robot. For rehabilitation, the timing of the motion is very important. However, a larger than necessary power–assist should not be employed, since the muscles of the patient may become weak. One example is a knee assistive instrument for walking rehabilitation [3]. The third type is a wearable robot for holding the posture. In fact, maintaining a posture such as the raised arm posture [4] and half–sitting posture [5], represent heavy tasks. As an example, workers raise their arms for an extended period time for maintenance of the wires on a ceiling or to prune a plant. Generally, the power–assist robot and the timing–assist robot are very costly to use for simple tasks. By contrast, the assist robot that holds the posture is not expensive because of its simple mechanism. Therefore, the development of the assist robot for holding posture is desirable if many users employ the wearable robot.

D. Wang et al. (Eds.): MeTrApp 2019, MMS 79, pp. 81–90, 2020.
https://doi.org/10.1007/978-981-15-0142-5_9

This study focuses on the mechanism of holding posture. There are important aspects involved in the mechanism of holding posture. That is, the system needs to be lightweight, to have a long continuous driving time, and to have an easy releasable locking mechanism. To realize the lightweight and long driving time attributes, actuators that need large currents or an air tank, (i.e., a motor or air cylinder has not), are to avoided. Serval robots were proposed and developed to realize the lightweight and long continuous driving time attributes [6]. Moreover, the various mechanisms to realize the robot mechanisms have been proposed and developed [7]. In these systems, it is important that the posture is not held completely and rigidly. Complete and rigid holding means that the user cannot move his or her body parts unless the release motion or a special procedure is executed. If a large force is exerted while holding the body parts, the user risks an injury since the overload acts on the body parts. However, this concern is not addressed in many studies.

This study proposes a mechanism to hold posture using friction and a cuneus such that the user is able to alter his or her posture without special motions and procedures. In Sect. 2, structure of the proposed mechanism and the principle of hold and release are discussed. Furthermore the mechanism is executed experimentally to confirm the motion. The experimental conditions and results are shown in Sect. 3. Finally, the conclusion is provided in Sect. 4.

2 Mechanism of Holding a Posture

In this section, the proposed mechanism for holding a posture is described. First, some target postures are illustrated to demonstrate the necessity for the mechanism for holding these postures. Second, the principle of the mechanism employing friction and a cuneus is described. Finally, the motion of the developed mechanism is analyzed and the forces used to hold the posture are calculated.

2.1 Target Postures

In this research, target postures are analyzed with respect to their maintenance, without assistance in motion. Maintaining the posture for a long time is hard work. In an example, the worker pruning the tree keeps his hands raised for a long time, as shown in Fig. 1. In another example, the caregiver or nurse sometimes changes the position of a person confined to a bed in order to prevent bedsores. In this act, the caregiver holds the position of his or her own upper arm, as shown in Fig. 2.

To realize the assisting instrument for the maintenance of the posture, a stopper, such as a bar, is useful in holding the motion. If the joint is fixed rigidly using the stopper bar, the following risky events might happen.

- The worker becomes injured in the case when overload is added.
- The worker cannot move to avoid the slip, even if he or she falls.

Therefore, it is desirable to release the stopper automatically if overload is added.

Fig. 1. Worker raising one arm to prune the plant.

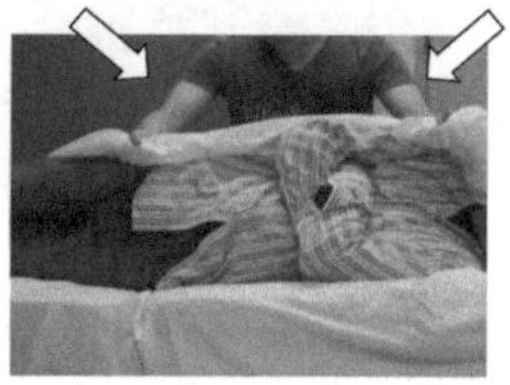

Fig. 2. Worker holding out both arms to change the patient's position.

2.2 Proposed Mechanism Using Friction and a Cuneus

In this study, a cuneus and friction are employed to hold the joint. In general, the cuneus is not released when strong forces are applied. We propose a new method using a cuneus to hold the joint while the torque is beneath a specified threshold value, and to release the joint if the torque becomes larger. The proposed mechanism is shown in Fig. 3. This mechanism consists of a friction disk with V-shaped grooves, a locking piece, and a guide block with a gap.

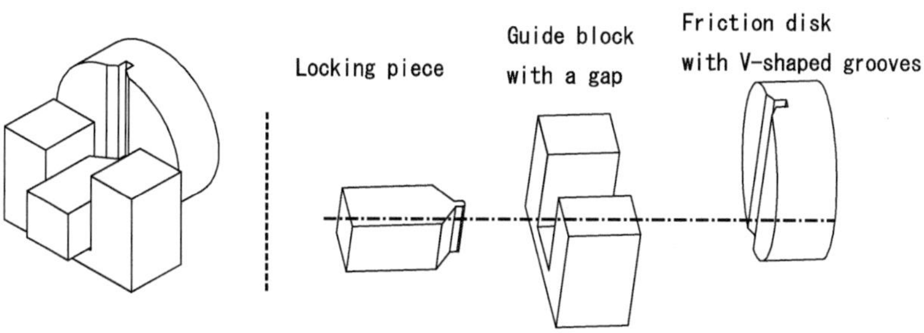

Fig. 3. Structure of the proposed mechanism.

The procedure of holding the motion of the friction disk is described. For the sake of simplicity of the argument, the motion is considered in two–dimensions, as shown in Fig. 4. The friction disk, the locking piece, and the guide block are arranged. The locking piece slides along the guide block. The gap distance in the guide block is slightly larger than the width of the locking piece.

First, the locking piece is placed on the gap of the guide block, as shown in Fig. 4(a). The locking piece does not contact the friction disk, such that the friction disk can move freely.

Second, if the locking piece is pushed, it moves toward the friction disk, as shown in Fig. 4(b). The locking piece collides with the friction disk. The pushing motion is controlled, for example, using a microcomputer.

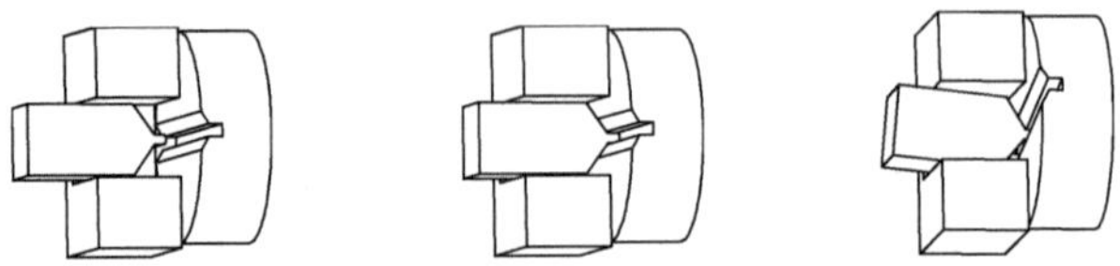

(a)release position (b)contact position (c)lock position

Fig. 4. Principle of the mechanism used to hold the motion.

Subsequently, the locking piece is inclined upon movement of the friction disk. Figure 4(c) shows the arrangement of the stopped state by friction and the cuneus. Three points of the locking piece are in contact with to the friction disk and the guide block. Consequently, the locking piece cannot move. In this case, if the force to change the position of the friction disk is larger than the static friction force, the friction force is shifted to dynamic friction force from static friction force. The balance of the three points of the forces of the locking piece is not maintained in case where the static friction force shifts to dynamic friction. Therefore, the position of the locking piece returns to its initial position shown in Fig. 4(a).

2.3 Calculation of Maximum Torque

In this mechanism, the gap distance of the guide block influences the inclination angle of the locking piece. The maximum friction force between the locking piece and guide block is changed according to the inclining angle. In this section, the maximum torque force is calculated.

First, the relation between the length and the angle is illustrated in Fig. 5. The descriptions of these parameters are listed in Table 1. x denotes the difference between the width of the locking piece (L_2) and the gap distance of the holding block (D_2).

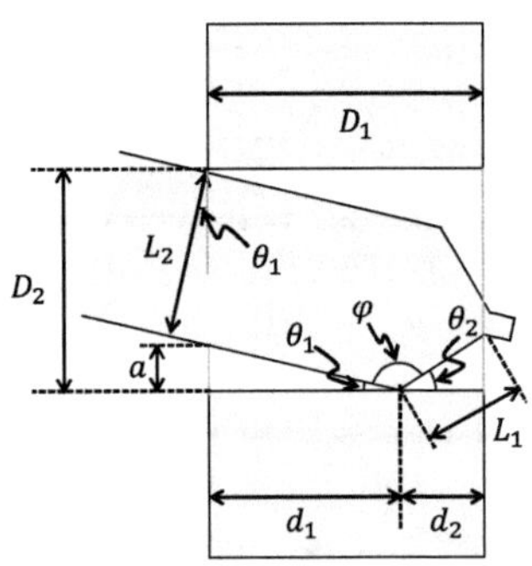

Fig. 5. Relation of the positions of the piece and the guide block.

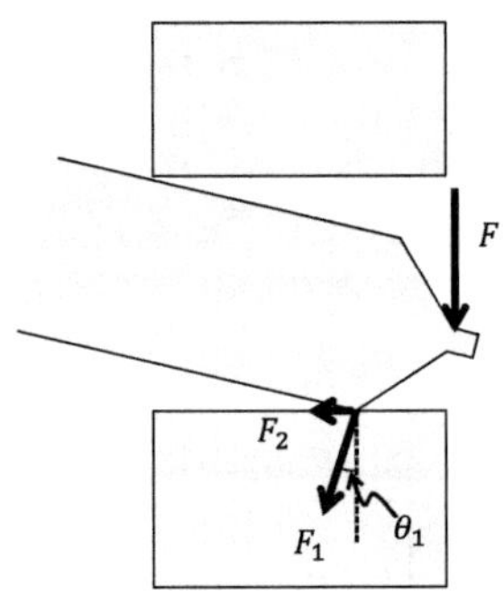

Fig. 6. Relation of the forces of the locking piece and the guide block.

Table 1. Parameters of the mechanism to calculate the maximum friction force.

Name	Meaning
φ	Angle of the locking piece
θ_1, θ_2	Angles between the locking piece and the guide block
D_1	Width of the guide block
D_2	Gap distance of the guide block
d_1, d_2	Length from the contact point between the locking piece and the guide block to the edge of the guide block
L_1	Length of the sharp shape of the locking piece
L_2	Width of the locking piece

The geometrical relation is obtained as follows;

$$\begin{aligned} \theta_1 + \theta_2 + \varphi &= \pi \\ d_1 + d_2 &= D_1 \\ d_1 \tan\theta_1 &= a \\ d_2 &= L_1 \cos\theta_2 \\ a + \frac{L_2}{\cos\theta_1} &= D_2 \\ D_2 &= x + L_2. \end{aligned}$$

The relation between x and θ_1 is given by Eq. (1):

$$x = L_2\left(\frac{1}{\cos\theta_1} - 1\right) + \{D_1 - L_1\cos(\pi - \varphi - \theta_1)\}\tan\theta_1. \quad (1)$$

Second, the friction between the locking piece and the guide block is calculated. The relation between these forces is illustrated in Fig. 6. F denotes the input force to move the friction disk, which is converted from the input torque.

$$\begin{aligned} F_1 &= F\cos\theta_1 \\ F_2 &= F_1 \sin\theta_1 \end{aligned}$$

The locking piece slips if the friction force F_2 is larger than the maximum friction force. It is difficult to calculate the maximum friction force, since the corner of the locking piece is in contact with the plane of the guide block and both materials are metal. For simplicity of the argument, we assume that the maximum friction force is constant. The maximum friction force is denote as F_{max}. The relation between the maximum friction force F_{max} and θ_1 is given by Eq. (2):

$$F_{max} \geq F_2. \quad (2)$$

The relation between F and θ_1 is given by Eq. (3):

$$F \leq \frac{F_{max}}{\cos\theta_1 \sin\theta_1}. \tag{3}$$

The maximum input force is obtained, when Eq. (3) satisfies the equal sign. The relation between the gap x and the maximum input force F is calculated using Eqs. (1) and (3). However, obtaining the equation of the relation directly is difficult, it is calculated by numerical simulation. The values of the parameter are follows; $\varphi = 120°$, $D_1 = 6.0$ mm, $D_2 = 6.5$ mm, $L_1 = 3.0$ mm, $L_2 = 6.5$ mm. These values were likewise applied in the experimental mechanism in Sect. 3.

The simulation results are shown in Fig. 7. The vertical axis depicts the maximum input force, whereas the horizontal axis specifies the gap distance between the width of the locking piece and the gap distance of the guide block. The maximum input force is normalized by the maximum friction force. It is clear from the graph that the relation is approximately inversely proportional.

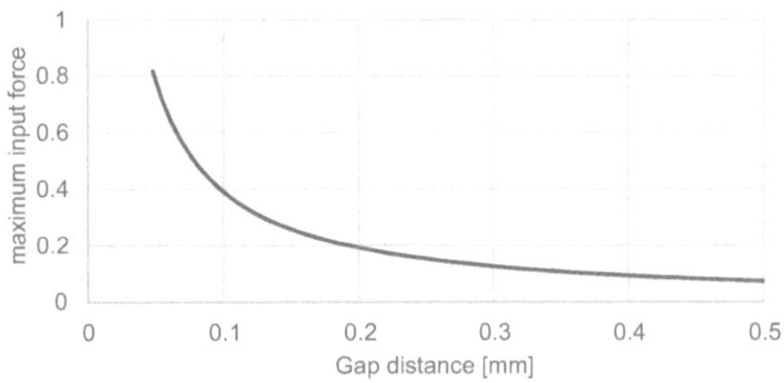

Fig. 7. Result of the numerical simulation.

3 Experiment

In this section, the experimental instrument and the results obtained with it are described. Two kinds of results (time variation and gap influence) are shown.

3.1 Experimental Environment

In this subsection, the actual mechanism of fabrication by metal parts is explained, and the experimental equipment used to measure the maximum friction torque is described.

First, the friction disk with V-shaped grooves, the locking piece, and the guide block with a gap are described. Figure 8 illustrates these parts. The shape of the opposite edge of the locking piece is added the stopper, which has a triangular shape. This shape does not influence the performance of the mechanism, and the role of the triangular part is to easily execute the experiments because the locking piece is not dropped. The guide block consists of three elements: a base block and two wall plates. The wall plates are fixed to the base block by a screw. When the gap distance of the block is modified, some numbers of thin plate whose thickness is strictly specified, are inserted between the wall and the base. We employ a thin plate of 0.05 mm thickness in these experiments.

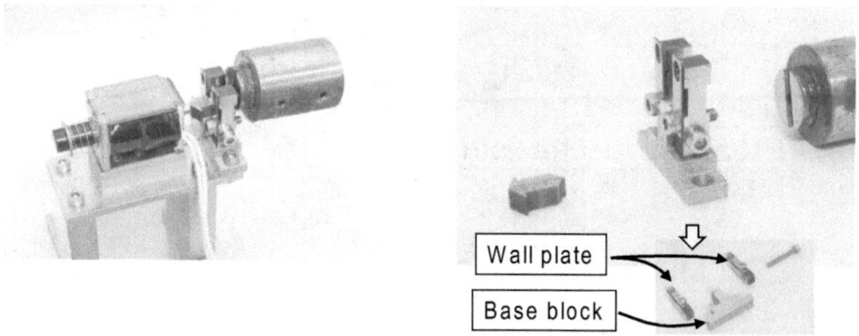

Fig. 8. Actual parts of the experimental instrument.

Second, a measurement system for the maximum frictional torque is described. The measurement system is composed of the proposed instrument, a torque measurement device, a handle to input the torque, and a solenoid to push the locking piece. Figure 9 shows a photograph of the measurement system.

The measurement procedure is explained as follows.

Step 1

The locking piece is pushed using the solenoid bar, and the tip of the locking piece enters the groove at the friction disk using the solenoid bar.

Step 2

The user gently rotates the handle attached to the torque measurement device. The tip of the locking piece enters the groove, and the locking piece is inclined. The user is pushing the solenoid bar during step 2.

Step 3

The solenoid bar is released. The locking piece is fixed, if the three points are contacted to the wall plates of the guide block and groove on the friction disk.

Step 4

The user adds to the torque gradually and slowly. The input torque is measured.

Step 5

The locking piece disengages and fries out if the input torque is more than a certain torque. Subsequently, the user returns the handle to its initial position.

This experiment is performed five times using the same conditions. The gap distance is modified by using the thin metal plate.

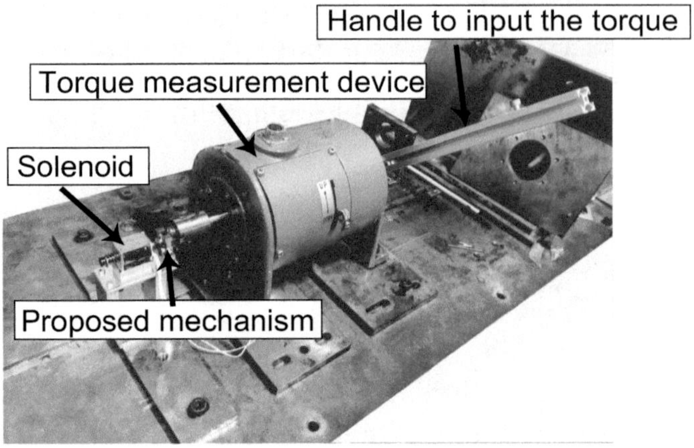

Fig. 9. Measurement system for torque.

3.2 Experimental Results

The experimental results are described in this subsection.

First, Fig. 10 shows the time variant of the torque. In this figure, the results of five experiments are included in case that the gap distance is 0.3 mm. These experiments were performed using the same parameters. The horizontal axis depicts time, and the vertical axis depicts torque. The input torque is manually given by the experimenter. Therefore, the timing is different from each other. The friction torque is increased gradually and is decreased suddenly. It is confirmed that the friction force is increased, when the input torque is increased. Moreover, when it is decreased, the locking piece is released. The values of the maximum friction torque are measured in this graph.

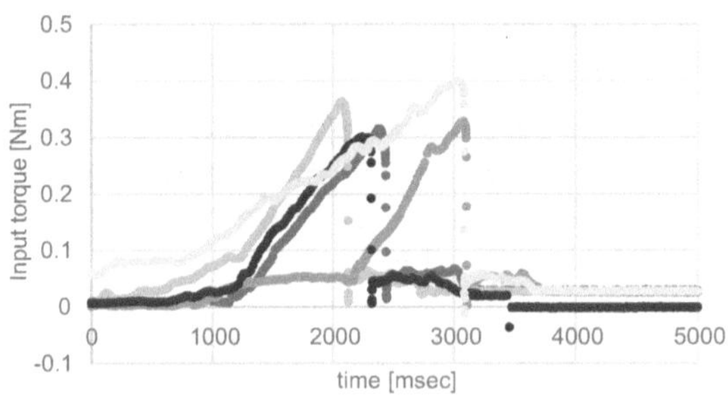

Fig. 10. Time variant of the friction torque.

Second, we obtained the maximum friction torque at various gap distances of the guide block. The results are shown in Fig. 11. This graph is an approximation inversely proportional to the gap distance, and the relation is in agreement with the results shown in Sect. 2. Consequently, it is clear that the friction force is controlled by the gap distance.

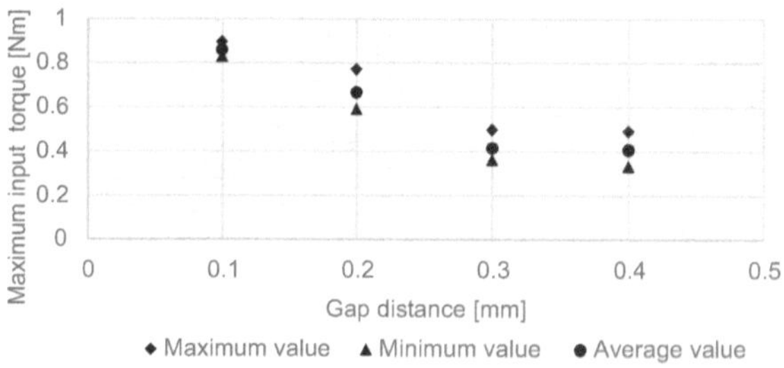

Fig. 11. Maximum torque at various gap distances

4 Conclusion

This study proposes a new releasable locking mechanism for holding a posture using friction and a cuneus. In general, the cuneus is not released even if strong forces are applied. The advantage of the proposed mechanism is the automatic release mechanism upon addition of a large torque is added. The maximum torque is calculated during posture maintenance. The relation between the maximum friction force and the gap distance is inversely proportional to an approximation. It is thus evident that the maximum torque to hold the posture can be controlled. We performed some experiments to confirm the motion during the mechanism and the values obtained by the calculation. We confirm that the locking piece is indeed released if the input force surpasses the threshold value. This behavior is in agreement with the predictions given by the calculation.

In a future work, this mechanism will be installed to our developed wearable type robot to hold the posture of the shoulder, as shown in Fig. 12. Moreover, we aim to investigate the influence of human motion on the mechanism.

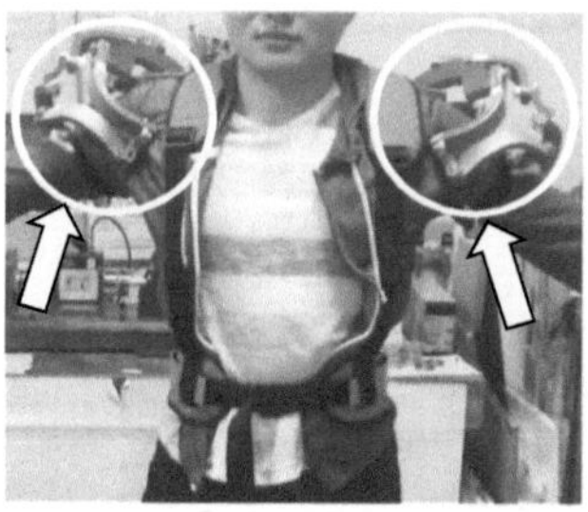

Fig. 12. Developing wearable robot enabling maintenance of the shoulder posture of the user.

Acknowledgment. This work was supported by JSPS KAKENHI Grant Number JP 17K01559.

References

1. Ruwan, G., Kazuo, K., Sanjaya, B.: A brief review on upper extremity robotic exoskeleton systems. In: 6th IEEE International Conference on Industrial and Information Systems (ICIIS), pp. 346–351 (2011). https://doi.org/10.1109/ICIINFS.2011.6038092
2. Viteckova, S., Kutilek, P., Jirina, M.: Wearable lower limb robotics: a review. Biocybern. Biomed. Eng. **33**(2), 96–105 (2013). https://doi.org/10.1016/j.bbe.2013.03.005
3. Zhu, Y., Nakamura, M., Ito, N., Fujimoto, H., Horiguchi, K., Wakabayashi, S., Takahashi, R., Terada, H., Haro, H.: Study of wearable knee assistive instruments for walk rehabilitation. J. Adv. Mech. Des. Syst. Manuf. **6**(2), 260–273 (2012). https://doi.org/10.1299/jamdsm.6.260
4. Araie, T., Nishizawa, U., Ikeda, T., Kakimoto, A., Toyama, S.: Evaluation of labor burden reduction achieved through wearing an agricultural power assist suit. Int. J. Model. Optim. **7**(4), 202–206 (2017). https://doi.org/10.7763/IJMO.2017.V7.584
5. Nwe, Y.Y., Toyama, S., Akagawa, M., Yamada, M., Sotta, K., Tanzawa, T., Kikuchi, C., Ogiwara, I.: Workload assessment with ovako working posture analysis system (OWAS) in Japanese vineyards with focus on pruning and berry thinning operations. J. Jpn. Soc. Hortic. Sci. **81**(4), 320–326 (2012). https://doi.org/10.2503/jjshs1.81.320
6. Ikemata, Y., Yasuhara, K., Sano, A.: A study of the leg-swing motion of passive walking. In: IEEE International Conference on Robotics and Automation, vol. 74, no. 738, pp. 1588–1593 (2008). https://doi.org/10.1109/ROBOT.2008.4543428
7. Plooij, M., Mathijssen, G., Cherelle, P., Lefeber, D., Vanderborght, B.: Lock your robot: a review of locking devices in robotics. IEEE Robot. Autom. Mag. **22**(1), 106–117 (2015). https://doi.org/10.1109/MRA.2014.2381368

Optimization Design of Flange Connection Bolt Set in Large-Scale Rolling Bearings Under Complicated Working Conditions

Huimin Dong(✉), Jing Wang, and Delun Wang

School of Mechanical Engineering,
Dalian University of Technology, Dalian 116024, China
donghm@dlut.edu.cn

Abstract.. This paper presents an optimal design method for flange connection bolt set under complicated working conditions in large-scale rolling bearings. In this method, the minimum mass of bolt set is used as optimization objective, and the diameter, length of bolt and the number of bolts are used as optimization variables. The reliability and installation space of bolts are taken as constraints, and the final solution is realized by genetic algorithm. Using this method, the flange connection bolt set for a pitch bearing in a wind turbine with consideration of the fatigue load and ultimate load is optimized. Comparing the optimization result with the bolt set originally used in the pitch bearing, it shows that the method can greatly reduce the mass of bolt set.

Keywords: Optimization design · Connection bolt set · Complicated working condition · Reliability constraint

1 Introduction

Bolts, as the important connection element, are widely used in engineering devices. The geometrical parameters of the bolted connection structure severely impact the safety and service life of the whole mechanism [1]. Therefore, how to realize the optimization design of the bolted connection structure under complicated working conditions is of great significance.

At present, many scholars have done relevant research on the optimization design of flange connection bolt set. He et al. [2] optimized the diameter of the bolt based on the MATLAB optimization toolbox. Chen et al. [3] optimized the reliability of the pressure vessel bolt set connection subjected to axial variable load with the economic cost of bolt set as the optimization target. Liu et al. [4] considered the geometrical parameter combination of bolt-flange structure joints as the design variable under static conditions, and carried out the multi-objective optimization design with the aim of strength and mass. Sun et al. [5] combined the fuzzy design with the optimization design theory and introduced the entropy theory to collate and derive the mathematical model for the reliability optimization design of the cylinder head bolts of diesel engines. Xia et al. [6] carried out an optimized design considering the fatigue life constraint on the spar bolt connection, and then carried out a series of fatigue tests to

D. Wang et al. (Eds.): MeTrApp 2019, MMS 79, pp. 91–98, 2020.
https://doi.org/10.1007/978-981-15-0142-5_10

verify the optimization results. Zhang et al. [7] established a bolt-connected finite element model to verify the feasibility of optimizing the bolt mass as the optimization goal.

In summary, the current optimization design of the flange connection bolt set only considers the load of the bolt set as the fatigue load or the ultimate load separately, and does not comprehensively consider the combined action of the fatigue load and the ultimate load. At the same time, the length of the bolt in the current bolt set connection optimization design is roughly determined by the thickness of the connected parts, and no clear optimization method is given. In a practical working environment, the bolt may be subjected to six degrees of freedom forces simultaneously. In this paper, an optimal design method of flange connection bolt set is developed under the joint action of fatigue load and ultimate load, the bolt diameter, length of the bolt and the number of bolts are used as optimization variables. The pitch bearing connection bolt set of a wind turbine is optimized.

2 Mathematical Model of Flange Connection Bolt Set Optimization Design

The optimal design of flange connection bolt set is to optimize the structural parameters of bolt set under certain constraints. The mathematical form is as follows.

$$\begin{cases} \min & f(x) \\ st. & g_i(x) \leq 0 \quad (i = 1, 2.\ldots..) \\ & G_j(x) \leq 0 \quad (j = 1, 2.\ldots..) \end{cases} \tag{1}$$

Where f(x) is the objective function, $g_i(x)$ is the reliability constraint conditions, $G_J(x)$ is the installation space constraint conditions.

2.1 Optimization Variable

The design of the flange connection bolt set needs to optimize the bolt diameter *d*, the number of bolts *n* and the length *L*, so the design variables are as follows.

$$x = [x_1, x_2, x_3]^{\mathrm{T}} = [d, n, L]^{\mathrm{T}} \tag{2}$$

2.2 Objective Function

In order to reduce the mass of the flange connection bolt set, the optimal design is based on the principle of the lightest total mass of the bolt set. For a selected material, the bolt density is fixed, and the lightest total mass is equivalent to the smallest total volume. The objective function can be equivalent to Eq. (3).

$$f(x) = x_1^2 \bullet x_2 \bullet x_3 \tag{3}$$

2.3 Optimization Constraints

In the optimization design of flange connection bolt set, the constraint conditions include bolt reliability and installation space requirements.

Flange Connection Bolt Set Load Distribution Calculation
The load on the most dangerous bolt should be calculated before setting constraints. Figure 1(a) shows the connection structure distribution diagram of the flange connection bolt set. The direction of the force coordinate system of the flange connection bolt set is shown in Fig. 1(b): X, Y - parallel to the flange connection bolt plane, Z - coincide with the flange axis. Under complicated working conditions, the flange connection bolt set is subjected to loads as follows: radial load F_X, F_Y, axial load F_Z, torque M_Z around the Z axis, overturning moment M_X, M_Y around the X and Y axes, equivalent fatigue load M_{ca}, F_{ZA}.

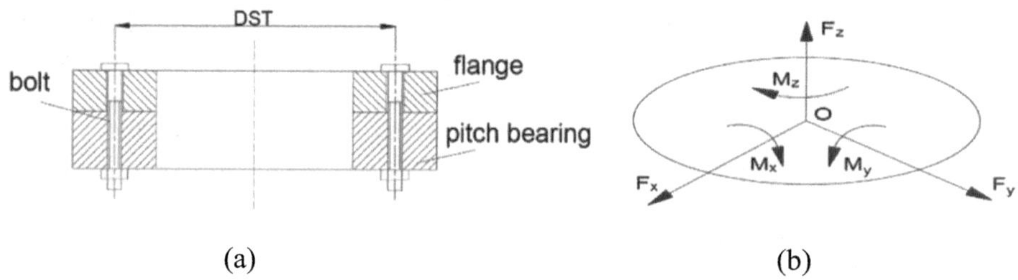

Fig. 1. (a) flange connection bolt; (b) coordinate system

Before analyzing the force condition of the flange connection bolt set, it is assumed that the bolt is an elastic body, the deformation is in the elastic range, and the pre-tightening force of each bolt is the same, the connected part is a rigid body. The axial force and radial force of single bolt are calculated according to the static balance equation.

The maximum radial load on the most dangerous bolt in the flange connection bolt set under the ultimate load is calculated by Eq. (4).

$$F_{Q\max} = \frac{\sqrt{F_X^2 + F_Y^2}}{n} + \frac{2|M_Z|}{D_{ST} \bullet n} \tag{4}$$

Where n represents the number of bolts and D_{ST} is the diameter of the bolt distribution circle.

The maximum axial load on the most dangerous bolt in the flange connection bolt set under ultimate load [8] is calculated by Eq. (5).

$$F_{A\max} = 4\frac{\sqrt{M_X^2 + M_Y^2}}{D_{ST} \bullet n} + \frac{F_Z}{n} \tag{5}$$

The maximum and minimum axial equivalent loads of the single bolt are calculated under the equivalent fatigue load by Eqs. (6) and (7).

$$F_{A\mathrm{dmax}} = 4\frac{M_{ca}}{D_{ST} \bullet n} + \frac{F_{ZA}}{n} \tag{6}$$

$$F_{A\mathrm{dmin}} = -F_{A\mathrm{dmax}} \tag{7}$$

Reliability Constraint

Bolt pre-tightening force must be a reasonable range. If the pre-tightening force is too large, the bolt may cause large deformation. If the pre-tightening force is too small, the connected parts may loosen or separate. Constraint 1 ensures that the bolt pre-tightening force meets the requirements during installation.

$$g_1(x) = F_{M\mathrm{max}} - F_{Mul}/S_Z \leq 0 \tag{8}$$

Where F_{Mul} is the allowable installation pre-tightening force, F_{Mmax} is the maximum installation pre-tightening force, and S_Z is the safety margin coefficient. The specific calculation method is based on the standard [9].

In the working state, the bolt is affected by the residual pre-tightening force and the external load. If the tensile stress of the bolt is too large, large deformation will occur and the connection will be invalid. Constraint 2 ensures that the tensile stress of the bolt meets the requirements under working conditions.

$$g_2(x) = \sigma_{redB} - R_{P02\mathrm{min}}/S_F \leq 0 \tag{9}$$

Where σ_{redB} is the tensile stress in the working state, R_{p02min} is allowable tensile stress, and S_F is the safety margin coefficient. The specific calculation method is based on the standard [9].

Under the fatigue load, the bearing capacity of the bolt connection is significantly lower than that under the static load, and the constraint 3 ensures that the tensile stress of the bolt meets the requirements under the fatigue load working condition.

$$g_3(x) = \sigma_{ab} - \sigma_{AS}/S_D \leq 0 \tag{10}$$

Where σ_{ab} is the continuous alternating stress acting on the bolt, σ_{AS} is the stress amplitude of the fatigue limit, and S_D is the safety margin coefficient. The specific calculation method is based on the standard [9].

After the connected parts are pre-tightened by the bolts, the pre-tightening force at the connection surface can generate friction. When the system is subjected to the radial load, the friction force can hinder the relative displacement of the connected parts and generate the slide. The constraint 4 ensures that the bolt and the connected parts do not slide each other.

$$g_4(x) = F_{Kerf} - F_{KR\mathrm{min}}/S_G \leq 0 \tag{11}$$

Where F_{Kerf} is the minimum tightening force required, F_{KRmin} is the minimum residual clamping load, and S_G is the anti-skid safety margin coefficient. The specific calculation method is based on the standard [9].

The thread connection must have sufficient engagement length, otherwise it will cause disengagement failure. The constraint 5 ensures that the minimum meshing length of the bolt meets the requirements.

$$g_5(x) = meff - mvorheff/S_X \leq 0 \quad (12)$$

Where *meff* is the minimum allowable screwing length of the bolt, *mvorheff* is the actual screwing length of the bolt, and S_X is the safety margin coefficient. The specific calculation method is based on the standard [9].

Installation Space Constraint

In order to ensure the convenience of bolt installation and disassembly, the constraints 6, 7 limit the space between the bolts in the bolt set.

$$G_6(x) = \pi \bullet D_{ST}/n - S_A \bullet d \leq 0 \quad (13)$$

$$G_7(x) = S_B \bullet d - \pi \bullet D_{ST}/n \leq 0 \quad (14)$$

Where S_A and S_B are bolt space coefficients.

2.4 Optimization Algorithm and Design Flow

In this method, there are many optimization constraints and the corresponding function calculation is complicated. The calculation of some constraints belongs to the solution of the discontinuity function, and the general optimal algorithm tends to fall into the local optimum. In the solution, the genetic algorithm has a good global search ability by cross mutation to find the optimal solution of the optimization model.

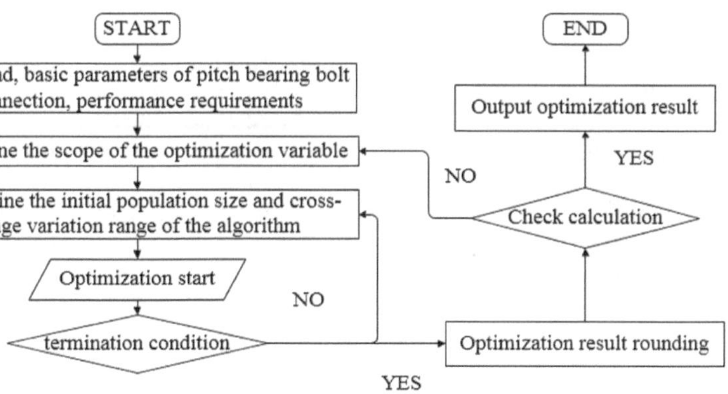

Fig. 2. Optimization process

In the whole optimization process, when all constraints meet the requirements and the mass of bolt set is minimum, the optimization result is rounded and the rounded result is brought into the check program. If the check result meets the requirements of the VDI 2230 bolt calculation standard, the optimization result is output. The block diagram of the optimized design of this algorithm is shown in Fig. 2.

3 Design Example

In the wind turbine, the hub, the pitch bearing and the impeller are bolted, and the forces acting on the bolts are very complicated. Therefore, this method is used to optimize the bolt connection of a specific wind turbine [10] pitch bearing (see Fig. 3).

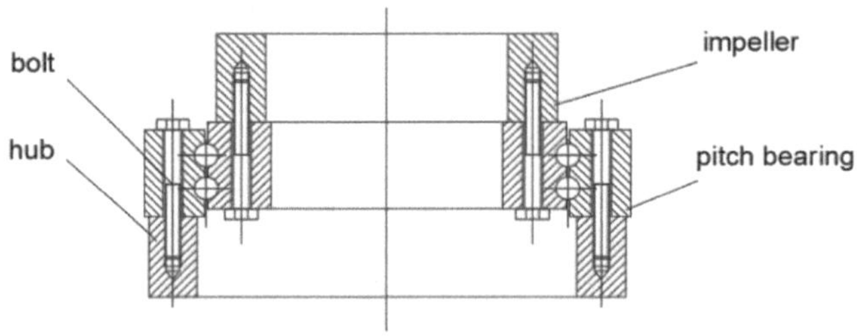

Fig. 3. Pitch bearing bolt connection

The basic parameters of the pitch bearing and the hub are shown in Table 1. The force of the pitch bearing connection bolt set under the ultimate load is $F_X = 465kN$, $F_Y = 520kN$, $F_Z = 824kN$, $M_X = -15699kNm$, $M_Y = 7326kNm$, equivalent fatigue load $M_{ca} = 21105kNm$, $F_{ZA} = 1732.4kN$, and the pitch bearing connection bolt set does not bear the torque M_z.

Table 1. Basic parameters

Parameter	Bearing/Hub	Parameter	Bearing/Hub
Bearing thickness (mm)	259—280	Young's modulus of bearing material (Mpa)	210000
Outer diameter at the joint surface (mm)	2642.5—2920	Young's modulus of Hub material (Mpa)	169000
Inner diameter at the joint surface (mm)	2420—2687	Hub thickness (mm)	205
Radial width at the joint surface (mm)	111.25—119.5		

The pitch bearing is connected by high strength bolt which Young's modulus is 210000 MPa, the bolt strength grade is 10.9. The optimized variables for the pitching connection bolt set are shown in Table 2.

Table 2. Optimization variable

Parameter	Bolt diameter d(mm)	Bolt number n	Bolt length L(mm)
Minimum value	20	20	200
Maximum value	64	120	400

The appropriate safety factor should be selected when optimizing the flange connection bolt set. If the safety factor is too small, the reliability of the optimized result will be reduced, and the safety factor is too large, the result will waste resources. The safety factors are selected according to the standard [9] as shown in Table 3.

Table 3. Safety factor

Parameter	Safety factor	Parameter	Safety factor
S_Z	1	S_D	1.2
S_F	1	S_G	1.2
S_X	1	S_A	1
S_B	2.5		

The results of the optimization of the method are compared with the parameters of the original pitch bearing bolt set. The comparison results are shown in Table 4.

Table 4. Optimization result

Parameter	Before optimization	Optimization result	Unit
Bolt number	100	120	
Bolt diameter	36	30	mm
Bolt length	358	313	mm
Maximum tensile stress	777.83	776.90	Mpa
Maximum torsional stress	279.70	282.79	Mpa
Actual yield strength	814.67	814.59	Mpa
Continuous alternating stress amplitude	17.24	18.48	Mpa
Bolt set volume	36.44	26.55	dm^2

Compared with the results before optimization, it can be seen that the values of the optimized bolt set working stress are similar and meet the requirements. The volume of the optimized bolt set is $26.55dm^2$, which is reduced 27.14% than before optimization.

4 Conclusions

This paper proposed a flange connection bolt set optimal design method under complicated working conditions with bolt minimum mass as the optimization target is proposed. The design problem of the bolt set with multiple load can be solved by genetic algorithm. Using the proposed method, a real flange connection bolt set has been optimized. The result shows that the method can perfectly deal with the design problem of flange connection bolt set with complicated loads.

The mass of the bolt set was reduced 27.14% under the almost identical working stress condition, which indicated that the optimal design method is effective and feasible.

Acknowledgements. This work was financially supported by the National Natural Science Foundation of China (No. U1808213)

References

1. Tian, J., Ai, T., Zhao, D., Wu, W.: Optimization design of bending stiffness of bolted joint structure. Aeronaut. Manuf. Technol. **03**, 86–89 (2014)
2. He, J., Han, W., Li, Z.: Optimization design scheme of bolt set based on MATLAB. Electromech. Inf. **21**, 145–146 (2011)
3. Chen, Y., Zhan, S., Huo, J.: Optimization design of bolt group connection based on MATLAB optimization toolbox. Min. Process. Equipment **33**(3), 28–29 (2012)
4. Liu, Y., Yin, Y.: Multi-objective optimization of a bolt-flange structure. Adv. Mat. Res. **5**, 233–235 (2011)
5. Sun, J., Luo, Y.: Reliability-optimization design based on fuzzy entropy for cylinder head bolts. J. Appl. Sci. **13**, 5189–5203 (2013)
6. Xia, T., Yao, W., Xu, L.: Metamodel-based optimization of the bolted connection of a wing spar considering fatigue resistance. Proc. Inst. Mech. Eng., Part G: J. Aerosp. Eng. **230**(5), 805–814 (2016)
7. Zhang, X., Wang, X., Zhang, M., Li, H., Zhang, J.: Optimization design and finite element verification of bolt set connection based on MATLAB. Machinery **55**(07), 68–71 (2017)
8. Xu, L.: Slewing Ring. Anhui Science & Technology Publishing House, Hefei (1998)
9. VDI 2230 Part 1, Systematic calculation of high duty bolted joints, Joints with one cylindrical bolt. Verein Deutsche Ingenieure, Dusseldorf, Germany (2003)
10. Sun, Z., Zhang, A.: Strength analysis of connecting bolts for wind turbine pitch bearings Most Influential Chinese Wind Issue in the World (11), 94–98 (2016)

Modal Analysis of Nutation Drive with Double Circular Arc Spiral Bevel Gear

Simin Ouyang, Ligang Yao(✉), Yongwu Cai, Jun Zhang, and Zhiyu Xie

School of Mechanical Engineering and Automation, Fuzhou University, Fuzhou, Fujian, China
ylgyao@fzu.edu.cn

Abstract. Nutation gear transmission combines nutation drive mode with double arc tooth profile, which has the advantage of large deceleration ratio and solves various problems in the traditional involute gear profile transmission. In his pa-per, the numerical solution and MATLAB software are used to solve the free vibration equation of the nutation system. And the undamped free vibration natural characteristics of the nutation transmission system of bevel gears are obtained. Then it reveals the influence of different physical parameters on the natural frequency of the system. The study and analysis mode shape of double circular spiral bevel gears under the natural frequency are carried out by using finite element software ANSYS. The natural frequencies of the system and each double circular arc bevel gear are compared and analyzed. After it the natural frequencies of the nutation drive system with 150%, 200% and 250% average meshing stiffness and 150%, 200% and 250% nutation gear support stiffness are calculated and compared. At last chapter, we get the orders vibration pattern and know the 7th order is the most influential one by finite element analysis in the ANASY. The natural frequencies of the system and each double arc bevel gear are compared and analyzed.

Keywords: Nutation drive · Double circular arc · Spiral bevel gear · Modal analysis

1 Introduction

Nutation drive of double circular spiral bevel gears is a new type of coaxial transmission. It has the advantages of large transmission ratio, simple and compact structure, small volume, small noise, large bearing capacity, stable transmission and high efficiency [1, 2]. It can be widely used in machine tools, instruments and petrochemical industry. Modal is the inherent vibration characteristic of mechanical structure [3–6]. Every modal has a specific natural frequency, damping ratio and modal mode. Therefore, modal analysis can be used to describe the motion process of gear transmissive system.

At present, there are many researches on modal analysis of gear transmission system, and great achievements have been made. Zhang established the elastic kinematics equation of the planetary gear train with small tooth difference and carried out

D. Wang et al. (Eds.): MeTrApp 2019, MMS 79, pp. 99–109, 2020.
https://doi.org/10.1007/978-981-15-0142-5_11

modal analysis on it, which can obtain more accurate inherent characteristics [7]. Zhang studied and analyzed the influence of spiral angle of double circular arc gear on modal analysis [8]. Zhang analyzed the influence of the modulus of hyperbolic normal circular arc gear, tooth number and transmission ratio on modal analysis [9]. After that, Deng and Zhang analyzed the effects with modulus and machining tool diameter for circular tooth line of cylindrical gears on modal analysis [10]. These provide a theoretical basis for the optimization design of gear transmission.

However, the research on the modal analysis of nutation gear transmission is still blank after literature review and retrieval. Due to the influence of time-varying meshing stiffness excitation and meshing impact, it is difficult to distribute the load equally and reasonably in the nutation transmission system of bevel gears, and the uneven distribution of load will lead to the increase of vibration and noise. Furthermore, the reliability, stability, transmission precision and life of the nutation drive will be seriously affected. So, modal analysis of nutation gear transmission system has important significance for vibration reduction and noise reduction.

2 Theoretical Principle of Modal Analysis

Modal analysis is a modern method which applies the system discrimination method to the engineering vibration field to research the structural dynamic characteristics. The research is mainly on the inherent vibration characteristics of mechanical structures. Generally, modal parameters can be obtained by calculation or experimental analysis, that is, modal parameters can be obtained by finite element method or the system input and output signals collected by experiments [11, 12].

The vibration differential equation of nutation reducer system can be expressed as follows:

$$M\ddot{x} + C\ddot{x} + Kx = f(t) \tag{1}$$

In Eq. (1), M, C and K are equivalent mass matrix, damping coefficient matrix and stiffness matrix of nutation drive system. The variable parameters x, $\dot{x}$ and $\ddot{x}$ are vibration displacement, vibration velocity and vibration acceleration. Function f (t) is external excitation load vector.

The theoretical modal analysis is based on the static finite element model. In view of the inherent characteristics (natural frequency, modal mode) of the system, the influence of external load can be ignored, and the damping term can be ignored because the damping effect is small. Thus, the equations of undamped free vibration can be established as follows:

$$M\ddot{x} + K\ddot{x} = 0 \tag{2}$$

The non-diagonal factor of stiffness matrix is not 0, so it is a non-diagonal matrix. Therefore, if there is elastic coupling in the system, then the dynamic equation of the system is a nonlinear coupled differential equation system. It is assumed that the mass points in the gear system are simply harmonic motion, the vibration i of harmonic motion is ω_i. Initial phase is ω_{i}. The corresponding mode shape is X_i. Take the main vibration $x = X_i \sin(\omega_i t + \varphi_i)$ and $\ddot{x} = \omega_i^2 X_i \sin(\omega_2 + \varphi_2)$ into Eq. (2). The corresponding characteristic equation can be obtained as follows.

$$\left(K - \omega_i^2 M\right) X_i = 0 \tag{3}$$

Set

$$S = M^{-1} K \tag{4}$$

Regular matrix S is called dynamic matrix of the system, which is also called the system matrix.

For such a n-order linear and homogeneous system of Eq. (3), it is necessary to get a non-zero solution that the determinant of the coefficient matrix is zero.

$$det\left(K - \omega_i^2 M\right) = 0 \tag{5}$$

This is a real coefficient equation of degree n about ω_i^2, with n pairs of roots. The n pairs of roots are named feature pairs, feature vector X_i and feature value ω_i^2. The ω_i^2 and X_i, which meet Eq. (5), are called modal parameter. It can be used to characterize the vibration characteristics of the system. Feature vector can reflect the spatial form of gear system vibration according to natural frequency ω_i^2, so it can also be called mode vector or mode.

For the actual system structure, n vibration frequencies (from small to large) can be obtained by formula (5): $\omega_1 < \omega_2 < \omega_3 < \omega_4 < \ldots < \omega_{\mathrm{n}}$.

For each natural frequency, the amplitude of a set of nodes can be determined according to Eq. (3), X_i (i = 1, 2, 3, …, n). The ratio of each component is kept constant. The numerical value is arbitrary, which is called eigenvector. In vibration theory, it is often referred to as the vibration mode of free vibration of structure.

Based on the above theory, this paper uses modal analysis to solve the natural frequency of the nutation gear transmission system.

The nutation gear transmission system is shown in Fig. 1.

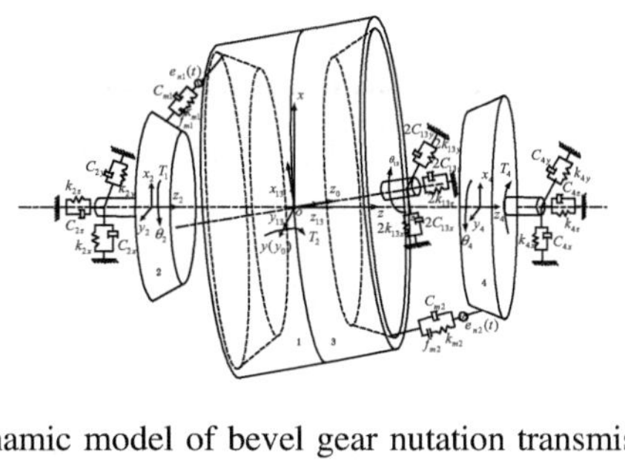

Fig. 1. Dynamic model of bevel gear nutation transmission system

Figure 1 is the dynamic model of the nutation transmission system of bevel gears [13]. It is described as follows:

Lateral bending vibration: each bevel gear is described by the translational degrees of freedom x_i, y_i (i = 13, 2, 4) in the x, y-directions. Axial vibration: each bevel gear is described translational degrees of freedom z_i in the z-direction. Torsional vibration: each bevel gear is described by the rotation of degrees of freedom θ_i around the z axis. Therefore, the outer bevel gear 1 and 3 are regarded as one whole and the system is converted into three gear parts, each gear considers 4 degrees of freedom for a total of 12 degrees of freedom. So, only the conditional parameters of three components (the numbers of them are 2, 13, 4) need to be considered.

From this dynamic model, the stiffness matrix K, Quality matrix M, and displacement vector X in the Eq. (2) can be obtained. X is $[x_2, y_2, z_2, \theta_2, x_{13}, y_{13}, z_{13}, \theta_{13}, x_4, y_4, z_4, \theta_4]$ and M is $[m_2, m_2, m_2, J_2, m_{13}, m_{13}, m_{13}, J_{13}, m_4, m_4, m_4, J_4]$. The stiffness matrix K is a matrix about K_m and K_d, which is too complicate to make detailed elaboration.

3 Modal Analysis of Nutation Transmission System of Double Arc Spiral Bevel Gear

This chapter uses MATLAB to calculate the natural frequency. And the factors affecting this natural frequency are analyzed in this chapter.

3.1 Solution of Modal Analysis About Bevel Gear Nutation Transmission System

On the previous chapter, a brief introduction of theoretical principle of modal analysis and dynamic model of the nutation transmission system. Then using these equations, we can get the ω_i we need in Eq. (5).

So, take K and M of Eq. (5) into MATLAB to calculate the dynamic differential equation. The parameters we need know are shown in Table 1. In this system, Support stiffness (K_i) is no difference in each direction, and bearing damping (C_i) is the same, which means $k_{ix} = k_{iy} = k_{iz} = k_i$ and $C_{ix} = C_{iy} = C_{iz} = C_i$.

Table 1. Parameters for bearing stiffness and support damping.

Essential parameter (k_i)	Numerical value (N/m)	Essential parameter (C_i)	Numerical value (N·m/s)
Support stiffness k_2	2.2 × 108	Bearing damping C_2	787
Support stiffness k_{13}	1.7 × 108	Bearing damping C_{13}	1803
Support stiffness k_4	2.9 × 108	Bearing damping C_4	2951

Table 2. Natural frequency of bevel gear nutation transmission system.

Order	1	2	3	4	5	6
Natural frequencies/Hz	0	6600	7870	7880	7880	16160
Order	7	8	9	10	11	12
Natural frequencies/Hz	16330	296300	35970	70210	107120	285490

From Table 1, the stiffness matrix K has been defined. And the quality matrix M can be obtained by measurement. According to the Eq. (5), the natural frequencies ω_i of the bevel gear nutation transmission system are obtained as shown in Table 2.

As can be seen, the minimum natural frequency of the nutation drive system is 0 Hz, which indicates that there is a rigid body motion at the beginning. Except the initial rigid body motion, the first 5 orders' natural frequencies of the whole nutation drive system are all small. After 6th order, the natural frequency is much large and the natural frequency span of the whole nutation transmission system is also large. Generally, the first ten natural frequencies have main influence on the system.

3.2 Modal Analysis of Nutation Transmission System with Different Mean Meshing Stiffness

The influence of different parameters on the natural frequencies of the system is very important for modal analysis. As mentioned in Sect. 2, the stiffness matrix K is a matrix about the mean meshing stiffness K_m and bearing support stiffness K_d. And in Eq. (5), quality matrix is determined with the determination of gears. Therefore, the mean meshing stiffness K_m and bearing support stiffness K_d are chosen to do comparative analysis.

Make the mean meshing stiffness of the two pairs of gears 150%, 200% and 250% times of the original one. The natural frequencies of different meshing stiffness are calculated as presented in Fig. 2.

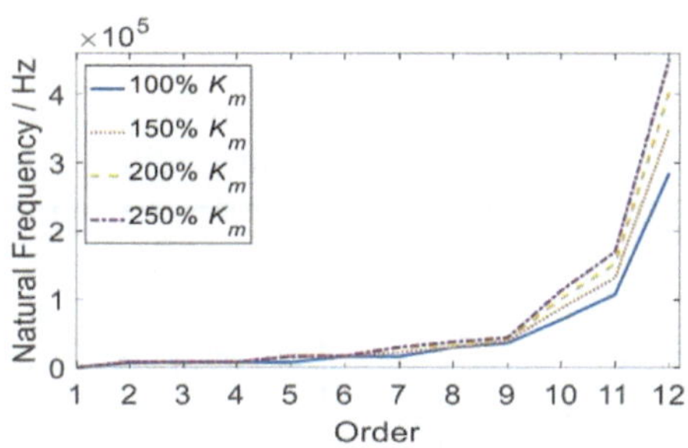

Fig. 2. Natural frequencies under different mean meshing stiffness K_m.

From Fig. 2, the trends of the natural frequency are similar. While, the natural frequency is larger when the mean meshing stiffness K_m growing. Specifically, this phenomenon is more obvious in the 5th order, 7th order and last forth orders. We can get the conclusion that the mean meshing stiffness K_m has slight effect on first 5 orders, but has more effect on the higher orders, especially, after 9th order. In order to see the law of influence more clearly, we've done a further deal with the data, which is shown in Fig. 3. The rate is relative errors compared with original meshing stiffness (100% K_m) under different meshing stiffness (150% K_m, 200% K_m and 250% K_m).

Figure 3 describes the rate of relative errors about natural frequency with growth of mean meshing stiffness K_m. The mean meshing stiffness K_m has the biggest influence on the natural frequency, but the magnitude makes little difference. After 7th order, the change of natural frequency has a positive correlation with mean meshing stiffness K_m. And the influence is unstable until the 9th order. Therefore, the meshing stiffness K_m has biggest effect on 5th order, and show more effect on the bigger order. Actually, the high order natural frequency is too high to make difference on gear transmission. So, when design a gear reducer, the meshing stiffness K_m can be ignored except on 5th order.

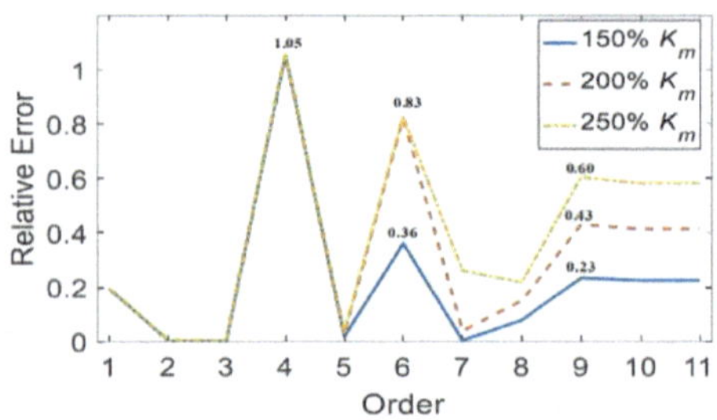

Fig. 3. The relative error of the natural frequency under different mean meshing stiffness K_m.

3.3 Modal Analysis of Nutation System Under Different Bearing Support Stiffness

In the same way, we can calculate the natural frequency of nutation transmission system when the bearing support stiffness K_d of nutation gear is 150%, 200% and 250% times of the original bearing support stiffness K_d. By comparing the natural frequency of bevel gear nutation drive system under different support stiffness K_d, the influence of bearing support stiffness K_d of nutation gear on this system is analyzed.

Through the natural frequency results of 150%, 200% and 250% of the original support stiffness K_d (Fig. 4), it can be seen that the support stiffness K_d of nutation gear has a greater effect on the low-order natural frequency, but less influence on the higher order. The natural frequencies of 10th–12th orders remain constant under different nutation gear support stiffness K_d. The bearing stiffness of nutation gear has great influence on the 6th and 7th order. The specific impact of natural frequency is shown in Fig. 5 in the same way.

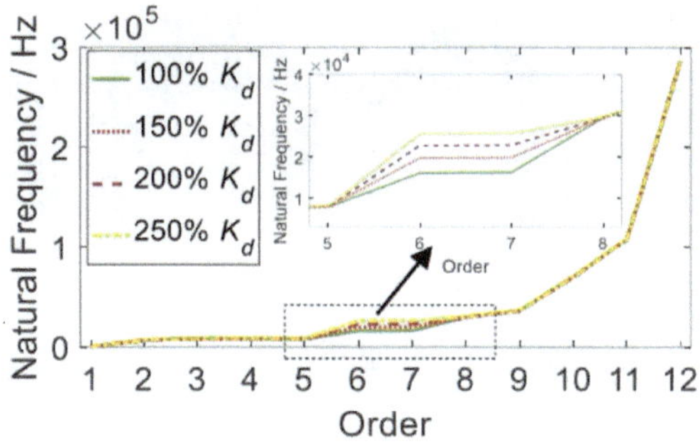

Fig. 4. Natural frequency under different bearing support stiffness K_d.

In this part, we using different support stiffness K_d to replace the mean meshing stiffness K_m as the contrast variables. Its influence is mainly on the 5th order and 6th order. And the bigger the support stiffness is, the greater the effect is. As mentioned before, influence on the lower order should be noticed more when we design such a nutation reducer, the support stiffness is necessary to check.

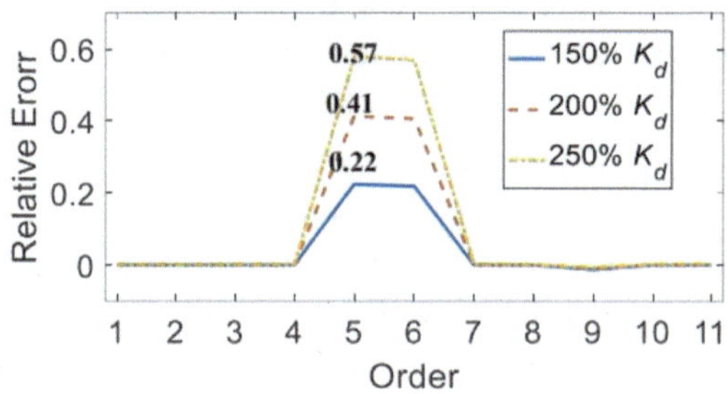

Fig. 5. Relative error of natural frequency under different support stiffness.

3.4 Modal Analysis of Nutation Transmission System of Double Arc Bevel Gear

After obtaining the natural frequency of nutation transmission system and the influence of parameters on the natural frequencies, it is necessary to complete the validation of natural frequency and vibration pattern to explore the natural frequency of the whole and the risk of resonance by using finite element method.

The finite element model is shown in Fig. 6. The material properties in this model of each double circular spiral bevel gear are material, modulus of elasticity E, Poisson's ratio μ and mass density. The material of gears is structural steel and the mass density is 7850 kg/m^3, and E = 206 GPa, μ = 0.29. The resulting natural frequency is shown in Fig. 7.

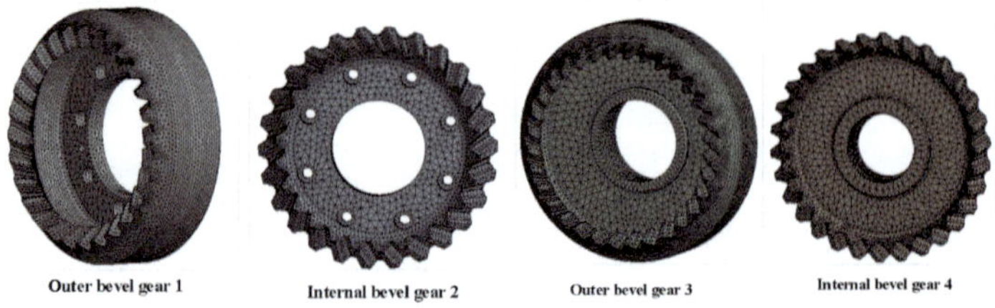

Fig. 6. Meshing diagram of gears.

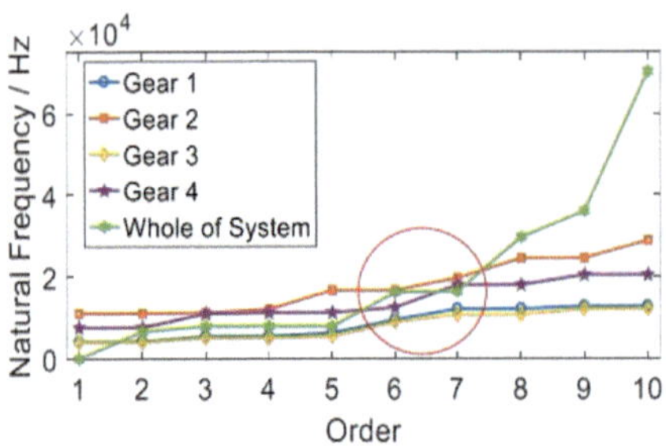

Fig. 7. Natural frequency of bevel gears and nutation transmission system

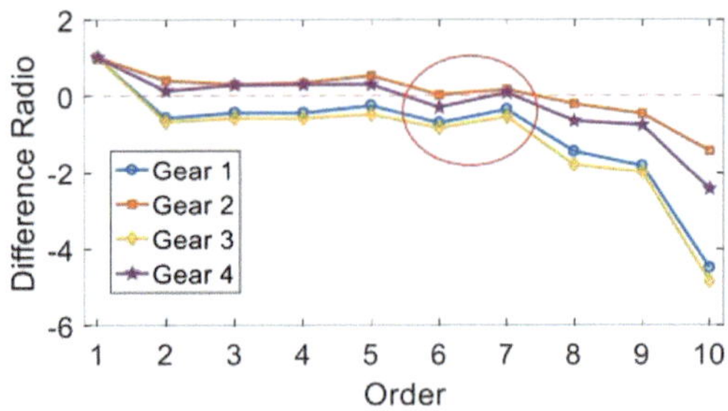

Fig. 8. Difference ratio of gears and nutation transmission system

In Figs. 7 and 8, there is no intersection between the gears and the system at the natural frequency. So, resonance does not occur under the current system. But we should notice the support stiffness to prevent resonance on 6th and 7th order, which each has a natural frequency of gear 2 and gear 4 that approaching the natural frequency of the system. There is a great possibility of resonance, especially when the support stiffness k_d is bigger as shown in Fig. 5.

Table 3. Vibration types of gears 1–4

Order	Type of gear 1	Type of gear 2	Type of gear 3	Type of gear 4
1	Torsional	Torsional	Torsional	Torsional
2	Torsional	Torsional	Torsional	Torsional
3	Folding	Circumferential	Folding	Folding
4	Folding	Folding	Folding	Folding
5	Umbrella	Folding	Umbrella	Bending
6	Circumferential	Bending	Circumferential	Circumferential
7	Bending	Bending	Bending	Bending
8	Bending	Circumferential	Bending	Bending
9	Radial	Bending	Radial	Radial
10	Radial	Bending	Radial	Radial

Combining with the animation and mode diagram of each order vibration process of double circular spiral bevel gear displayed by ANSYS Workbench post-processing program, the low order array of double circular spiral bevel gear can be classified into torsional vibration, diagonal vibration, umbrella vibration, circular vibration, and so on. Bending vibration and radial vibration are the types of modes of vibration (Table 3).

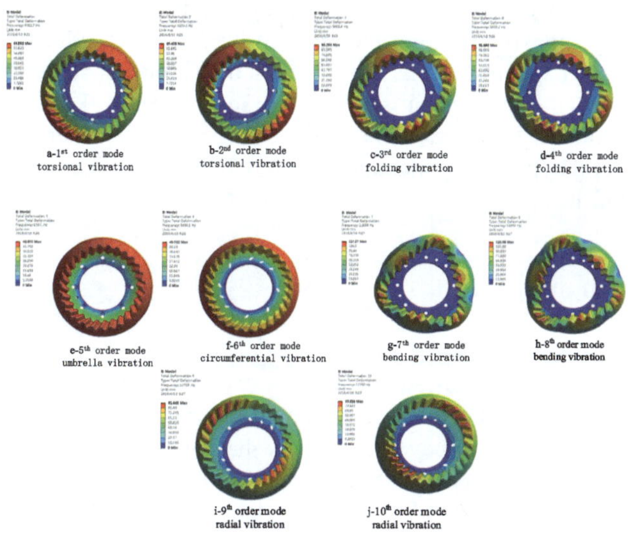

Fig. 9. First ten orders vibration patterns of gear 1.

Take the mode diagram of bevel gear 1 as an example, the shapes of each order of gear 1 are shown in the Fig. 9. The main results are as follows:

(1) The torsional vibration of bevel gear is as follows: the face of bevel gear swings forward and backward along a line of radial axis, as shown in Fig. 9(a) and (b).

(2) The main forms of counter-bending vibration are as follows: the axial direction of bevel gear is V-shaped, and it is regular polygon on the end face, as shown in Fig. 9 (c) and (d).

(3) The shape of umbrella vibration is as follows: the tooth of bevel gear retracts along the axis to the big end to form an umbrella shape, as shown in Fig. 9(e).

(4) The form of circular vibration is that there is no vibration and displacement in the axial direction of bevel gear, as shown in Fig. 9(f).

(5) Bending vibration: the main form is the regular wave array in the axial direction and the regular polygon on the end face, as shown in Fig. 9(g) and (h).

(6) Radial vibration: the main manifestation is that the bevel gear expands along the radial direction. The elliptical shape appears on the end of the bevel gear, and the axial vibration is basically non-vibrating, as shown in Fig. 9(i) and (j).

As illustrated in Fig. 9, the minimum natural frequency of outer bevel gear 1 is 4162.7 Hz and maximum natural frequency is 12750 Hz. The maximum deformation is in the 7th order. There are some natural frequencies of some gears are similar, such as 1st order and 2nd order, the 7th order and 8th order, the 9th order and 10th order. The main vibration modes of bevel gear 1 are torsional vibration, bending vibration, bevel vibration, circumferential vibration, bending vibration and radial vibration. It can be seen that the first and second order vibration modes of double circular spiral bevel gear are torsional vibration, the middle order vibration is umbrella type vibration, for the folded vibration and circular vibration, while the higher order vibration is bending vibration and radial vibration, and the frequency of adjacent order is close.

Because of the maximum deformation is in the 7th order, which is same to each gear, the resonance happened in this order is most dangerous. When design a reducer of bevel gear nutation transmission system, the resonance in 7th order needs more attention. To avoid large deformations, we need get the 7th order of nutation transmissive system and bypass it by making sure the vibration frequency at stable rotational speed away from it.

4 Conclusion

In this paper, the modal analysis of the nutation transmission system about double circular spiral bevel gear is carried out. Then it calculates the natural frequency of the nutation transmission system by using dynamic differential equation of nutation drive with double circular arc spiral bevel gear. After that, this paper studies the influence of different mean meshing stiffness and support stiffness on the natural frequencies. The comparing analysis of the natural frequencies under 150%, 200% and 250% times of meshing stiffness and support stiffness are presented. The results show that the average meshing stiffness mainly affects the higher order natural frequency of nutation transmission system, while the bearing support stiffness of the nutation gear mainly affects the low-order natural frequency of nutation drive system. So, it is better to paid more attention to the support stiffness, when design a nutation transmission system about double circular spiral bevel gear. At last, based on the ANSYS Workbench platform, the modes of each double circular spiral bevel gear in nutation drive reducer are analyzed, the first 10 modes are extracted, and the first 10 modes of double circular spiral bevel gear are judged. And find out the most influential natural frequency, which we should pay more attention on in further studies.

Acknowledgments. The authors acknowledge the support of the NSFC (Grant Nos. 51775114 and 51875105).

References

1. Zhang, J., Yao, L.G., Huang, S.J., Gu, B.: Precise tooth profile modeling of internal meshing double circular arc spiral bevel gear. J. Mechanical Manufacturing Engineering (1), 119–124 (2015)
2. Kemper, Y.: Nutating traction drive. J. Eng. Power Trans. ASME **103**(1), 154–157 (1981)
3. Lin, Z., Yao, L.G.: Mathematical model and 3D modeling of involute spiral bevel gears for nutation drive. J. Adv. Mat. Res. **694**, 503–506 (2013)

4. Lin, Z., Yao, L.G.: General mathematical model of internal meshing spiral bevel gears for nutation drive. J. Appl. Mech. Mat. **101**, 708–712 (2011)
5. Chen, A.H., Liu, D.S., Luo, S.M.: Numerical study on dynamic transmission error and vibration stability of gear system. J. Acta Mech. Eng. **4**, 21–25 (2004)
6. Zhang, J., Guo, F., Xie, S.L.: Elastic dynamic modeling and modal analysis of small tooth difference star gear reducer. J. Vibr. Shock **34**(12), 169–175 (2015)
7. Ye, Y.D., Zhou, Z.B.: Finite element modal analysis of spur gear based on ANSYS. J. Mech. Drive **30**(5), 63–65 (2006)
8. Lin, Z., Yao, L.G.: General mathematical model of internal meshing spiral bevel gears for nutation drive. J. Appl. Mech. Mat. **101**, 708–712 (2011)
9. Zhang, X.P., Shi, D.H., Chen, H.J.: Modal analysis of hyperbolic normal circular arc gears. J. Mech. Drive **2**, 62–65 (2017)
10. Tang, L.B., Hu, L., Yang, J.L.: Modal analysis of helical gear with input shaft of a gearbox. J. Noise Vibr. Control **38**(S2), 14–17 (2018)
11. Deng, F.Y., Zhang, Q.: Mathematical model and modal analysis of new circular cylindrical gear. J. Mechanical Drive (1) (2018)
12. Zhang, L.: Numerical modal analysis of double arc gear. J. Mech. Strength **38**(2), 425–428 (2016)
13. Cai, Y.W., Yao, L.G., Xie, Z.Y.: Influence analysis of system parameters on characteristics of the nutation drive with double circular arc spiral bevel gears. J. Forschung im Ingenieurwesen **81**(2–3), 125–133 (2017)

Engineering Applications

Simulation Analysis on Key Mechanisms of the Automatic Tobacco Harvester Based on ADAMS

Jie Yu, Jinkui Chu(✉), and Yang Li

School of Mechanical Engineering,
Dalian University of Technology, Dalian 116023, China
{yujie1217, liydut}@mail.dlut.edu.cn,
chujk@dlut.edu.cn

Abstract. Tobacco is widely cultivated in many countries. However, manual harvesting of tobacco leaves cannot meet the requirements of modern agriculture. Therefore, it is essential to develop the mechanized tobacco harvester. In this paper, an automatic tobacco harvester was designed and analyzed. The virtual prototype model was established using ADAMS. The traveling performance of the automatic tobacco harvester was analyzed using ADAMS/Tire. It can be found that the harvester work well on a flat sandy loam soil road. In addition, kinematic and dynamic simulation analysis on lifting mechanisms of the automatic tobacco harvester based on ADAMS were carried out. The results showed that the kinematic trajectory can meet the design requirements. Meanwhile, the dynamic parameters of key mechanisms were obtained, which would provide basic parameters for the design of the subsequent hydraulic system.

Keywords: Automatic tobacco harvester · Simulation analysis · Virtual prototype · ADAMS

1 Introduction

Tobacco leaves harvesting is a basic job in the process of tobacco production. The harvesting period and harvesting method of tobacco leaves are related to yield and quality of tobacco leaves, therefore scientific and timely harvesting is essential to ensure the quality of tobacco leaves [1–3]. Manual harvesting requires a large amount of labor and time. Therefore, when harvesting on large areas of tobacco fields, manual harvesting cannot meet agricultural requirements. The mechanized tobacco harvester has high harvesting efficiency. Therefore, it is necessary to develop the mechanized tobacco harvester.

ADAMS (Automatic Dynamic Analysis of Mechanical Systems) is a virtual prototype analysis software with powerful dynamic simulation analysis functions [4]. Static, kinematic and dynamic analysis of virtual mechanical systems can be carried out on the basis of ADAMS. In recent years, virtual prototyping technology has been widely used in the development of mechanical systems.

In this paper, the three-dimensional (3D) model of automatic tobacco harvester was established with SolidWorks, the virtual prototype model was established using

D. Wang et al. (Eds.): MeTrApp 2019, MMS 79, pp. 113–122, 2020.
https://doi.org/10.1007/978-981-15-0142-5_12

ADAMS. And simulation analysis on key mechanisms of automatic tobacco harvester based on ADAMS was carried out, which would provide technical support for further design and development of automatic tobacco harvester.

2 Mechanical Structure of the Automatic Tobacco Harvester

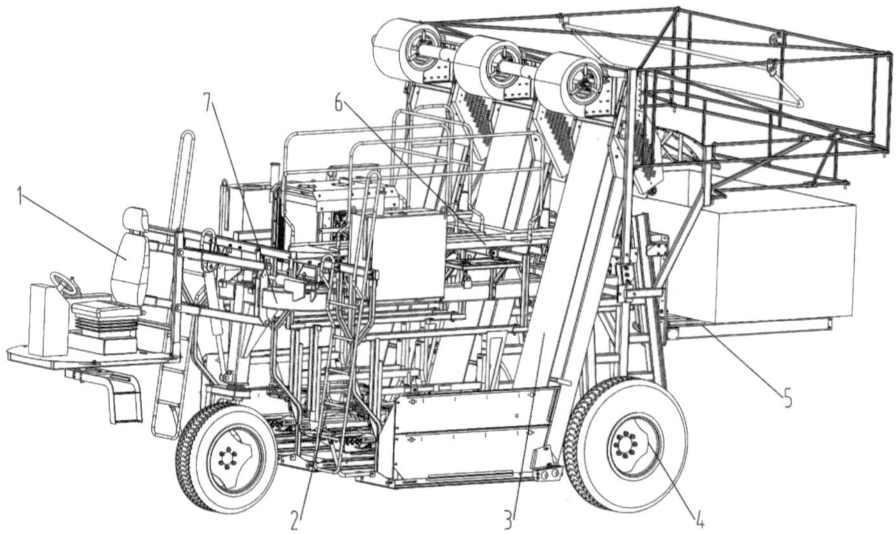

1. Cab 2. Harvesting mechanism 3. Conveying mechanism 4. Traveling mechanism 5. Tobacco basket lifting mechanism 6. Harvesting platform lifting mechanism 7. Frame

Fig. 1. Structure of the automatic tobacco harvester

According to the requirements of tobacco leaves harvesting, an automatic tobacco harvester should have three major functions: traveling, harvesting and storage. Tobacco leaves mature from bottom to top successively, so the tobacco leaves are divided into upper, middle and lower parts, which are harvested at different times. Therefore, it is necessary for the automatic tobacco harvester to design a harvesting platform lifting mechanism. And Fig. 1 shows the structure of the automatic tobacco harvester. It is mainly composed of the cab, the frame, the harvesting mechanism, the conveying mechanism, the harvesting platform lifting mechanism, the tobacco basket lifting mechanism, and the traveling mechanism. In addition, compared with electric drive and mechanical drive, hydraulic drive is more suitable for large agricultural harvester with poor working conditions. Therefore, hydraulic drive is applied in the automatic tobacco harvester.

3 Kinematic and Dynamic Simulation Analysis of Key Mechanisms

3.1 The Traveling Mechanism of the Harvester

It is important to study the vehicle trafficability of large harvesting machines on soft soil roads. The automatic tobacco harvester weighs about 6 ton, and three wheels are driven by three hydraulic motors. In order to obtain the dynamic parameters of the traveling mechanism and the vehicle trafficability on a soft soil road, virtual prototyping technology is used in this section.

The soft-soil tire model in ADAMS/Tire is used to analyze the traveling performance of the automatic tobacco harvester on a soft soil road. It should be noted that two classical soil elastoplastic theories are applied to analyze the wheel-soil interaction in the soft-soil tire model.

Firstly, the pressure-sinkage model proposed by Bekker is applied to calculate the normal stress in the soft-soil tire model. The formula of normal stress is as follows [6, 7]

$$\sigma = \left(K_c/b + K_\phi\right)h^n \tag{1}$$

in which $\mathbf{K_c}$ and $\mathbf{K_\phi}$ are the cohesive and frictional moduli respectively, **h** and **b** are the sinkage and wheel width respectively, **n** is the sinkage exponent.

Secondly, the soil shear model proposed by J.Janosi is applied to calculate the shear stress in the soft-soil tire model. The formula of shear stress is as follows [8]

$$\tau = (c + \sigma \tan\phi)\left(1 - e^{-j/k}\right) \tag{2}$$

in which **c** is the cohesion stress of the soil, $\boldsymbol{\phi}$ is the friction angle of the soil, **j** is the shear displacement, **k** is the shear deformation moduli.

The 3D model of the automatic tobacco harvester is imported into ADAMS. And then the tire model and ground model are established. The flexible tire which takes the elastic deformation into account is used in this model. The vertical stiffness of the tire is set to 310 N/mm, and rolling resistance is set to 0.03. Since the automatic tobacco harvester works in the field, the sandy loam soil is applied in the ground model. The physical parameters of the sandy loam soil are listed in Table 1. And Fig. 2 shows the virtual prototype model of the traveling mechanism on a flat sandy loam soil road in ADAMS.

Table 1. Physical parameters of the sandy loam soil

ϕ(rad)	c(N/m^2)	K_c(N/m$^{(n+1)}$)	K_ϕ(N/m$^{(n+2)}$)	n
0.506	1720	5270	1.52×10^6	0.7

Then three revolute motions are added to three wheels. The traveling speed of the harvester is set to 500 mm/s. It is worth noting that the radius of the front and rear wheels are different, therefore the values of the revolute motions are set respectively.

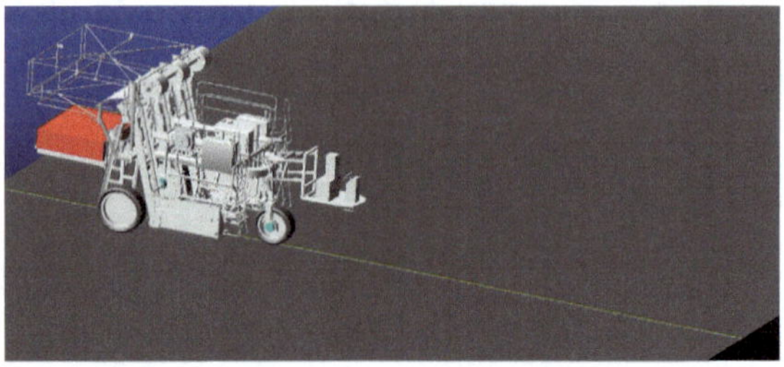

Fig. 2. Virtual prototype model of the traveling mechanism

The unloaded radius of the front wheel is 492.5 mm, therefore the revolute motion of the front wheel is set as follows

$$motion1 = \frac{v}{2\pi R_1} \cdot 360^\circ = \frac{500}{2\pi \cdot 492.5} \cdot 360^\circ \approx 58.17^\circ/s \tag{3}$$

The unloaded radius of each rear wheel is 605 mm, therefore the revolute motions of the rear wheels are set as follows

$$motion2 = motion3 = \frac{v}{2\pi R_2} \cdot 360^\circ = \frac{500}{2\pi \cdot 605} \cdot 360^\circ \approx 47.35^\circ/s \tag{4}$$

From the post-processing results, the driving torques of three wheels can be obtained. It can be observed from Fig. 3 that the driving torque of front wheel is finally stabilized at 145.85 N·m. And the driving torques of rear wheels are finally stabilized at 989.15 N·m and 998.28 N·m. These will provide the basic dynamic parameters for the design of the subsequent hydraulic system.

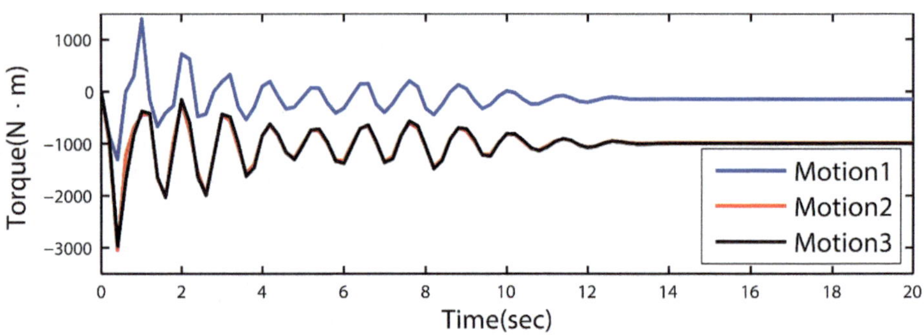

Fig. 3. Driving torque curves of three hydraulic motors

Figure 4 shows the actual traveling speed of the harvester. And it is finally stabilized at 472 mm/s. It should be noted that the actual speed is less than the set value (500 mm/s) due to the elastic deformation of the tire. Figure 5 shows the displacement of the harvester in vertical direction during the simulation. It can be observed that the sinkage is finally stabilized at 99 mm. Figure 6 shows the slip ratio of the front and rear wheels. It can be seen that the slip ratio of three wheels are very small.

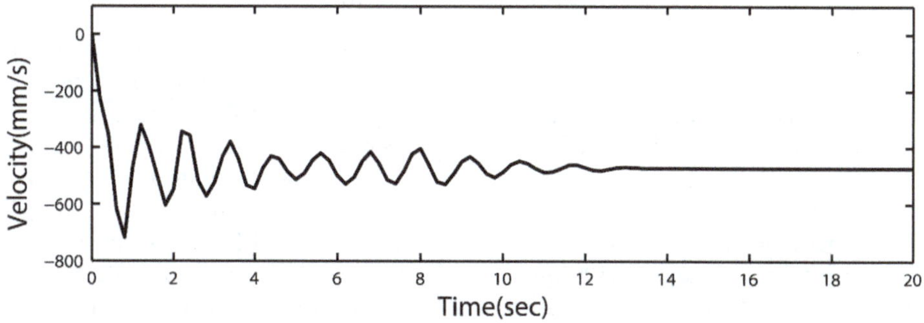

Fig. 4. Traveling speed of the harvester

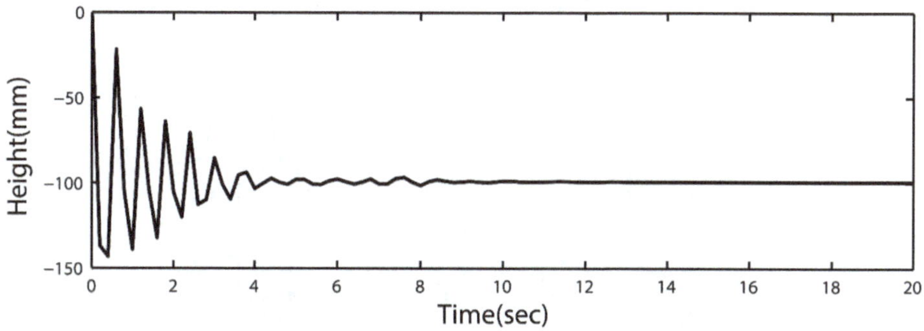

Fig. 5. Displacement of the harvester in vertical direction

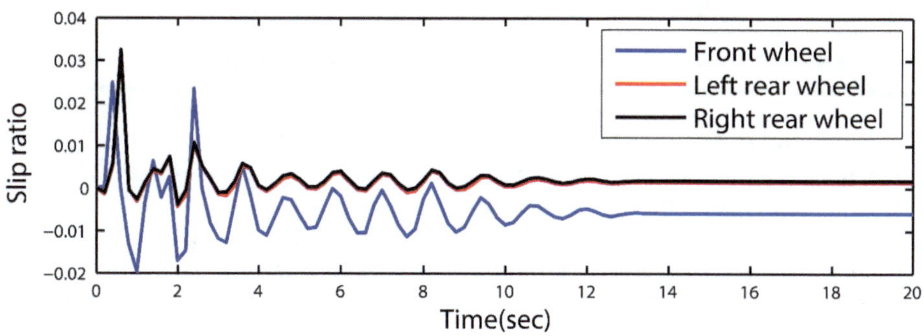

Fig. 6. Slip ratio of the front and rear wheels

The simulation results show that the automatic tobacco harvester has better trafficability on a flat sandy loam soil road and can meet the requirements for tobacco leaves harvesting.

3.2 The Harvesting Platform Lifting Mechanism

Tobacco leaves are harvested in batches. Two or three pieces of tobacco leaves are harvested at each time. Therefore, a lifting mechanism is designed to adjust the height of harvesting mechanism. Meanwhile, in order to make the harvested tobacco leaves fall on the conveyor belt smoothly, a large distance between the harvesting mechanism and the conveying mechanism is not allowed. Therefore, the lifting mechanism is designed to lift the harvesting mechanism and the conveying mechanism at the same time.

The basic design requirements of the harvesting platform lifting mechanism is that the lifting height is at least 300 mm, and the lifting process is stable. Therefore, a parallel four-bar mechanism is applied to lift the harvesting mechanism and the conveying mechanism. And the lifting mechanism is powered by a hydraulic cylinder. The specific structure of the harvesting mechanism and the conveying mechanism can be found in Ref. [5]. In order to obtain the kinematic and dynamic parameters of the harvesting platform lifting mechanism, virtual prototyping technology is used in this section.

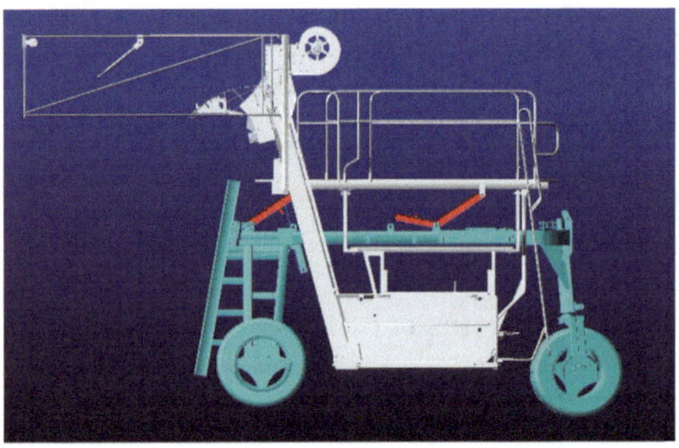

Fig. 7. Virtual prototype model of the harvesting platform lifting mechanism

Firstly, import the 3D model into ADAMS. The model consists of four parts, the frame, the parallel four-bar mechanism, the harvesting and conveying mechanism, and the hydraulic cylinder. Figure 7 shows the imported model in ADAMS.

Secondly, set material parameters and add constraints in ADAMS. The frame is fixed to the ground. The harvesting and conveying mechanism is fixed to the link rod of the parallel four-bar mechanism. Revolute joints are added to each hinge of the four-bar mechanism. A translation joint and an impact force are added between cylinder and piston of the hydraulic cylinder.

Thirdly, set a motion and solve the model. A translation motion is added to the translation joint. To represent the reciprocating motion of the hydraulic cylinder, a

STEP function is used in this paper. It should be noted that the hydraulic cylinder stroke is designed to be 290 mm. The value of the motion is set as follows

STEP(time, 0, 0, 4.7, 141)+STEP(time, 4.7, 0, 14.3, −290)+STEP(time, 14.3, 0, 20, 170).

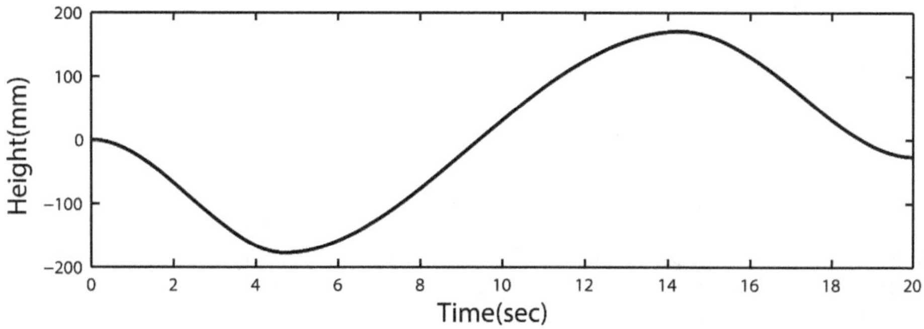

Fig. 8. Lifting curve of the centroid of harvesting and conveying mechanism

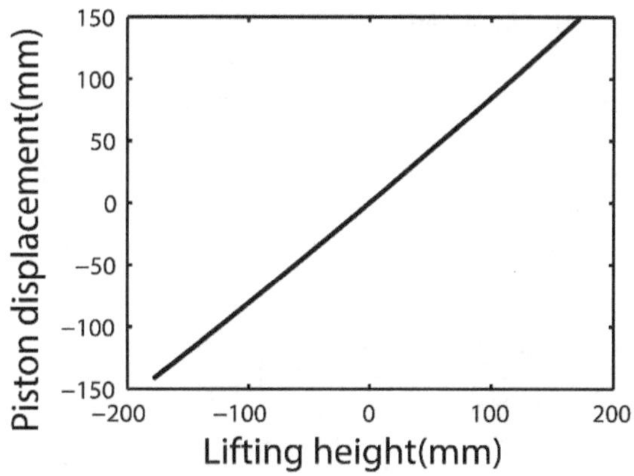

Fig. 9. Relation between the lifting height and the piston displacement

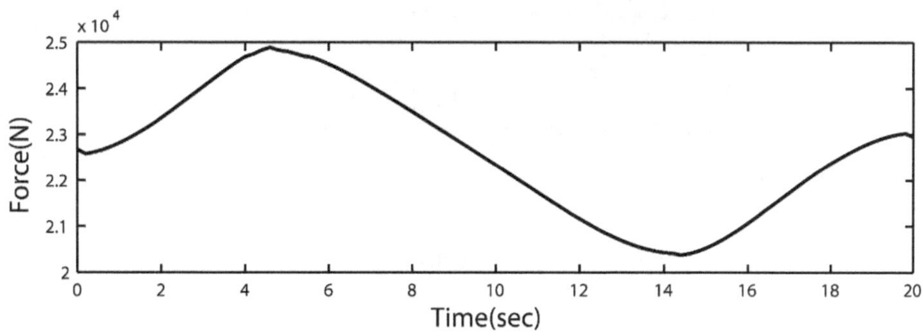

Fig. 10. Driving force curve of the hydraulic cylinder

From the post-processing results, the lifting height and dynamic parameters of the harvesting platform lifting mechanism can be obtained. Figure 8 shows the lifting curve of the centroid of harvesting and conveying mechanism. It can be obtained that the maximum value of the lifting height is approximately 348 mm, which fully meets the agronomic requirements for tobacco leaves harvesting. It can be obtained from Fig. 9 that the lifting height is linear with the piston displacement. Therefore, the lifting process is stable with the working of the hydraulic cylinder. Figure 10 shows the driving force curve of the hydraulic cylinder. It can be obtained that the range of driving force is 20377 N–24890 N during the lifting process, which will provide basic dynamic parameters for the design of the subsequent hydraulic system.

3.3 The Tobacco Basket Lifting Mechanism

During the harvesting process, the tobacco leaves are transferred from the conveying system to the tobacco basket. The tobacco basket is sent to the flue-cured tobacco room for further processing of tobacco leaves after harvesting. Therefore, a tobacco basket lifting mechanism is designed to achieve the loading and unloading of tobacco leaves. In this paper, two hydraulic cylinders are applied to lift the tobacco basket. According to the design requirements, the tobacco basket needs to carry a load of 1 ton.

In order to obtain the kinematic and dynamic parameters of the tobacco basket harvesting lifting mechanism, virtual prototyping technology is used in this section. Similar to the previous section, the 3D model of the tobacco basket lifting mechanism is imported into ADAMS. The model consists of four parts, the frame, the tobacco basket lifting frame, two hydraulic cylinders, and the tobacco basket. The frame and the tobacco basket lifting frame are connected by slider-guide rail mechanisms. Figure 11 shows the imported model in ADAMS.

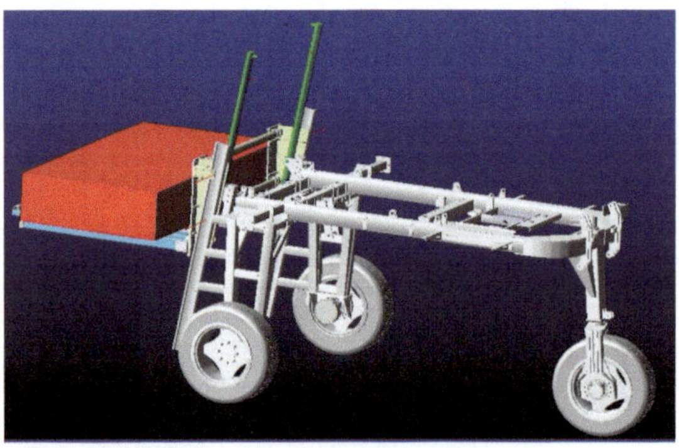

Fig. 11. Virtual prototype model of the tobacco basket lifting mechanism

Then set material parameters and add constraints in ADAMS. Finally set motions and solve the model. Two translation motions are added to two hydraulic cylinders, and the value of each motion is set as follows

STEP(time, 0, 0, 8, 240)+STEP(time, 8, 0, 48, −1200)+STEP(time, 48, 0, 88, 1200).

From the post-processing results, the lifting height and dynamic parameters of the tobacco basket lifting mechanism can be obtained. Figure 12 shows the distance from the bottom of tobacco basket to the ground during the simulation. It can be obtained that the minimum value is approximately 569 mm, the maximum value is approximately 1753 mm, which fully meets the loading and unloading requirements of the tobacco basket. Figure 13 shows the driving force curves of two hydraulic cylinders. It can be found that the two curves are similar, which is in line with the actual situation. And there are fluctuations in the two curves, which may be caused by the gap between the slider and the guide rail. During the simulation, the maximum value of the driving force is 8577 N, which will provide basic dynamic parameters for the design of the subsequent hydraulic system.

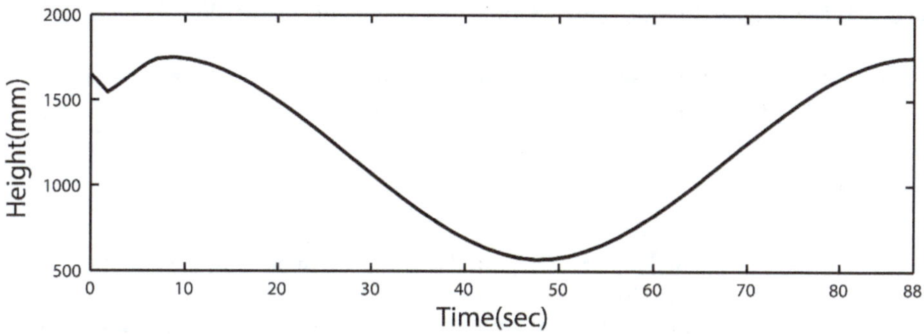

Fig. 12. Lifting curve of the tobacco basket

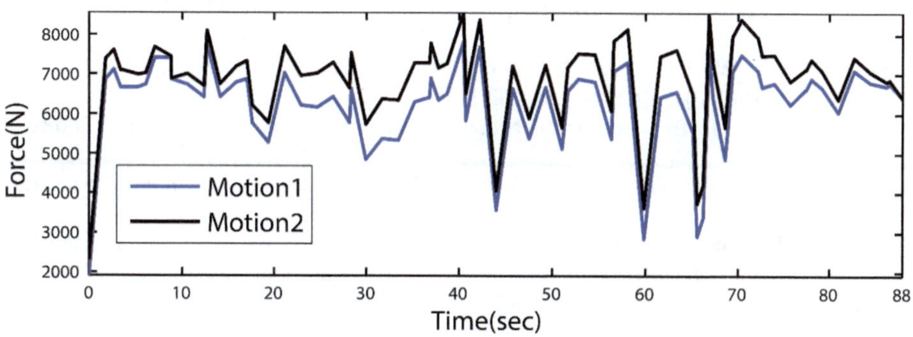

Fig. 13. Driving force curves of two hydraulic cylinders

4 Conclusion

In this paper, simulation analysis on key mechanisms of the automatic tobacco harvester based on ADAMS were carried out. Firstly, the trafficability of the automatic tobacco harvester on a soft soil road was analyzed. The driving torques of three wheels were obtained. And it can be observed from the post-processing results that the automatic tobacco harvester has better trafficability on a flat sandy loam soil road.

Secondly, kinematic and dynamic simulation analysis of the harvesting platform lifting mechanism was carried out. From the post-processing results, it can be observed that the lifting height meets the agronomic requirements for tobacco leaves harvesting and the lifting process is stable. The driving force which is powered by hydraulic cylinder can also be obtained. This will provide basic dynamic parameters for the design of the subsequent hydraulic system.

Thirdly, the tobacco basket lifting mechanism was analyzed based on ADAMS. The lifting height and the driving force were obtained. And the lifting height fully meets the loading and unloading requirements of the tobacco basket.

Acknowledgements. This research was funded by National Natural Science Foundation of China and The Fundamental Research Funds for the Central Universities.

References

1. Yu, Y., Wei, F., Qin, W., Liu, T., Zeng, M., Ning, W.: Technology development of tobacco harvester. J. Agric. Mech. Res. **38**(2), 255–262 (2016). (in Chinese)
2. Wells, L.G., Day, G.B., Smith, T.D.: Automated harvesting of burley tobacco I. system development. Trans. Am. Soc. Agric. Eng. **33**(4), 1033–1037 (1990)
3. Casada, J.H., Bader, M.J., Walton, L.R., Swetnam, L.D., Fiedeldey, M.E.: Mechanical harvesting system for burley tobacco. Appl. Eng. Agric. **3**(1), 95–98 (1987)
4. Du, Y., Zhu, Z., Mao, E., Song, Z., Gao, J., Zhu, K.: Simulation on small-scale corn harvester for hilly area based on ADAMS. Trans. Chin. Soc. Agric. Mach. **42**(A1), 1–5 (2011). (in Chinese)
5. Yu, J., Chu, J., Wu, Z., Wang, Y.: Scheme design of automatic tobacco harvester. J. Agric. Mech. Res. **41**(4), 70–74 (2019). (in Chinese)
6. Zhang, K.: Vehicle-Terramechanics. National Defense Industry Press, Beijing (2002). (in Chinese)
7. Bekker, M.G.: Introduction to Terrain-Vehicle. The University of Michigan Press, Ann Arbor (1969)
8. Janosi, Z., Hanamoto, B.: The analytical determination of drawbar pull as a function of slip for tracked vehicle in deformable soils. In: 1st International Conference on the Mechanics of Soil-Vehicle Systems, vol. 41, pp. 707–736. Edizioni Minerva Tecnica, Torino, Italy (1961)

Design and Analysis of a Novel Primary and Secondary Embedded Deployable Offshore Cage

Zi-Yu Liu, Lu-Bin Hang(✉), Le-Le Bai, Xiao-Bo Huang, Peng-Cheng Zhang, Su-Long Li, Shi-Sen Lin, Jun-Rong Chen, and Ming-Yuan Wang

Shanghai University of Engineering Science, Shanghai, China
hanglb@126.com

Abstract. Cage aquaculture is an important strategic direction for marine aquaculture, which plays an important role in expanding aquaculture space and protecting the offshore ecological environment. First, based on the concept of embedded deployable mechanisms, a novel embedded deployable offshore cage driven by cables is proposed. Over-constrained spatial mechanisms and planar compliant mechanisms are respectively adopted as the primary and secondary cages while the secondary cages are embedded in the primary cage. A closed-loop cable-driven module is designed to deploy the cage through the pull and release of the cable. Then, due to the kinematic characteristics of threefold-symmetric Bricard mechanism with three symmetric planes intersecting at a determined line, the determined line is regarded as the central line of the driven mechanism. Finally, the relationship between the input angle and the output angle is obtained from the kinematic equations of the primary cage frame, then the relationship between the driven angle and corresponding open state of the cage frame can be derived by geometric projection.

Keywords: Offshore cage · Deployable mechanism · Primary and secondary cage · Kinematic characteristics · Geometric projection

1 Introduction

As the near-shore spaces of today can hardly satisfy the increasing demand for marine foods, the fish farming industry has the tendency to move to offshore area [1]. Offshore cages represent the advanced productivity of marine aquaculture and play important roles in expanding sea-culture space and protecting the coastal ecological environment [2]. According to the material of the frame, offshore cages can be divided into rigid structure cages and flexible structure cages [3]. Rigid structure cages have the advantages of larger capacity, stronger wave resistance and higher degrees of automation, which are the developing directions of the marine farms [4]. At present, there are many advanced rigid cages in the world such as Hex Box Ocean Aqua-farms in Norway (see Fig. 1a), the "Deep Blue NO.1" full-submarine deep-sea fishing grounds in China (see Fig. 1b) and the Ocean Spar Sea Station in the United States

D. Wang et al. (Eds.): MeTrApp 2019, MMS 79, pp. 123–132, 2020.
https://doi.org/10.1007/978-981-15-0142-5_13

(see Fig. 1c). However, the structure of the rigid cages is fixed after assembly, which increases transport difficulties and limits the application of rigid structure cages.

(a) Hex Box Ocean Aqua-farms [5]

(b) "Deep Blue NO.1" full-submarine fishing grounds[6]

(c) The Ocean Spar Sea Station[3]

Fig. 1. Advanced offshore cages.

Deployable mechanisms refer to a type of mechanisms which can expand to deployed configuration for working and collapsed to a small, closed, or folded configuration for storage [7]. At present, deployable mechanisms have been widely used in aerospace industry and architecture, such as Folding Articulated Square Truss, Solid Surface Deployable Antenna and a retractable roof designed by Luo [8, 9]. Deployable mechanisms have small enveloping space and are convenient for transportation, which makes them have prospects in the design of offshore cages.

A novel embedded deployable offshore cage is designed in this paper. The cage is designed with a deployable secondary-primary mechanism. Through the pull and release of the cable, the cage can be deployed and folded. The rest of the paper is arranged as follows. In Sect. 2, the scheme of the deployable cage system is described and the states of the cage during the process of deployment are determined. In Sect. 3, the mechanical design of the driven system is described in detail. In Sect. 4, based on the geometric invariance of the threefold-symmetric Bricard mechanism, the deployable cage driven mechanism with determined axis as folding center is designed. Then the relationship between the driven angle and the open state of the cage can be obtained.

2 Novel Primary and Secondary Embedded Deployable Cage System

2.1 Scheme of the Embedded Deployable Cage System

Based on the concept of deployable mechanisms, a novel embedded deployable offshore cage system is proposed. As shown in Fig. 2a, the new cage system is composed of a deployable primary and secondary cage box, floating devices, and anchoring devices [10]. The deployable primary and secondary cage box contains the cage frame and fishing nets. The cage frame includes a primary frame and several secondary frames, in which the primary frame is a spatial deployable mechanism and the secondary frames are planar flexible deployable mechanisms. The floating devices are connected with the upper frame to provide buoyancy and the anchoring devices are installed at the bottom to anchor the cage.

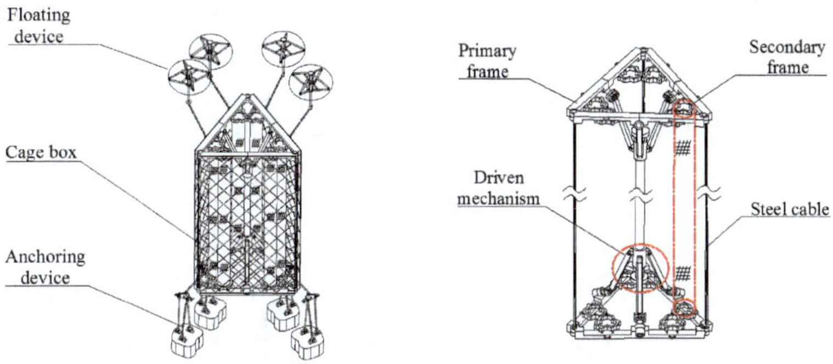

(a)Model of the deployable cage system (b) Model of the cage frame

Fig. 2. The novel embedded deployable offshore cage.

2.2 States of the Cage During the Process of Deployment

The deployable process can be divided into three states as shown in Fig. 3. When the frame is fully deployed, the primary frame is deployed to the planar state. Lower frame moves downward by releasing the steel cables to deploy vertically. When the frame is folded along the vertical direction, the winches retract the steel cable to realize the vertical folding. When the frame is fully folded, the cage driven mechanism drives the primary frame to fold from the plane state to the six-bar parallel state.

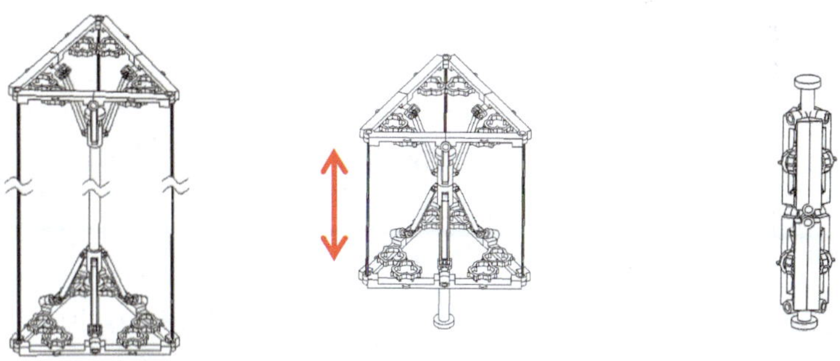

(a) Fully deployed state (b) Vertical-folded state (c) Fully folded state

Fig. 3. Three states of the cage frame.

2.3 Key Configurations of the Primary and Secondary Frame Mechanism

The cage mechanism is shown Fig. 4c, which is composed of the primary frame and the secondary frames. Threefold-symmetric Bricard mechanism is adopted as the primary frame and the process of deployment is shown in Fig. 4a. Six bars are parallel into a bundle to achieve the most compact configuration when fully folded while six bars move to a plane when fully deployed [11]. Threefold-symmetric Bricard mechanism has advantages of simple structure, good rigidity, and high reliability [12]. Adopting threefold-symmetric Bricard mechanism as the primary frame not only improves the reliability of cage deployment but also reduces the driven difficulty.

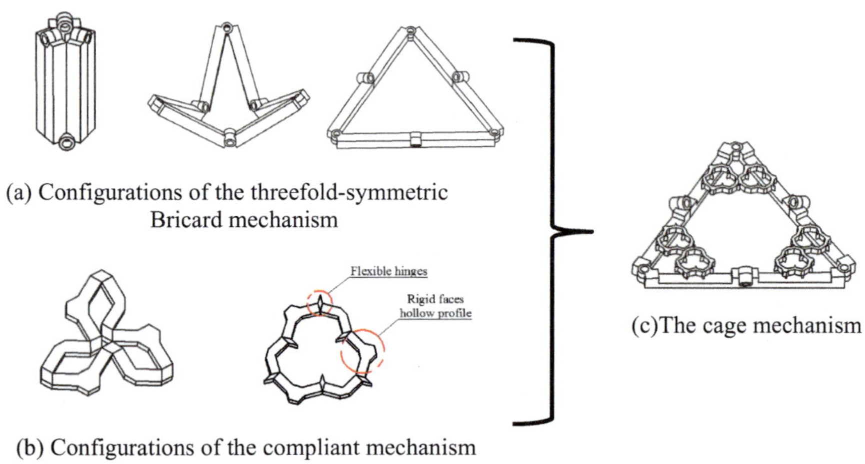

Fig. 4. The model of the cage mechanism.

As shown in Fig. 4b, the secondary frame is a new kind of compliant mechanism proposed by Overvelde et al. [13]. It is composed of rigid faces and flexible hinges. The secondary frames are embedded in the primary frame and deployed independently. The secondary frames are compatible with the folding of the primary frame by deformation while deployed by the rebound of the hinges.

3 Cage Driven Mechanism and Closed-Loop Cable-Driven Module

As shown in Fig. 5a, cage driven mechanism consists of a central column, a slider that moves along the column, driven links, and cross hinges. When the primary frame is deployed to a plane, cross hinges are connected to the revolute joints and the driven links. The ends of the hinges are restrained in the same plane with the revolute joints. The primary frame can be deployed by the rotation of the driven links.

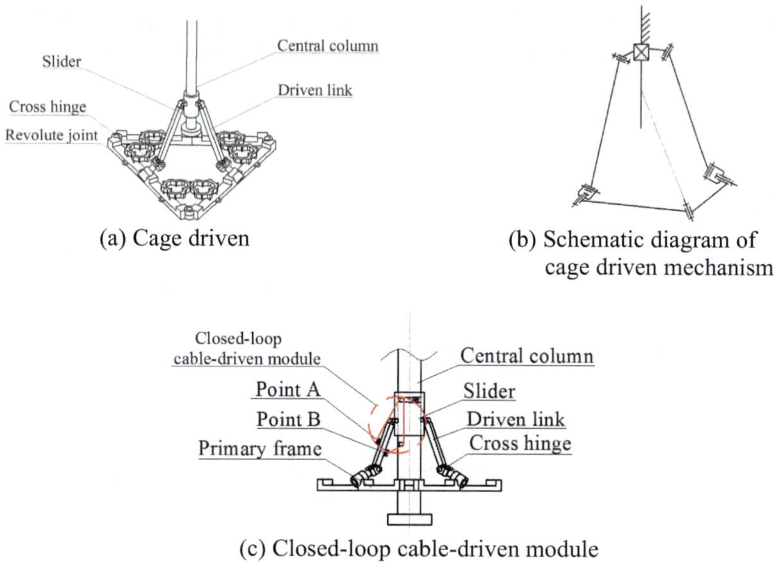

Fig. 5. Cage driven mechanism and closed-loop cable-driven module

Since offshore cages work in the marine environment, the driven system is required to be simple and light. Cable-driven module satisfies the requirements to deploy the frame with the characteristics of good driving flexibility, light mass, and long transmission distance [14].

The closed-loop cable-driven module is proposed in Fig. 5c. Two ends of the cable are connected to the points A and B. The cable is wound around the motor to form a close-loop. The driven links will rotate to deploy the frame by the rotation of the motor. When the motor rotates clockwise, the cable is pulled and the primary frame is folded. On the contrary, the cable is released and the primary frame is deployed.

4 Kinematic Analysis of the Cage Mechanism

4.1 Analysis of Kinematic Characteristic of the Cage Mechanism

The primary frame is composed of threefold-symmetric Bricard mechanism with twist angle $\alpha = \pi/3$ to achieve the most compact folded configuration [15]. It has been known that three planes of symmetry intersect at a determined line in any position for any threefold-symmetric Bricard mechanism.

As shown in Fig. 6a, O_1O_2 is the determined line of three symmetric planes. When the frame is fully deployed, the central axis of the column coincides with the line O_1O_2. At this time, the axes of joints B, D, F are parallel to each other and the axes of joints A, C, and E intersect at point M, which is on the line O_1O_2. As shown in Fig. 6b, during the process of deployment the axes of joint A, C, and E intersect at point M, the axes of joint B, D and F intersect at point N, points M and N are on the line O_1O_2. As shown in Fig. 6c, joints A, C, E and joints B, D, and F move into two parallel planes, the axes of the joints intersect at points M and N when the frame is fully folded.

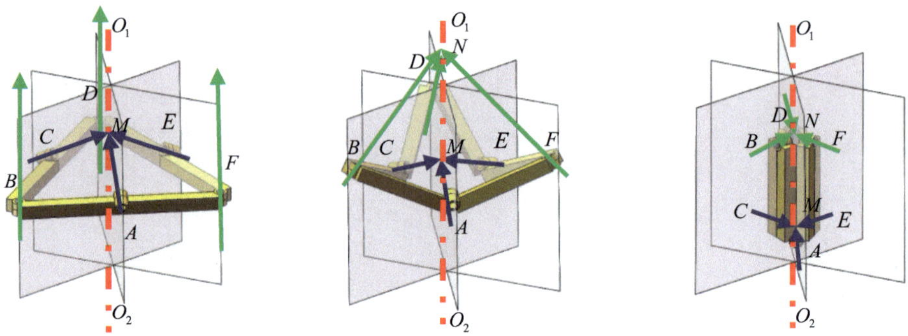

(a)Fully deployed configuration

(b) Configuration during the process of deployment

(c) Fully deployed configuration

Fig. 6. Axis relationship of threefold-symmetric Bricard mechanism.

It can be seen from the above analysis that the axes of joints A, C, E and joints B, D, F always intersect at points M and N which move along line O_1O_2. Each joint always moves on a symmetry plane and rotates around the axis in the plane, that is, the movements of each joint in the plane of symmetry are two translations and one rotation. According to the above analysis, the driven mechanism needs to satisfy the following conditions:

1. There are always three symmetric planes which coincide with threefold-symmetric Bricard mechanism.
2. By taking the line O_1O_2 as the frame, the driven links rotate in the symmetric planes to drive the primary frame.

As shown in Fig. 7, taking the center column as the frame, three driven links are evenly installed on the slider. The cross hinges are connected to the joints A, C, and E. Three symmetric planes always exist in the driven mechanism through the constraint of the slider and the driven link. When the driven link rotates in the symmetrical plane, it drives the cross hinge to move in the symmetrical plane. The end of the cross hinge and the end of the revolute joint are restrained in the same plane. Through the cross hinge, two translations and one rotation of the joint in the symmetrical plane can be realized. Since the threefold-symmetric Bricard mechanism is one degree of freedom, the primary frame can be driven by rotating any link.

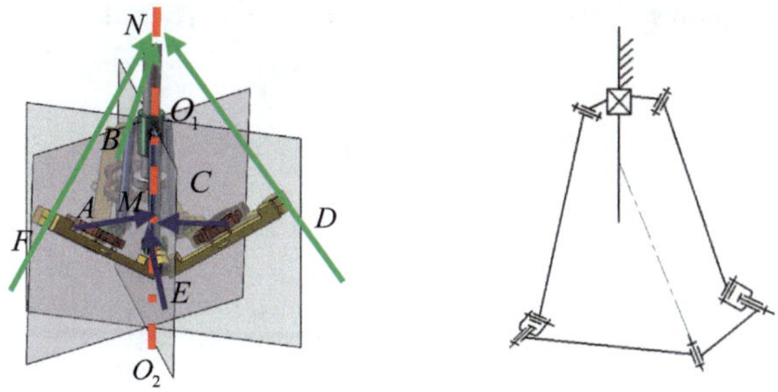

(a) Axis relationship of cage driven mechanism

(b) Schematic diagram of cage driven mechanism

Fig. 7. Model of the cage driven mechanism

4.2 Relationship Between the Driven Angle and the Open State of the Cage Frame

The relationship between the driven angle of the driven link and the corresponding open state of the frame is derived in this section. First, the input angle and the output angle can be obtained by establishing the kinematic equations of the primary frame. Then, the relationship between the driven angle and the displacement relative to the central column is analyzed through the projection in symmetry planes and the vertical plane, and the open state is further obtained.

The D-H model of the threefold-symmetric Bricard mechanism is shown in Fig. 8, whose geometric parameters can be expressed as [15]:

$$\begin{aligned} & a_{12} = a_{23} = a_{34} = a_{45} = a_{56} = a_{61} = l \\ & \alpha_{12} = \alpha_{34} = \alpha_{56} = \alpha, \alpha_{23} = \alpha_{45} = \alpha_{61} = 2\pi - \alpha \\ & R_i = 0 (i = 1, 2, \ldots, 6) \end{aligned} \tag{1}$$

where $a_{i(i+1)}$ is the length of the link $i(i+1)$, $\alpha_{i(i+1)}$ is the twist angle between the axes of joints i and $i+1$, R_i is the offset along joint i.

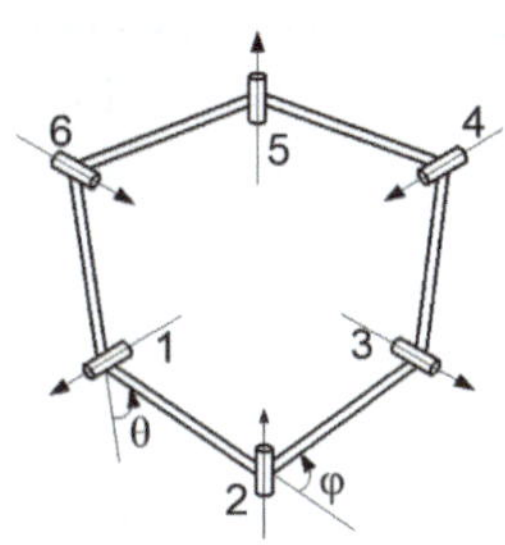

Fig. 8. Threefold-symmetric overconstrained mechanism.

Because of symmetry, only two revolute variables are considered in detail:

$$\begin{aligned}\theta_1 = \theta_3 = \theta_5 = \theta \\ \theta_2 = \theta_4 = \theta_6 = \varphi\end{aligned} \tag{2}$$

The mobility condition of the mechanism in Fig. 8 is:

$${}^{3}_{4}T\,{}^{2}_{3}T\,{}^{1}_{2}T = {}^{5}_{4}T\,{}^{6}_{5}T\,{}^{1}_{6}T \tag{3}$$

By using the parameters in Eqs. (1), (2) and (3) we can obtain the expression of the closure equation when $\alpha = \pi/3$:

$$1 + 3(\cos\,\theta + \,\cos\,\varphi) + 5\cos\,\theta\,\cos\varphi - 4\sin\,\theta\,\sin\,\varphi = 0 \tag{4}$$

Projecting the cage frame to the vertical plane and the symmetry plane, the projection views are shown in Fig. 9.

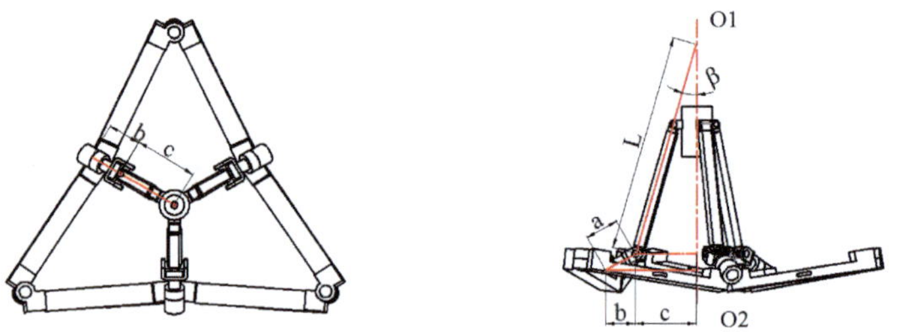

(a)Vertical projection view (a)Symmetric projection view

Fig. 9. Projection views of the frame mechanism.

As shown in Fig. 10, the length of the driven links and the cross hinges are respectively known as L and a. The design work can be done based on the parameters of β, b, c, where β is the rotating angle between the driven link and the line O_1O_2, c is the vertical projection length of the driven link, b is the vertical projection length of the cross hinge. The projection length from the frame to the line O_1O_2 is $b + c$. Given any input angle θ, $b + c$ is determined and the rotation angle of the driven link can be expressed as followed:

$$\beta = \arccos\frac{c}{L} \tag{5}$$

The input angle θ corresponds to the length of the frame from the axis, and the projection length b, c is fixed for the given input angle, so there is a mapping relationship between the rotation angle of the driving link and the input angle.

For example, when $L = 400$ mm, $a = 148$ mm, $\beta = 21.5°$, the input-output angle is $(\theta, \varphi) = \left(-\frac{\pi}{3}, \frac{4\pi}{5}\right)$ and the open state of the cage is shown as Fig. 10.

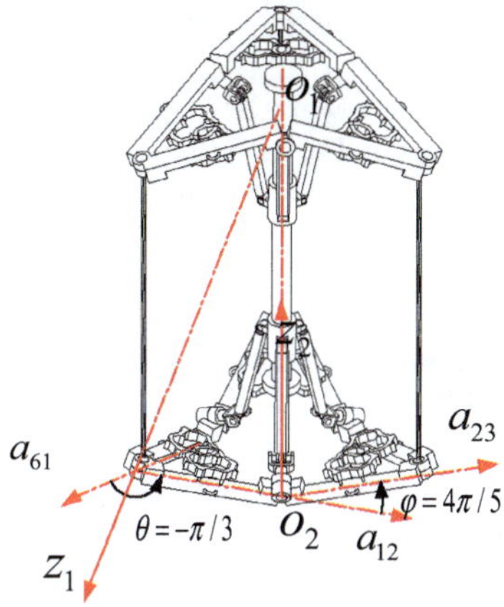

Fig. 10. The open state of the cage frame when $\beta = 21.5°$

5 Conclusions

(1) In this paper, a novel embedded primary and secondary deployable offshore cage is designed based on the concept of deployable mechanisms. The cage has a small folded space to overcome the inconvenience of transportation. A closed-loop cable-driven module is designed to deploy the cage through the pull and release of the cable.

(2) Based on the kinematic characteristics of threefold-symmetric Bricard mechanism, this paper analyses the deployment conditions of the cage frame, and designs the deployable cage driven mechanism with determined axis as folding center.

(3) The input-output angle can be obtained by the kinematic equations of the primary frame. Then, the relationship between the driven angle and displacement of the driven link relative to the central column is analyzed through the projection in symmetry plane and the vertical plane, and the open state is further determined.

References

1. Li, L., Fu, S., Xu, Y., et al.: Dynamic responses of floating fish cage in waves and current. Ocean Eng. **72**, 297–303 (2013)
2. Guo, G.X., Tao, Q.Y., Huang, X.H., et al.: Progress on frontier of equipment technology for sea-cage aquaculture. J. Agric. Sci. Technol. **13**(5), 44–49 (2011)
3. Loverich, G., Forster, J.: Advances in offshore cage design using spar buoys. Marine Technol. Soc. J. **34**(1), 18–28 (2000)
4. Scott, D.C.B.: Offshore cage systems: a practical overview. Options Méditerranéennes Série B Etudes Et Recherches (2000)
5. Hex Box Norway. http://www.oceanaquafarms.com/product/hex-box-global/. Accessed 04 June 2019

6. “Deep Blue 1” attracts worldwide attention. http://www.sohu.com/a/251994899_421212. Accessed 04 June 2019
7. Yi, Y., Yan, P., Hua, Y.P., et al.: Deployable parallel lower-mobility manipulators with scissor-like elements. Mech. Mach. Theory **135**, 226–250 (2019)
8. Liu, R.Q., Tian, D.K., Deng, Z.Q.: Research actuality and the prospect of structure for space deployable antenna. J. Mach. Des. **27**(9), 1–10 (2010)
9. Luo, Y., Yu, Y., Liu, J.: A retractable structure based on Bricard linkages and rotating rings of tetrahedra. Int. J. Solids Struct. **45**(2), 620–630 (2008)
10. Shainee, M., Ellingsen, H., Leira, B.J., et al.: Design theory in offshore fish cage designing. Aquaculture **392**(10), 134–141 (2013)
11. Chen, Y.: Design of Structural Mechanisms. Ph.D. thesis. University of Oxford (2003)
12. Guo, S., Fang, Y.F., Yue, C.: Structure synthesis of single closed-loop multi-degree of freedom of over-constrained mechanism based on screw theory. J. Mech. Eng. **27**(9), 1–10 (2010)
13. Overvelde, J.T.B., Weaver, J.C., Hoberman, C., et al.: Rational design of reconfigurable prismatic architected materials. Nature **541**(7637), 347–352 (2017)
14. Mao, Y., Agrawal, S.K.: Design of a Cable-Driven Arm Exoskeleton (CAREX) for neural rehabilitation. IEEE Trans. Rob. **28**(4), 922–931 (2012)
15. Chen, Y., You, Z., Tarnai, T.: Threefold-symmetric Bricard linkages for deployable structures. Int. J. Solids Struct. **42**(8), 2287–2301 (2005)

An Active/Passive Joint for Reconfiguration Applications

Mykhailo Riabtsev(✉), Victor Petuya, Aitor Riera, and Mónica Urízar

Department of Mechanical Engineering, University of the Basque Country UPV/EHU, Plaza Torres Quevedo 1, 48013 Bilbao, Spain
mriabtsev001@ikasle.ehu.eus

Abstract. In this paper, an active/passive reconfigurable 2R joint is presented. Direct and inverse kinematic problems and the force transmission are considered. The advantages of the current design show its superiority with respect to previous designs. A prototype of the joint was built and tested, proving the viability of the concept. The proposed joint can be applied in reconfigurable manipulators, serial robot arm wrists, humanoid robot shoulder and waist, where 2R active joints are required.

Keywords: Active/passive joint · Manipulator · Reconfigurable · Prototype

1 Introduction

The process of readjusting an assembly line for a new product is very costly if it involves equipment replacement. One way of reducing cost is to use reconfigurable mechanisms, which can be readjusted for a new task and do not require complete replacement.

There are different reconfiguration strategies. The example of the *offline* reconfiguration approach is modular reconfiguration [1], when mechanisms are made from standard modules which can be easily reconnected in the required way. Other methods lay on additional actuation [2], or on the contrary, locking degrees of freedom [3]. Some research is focused on the reconfiguration by changing the dimensions of the base or the end-effector platforms [4] or other links of the mechanism [5]. Another widely used way of changing the mechanism's topology is the application of reconfigurable joints. These joints can combine different types of motion [6], change orientation of their axis [7] or involve other strategies [8–10].

In the current paper, a 2R active/passive reconfigurable joint is presented. The design of the joint is described in Sect. 2, kinematics and force transmission analysis is provided in Sect. 3, prototype manufacturing is described in Sect. 4. Finally, possible applications and future work are highlighted.

D. Wang et al. (Eds.): MeTrApp 2019, MMS 79, pp. 133–140, 2020.
https://doi.org/10.1007/978-981-15-0142-5_14

2 Design Description

The joint proposed in this paper (Fig. 1) consists of a central ball with a slot crossing the sphere. The red input is placed in this slot and defines the first rotational input. The blue shaft transmits rotation to the ball through the bevel gear to control the second rotational degree of freedom. If two inputs are used (Fig. 1) in order to increase the accuracy and balance the joint, the input command for the two light blue shafts should have the same value but in opposite directions.

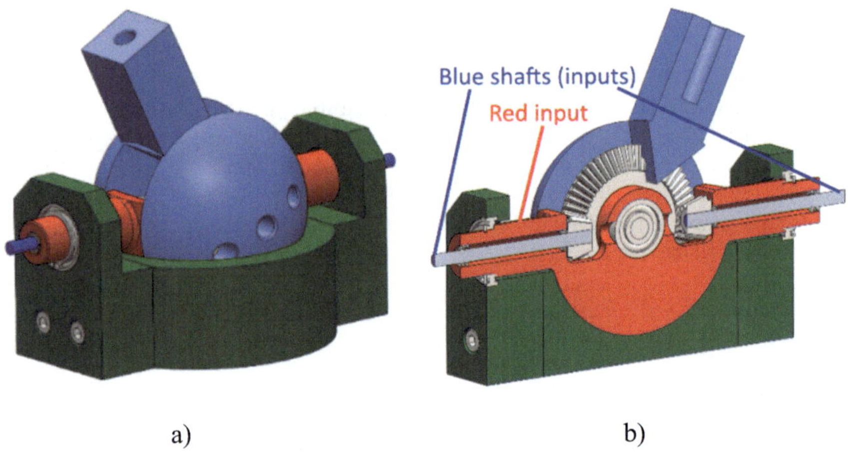

Fig. 1. CAD model of the proposed joint: (a) general view; (b) cross section.

The proposed joint is kinematically a cardan joint with the capacity to be actuated in both axes. The design proposed in this paper, aims at increasing the load carrying capacity of the joint compared to the conventional universal joints. Thus, the contact area between the central ball and the base part allows a good distribution of the load.

The proposed joint represents an evolution of a previous design shown in Fig. 2 [11], which was based in the concept of the spherical five bar linkage. This previous design had several drawbacks. Firstly, the relation between the input commands in the blue shafts depends on the input command in the red one. In addition, the clearances between the sliding parts are hard to compensate and the value of transmitted torque depends on the dimensional parameters of the joint (ball radius, link lengths). Finally, the structural elements of the joint make workspace extension impossible.

In the proposed joint, most of these drawbacks have been eliminated. The workspace volume is now 73.9% bigger due to the absence of the top ring and elimination of the singularity position inherent to the previous design. The joint has increased the percentage of the ball-to-base contact (from 60.87% to 93.22%) because of the open slots elimination. The actuation is now uniform and symmetrical. The use of two actuators for the blue shafts creates a pretension in the gear that eliminates the clearances and improves the accuracy of the joint. In addition, the kinematics of the joint are very simple and there is the possibility of one-side actuation that can be useful for some applications.

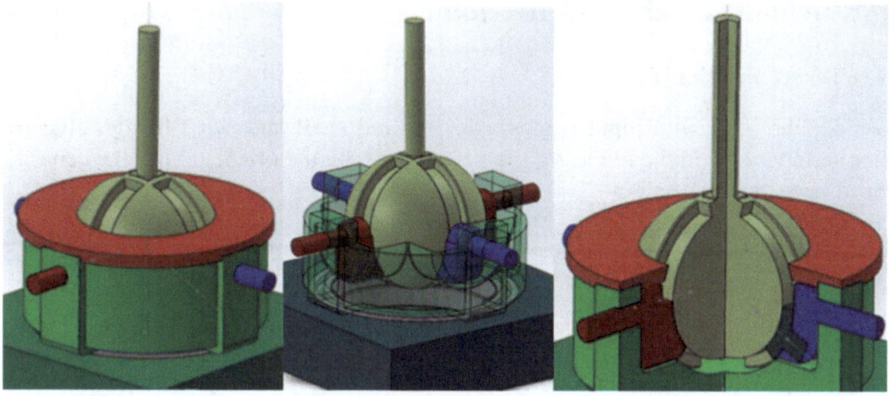

Fig. 2. Initial design [11]

Between the initial design shown in Fig. 2 and the presented design shown in Fig. 3, other alternatives have been proposed. For example, in the joint of Fig. 3 the main design goals were the kinematics simplification and the increase of the workspace volume. In this design, the workspace volume is 236,4% bigger in comparison to the design shown on Fig. 2, due to the actuators arrangement and the elimination of the elements, which can limit the joint movement. Unfortunately, this solution implied the elimination of the joint ball, thus affecting the load capacity. The other disadvantage of this design is the clearance in the gear, which is hard to compensate and affects the accuracy of the joint movement.

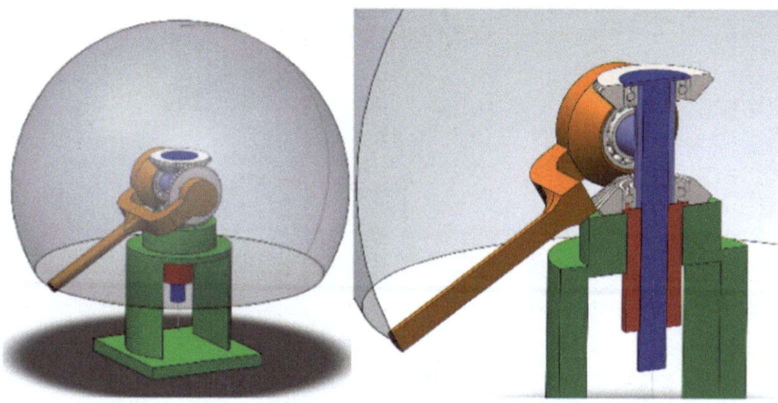

Fig. 3. Intermediate design: joint with extended workspace

As explained, the new joint can be simultaneously actuated to control the two rotational degrees of freedom. Additionally, the joint can be implemented in a reconfigurable manipulator as an active/passive rR joint. In this case, the red input defines de orientation of the rotation axis and once it is blocked, a reconfigurable revolute joint is obtained.

3 Kinematic Analysis of the Joint

3.1 Direct Position Problem

Consider the joint with input angles α for the red shaft and β_{input} for the blue shaft. After an input of α in the red shaft, the vector defining the orientation of the central rod (Fig. 4) is,

$$n' = \begin{Bmatrix} 0 \\ -\sin\alpha \\ \cos\alpha \end{Bmatrix} \tag{1}$$

The input β_{input} causes a rotation of β around the V' axis being,

$$\beta = \left(\frac{Z_1}{Z_2}\right)\beta_{\text{input}} \pm \alpha \tag{2}$$

where Z_1 and Z_2 are the number of teeth in the pinion and in the gear.

Thus, the orientation vector of the central rod is calculated as follows,

$$n = \begin{bmatrix} \cos\beta & 0 & \sin\beta \\ 0 & 1 & 1 \\ -\sin\beta & 0 & \cos\beta \end{bmatrix} \begin{Bmatrix} 0 \\ -\sin\alpha \\ \cos\alpha \end{Bmatrix} = \begin{Bmatrix} \sin\beta\cos\alpha \\ -\sin\alpha \\ \cos\beta\cos\alpha \end{Bmatrix} \tag{3}$$

which normalized, has the form,

$$n = \begin{Bmatrix} \tan\beta \\ -\tan\alpha/\cos\beta \\ 1 \end{Bmatrix} \tag{4}$$

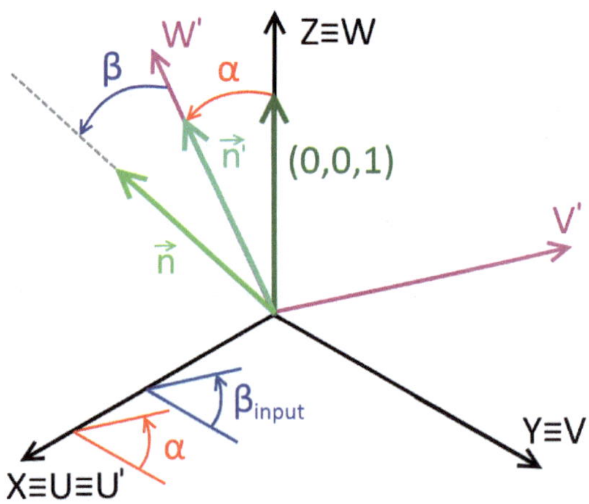

Fig. 4. Position parameters of the joint.

3.2 Inverse Position Problem

In this case, the components of vector $n = \{u \quad v \quad 1\}^T$ are known. Thus, the values for the red and blue inputs can be directly calculated as,

$$\begin{aligned} \beta &= \tan^{-1} u \\ \alpha &= -\tan^{-1}(v \cos\beta) \end{aligned} \tag{5}$$

being (2) the relation between β and β_{input}.

3.3 Force Transmission in the Joint

As it is shown subsequently, the computation of the force transmission becomes rather simple for the proposed joint. As the joint has two inputs, two input torques should be considered. Red input torque T_r creates red force component F_r:

$$F_r = \frac{T_r}{d_r} \tag{6}$$

where d_r is the shortest distance between the point where the force F_r is applied and the axis of rotation of red input (Fig. 5). This distance is calculated as,

$$d_r = R\cos\beta \tag{7}$$

being, R the distance between the application point and the center of the joint ball. The direction of the force F_r is $(-V')$.

Additionally, blue input torque T_b creates red force component F_b:

$$F_b = \left(\frac{Z_2}{Z_1}\right) T_b R \tag{8}$$

This feature provides additional force amplification in the joint and can be considered as an advantage for certain applications. The direction of the force F_r is on plane *W'U'* and perpendicular to vector ***n*** (Fig. 4).

The two force components are perpendicular to each other, thus the module of the resultant force yields:

$$F = \sqrt{F_r^2 + F_b^2} \tag{9}$$

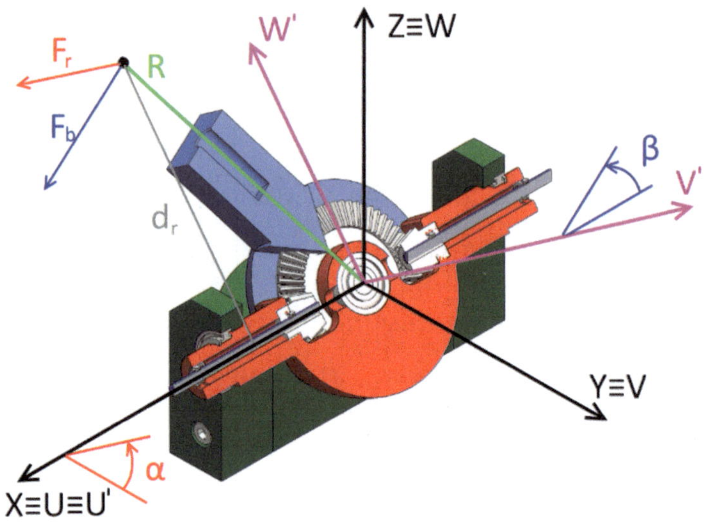

Fig. 5. Force transmission in the joint

4 Prototype Manufacturing

For testing the proposed joint, a demonstrative prototype was manufactured (Fig. 6). The base of joint, the ball and the red input were printed with Stratasys' Dimension sst1200. The bevel gear and the pinions were printed with Markforged Mark Two 3D printer, which provides better surface quality.

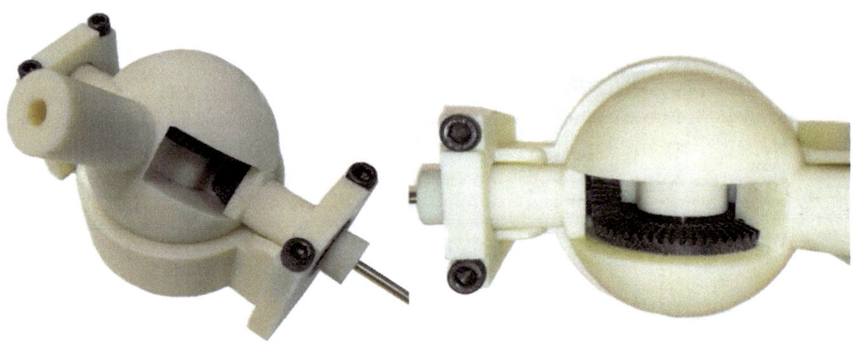

Fig. 6. Demonstrative prototype

The mechanical properties of the printed gears and their accuracy do not allow implementing this prototype directly in a manipulator, but, overall, it becomes an effective way for analyzing the concept of the proposed joint.

The testing of the prototype demonstrated the lack of durability of the gear teeth. Currently, the design of the joint is being improved with the aim of building a practical joint using steel/aluminum, which will be able to be used in a reconfigurable mechanism. This process requires modification of each part of the joint for the conventional machining methods, because 3D printing cannot provide satisfactory tolerances and the surface quality.

Among the possible applications, the joint presented in this paper can be incorporated in parallel robots, act as a wrist of a serial robotic arm, shoulder joint or waist of a humanoid robot, walking robot leg joints, etc.

5 Conclusions

In this paper, a 2R active/passive joint has been presented. The proposed joint is an evolution of previous designs and it demonstrates improved workspace, simpler input control and higher load carrying capacity and accuracy. In addition, there is the possibility of one-side actuation that can be useful for some applications. The joint can be applied as a reconfigurable rR joint which is able to change the orientation of its axis, or as a lockable universal joint.

Currently we are working on the design improvement in order to build a prototype in steel/aluminum. In addition, we are working on finite element structural analysis of the joint to find out the weaknesses and limitations of the current design.

Acknowledgments. The authors wish to acknowledge the financial support received from the Spanish Government through the Ministerio de Economía y Competitividad (Project DPI2015-67626-P (MINECO/FEDER, UE)) and the financial support given to the research group through the project with Ref. IT949-16, given by the Departamento de Educación, Política Lingüística y Cultura of the Regional Government of the Basque Country.

References

1. Chen, I.-M.: Rapid response manufacturing through a rapidly reconfigurable robotic workcell. Robot. Comput. Integr. Manuf. **17**, 199–213 (2001)
2. Wohlhart, K.: Kinematotropic linkages. In: Lenarčič, J., Parenti-Castelli, V. (eds.) Recent Advances in Robot Kinematics, pp. 359–368. Springer, Dordrecht (1996)
3. Grosch, P., Di Gregorio, R., López, J., Thomas, F.: Motion planning for a novel reconfigurable parallel manipulator with lockable revolute joints. In: Proceedings of the IEEE 2010 International Conference on Robotics and Automation (ICRA), Anchorage, AK, USA, 3–7 May 2010, pp. 4697–4702 (2010)
4. Viegas, C., Tavakoli, M., de Almeida, A.T.: A novel grid-based reconfigurable spatial parallel mechanism with large workspace. Mech. Mach. Theory **115**, 149–167 (2017)
5. Nansai, S., Rojas, N., Elara, M., Sosa, R.: Exploration of adaptive gait patterns with a reconfigurable linkage mechanism. In: Proceedings of the 2013 IEEE/RSJ International Conference on Intelligent Robots and Systems (IROS 2013), Tokyo, 2013, pp. 4661–4668 (2013)

6. Yan, H.S., Kuo, C.H.: Topological representations and characteristics of variable kinematic joints. J. Mech. Des. **128**, 384–391 (2006)
7. Gan, D., Dai, J.S., Liao, Q.: Mobility change in two types of metamorphic parallel mechanisms. J. Mech. Robot. **1**, 041007-1–041007-9 (2009)
8. Wei, Y., Fang, Y., Guo, S.: Reconfigurable parallel mechanisms with three types of kinematotropic chains. In: Proceedings of the 14th International Federation for the Promotion of Mechanism and Machine Science (IFToMM) World Congress, Taipei, Taiwan, 25–30 October, 2015 (2015)
9. Halverson, P.A., Howell, L., Magleby, S.: Tension-based multi-stable compliant rolling-contact elements. Mech. Mach. Theory **45**(2), 147–156 (2010)
10. Lee, C., Hervé, J.: Various types of RC//RC -like linkages and the discontinuously movable Koenigs joint. Mech. Mach. Theory **96**, 255–268 (2015)
11. Riabtsev, M., Petuya, V., Riera, A., Macho, E.: Design of an active reconfigurable 2R joint. In: Proceedings of the 15th International Federation for the Promotion of Mechanism and Machine Science (IFToMM) World Congress, Krakov, 30–4 July, 2019 (2019)

Kinematics Analysis and Dynamics Simulation Optimization of Lower Limb Rehabilitation Robot for the Stroke

Li Dong[1], Lu Li[2(✉)], and Shouqi Chen[1]

[1] University of Science and Technology of China, Hefei 230026, China
{dlustc, chensq09}@mail.ustc.edu.cn
[2] Institute of Advanced Manufacturing Technology Hefei Institutes of Physical Science, Chinese Academy of Sciences, Changzhou 213164, China
lli@iamt.ac.cn

Abstract. For a new type of lower limb rehabilitation robot for the stroke with 4 degrees of freedom(DOF), the D-H method was used to analyze the kinematics, the kinematic equations were established, and the workspace of the rehabilitation robot was simulated. Based on the Lagrange equations, the dynamic equations of the rehabilitation robot were deduced and the torque of each joint was simulated. Finally, the gravity compensation mechanism was designed to optimize the starting torque and maximum torque of the flexion and extension joint of hip, the sum of which was reduced by 72.51%.

Keywords: Lower limb rehabilitation robot · Kinematics analysis · Dynamics analysis · Structural optimization · Gravity compensation mechanism

1 Introduction

Acute cerebrovascular disease, also known as stroke, is a common disease worldwide, with high morbidity, high mortality and high disability rate. According to the *China Cardiovascular Disease Report* 2018, China's morbidity and mortality rate are still on the rise, and stroke has become the first cause of death for Chinese residents [1]. Therefore, the rehabilitation treatment of hemiplegia after stroke has become the research focus and hot spot of modern rehabilitation medicine and rehabilitation engineering [2]. Lower limb rehabilitation robots help the stroke patients with rehabilitation training, which have the characteristics of high efficiency and repeatability.

The research of medical rehabilitation robot started in the 1980s. For example, the HAL exoskeleton robot developed by the University of Tsukuba in Japan can assist elderly and lower limb disabled people to walk [3]. Lokomat, an exoskeleton lower limb rehabilitation robot developed by Swiss medical device company HOCOMA in collaboration with the rehabilitation center of the Zurich Medical College, can assist the lower limb disabled people to perform rehabilitation on the treadmill [4]. Hefei University of Technology proposed a sitting lower limb rehabilitation mechanism based on five-bar adjustable mechanism, which has the advantages of simple structure, complete functions and suitable for family rehabilitation [5]. Others include active leg

D. Wang et al. (Eds.): MeTrApp 2019, MMS 79, pp. 141–151, 2020.
https://doi.org/10.1007/978-981-15-0142-5_15

exoskeleton (ALEX) at the University of Delaware in the United States [6], lower extremity powered exoskeleton (LOPES) from the University of Twente in the Netherlands [7], etc.

At present, the lower limb rehabilitation robots that have been put into production on the market are mostly assisted upright walking mode or suspension weight reduction mode. The former requires the patients with sufficient upper limb strength and has a risk of balance. The latter is not popular because of the high installation requirements and the overall size. In order to make the application range of rehabilitation robots more extensive, the rehabilitation robot involved in this paper adopts sitting/lying structure, which makes the robot small in size, light in weight and suitable for ordinary household use. In order to maximize the imitative function of normal people's legs, the rehabilitation robot involved in this paper has 4 DOF in one leg, in which the hip joint has 2 DOF, the knee joint has 1 DOF, and the ankle joint has 1 DOF. The 4 DOF are driven by 4 DC motors, and the 4 DC motors can be driven separately for targeted training or simultaneously.

2 Kinematics Analysis

2.1 Kinematics Model and Equation Establishment

The kinematics analysis of robots mainly studies the motion characteristics, that is, the relationship between the position, velocity, acceleration and position of the mechanism. In this paper, D-H method is used to analyze the kinematics of robot system, this method fixes a coordinate system on each connecting rod of the robot, and then uses a 4×4 homogeneous transformation matrix to describe the spatial relationship between two adjacent connecting rods. By transforming sequentially, the position of the end actuator relative to the base coordinate system is deduced, and the kinematic equations of the robot are established [8].

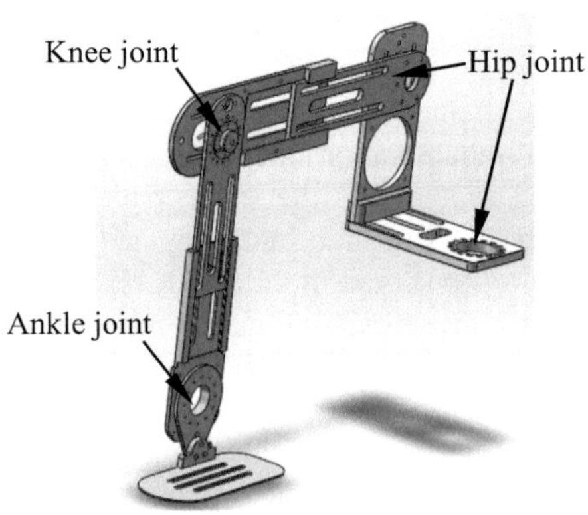

Fig. 1. Structure diagram.

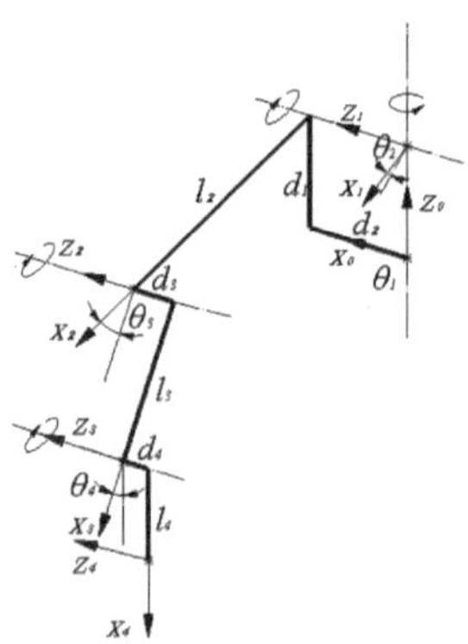

Fig. 2. D-H coordinate system diagram.

In order to facilitate the kinematics analysis, the robot is simplified into a rigid linkage mechanism. The structure diagram of the structure is shown in Fig. 1, the D-H coordinate system is shown in Fig. 2, and Table 1 is the D-H parameters.

Table 1. D-H parameters table.

Connecting rod number	$a_i/(mm)$	$\alpha_i/(°)$	$d_i/(mm)$	$\theta_i/(°)$	Range of variable $\theta_i/(°)$
1	$l_1 = 0$	90	$d_1 = 230$	θ_1	−20–10
2	$l_2 = 500$	0	$d_2 = 150$	θ_2	−80–30
3	$l_3 = 420$	0	$d_3 = 30$	θ_3	−135–0
4	$l_4 = 80$	0	$d_4 = 15$	θ_4	−45–30

Where a_i represents the length of the connecting rod, α_i represents the torsion angle of the connecting rod, d_i represents the connecting rod distance, and θ_i represents the connecting rod angle [9].

The spatial transformation matrix between adjacent connecting rods of the D-H method is

$$ {}^{i-1}_{\ \ i}T = \begin{bmatrix} cos\theta_i & -sin\theta_i cos\alpha_i & sin\theta_i sin\alpha_i & a_i cos\theta_i \\ sin\theta_i & cos\theta_i cos\alpha_i & -cos\theta_i sin\alpha_i & a_i sin\theta_i \\ 0 & sin\alpha_i & cos\alpha_i & d_i \\ 0 & 0 & 0 & 1 \end{bmatrix} \tag{1} $$

Bring the parameters in Table 1 into Eq. (1), and calculate ${}^0_1T$, ${}^1_2T$, ${}^2_3T$, and ${}^3_4T$, then the kinematic equation of the robot is

$$ {}^0_4T = {}^0_1T\,{}^1_2T\,{}^2_3T\,{}^3_4T = \begin{bmatrix} c_1c_{234} & -c_1s_{234} & s_1 & (d_2+d_3+d_4)s_1 + l_2c_1c_2 + l_4c_1c_{234} + l_3c_1c_{23} \\ s_1c_{234} & -s_1s_{234} & -c_1 & l_2s_1c_2 - (d_2+d_3+d_4)c_1 + l_4s_1c_{234} + l_3s_1c_{23} \\ s_{234} & c_{234} & 0 & d_1 + l_2s_2 + l_3s_{23} + l_4s_{234} \\ 0 & 0 & 0 & 1 \end{bmatrix} \tag{2} $$

Where s_i represents $sin\theta_i$, c_i represents $cos\theta_i$, s_{ij} represents $sin(\theta_i + \theta_j)$, the rest can be deduced by analogy.

2.2 Workspace Simulation

The workspace refers to the movement space at the end of the robot and is an important index to measure the performance of the robot. In the process of robot design, it is necessary to consider not only the structure and control system of the robot, but also the workspace of the robot. The size of the workspace is related to the size of the robot's connecting rod and the overall structure of the robot [10]. Figure 3 shows the results of simulation based on the design parameters of the robot in Table 1.

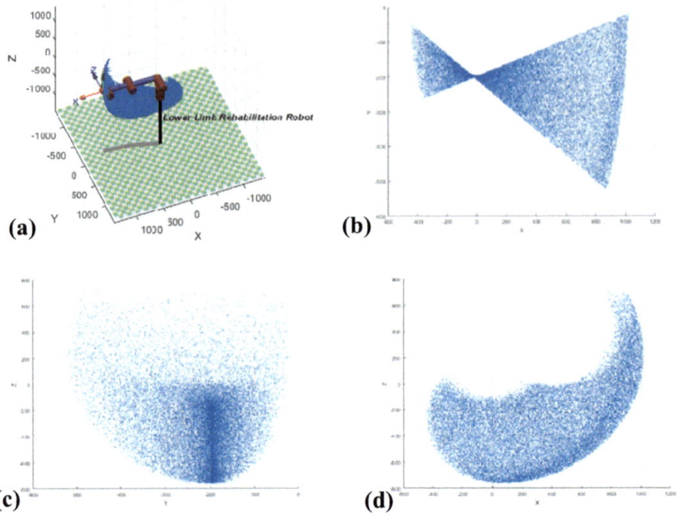

Fig. 3. (a) 3D diagram of the foot workspace (b) xoy plane projection diagram (c) yoz plane projection diagram (d) xoz plane projection diagram.

It can be seen from Fig. 3 that the robot has a large space in the direction of the leg flexion and extension, and the movement space on both sides is small, which is consistent with the movement pattern of the normal human leg.

3 Dynamics Analysis

3.1 Dynamics Modeling and Solution

Dynamics studies the relationship between motion and force. The dynamic equations are equations describing the mechanical system of a robot, which represent the relationship between the position, velocity, acceleration and driving torque of each joint. In this paper, Lagrange method was used to analyze the dynamics of the rehabilitation robot system. The Lagrange equation form is as follows

$$L(q_i, \dot{q}_i) = K - P \tag{3}$$

$$F_i = \frac{d}{dt}\left(\frac{\partial L}{\partial \dot{q}_i}\right) - \frac{\partial L}{\partial q_i} \quad , \quad i = 1, 2, \cdots n \tag{4}$$

Where L is the Lagrange function, K is the sum of kinetic energy, P is the sum of potential energy, F_i is the generalized force or generalized moment, q_i is the generalized coordinate, $\dot{q}_i$ is the generalized velocity, and n is the number of DOF [11].

A simplified model of the robot dynamics is shown in Fig. 4, m_{ri} represents the motor rotor mass, J_i represents the motor rotor's moment of inertia, q_{mi} represents the angle before the reducer, N represents the reduction ratio, and m_{li} represents mass of the connecting rod, I_i represents the moment of inertia of the connecting rod, q_i represents the angle after passing the reducer, l_i represents the length of the connecting rod, and d_i represents the length of the connecting rod centroid.

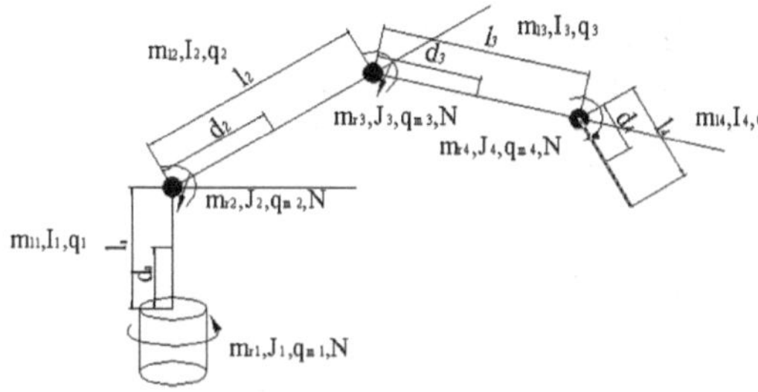

Fig. 4. Simplified model of 4 DOF lower limb rehabilitation robot.

Calculate the kinetic energy and potential energy of each motor rotor and connecting rod according to Fig. 4, and bring them into Eq. (4) to calculate the joint torque.

$$\tau_i = \frac{d}{dt}\left(\frac{\partial K}{\partial \dot{q}_i}\right) - \frac{\partial K}{\partial q_i} + \frac{\partial P}{\partial q_i} \tag{5}$$

Where τ represents the joint torque, and the final dynamic equation is

$$\boldsymbol{H}(\boldsymbol{q})\ddot{\boldsymbol{q}} + \boldsymbol{C}(\boldsymbol{q},\dot{\boldsymbol{q}})\dot{\boldsymbol{q}} + \boldsymbol{G}(\boldsymbol{q}) = \boldsymbol{\tau} \tag{6}$$

Where $\boldsymbol{H(q)}$ is the inertia matrix, $\boldsymbol{C}(\boldsymbol{q},\dot{\boldsymbol{q}})$ is the matrix of centrifugal force and Coriolis force, and $\boldsymbol{G(q)}$ is the gravity matrix.

$$\boldsymbol{H}(\boldsymbol{q}) = \begin{bmatrix} H_{11} & H_{12} & H_{13} & H_{14} \\ H_{21} & H_{22} & H_{23} & H_{24} \\ H_{31} & H_{32} & H_{33} & H_{34} \\ H_{41} & H_{42} & H_{43} & H_{44} \end{bmatrix}, \boldsymbol{C}(\boldsymbol{q},\dot{\boldsymbol{q}}) = \begin{bmatrix} C_{11} & C_{12} & C_{13} & C_{14} \\ C_{21} & C_{22} & C_{23} & C_{24} \\ C_{31} & C_{32} & C_{33} & C_{34} \\ C_{41} & C_{42} & C_{43} & C_{44} \end{bmatrix}, \boldsymbol{G}(\boldsymbol{q}) = \begin{bmatrix} G_1 \\ G_2 \\ G_3 \\ G_4 \end{bmatrix}, \tau = \begin{bmatrix} \tau_1 \\ \tau_2 \\ \tau_3 \\ \tau_4 \end{bmatrix}$$

$$\begin{aligned} H_{11} = {} & J_1N^2 + I_1 + m_{l2}d_2^2c_2^2 + m_{r3}l_2^2c_2^2 + m_{l3}(l_2c_2 + d_3c_{23})^2 + m_{r4}(l_2c_2 + l_3c_{23})^2 \\ & + m_{l4}(l_2c_2 + l_3c_{23} + d_4c_{234})^2. \end{aligned}$$

$$H_{12} = H_{13} = H_{14} = H_{21} = H_{31} = H_{41} = 0.$$

$$\begin{aligned} H_{22} = {} & J_2N^2 + I_2 + m_{l2}d_2^2 + J_3 + m_{r3}l_2^2 + I_3 + m_{l3}(l_2^2 + d_3^2 + 2l_2d_3s_3) + J_4 + m_{r4}(l_2^2 + l_3^2 + 2l_2l_3s_3) + I_4 + \\ & m_{l4}(l_2^2 + l_3^2 + d_4^2 + 2l_2l_3s_3 + 2l_2d_4s_{34} + 2l_3d_4s_4). \end{aligned}$$

$$\begin{aligned} H_{23} = H_{32} = {} & J_3N + I_3 + m_{l3}(d_3^2 + l_2d_3s_3) + J_4 + m_{r4}(l_3^2 + l_2l_3s_3) + I_4 + m_{l4}(l_3^2 + d_4^2 + l_2l_3s_3 + l_2d_4s_{34} + \\ & 2l_3d_4s_4) \end{aligned}$$

$$H_{24} = H_{42} = J_4N + I_4 + m_{l4}(d_4^2 + l_2d_4s_{34} + l_3d_4s_4).$$

$$H_{33} = J_3N^2 + I_3 + m_{l3}d_3^2 + J_4 + m_{r4}l_3^2 + I_4 + m_{l4}(l_3^2 + d_4^2 + 2l_3d_4s_4).$$

$$H_{34} = J_4N + I_4 + m_{l4}(d_4^2 + l_3d_4s_4 + l_3d_4c_4).$$

$$H_{43} = J_4N + I_4 + m_{l4}(d_4^2 + l_3d_4s_4).$$

$$H_{44} = J_4N + I_4 + m_{l4}d_4^2.$$

$$C_{11} = -2\dot{q}_2[m_{l3}(l_2c_2 + d_3c_{23})(l_2s_2 + d_3s_{23}) + m_{r4}(l_2c_2 + l_3c_{23})(l_2s_2 + l_3s_{23}) + m_{l2}d_2^2c_2s_2 + m_{r3}l_2^2c_2s_2 + m_{l4}(l_2c_2 + l_3c_{23} + d_4c_{234})(l_2s_2 + l_3s_{23} + d_4s_{234})].$$

$$C_{12} = C_{33} = C_{44} = 0.$$

$$C_{13} = -2\dot{q}_1[m_{l3}d_3s_{23}(l_2c_2 + d_3c_{23}) + m_{r4}l_3s_{23}(l_2c_2 + l_3c_{23}) + m_{l4}(l_2c_2 + l_3c_{23} + d_4c_{234})(l_3s_{23} + d_4s_{234})].$$

$$C_{14} = -2\dot{q}_1[m_{l4}d_4s_{234}(l_2c_2 + l_3c_{23} + d_4c_{234})].$$

$$C_{21} = \dot{q}_1[m_{l2}d_2^2c_2s_2 + m_{l3}(l_2c_2 + d_3c_{23})(l_2s_2 + d_3s_{23}) + m_{r4}(l_2c_2 + l_3c_{23})(l_2s_2 + l_3s_{23}) + m_{r3}l_2^2c_2s_2 + m_{l4}(l_2c_2 + l_3c_{23} + d_4c_{234})(l_2s_2 + l_3s_{23} + d_4s_{234})].$$

$$C_{22} = \dot{q}_3(2m_{l3}l_2d_3c_3 + 2m_{r4}l_2l_3c_3 + 2m_{l4}l_2l_3c_3 + 2m_{l4}l_2d_4c_{34}) + \dot{q}_4(2m_{l4}l_2d_4c_{34} + 2m_{l4}l_3d_4c_4)$$

$$C_{23} = \dot{q}_3(m_{r4}l_2l_3c_3 + m_{l4}l_2l_3c_3 + m_{l4}l_2d_4c_{34}) + \dot{q}_4(2m_{l4}l_2d_4c_{34} + 2m_{l4}l_3d_4c_4).$$

$$C_{24} = \dot{q}_4(m_{l4}l_2d_4c_{34} + m_{l4}l_3d_4c_4).$$

$$C_{31} = \dot{q}_1[m_{l3}d_3s_{23}(l_2c_2 + d_3c_{23}) + m_{l4}(l_2c_2 + l_3c_{23} + d_4c_{234})(l_3s_{23} + d_4s_{234}) + m_{r4}l_3s_{23}(l_2c_2 + l_3c_{23})].$$

$$C_{32} = -\dot{q}_2(m_{l3}l_2d_3c_3 + m_{r4}l_2l_3c_3 + m_{l4}l_2l_3c_3 + m_{l4}l_2d_4c_{34}).$$

$$C_{34} = 2m_{l4}l_3d_4c_4(\dot{q}_2 + \dot{q}_3).$$

$$C_{41} = \dot{q}_1m_{l4}d_4s_{234}(l_2c_2 + l_3c_{23} + d_4c_{234}).$$

$$C_{42} = -\dot{q}_2m_{l4}d_4(l_3c_4 + l_2c_{34}).$$

$$C_{43} = -m_{l4}l_3d_4c_4(2\dot{q}_2 + \dot{q}_3).$$

$$G_2 = g[m_{l2}d_2c_2 + m_{l3}(l_2c_2 + d_3c_{23}) + m_{l4}(l_2c_2 + l_3c_{23} + d_4c_{234}) + m_{r3}l_2c_2 + m_{r4}(l_2c_2 + l_3c_{23})].$$

$$G_3 = g[m_{l3}d_3c_{23} + m_{l4}(l_3c_{23} + d_4c_{234}) + m_{r4}l_3c_{23}].$$

$$G_4 = m_{l4}d_4gc_{234}.\ G_1 = 0.$$

3.2 Dynamics Simulation

In the dynamic simulation, the motion of each joint of the rehabilitation robot can be intuitively seen, such as output displacement, velocity, acceleration and force. By observing whether there is any mutation in the output curve, we can judge whether the designed structure is reasonable [12]. It is important to note that the lower limb rehabilitation robot needs to drive the patient's legs to move together, so the weight of the patient's legs needs to be taken into account when modeling.

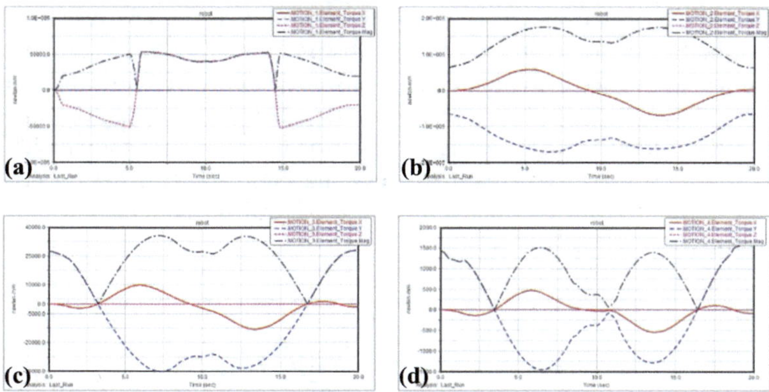

Fig. 5. Torque diagram of joints (a) hip abduction joint (JOINT_1) (b) flexion and extension joint of hip (JOINT_2) (c) flexion and extension joint of knee (JOINT_3) (d) flexion and extension joint of ankle (JOINT_4).

The torque curves of each joint in a motion cycle can be seen from Fig. 5. The maximum torque determines the selection of the motor, which is the first factor to be considered. The starting torque is also important because of the high current of the motor at startup, which can overheat components and potentially destabilize the system or damage related equipment. The starting torque and maximum torque of each joint are shown in Table 2.

Table 2. Starting torque and maximum torque of 4 joints.

Joint	Starting torque $(N \cdot m)$	Maximum torque $(N \cdot m)$
JOINT_1	1.21	54
JOINT_2	65.8	176.8
JOINT_3	27.6	35.63
JOINT_4	1.43	1.61

It can be seen from Table 2 that both the starting torque and the maximum torque at JOINT_2 are much larger than the other three joints, which put a high demand on the performance of the drive equipment. The motor torque at this joint is required to be large, weight and size of the motor will also increase as well as the energy consumption increase. In order to reduce the starting torque and maximum torque of JOINT_2 and considering that the rehabilitation robot requires a small overall size and light weight, the gravity compensation mechanism is designed in this paper.

4 Structural Optimization

The compensation arm in the gravity compensation mechanism is fixed with the thigh connecting rod and extends in reverse, rotating with the rotation of the thigh connecting rod. The upper end of the spring is attached to the ends of the compensation arm, and the bottom is attached to the base.

As can be seen from Fig. 5, the joint torque has increased and reached the maximum from the beginning to about 6.5 s. In this process, JOINT_2 is in the ascending phase, and the spring needs to provide a downward pulling force to reduce the torque. At this time, the larger the pulling force, the better the effect, but similarly, in the falling phase of JOINT_2, the spring force is the resistance and the joint torque will increase. The spring force is related to spring stiffness and preload force, so it is necessary to select suitable spring stiffness and preload force, the method of simulation test is adopted in this paper (Fig. 6).

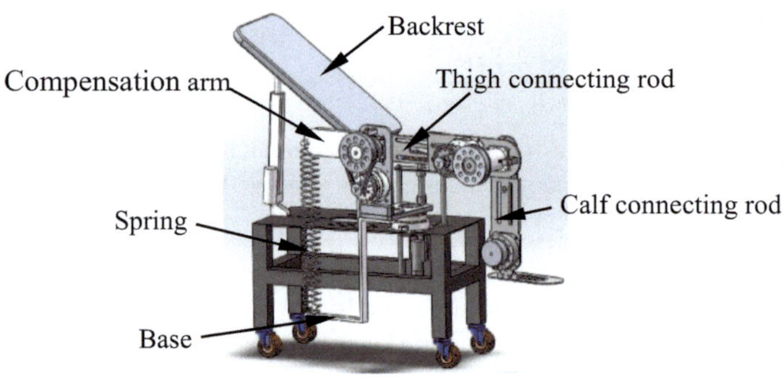

Fig. 6. Structure diagram after adding gravity compensation mechanism.

Select 10 kinds of springs, the stiffness is 0.2 N/mm, 0.4 N/mm, 0.6 N/mm up to 2 N/mm. The preload force is calculated by defining the spring deformation as zero when the compensation arm is at the lowest position, and select 6 simulation points of 1.0 times, 1.2 times, 1.4 times, 1.6 times, 1.8 times and 2.0 times in turn, the maximum is only 2 times because of space limitation.

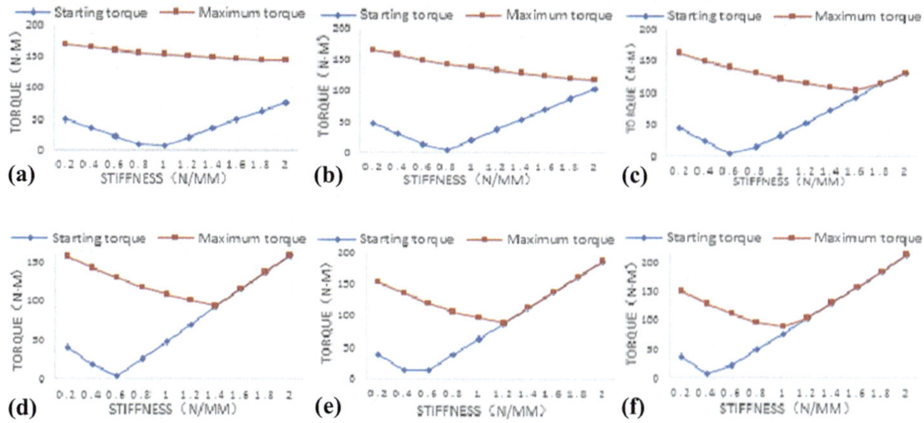

Fig. 7. Relationship between spring stiffness and torque in different preload forces (a) 1.0 times preload (b) 1.2 times preload (c) 1.4 times preload (d) 1.6 times preload (e) 1.8 times preload (f) 2.0 times preload.

The following conclusions can be obtained from Fig. 7: (1) Under the preload force of the same multiple, with the increase of spring stiffness, the maximum torque shows a decreasing trend, and the greater the preload force, the more obvious the decrease will be. However, when the preload is 1.4 times and above, the maximum torque of the spring with large stiffness reaches the lowest point and bounces up, and the greater the stiffness, the greater the rebound, which shows that the spring stiffness should not be too large. (2) Under the preload force of the same multiple, with the increase of spring stiffness, the starting torque shows the trend of first falling and then rising, so the spring stiffness should not be selected too small. And when the preload is 1.4 times and above, the starting torque of the spring with large stiffness has become the maximum torque, which also indicates that the spring stiffness should not be too large.

It can be seen from Fig. 7 that the turning point of the starting torque is around 0.8 N/mm, so the stiffness is selected to be 0.6 N/mm, 0.8 N/mm and 1.0 N/mm for further analysis. In order to select the appropriate preload force, the preload is subdivided again, and each increase of 0.1 times to take a data, the simulation results are shown in Fig. 8. In order to observe the gap of simulation results more intuitively, their torque reduction rate is calculated respectively, and the total reduction rate is calculated as a reference.

It can be seen from Fig. 8 that when $k = 0.6$ N/mm and the preload force multiple is 1.5 times, the total reduction rate is the largest, the maximum is 123.35%, but the maximum torque reduction rate is only 24.26% at this time, according to Fig. 8, the maximum torque is 133.9 $N \cdot m$. As mentioned before, the maximum torque is more important for motor selection than the starting torque, so the reduction rate of both should be comprehensively considered. Under comprehensive consideration, when $k = 0.8$ N/mm and the preload force multiple is 2 times, it is the optimal value. At this time, the maximum torque reduction rate is 45.31%, the value is 96.7 $N \cdot m$, the starting torque reduction rate is 27.20%, the value is 47.9 $N \cdot m$, and the total reduction rate is 72.51%.

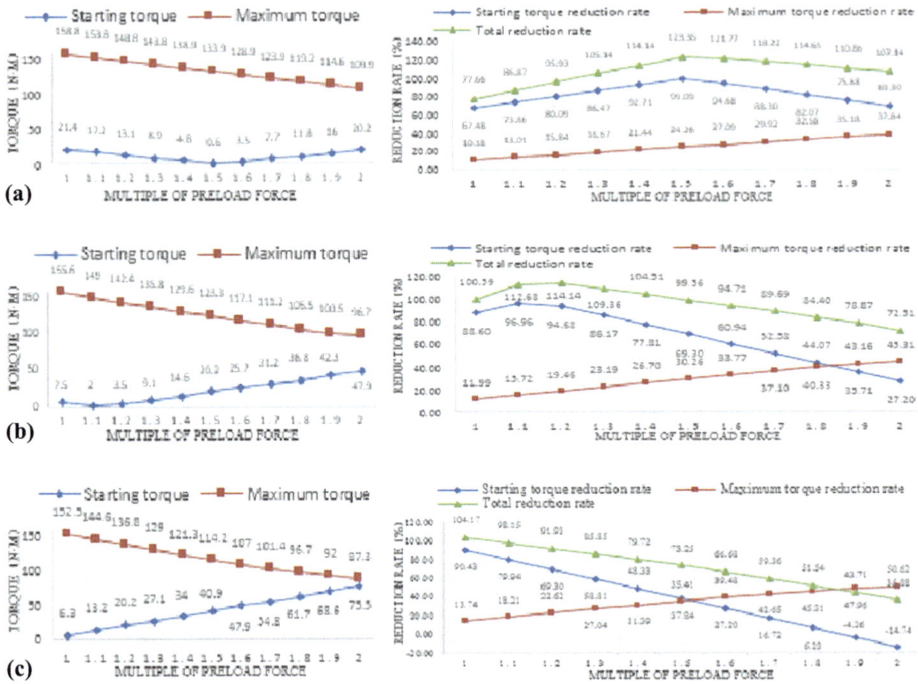

Fig. 8. Relationship between preload force and torque in different spring stiffness (a) k = 0.6 N/mm (b) k = 0.8 N/mm (c) k = 1.0 N/mm.

5 Conclusions

(1) According to the kinematic and dynamic analysis of the lower limb rehabilitation robot for the stroke, there is no mutation in the system during the movement cycle, and the overall structure design of the system is reasonable.

(2) By designing the gravity compensation mechanism and selecting the appropriate parameters, the mechanical structure is optimized and the torque value of the JOINT_2 is reduced by 72.51%, which provides experience for further optimization in the future.

Acknowledgements. This project is supported by the National Key Research and Development Program of China (No. 2017YFB1303200), and National Natural Science Foundation of China (Grant No. 51705500).

References

1. Hu, S., Gao, R., Liu, L., Zhu, M., Wang, W., Wang, Y., Wu, Z., Li, H., Gu, D., Yang, Y., Zheng, Z., Chen, W.: Summary of the 2018 report on cardiovascular diseases in China. Chin. Circ. J. **34**(03), 209–220 (2019)

2. Ward, N.S., Cohen, L.G.: Mechanisms underlying recovery of motor function after stroke. Arch. Neurol. **61**(12), 1844–1848 (2004)
3. Suzuki, K., Mito, G., Kawamoto, H., Hasegawa, Y., Sankai, Y.: Intention-based walking support for paraplegia patients with robot Suit HAL. Adv. Robot. **21**(12), 1441–1469 (2007)
4. Riener, R., Lünenburger, L., Maier, I.C., Colombo, G., Dietz, V.: Locomotor training in subjects with sensori-motor deficits: an overview of the robotic gait orthosis lokomat. J. Healthc. Eng. **1**(2), 197–216 (2010)
5. Wu, B.: Design and research of sitting rehabilitation robot of limbs coordination training, MS thesis, Hefei University of Technology, Hefei (2015)
6. Banala, S.K., Agrawal, S.K., Scholz, J.P.: Active Leg Exoskeleton (ALEX) for gait rehabilitation of motor-impaired patients. In: IEEE 10th International Conference on Rehabilitation Robotics, pp. 401–407. IEEE, Netherlands (2007)
7. Veneman, J.F., Kruidhof, R., Hekman, E.E., Ekkelenkamp, R., Van Asseldonk, E.H., Van Der Kooij, H.: Design and evaluation of the LOPES exoskeleton robot for interactive gait rehabilitation. IEEE Trans. Neural Syst. Rehabil. Eng. **15**(3), 379–386 (2007)
8. Shao, Z.: Structural design and simulation analysis of external bone robot of lower extremities, MS thesis, Yangzhou University, Yangzhou (2016)
9. Cui, Y., Shi, P., Hua, J.: Kinematics analysis and simulation of a 6-dof humanoid robot manipulator. In: 2nd International Asia Conference on Informatics in Control, Automation and Robotics, pp. 246–249. IEEE, China (2010)
10. Zhao, H.: Design and analysis of leg mechanism for based on flexible bionic joint for quadruped robot, MS thesis, Hebei University of Technology, Tianjin (2015)
11. Zhang, D.: Dynamic Modeling and Control of Flexible Joint Robot, MS thesis, Nanjing University of Science and Technology, Nanjing (2017)
12. Li, Z., Kota, S.: Virtual prototyping and motion simulation with ADAMS. J. Comput. Inf. Sci. Eng. **1**(3), 276–279 (2001)

A Configuration Synthesis Method of Deployable Mechanism Cells Used for Large Supporting Structures in Space

Chenhan Guang, Zilu Chen, and Yang Yang(✉)

School of Mechanical Engineering and Automation, Beihang University, XueYuan Road No. 37, HaiDian District, Beijing 100191, China
yang_mech@126.com

Abstract. More large payloads will be launched with the development of the aerospace industry, which means supporting structures for large payloads are also needed. Meanwhile, deployable supporting structures that can be synchronously packaged and deployed have been studied. However, if much larger or more complex structures are needed, building structures with identical deployable cells in space could be another choice. Therefore, we synthesize deployed configurations of deployable mechanism cells for large supporting structures in space with the help of finding the basic unit for a certain three-dimensional array. Then we present several examples of deployable mechanism cells based on the synthesized deployed configurations.

Keywords: Configuration synthesis method · Deployable mechanism cells · Large supporting structures

1 Introduction

Much larger space platforms, such as huge space observatories, space stations, etc. will be launched in space with the development of the aerospace industry, and large key devices on them need supporting structures. The supporting structures are also large, so that they cannot be directly loaded into launch vehicles. Hence the research on the structures which can be packaged or folded before launching and then deploy in orbit is necessary.

Up to now, many researchers have deeply studied this kind of structure or mechanism. Furuya (see [1]) presented various types of deployable tensegric modules and used them to build types of one-dimensional truss masts. A modular cable-mesh deployable structure for large-scale satellite communication antennas was presented by Meguro et al. (see [2]). Tan and Pellegrino (see [3]) developed a cable-stiffened deployable pantographic structure. Chen (see [4]) introduced a general approach to constructing large structural mechanisms with Bennett linkages and hybrid Bricard linkages. Using the plane-symmetric Bricard linkage, Qi et al. (see [5]) constructed a kind of deployable plane truss containing

D. Wang et al. (Eds.): MeTrApp 2019, MMS 79, pp. 152–160, 2020.
https://doi.org/10.1007/978-981-15-0142-5_16

quadrangular modules and a kind of n-sided regular polygon deployable ring. Huang et al. (see [6]) treated spatial four-bar linkages, five-bar linkages and six-bar linkages as modules to construct large deployable mechanisms, and summarized three methods to connect modules. Constructed by scissor linkages and Sarrus linkages, a family of mechanisms which could be deployed in two or three directions were presented by Lu et al. (see [7]). Guang and Yang (see [8]) constructed a family of deployable mechanisms which were transformed from a kind of origami pattern with thickness. You and Pellegrino (see [9]) developed a pantographic ring serving as the supporting structure of a mesh reflector. Deng (see [10]) systematically introduced designing methods of deployable mechanisms in space.

The previous studies are focused on using modules to construct large deployable structures on which all modules are pre-connected and deployed synchronously. Meanwhile, shapes of deployed structures are fixed and relatively simple. If more complex or much larger structures are needed, the existing method is not suitable.

To solve this problem, assembling truss structures in space is one of the promising approaches. Deployable mechanism cells acting as building blocks are loaded into launch vehicles, and then they deploy automatically or passively in orbit. After deploying, the cells are to be assembled by robot arms or astronauts to construct target truss structures. Maybe the cells can also be used as building blocks when human beings land on the moon or other planets in the future.

Therefore, we intend to explore the method to synthesize the building blocks. With the help of finding the basic unit for a certain three-dimensional array, we introduce a method to synthesize deployed configurations of mechanism cells for large supporting structures in space. Then we present several examples of deployable mechanism cells based on the synthesized deployed configurations.

2 Method

Assembled truss is essentially a geometric three-dimensional array, and any array can be achieved by repeating its basic unit. Hence, the basic unit can be used as the deployed configuration for the mechanism cell. The method to synthesize deployed configurations of mechanism cells for large supporting structures is to find out the basic unit for the truss array.

In this study, we introduce and explain the method through the arrays constructed by cubic units (trusses). For other arrays, this method can be used similarly.

Apparently, cubic trusses can be used to construct larger trusses, as shown in Fig. 1. In this figure, $\mathrm{A}_{i,j,k}, \mathrm{B}_{i,j,k}, ..., \mathrm{H}_{i,j,k}$ are the eight vertices of a cubic truss, and the subscripts i, j and k denote the serial numbers of cubic trusses along the axis x, the axis y and the axis z, respectively. Here, the axis x is parallel to $\overrightarrow{\mathrm{A}_{1,1,1}\mathrm{H}_{1,1,1}}$ vector, the axis y is parallel to $\overrightarrow{\mathrm{A}_{1,1,1}\mathrm{B}_{1,1,1}}$ vector, and the axis z is parallel to $\overrightarrow{\mathrm{A}_{1,1,1}\mathrm{D}_{1,1,1}}$ vector.

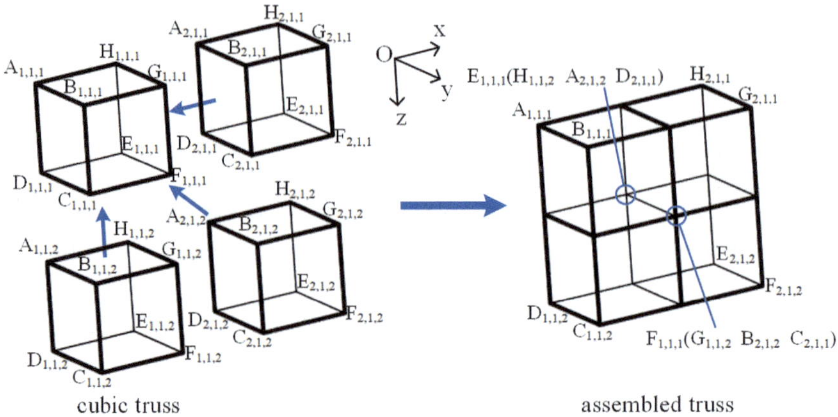

Fig. 1. Building larger truss with cubic trusses

Then we find that an inner rod of the assembled truss consists of four rods from its four adjacent cubic trusses. For example, the inner rod $\overline{E_{1,1,1}F_{1,1,1}}$ in the assembled truss is composed of the rod $\overline{E_{1,1,1}F_{1,1,1}}$, the rod $\overline{D_{2,1,1}C_{2,1,1}}$, the rod $\overline{H_{1,1,2}G_{1,1,2}}$ and the rod $\overline{A_{2,1,2}B_{2,1,2}}$. Assuming that all the rods of one cubic truss satisfy all the mechanical requirements in space, the combination of four rods for an inner rod is redundant, and this will lead to waste of payload capacity for launch vehicles. Therefore, we tend to eliminate the redundance of inner rods.

To solve this problem, we introduce a reverse method from the perspective of the geometry, which is to find the basic unit for an array.

Begin with the simplest case. Actually, an array can be assembled by units each of which is constructed by three orthogonal line segments meeting at one point, as shown in Fig. 2.

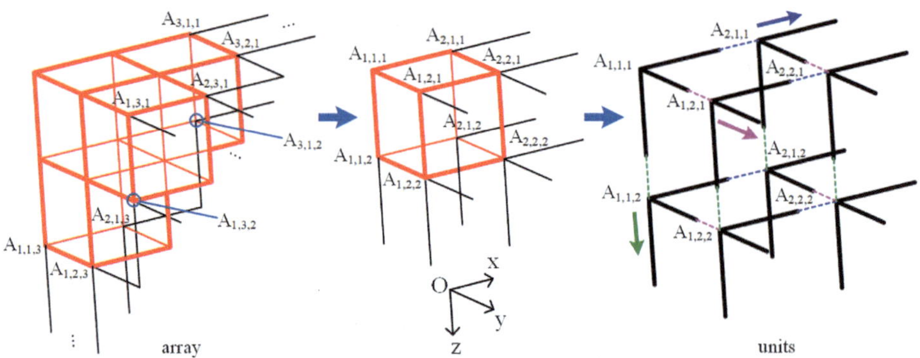

Fig. 2. Larger truss built by units, each of which consists of three orthogonal line segments

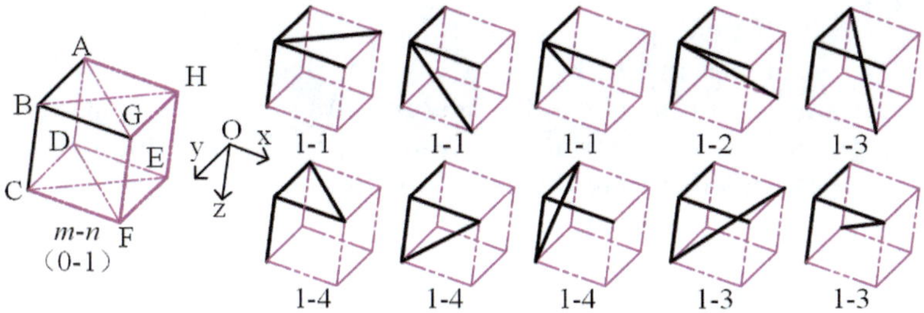

Fig. 3. Trusses with one diagonal rod

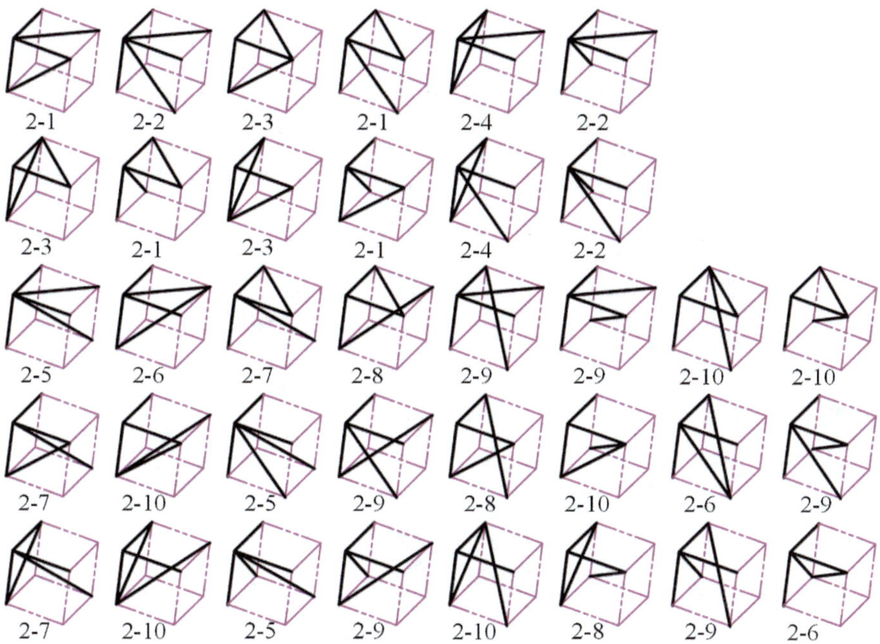

Fig. 4. Trusses with two diagonal rods

In Fig. 2, $A_{i,j,k}$ denotes the vertex of one unit, where the three segments meet. The other endpoints of the three rods (segments) in one unit need not to be signed since they are to be connected with the vertices of adjacent units in assembled array. So the eight vertices of any cubic unit in the array can be denoted by $A_{i,j,k}$, $A_{i+1,j,k}$, $A_{i,j+1,k}$, $A_{i+1,j+1,k}$, $A_{i,j,k+1}$, $A_{i+1,j,k+1}$, $A_{i,j+1,k+1}$ and $A_{i+1,j+1,k+1}$, respectively.

For the array as shown in Fig. 2, its inner rod $\overline{A_{2,1,2}A_{2,2,2}}$ is not a combination of four rods, but just itself. If so, the deployable mechanism cells transformed from this unit contain less rods, and more cells can be launched in orbit.

Then similar basic units can be derived from the simplest unit. Diagonal rods can be added on units. The unit with no diagonal rod is related with five planes which are ABCD plane, ABGH plane, BCFG plane, BCEH plane and AGFD plane as shown in Fig. 3. When any two diagonal points are connected with straight line, there is to be a diagonal rod. As long as the diagonal rods are on the five planes, the units formed are also basic units. In addition, the sign in this section, m-n denotes that this unit is with m rods on it and it is the nth type of the deployed configuration.

Table 1. Numbers of basic units and deployed configurations with different numbers of diagonal rods

Number of rods	0	1	2	3	4	Sum
Number of basic units	1	10	36	56	32	135
Number of deployed configurations	1	4	10	14	10	39

Fig. 5. Trusses with three diagonal rods

Usually, there is only one diagonal rod on each of the five planes for the truss structure, and BCEH plane and AGFD plane are not to contain diagonal rods at the same time. According to the principles, we illustrate the basic units whose numbers of diagonal rods are one, two, three and four in Figs. 4, 5 and 6, respectively.

When the number of diagonal rods is one, the number of basic units can be calculated and equals 10 ($C_2^1 + C_2^1 + C_2^1 + C_2^1 + C_2^1 = 10$). When the number of diagonal rods is two, the number of basic units can be calculated and equals 36 ($C_3^2 \times C_2^1 \times C_2^1 + C_3^1 \times C_2^1 \times C_2^1 \times C_2^1 = 36$). When the number of diagonal rods is three, the number of basic units can be calculated and equals 56 ($C_3^3 \times C_2^1 \times C_2^1 \times C_2^1 + C_3^2 \times C_2^1 \times C_2^1 \times C_2^1 \times C_2^1 = 56$). When the number of diagonal rods is four, the number of basic units can be calculated and equals 32 ($C_3^3 \times C_2^1 \times C_2^1 \times C_2^1 \times C_2^1 \times C_2^1 = 32$).

From Figs. 4, 5 and 6, although all the units can construct different arrays, some units are the same for deployed configurations of deployable mechanism cells. We select and sign the units, and if the configurations are identical they are signed with the same sign (they are of the same n). The numbers of units and the numbers of distinct deployed configurations with different numbers of diagonal rods are listed in Table 1.

The above-mentioned units are parts of all the units to construct large arrays. The other complex units as shown in Fig. 7 are not to discussed, because they can be found out by the same method.

After obtaining deployed configurations, we can select some of them and transform them into deployable mechanism cells.

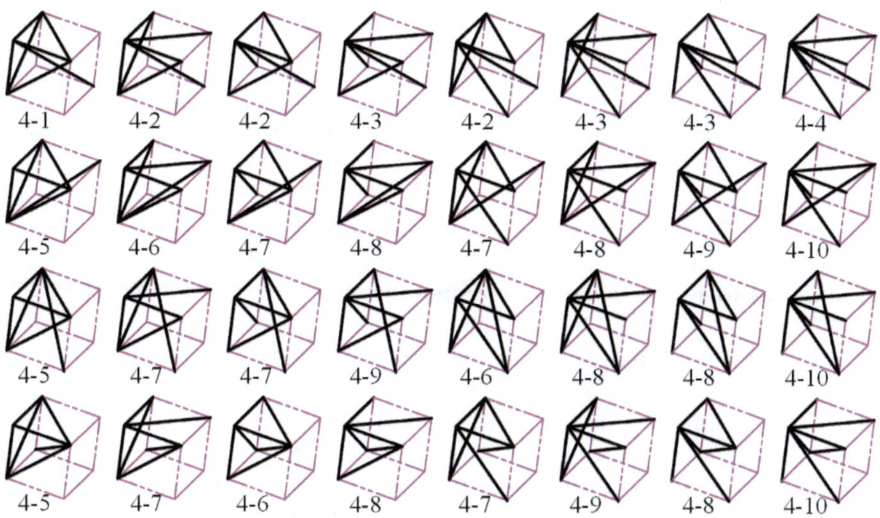

Fig. 6. Trusses with four diagonal rods

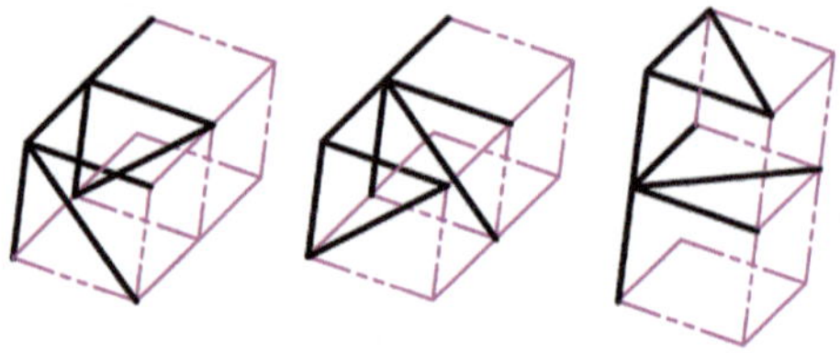

Fig. 7. Complex units

3 Examples

In this section, we choose 0-1 (in Fig. 3) configuration and 3-1 (in Fig. 5) configuration, and transform them into deployable mechanism cells to assemble large truss.

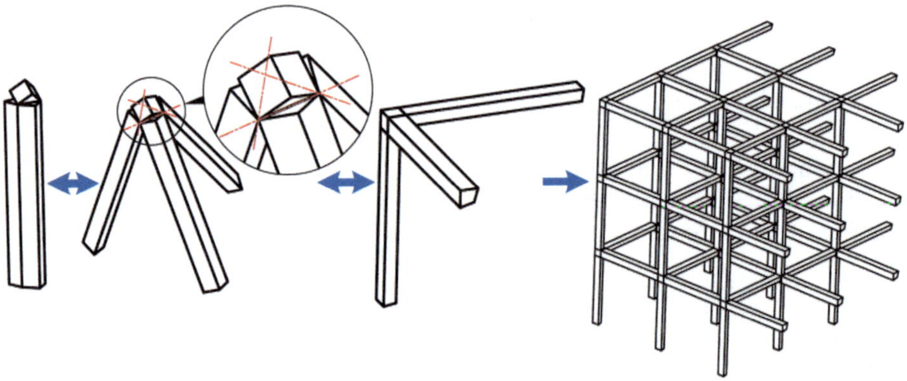

Fig. 8. Transformed mechanism cell from 0-1 configuration and assembled truss

For 0-1 configuration, the deployable mechanism cell can be constructed by a connecting block and three rods, and the three rods are jointed with the connecting block by three revolute joints (the axes of joints are denoted by fine dot dash lines) as shown in Fig. 8. This cell has three DoFs (degrees of freedom).

The configuration 0-1 can also be transformed by the spatial 6R linkage (see [10]) whose DoF is one, as shown in Fig. 9.

For 3-1 configuration, the deployable mechanism cell is constructed by three linkage mechanisms which are one spatial 5R linkage mechanism (see [10]) and two identical planar 4R linkage mechanisms, as shown in Fig. 10. This cell has only one DoF, and after deployment it forms a stable tetrahedral structure.

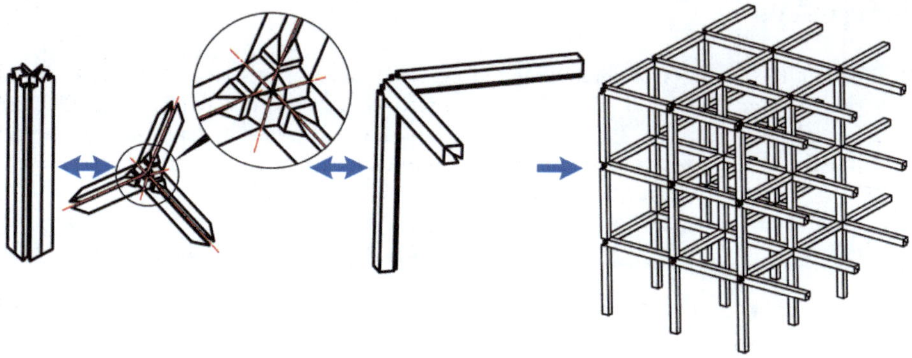

Fig. 9. Another transformed mechanism cell from 0-1 configuration and assembled truss

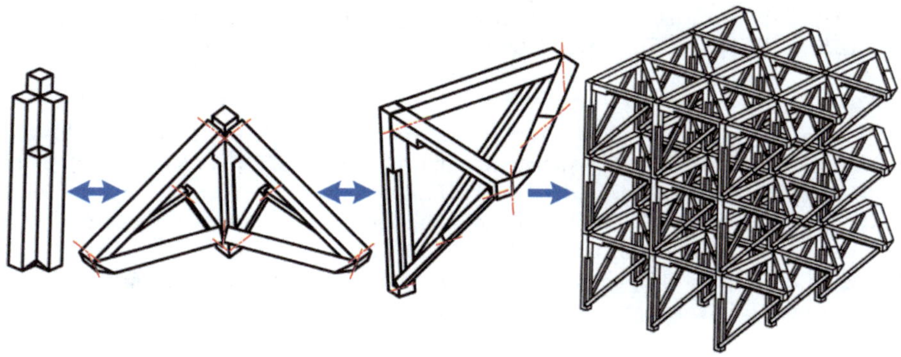

Fig. 10. Transformed mechanism cell from 3-1 configuration and assembled truss

4 Conclusions and Future Work

In this study, based on finding the basic units of arrays, the configuration synthesis method of deployable mechanism cells is introduced. Then the basic units with or without diagonal rods are listed, and all the distinct configurations are selected and marked.

Some examples of deployable mechanism cells which are transformed from two of the selected configurations are illustrated to confirm the validity of this method.

In the future, we will transform more configurations into mechanism cells, and test the mechanical properties of the assembled trusses to find the better cells. The connecting method of deployed cells will also be studied.

Acknowledgements. This work is supported by Aerospace Science and Technology Fund of China Aerospace Science and Industry Corporation (KZ37019701).

References

1. Furuya, H.: Concept of deployable tensegrity structures in space application. Int. J. Space Struct. **7**(2), 143–151 (1992). https://doi.org/10.1177/026635119200700207
2. Meguro, A., Mitsugi, J., Ando, K.: A modular cable-mesh deployable structure for large-scale satellite communication antennas. Electron. Commun. Jpn. Part I **77**(8), 90–100 (1994). https://doi.org/10.1002/ecja.4410770809
3. Tan, G.E.B., Pellegrino, S.: Nonlinear vibration of cable-stiffened pantographic deployable structures. J. Sound Vib. **314**, 783–802 (2008). https://doi.org/10.1016/j.jsv.2008.01.022
4. Chen, Y.: Design of structural mechanisms. Ph.D. thesis, University of Oxford, Oxford, UK (2003)
5. Qi, X., Huang, H., Miao, Z., Li, B., Deng, Z.: Design and mobility analysis of large deployable mechanisms based on plane-symmetric Bricard linkage. ASME J. Mech. Des. **139**(2), 022302 (2016). https://doi.org/10.1115/1.4035003
6. Huang, H., Deng, Z., Li, B.: Mobile assemblies of large deployable mechanisms. JSME J. Space Eng. **5**(1), 1–14 (2012). https://doi.org/10.1299/spacee.5.1
7. Lu, S.N., Zlatanov, D., Ding, X.L., Molfino, R., Zoppi, M.: Novel deployable mechanisms with decoupled degrees-of-freedom. ASME J. Mech. Rob. **8**(2), 021008 (2016). https://doi.org/10.1115/1.4031639
8. Guang, C., Yang, Y.: Single-vertex multicrease rigid origami with nonzero thickness and its transformation into deployable mechanisms. ASME J. Mech. Robot. **10**(1), 011010 (2018). https://doi.org/10.1115/1.4038685
9. You, Z., Pellegrino, S.: Cable-stiffened pantographic deployable structures. Part 2: mesh reflector. AIAA J. **35**, 1348–1355 (1997)
10. Deng, Z.: Design of Space Deployable and Foldable Mechanisms, Chap. 5. Harbin Institute of Technology Press, Harbin (2013). (in Chinese)

Parameter Analysis and Optimal Design of Energy Storage Device for Non-powered Walking Aid

Fanyu Meng, Wenxiang Zhao, Rui Wang, and Zhixiu Hao(✉)

Tsinghua University, Beijing, China
THUme51mfy@163.com

Abstract. This paper optimizes the energy storage device of a non-powered walking aid. Through the use of gait analysis and linear fitting method, we have found the approximate relationship and specific values between the hip rotation angle and the hip rotation torque. Through the method of mechanical analysis, we find out how the geometric parameters of the cam and spring affect the relationship between the torque of hip joint and its angle. Furthermore, all suitable design parameters are obtained by polar coordinate operation and quadratic polynomial fitting analysis of cam profile. Finally, a 3D model is designed and the actual device is assembled.

Keywords: Non-powered walking aid · Energy storage device · Optimal design

1 Introduction

Hemiplegia is clinically manifested as partial or complete loss of free movement of the upper and lower limbs. The main cause of hemiplegia is stroke [1]. Brain and nervous system functional damages in stroke patients may lead to random dyskinesia or even paralysis of local skeletal muscle [1]. Using exercise aids to help and replace the movements of the limbs, the patient can recover some of the original motor functions on the basis of physiological "structural rehabilitation" and achieve the "functional rehabilitation" state through training on long-term rehabilitation exercise [2].

Currently, there are many different forms of lower extremity exoskeleton robots that help patients with hemiplegia to improve their walking ability. Some of the influential robots include Locomat [3] from Swiss Hocma Medical Devices Company, H2 [4] designed by Technaid et al., LOPES designed by Veneman of Institute of Biomedical Technology of Twente University [5] and Ekso exoskeleton rehabilitation robot developed by robot Ekso Bionics, etc. [6]. However, these devices are all active devices, which have many problems in applications such as excessive weight, limited drive, short battery power supply time, uncomfortable wearing, high price, etc. Therefore, scholars at Nagoya University have developed a non-powered walking aid ACSIVE based on the principle of "passive dynamic walking [7]", which greatly reduces the cost and weight of the device, hence making it possible for one to wear the walking assistance device on a daily basis. However, user feedback on the effectiveness

D. Wang et al. (Eds.): MeTrApp 2019, MMS 79, pp. 161–171, 2020.
https://doi.org/10.1007/978-981-15-0142-5_17

of the device effect is not very good. Cadence Biomedical released its Kickstart Walking System back in August 2012, based on the body structure of large terrestrial animals for having long tendons that span across multiple joints, which makes it efficient for walking and running. However, Kickstart only used unilateral energy storage and release. For patients with moderate to severe hemiplegia, the affected side has limited energy storage capacity, hence, the effect from Kickstart design is not very significant.

In this paper, we propose an energy storage device based on the spring-cam mechanism for double-sided energy transfer on non-powered walking aids. For patients with hemiplegia who are unable to perform complete hip flexion and extension, we can design an energy storage device for the purpose of helping them to walk. To do so, using energy storage device transfers the energy provided by the healthy leg to the affected leg. Compared to other energy storage methods, the spring-cam mechanism can maximize energy harvesting efficiency, so we will analyze the gait and use mathematical methods to design the spring-cam mechanism. Finally, the design and physical construction of the 3D model are completed.

2 Design and Optimization of Energy Storage Devices

2.1 Gait Analysis

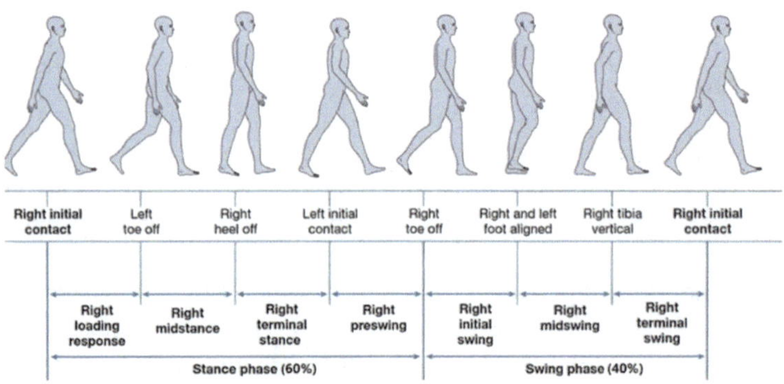

Fig. 1. Normal gait cycle diagram of human walking in a longitudinal plane [8]

The periodic movement of the feet and the sufficient reaction force on the ground support the body to form the typical walking behavior of the human body. For each leg, one gait cycle is divided into a stance phase and a swing phase. In stance phase, the foot is in contact with the ground, accounting for about 60% of the entire cycle for a normal person's gait; the swing phase is when the foot is off from the ground, accounting for about 40% of the cycle.

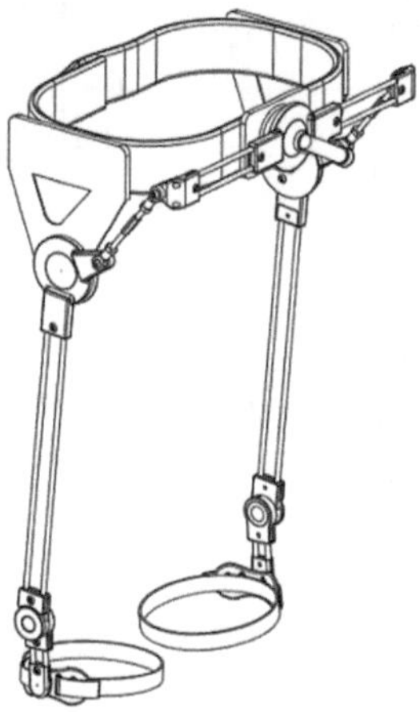

Fig. 2. Mechanism diagram of symmetrical four-bar linkage

Assume that the left leg in Fig. 1 is the healthy side leg of the hemiplegic patient, and the right leg is the paralyzed side leg of the hemiplegic patient. Figure 2 displays a Mechanism diagram of symmetrical four-bar linkage. Since both the linkage mechanism and the human gait feature are symmetrical, the energy storage device should storage energy from 0% to 50% and release energy from 50% to 100%. According to the analysis above, the state of the energy storage device and the performance of the assisting can be illustrated in Table 1 below.

Table 1. The state of the energy storage device

Gait cycle	0–50%	50%–60%	60%–100%
Energy storage device	Spring compression increases elastic potential energy	Spring release elastic potential energy	Spring release elastic potential energy
Performance of the assisting task	Healthy side leg is more laborious	Both legs stand on the ground	Legs with hemiplegia are more relaxed

2.2 Kinetic Parameters of the Hip Joint

In order to calculate the torque that the energy storage device needs to provide, we need to first understand how much force a hemiplegic patient would need to assist with walking. So we should try to get the relationship between the hip torque and angle of ordinary people. Research by Hao Zhixiu and colleagues from Tsinghua University [9] have illustrated the angle curve of the hip joint in a sagittal plane and the torque of the hip joint bending for ordinary people, as shown in Fig. 3 [9].

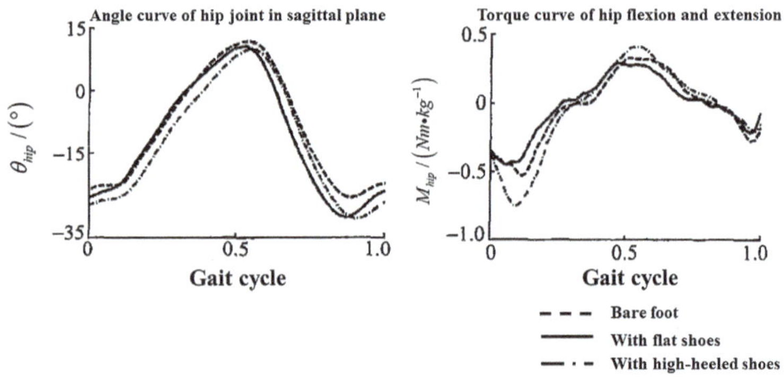

Fig. 3. Two important kinematic parameters of a hip joint

Figure 3 shows us the changes in hip angle and hip torque for ordinary people in one gait cycle and we find that hip angle and hip torque are dependent on gait cycle (time), which is an independent variable. So we can get the relationship between two dependent variables by eliminating the independent variable. In particular, we first digitize the two curves respectively in Fig. 3 with Origin. As data points (tuples) obtained this way on the hip angle curve (*gait cycle percentage*, *hip angle*) and the torque curve (*gait cycle percentage*, *hip torque*) do not correspond to the same position on a gait cycle, we use linear interpolation to match them. With angle and torque data corresponding to the same gait cycle percentages, we can eliminate the gait cycle variable, and scatter plot the hip torques with respect to the hip angles in Fig. 4. Notice that 50% to 100% of the gait cycle is the swing phase of paralyzed leg of a hemiplegia patient. In order to simplify the mechanical design, we first use a least squared method to linearly fit these points, the resulting curve being shown in Fig. 4. Since r^2 is close to 0.9, the degree of linear fit is relatively high. The linear approximation will guide the design of a cam mechanism in the walking assistive device. We will further validate this approximation in Sect. 3.

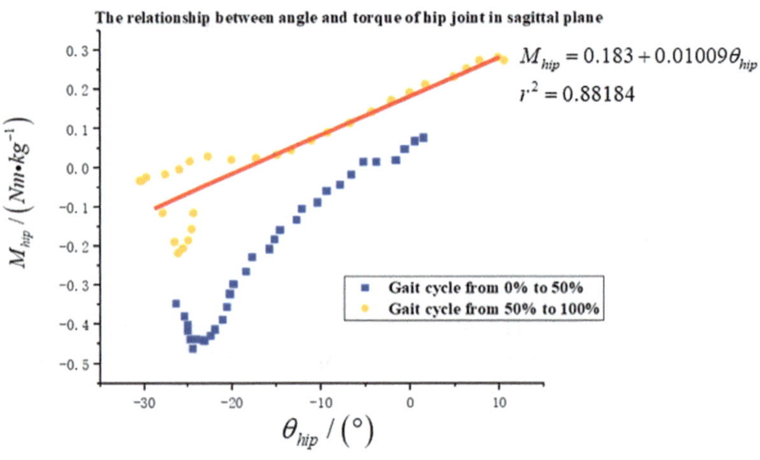

Fig. 4. Relationship between θ_{hip} and M_{hip}

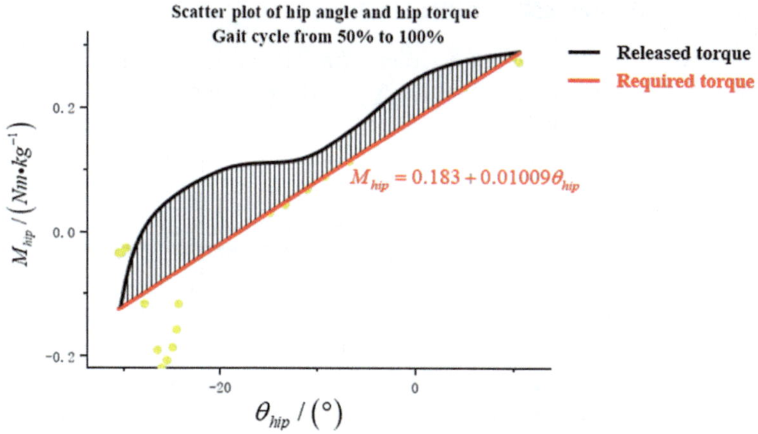

Fig. 5. Scatter plot of θ_{hip} and M_{hip}(gait cycle from 50% to 100%)

The area enclosed by the curve and the horizontal axis in Fig. 5 represents the energy for hip joint movement in sagittal plane per unit weight. In particular, for the cam-spring mechanism, the geometry of the cam can be designed reasonably so that the torque released by the energy storage device is as close as possible to the required torque of the patient's hip flexion and extension (red straight line in Fig. 5), thereby avoiding inefficient energy storage and release (The shaded area in Fig. 5) and achieving its highest energy harvesting efficiency.

Figure 5 also shows why the cam-spring mechanism has higher energy harvesting efficiency than other energy storage devices such as tension springs and torsional springs. This is because we can control the relationship between its released torque and angle of hip joint by designing its cam profile, but this relationship is fixed in other energy storage methods.

For a simulated patient weighing 60 kg, in the cam plane, when the spring is compressed to its limit, the energy storage device provides the maximum torque, which is 60% of the patient's required torque. We can get:

$$M_{cam-\max} = 60\% \times 0.38 \times 60 = 13.68\,\mathrm{N} \bullet \mathrm{m} \quad (1)$$

In Eq. (1), 0.38 is maximum variation of M_{hip}. After rod length optimization of the symmetrical four-bar linkage mechanism, the proportional coefficient of the cam rotation angle and the hip joint rotation angle is always 1, so we have:

$$\theta_{cam} = \theta_{hip} \quad (2)$$

Assuming that clockwise is the positive direction of θ_{cam}, as shown in Fig. 7.

2.3 Simplified Mechanical Analysis

When we get $M_{cam-\max}$, θ_{cam} and the straight line shown in Figs. 4 and 5, we have the theoretical conditions for mechanical analysis. Through mechanical analysis, we can establish a connection between M_{cam} and cam parameters to prepare for further design of the cam profile. When performing mechanical analysis on the roller, two frictional forces are easily ignored. The first friction is the frictional force F_{f1} between the spring and the outer wall. The second is the rolling friction F_{f2} on the contact point of the cam and the roller. For the energy loss of frictional heat generation, we can use a square groove linear guide to reduce the friction, which is shown in Fig. 7. After consulting the data, we get: the rolling friction is negligible; the friction coefficient of the linear slide is generally between 0.1 and 0.2 due to the installation and actual processing, as the main reason for reducing the energy conversion efficiency.

In the calculation we used $\mu = 0.15$, assuming that the roller moves in a uniform linear motion during the whole process, the mechanical analysis of the roller is shown in Fig. 6 (the meanings of notations are explained below).

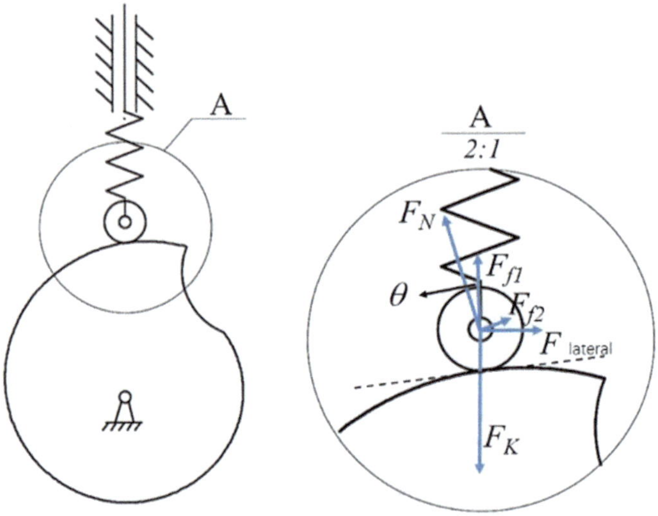

Fig. 6. Roller diagram of all of forces

From Fig. 6, we can get the following equations:

$$F_K = k\Delta x \tag{3}$$

$$F_{lateral} = F_N \sin\theta \tag{4}$$

$$F_{f2} \approx 0 \tag{5}$$

$$F_{f1} = F_{lateral}\mu \tag{6}$$

$$F_N \cos\theta + F_{f1} = F_K \tag{7}$$

$$F_N(l_0 + \Delta x)\sin\theta = M_{cam} \tag{8}$$

Solving equations to get:

$$M_{cam} = \frac{k\Delta x(l_0 + \Delta x)\tan\theta}{1 + 0.15\tan\theta} \tag{9}$$

In the equation group, the mechanical variables are explained as follows. F_K is the pressure of the spring on the roller. F_N is the pressure of the cam on the roller. $F_{lateral}$ is the pressure of the outer wall on the roller. M_{cam} is the torque stored by the cam. Variables of mechanism design parameters are explained as follows. k is the stiffness coefficient of the spring. Δx is the compression of the spring. l_0 is the minimum distance between the center of the roller and the center of rotation of the cam. θ is the angle between the normal of the cam profile and the line where the linear slide is located at the point of contact between the cam and the roller. μ is the friction coefficient of the sliding guide.

$\Delta x_{\max}$ is the max compression of the spring, $\theta_{cam-\max}$ is the max degree of θ_{cam}. $\Delta x_{\max}$, l_0 and $\theta_{cam-\max}$ decide on the size of the cam; θ, θ_{cam}, Δx decide on the contour line of the cam; k and $\Delta x_{\max}$ determine the choice of spring. These are shown visually in Fig. 7.

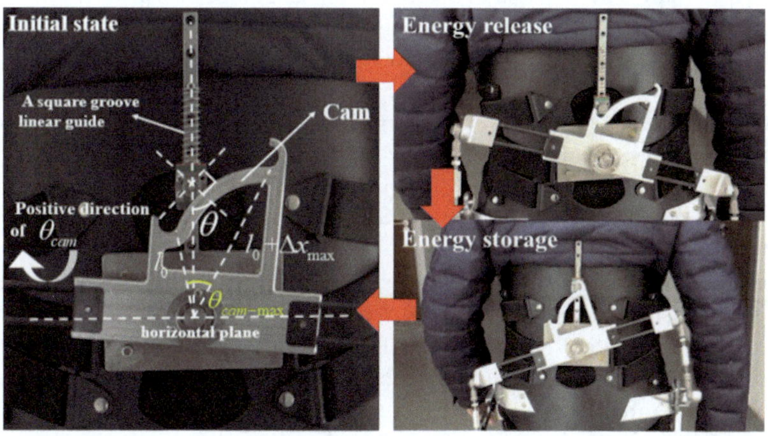

Fig. 7. Parameters labeling and three states of the energy storage device

We can calculate $\theta \in [0°, 90°]$, because cam profile is convex curve by simple mathematical operations. In order to get the relationship between Δx, θ and M_{cam}, we assume $k = 0.1\,\text{N} \bullet \text{mm}^{-1}$, we get the trend graph shown in Fig. 8 through MATLAB.

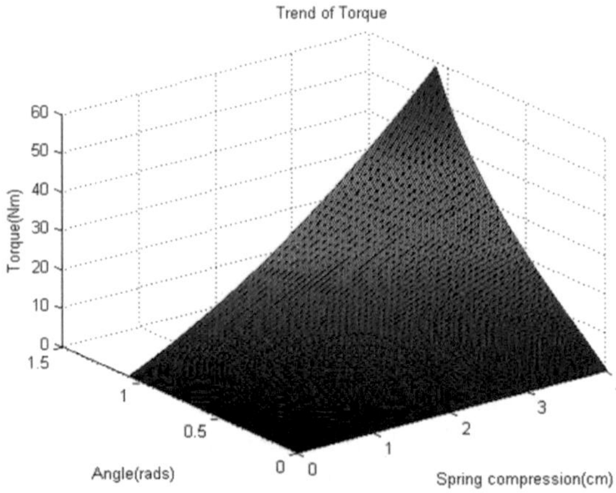

Fig. 8. Relationship between Δx, θ and M_{cam}

From Fig. 8, we find when θ is fixed, M_{cam} increases as Δx increases; when Δx is fixed, M_{cam} increases as θ increases.

3 Results and Discussion

Because there are many geometric parameters of energy storage device, we need to determine some parameters first before deducing other parameters based on their relationship. Finally, we are able to determine the design of the spring-cam mechanism and create a physical model as shown in Fig. 7.

Firstly, three fixed parameters should be determined, these are $\Delta x_{\max}$, l_0 and k. Based on the size of the volunteer and the size of the existing products in the market, we determined $\Delta x_{\max} = 40\,\text{mm}$ and $l_0 = 60\,\text{mm}$. For selection of parameter k, specifically, we brought in two boundary conditions into Eq. (3). When the spring is in its original length, $\Delta x = 0\,\text{mm}$, $M_{cam} = 0\,\text{N} \bullet \text{m}$; when the spring is at maximum compression, $\Delta x = 40\,\text{mm}$, $M_{cam} = 13.68\,\text{N} \bullet \text{m}$, make $\theta = 45°$, get the result of $k = 3.94\,\text{N} \bullet \text{mm}^{-1}$.

Then, according to the three known parameters, a polar coordinate system and cam profile is established as shown in Fig. 9. We find that only when the polar coordinate equation of cam profile is in the form of $\rho = a\alpha + b$, can we obtain the analytical solution by the equality of Eq. (3) and the polar coordinate equation. Therefore, in order to facilitate the calculation, we assume the cam contour as Archimedes spiral whose equation is in the form of $\rho = a\alpha + b$. We can get the following equations:

$$M_{cam} = \frac{k\Delta\rho\rho\tan\theta}{1 + 0.15\tan\theta} \tag{10}$$

$$\frac{\theta_{cam} + 30}{360} = \frac{\alpha}{2\pi} \tag{11}$$

$$\rho = l_0 + \Delta\rho \tag{12}$$

$$y = \rho \sin\alpha \tag{13}$$

$$x = \rho \cos\alpha \tag{14}$$

$$\tan\beta = \frac{dy}{dx} \tag{15}$$

$$\rho = a\alpha + b \tag{16}$$

$$f(0) = \rho_0 = b = 0.06 \tag{17}$$

$$\theta = \alpha + (\frac{\pi}{2} - \beta) \tag{18}$$

$$\tan(\theta) = \tan(\alpha + (\frac{\pi}{2} - \beta)) \tag{19}$$

Solving equations, we obtain:

$$M_{cam} = 3.94 \times 10^3 \times a^2 \times \frac{a\alpha^2 + 0.06\alpha}{a\alpha + 0.06 + 0.15a} \tag{20}$$

We change the value of a to make the curve represented by Eq. (4) get as close to the linear fitting curve in Fig. 5 as possible. Now we can calculate the approximate solution of $a = 1.211 \times 10^{-2}$ and then the optimized cam profile has been obtained whose polar coordinate equation is $\rho = 1.211 \times 10^{-2}\alpha + 0.06$. By optimizing the cam profile, we get the relationship between optimum torque and angle as shown in Fig. 10. From Fig. 10, we can see that the curve we obtained by designing the cam profile

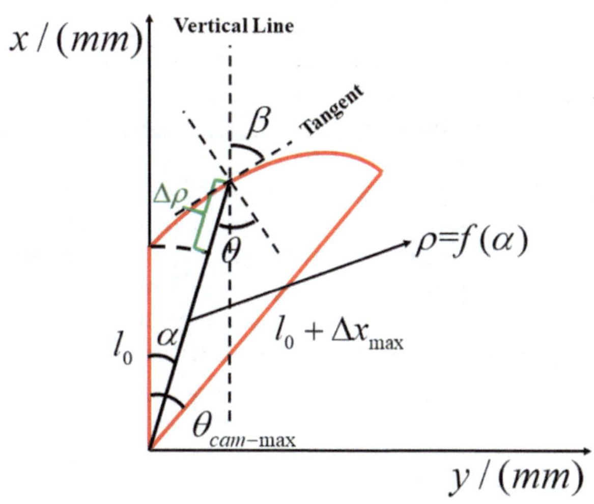

Fig. 9. cam profile in a polar coordinate system

(black curve) can cover all the yellow scatters and is very close to the target curve (red curve), which also proves that our optimization is effective. So far, all the geometric parameters of the cam spring mechanism have been calculated.

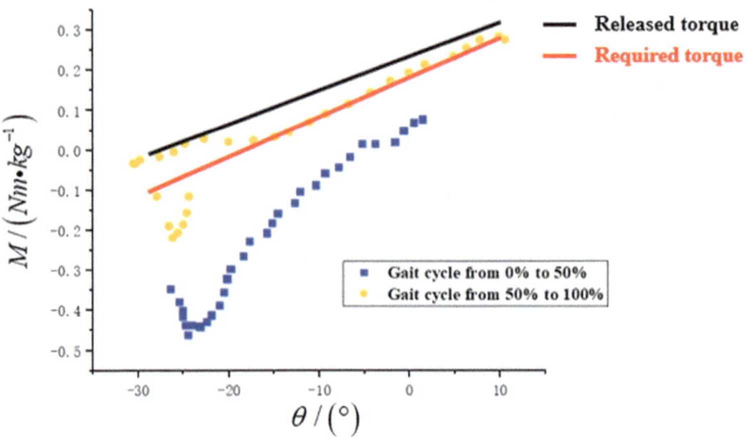

Fig. 10. Relationship between θ_{cam} and M_{cam}

In summary, we have decided all the design parameters of the cam-spring energy storage device.

4 Conclusion

In this paper, we design an energy storage device for non-powered walking assistance. Through gait analysis and linear fitting method, we find the specific values and the approximate relationship between the hip rotation angle and the hip rotation torque. Through the method of mechanical analysis, we find out how the geometric parameters of the cam and spring affect the relationship between the torque of hip joint and its angle. Furthermore, all suitable design parameters are obtained by polar coordinate operation and quadratic polynomial fitting analysis of cam profile. Finally, a 3D model is designed and the actual device is assembled.

The innovation of this paper is the discussion of the influence of all geometrical parameters on the mechanical properties of the spring-cam mechanism through detailed mathematical and mechanical methods, which provides a reference for designing the energy storage device in the form of a spring-cam mechanism.

One of the deficiencies to be considered in future research is that the approximations made during the experiment could to some extent affect the accuracy of the final result.

References

1. Esquenazi, A., Talaty, M., Jayaraman, A.: Powered exoskeletons for walking assistance in persons with central nervous system injuries. A narrative review. PM R **9**(4), 46–62 (2016)
2. Liu, M., Wu, B., Wang, W.Z., Lee, L.M., Zhang, S.H., Kong, L.Z.: Stroke in China: epidemiology, prevention, and management strategies. Lancet Neurol. **6**(5), 456–464 (2007)
3. Jezernik, S., Colombo, G., Keller, T., et al.: Robotic orthosis lokomat: a rehabilitation and research tool. Neuromodulation Technol. Neural Interface **6**(2), 108–115 (2010)
4. Bortole, M., Ama, A.D., Rocon, E., et al.: A robotic exoskeleton for overground gait rehabilitation. In: IEEE International Conference on Robotics and Automation. IEEE (2013)
5. Veneman, J.F., Kruidhof, R., Hekman, E.E.G., et al.: Design and evaluation of the LOPES exoskeleton robot for interactive gait rehabilitation. IEEE Trans. Neural Syst. Rehabil. Eng. **15**(3), 379–386 (2007)
6. Strausser, K.A., Swift, T.A., Zoss, A.B., et al.: Prototype medical exoskeleton for paraplegic mobility: first experimental results. In: ASME Dynamic Systems and Control Conference (2010)
7. McGeer, T.: Passive dynamic walking. Int. J. Robot. Res. **9**(2), 62–82 (1990)
8. Vaughan, C.L., Davis, B.L., O'Connor, J.C.: Dynamics of Human Gait, 2nd edn. Kiboho Publishers, Cape Town (1992)
9. Zhixiu, H., Jibin, Z., Dewen, J., et al.: How footwear affects the kinematics and dynamics of three-dimensional human gait. J. Tsinghua Univ. (Sci. Technol.) **46**(8), 1388–1392 (2006)

The Design and Analysis of Autonomous Underwater Pipeline Inspection Robots

Fujie Yu and Yuan Chen(✉)

School of Mechanical, Electrical and Information Engineering, Shandong University at Weihai, Weihai, China
cyzghysy@sdu.edu.cn

Abstract. In recent years, some research institutes have proposed the use of cable-free autonomous underwater vehicle-manipulator system (UVMS) for pipeline inspection. However, there are still some problems in applying conventional UVMS to the pipeline inspection and maintenance. According to the requirements of pipeline inspection mission, we have proposed a novel UVMS with adsorption capacity. Compared with the conventional UVMS, the UVMS can be adapted to a pipeline having a diameter of Φ500–1000 mm when adsorbed, so the end effector has higher operational precision compared to the floating state. Furthermore, the kinematics and the dynamic models of UVMS are established. Among them, the dynamic model of UVMS is established by the SimMechanics toolbox under MATLAB/Simulink, which overcomes the complexity of high-dimensional and dynamic interactions of UVMS.

Keywords: UVMS · SimMechanics · Precise operation

1 Introduction

At present, marine oil pipelines around the world stretch for tens of thousands of kilometers. According to statistics, there are 37,000 km of submarine pipelines in the Gulf of Mexico. According to the survey, 90% of the submarine pipeline and related facilities have never carried out maintenance activities since they were built. The main reason is that the maintenance technology of submarine oil pipelines is difficult and costly [1]. However, once the accident occurs, it will not only cause huge economic losses, but also cause serious pollution to the marine environment. In order to ensure the safe of the submarine pipeline, it is necessary to carry out routine inspection and maintenance. ROV has a relatively good working ability, but it is difficult to achieve a wide range of operations due to the umbilical cable. AUV is not limited by the umbilical cable, and it is suitable for carrying a wide range of scanning with the corresponding sensor, but the working ability is weak. Based on the above, some research institutes have proposed the use of cable-free autonomous underwater vehicle-manipulator system (UVMS) for pipeline inspection.

However, there are still some problems in applying conventional UVMS to the pipeline inspection and maintenance tasks. Due to the influence of ocean currents and complex hydrodynamic interference, UVMS still has the problems of insufficient operation accuracy and poor stability, which will lead to some high precision

D. Wang et al. (Eds.): MeTrApp 2019, MMS 79, pp. 172–181, 2020.
https://doi.org/10.1007/978-981-15-0142-5_18

underwater maintenance tasks can not competent [2–4]. What's more, UVMS is susceptible to collision with pipelines or related facilities during maintenance due to other external forces. Therefore, some research institutes have proposed a novel UVMS configuration. For example, in the ALIVE project led by Ocean Systems Lab in Heriot-Watt University, the heavy-duty underwater robot is equipped with two hydraulic grippers and a 7-function working manipulator. When working, it first grabs the work target with two hydraulic grippers and then the 7-function manipulator independently completes the switching operation of the valve (UVMS for pipeline patrol tasks should be light and flexible, it is difficult for the two hydraulic grippers to grasp the cylindrical target) [5]; Seoul National University in South Korea proposed a two-manipulator UVMS for up and down configuration in order to switch the valves on the pipeline. The principle is that first using the lower manipulator to grasp the pipe rod to increase the stability of UVMS in water, and then using the upper manipulator to hold the valve for rotation (the UVMS is only suitable for valves with small pipe diameter) [6]. In addition, most of the manipulator are mounted in the lower part of the carrier of UVMS, which makes UVMS very inconvenient to place on land or docking on the seabed, especially when it is necessary to stop for a long time or have to land when the power source fails.

For the UVMS, the kinematics and dynamics are the basis and premise for its trajectory planning and motion control [7]. In this paper, the kinematics model of the UVMS is first established. For the kinetic model, considering the high dimensionality of the UVMS and the complexity of the hydrodynamic parameters, it is difficult to obtain accurately model through rigorous mathematical derivation. Therefore, the kinetic model of UVMS is established by the SimMechanics toolbox under MATLAB/Simulink, which provides a dynamic simulation platform with visualization function. Finally, the simulation result shows that the kinematics model and the dynamics visualization platform are accurate and feasible.

2 Functional Requirements and Design of UVMS

2.1 Functional Requirements of UVMS

According to the requirements of pipeline inspection mission, the movement of UVMS is divided into three different states. The corresponding tasks in these three states are described as follows:

(1) Suspension operation: rapid inspection of submarine oil pipelines and related facilities; maintenance tasks without adsorption conditions and low operational accuracy requirements, such as sampling, pipeline inspection, etc.

(2) Vertical adsorption operation: pipe with an angle between the axis and the vertical line of the ground is less than $\pm 30°$; the operation accuracy of the maintenance object is high, such as the insertion and removal of the underwater cable, the opening and closing of the valve, etc.; It has adsorption conditions and has energy saving requirements.

(3) Lateral docking operation: Pipe with an angle between the axis and the horizontal line is less than ±30°; the operation accuracy of the maintenance object is high; It has adsorption conditions and has energy saving requirements.

2.2 UVMS Structural Design

Innovatively designed UVMS for pipeline inspection is shown in Fig. 1. The electromagnet adsorption module, the configuration and distribution of the propeller, and the installation position of the manipulator are mainly analyzed.

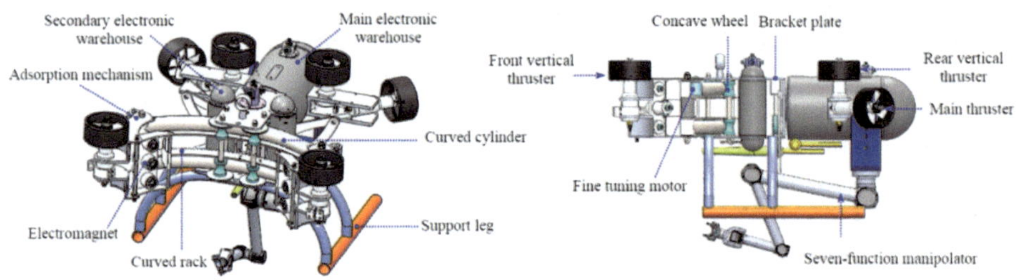

Fig. 1. The structure of UVMS for pipeline inspection

Firstly, considering that compared with other methods, the electromagnet adsorption method has the advantages of simple control, fast response, and small space occupation. Therefore, the electromagnet is selected as the adsorption device for fixing the UVMS to the pipeline or other cylindrical adsorbable objects. Among them, the electromagnet bracket is designed as an adjustable mechanism, so that UVMS can adapt to different pipe diameters when adsorbed. The curved cylinder is clamped by four concave wheels at the front and the rear, and the curved rack is fixed with the curved cylinder and has the same radius. When the UVMS is adsorbed on the pipeline, the curved rack will be in a fixed state. At this time, by controlling the gear meshing with the curved rack to achieve the movement of the UVMS along the bend of the bobbin, the purpose of expanding the working space of the UVMS in the fixed state is achieved.

Secondly, considering that the pipeline or other cylindrical adsorbable objects are not completely vertical or horizontal, the carrier of UVMS needs to have the ability to adjust the pitch and roll angle to complete the adsorption tasks. To this end, four vertical high-power thrusters are arranged to be distributed around the outer edge of the UVMS. This arrangement is beneficial to the stability of the UVMS under normal operation, and also facilitates the adjustment of the UVMS to the posture required to butt adsorption when adjacent to the pipeline.

Finally, the installation position of the manipulator should consider the flexible working space of the manipulator and the impact of the pipeline on the working space of the manipulator after UVMS adsorption. To sum it up, the bottom of the tail of the UVMS is selected as the installation position. In addition, the support legs are fixed to the lower bracket plate, so that the relative positions of the legs with the manipulator during the circumferential movement of the UVMS are unchanged, and the movement interference between the legs and the manipulator is prevented.

As shown in Fig. 2(a), the UVMS can be adapted to pipelines with a diameter range of Φ500–1000 mm. Figure 2(b) is the adjustment mechanism of the electromagnet, the principle of the mechanism is that when the UVMS is adjacent to the pipeline for docking adsorption, the fine-tuning motor drives the gears to mesh to rotate the electromagnet holder around the electromagnet shaft in the spring piece. Under the adaptive support force provided by the spring piece, the electromagnet adsorption module will be attached to the surface of the pipeline, the UVMS can be fixed to the pipeline even without propeller provides thrust.

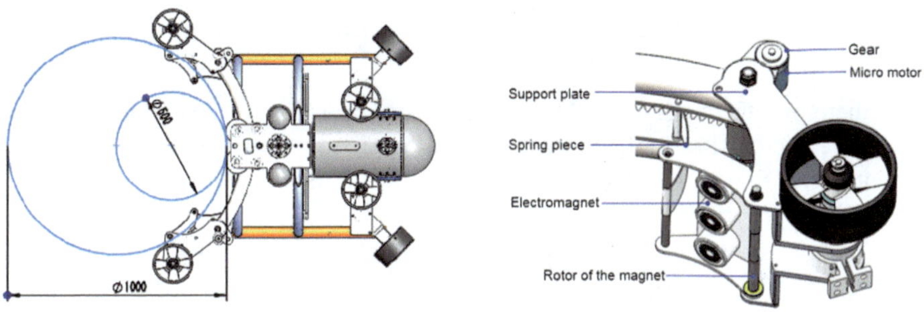

(a) The range that UVMS can adapt (b) Adjustment mechanism

Fig. 2. Electromagnet adjustment module of UVMS

3 Kinematic Model of UVMS

The model and the coordinate frame arrangement of UVMS are shown in Fig. 3. $\pounds_o$ is the inertial frame, $\pounds_b$ is the vehicle frame, $\pounds_i$ is the joint frame, and $\pounds_e$ is the end effector frame. The kinematic equation of UVMS can be expressed as follows [9]:

$$\dot{\xi} = J_{a,s}\,\zeta \tag{1}$$

where $\xi = [\eta\ q]^{\mathrm{T}}$ is the vector of absolute positions, euler angles and the manipulator joint angles of UVMS; $\eta = [x\ y\ z\ \Phi\ \theta\ \Psi]^{\mathrm{T}}$ is the absolute positions, euler angles of the vehicle, x-surge position, y-sway position, z-heave position, ϕ-roll angle, θ-pitch angle, ψ-yaw angle respectively; $q = [q_1\ q_2\ q_3\ q_4\ q_5\ q_6]^{\mathrm{T}}$, and q_i is the joint position of the link i; The velocity of UVMS in the vehicle frame is $\zeta = [v\ \dot{q}]^{\mathrm{T}}$, where $v = [u\ v\ w\ p\ q\ r]^{\mathrm{T}}$, u, v, w representing the linear velocity of x, y, and z respectively; p, q, r representing the angular velocity of roll, pitch and yaw respectively; $J_{a,s}$ is the Jacobian matrix of the system velocity converted to the inertial frame velocity.

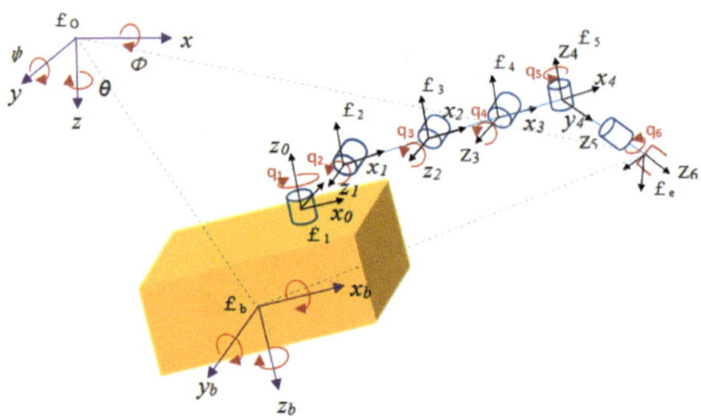

Fig. 3. UVMS and its coordinate frame arrangement

Furthermore, the tasks of UVMS are often accomplished by the end effector, so the states of the end effector in the inertial frame $\pounds_o$ needs to be considered. The velocity of the joint frame $\pounds_i$ with respect to the inertial frame $\pounds_o$ is expressed as:

$$V_{oi} = J_{io}\zeta = \left[Ad_{g_{bi}}^{-1} \, Ad_{g_{bi}}^{-1} J_i \right] \zeta \tag{2}$$

By using Eq. (2) and take $i = e$, the end effector velocity can be obtained as follows:

$$V_{oe} = J_{eo}\zeta = \left[Ad_{g_{be}}^{-1} \, Ad_{g_{be}}^{-1} J_n \right] \zeta \tag{3}$$

For the manipulator, the D-H parameters, radius, length and joint limits are at Table 1.

Table 1. D-H parameters [m, rad], radius [m] and length [m] of the manipulator.

Parameters	a	d	θ	α	Radius	Length	limits
Link1	0.1	0	q_1	π/2	0.05	0.15	−π/6–π/6
Link2	0.5	0	q_2	0	0.05	0.6	−π/2–π/10
Link3	0.4	0	q_3	0	0.04	0.4	−2.8–2.8
Link4	0.2	0	q_4	−π/2	0.03	0.2	−2.8–2.8
Link5	0	0	q_5	−π/2	0.03	0.1	−4–0.87
Link6	0	0.1	q_6	0	0.03	0.1	−2π–2π

Figure 4 shows the working space of the manipulator obtained by Monte Carlo method [8]. Figure 4(a) shows the allowed working space in the suspended state, and Fig. 4(b) shows the allowed working space when UVMS is adsorbed on the pipeline.

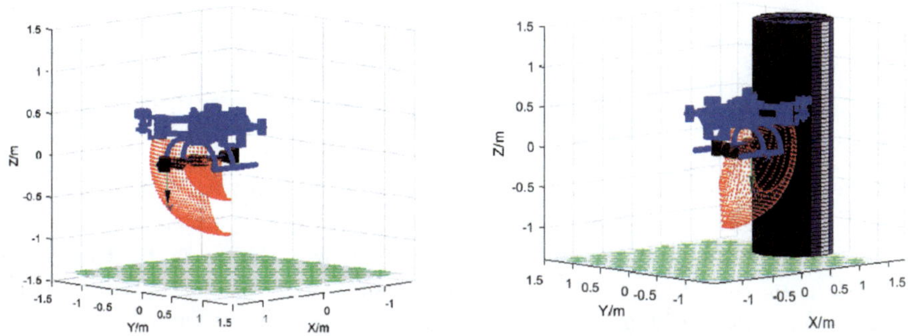

(a) UVMS is in suspension (b) UVMS is adsorbed on the pipeline

Fig. 4. Allowable workspace for the manipulator of UVMS

4 UVMS Dynamics Modeling Based on SimMechanics

SimMechanics is a toolbox for the Simulink catalog containing linkages, sensors, actuators, constraints and force elements. With the toolbox, the dynamics modeling complexity of UVMS can be simplified and the motion process of UVMS can be visualized. It should be noted that the SimMechanics is generally used to describe the rigid body dynamics and the interaction between them of the system. Therefore, in the SimMechanics, the hydrodynamics of the UVMS should be taken into account.

The dynamics equation of UVMS is as follows:

$$M\dot{\zeta} + C(\zeta)\zeta + D(\zeta)\zeta + g(\eta_{ie}) = \tau \tag{4}$$

Where M (ξ) is the inertia matrix of the system. C (ξ, ζ_r) is the coriolis and centripetal matrix of the system. D (ξ, ζ_r) is the damping matrix of the system. τ is the control force and torque of UVMS.

For the visualization model of UVMS, the input force and torque should satisfy:

$$M\dot{\zeta} = \tau - C(\zeta)\zeta - D(\zeta)\zeta - g(\eta_{ie}) = \tau' \tag{5}$$

Figure 5 shows the UVMS dynamics module based on SimMechanics, it consists of force/torque calculation module, visualization module, and feedback variable number calculation module (from left to right).

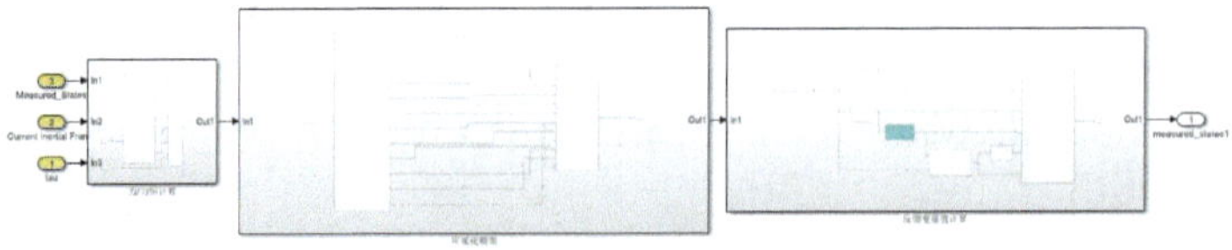

Fig. 5. Internal diagram of the UVMS dynamics module

The internal composition of visualization model based on SimMechanics is shown in Fig. 6. The green line connects the SimMechanics part and the purple line connects the Simulink content. In1 is the input port of force and torque, which is a 12×1 vector. For a submodule, the port B is the base coordinate system, the port F is the follower coordinate system, and the modules are connected by coordinates to achieve the construction of the system model. Figure 7 shows the 3D display window of the Mechanics Explorers during simulation. After the system simulation is completed, the overall motion animation of the UVMS is automatically generated.

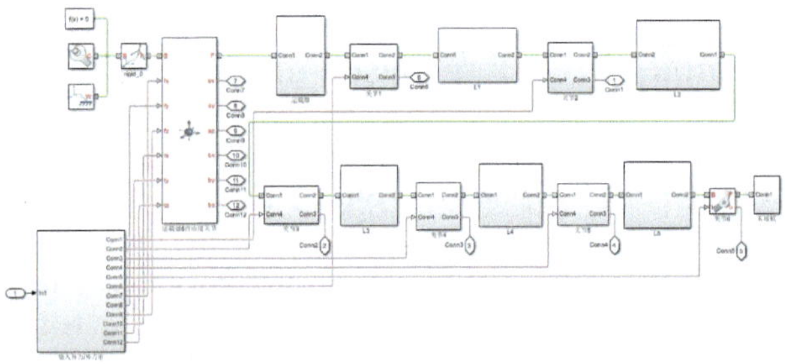

Fig. 6. The internal composition of the visualization module of UVMS

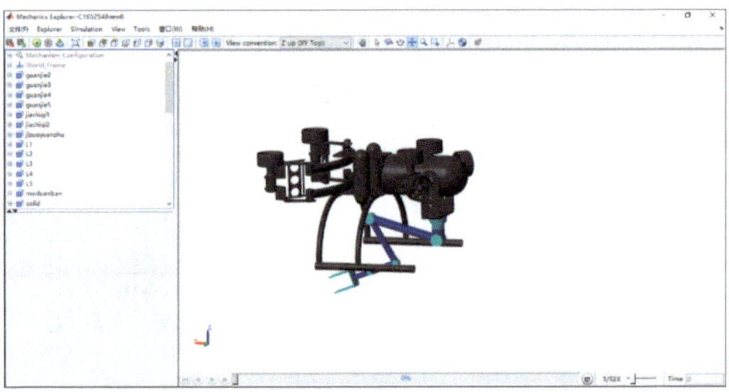

Fig. 7. Mechanics Explorers-SimMechanics simulation platform interface

Considering that the hydrodynamic coefficient of a real carrier is difficult to determine, an object with a regular shape can be estimated by an empirical formula. For this purpose, the shape of the carrier is simplify as a rectangular body of 2.3 × 2 × 1 m cuboid. In the simulation, the calculation step is set to variable step, and the solution mode is ode45. In order to detect the correctness of the UVMS dynamics platform based on SimMechanics, it was replaced and simulated with the dynamic part of the general UVMS overall control system. The simulation time is 40 s, the simulation step is 0.008 s, and the initial pose of the end effector is: $(x_e\ y_e\ z_e) = [1.35\text{ m}\ 0\text{ m}\ -0.2\text{ m}]$, $(\Phi_e\ \theta_e\ \Psi_e) = [-110°\ 0°\ -90°]$. The trajectory equation of the task performed by UVMS are given as follows:

$$\xi_d(t) = \left. \begin{bmatrix} 7.6 \times 10^{-6}t^4 - 6.7 \times 10^{-4}t^3 + 0.017t^2 - 0.028t + 0.038 \\ -7.6 \times 10^{-6}t^4 + 5.5 \times 10^{-4}t^3 - 0.009t^2 + 0.028t - 0.048 \\ 0 \end{bmatrix} \text{m} \right\} 0 \leq \text{t} \leq 40\,\text{s} \tag{6}$$

Figure 8 is the trajectory tracking curve of the position and orientation of the end effector. It can be seen that the measured trajectory of the end effector is almost completely coincident with the expected trajectory. The position tracking error is within ±0.02 m, and the orientation tracking error is within ±2°. The Screenshot of motion animation is shown in Fig. 9. Due to SimMechanics' visualization function, the movement of each joint and carrier can be clearly seen.

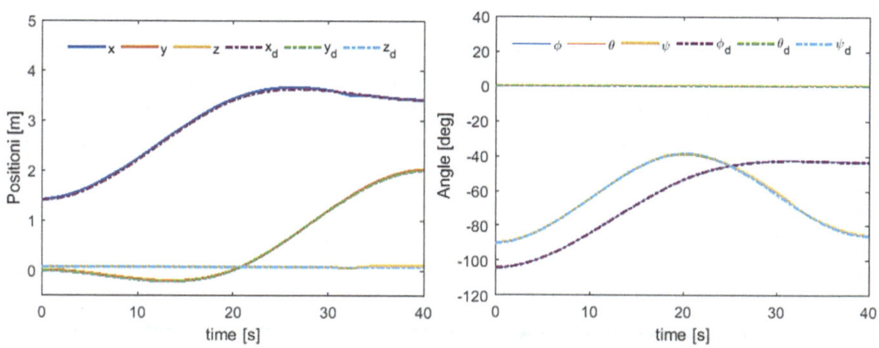

Fig. 8. Position and orientation tracking curve of UVMS

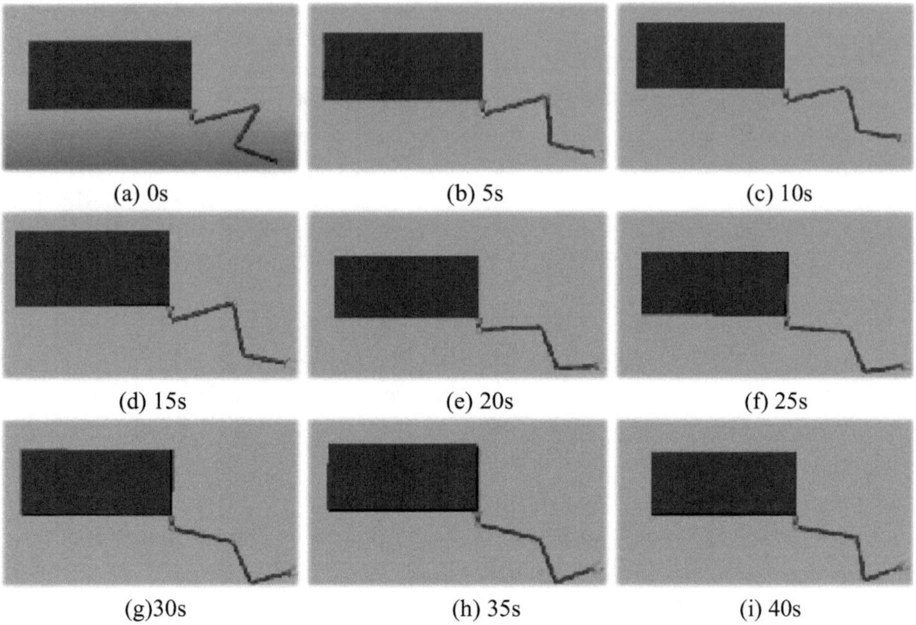

Fig. 9. Trajectory tracking process of UVMS

5 Conclusion

In this paper, the inspection of underwater pipelines are taken as research tasks, and the UVMS with pipeline adsorption capacity is designed. Compared with conventional UVMS, the UVMS can be adapted to a pipeline having a diameter of Φ500–1000 mm when adsorbed, so the end effector has higher operational precision. For the kinetic model, the dynamic model of UVMS is established by the SimMechanics toolbox under MATLAB/Simulink, which overcomes the complexity of high-dimensional and dynamic interactions of UVMS. For future studies, this provides a good dynamic simulation platform with visualization functions.

Acknowledgements. Grateful acknowledgement is given to the National Natural Science Foundation of China with Grant No. 51375264, Major Scientific and Technological Innovation Project of Shandong Province with Grant No. 2017CXGC0923, Key Research and Development Program of Shandong Province with Grant No.2018GGX103025, the Fundamental Research Funds for the Central Universities with Grant No. 2019ZRJC006, and Natural Science Foundation of Shandong Province with Grant No. ZR2019MEE019.

References

1. Zhang, M.S.: China's submarine oil and gas pipelines have serious hidden dangers. China Energy News, p. 2 (2013)

2. Mohan, S.: Coordinated motion control in task space of an autonomous underwater vehicle–manipulator system. Ocean Eng. **104**(262), 155–167 (2015)
3. Han, J., Park, J., Chung, W.K.: Robust coordinated motion control of an underwater vehicle-manipulator system with minimizing restoring moments. Ocean Eng. **38**, 1197–1206 (2011)
4. Wei, Y.H.: Technology of Control UVMS. Harbin Engineering University Press, Harbin (2016)
5. Evans, J., Redmond, P., Plakas, C., et al.: Autonomous docking for Intervention-AUVs using sonar and video-based real-time 3D pose estimation. In: Proceedings of OCEANS 2003. IEEE (2003)
6. Bae, J., Bak, J., Jin, S., et al.: Optimal configuration and parametric design of an underwater vehicle manipulator system for a valve task. Mech. Mach. Theory **123**, 76–88 (2018)
7. Cui, Y., Sarkar, N.: A unified force control approach to autonomous underwater manipulation. Robotics **19**, 255–266 (2001)
8. Rastegar, J., Fardanesh, B.: Manipulator workspace analysis using the Monte Carlo method. Mech. Mach. Theory **25**(2), 233–239 (1990)
9. Antonelli, G.: Underwater Robots. Springer Tracts in Advanced Robotics. Springer, Cham (2013)

Novel Solution for the Mechanism Used in the Actuation of Intelligent Roof Windows

Carmen Sticlaru, Petrisor Laurentiu Tuca, Erwin-Christian Lovasz, Corina-Mihaela Gruescu(✉), Inocenţiu Maniu, and Cristian Emil Moldovan

University Politehnica Timisoara, 300002 Timisoara, Romania
corina.gruescu@upt.ro

Abstract. The window roofs are available in the variants top-hung and center pivot roof window with actuation based on rack or chain mechanisms. The new solution proposed for center pivot roof windows is based on a geared linkage. Its scheme is explained and its analysis is performed in order to get functional parameters. The CAD model was generated and used to simulate the kinematic behavior, then a prototype was assembled to test the practical functioning. The mechanical system is actuated by an intelligent control system, which includes an Arduino board and sensors for light, temperature, pressure and humidity.

Keywords: Roof window · Intelligent actuation · Geared linkage

1 Introduction

Since ancient times, the natural lighting and ventilation of interior spaces was important. The gaps in the wall brought a series of problems, such as the penetration of cold air, rain, wind, heat, etc. Later, these gaps in the wall were called windows. The windows are among the most expressive and essential accessories of our dwellings, their design varies in terms of size, structure or even shape. The windows as they are today have appeared in the eighteenth century along with the industrial revolution, when new methods of glass processing on large scale were established.

A first mention of the attic windows appears in an illustration of the room belonging to the famous Danish poet and writer Hans Christian Anderson (n: 1805 - d: 1875). [1] Another mention of the attic window is found in the work of the famous painter Vincent van Gogh (n: 1853 - d: 1890) entitled "Farms near Auvers" [2]. The popularity of the attic windows began to grow around 1970, when automated operation was proposed by Velux [3] in Denmark.

There are two main solutions on the market, regarding the opening of the roof window:

- top-hung roof window (Fig. 1 [3]), suitable for roof pitches between 15° and 55°
- center pivot roof window (Fig. 2 [3]), suitable for roof pitches between 15° and 90°.

The command of the mechanisms which open and close the window is either manual or automated, depending on the accessibility to the mechanism and on the choice of the customer.

D. Wang et al. (Eds.): MeTrApp 2019, MMS 79, pp. 182–191, 2020.
https://doi.org/10.1007/978-981-15-0142-5_19

Fig. 1. Top-hung roof window

Fig. 2. Center pivot roof window

There are two main types of mechanisms available for the roof windows offered on the present market:

- single or double rack mechanism (Fig. 3) [4]
- single or double chain mechanism (Fig. 4) [4].

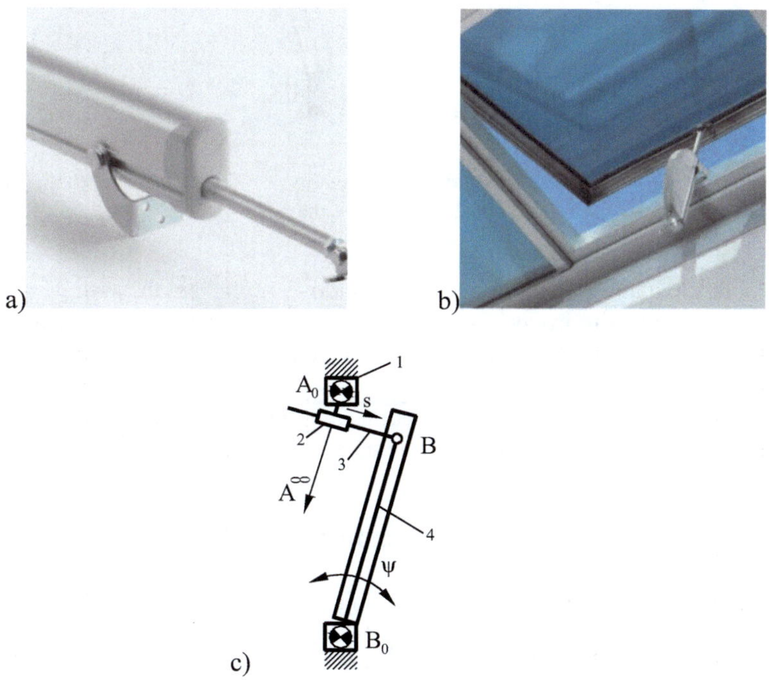

Fig. 3. Single rack actuation. a. Actuation assembly b. Window with rack opening c. Scheme of the equivalent linkage

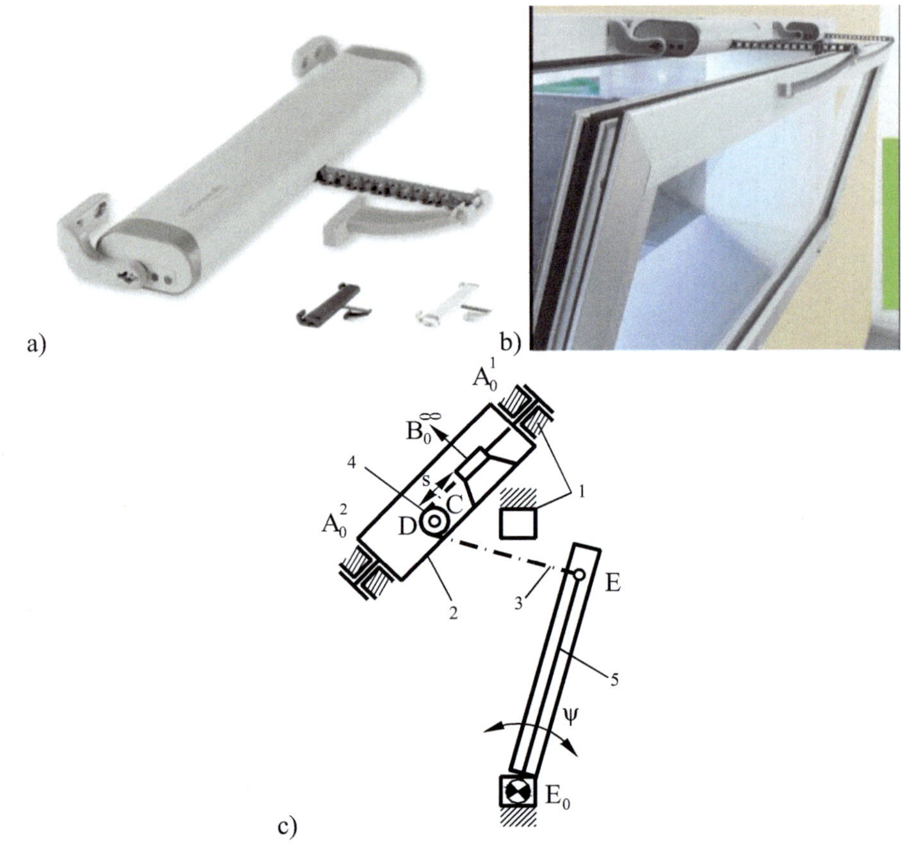

Fig. 4. Single chain actuation. a. Actuation assembly b. Window with chain opening c. Kinematic scheme (1 – fixed element, 2 – case, 3 – chain, 4 – wheel, 5 –window)

The both actuation systems connect the windows frame and the sash preventing the access for mounting the insects net, for instance, and introduce an undesired visual effect. The actuation solutions use complex mechanisms with friction loss and low amplitude of opening.

Considering all above aspects, the authors propose a novel solution based on a geared linkage with linear actuation mounted on the exterior side of the frame.

2 Kinematic Schema and Kinetostatic Analysis

2.1 Kinematic Schema

The proposed mechanism for the window actuation contains an inverted slider crank connected with a cycloid gear train (Fig. 5). The driving element (3) performs a linear movement and the driven element is the gear (5), whose shaft and bearings are mounted in the window sash. This mechanism known as geared linkage with linear actuation allows a large swing angle χ of the driven element with favorable transmission angle of the base mechanism and approximately linear transmission function $\chi(s)$.

The transmission function $\chi(s)$ of the geared linkage with linear actuation depends on the transmission functions of the base mechanisms, denoted $\varphi(s)$ and $\psi(s)$, inverse of the transmission ratio $\rho = -r_3/r_5$, and the input stroke s. This is [5–8]:

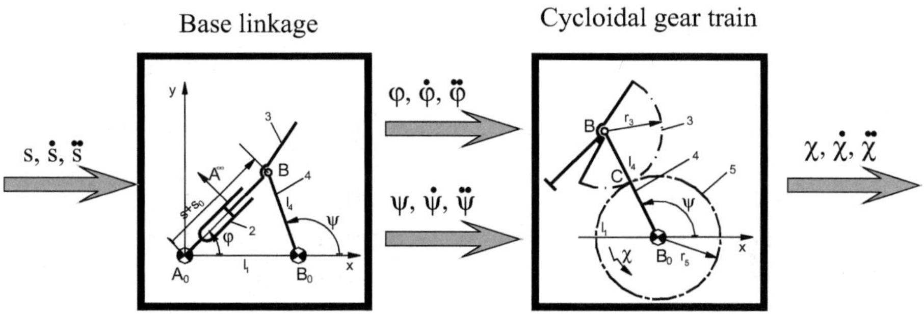

Fig. 5. Kinematic schema and parameters

$$\chi(s) = (1 - \rho) \cdot \psi(s) + \rho \cdot \varphi(s), \quad (1)$$

The first and second order transmission functions are:

$$\chi'(s) = (1 - \rho) \cdot \psi'(s) + \rho \cdot \varphi'(s), \quad (2)$$

$$\chi''(s) = (1 - \rho) \cdot \psi''(s) + \rho \cdot \varphi''(s). \quad (3)$$

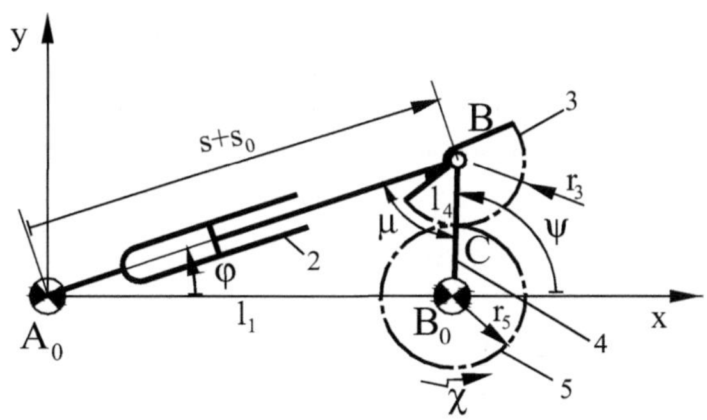

Fig. 6. Kinematic scheme of the geared linkage with linear actuation

By writting the vectorial equation for closing the loop of the base mechanism shown in Fig. 6:

$$(s_0 + s) \cdot e^{i \cdot \varphi} = l_1 + l_4 \cdot e^{i \cdot \psi}, \tag{4}$$

can be computed the transmission functions of 0-order $\varphi(s)$ and $\psi(s)$ as follows:

$$\varphi(s) = 2 \arccos \frac{l_4^2 - l_1^2 - (s_0 + s)^2}{2 \cdot l_1 \cdot (s_0 + s)}, \tag{5}$$

and

$$\psi(s) = \arccos \left(\frac{(s_0 - s)^2 - l_1^2 - l_4^2}{2 \cdot l_1 \cdot l_4} \right). \tag{6}$$

For the geared linkage used for actuating the window were chosen the linkage dimensions as shown in Table 1.

Table 1. Geometrical parameters of the geared linkage with linear actuation

Nr	Parameter	Dimension	Nr	Parameter	Dimension
1.	Frame length	l_1 = 208 mm	4.	Gear ratio (m = 1, z_1 = 20, z_2 = 40)	ρ = 0.5
2.	Crank length	l_4 = 27 mm	5.	Stroke	s = 54 mm
3.	Initial stroke	s_0 = 181 mm	6.	Minimum transmission angle	μ_{min} = 50°

The variations of transmission function 0-ord and 1-ord are shown in Fig. 7. For an optimized approximated linear transmission function, which is recommended for an easier control of the motion, the following condition is to be satisfied in the flat domain of the transmission function 1-ord:

$$\chi'(s_1) = \chi'(s_2). \tag{7}$$

As shown in Fig. 7b, in order to satisfy the condition (5) the initial stroke should be changed to s_0 = 200 mm and the stroke to s = 35 mm (s_1 = 11 mm and s_2 = 46 mm). The angular displacement of the window is limited to about χ_{max} = 121°.

The current transmission angle of the base linkage will be computed with the relationship:

$$\mu(s) = \psi(s) - \varphi(s). \tag{8}$$

Figure 8 shows that the variation of the of the transmission angle within the range of the stroke s = [s_1, s_2] is satisfied:

$$\mu(s) \geq \mu_{\min}. \tag{9}$$

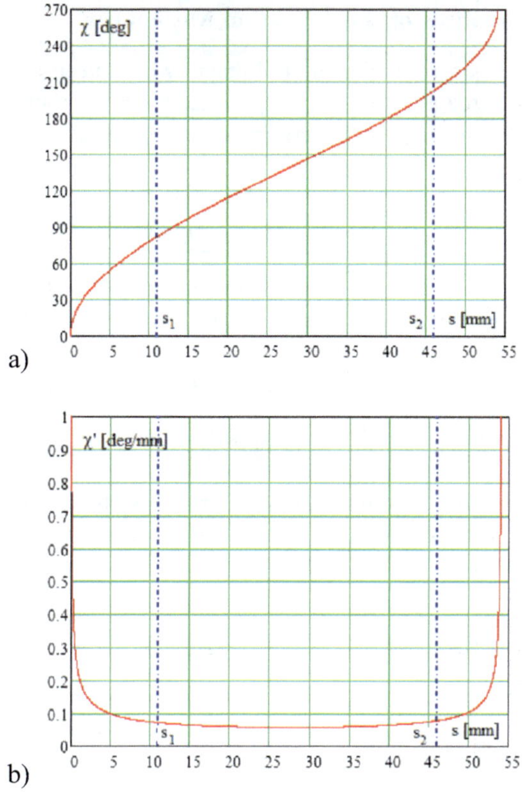

Fig. 7. Transmission functions 0-ord. (a) and 1-ord (b) of the geared linkages

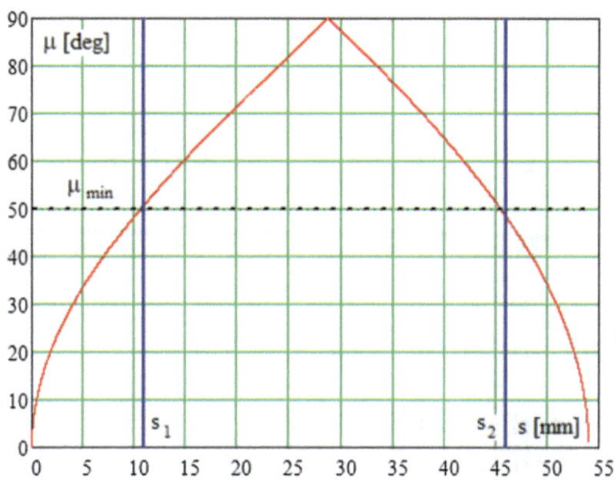

Fig. 8. Transmission angle of the base linkages

2.2 Kinetostatic Analysis of the Actuation Mechanism

The kinetostatic analysis of the geared linkage with linear actuation develops by computing the drive force and the reactions variation in the joints. The drive force of the actuator, according to [6], will be computed with the relationship:

$$F_A = \frac{M_R \cdot [s_0(1+\rho) - \rho \cdot l_4 \sin(\varphi(s) - \psi(s))]}{l_4 \cdot s_0 \cdot \cos((\varphi(s) - \psi(s))}, \tag{10}$$

where: M_R is the friction moment resulted by acting the center pivot roof window, while the weight of the window is symmetrically distributed to the central pivot:

$$M_R = \mu \cdot G \cdot r, \tag{11}$$

with: μ - friction coefficient, G - the weight of the window, r - the radius of the central pivot.

By considering the weight of the window G = 20 N, the friction coefficient μ = 0.1 and the radius of the central pivot r = 6 mm, the reacting moment is M_R = 12 Nmm. The variation of the drive force is shown in Fig. 9. The maximum value is about 1.8 N.

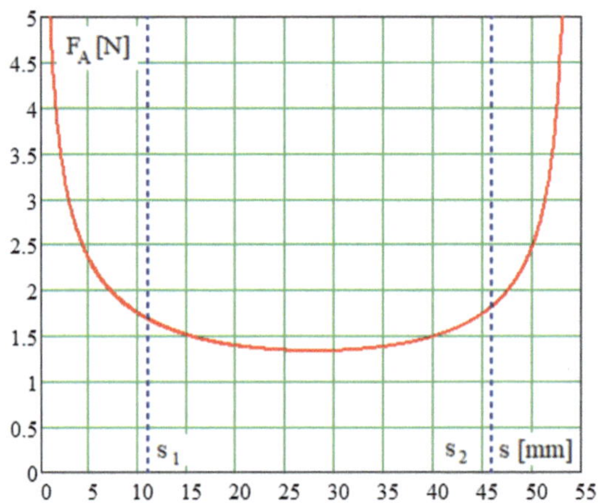

Fig. 9. Variation of the drive force of the linear actuator

3 CAD Modeling and Prototype

The mechanism was modelled in CREO 2.0.

Figure 10 renders a general view of the mechanism (a) and the side-views of the extreme positions of the mechanism (b – initial position; c – final position).

The CAD model was used to simulate the functioning of the window mechanism. The result of the simulation given in Fig. 11 validates the analytical approach of the kinematic and kinetostatic analysis.

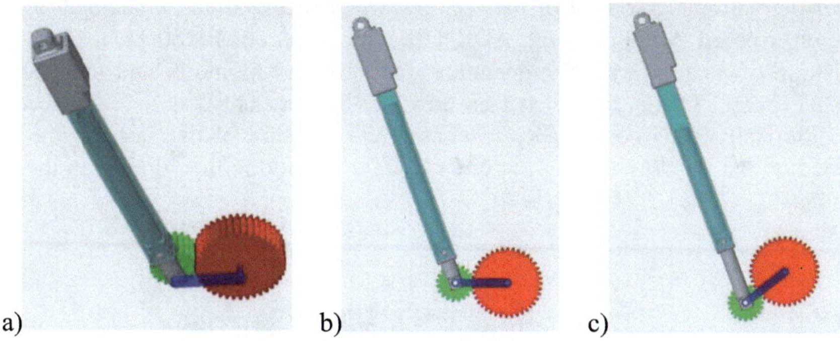

Fig. 10. CAD model of the mechanism (a – general view, b – initial position, c – final position)

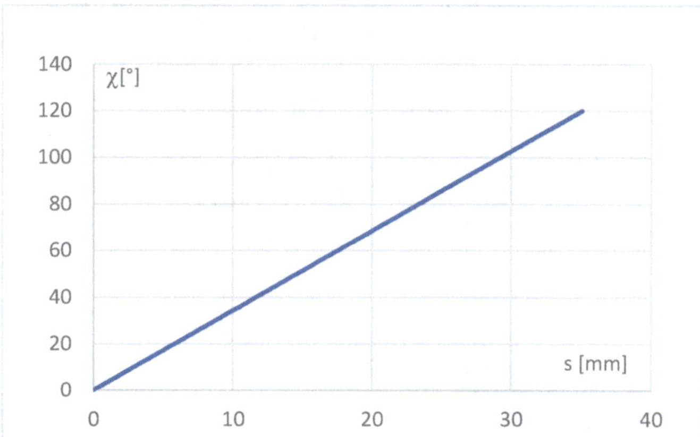

Fig. 11. Angular displacement of the window in respect to actuator stroke

The actuation is performed by means of the linear actuator type Actuonix L 12, which allows a maximum force of 80 N with the maximum stroke of 100 mm [9] (Fig. 12).

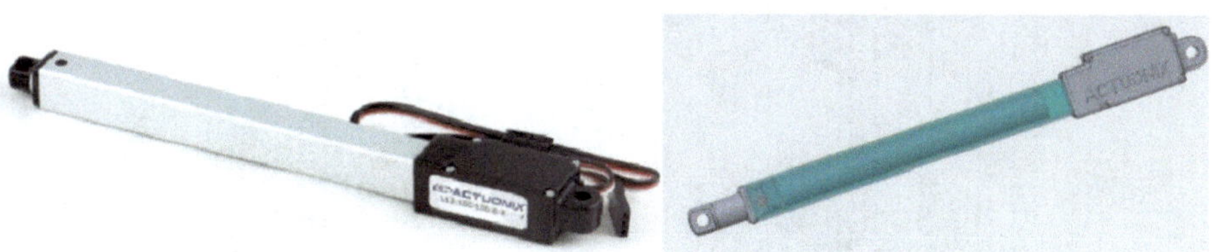

Fig. 12. Linear actuator type Actuonix L 12

For the automatic control of the window was conceived and assembled a system based on a board Arduino Mega ADK [10], the sensor BME280 [11], used for the acquisition of data regarding temperature, pressure and humidity and the phototransistor PT15-21C/TR8 [12], which measures the light intensity.

Additionally, it is compulsory to control the pressure of the sash on the gasket mounted on the window's frame, in order to tight sealing on the window in the closed position.

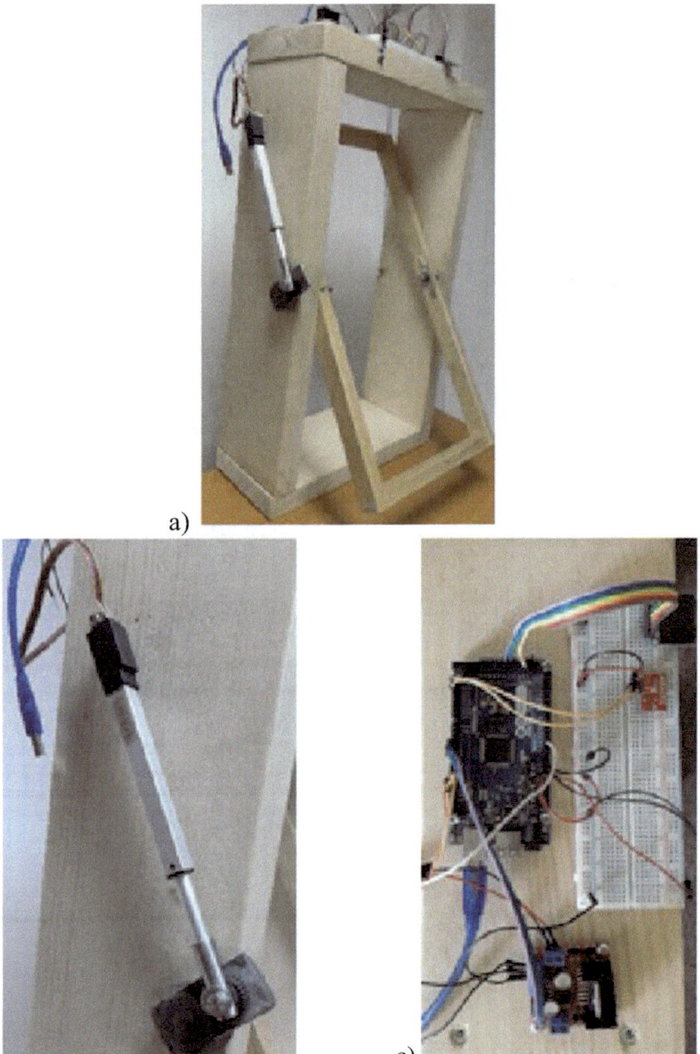

a) b) c)

Fig. 13. Prototype of roof window. a. Assembly b. Detail (the mechanism) c. Detail (the command parts)

An appropriate code was written in order to control the closing and opening of the window at given values of ambient light intensity, humidity, temperature and pressure.

The prototype (Fig. 13) resulted by mounting the mechanism and control parts on a frame and sash made of wood. The assembly was tested in regard with its functioning at different values of parameters.

4 Conclusions

The opening/closing mechanism based on a geared linkage with linear actuator is novel as solution for the acting of roof-windows. The mechanism is suited for the pivot type of sash mobility.

The intelligent control system of the opening or closing the window use a micro-controller, a sensor for measuring the temperature, pressure and humidity and an phototransistor for measuring the light intensity.

The solution is advantageous because of its small size, high efficiency, linear characteristic and high mechanical reliability.

References

1. https://www.fakro.ro/stiri/istoria-ferestrelor-de-mansarda/. Accessed 10 July 2019
2. https://www.tate.org.uk/art/artworks/gogh-farms-near-auvers-n04713. Accessed 10 July 2019
3. https://www.velux.co.uk. Accessed 01 June 2019
4. https://kadra.ro/. Accessed 01 June 2019
5. Lovasz, E.-C., Modler, K.-H., Pop, C., Pop, F., Mărgineanu, D.T., Maniu, I.: Type synthesis and analysis of geared linkages with linear actuation. Mechanika **24**(1), 108–114 (2018)
6. Lovasz, E.-C., Margineanu, D.T., Ciupe, V., Maniu, I., Gruescu, C.M., Zabava, E.S., Stan, S.D.: Design and control solutions for haptic elbow exoskeleton module used in space telerobotics. Mech. Mach. Theory **107**, 384–398 (2017)
7. Modler, K.-H., Lovasz, E.-C., Perju, D., Hollmann, Ch.: Geared linkages with linear displacement actuator used as function generating mechanisms. In: 11th World Congress in Mechanism and Machine Science, Tianjin, vol. 3, pp. 1254–1259 (2004)
8. Lovasz, E.-C., Sticlaru, C., Suciu, C., Gruescu, C.M., Ceccarelli, M., Maniu, I., Moldovan, C.E.: Novel actuation design of an active elbow orthosis. In: 15th World Congress in Mechanism and Machine Science, Krakow (2019)
9. https://www.actuonix.com/L12-Micro-Linear-Actuators-s/1821.htm. Accessed 01 June 2019
10. https://www.arduino.cc/en/Main/ArduinoBoardMegaADK?from=Main. Arduino BoardADK. Accessed 01 June 2019
11. https://www.bosch-sensortec.com/bst/products/all_products/bme280. Accessed 01 June 2019
12. https://www.digikey.com/product-detail/en/everlight-electronics-co-ltd/PT15-21C-TR8/1080-1380-1-ND/2676114. Accessed 01 June 2019

Design and Kinematic Analysis of a Novel Wire-Driven Spherical Scissors Mechanism

Ruiqin Li[1(✉)], Ziwei Liu[1], Wenbo Mi[1], Shaoping Bai[2], and Jianwei Zhang[3]

[1] School of Mechanical Engineering, North University of China, Taiyuan 030051, China
liruiqin@nuc.edu.cn
[2] Department of Materials and Production, Aalborg University, 9220 Aalborg, Denmark
shb@mp.aau.dk
[3] Department of Informatics, University of Hamburg, 22527 Hamburg, Germany

Abstract. A novel wire-driven shoulder exoskeleton is presented. The exoskeleton is designed with spherical scissors mechanism to achieve a compact and light weight design. Wire-driven approach is adopted to actuate the mechanism effectively without increasing its weight and size. Kinematic simulation verifies the feasibility of the wire-driven system. The structural stiffness is evaluated through FEA.

Keywords: Wire-driven mechanism · Spherical scissors mechanism · Kinematics · Shoulder joint · Rehabilitation

1 Introduction

Shoulder joint is one of the most complex joints in human body. In recent years, shoulder joint dyskinesia occurs frequently due to the increase of motor injury and nerve diseases, which requires innovative shoulder mechanisms for rehabilitation purpose. Many shoulder mechanisms have been reported in literature. Christensen et al. [1] presented a 3-DOF spherical mechanism with a double parallelogram linkage for a shoulder exoskeleton joint. The shoulder joint was designed, constructed, and integrated in a 4-DOF wearable upper-body exoskeleton. Hunt et al. [2] designed a 5-DOF shoulder exoskeleton using parallel actuation and an integrated passive slip interface. The parallel mechanism is of low inertia and high stiffness. Klein et al. [3] designed a parallel shoulder mechanism. The mechanism can produce large forces through a substantial portion of the range of motion of the human arm while remaining lightweight, for utilizing the push rod motors to drive the exoskeleton. Kumar et al. [4] designed a dual-arm exoskeleton named RECUPERA, which features a lightweight design, high level of modularity, decentralized computing, and various levels of safety implementation.

Cable-driven mechanism has a simple and light weight mechanical structure that matches perfectly with shoulder exoskeletons. Many cable-driven mechanisms have been designed. Cui et al. [5] reported the design and control of a 7-DOF cable-driven

D. Wang et al. (Eds.): MeTrApp 2019, MMS 79, pp. 192–200, 2020.
https://doi.org/10.1007/978-981-15-0142-5_20

arm exoskeleton (CARXE-7). Eight cables are routed through the exoskeleton cuffs to drive the whole-arm motion. While the exoskeleton can help the subjects to follow the path more closely, the cable-driven mechanism is too large and the cable layout is complex. Ball et al. [6] presented a 5-DOF wire-driven shoulder rehabilitation robot named MEDARM, but its pulley has a large volume and makes it difficult to be wore. Salah [7] presented a wire-driven mechanism, which used a pair of wires to control end effector. Lee et al. [8] proposed a wire-driven series elastic actuator for lower limb rehabilitation, which uses pulleys to lift lower limb. Kim et al. [8–10] designed a low-inertia high-stiffness manipulator, whose joint uses two pairs of pulleys to rotate elbow joint. Dehez et al. [11] reported a two-DOF rehabilitation robot, using poly-articulated structure and Bowden cables.

Shoulder joints are essentially spherical mechanisms. It is noted that spherical mechanisms are being increasingly used in different applications, including orientating devices and exoskeletons [12]. Lo et al. [13] presented a redundant 4R spherical wrist mechanism for a shoulder exoskeleton. It can move throughout the entire shoulder workspace with feasibly low joint velocities. Hsieh et al. [14] reported a lightweight shoulder exoskeleton, which consisted of two spherical mechanisms with two slider crank mechanisms. The actuators were attached side-by-side, close to a human body. Castro et al. [15] proposed a spherical scissors mechanism for shoulder rehabilitation, which achieved a very compact design. Lum et al. [16] introduced a spherical mechanism for a minimally invasive surgical robot. Maden et al. [17] analyzed the geometric design of the planar scissor structural mechanisms in terms of their possible configurations and then developed trigonometric calculation methods for different types of the planar scissor structural mechanisms. Wu et al. [18] presented an upper limb exoskeleton which is powered by wire-driven system.

It's noted that, for these mechanisms mentioned, suitable actuations are rarely designed to drive these spherical mechanisms. If the mechanism is driven directly by motors, the compactness and lightweight features of the spherical mechanism will disappear. An effective driving approach is thus required for shoulder exoskeletons.

This paper presents a novel wire-driven spherical scissors mechanism for shoulder joint rehabilitation. This driving system consists of pulleys, holders and steel wires. The spherical scissor mechanism can stretch and shrink and rotates around remote spherical center. By this way, a system of lightweight and compact shoulder exoskeleton is obtained.

2 Wire-Driven Spherical Scissors Mechanism

As shown in Fig. 1, the spherical scissors mechanism has a common remote spherical center. Three sloped holders with the pulley in same size are fixed on the surfaces of arc-shape links using rivets in the revolute joints. The pulleys are mounted on the mandrel on the plane of the holders. Two steel wires, fixed with Pulley I and Pulley III, respectively, pass through pulley II and change their directions via pulley IV. Two output wires are wrapped by brake hose and linked to two motors, respectively.

Two motors are fixed on a back plate, as shown in Fig. 2. By controlling the length of two steel wires, the single-DOF spherical scissors mechanism can stretch or shrink.

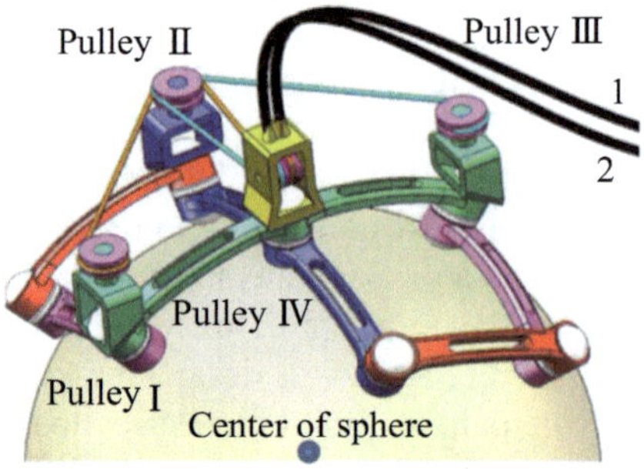

Fig. 1. The wire-driven spherical scissors mechanism

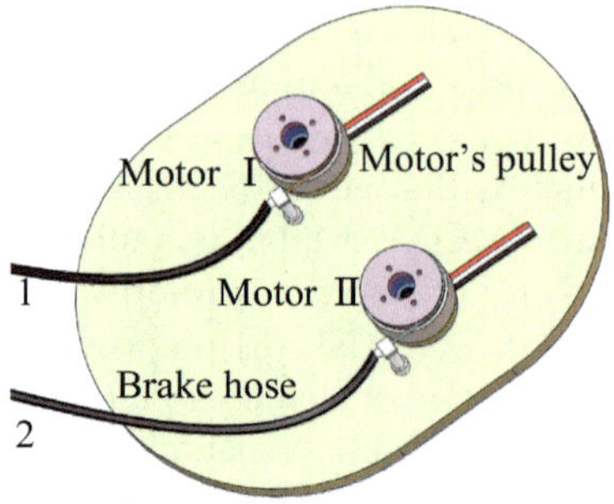

Fig. 2. The driving system

3 Analysis of the Steel Wires

While steel wires in the driving system are elastic, to facilitate the calculation, we neglect their deformation. Figures 3 and 4 show geometric models of the wire-driven mechanism.

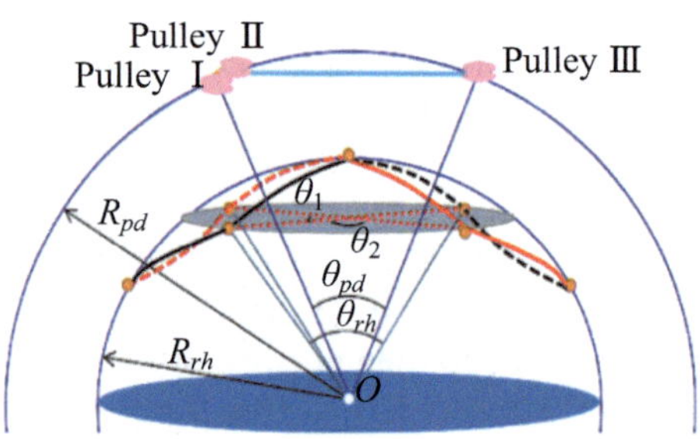

Fig. 3. Geometric model of the wire-driven mechanism

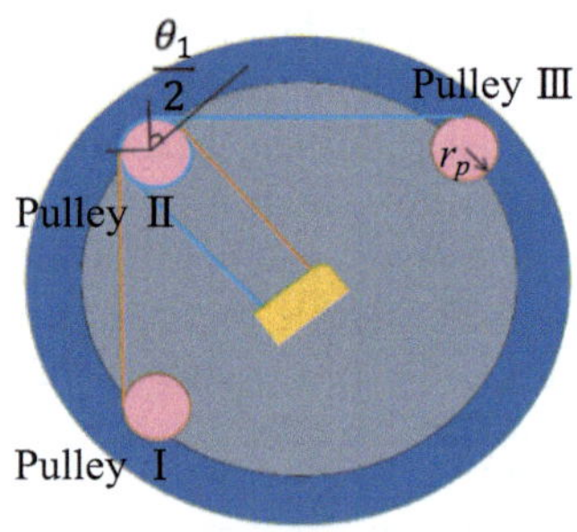

Fig. 4. The layout of the wires in the wire-driven mechanism

The distance R_{pd} from spherical center O to the pulley should always satisfy:

$$R_{pd} \geq \frac{R_{rh}}{\cos \theta_{pd}/2} \tag{1}$$

where R_{rh} is the radius of the spherical scissors mechanism. r_p is the radius of pulley.

The total length l_{wire} of the steel wires is found by

$$l_{wire} = l_1 + l_2 \tag{2}$$

where, l_1 and l_2 present two wires' length, respectively.

$$l_1 = R_{pd} \cdot \sin\frac{\theta_{pd}}{2} \cdot \left(1 + 2 \cdot \sin\frac{\theta_1}{2}\right) + \left(\frac{\pi}{2} + \frac{\theta_1}{2}\right) \cdot r_p \tag{3}$$

$$l_2 = R_{pd} \cdot \sin\frac{\theta_{pd}}{2} \cdot \left(1 + 2 \cdot \sin\frac{\theta_2}{2}\right) + \left(\frac{\pi}{2} + \frac{\theta_2}{2}\right) \cdot r_p \tag{4}$$

$$\theta_2 = \pi - \theta_1 \tag{5}$$

where θ_{pd} is angle of pulley relative to spherical center. θ_1 is the joint angle of the mechanism. θ_2 is the auxiliary angle of θ_1.

It can be known from Eqs. (2)–(5) that the total length of the steel wires l_{wire} is related to the angle θ_1. Figure 5 shows the variation of the total wire length l_{wire} with respect to θ_1.

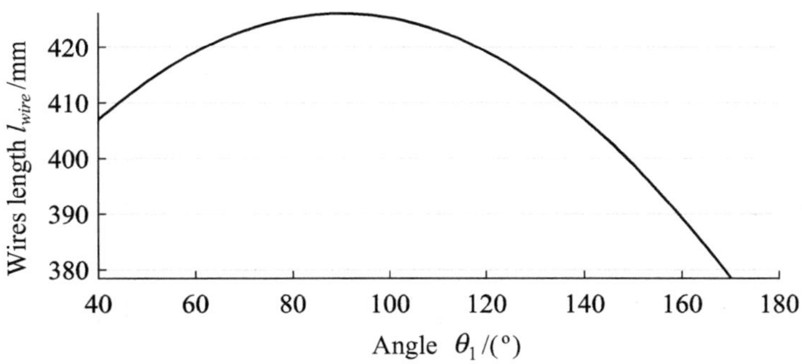

Fig. 5. Variation of the wire length for rotation angle θ_1

Thus, controlling the movement of the spherical scissors mechanism by pulling and stretching steel wires of the motors is feasible.

4 Kinematic Simulation

A model for simulation was developed in SolidWorks software as shown in Fig. 6, the range of motion for the mechanism θ_m is set to 0°–130°.

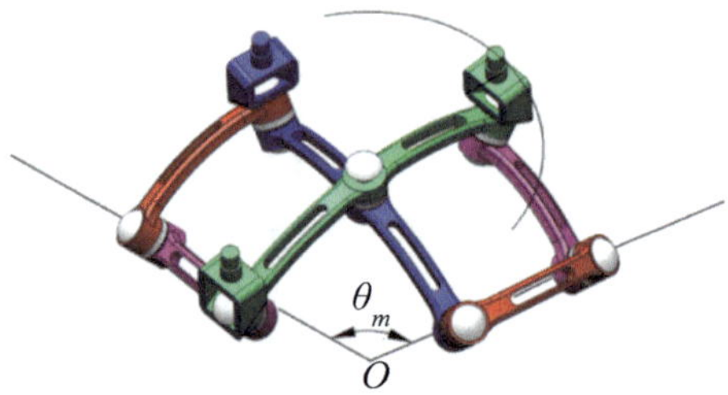

Fig. 6. The motion of the wire-driven mechanism

To verify the wires's change law obtained from Sect. 3, a simplified wire-driven mechanism model is moving by Adams, as shown in Fig. 7. Let the wire's end moving at a speed of 50 mm/s and apply payload of 0.1 N/mm for 0.5 s. The motion of the mechanism is displayed in Figs. 8 and 9.

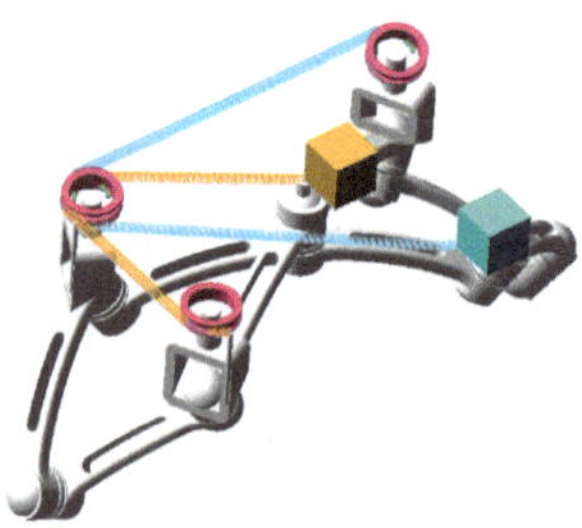

Fig. 7. The simplified mechanism model

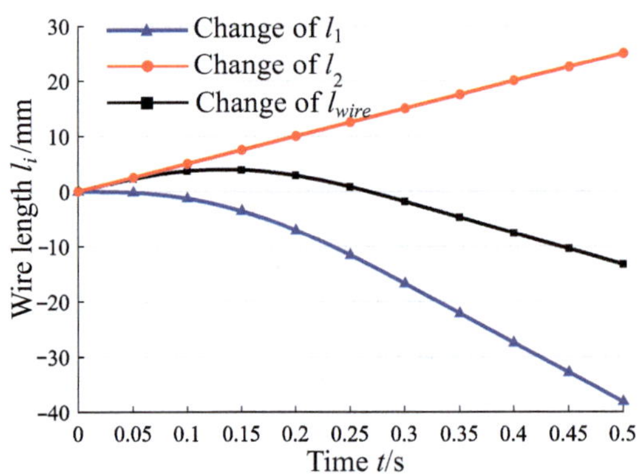

Fig. 8. Changes of the wire lengths

In Fig. 8, the changes of l_1 and l_2 indicate the retracted length and the pull-out length in the driving system, respectively. Thus, they are in the opposite directions. In the simulation, the wire length l_2 is pulled out at a uniform velocity, and the change trend of the wire length l_1 is shown as non-uniform retraction. In this process, the change trend of the total wire length l_{wire} first becomes longer and then shortens. The total wire length l_{wire} at the end is less than that at the beginning, which is in accordance with the change law calculated expressed in Eqs. (2)–(5).

Figure 9 shows the changes of the angular displacement, angular velocity of the angle θ_1 of the spherical scissors mechanism.

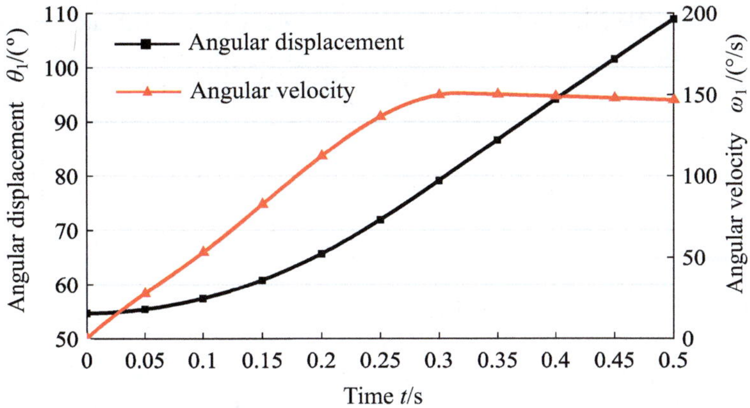

Fig. 9. Changes of the angular displacement and angular velocity of the angle θ_1 of the mechanism

5 The Stiffness Analysis of the Joint

Assuming the mechanism has enough stiffness and strength to be seen as rigid body. The mechanism's stiffness is mainly determined by wires.

Using energy conservation principle equation, we have

$$\tau_{joint}\delta\theta = T_{pull}\delta(l_1) + T_{loosen}\delta(l_2) \tag{6}$$

where τ_{joint} is the resultant torque. $\delta\theta$, $\delta(l_1)$ and $\delta(l_2)$ are variations of three variables.

$$\delta(l_1) = \left(R_{pd} \cdot \sin\frac{\theta_{pd}}{2} \cdot \cos\frac{\theta_1}{2} + \frac{r_p}{2}\right) \cdot \delta\theta \tag{7}$$

$$a = R_{pd} \cdot \sin\frac{\theta_{pd}}{2} \tag{8}$$

$$\delta(l_2) = \left(-R_{pd} \cdot \sin\frac{\theta_{pd}}{2} \cdot \sin\frac{\theta_1}{2} - \frac{r_p}{2}\right) \cdot \delta\theta \tag{9}$$

Applying infinitesimal torque δ_τ to the mechanism and the steel wires generate tensions T_{pull} and T_{loosen}, l_p is the steel wire's length in preloaded state.

$$T_{pull} = k_{wire}\left(l_p + \delta(l_1)\right) \tag{10}$$

$$T_{loosen} = k_{wire}\left(l_p + \delta(l_2)\right) \tag{11}$$

where k_{wire} is steel wire stiffness.

The joint stiffness k_{joint} of the wire-driven system can be calculated as follows:

$$k_{joint} = \frac{\delta\tau}{\delta\theta} = k_{wire} \cdot \left[\left(a \cdot \cos\frac{\theta_1}{2} + \frac{r_p}{2} \right)^2 + \left(a \cdot \sin\frac{\theta_1}{2} + \frac{r_p}{2} \right)^2 - l_p \cdot a \cdot \left(\sin\frac{\theta_1}{2} + \cos\frac{\theta_1}{2} \right) \right] \quad (12)$$

The joint stiffness k_{joint} can be used in wire selection, finding suitable pulley's radius, etc.

6 Structural Stiffness Evaluation

The structural stiffness of the wire-driven spherical scissors mechanism is analyzed using FEA software. The mechanism, made of aluminum alloy, is divided into 22941 nodes and 10118 elements and smallest gridding is 5 mm. A force of 800 N is applied to the moving pulley, which imitates the tension force applied by steel wires. The deformations of the holder are shown in Fig. 10. The highest deformation of three pulleys is found as 0.6 mm, which is acceptable.

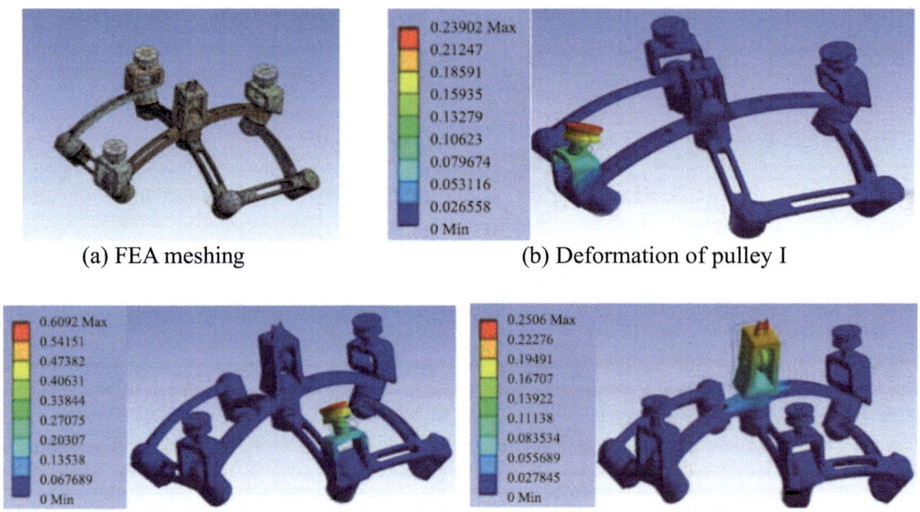

(a) FEA meshing (b) Deformation of pulley I

(c) Deformation of pulley II (d) Deformation of pulley III

Fig. 10. The deformation of the three pulleys

7 Shoulder Joint Rehabilitation Mechanism

Using wires-driven mechanism to drive human's shoulder makes exoskeleton light and flexible, which reduces the power requirement for actuators.

Users can wear exoskeleton with such a driving system under their clothes. Figure 11 shows a scheme of a shoulder rehabilitation device with double spherical scissors mechanisms. Four actuators are located in a back plate. The device delivers the shoulder joint through black brake hoses to help shoulder rehabilitation.

The mechanism is made of aluminum alloy AL6061. Wires are manufactured in high carbon steel, while the brake hose are made of organic polymer material. The total mass of the mechanism is only 0.462 kg. Compared with other mechanism used for shoulder rehabilitation of the same function, the wire-driven system has a very low mass and inertia.

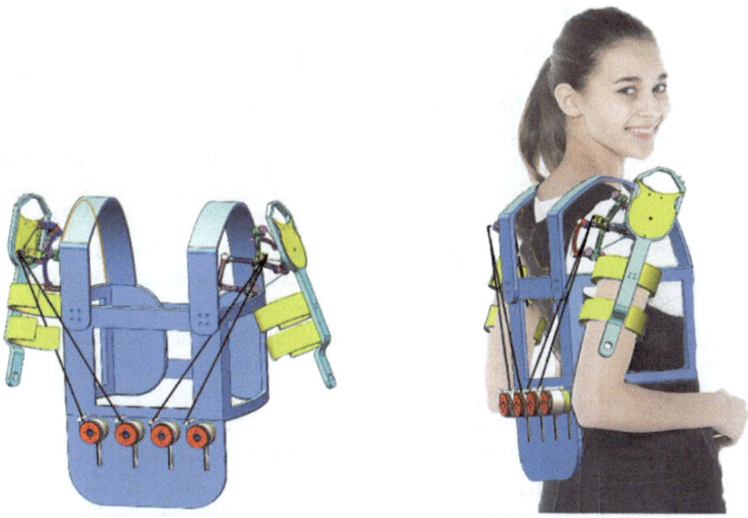

(a) The mechanism (b) Mounted on human

Fig. 11. Shoulder exoskeleton

8 Conclusions

A wire-driven spherical scissors mechanism is presented, which features of low mass and inertia. Kinematic models are developed to calculate the length change of steel wires for the motion control. The stiffness of the mechanism is verified by FEA method.

In future work, we will optimize the mechanism structure and build a prototype for testing and experiments.

Acknowledgements. This research was funded by the key research and development project of Shanxi province (Grant Nos. 201803D421027, 201803D421028), the foundation of Shanxi key laboratory of advanced manufacturing technology (Grant No. XJZZ201702).

References

1. Christensen, S., Bai, S.: Kinematic analysis and design of a novel shoulder exoskeleton using a double parallelogram linkage. ASME J. Mech. Robot. **10**, 041008_1–10 (2018)
2. Hunt, J., Lee, H., Artemiadis, P.: A novel shoulder exoskeleton robot using parallel actuation and a passive slip interface. ASME J. Mech. Robot. **9**, 011002_1–7 (2017)
3. Klein, J., Spencer, S., Allington, J., et al.: Optimization of a parallel shoulder mechanism to achieve a high-force, low-mass, robotic-arm exoskeleton. IEEE Trans Robot. **26**(4), 710–715 (2010)
4. Kumar, S., Wöhrle, H., Trampler, M., et al.: Modular design and decentralized control of the RECUPERA exoskeleton for stroke rehabilitation. Appl. Sci. **29**(626), 1–23 (2019)
5. Cui, X., Chen, W.H., Jin, X., et al.: Design of a 7-DOF cable-driven arm exoskeleton (CAREX-7) and a controller for dexterous motion training or assistance. IEEE/ASME Trans. Mechatron. **22**(1), 161–172 (2016)
6. Ball, S.J., Brown, I.E., Scott, S.H.: MEDARM: a rehabilitation robot with 5DOF at the shoulder complex. IEEE/ASME International Conference on Advanced Intelligent Mechatronics, Zurich, Swiss, 4–7 September 2007 (2007)
7. Salah, B., Janeh, O., Noche, B., et al.: Design and simulation based validation of the control architecture of a stacker crane based on an innovative wire-driven robot. Robot. Cim. Int. Manuf. **44**, 117–128 (2017)
8. Lee, C., Kim, J.Y., Kim, S.Y., et al.: Human force observation and assistance for lower limb rehabilitation using wire-driven series elastic actuator. Mechatronics **55**, 13–26 (2018)
9. Kim, Y.J.: Anthropomorphic low-inertia high-stiffness manipulator for high-speed safe interaction. IEEE Trans. Robot. **33**(6), 1358–1374 (2017)
10. Lee, J., Kim, Y.J., Roh, S., et al.: Tension propagation analysis of novel robotized surgical platform for transumbilical single-port access surgery. In: 2013 IEEE/RSJ International Conference on Intelligent Robots and Systems, Tokyo, Japan, 3–7 November 2013 (2013)
11. Dehez, B., Sapin, J.: ShouldeRO, an alignment-free two-DOF rehabilitation robot for the shoulder complex. In: 2011 IEEE International Conference on Rehabilitation Robotics, Rehab Week Zurich, ETH Zurich Science City, Switzerland, 29 June–1 July 2011 (2011)
12. Bai, S., Li, X., Angeles, J.: A review of spherical motion generation using either spherical parallel manipulators or spherical motors. Mech. Mach. Theory **140**, 377–388 (2019)
13. Lo, H.S., Xie, S.Q.: Optimization and analysis of a redundant 4R spherical wrist mechanism for a shoulder exoskeleton. Robotica **32**, 1191–1211 (2014)
14. Hsieh, H.C., Chien, L., Lan, C.C.: Dimensional synthesis of a lightweight shoulder exoskeleton. In: 2015 IEEE International Conference on Advanced Intelligent Mechatronics, Busan, Korea, 7–11 July 2015 (2015)
15. Castro, M.N., Rasmussen, J., Andersen, M.S., et al.: A compact 3-DOF shoulder mechanism constructed with scissors linkages for exoskeleton applications. Mech. Mach. Theory **132**, 264–278 (2019)
16. Lum, M.J.H., Rosen, J., Sinanan, M.N., et al.: Optimization of a spherical mechanism for a minimally invasive surgical robot: theoretical and experimental approaches. IEEE Trans. Bio-Med. Eng. **53**(7), 1440–1445 (2006)
17. Maden, F., Korkmaz, K., Akgün, Y.: A review of planar scissor structural mechanisms: geometric principles and design methods. Archit. Sci. Rev. **54**, 246–257 (2011)
18. Wu, Q.C., Wang, X.S., Chen, B., et al.: Patient-active control of a powered exoskeleton targeting upper limb rehabilitation training. Front. Neurol. **9**, 817_1–13 (2018)

Machinery Dynamics

Numerical Simulations of the Dynamic Pressure Oil-Gas Separator

Yutao Yan(✉), Yao Yu, Zhili Sun, and Xulei Su

School of Mechanical Engineering and Automation, Northeastern University, Shenyang 110819, People's Republic of China
ytyan@mail.neu.edu.cn

Abstract. The movement of fluid in oil-gas separator and its separation efficiency were investigated by the computational fluid dynamics method. The results show that the fluid velocity is decreasing along a helical trajectory and axis direction in the oil-gas separator. The central velocity is a backward motion tendency. The velocity vector moves from the bottom of the separator to overflow gate owing to the effects of the strongly inner swirling flow. The rotating center shape of the inner swirling flow is similar to the tornado. The rotation direction of the velocity vector is same as the outer swirling flow in central region. In the central region of the separator, there is the subatmospheric pressure which results in the inner swirling flow. The gas bubbles do downward helical movement, and gradually move the center of the separator, the last overflow the entire flow field from the inner fluid level. The gas bubbles diameter and inlet velocity have remarkable effects on separation efficiency. This study has revealed the flow mechanism of oil-gas separators.

Keywords: Oil-gas separator · Turbulence · Velocity field · Pressure field · Separation efficiency

1 Introduction

The dynamic pressure oil-gas separator is a device which is used to separate oil droplets from oil-gas two phase mixed fluid by centrifugal force. It has been widely applied in industry due to its simple structure, high separation efficiency, fast separation speed and low maintenance requirements. In a dynamic pressure separator, the swirling turbulence flow forces oil droplets towards the wall surface, separating oil droplets. Therefore, the flow field significantly influences the separating efficiency [1]. The separation process is an extremely complex three-dimensional strongly swirling and high strength turbulence two-phase flow, and its flow behavior is very complex in separators. So it is very difficult to predict internal flow field status by means of the experiments or analytical approach. With the rapid development of numerical calculation technique, the separator's internal flow field has been studied by numerical simulation methods [2, 3].

Mousavian et al. [4] compared three models, the Reynolds stress model without considering the air-core, the κ-ε model, and Reynolds stress turbulence model with the volume of fluid multiphase model for simulating the air-core in terms of their

D. Wang et al. (Eds.): MeTrApp 2019, MMS 79, pp. 203–212, 2020.
https://doi.org/10.1007/978-981-15-0142-5_21

predications of velocity, axial and tangential velocity distributions, and separation proportion. The results showed the Reynolds stress model with air-core simulation model clearly predicated the experimental data. Jawarneh et al. [5] simulated two-phase, strongly swirling flow in cylindrical separator with double vortex generators to predict separation efficiency by using the mixture-granular multiphase and the RNG κ-ε turbulence models, and found that the predications obtained with the RNG κ-ε model are reasonably good, the flow field is of a forced-vortex inside the core and a free-vortex outside the core. Yang et al. [6] studied the flow field and separation efficiency in a hydrocyclone by a three dimensional simulation, and computed the particle trajectories by integrating the force balance equations. The simulation results agreed favorably with experimental data. It was found that the unsteady RANS-based simulation that used a relatively coarse grid could provide reasonable and industrially relevant results with limited computational effort by comparing the finite volume RANS model with two LES approaches [7]. Sun et al. [8] studied the motion of oil droplet in a separator by numerical simulation, and the results showed that oil droplets' locus is mostly affected by the number of vanes, the rotating speed of impeller and the diameter of oil droplet. Xu et al. [9] simulated the effects of low-holdup solid particles on the bubbling behavior in a gas-liquid-solid flow system with single gas nozzle by discrete particle model and volume of fluid method, and numerically investigated the influences of the particles properties and fluid properties. The results showed that the simulations agree well with experiments.

This paper uses numerical simulation methods to analysis the velocity field, pressure field, flow trajectory and separation efficiency of the dynamic pressure oil-gas separator. The objective is to reveal the flow mechanism of the dynamic pressure oil-gas separator.

2 Mathematical Models

2.1 Turbulence Models

Governing Equation. In the dynamic pressure oil-gas separator, the internal fluid flow was considered to be three-dimensional, viscosity, steady, axis-symmetric and strong swirling turbulent flow. The fluid flow characteristics can be depicted by mass equation and momentum equation, which are the basis for the calculation and simulation [2, 6], the equations for mass and momentum in a general form are as follows:

Continuity equation:

$$\frac{\partial \rho}{\partial t} + \frac{\partial}{\partial x_{\mathrm{i}}}(\rho u_{\mathrm{i}}) = 0 \tag{1}$$

Momentum equation:

$$\frac{\partial}{\partial t}(\rho u_{\mathrm{i}}) + \frac{\partial}{\partial x_{\mathrm{j}}}(\rho u_{\mathrm{i}} u_{\mathrm{j}}) = -\frac{\partial p}{\partial x_{\mathrm{i}}} + \frac{\partial}{\partial x_{\mathrm{j}}}\left[\mu\left(\frac{\partial u_{\mathrm{i}}}{\partial x_{\mathrm{j}}} + \frac{\partial u_{\mathrm{j}}}{\partial x_{\mathrm{i}}}\right) - \rho\overline{u'_{\mathrm{i}} u'_{\mathrm{j}}}\right] + \rho g_{\mathrm{i}} \tag{2}$$

Where the subscripts i, j = 1, 2, 3 indicate the components in the Cartesian coordinate system, u_i, u_j are the fluid mean velocity component, x_i, x_j are the Cartesian coordinate component, ρ is the fluid density of the fluid, t is the time, p is the fluid mean pressure, μ is the fluid viscosity, g_i is the component of gravitational acceleration, u'_i, u'_j are the ith fluid fluctuating velocity component, and the $-\rho\overline{u'_i u'_j}$ is known as the Reynolds stress tensor which represents the effects of the turbulent fluctuations on the fluid flow [10].

RNG κ-ε Model. The RNG κ-ε turbulence model is derived from the instantaneous Navier-Stokes equations using a mathematical technique called ReNormalization Group (RNG) methods [10]. To high Reynolds number flows, the RNG κ-ε model is similar in form to the standard κ-ε model, but an additional terms and functions appear in the transport equations, and it has shown substantial improvements over the standard κ-ε model. So, the RNG κ-ε model includes the effects of swirl on turbulence and enhances the solution accuracy for swirling flows. The Reynolds stress tensor of the RNG κ-ε model based on Boussinesq hypothesis is expressed by

$$-\rho\overline{u'_i u'_j} = -\frac{2}{3}\rho\kappa\delta_{ij} + \mu_t\left(\frac{\partial u_i}{\partial x_j} + \frac{\partial u_j}{\partial x_i}\right) \tag{3}$$

Where κ is the turbulent kinetic energy, δ_{ij} is the Kronecker delta, and μ_t is the eddy viscosity. To high Reynolds number flows, the μ_t is given by

$$\mu_t = C_\mu \rho \kappa^2/\varepsilon \tag{4}$$

Where ε is the turbulent dissipation rate, and C_μ is the constant, i.e. C_μ = 0.0845 [10].

The turbulent kinetic energy κ and the dissipation rate ε in the RNG κ-ε model are modeled by the following two transport equations:

Kinetic energy equation:

$$\frac{\partial}{\partial t}(\rho\kappa) + \frac{\partial}{\partial x_i}(\rho u_i \kappa) = \frac{\partial}{\partial x_i}\left(\alpha_\kappa \mu_{eff}\frac{\partial \kappa}{\partial x_i}\right) + G_\kappa - \rho\varepsilon \tag{5}$$

Dissipation rate equation:

$$\frac{\partial}{\partial t}(\rho\varepsilon) + \frac{\partial}{\partial x_i}(\rho u_i \varepsilon) = \frac{\partial}{\partial x_i}\left(\alpha_\varepsilon \mu_{eff}\frac{\partial \varepsilon}{\partial x_i}\right) + C_{\varepsilon 1}\frac{\varepsilon}{k}G_\kappa - C_{\varepsilon 2}\frac{\varepsilon}{k} - R_\varepsilon \tag{6}$$

Where μ_{eff} is the effective viscosity defined as the summation of μ and μ_t, namely μ_{eff} = μ+μ_t, α_κ and α_ε are the inverse effective Prandtl numbers for κ and ε, respectively and for high Reynolds number their values are $\alpha_\kappa = \alpha_\varepsilon \approx 1.393$ [5], G_κ is the production of turbulent kinetic energy defined by

$$G_\kappa = 2\mu_t \bar{S}_{ij}\bar{S}_{ij} \tag{7}$$

Where the mean strain rate tensor $\bar{S}_{ij}$ is expressed as

$$\overline{S_{ij}} = \frac{1}{2}\left(\frac{\partial u_i}{\partial x_j} + \frac{\partial u_j}{\partial x_i}\right) \tag{8}$$

The term R_ε in Eq. (6) is the rate of strain which plays an important role in the anisotropy of the large scale eddies. In the RNG κ-ε model, the R_ε is given by

$$R_\varepsilon = \frac{C_\mu \rho \eta^3 (1 - \eta/\eta_0)\varepsilon^2}{1 + \beta\eta^3}\frac{\varepsilon^2}{\kappa} \tag{9}$$

Where

$$\eta = \frac{\kappa}{\varepsilon}\sqrt{2\overline{S_{ij}}\,\overline{S_{ij}}} \tag{10}$$

The values of the model constants which appear in the RNG theory are basically theoretically obtained and usually take the values $C_{\varepsilon 1}$ = 1.42, $C_{\varepsilon 2}$ = 1.68, β = 0.012, η_0 = 4.38 [10].

2.2 Multi-phase Flow Model

In order to trace the axis motion, radius motion and tangential motion of the gas droplet or bubbles in the oil-gas separator, the stochastic trajectory model is employed. The resistance force of gas bubbles withstood includes the mean flow field resistance force and fluctuation flow field resistance force. It was assumed that the effect of the motion of the gas bubbles on the flow and the gas bubbles/gas bubbles interactions are negligible in the gas and oil separator, due to the concentration of the gas bubbles is commonly very small (less than 10%). The gas bubbles were regarded as spheres owing to the bigger surface tension of gas phase. Then at the same time, the buoyant forces, inertial forces, gravitational forces, additional mass forces, Basset forces, Magness forces and Saffman forces were neglected due to their forces on the magnitude are much less than the drag forces in this study. So, the drag forces could become a domination mechanism [8].

The trajectory of a discrete phase particle was predicted by integrating force balance on the discrete phase without mutual interaction, which was written in a Lagrangian reference frame. This force balance equates the particle inertia with the forces acting on particle, and can be written (for the X direction in Cartesian coordinates) as:

$$\frac{du_p}{dt} = F_D(u - u_p) \tag{11}$$

Where u_p is the velocity of the discrete phase, u is the velocity of the continuous phase, $F_D(u\text{-}u_p)$ is the drag force per unit mass, where:

$$F_D = \frac{18\mu}{\rho_P d_p^2}\frac{C_D \mathrm{Re}}{24} \tag{12}$$

Here, ρ_p is the density of the discrete phase, d_p is the gas bubble diameter, μ is the viscosity of the continuous phase, Re is the Reynolds number, which is defined as:

$$\mathrm{Re} = \frac{\rho d_p |u_p - u|}{\mu} \tag{13}$$

The drag coefficient C_D can be obtained from:

$$C_D = a_1 + \frac{a_2}{\mathrm{Re}} + \frac{a_3}{\mathrm{Re}^2} \tag{14}$$

Where a_1, a_2, a_3 are constants that apply to smooth spherical particles over several ranges of Re given by Morsi and Alexander [11].

The kinetic trajectory equations of the discrete phase in the two-phase fluid flow are obtained by integrating the moment equation, and expressed (for the X direction in Cartesian coordinates) as:

$$x_p = \int u_p \, dt \tag{15}$$

The trajectory component for the Y and Z direction in Cartesian coordinates could be obtained by the same method, and the geometric locus of the discrete phase ultimately obtained.

3 Simulation Methods

3.1 Physical Model

The dynamic pressure oil-gas separator was investigated by numerical simulation method. The physical dimension of separator includes nominal diameter, height, inlet diameter, inlet radial position and overflow vent position, and their values are 100 mm, 200 mm, 20 mm, 20 mm and 30 mm, respectively. The physical model of the oil-gas separator was built by the commercial package code Pro/E, and the separator model was meshed with GAMBIT, which is the pre-processor of the commercial code FLUENT. The unstructured tetrahedral mesh was used to mesh the model for the numerical computation, as shown in Fig. 1.

Fig. 1. Mesh methodology.

3.2 Boundary Conditions

Before any progress could be made in the numerical computations, the boundary conditions which specify the fluid flow in the separator had to be addressed. According to actual operating conditions and the flow characteristics, the simulation boundary conditions were made. A constant inlet velocity at the inlet boundary surface and a constant pressure at the outlet boundary surface were used. The oil-gas mixed fluid at a constant velocity along cross section perpendicularly got into separator. At the outlet boundary, there is no information about the variables and some assumptions have to be made. The diffusion fluxes in the direction normal to the exit plane are assumed to be zero. At the solid walls, the no slip boundary condition was applied where the fluid velocities at the walls were specified to be zero, and that no the matter and energy interchanged. The centerline boundary was considered axis of symmetry.

3.3 Numerical Schemes

The gas bubbles phase was treated as a dilute phase owing to the volume fraction of the gas bubble less than 10%. The split-phase flow model was used to simulate two-phase flow of the dynamic pressure oil-gas separator. In this study, the semi-implicit pressure linked equations algorithm for pressure-velocity coupling technique was adopted to solve the discrete governing fluid flow equations over the computational domain. The second order upwind schemes were used for the turbulence kinetic energy, turbulence dissipation rate and volume fraction, and the momentum discretization method was quadratic upwind interpolation of convective kinematic scheme [10]. The discrete phase model technique was adopted to achieve the discrete phase gas bubble separation behavior inside the separator [4].

4 Results and Discussions

4.1 Velocity Field Analysis

The internal flow field velocity distribution about the plane $X = 0$ are shown in the Fig. 2. The fluid velocity is gradually decreases with the fluid velocity decreasing along axis direction, and a backward motion tendency at the centrality in Fig. 2(a). The tangential velocity distribution of the origination and middle section is shown in Fig. 2(b), it can be seen that the tangential velocity become bigger and bigger along radial direction, but suddenly decrease near the wall surface. The tangential velocity distribution has obviously axial symmetry and swirling flow characteristics. To dynamic pressure separator, the inlet velocity of the oil-gas mixture fluid has obviously influence to the velocity distributions.

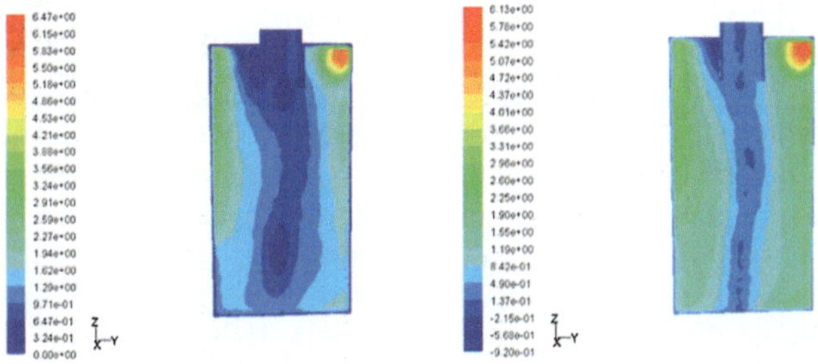

(a) Contours of velocity magnitude (b) Contours of the tangential velocity

Fig. 2. Velocity distribution on *Z-Y* plane ($X = 0$).

The velocity magnitude plot of the velocity vectors is presented in Fig. 3. The fluid velocity of inlet is maximum, and all decreasing along a helical trajectory and axis direction in the oil-gas separator. The velocity vector in $X = 0$ axial section is shown in Fig. 4. The arrowhead represents the direction of the velocity vector, and the length represents the size of the velocity vector. It can be seen that the velocity moves from the bottom of the separator to overflow gate owing to the effects of the strongly inner swirling flow, the rotating center shape of the inner swirling flow is similar to the tornado, the rotation direction of the velocity vector is same as the outer swirling flow in central region. The general trend is that the velocity vector near the internal face of separator is downward trend along rotation direction, but the velocity vector in the central region is upward trend and the rotation direction is similar to the outer swirling flow. The central velocity is backflow tendency, and the velocity vectors point to overflow port from the bottom of the separator, and the phenomenon is identical with the theory. The inner swirling flow leads to upward velocity vector, and is the major power of the gas bubble separated.

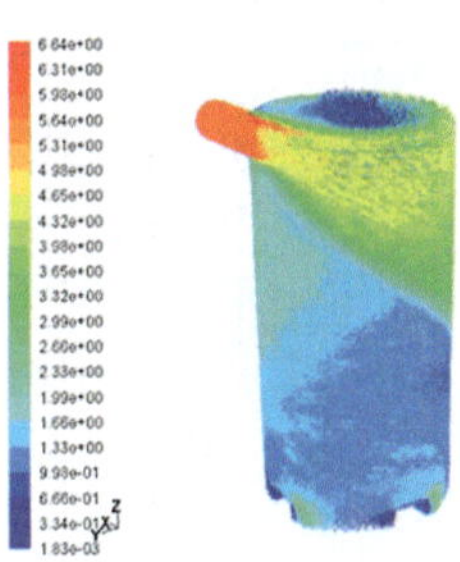

Fig. 3. Velocity vector magnitude.

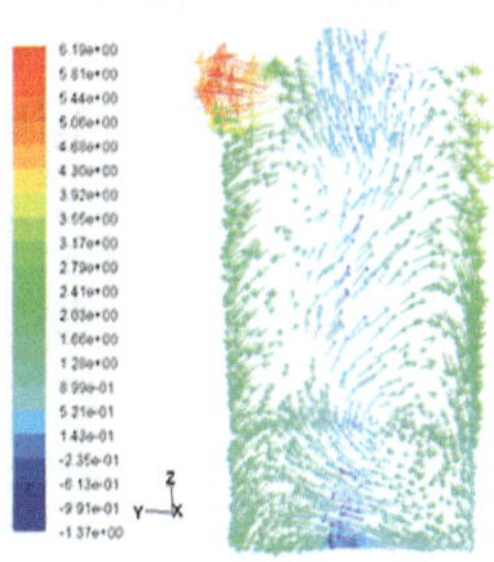

Fig. 4. Velocity vector on $X = 0$ axial section.

The axial velocity only provides the motive force from inlet to outlet, decides the detention time of fluid, and does not play a part to separating. The tangential velocity vectors distribution of the different cross section is shown in Fig. 5. It can be seen that the fluid swirling center is different at different height along Z axis, and gradually moves to separator's geometrical center. At $Z = 100$ cross section, the swirling center deviates separator's geometrical center, the tangential velocity gradually decreases, and is asymmetrical distribution from outside to inside in the radial direction in Fig. 5(a). At $Z = 50$ cross section, the swirling center substantially coincides with separator's geometrical center, the tangential velocity slowly decreases, and its distribution is relatively symmetrical from outside to inside in the radial direction in Fig. 5(b). These show that the separator's internal fluidic field is a strong complex swirling one. The swirling center has a certain curvature, and its shape is similar to tornado.

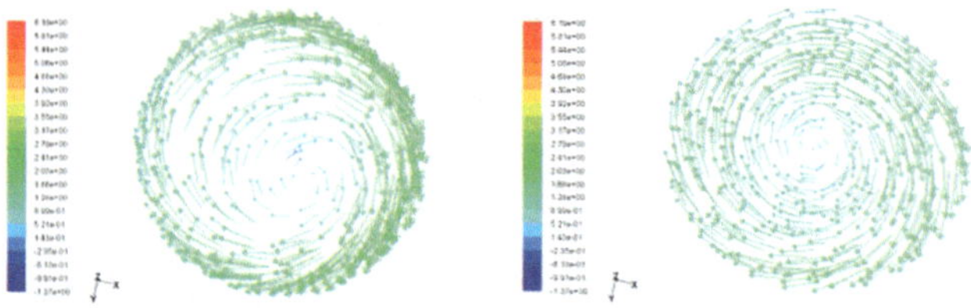

(a) Z=100mm (b) Z=50mm

Fig. 5. Tangential velocity vector distribution.

4.2 Pressure Field Analysis

The pressure field distribution is affected by fluid velocity and kinetic energy. The pressure field distribution is shown in Fig. 6. It can be seen that the maximum pressure is at the separator's inlet, and become gradually decreasing along axial direction, there is subatmospheric pressure at the bottom outlet in Fig. 6(a). Because the kinetic energy become gradually decreasing along axial direction. In Fig. 6(b), the pressure become increasing gradually along radial direction. The pressure depends on tangential velocity, while the tangential velocity also increasing gradually along radial direction. The subatmospheric pressure appears in the separator's central region and results to the inner swirling flow.

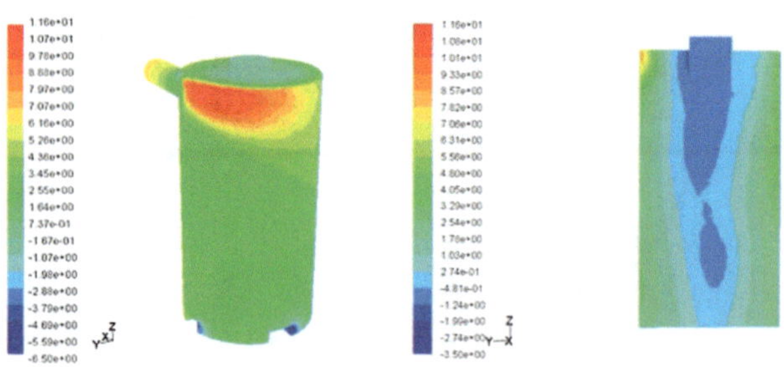

(a) Pressure distribution (b) X=0 axial section

Fig. 6. Static pressure distribution.

4.3 Trajectory Analysis

The gas bubbles trajectory in the separator is shown in Fig. 7. It can be seen that the gas bubbles do downward helical movement with fluid rotation along axial direction, gradually move toward the separator's center, and overflow the inner fluid level of the entire flow field. And then, the gas bubbles are separated from fluid, and flow by separator's center. Figure 8 shows the oil droplet trajectory in the separator. It can be seen that the oil droplet do downward helical movement with fluid rotation along axial direction, gradually move toward the separator's wall surface, and flow the entire flow field from bottom outlet. The influencing factors on separating property mainly includes inlet velocity and particle diameter.

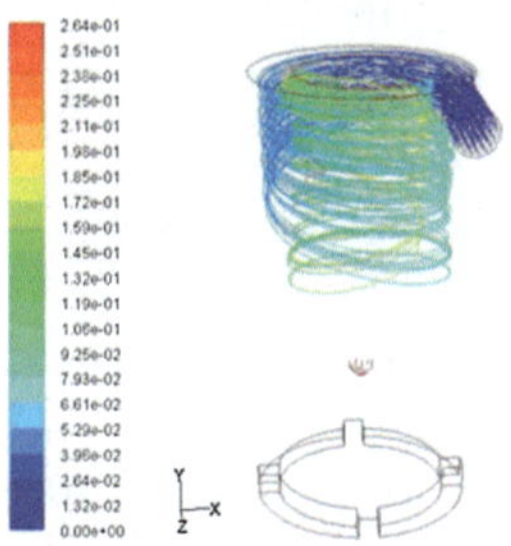

Fig. 7. Gas bubbles trajectory.

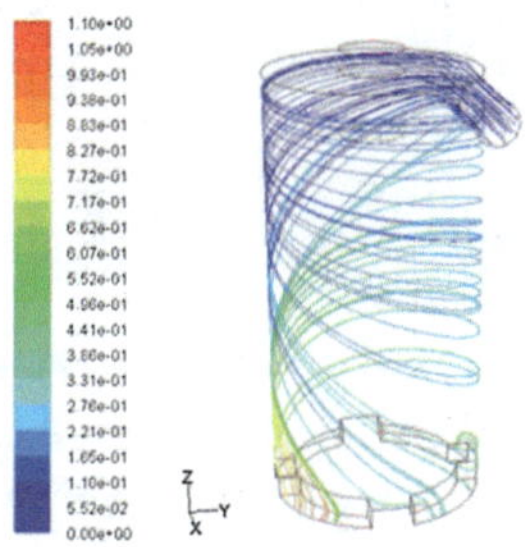

Fig. 8. Oil droplets trajectory.

4.4 Separation Efficiency Analysis

Figure 9 shows the numerically obtained separation efficiency curves with inlet velocity and gas bubbles diameter. To identical diameter gas bubbles, the separation efficiency increases with inlet velocity increasing. The separation efficiency also increases with gas bubbles diameter increasing. So, the gas bubbles diameter and inlet velocity have remarkable effects on separation efficiency.

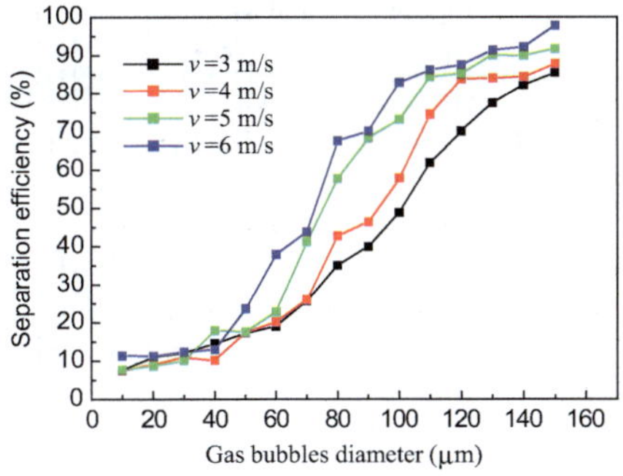

Fig. 9. Separation efficiency *vs.* inlet velocity and gas bubbles diameter.

5 Conclusions

The flow behavior of dynamic pressure oil-gas separator is analyzed by numerical simulation. The results indicated that the fluid velocity is all decreasing along axis direction, and central velocity is a backward motion tendency in the oil-gas separator. The rotating center shape of the inner swirling flow is similar to the tornado. The rotation direction of the velocity vector is same as the outer swirling flow in central region. There is the subatmospheric pressure which results in the inner swirling flow in the central region of the separator. The gas bubbles do downward helical movement, and gradually move the center of the separator. The gas bubbles diameter and inlet velocity have remarkable effects on separation efficiency. note that the first paragraph of a section or subsection is not indented. The first paragraphs that follows a table, figure, equation etc. does not have an indent, either.

References

1. Hoffmann, A.J., Stein, L.E.: Gas Cyclones and Swirl Channels: Principles, Design and Operation. Chemical Industry Press, Beijing (2004)
2. Gao, X., Chen, J.F., Feng, J.M., et al.: Numerical investigation of the effects of the central channel on the flow field in an oil-gas cyclone separator. Comput. Fluids **92**(3), 45–55 (2014)
3. Elsayed, K., Lacor, C.: Optimization of the cyclone separator geometry for minimum pressure drop using mathematical models CFD simulations. Chem. Eng. Sci. **65**(22), 6048–6058 (2010)
4. Mousavian, S.M., Najafi, A.F.: Numerical simulations of gas-liquid-solid flows in a hydrocyclone separator. Arch. Appl. Mech. **79**, 395–409 (2009)
5. Jawarneh, A.M., Tlilan, H., Al-Shyyab, A., et al.: Strongly swirling flows in a cylindrical separator. Miner. Eng. **21**(5), 366–372 (2008)
6. Yang, I.H., Shin, C.B., Kim, T.H., et al.: A three-dimensional simulation of a hydrocyclone for the sludge separation in water purifying plants and comparison with experimental data. Miner. Eng. **17**(5), 637–641 (2004)
7. Gronald, G., Derksen, J.J.: Simulating turbulent swirling flow in a gas cyclone: a comparison of various modeling approaches. Power Technol. **205**(1–3), 160–171 (2011)
8. Sun, X.J., Sun, H.O., Jiang, R.Q.: Numerical simulation of locus of oil droplet in oil-air separator. Appl. Sci. Technol. **33**(10), 69–72 (2006)
9. Xu, Y.G., Liu, M.Y., Tang, C.: Three-dimensional CFD-VOF-DPM simulations of effects of low-holdup particles on single-nozzle bubbling behavior in gas-liquid-solid systems. Chem. Eng. J. **222**(4), 292–306 (2013)
10. Ma, L., Ingham, D.B., Wen, X.: Numerical modelling of the fluid and particle penetration through small sampling cyclones. J. Aerosol Sci. **31**(9), 1097–1119 (2000)
11. Chen, X.L., Zhong, W.Q., Sun, B.B., et al.: Study on gas-solid flow in an obstructed pulmonary airway with transient flow based on CFD–DPM approach. Power Technol. **217**(2), 252–260 (2012)

A New Dynamic Model to Predict Vibration of Cylindrical Roller Bearings with Clearance and Off-Sized Roller Defects

Huimin Dong(✉), Kui Zhang, and Chu Zhang

School of Mechanical Engineering,
Dalian University of Technology, Dalian 116024, China
donghm@dlut.edu.cn

Abstract. In this paper, a new dynamic model to predict the vibration of cylindrical roller bearings with clearance and off-size roller defects is presented. The contact between rollers and rings is modeled as gap-activated non-linear spring-damper elements. The motion equations of a bearing are established by means of Newton's Second Law with consideration of the nonlinear contact forces. The solvable algebraic equations in time domain are derived by Newmark scheme. With a simple transformation, the equations can be converted into a set of linear complimentary equations. By employing Lemke algorithm, the accurate results can be obtained. Case studies, a 10-roller cylindrical bearing of having various clearance and single off-size roller defects, show that the proposed method is feasible and effective.

Keywords: Cylindrical roller bearing · Clearance · Off-size roller defects · Linear complimentary equation

1 Introduction

Roller bearings, as the important supporting elements of rotor system, are widely used in machines. The bearing dynamic performance significantly affects vibration and noise of the supported rotor system. An analysis of bearing dynamic behavior is essential to predict vibration responses of rotors and machines [1]. Bearing vibration always exists in operating, even it is free of defect [2]. It is impossible to manufacture perfect surface or contour, which means that radial clearance and off-sized roller defects inherently exist in a bearing. The vibration analysis is very difficult with taking account for nonlinear contact force and errors [3]. Some researchers had conducted about the radial clearance or off-size roller defects influence on bearing vibration. Wang developed a real engine rotor bearing model by combining rotor finite element model with bearing lumped mass model to research the effect radial clearance of ball bearing on the vibration of whole engine rotor [4]. It pointed out radial clearance severely influences the acceleration amplitude and frequency of rotor system and reducing the radical clearance can reduce rotor vibration and increase its stability. Aktürk established a ball bearing lumped parameter model and investigated the influence of off-size ball on bearing vibration [5]. It was found that the vibrations due to off-size balls are at cage

D. Wang et al. (Eds.): MeTrApp 2019, MMS 79, pp. 213–222, 2020.
https://doi.org/10.1007/978-981-15-0142-5_22

speed and its harmonic. Harsha [1] presented an analytical model with considering off-size ball to investigate the effect of ball elastic deformation on rotor bearing system. It points out that the highest radial vibrations due to ball elastic deformation are at the speed of the number of balls times the cage speed and the vibration due to single off-size roller occurs at the cage speed. Tong presented a 5-DOF tapered roller bearing model to study the vibration characteristics of tapered roller bearings with roller diameter error [6]. And the results show the roller diameter error significantly affect the vibration amplitude of bearing. Previous studies have proved radial clearance and off-size roller detects severely impact the vibration of bearing and rotor system. This paper deals with the dynamic modeling to predict the vibration response of cylindrical roller bearings focusing on the radial clearance and off-sized roller defects.

The rest of the paper is organized as follows. In Sect. 2, the dynamic model and solving method are introduced in detail. In Sect. 3, the impact of radial clearance and off-size roller defects on bearing vibration is investigated. Eventually, some conclusions are given in Sect. 4.

2 Model Description

A roller cylindrical bearing with clearance and single off-size roller, shown in Fig. 1, whose outer ring is supported by a housing unit. The inner ring is attached to a shaft through interference fit. A single external load is assumed to act at the shaft center. We assume that (1) the cross sections of the rollers, inner and outer raceways are perfectly round, (2) rollers only roll on the raceway without sliding, (3) the pressure distribution along the contact lines are uniform in the bearing axial direction, (4) rollers evenly distribute between rings, and (5) rollers distribute on the pitch circle when external load has not applied.

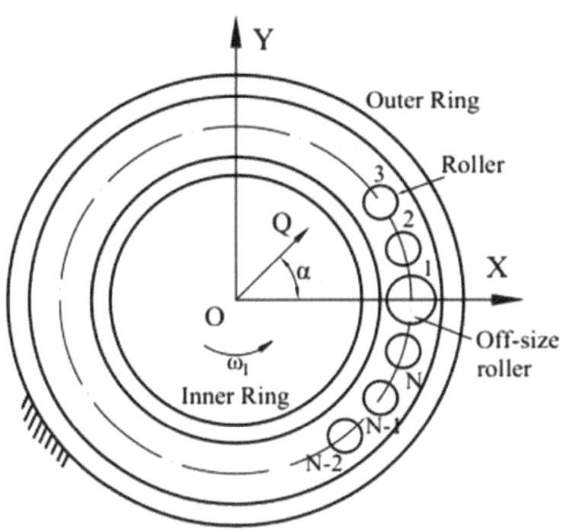

Fig. 1. An illustration of a cylindrical roller bearing with clearance and single off-size roller.

2.1 Contact Model of Between a Roller and Rings

According to above assumptions, each roller only vibrates in the bearing radial direction. And the contact between j-th roller and rings can be simply described as a 2D model shown in Fig. 2.

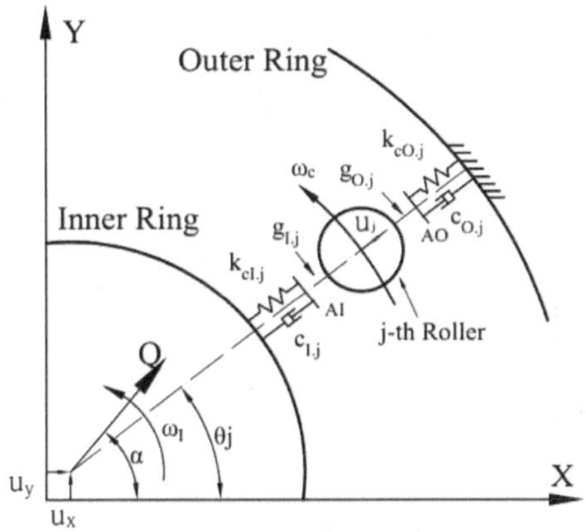

Fig. 2. An illustration of the contact between j-th roller and rings.

The contact between rollers and rings is modeled as gap-activated non-linear spring-damper elements as shown in Fig. 2. The deformation δ_{ij} of the equivalent spring-damper element between j-th roller and inner ring can be expressed as:

$$\delta_{ij} = u_x \cos\theta_j + u_y \sin\theta_j - u_{AI} \quad (1)$$

where u_x and u_y are the displacement of inner ring, u_{AI} is the displacement of AI.

The deformation δoj of the equivalent spring-damper element between j-th roller and outer ring can be expressed as:

$$\delta_{oj} = u_{AO} \quad (2)$$

where u_{AO} is the displacement of AO.

From the geometric consideration, the radial gap between the j-th roller and inner ring, $g_{i.j}$, and the radial gap between the j-th roller and outer race, $g_{o.j}$, are related to the displacement as follows:

$$g_{i.j} - g_{i.j0} = u_j - u_{AI}; g_{o.j} - g_{o.j0} = u_{AO} - u_j \quad (3)$$

where u_j is the radial displacement of j-th roller, $g_{i.j0}/g_{o.j0}$ is the initial gap between j-th roller and inner/outer ring.

Substituting the Eq. (3) into Eqs. (1 and 2), the equation of deformation coordination including gap can be derived as:

$$\delta_{ij} = u_x \cos\theta_j + u_y \sin\theta_j - u_j + g_{ij} - g_{ij0}; j = 1, \ldots, N \quad (4)$$

$$\delta_{oj} = u_j + g_{oj} - g_{ojo}; j = 1, \ldots, N \quad (5)$$

The contact force between rollers and rings can be determined by Hertz contact theory. The equation of contact force F can be written as:

$$F = k\delta^{\frac{10}{9}} + 1.5\alpha_e k \delta^{\frac{10}{9}} \dot{\delta} \quad (6)$$

where δ is the amount of the relative compression induced by load F, $\dot{\delta}$ is the differential of the relative compression, α_e = 0.08–0.32 s/m is restitution coefficient for steel [7], k Hertz stiffness coefficient [8].

$$k = \left[4.7965 \left(\frac{1 - \upsilon_r^2}{\pi E_r} + \frac{1 - \upsilon_{i/o}^2}{\pi E_{i/o}} \right)^{0.9} \frac{\left(1 \pm \kappa_{ri/ro}\right)^{0.1}}{l^{0.74} D_r^{0.1}} \right]^{-\frac{10}{9}}$$

where $\kappa_{ri} = D_r/D_i$; $\kappa_{ro} = D_r/D_o$; $E_r/E_i/E_o$ is the moduli of elasticity of roller/inner ring/outer ring; $v_r/v_i/v_o$/is Poisson's ratios of roller/inner ring/outer ring; $\pm$ is for contact between roller and inner/outer ring.

Considering the damping force is much less than the contact force, the non-linear contact force can be linearized as:

$$F = k_c\delta + c\dot{\delta} \quad (7)$$

where k_c/c is equivalent contact stiffness/damping $k_c = k^{0.9}F^{0.1}/c = 1.5\alpha_e F$, k_c and c can be determined by iteration scheme.

2.2 Motion Equation and Solution

With the determination of interaction force between roller and rings, the motion equations of bearing can be derived by Newton's Second Law.

(1) Dynamic differential equations of inner ring

$$\begin{aligned} m_i \ddot{u}_c &= Q_x - \sum_{j=1}^{N} F_{I.j} \cos\theta_j \\ m_i \ddot{u}_y &= Q_y - \sum_{j=1}^{N} F_{I.j} \sin\theta_j - F_{w.i} \end{aligned} \quad (8)$$

(2) Dynamic differential equations of inner roller

$$m_j \ddot{u}_j = F_{I.j} - F_{O.j} - F_{w.j} \sin\theta_j; j = 1, \ldots, N \quad (9)$$

(3) Contact forces between inner ring and rollers

$$k_{I.j}\delta_{I.j} + c_{I.j}\dot{\delta}_{I.j} = F_{I.j}; j = 1, \ldots, N \tag{10}$$

(4) Contact forces between outer ring and rollers

$$k_{O.j}\delta_{O.j} + c_{O.j}\dot{\delta}_{O.j} = F_{O.j}; j = 1, \ldots, N \tag{11}$$

where m_i/m_j is the mass of inner ring/j-th roller; $k_{I.j}/k_{O.j}$ is the equivalent stiffness between j-th roller and inner/outer ring; $c_{I.j}/c_{O.j}$ is the equivalent damping between j-th roller and inner/outer ring, $F_{w.i}/F_{w.j}$ is the gravity of inner ring/j-th roller.

Substituting the Eqs. (4, 5) into Eqs. (8–11), the followed matrix can be gained:

$$\boldsymbol{M\ddot{x}} + \boldsymbol{C\dot{x}} + \boldsymbol{Kx} = \boldsymbol{Q} + \boldsymbol{F} \tag{12}$$

where $\boldsymbol{x}$ is the displacement vector; $\boldsymbol{Q}$ is the generalized force vector; $\boldsymbol{F}$ is the interaction force vector.

By using Newmark scheme, solution equation of Eq. (12) in time domain can be obtained at each time step, as Eq. (13).

$$\bar{\boldsymbol{K}}\boldsymbol{x}_{k+1} = \bar{\boldsymbol{Q}}_{k+1} + \boldsymbol{F}_{k+1} \tag{13}$$

where:

$$\bar{\boldsymbol{K}} = \frac{1}{\beta dt^2}M + \frac{\gamma}{\beta dt}\boldsymbol{C} + \boldsymbol{K}$$

$$\begin{aligned}\bar{\boldsymbol{Q}}_{k+1} = \boldsymbol{Q}_{k+1} &+ \boldsymbol{M}\left(\frac{1}{\beta dt^2}\boldsymbol{x}_k + \frac{1}{\beta dt}\dot{\boldsymbol{x}}_k + \left(\frac{1}{2\beta} - 1\right)\ddot{\boldsymbol{x}}_k\right)\\ &+ \boldsymbol{C}\left(\frac{\gamma}{\beta dt}\boldsymbol{x}_k + \left(\frac{\gamma}{\beta} - 1\right)\dot{\boldsymbol{x}}_k + \left(\frac{\gamma}{2\beta} - 1\right)\ddot{\boldsymbol{x}}_k\right)\end{aligned}$$

where dt is time step; β and γ are the Newmark coefficient, $\gamma \geq 0.5$, $\beta \geq 0.25(0.5 + \gamma)^2$ $\gamma = 0.5$, $\beta = 0.25$, $dt = 10^{-5}$ in this article.

It is noted that Eq. (13) cannot be solved directly because $\boldsymbol{F}_{k+1}$ are unknown a priori. But when the Eq. (13) is partitioned like (14), we fortunately find contact force $\boldsymbol{F_r}$ and clearance $\boldsymbol{g}$ are complementary. Therefore, some mathematical transformation can be done to convert the Eq. (14) into a complementary equation about $\boldsymbol{F_r}$ and $\boldsymbol{g}$.

$$\begin{bmatrix} \bar{\boldsymbol{k}}_{11} & \bar{\boldsymbol{k}}_{12} \\ \bar{\boldsymbol{k}}_{21} & \bar{\boldsymbol{k}}_{22} \end{bmatrix} \begin{Bmatrix} \boldsymbol{u} \\ \boldsymbol{g} \end{Bmatrix}_{k+1} = \begin{Bmatrix} \bar{\boldsymbol{q}} \\ \bar{\boldsymbol{q}}_g \end{Bmatrix}_{k+1} + \begin{Bmatrix} \boldsymbol{0} \\ \boldsymbol{F}_r \end{Bmatrix}_{k+1} \tag{14}$$

It is obvious that the first set of equation does not contain the unknown contact forces $\boldsymbol{F_r}$. As a result, displacement vector $\boldsymbol{u}$ can be determined from the vector $\boldsymbol{g}$ using the following equation:

$$\boldsymbol{u}_{k+1} = \bar{\boldsymbol{k}}_{11}^{-1}(\bar{\boldsymbol{q}} - \bar{\boldsymbol{k}}_{12}\boldsymbol{g}_{k+1}) \tag{15}$$

It can be shown that $\bar{\boldsymbol{k}}_{11}^{-1}$ exists for $m_i/m_j > 0$, $j = 1, \ldots, N$.

Substituting the Eq. (15) into the second set of equations in Eq. (14), we obtain:

$$\boldsymbol{F}_{r.k+1} - \boldsymbol{K}^{*}\boldsymbol{g}_{k+1} = \boldsymbol{Q}^{*} \tag{16}$$

where:

$$\boldsymbol{K}^{*} = \bar{\boldsymbol{k}}_{22} - \bar{\boldsymbol{k}}_{21}\bar{\boldsymbol{k}}_{11}^{-1}\bar{\boldsymbol{k}}_{12};\ \boldsymbol{Q}^{*} = \bar{\boldsymbol{k}}_{21}\boldsymbol{k}_{11}^{-1}\bar{\boldsymbol{q}}_{k+1} - \boldsymbol{q}_{g,k+1}$$

The contact force vector $\boldsymbol{F}_{r.k+1}$, and the gap vector $\boldsymbol{g}$, satisfy the following non-negative and complementary conditions:

$$\boldsymbol{F}_{r.k+1} \geq \boldsymbol{0};\ \boldsymbol{g}_{k+1} \geq \boldsymbol{0};\ \boldsymbol{F}_{r.k+1}^{T} \bullet \boldsymbol{g}_{k+1} = \boldsymbol{0} \tag{17}$$

It can be easily verified that matrix $\boldsymbol{K}^{*}$ is symmetric and positive-definite. According to Bazaraa [9], a solution to Eq. (16), subjected to the conditions in Eq. (17), may be obtained using the Lemke algorithm.

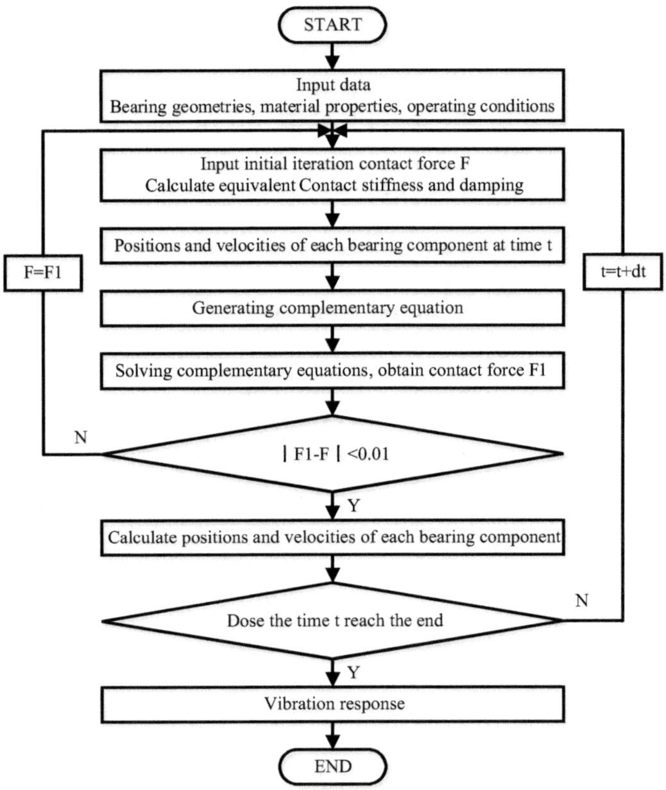

Fig. 3. Solution flow chart of bearing vibration.

It can be noted that Eq. (16) cannot be solved directly because matrix $\boldsymbol{K}^*$ contains terms that are dependent on the unknown contact forces. Newton-Raphson iterative scheme is introduced to handle the load-dependent contact stiffness. The initial contact stiffness between the rollers and the inner or the outer ring at each time step can be determined by the method reported by Yu [10]. Then the dynamic model mentioned above can be solved according to the process shown in Fig. 3.

3 Case Study

Clearance and off-size roller defect are inevitable in a real cylindrical roller bearing. And these errors will result in severe vibration, noise and even lead to the decline of bearing precision and premature failure. By investigating the effect of these errors on bearing vibration, bearing vibration can be controlled as much as possible in the design or assembly stage.

The cylindrical roller bearing dynamic model presented in the previous section is implemented into a computer procedure. A real cylindrical roller bearing with various clearance, off-size roller defects are simulated and the vibration spectrum is obtained by FFT. Then, three cases have been done to verify the validity of the proposed model and investigate the effect of clearance and off-size roller defects on cylindrical roller bearing vibration. It is assumed that bearing subject to a 5000 N ($Q = 5000$ N, $\alpha = 0$) load and the cage rotating frequency fc, and the roller passage frequency f are 150 Hz, 1500 Hz. The parameters of a cylindrical bearing are in Table 1.

Table 1. Parameters of a cylindrical bearing.

Terms	Parameter	Terms	Parameter
Diameter of inner ring	53 mm	Mass of inner ring	0.245 kg
Diameter of outer ring	83 mm	Poisson's ratio of rollers/rings	0.3
Roller diameter	15 mm	Elasticity modulus of rollers/rings	206 Gpa
Roller length	15 mm	Number of rollers	10

In case1, the vibration of bearing with none errors is simulated. The time vibration and its spectrum are shown in Figs. 4 and 5. They shows the vibrations of the bearing with none errors mainly occur at the roller passage frequency and its harmonic. This is completely consistent with Harsha's research.

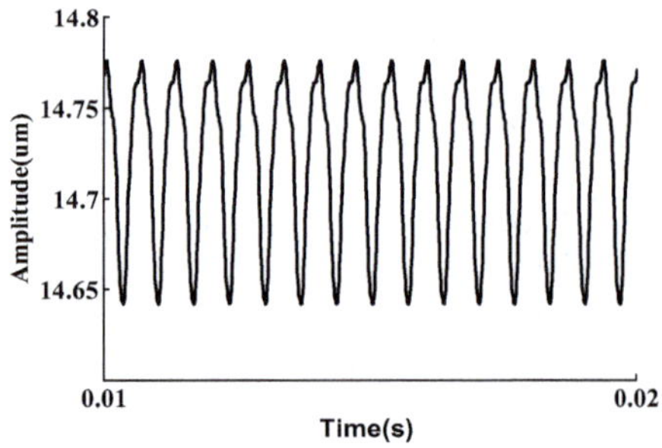

Fig. 4. The time vibration response of cylindrical roller bearing with none errors.

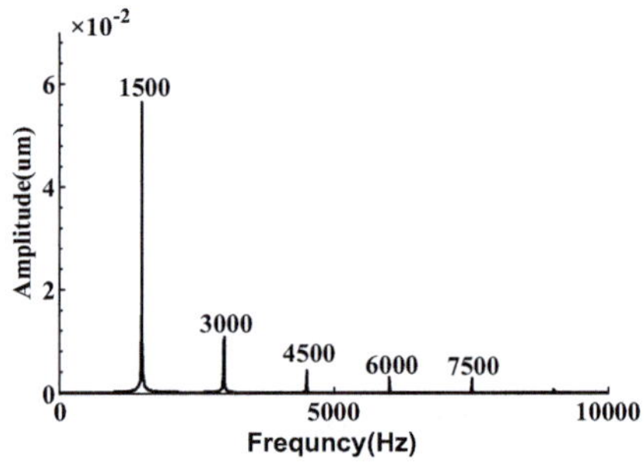

Fig. 5. The vibration spectrum of cylindrical roller bearing with none errors.

In case2, the vibrations of bearing with ±4 um, ±3 um, ±2 um, ±1 um, 0 um clearance have been simulated, and the results are shown in Fig. 4. It shows the variation of clearance mainly affects the vibration amplitude at f and 4f. With the clearance increasing, the vibration at 4f will replace it at f as the highest vibration. And small negative clearance can reduce the vibration (Fig. 7).

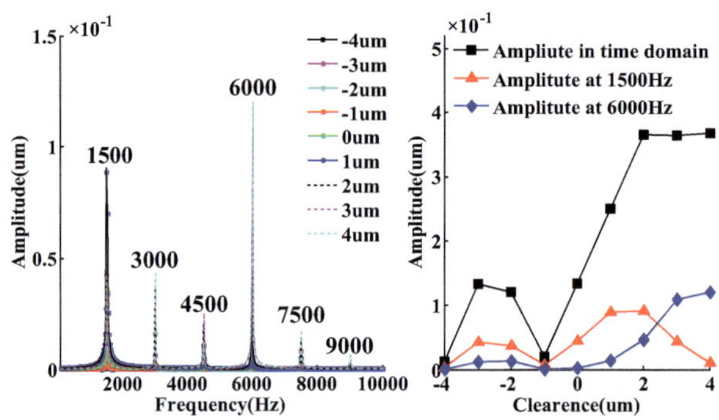

Fig. 6. Bearing vibration with different clearance.

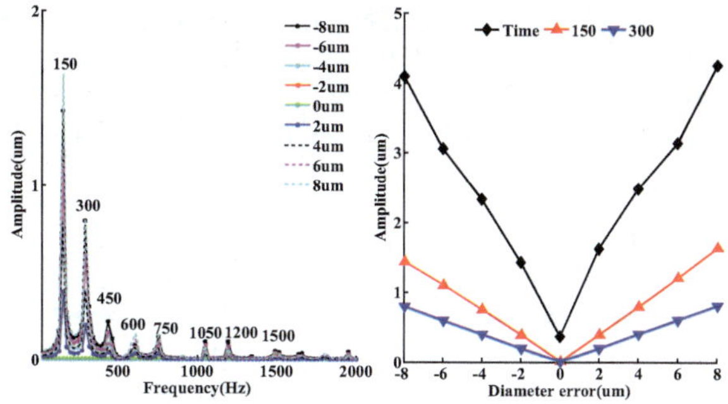

Fig. 7. Bearing vibration with different single off-size roller.

In case3, the vibrations of bearing with ±8 um, ±6 um, ±4 um, ±2 um, 0 um single off-size roller and +4 um clearance have been simulated. The results are shown in Fig. 6. It shows the vibration of bearing with clearance and single off-size roller mainly occur at the cage rotating frequency and its harmonic and the highest is at the frequency of fc. And when the rollers has any diameter error, the vibration amplitude of bearing will increase.

4 Conclusion

This paper develops a new lumped parameter model of cylindrical roller bearings to predict vibration with consideration of radial clearance and off-sized roller defects, wherein the complex nonlinear vibration problem is converted into a simple standard linear complementary problem. And the nonlinear vibration response of a cylindrical roller bearing can be solved efficiently by means of Lemke algorithm. With this model, the effect of radial clearance and off-size roller on the vibration of cylindrical roller bearing can be investigated. The case studies exhibited the effectiveness of this new model. The numerical results show that radial clearance mainly influences the vibration amplitude at roller pass frequency and its harmonic. And off-size roller defects influence the vibration amplitude at cage rotating frequency. It also can be found that off-size roller defects severely influence the bearing vibration amplitude. Thus, it is vital to control the diameter precision of the rollers.

Acknowledgements. This work was supported by the National Key R&D Program of China (No. 2018YFB2001504).

References

1. Harsha, S.: The effect of ball size variation on nonlinear vibrations associated with ball bearings. Proc. Inst. Mech. Eng. Part K J. Multi-Body Dyn. **218**(4), 191–210 (2004)
2. Meyer, L., Ahlgren, F., Weichbrodt, B.: An analytic model for ball bearing vibrations to predict vibration response to distributed defects. ASME **102**(2), 205–210 (1980)

3. Aktürk, N., Uneeb, M., Gohar, R.: The effects of number of balls and preload on vibrations associated with ball bearings. ASME **119**(4), 747–753 (1997)
4. Wang, H., Gong, J., Chen, G.: Characteristics analysis of aero-engine whole vibration response with rolling bearing radial clearance. J. Mech. Sci. Technol. **31**(5), 2129–2141 (2017)
5. Aktärk, N., Gohar, R.: The effect of ball size variation on vibrations associated with ball-bearings. Proc. Inst. Mech. Eng., Part J J. Eng. Tribol. **212**(2), 101–110 (1998)
6. Tong, V., Hong, S.: Vibration characteristics of tapered roller bearings with roller diameter error. MATEC Web Conf. **51**, 1004 (2016)
7. Jin, D., Hu, H.: Impact Vibration and Control. Science Press, Beijing (2005)
8. Harris, T., Kotzalas, M.: Advanced Concepts of Bearing Technology: Rolling Bearing Analysis, 5th edn. Taylor & Francis, Boca Raton (2007)
9. Bazaraa, M., Sherali, H., Shetty, C.: Nonlinear programming: theory and algorithms. Technimetrics **49**(1), 105 (1994)
10. Yu, S., Wang, D., Dong, H., Wang, B.: A new method for determining load distributions among rollers of bearing with manufacturing errors. Proc. Inst. Mech. Eng. Part C J. Mech. Eng. Sci. **227**(11), 2402–2415 (2013)

The Analysis for Stress Intensity Factor of Three-Dimensional Crack Under Combined Action of Friction Force and Tension

Yutao Yan(✉), Yibo Zhang, Xin Wu, and Wendong Li

School of Mechanical Engineering and Automation, Northeastern University, Shenyang 110819, People's Republic of China
ytyan@mail.neu.edu.cn

Abstract. The active system model was established under dry sliding friction in reciprocating motion, and the three-dimensional crack was established in the friction surface center of cuboid model. The characteristics of stress intensity factors were investigated by fracture mechanics and finite element method. The variation regulations of the stress intensity factors were obtained under different contact pressure, friction coefficient, ellipse shape ratio and ellipse parameter angle. The stress intensity factor K_{I} increases with increasing of contact pressure in away from crack, and decreases with increasing of contact pressure in near crack. The stress intensity factor K_{II} increases with increasing of contact pressure during a working cycle. During sliding block moving toward crack, the K_{I} decreases with increasing of friction coefficient. The K_{I} increases with increasing of friction coefficient during sliding block moving away from crack. The influences of friction coefficient on variation of K_{II} nearly coincide with contact pressure. The K_{I} and K_{II} increase with increasing of ellipse shape ratio. The K_{I} decreases, and then increases, the average value K_{II} decreases with increasing of ellipse parameter angle. The fatigue cracks are mainly mode I-II composite cracks. The squeeze or tear effects generated by combined action of friction force and tension have obvious influences on the K_{I}. The influences of the relative sliding between crack faces generated by contact pressure on the K_{II} are dominant.

Keywords: Stress intensity factor · Three-dimensional crack · Active system · Squeeze effect · Tear effect

1 Introduction

The active system is any mechanical system that transmits alternating workloads with simultaneous occurrence of the friction and wear process. The system exists widely in modern mechanical system, such as the shaft/hub with stationary or running fit, the crankpin/connecting rod with the sliding bearing, wheel/rail etc. [1, 2]. The valid evaluation on damage, durability and limiting states of the active system is impossible from the viewpoint of either the sole tribology or the exclusive mechanical fatigue [1]. The major research direction in the tribo-fatigue is influences of combined action of friction/wear damage and fatigue damage on surface crack. The stress intensity factors

D. Wang et al. (Eds.): MeTrApp 2019, MMS 79, pp. 223–232, 2020.
https://doi.org/10.1007/978-981-15-0142-5_23

reflect the strength of stress field at the crack tip. The common numerical analysis methods include weight function method [3, 4], energy release rate method [5], boundary collocation method [6] and finite element method [7, 8] etc.

A hybrid experimental-finite element method of the fatigue crack propagation rate was presented, the relation between fatigue crack growth rate and stress intensity factors was obtained by Yue [9]. The displacement field and stress intensity factors of three-dimensional cracks were determined by a hybrid analytic-numerical method by Ju [10]. Dündar et al. [11] assessed the variation of stress intensity factors and fatigue crack propagation characteristics of structures with multiple three-dimensional cracks by finite element method. He et al. [12] investigated the wear and rolling contact fatigue characteristics of the wheel/rail, the damage mechanism was to transform at different speed, the angle of fatigue crack propagation and the crack depth increased at low wear rate. The influences of tangential force on wear and rolling contact fatigue under normal, tangential and lateral force was explored, the wear loss of material increased with the tangential force increasing, the wear and fatigue damage behaviors changed with variation of the tangential force [13]. Abdelbary et al. [14] explored the influence of crack orientation on wear properties of polyamide under static and cyclic loads, indicated the wear rate change due to surface crack existence. Although these authors have done a lot of work, they may only consider the effect of one or more parameters on stress intensity factors, which leads to incomplete analysis.

The aim of this study is to investigate variation of stress intensity factors with working parameters and structure parameters based on fracture mechanics and finite element method under dry sliding friction. In the simulation analysis, the cuboid model containing the three-dimensional surface crack is stationary, while the sliding block is sliding in reciprocating motion. The results are to afford foundation for the further research of tribo-fatigue.

2 Models and Theoretical Background

The stress intensity factors of the three-dimensional surface crack were investigated by the finite element analyses under combined action of the friction and the tension. The sliding block was established with the size of 4 mm × 2 mm × 2 mm, and without mass. The cuboid model containing three-dimensional surface crack was established with the size of 10 mm × 10 mm × 3 mm, and the three-dimensional surface crack was established in the center of cuboid model with a = 1.5 mm in Fig. 1. The material is assumed to be linear elastic, the Young's modulus E is 2.06×10^5 MPa, and the Poisson's ratio υ is set to 0.3. A dry sliding friction pair was formed by the bottom face of the sliding block and the top face of cuboid model. The sliding block reciprocates, and its velocity is set to 1 mm/s. A cycle of reciprocating motion consists of two strokes. The contact pressure σ_P was applied to the top face of the sliding block, and the tension σ_F was applied to the right face of the cuboid model. The leftmost face of the cuboid model was fixed with all freedoms, while the bottom face was fixed in the x-direction but kept deformable in the y-direction. The surface to surface contact type with friction was used. The finite elements analyses were carried out with ANSYS code. The 1/2 model was applied in analyzing considering model symmetry and computational efficiency. The SOLID186 element type was chosen. A mesh refinement employed the three-dimensional wedge shape elements was applied across the crack front in order to deal with singularity of stress field and strain field, and improved the calculation accuracy in Fig. 2. The effects of contact pressure σ_P, friction coefficient f, ellipse shape ratio e ($e = b/a$) and ellipse parameter angle φ on stress intensity factors was analyzed. The examined values of various parameters and implication are reported in Table 1 and Fig. 1.

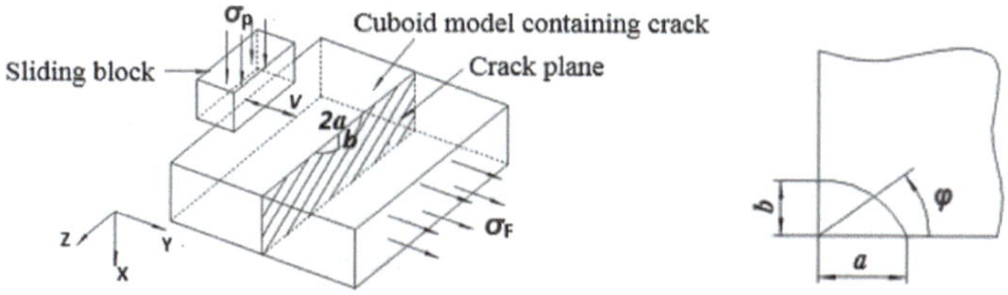

Fig. 1. Active system.

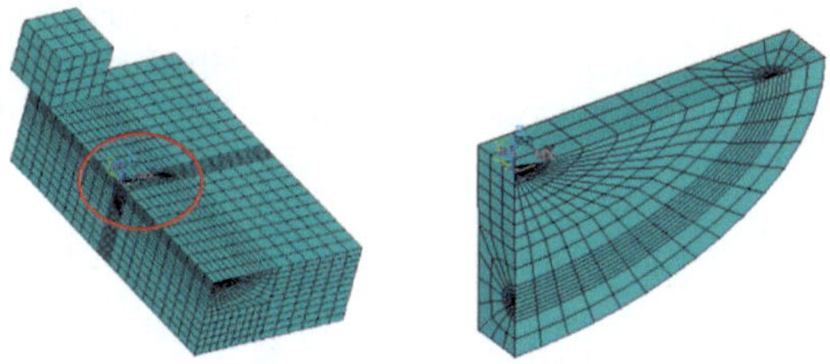

(a) Finite element model (b) Partial enlarged drawing

Fig. 2. Analysis model.

Table 1. Parameters setting.

Symbol	Parameter	Value		
σ_F/MPa	Tension	200		
σ_P/MPa	Contact pressure	100	200	300
f	Friction coefficient	0.10	0.15	0.20
φ/°	Ellipse parameter angle	30	60	90
e	Ellipse shape ratio	0.5	0.7	0.9

Compared with other methods, such as stress extrapolation method, the displacement method is a relatively stable method to calculate the stress intensity factors of the three-dimensional crack front. Therefore, the displacement method was applied to calculate the stress intensity factors of the three-dimensional crack front in finite element analysis. The singularity of the stress field and strain field for the three-dimensional crack front was described by using the evolutive 20-node or 12-node isoparametric elements [15]. The singularity of elements was obtained by moving the mid-side node to the quarter point of near the three-dimensional crack front. The stress intensity factors can be calculated by the crack front point and the quarter point of crack front [15, 16]. The displacement components of the crack front point is as follow

$$u = \frac{K_I}{4G}\sqrt{\frac{r}{2\pi}}\left[(2k-1)\cos\frac{\theta}{2} - \cos\frac{3\theta}{2}\right] - \frac{K_{II}}{4G}\sqrt{\frac{r}{2\pi}}\left[(2k+3)\sin\frac{\theta}{2} + \sin\frac{3\theta}{2}\right] \quad (1)$$

$$v = \frac{K_I}{4G}\sqrt{\frac{r}{2\pi}}\left[(2k-1)\sin\frac{\theta}{2} - \sin\frac{3\theta}{2}\right] - \frac{K_{II}}{4G}\sqrt{\frac{r}{2\pi}}\left[(2k+3)\cos\frac{\theta}{2} + \cos\frac{3\theta}{2}\right] \quad (2)$$

$$w = \frac{2K_{III}}{G}\sqrt{\frac{r}{2\pi}}\sin\frac{\theta}{2} \quad (3)$$

Where r is the distance from the crack front point. u, v and w is the displacement component in the x, y and z-directions, respectively. θ is the angular co-ordinate around crack front in Fig. 3. G is shear modulus. k is material parameter, $k = 3 - 4\upsilon$ in the plane strain state, and $k = (3 - \upsilon)/(1 + \upsilon)$ in the plane stress state. K_I, K_{II} and K_{III} is the stress intensity factors of mode I, mode II and mode III, respectively.

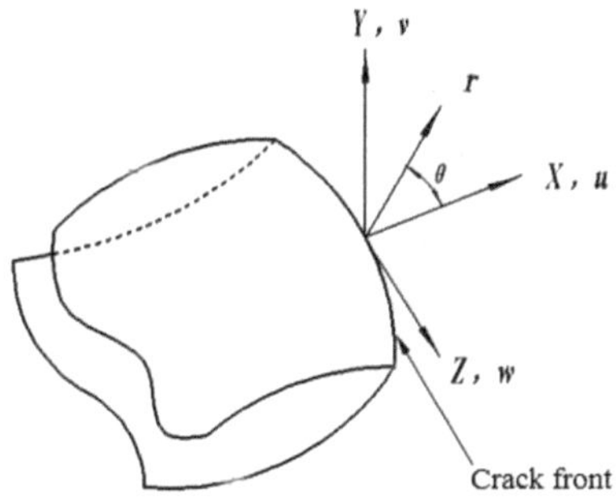

Fig. 3. Local coordinate of the crack front.

The stress intensity factors are respectively obtained by solving the displacement of the crack front point and the quarter point of crack front on the plane at $\theta = \pi$. Then, the stress intensity factors K_I, K_{II} and K_{III} can be described as

$$K_I = \frac{2G}{k+1}\sqrt{\frac{2\pi}{r}}(v_{(1/4)} - v_0) \quad (4)$$

$$K_{II} = \frac{2G}{k+1}\sqrt{\frac{2\pi}{r}}(u_{(1/4)} - u_0) \quad (5)$$

$$K_{\mathrm{III}} = \frac{2G}{k+1}\sqrt{\frac{2\pi}{r}}\left(w_{(1/4)} - w_0\right) \tag{6}$$

Where u_0 and $u_{(1/4)}$ are the displacement of the crack front point and the quarter point of crack front in the *x*-direction, respectively; v_0 and $v_{(1/4)}$ are the displacement of the crack front point and the quarter point of crack front in the *y*-direction, respectively; w_0 and $w_{(1/4)}$ are the displacement of the crack front point and the quarter point of crack front in the *z*-direction, respectively.

3 Results and Discussions

Figure 4 shows the variation of stress intensity factors with the contact pressure in one period for $f = 0.1$, $\varphi = 30°$, $e = 0.5$ and $\sigma_F = 200$ MPa. It can be seen from the Fig. 4a that the K_{I} decreases with increasing of contact pressure in near the crack zone ($3\ \mathrm{s} \le t \le 6\ \mathrm{s}$ or $13\ \mathrm{s} \le t \le 16\ \mathrm{s}$), and increases with increasing of contact pressure in away from the crack zone (in other times). The value of the K_{I} is the minimum, when the sliding block locates in directly above the three-dimensional crack in the stroke or the reverse stroke. This result is good agreement with the results in Reference [17]. Either the stroke or the reverse stroke, the friction force causes the crack closure while the sliding block moves to the crack, and causes the crack opening with the sliding block moves away from the crack. The squeeze effect generated by the friction force and contact pressure is obvious, the bigger the contact pressure and the friction force, the smaller the K_{I} in near the crack zone. The tear effect generated by friction force and tension is also obvious, the bigger the contact pressure and the friction force, the bigger the K_{I} in away from the crack zone. In a reciprocating cycle, the variation of the K_{I} is decreasing-increasing in the stroke, and is decreasing-increasing-decreasing in the reverse stroke due to the squeeze effect or tear effect generated by the friction force. The further away the sliding block from the crack, the less the influence of the friction force on K_{I}. The friction force and tension are in the same direction in the stroke, the crack closure or crack opening are enhanced due to the action of the friction force. However, the friction force and tension are in opposite direction in the reverse stroke, the crack closure or crack opening are reduced because of the action of friction force. It can be seen from the Fig. 4b that the K_{II} increases with increasing of contact pressure in the reciprocating period. The value of K_{II} for $\sigma_P =$ 300 MPa is improved 16% on average than that of $\sigma_P = 100$ MPa. The larger the contact pressure, the more the fluctuation of K_{II}. The influence of contact pressure on K_{II} is very obvious. The reason of the variation may come from the relative motion between the crack faces due to the contact pressure. The variation of K_{II} is increasing-decreasing- increasing in the stroke. In the reverse stroke, the variation of K_{II} is the same as that in the stroke, but the value of K_{II} is smaller. It can be analyzed that the combined action on the friction force and tension has also definite effect on K_{II}. In the stroke, the directions of the friction force and tension is the same, the friction force is positive effects on K_{II}. The directions of friction force and tension is opposite in the reverse stroke, and is negative effects on K_{II}. The minimal value of K_{II} is observed, while the sliding block locates in directly above the three-dimensional crack in the stroke or the reverse stroke. There is no relative motion between crack faces caused by contact pressure.

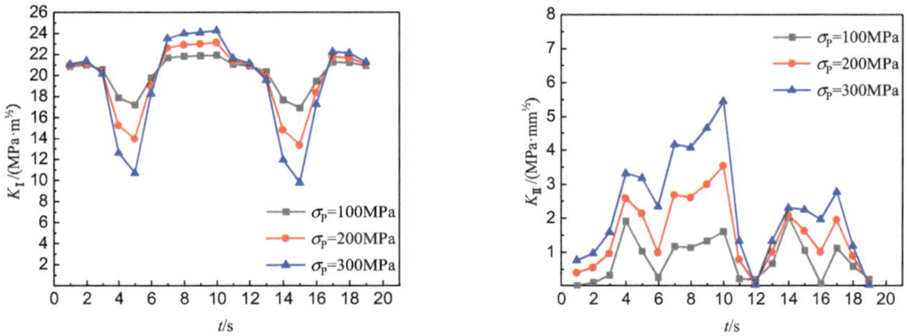

(a) Variation of K_I (b) Variation of K_{II}

Fig. 4. Variation of the stress intensity factors with the contact pressure ($f = 0.1$, $\varphi = 30^o$, $e = 0.5$, $\sigma_F = 200$ MPa).

Figure 5 shows the variation of stress intensity factors with the friction coefficient in one period for $\sigma_P = 200$ MPa, $\varphi = 30^o$, $e = 0.5$ and $\sigma_F = 200$ MPa. In the working period, the K_I decreases with increasing of the friction coefficient while the sliding block moves toward the crack. The K_I increases with increasing of the friction coefficient under the sliding block moving away from the crack. When the sliding block moves toward the crack, the friction force results in the crack closure due to its squeeze effect. The squeeze effect increases with increasing of the friction coefficient. However, the friction force causes the crack opening due to its tear effect, while the sliding block moves away from the crack. The tear effect increases with increasing of the friction coefficient. The Friction coefficient has great influence on the K_I. In a reciprocating cycle, either in the stroke or the reverse stroke, the value of K_I decreases, and then increases. The variation of K_I is much small at the stroke center. The variation of K_I in the reverse stroke is the same as that in the stroke. The variation gradient of K_I is very obvious in the central zone of the cuboid model. The effects of the friction coefficient on the variation of K_I are the same as that of the contact pressure. It can be seen from the Fig. 5b that the value of K_{II} increases with increasing of the friction coefficient. The K_{II} of $f = 0.2$ is two times as that of $f = 0.1$ on average. In the stroke, the variation of K_{II} is quick increasing, decreasing, and then increasing. The variation of K_{II} is quick increasing, decreasing, slight increasing, and then quick decreasing in the reverse stroke. The minimal value of K_{II} is observed in the beginning of the stroke or the reverse stroke. The reason may be the little influence of the contact pressure on the relative motion of crack faces due to sliding block far from the crack. The average value of K_{II} in the reverse stroke is smaller than that in the stroke. The combined action of friction force and tension also affects the value of K_{II}. The relative motion between the crack faces is affected by the crack closure or the crack opening.

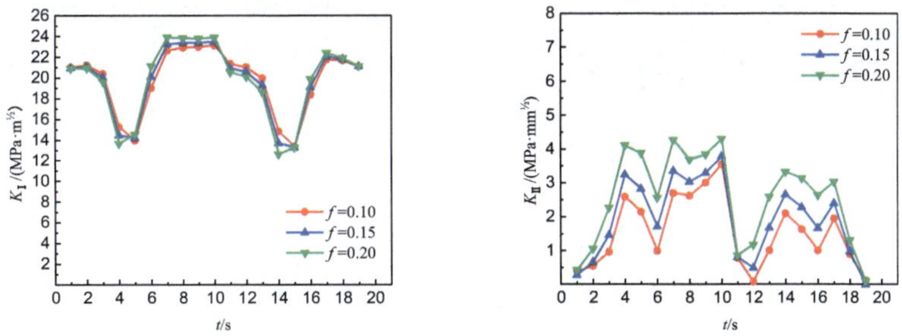

(a) Variation of K_{I} (b) Variation of K_{II}

Fig. 5. Variation of the stress intensity factors with friction coefficient (σ_P = 200 MPa, φ = 30°, e = 0.5, σ_F = 200 MPa).

The variation of stress intensity factors with ellipse shape ratio is shown in Fig. 6 for σ_P = 200 MPa, f = 0.1, φ = 30° and σ_F = 200 MPa. The K_{I} increases with increasing of ellipse shape ratio in Fig. 5a. The value of K_{I} for e = 0.7 is improved 16% on average than that of e = 0.5, with a maximum of 26%. The value of K_{I} for e = 0.9 is only improved 11% on average than that of e = 0.7, with a maximum of 23%. The increasing gradient of value of K_{I} is weakened with increasing of ellipse shape ratio. With increasing ellipse shape ratio, the crack depth is bigger, *i.e.* b is greater, under a maintaining constant. The crack depth has obvious effects on K_{I}. In a reciprocating cycle, the variation of the three curves nearly coincides with each other (see Fig. 6a). During the sliding block moving toward the crack, the value of K_{I} gradually decreases because of the squeeze effect of the friction force. The squeeze effect causes the crack closure. The effect of tension on K_{I} is reduced due to the friction force. On the contrary, the value of K_{I} gradually increases due to the tear effect of the friction force during the sliding block moving away from the crack. The tear effect causes the crack opening. The effect of tension on K_{I} is enhanced due to the friction force. The value of K_{I} is the minimum, when the sliding block locates in center of the cuboid model, namely directly above the crack. At this position, the crack is in a closed state. The tension is dominant to K_{I}, the effect of the friction force and the contact pressure is negligibly small. It can be seen from the Fig. 6b that the K_{II} increases with increasing of ellipse shape ratio. The reason of the variation may come from the enlarging of the crack depth and the enlarging of sliding effect between the crack faces. The value of K_{II} for e = 0.7 is improved 68% on average than that of e = 0.5. The value of K_{II} for e = 0.9 is improved 35% on average than that of e = 0.7. The fluctuations of K_{II} are drastic and there is a distinct enhancement of K_{II} at t = 4 s, 6 s, 14 s, 16 s. The variation of K_{II} nearly coincides with each other in the stroke or in the reverse stroke. But the value of K_{II} is greater in the stroke than that in the reverse stroke. The reason may come from the influence of the combined action of the friction force and the contact pressure on the relative motion between the crack faces.

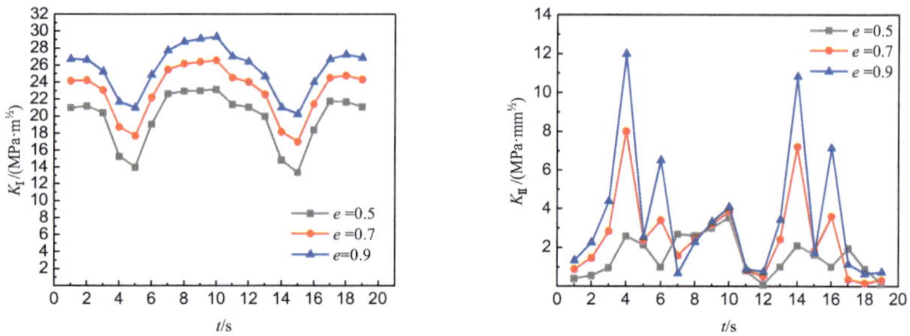

(a) Variation of K_{I} (b) Variation of K_{II}

Fig. 6. Variation of the stress intensity factors with ellipse shape ratio (σ_{P} = 200 MPa, f = 0.1, φ = 30°, σ_{F} = 200 MPa).

Figure 7 shows the variation of the stress intensity factors with the ellipse parameter angle. It can be seen from the Fig. 7a that the K_{I} slightly decreases and then increases. The average value of K_{I} for φ = 0° is nearly coincident with that of φ = 30°, and then the variation of K_{I} is only 5% on average from φ = 60° to φ = 90°. But the value of K_{I} for φ = 30° is improved 15% than that of φ = 60°. The K_{I} of different positions of the crack front is different. The reason of variation of K_{I} may be from the different crack depth due to the different position of the crack front [18]. The variation of K_{I} is decreasing, and then increasing in the stroke or the reverse stroke. The reason of variation comes from the squeeze effect or tear effect caused by the friction force. The variation of K_{II} can be seen from the Fig. 7b. The K_{II} decreases with increasing of ellipse parameter angle. During φ = 30°, 60° and 90°, the variation of the three curves nearly coincides with each other. The fluctuations of K_{II} are drastic at φ = 0°. Especially, at t = 4 s, 6 s, 14 s and 16 s, the value of K_{II} is extremely high. The ellipse parameter angle reflects the position of the crack front, and indicates the crack depth. The reasons of variation of stress intensity factor may come from the influences of all loads on crack depth.

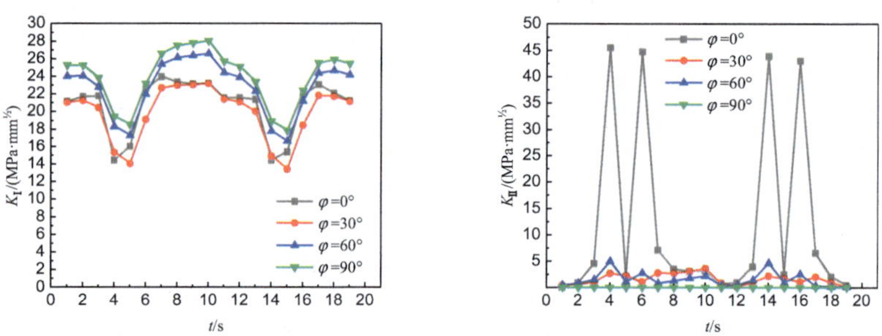

(a) Variation of K_{I} (b) Variation of K_{II}

Fig. 7. Variation of the stress intensity factors with ellipse parameter angle (σ_{P} = 200 MPa, f = 0.1, e = 0.5, σ_{F} = 200 MPa).

In this study, the combined action of the friction force and tension causes crack to closure or opening, and the contact pressure results in the relative movement between the crack faces. The fatigue crack is I-II mixed mode crack. The squeeze effect or tear effect results in the variation of the stress intensity factors. The plastic slippage and plastic deformation regions is formed due to the combined action of all loads. The passivation of the crack front appears. There is a residual stress field in the crack front zone after the sliding block moves across crack region in the stroke, which accelerates crack opening. In the reverse stroke, the hysteresis effect of crack opening affects the variation of stress intensity factors.

4 Conclusions

(1) The K_{I} increases with increasing of contact pressure in faraway from crack zone, and decreases with increasing of contact pressure in near the crack zone. The value of K_{II} increases with increasing of contact pressure. The K_{I} decreases with increasing of friction coefficient during t = 1–5 s and t = 11–15 s due to the squeeze effect, and increases with increasing of friction coefficient during t = 6–10 s and t = 16–19 s due to the tear effect. The K_{II} increases with increasing of friction coefficient. With increasing of ellipse shape ratio, the K_{I} and K_{II} all increase. The K_{I} slightly decreases, and then increases with increasing of ellipse parameter angle. The K_{II} decreases with increasing of ellipse parameter angle. The influences of all loads on K_{I} and K_{II} are different.

(2) In this study, the fatigue crack is mode I-II composite crack. The squeeze effect caused by friction force results in the decreasing of K_{I} during sliding block moving toward crack, and then the tear effect caused by friction force results in the increasing of K_{I}. The influences of contact pressure and crack depth on K_{II} are dominant. The plastic slippage and plastic deformation appear in the crack front, and result in crack passivation. The existence of residual stress and hysteresis effect of crack affect the variation of the stress intensity factors.

Acknowledgement. This project was supported by the National Natural Science Foundation of China (No. 51875095).

References

1. Sosnovskiy, L.A.: Tribo-Fatigue: Wear-Fatigue Damage and Its Prediction. Springer, Berlin (2005)
2. Sherbakov, S.S., Zhuravkov, M.A.: Interaction of several bodies as applied to solving tribo-fatigue problems. Acta Mech. **224**(7), 1541–1553 (2013)
3. Kim, J.H., Lee, S.B.: Prediction of crack opening stress for part-through cracks and its verification using a modified strip-yield model. Eng. Fract. Mech. **66**(1), 1–14 (2000)
4. Benuzzi, D., Bormetti, E., Donzella, G.: Stress intensity factor range and propagation mode of surface cracks under rolling-sliding contact. Theor. Appl. Fract. Mech. **40**(1), 55–74 (2003)

5. Zambelis, G., Botelho, T.D.S., Klinkova, O., et al.: Evaluation of the energy release rate in mode I of asymmetrical bonded composite/metal assembly. Eng. Fract. Mech. **190**, 175–185 (2018)
6. Mohammad, F.M., Dehghani, I.: Kinked crack analysis by a hybridized boundary element/boundary collocation method. Int. J. Solids Struct. **47**(7–8), 922–933 (2010)
7. Dai, D.N., Hills, D.A., Nowell, D.: Modelling of growth of three-dimensional cracks by a continuous distribution of dislocation loops. Comput. Mech. **19**(6), 538–544 (1997)
8. Zhang, Y.M., Yi, D.K., Xiao, Z.M., et al.: Elastic-plastic fracture analyses for pipeline girth welds with 3D semi-elliptical surface cracks subjected to large plastic bending. Int. J. Press. Vessels Pip. **105–106**, 90–102 (2013)
9. Yue, J.X., Dong, Y., Soares, C.G.: An experimental-finite element method based on beach marks to determine fatigue crack growth rate in thick plates with varying stress state. Eng. Fract. Mech. **196**(5), 123–141 (2018)
10. Ju, S.H., Hsu, H.H.: A hybrid analytic-numerical method for three-dimensional cracks. Theor. Appl. Fract. Mech. **85**, 262–275 (2016)
11. Dündar, H., Ayhan, A.O.: Three-dimensional fracture and fatigue crack propagation analysis in structures with multiple cracks. Comput. Struct. **158**, 259–273 (2015)
12. He, C.G., Guo, J., Liu, Q.Y., et al.: Experimental investigation on the effect of operating speeds on wear and rolling contact fatigue damage of wheel materials. Wear **364–365**, 257–269 (2016)
13. He, C.G., Huang, Y.B., Ma, L., et al.: Experimental investigation on the effect of tangential force on wear and rolling contact fatigue behaviors of wheel material. Tribol. Int. **92**, 307–316 (2015)
14. Abdelbary, A., Abouelwafa, M.N., Fahham, I.M.E., et al.: The influence of surface crack on the wear behavior of polyamide 66 under dry sliding condition. Wear **271**(9–10), 2234–2241 (2011)
15. Barsoum, R.S.: On the use of isoparametric finite elements in linear fracture mechanics. Int. J. Numer. Methods Eng. **10**(1), 25–37 (1976)
16. Henshell, R.D., Shaw, K.G.: Crack tip finite elements are unnecessary. Int. J. Numer. Method Eng. **9**, 495–507 (1975)
17. Zhou, Z.Y., Ji, B.H., Fu, Z.Q., et al.: Fatigue crack retrofitting by closing crack surface. Int. J. Fatigue **119**, 229–237 (2019)
18. Mirahmadi, H., Azimi, M., Mirjavadi, S.S.: Calculation of stress intensity factor for functionally graded cylinders with two radial cracks using the weight function method. Theor. Appl. Fract. Mech. **85**, 447–456 (2016)

A Directly Coupled Finite Element – Meshless Method with Varying Dilation Parameter

Yaqing Zhang, Wenjie Ge(✉), Ziang Zhang, Xiaojuan Mo, and Yonghong Zhang

Northwestern Polytechnical University, Xi'an 710072, China
gwj@nwpu.edu.cn

Abstract. Compliant mechanisms are flexible mechanisms that can transfer input forces or displacements through elastic body deformation. Thus, large deformation of structural analysis is necessary, especially for the topology optimization-based design. The meshless method has more advantages in simulating large deformation structural analysis, but the calculation is more expensive than the finite element method. Thus, a directly coupled method of finite element (FE) method and element-free Galerkin (EFG) method is presented in this study. A smooth cosine weight function is proposed to establish continuous shape functions in the whole design domain. To retain the feature of EFG method, a continuous varying dilation parameter is developed in the coupling domain to connect FE and EFG domains. Several benchmark problems are given to demonstrate the effectiveness of the proposed approach. Comparing with some other coupling methods, the proposal is quite succinct and tractable, shows higher precision or cost less computation comparing with other methods.

Keywords: FE method · EFG method · Directly coupled method · Smooth cosine weight function · Varying dilation parameter

1 Introduction

The FE method is quite sensitive to the element distortion and mesh entanglement, and often encounters difficulties in mesh rezoning progress. According to the study of Sandia National Laboratories, the geometry decomposition, meshing and mesh manipulation account for 52% of the overall analysis time [1]. As an alternative, meshless methods implement numerical simulation by a set of arbitrarily scattered field nodes. Without requiring any additional efforts to remain mesh connectivity, meshless methods make the description and refinement of the intricate design easily. However, the meshless methods cannot compare favorably with the FE method in many aspects, such as the computational efficiency and the exerting of boundary conditions.

Coupling the FE method and the meshless method together can be a complementary approach. The meshless method can be employed in the region where elements distortion or cracks region exists. And in other regions, remain the FE method to reduce the computation. Besides, applying the FE method on the displacement boundary can also circumvent the trouble of imposing displacement boundary conditions in many meshless methods. Several coupling methods have been proposed, such

D. Wang et al. (Eds.): MeTrApp 2019, MMS 79, pp. 233–243, 2020.
https://doi.org/10.1007/978-981-15-0142-5_24

as the interface elements-based coupling method [2, 3], the extended weak forms-based coupling method [4, 5], the continuous blending method [6, 7] and so on.

In the interface elements-based coupling method, a hybrid displacement approximation is adopted in the interface between FE and EFG domains. But the high-order compatibility can hardly be satisfied on interface elements boundary. A smooth blending function [8] is employed to underplay the discontinuity on the interface boundary. However, the improved method can only guarantee continuous derivatives along the meshless boundary. In the extended weak forms-based coupling methods, additional displacement constraints are introduced to connect the two domains by the Lagrange multiplier in weak forms and the same high-order compatibility problem also exists. A transitional region is introduced to obtain the higher-order compatibility by adding transitional nodes [5], which means more calculation of the external stiffness matrix the transitional region. In the continuous blending method, its polynomial basis is constructed for the mixed approximations in the interface region. In this method, the weight function is chopped off, which would lead the shape function hierarchical. The discontinuous shape functions in interface regions encumber the application of the coupled method in many situations where highly depends on the sensitivity analysis, such as the topology optimization [9] and the shape optimization [10].

In this study, a directly coupled FE-EFG method is proposed. In Sect. 2, the idea of the directly coupled method is put forward and a smooth cosine weight function is proposed. A varying dilation parameter is introduced in the coupling domain in Sect. 4. In Sect. 5, two examples are present to illustrate the feasibility of this approach.

2 Directly Coupled FE-EFG Method

2.1 Imitating FE Method by EFG Method

The main difference between FE and meshless methods is the approximate stratagem of the two methods. Namely, their shape functions are constructed in different ways. For both meshless and FE methods, a local approximate solution of $u_i(\mathbf{x})$ can be expressed by

$$u_i^h(\mathbf{x}, \bar{\mathbf{x}}) = \sum\nolimits_{j=1}^{m} p_j(\mathbf{x}) a_j(\bar{\mathbf{x}}) = \boldsymbol{p}^{\mathrm{T}}(\mathbf{x}) \boldsymbol{a}(\bar{\mathbf{x}}), \tag{1}$$

$$\boldsymbol{p}^{\mathrm{T}}(\mathbf{x}) = [p_1(\mathbf{x}), p_2(\mathbf{x}), \ldots, p_m(\mathbf{x})], \tag{2}$$

$$\boldsymbol{a}^{\mathrm{T}}(\bar{\mathbf{x}}) = [a_1(\bar{\mathbf{x}}), a_2(\bar{\mathbf{x}}), \ldots, a_m(\bar{\mathbf{x}})], \tag{3}$$

in which $p_j(\bar{\mathbf{x}})$ is the basis function and $a_j(\mathbf{x})$ is the unknown parameter. m is the number of the basis functions.

The shape functions of the FE method are obtained by the Lagrange interpolation of the element nodes. In FE methods, the number of the basis functions is equal to the nodal number in the element. While in meshless methods the construction of shape functions is based on some other methods, such as moving least square (MLS) method or kernel function. Particular nodes in an influence domain are adopted in the meshless approximation, as shown in Fig. 1. Unlike the FE approximation, the number of basis functions in meshless method does not need to be equal to the node number in the influence domain.

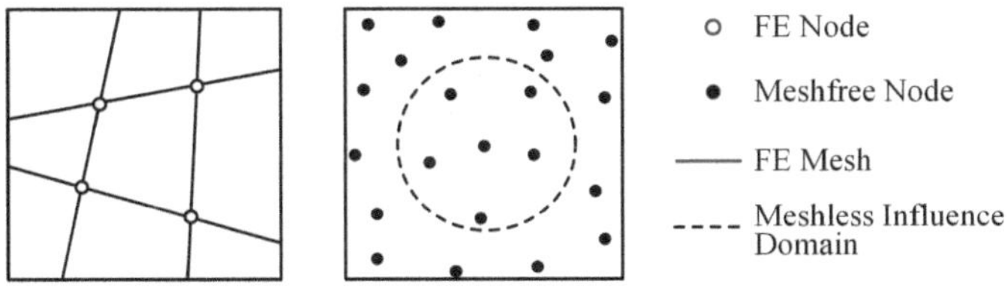

Fig. 1. Discrete model of FE and Meshless method.

Considering a moving least squares approximation satisfying the Kronecker delta criterion, this fitting method could be quite similar to the finite element approximation, under an appropriate arrangement of the nodes in the influence domain. In this situation, the EFG method would degenerate into the FE method. For example, considering a rectangle formed by the nodes in EFG method, suppose that a weight function $w(\mathbf{x}-\mathbf{x}_I)$ has the following properties: (1) positive inside the rectangle and zero outside; (2) zero on the sides which the nodes $\mathbf{x}_I$ faces. Let the basis function $\boldsymbol{p}^T(\mathbf{x}) = [1, x, y, xy]$. Then the MLSA would be a bilinear interpolation, which means the functions $\Phi_I(\mathbf{x})$ would also be bilinear on the rectangle, in conformity with the shape function of a 4-nodes rectangle element FE method. Hence, by choosing appropriate parameters of EFG method, the FE method can be considered as a subclass of EFG method (namely a vestigial EFG method), without constructing the interface shape functions or introducing the extended weak forms in the coupling region.

2.2 Weight Function Descriptions

The weight function plays a pivotal role in the directly coupled method. To imitate the FE method with the EFG method closely, an important ingredient is choosing an appropriate weight function associated with an appropriate influence domain. Several properties need to be satisfied when constructing the general weight function: (1) $w(\mathbf{x}-\mathbf{x}_I)$ should to be compact support, namely, zero on the area outside the influence domain of point $\mathbf{x}_I$; (2) the value of $w(\mathbf{x}-\mathbf{x}_I)$ should be maximal at point $\mathbf{x}_I$; (3) $w(\mathbf{x}-\mathbf{x}_I)$ should be smooth.

In the weight function, the support size of influence domain is dominated by a dilation parameter H. For a 2-dimensional problem with regularly distributed nodes, the support size can be taken as the distance to the nearest neighbor, namely 4 nodes would contribute to the approximation when setting $H = 1.0$. An ideal shape function of the directly coupled method should coincide with the FE shape function when small influence domain is adopted. Namely, when H is close to 1.0, the EFG method should degenerate into FE method. Note that, the nodes number, positions, and their dilation parameters cannot be taken arbitrarily in the EFG method. For an n-dimensional space, at least $n+1$ nodes are needed in the influence domain to form a non-degenerated n th rank simplex element [11]. If the above rule is not met in the program, H needs to be increased until the influence domain contains enough nodes.

The shape functions based on several extant weight functions are shown in the first 7 plots in Fig. 2, with different values of H in the one-dimension situation. The shape functions perform well when the H takes a relatively large value, typically 2.0–4.0 adopted in precedent studies. When taking $H \leq 1.5$, the derivatives of the shape functions may be strongly non-polynomial, which leads to numerical integration not convenient [4].

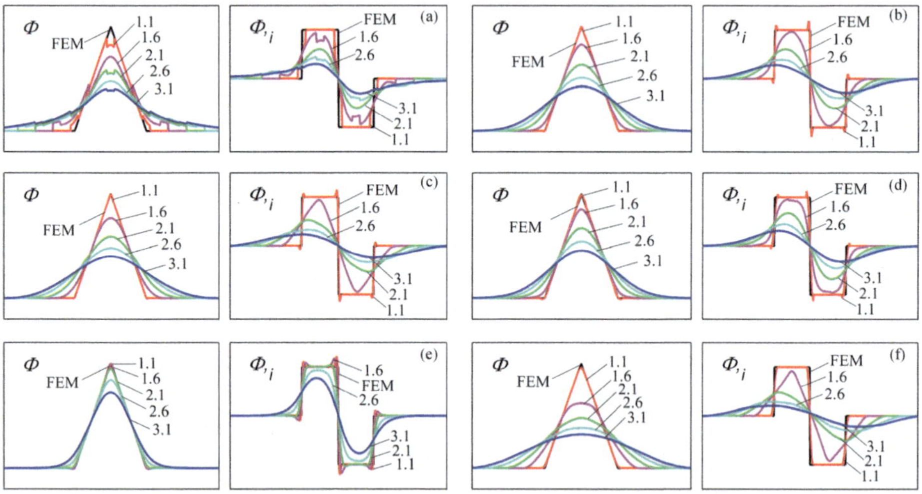

Fig. 2. Shape functions in different H with weight function of (a) quadratic weight function; (b) cosine weight function; (c) rational weight function; (d) cubic spline weight function; (e) quartic spline weight function; (f) Gaussian weight function; (g) exponential weight function; (h) smooth cosine weight function.

A much more continuous weight function is needed when taking small influence domain in the EFG method. A smooth cosine weight function is proposed in this study:

$$w(r) = \begin{cases} (\cos(\pi r) + 1)/2 & 0 \leq r \leq 1 \\ 0 & r > 1 \end{cases}, \tag{4}$$

with its derivative

$$\mathrm{d}w/\mathrm{d}x = \begin{cases} -\pi \sin(\pi r)\mathrm{sign}(x - x_I)/2 & 0 \leq r \leq 1 \\ 0 & r > 1 \end{cases}, \tag{5}$$

where r is calculated by $r = \|x - x_I\|/(H \cdot c_I)$, c_I is the neighbor nodes distance of x_I, H is the dilation parameter of nodal distance.

As shown in Fig. 2(h), the new shape function is nearly concordant with the FE shape function when the H takes a small value. And the proposed weight function can guarantee C^2 continuity, which means the second derivative of shape function is also continuous. The inconvenient of numerical integration is not existing anymore. Thus the proposed weight function is more suitable for the direct coupling method than its counterparts.

3 Varying Dilation Parameter for EFG Methods

As a small influence domain is necessary for the directly coupled method, that means some advantages of the EFG method are suppressed. A coupling domain Ω_C, with continuous varying dilation parameter $H(\mathbf{x}) \in (H_{min}, H_{max})$, is proposed in this method for the transition from FE domain to EFG domain, as displayed in Fig. 3. In the coupling domain, the dilation parameter varies from H_{min} on the FE boundary to H_{max} on the EFG boundary.

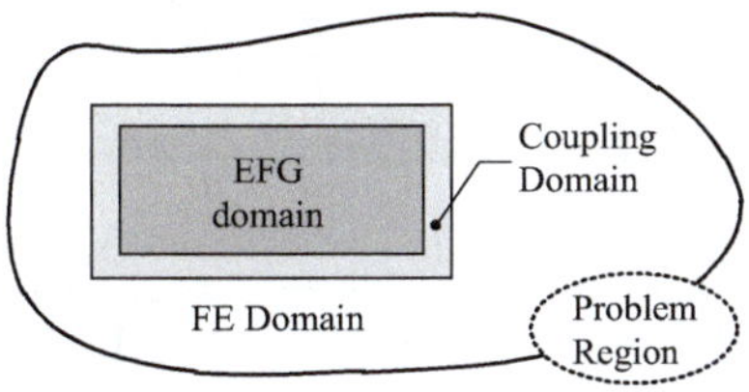

Fig. 3. Subdomains of the coupling method.

The shape functions of the interface elements-based coupling method and the directly coupled method are discussed in the following. And for the sake of better illustration, their coupling shape functions are shown in one-dimension problems. Taking the dilation parameter of $H = 3.0$, the shape functions of interface elements-based coupling method are continuous in the interface domain Ω_C, shown in Fig. 6. But the discontinuity of their derivatives is evident. Due to the discontinuity of the derivative, the force equilibrium condition in the interface domain cannot be satisfied, leading to the inaccuracy of stress analysis (Fig. 4).

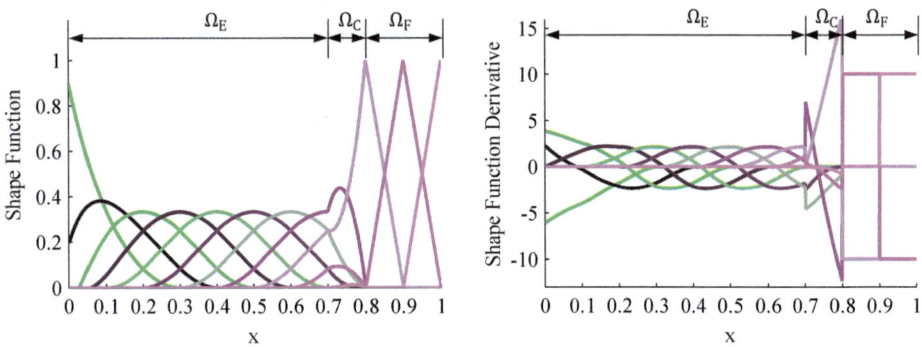

Fig. 4. Shape function and their derivative of interface elements based coupling method.

For the directly coupled method, the weight function of the EFG domain has to be zero on the FE domain, which means the weight function needs to keep a relatively small support radius. But a relatively large radius is needed in order to keep the advantage of the EFG method. Hence, the varying dilation parameter is taking $H_{min} = 1.0$ on the FE boundary and $H_{max} = 3.0$ on the EFG domain here. The shape functions of the proposed method are shown in Fig. 8. In this approach, the shape functions and their derivatives can transit from the FE domain to the EFG domain smoothly. The shape function of the node locating at 0.8 can satisfy the Kronecker delta criterion in the FE part so that the EFG method can be connected to the FE method directly. Also, the boundary conditions can be imposed on these nodes directly.

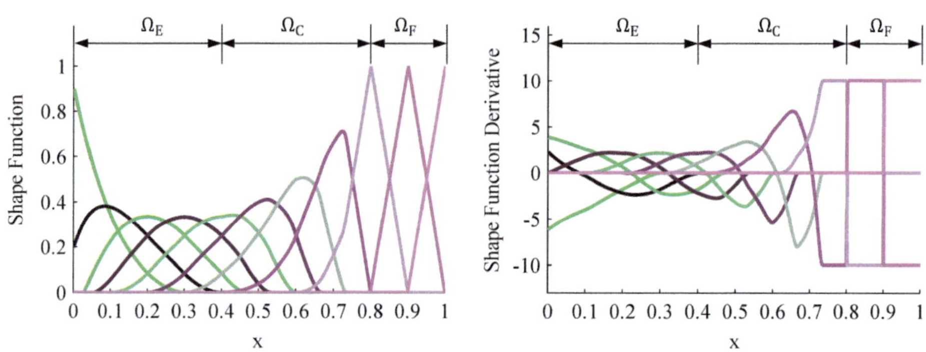

Fig. 5. Shape function and their derivative of directly coupled method.

It should be noted that the influence domain of EFG nodes cannot extend to the FE domain. Hence, a basic principle is that the influence domain of EFG nodes should be secluded from the FE domain, namely

$$L_C \geq H_{max} \cdot d_C \tag{6}$$

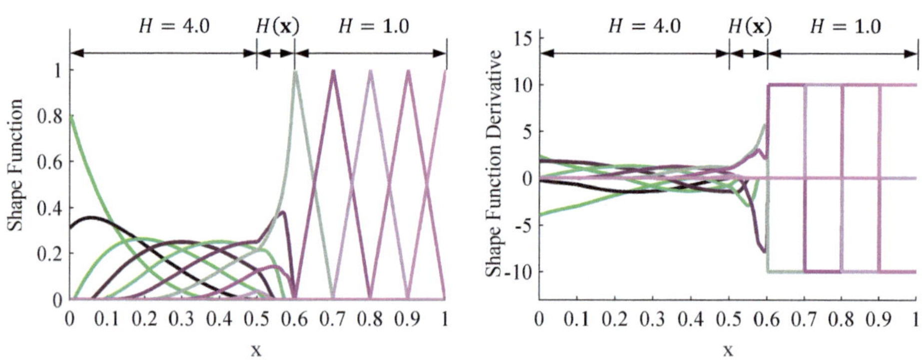

Fig. 6. Shape function and their derivative of EFG method with steep varying dilation parameter.

where L_C is the coupling domain width, d_C is equal to the nodal space in coupling domain. The shape functions, with an inappropriate combination of L_C and H_{max} in the coupled method, are shown in Fig. 9, taking $H(\mathbf{x}) \in (1.0, 4.0)$ in a small coupling domain between 0.5 and 0.6.

Noted that, the dilation parameter in this section varies linearly in the coupling domain. For the two-dimensional situation, the varying dilation parameter is interpolated by biharmonic spline interpolation.

4 Numerical Example

In this section, the proposed directly coupled method is implemented on several numerical examples. The computations are coded in MATLAB and performed by a personal computer that has E5-1630v3 CPU and 32 GB SDRAM.

4.1 Numerical Example

A cantilever beam is shown in Fig. 7, with a unidirectional tensile load of $P = 1000$, length $L = 48$, and width $D = 12$. The model is discretized with 81 points in the x-direction and 21 points in the y-direction. As shown in Fig. 5, the beam here is divided into three parts (FE domain, coupling domain, and meshless domain) with $L_E : L = 1 : 2$. The structural material is set to Young's modulus of $E = 30 \times 10^6$ and Poisson's ratio of $\nu = 0.3$.

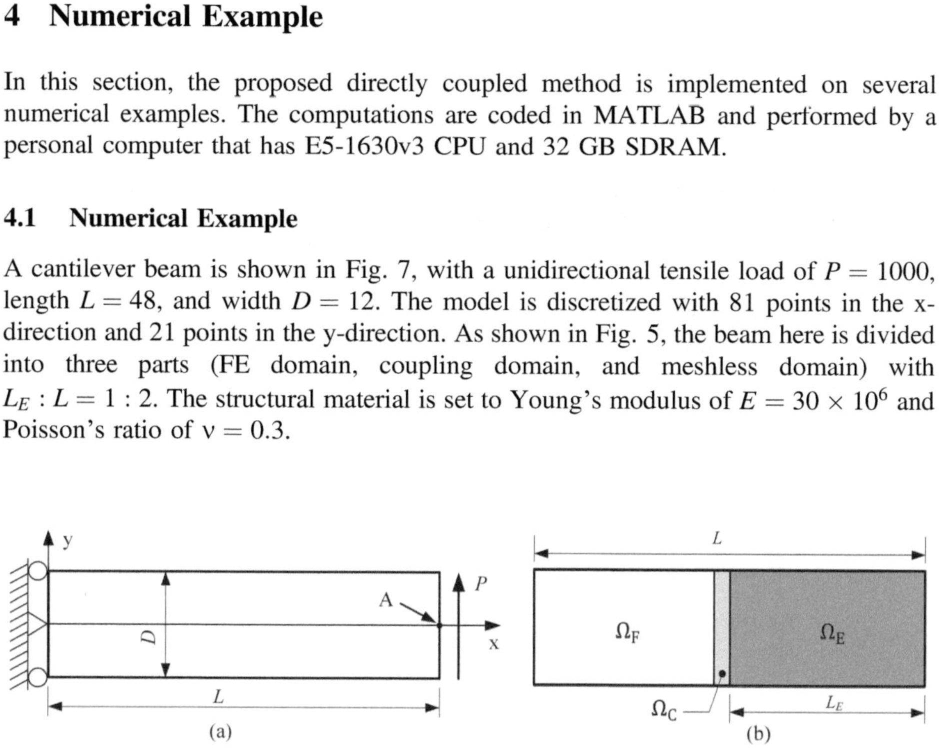

Fig. 7. (a) Cantilever beam model; (b) arrangement of FE and EFG domain.

The shear stresses of analytical solution, EFG solution, FE solution, proposed coupling solution, interface element-based coupling solution, and extended weak forms-based coupling solution are displayed in Fig. 6. The stress of the EFG method is closest to the analytical solution. And the solutions in the FE and EFG domains in the coupled method show great similarities to the FE solution and EFG solution respectively. In the proposed method, the stress transit from FE domain to EFG domain smoothly. As the derivatives of shape functions in the interface domain are discontinuous, the stresses are discontinuous in the interface elements-based method and extended weak forms-based method.

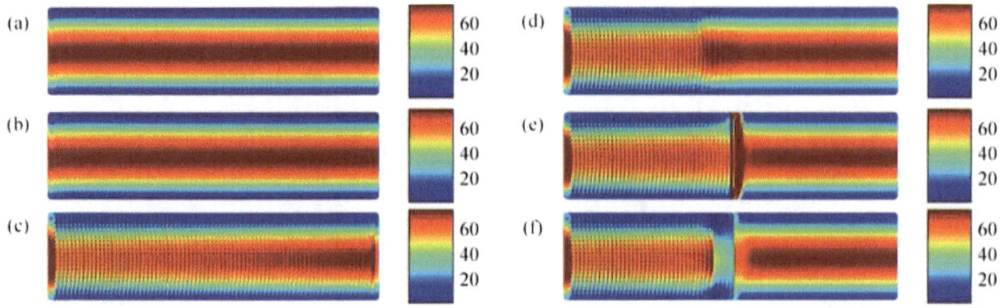

Fig. 8. Shear stresses: (a) analytical solution; (b) EFG solution; (c) FE solution; (d) proposed coupling method; (e) interface element-based coupling method; (f) extended weak forms-based coupling method.

Table 1 compares the vertical displacements at point A and time costs of the different methods. The EFG method gets the highest accuracy and the highest time cost while the FE method costs the least computation. The interface elements-based coupling method and the extended weak forms-based coupling method have lower accuracy, which restricts their applicability.

The error and time cost of the proposed coupling method are between the EFG method and the FE method, conforming to the anticipated result. And the solution of the proposed method is more accurate than the other two coupling methods. Meanwhile, the computational efficiency is slightly higher than the interface elements-based coupling method, as actually more EFG domain is adopted in the coupling domain. And it is much lower than the extended weak forms-based coupling method because no extra stiffness matrix needs to be constructed.

Table 1. Comparison of vertical displacement errors and time costs

	u_y (10^{-4})	Error (%)	Time cost (/s)
Analytical solution	89.00	/	/
EFG method	89.00	-8.91×10^{-4}	36.41
FE method	88.89	−0.12	1.7895
The proposed coupling method	88.92	-8.64×10^{-2}	18.02
Interface elements-based method	85.73	−3.67	15.4433
Extended weak forms-based method	86.56	−2.75	34.2048

The convergence of the five methods is shown in Fig. 8. The error in the y-axis is the energy error norm and the h in the x-axis is the mesh parameter (setting as the largest distance of adjacent nodes). The EFG method has the greatest convergence rate. The convergence rate of the directly coupled method is slightly greater than the FE method, and the other two coupling methods own the worst results.

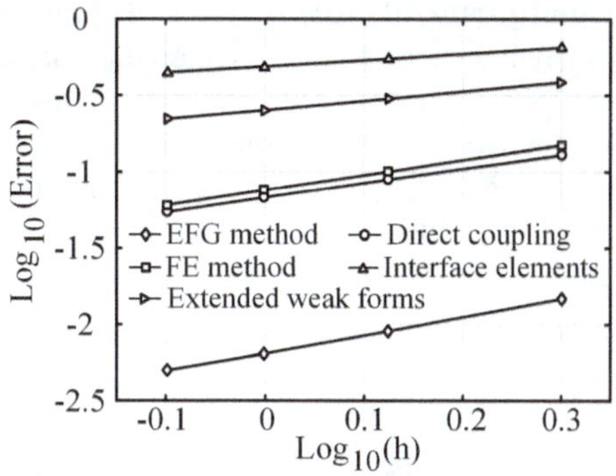

Fig. 9. Energy error norm of EFG method, FE method, proposed coupling method, interface element-based coupling method, and extended weak forms-based coupling method.

4.2 Direct Coupling Method with Different Stiffness

Consider a hyperelastic nonlinear problem here. A square shape structure with two different stiffness regions is shown in Fig. 2 (a), with $L = 100$, $D = 100$, traction $F = -3 \times 10^{-2}$, the Lamé parameters $\mu_1 = 1$ and $K_1 = 10$ in the high-stiffness region, and $\mu_1 = 1 \times 10^{-3}$ and $K_1 = 10 \times 10^{-3}$ in the low-stiffness region.

As the low-stiffness region has little influence on the total deformation, the linear FE domain is assigned to this region. Several triangular elements are mixed in the quadrilateral FE approximation, showing in Fig. 10(b). Incremental Newton-Raphson method is taken, with load increment of 10 steps and the iteration upper limit of 30 in each load increment.

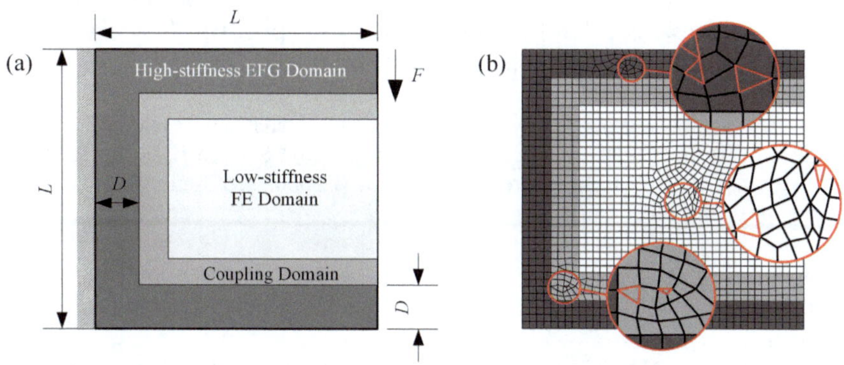

Fig. 10. (a) Schematic diagram of C-shape problem, (b) Low-quality FE meshes of C-shape problem.

The deformations of the EFG method, the directly coupled method, and the FE method are shown in Fig. 3. The vertical displacements of point P are −87.11 mm (EFG method), −87.07 mm (coupling method), and −60.32 mm (FE method).

The FE-based result is significantly different from the coupling and EFG results, due to the fact that the stiffness matrices of triangular elements are extremely large in the quadrilateral FE method. The solution of the low-stiffness region only has little effect on the whole deformation, hence the result of the coupled method is similar to the EFG method. Though the low-stiffness region deformation of the meshless-based method is the smoothest, it cost much computation cost: the EFG method cost 33.10 s in each iteration averagely while the coupling method cost only 18.32 s (Fig. 11).

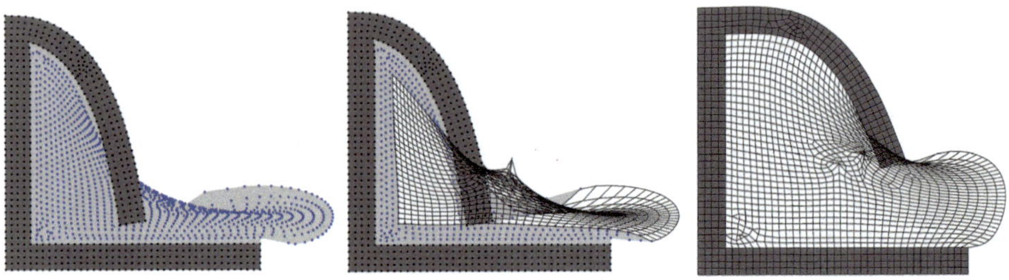

Fig. 11. Deformations of the C-shape structure: (a) EFG method; (b) Directly coupled FE-EFG method; (c) FE method.

5 Conclusion

This study developed a directly coupled FE–EFG method, by degenerating the EFG method to the FE method in a coupling domain. A smooth cosine weight function is presented to guarantee higher order compatibility in coupling domain. A coupling domain with variable dilation parameter is adopted to keep the advantage of the EFG method. The proposal is tested by two numerical examples, which shows higher precision or less computation cost comparing with other methods.

Funding. This work was supported by the National Key Research and Development Program of China (2017YFB1300102), the Project Supported by Natural Science Basic Research Plan in Shaanxi Province of China under Program No. 2017JM500, and National Natural Science Foundation of China under Grant Nos. 51375383.

References

1. Bazilevs, Y., Calo, V.M., Cottrell, J.A., Evans, J.A., Hughes, T.J.R., Lipton, S., Scott, M.A., Sederberg, T.W.: Isogeometric analysis using T-splines. Comput. Methods Appl. Mech. Eng. **199**, 229–263 (2010)
2. Belytschko, T., Organ, D., Krongauz, Y.: A coupled finite element-element-free Galerkin method. Comput. Mech. **17**, 186–195 (1995)
3. Rao, B.N.: Coupled meshfree and fractal finite element method for unbounded problems. Comput. Geotech. **38**, 697–708 (2011)

4. Hegen, D.: Element-free Galerkin methods in combination with finite element approaches. Comput. Methods Appl. Mech. Eng. **135**, 143–166 (1996)
5. Gu, Y.T., Zhang, L.C.: Coupling of the meshfree and finite element methods for determination of the crack tip fields. Eng. Fract. Mech. **75**, 986–1004 (2008)
6. Huerta, A., Fernandez-Mendez, S.: Enrichment and coupling of the finite element and meshless methods. Int. J. Numer. Methods Eng. **48**, 1615–1636 (2000)
7. Erkmen, E., Bradford, M.A.: Coupling of finite element and meshfree methods for locking-free analysis of shear-deformable beams and plates. Eng. Comput. **28**, 1003–1027 (2011)
8. Dolbow, J., Belytschko, T.: Volumetric locking in the element free Galerkin method. Int. J. Numer. Methods Eng. **46**, 925–942 (1999)
9. Zhang, Y.Q., Ge, W.J., Tong, X.X., Ye, M.: Topology optimization of structures with coupled finite element - element-free Galerkin method. Proc. Inst. Mech. Eng. Part C-J. Eng. Mech. Eng. Sci **232**, 731–745 (2018)
10. Gong, S.G., Xie, G.L., Zhang, J.P., Nie, S.H., Li, Y.M.: Sensitivity analysis and shape optimization based on FE-EFG coupled method. Res. Eng. Des. **20**, 117–128 (2009)
11. Liu, W.-K., Li, S., Belytschko, T.: Moving least-square reproducing kernel methods (I) Methodology and convergence. Comput. Methods Appl. Mech. Eng. **143**, 113–154 (1997)

Parallel Manipulators

Kinematics and Dynamics Simulation Analysis of a 3-DOF Parallel Mechanism for Application in Hybrid Machine

Haiqiang Zhang[1,2(✉)], Hairong Fang[1], Dan Zhang[2], Xueling Luo[2], and Fuqun Zhao[1]

[1] School of Mechanical, Electronic and Control Engineering, Beijing Jiaotong University, Beijing, People's Republic of China
{hqzhang, hrfang, 16116362}@bjtu.edu.cn
[2] Lassonde School of Engineering, York University, Toronto, ON M3J 1P3, Canada
{dzhang99, lxling07}@yorku.ca

Abstract. A novel parallel mechanism with one translational and two rotational coupling degrees of freedom, which can be utilized as the main body module for application in hybrid kinematic machine, is proposed. The inverse kinematics is systematically established based on the closed-loop vector method. In terms of the principle of virtual work and d'Alembert's formulation, the dynamics formulation is deduced in sequence. Finally, the illustrative simulation example is conducted to demonstrate the analytic results of the kinematics and dynamics, and the analytical solutions are verified by Simulink and Recurdyn collaborative simulation. Simultaneously, the dynamic dexterity is further conducted to evaluate the performance, and the results show that the proposed parallel mechanism has greater dynamic performance, which can be demonstrated practically by the alternative application of the parallel kinematic module for the hybrid machine tool.

Keywords: Parallel mechanism · Kinematics · Dynamics · Collaborative simulation · Dynamic dexterity

1 Introduction

Serial mechanisms and parallel mechanisms are two prime types of robots widely employed in various fields of both academy and practice such as kinematic machine tool, surgical robot, pick and place robot, and so on [1–3]. Serial mechanisms connected to each other in serial form have open loop structure which leads to lower precision, poor repeatability, as well as lower stiffness. In contrast, parallel mechanisms have taken more attention by domestic and foreign scholars due to their advantages such as high stiffness, lower inertia, high precision, and good dynamic characteristics over serial mechanisms. In general, the driving devices are installed at the fixed platform or a position close to the fixed platform, so that the moving components is light in weight and high in precision, good dynamic response, moreover, the fully symmetrical parallel mechanism has good isotropy. According to these characteristics, the parallel

D. Wang et al. (Eds.): MeTrApp 2019, MMS 79, pp. 247–258, 2020.
https://doi.org/10.1007/978-981-15-0142-5_25

mechanism has been widely used in fields requiring high stiffness and high precision. Unfortunately, parallel mechanisms have inherent limitations such as small and complex workspace, especially the parallel mechanisms with translational and rotational coupling degrees of freedom, which is a fatal disadvantage and limit its application. In which of the situations, combining the advantages of parallel mechanisms and serial mechanisms, hybrid mechanisms are being developed in manufacturing and engineering fields in recently years [4, 5].

Tsai et al. [6] introduced a new hybrid machine which typically consisted of a position mechanism and an orientation mechanism that can be arranged in series or in parallel to achieve many various machine centers. Fully parallel mechanisms have relatively very limited workspace, especially in terms of orientation characteristics. Fully serial machines have a problem of error accumulation, so a modular method to construct a post-process system for a novel hybrid parallel serial five axis machine tool is proposed in [7]. Gao et al. [8] developed a five-axis hybrid computer numerical control machine by integrating an X-Y gantry system on the basis of parallel manipulator, and the inverse kinematics and optimization of the parallel manipulator were mathematically analyzed. Chen et al. [9] proposed a hybrid machining center including a three degree of freedom parallel platform and traditional computer numerical control X-Y table, and pointed that it is a trend for machine to be reconfigurable hybrid machine for the fabrication of complex and precision products. Coppola et al. [10] addressed a 6-DOF reconfigurable hybrid parallel manipulator for sustainable manufacturing, and a systematical kinematics, orientation workspace, singularity, and stiffness are developed in detail. Xie et al. [11] presented a novel redundant three degrees of freedom parallel kinematic mechanism with high rotational capability, investigated the motion/force transmissibility performance, and applied the parallel module to construct a redundant hybrid machine for freedom surface. Sun et al. [12] investigated a novel 5-axis hybrid reconfigurable robot named Tricept-IV, including one four degrees of freedom hybrid module and one two degrees of freedom end effector, and demonstrated a method of workspace decomposition applied in the dimensional synthesis of a 4-DOF hybrid module. In general, the hybrid architecture contains several different configuration type such as parallel-serial, parallel-parallel, serial-parallel, and so on, therefore, the mechanism proposed in this paper is no exception.

The rest of this paper is organized as follows. Section 1 has provided the introduction. In Sect. 2, the proposed parallel mechanism is described in detail, and presented the candidate application in hybrid machine. In Sect. 3, the inverse kinematics including position, velocity and acceleration are derived. Next, the dynamics is deduced on the basis of the principle of virtual work in Sect. 4. Section 5 the simulation comparative analysis based on the Simulink and Recurdyn are conducted, some simulation results and dynamic dexterity performance are depicted. Finally, Sect. 6 provides contributions and suggestions for further work.

2 Architecture Description

The parallel mechanism proposed in this paper includes a fixed platform, a moving platform, and four limbs with variable lengths (two RPU configuration kinematic legs $l_i(i = 1, 3)$, and two SPR configuration kinematic legs $l_i(i = 2, 4)$), where R is rotational joint, P is prismatic joint, U is universal joint, and S is spherical joint. Herein the fixed platform has a central point **B** and four kinematic joints $\mathbf{B}_i$ which are uniformly distributed at the vertex of the square. Similarly, the moving platform has a central joint **A** and four kinematic joints $\mathbf{A}_i$ which are also uniformly located at the vertex of the square. Moreover, its circumcircle radii are r_a and r_b, respectively. It should be noted that the complicated spherical joints in here are replaced by three revolute joints, whose axis are not collinear in space, are equivalent to a spherical joint. And more importantly, the compound spherical joint has larger rotational angle than that of integrated spherical joint. The parallel mechanism possesses one translation and two rotations degrees of freedom, so it can be used to a candidate orientation mechanism and connected two long tracks in series to achieve five-axis hybrid machine tool (as depicted in Fig. 1), which can complete the complex surface machining for engineering application.

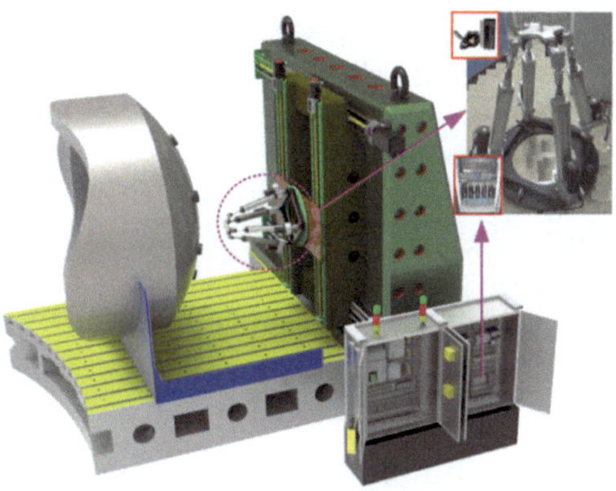

Fig. 1. A five-axis hybrid kinematic machine

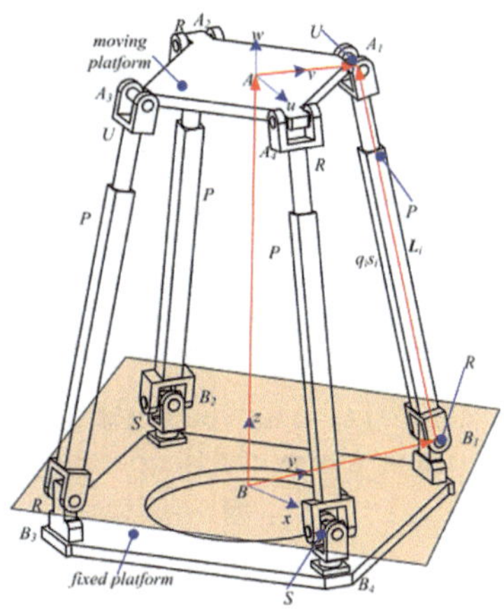

Fig. 2. Kinematic model of parallel manipulator

Figure 2 shows the kinematic schematic of the parallel mechanism. A fixed coordinate system *B-xyz* with the z-axis being vertically the base is attached to the fixed platform, whose *x*-axis is along the vector BB_1, and *y*-axis is determined by the right hand rule. Similarly, the moving coordinate system *A-uvw* is attached on the moving platform and *w*-axis is vertical the moving platform. The *u*-axis is collinear with the vector AA_1, and the *v* axis is determined by the right hand rule.

3 Kinematics Analysis

The moving platform can move along z direction and rotate around the x-axis and v-axis with consideration of mobility property. The homogeneous transformation matrix **T** of point **A** can be expressed as follows:

$$\mathbf{T} = \begin{bmatrix} \mathbf{R} & \mathbf{p} \\ 0 & 1 \end{bmatrix} = Rot(x, \alpha)Trans(z, d)Rot(v, \beta)$$
$$= \begin{bmatrix} \cos\beta & 0 & \sin\beta & x \\ \sin\alpha\sin\beta & \cos\alpha & -\sin\alpha\cos\beta & y \\ -\sin\beta\cos\alpha & \sin\alpha & \cos\alpha\cos\beta & z \\ 0 & 0 & 0 & 1 \end{bmatrix} \tag{1}$$

where $\mathbf{p} = [x,\ y,\ z]^{\mathrm{T}}$ is the position coordinates of point A in the fixed coordinate system, α is the rotation angle around the x-axis, and β is the rotation angle around the v-axis.

Additionally, once the rotation matrix **R** is obtained, then the rotational angels will be obtained

$$\begin{cases} \alpha = arc\tan 2(\mathbf{R}(3,2), \mathbf{R}(2,2)) \\ \beta = arc\tan 2(\mathbf{R}(1,3), \mathbf{R}(1,1)) \end{cases} \tag{2}$$

where $R(\bullet, \bullet)$ represents the row and column elements of the rotation matrix **R**.

Moreover, the aforementioned relations in (1) should also be summarized

$$\begin{cases} x = 0 \\ y = -ds\alpha \\ z = ds\alpha \end{cases} \tag{3}$$

where d is the movement distance along the z direction. So it is very easy to obtain the parasitic motion of the parallel mechanism

$$y = -z\tan\alpha \tag{4}$$

Referring to Fig. 2, the closed vector constraint equation of i-the limb can be written as

$$\mathbf{b}_i + d_i\mathbf{s}_i = \mathbf{p} + \mathbf{a}_i \tag{5}$$

where $\mathbf{b}_i$ is the position vector in the fixed coordinate system, and ${}^{A}\mathbf{a}_i$ is the position vector in the relative coordinate system, and $\mathbf{a}_i$ is the vector ${}^{A}\mathbf{a}_i$ in the fixed coordinate system, namely $\mathbf{a}_i = \mathbf{R} \cdot {}^{A}\mathbf{a}_i$.

The inverse kinematic solution of the parallel mechanism is determined as follows

$$d_i = \|\mathbf{p} + \mathbf{a}_i - \mathbf{b}_i\| \tag{6}$$

The unit vector of s_i can be obtained from (6)

$$s_i = \|\mathbf{p} + \mathbf{a}_i - \mathbf{b}_i\| / d_i \tag{7}$$

The velocity of the i-th active limb can be differentiated (5) with regard to time and written in matrix form

$$\dot{d}_i = \mathbf{J}_q \begin{bmatrix} v_p \\ w_p \end{bmatrix} \tag{8}$$

where $\mathbf{v}_p$ is the linear velocity and $\boldsymbol{\omega}_p$ is angular velocity of the mechanism, respectively.

In which $\mathbf{J}_q = \begin{bmatrix} \mathbf{s}_1^T & [\mathbf{a}_1 \times \mathbf{s}_1]^T \\ \mathbf{s}_2^T & [\mathbf{a}_2 \times \mathbf{s}_2]^T \\ \mathbf{s}_3^T & [\mathbf{a}_3 \times \mathbf{s}_3]^T \\ \mathbf{s}_4^T & [\mathbf{a}_4 \times \mathbf{s}_4]^T \end{bmatrix}$

Additionally, the velocity of the moving platform can be also expressed by the independent variables $\dot{\mathbf{X}} = \begin{bmatrix} \dot{\alpha} & \dot{\beta} & \dot{z} \end{bmatrix}^T$

$$\mathbf{v}_p = \begin{bmatrix} \frac{\partial x}{\partial \alpha} & \frac{\partial x}{\partial \beta} & \frac{\partial x}{\partial z} \\ \frac{\partial y}{\partial \alpha} & \frac{\partial y}{\partial \beta} & \frac{\partial y}{\partial z} \\ \frac{\partial z}{\partial \alpha} & \frac{\partial z}{\partial \beta} & \frac{\partial z}{\partial z} \end{bmatrix} = \mathbf{J}_{vp} \dot{\mathbf{X}} \tag{9}$$

where $\mathbf{J}_{vp} = \begin{bmatrix} 0 & 0 & 0 \\ -z\sec^2\alpha & 0 & -\tan\alpha \\ 0 & 0 & 1 \end{bmatrix}$

The angular velocity of the moving platform can be expressed as a linear superposition of the Euler angles

$$\boldsymbol{\omega}_p = Rot(\alpha)\dot{\alpha} + \mathbf{0}\dot{z} + Rot(\beta)\dot{\beta} = \mathbf{J}_{wp}\dot{\mathbf{X}} \tag{10}$$

where $\mathbf{J}_{wp} = \begin{bmatrix} 1 & 0 & 0 \\ 0 & c\alpha & 0 \\ 0 & s\alpha & 0 \end{bmatrix}$

Therefore, the velocity mapping relationship between actuator joints variables and the independent parameter variables can be expressed as

$$\dot{\mathbf{d}} = \mathbf{J}_0 \dot{\mathbf{X}} = \mathbf{J}_q \mathbf{J}_p \dot{\mathbf{X}} \tag{11}$$

where $\mathbf{J}_0$ is the velocity Jacobian matrix of the parallel mechanism, and $\mathbf{J}_p$ is the Jacobian matrix of the moving platform, i.e. $\mathbf{J}_p = \begin{bmatrix} \mathbf{J}_{vp} \\ \mathbf{J}_{wp} \end{bmatrix}$

By taking derivative of (11) with respect to time, the acceleration of the active limbs can de deduced as

$$\ddot{\mathbf{d}} = \dot{\mathbf{J}}_0\dot{\mathbf{X}} + \mathbf{J}_0\ddot{\mathbf{X}} \tag{12}$$

Where

$$\dot{\mathbf{J}}_0 = \begin{bmatrix} [\mathbf{w}_1 \times \mathbf{s}_1]^T & [(\mathbf{w}_p \times \mathbf{a}_1) \times \mathbf{s}_1 + \mathbf{a}_1 \times (\mathbf{w}_1 \times \mathbf{s}_1)]^T \\ [\mathbf{w}_2 \times \mathbf{s}_2]^T & [(\mathbf{w}_p \times \mathbf{a}_2) \times \mathbf{s}_2 + \mathbf{a}_2 \times (\mathbf{w}_2 \times \mathbf{s}_2)]^T \\ [\mathbf{w}_3 \times \mathbf{s}_3]^T & [(\mathbf{w}_p \times \mathbf{a}_3) \times \mathbf{s}_3 + \mathbf{a}_3 \times (\mathbf{w}_3 \times \mathbf{s}_3)]^T \\ [\mathbf{w}_4 \times \mathbf{s}_4]^T & [(\mathbf{w}_p \times \mathbf{a}_4) \times \mathbf{s}_4 + \mathbf{a}_4 \times (\mathbf{w}_4 \times \mathbf{s}_4)]^T \end{bmatrix} \cdot \mathbf{J}_p + \mathbf{J}_q \cdot \begin{bmatrix} 0 & 0 & 0 \\ -(\dot{z} \cdot \sec^2 \alpha + \frac{2z \bullet \sin \alpha}{\cos^3 \alpha} \bullet \dot{\alpha}) & 0 & (-1 - \tan^2 \alpha) \bullet \dot{\alpha} \\ 0 & 0 & 0 \\ 0 & 0 & 0 \\ 0 & -\sin \alpha \bullet \dot{\alpha} & 0 \\ 0 & \cos \alpha \bullet \dot{\alpha} & 0 \end{bmatrix} \tag{12}$$

In which w_i is the angular velocity of the i-th limb defined in the fixed coordinate frame, and it can be expressed as.

$$\boldsymbol{\omega}_i = \frac{\mathbf{s}_i \times \mathbf{v}_{ai}}{d_i} = \mathbf{J}_{wi}\dot{\mathbf{X}} \tag{13}$$

where the velocity $\mathbf{v}_{ai}$ of kinematic joints A_i connected to the moving platform can be obtained in the fixed coordinate system

$$\mathbf{v}_{ai} = \mathbf{v}_p + \boldsymbol{\omega}_p \times \mathbf{a}_i = \mathbf{J}_{ai} \cdot \mathbf{J}_p \cdot \dot{\mathbf{X}} \tag{14}$$

and $\mathbf{J}_{ai} = \begin{bmatrix} 1 & 0 & 0 & 0 & a_{iz} & -a_{iy} \\ 0 & 1 & 0 & -a_{iz} & 0 & a_{ix} \\ 0 & 0 & 1 & a_{iy} & -a_{ix} & 0 \end{bmatrix}$, $\mathbf{J}_{wi} = \frac{\tilde{\mathbf{s}}_i \mathbf{J}_{ai} \mathbf{J}_p}{d_i}$

Suppose e_1 is the distance between the centroid B_i and the center of mass of lower sub-limb, while e_2 is the distance of the centroid A_i and the center of mass of upper sub-limb. The position vector of the centroid of the i-th limb with respect to the fixed coordinate system can be written as

$$\mathbf{r}_{1i} = \mathbf{b}_i + e_1\mathbf{s}_i \tag{15}$$

$$\mathbf{r}_{2i} = \mathbf{b}_i + (d_i - e_2)\mathbf{s}_i \tag{16}$$

Taking the derivative of (15) and (16) with respect to time, the velocity of the center can be obtained with respect to the fixed coordinate system.

$$\mathbf{v}_{1i} = e_1\boldsymbol{\omega}_i \times \mathbf{s}_i = \mathbf{J}_{v1i}\dot{\mathbf{X}} \tag{17}$$

where $\mathbf{J}_{v1i} = -e_1 \cdot \tilde{\mathbf{s}}_i\mathbf{J}_{wi}$

$$\mathbf{v}_{2i} = (d_i - e_2)\boldsymbol{\omega}_i \times \mathbf{s}_i + \dot{d}_i \cdot \mathbf{s}_i = \mathbf{J}_{v2i}\dot{\mathbf{X}} \tag{18}$$

and $\mathbf{J}_{v2i} = (e_2 - d_i) \cdot \tilde{\mathbf{s}}_i\mathbf{J}_{wi} + \mathbf{s}_i\mathbf{J}_{di}\mathbf{J}_p$

Combining Eqs. (13), (17) and (18) yields

$$\mathbf{J}_{v\omega 1i} = \begin{bmatrix} \mathbf{J}_{v1i} \\ \mathbf{J}_{wi} \end{bmatrix}, \quad \mathbf{J}_{v\omega 2i} = \begin{bmatrix} \mathbf{J}_{v2i} \\ \mathbf{J}_{wi} \end{bmatrix} \tag{19}$$

where $\mathbf{J}_{v\omega 1i}$ and $\mathbf{J}_{v\omega 2i}$ are the Jacobian matrices of the *i*-th sub-limb.

Taking the derivative of (14) with respect to time, the acceleration of the point A_i can be obtained

$$\dot{v}_{ai} = \dot{\mathbf{v}}_p + \dot{\boldsymbol{\omega}}_p \times \mathbf{a}_i + \boldsymbol{\omega}_p \times (\boldsymbol{\omega}_p \times \mathbf{a}_i) \tag{20}$$

The angular acceleration velocity of the i-th sub-limb can be expressed as

$$\dot{\omega}_i = \frac{\mathbf{s}_i \times \dot{v}_{ai} - 2\dot{d}_i\boldsymbol{\omega}_i}{d_i} = \mathbf{J}_{wi}\ddot{\mathbf{X}} + \dot{\mathbf{J}}_{wi}\dot{\mathbf{X}} \tag{21}$$

where $\dot{\mathbf{J}}_{wi} = \frac{\tilde{\mathbf{s}}_i((\dot{\mathbf{J}}_{vp} - \tilde{\mathbf{a}}_i\dot{\mathbf{J}}_{wp}) - [(\mathbf{J}_{wp}\dot{\mathbf{X}})\times]\tilde{\mathbf{a}}_i\mathbf{J}_{wp}) - 2\mathbf{J}_{di}\mathbf{J}_O\dot{\mathbf{X}}\mathbf{J}_{wi}}{d_i}$

The acceleration of the center of mass of the *i*-th limb can be obtained by differentiating (17) and (18) with respect to time

$$\dot{v}_{1i} = e_1\dot{\boldsymbol{\omega}}_i \times \mathbf{s}_i + e_1\boldsymbol{\omega}_i \times (\boldsymbol{\omega}_i \times \mathbf{s}_i) = \mathbf{J}_{v1i}\ddot{\mathbf{X}} + \dot{\mathbf{J}}_{v1i}\dot{\mathbf{X}} \tag{22}$$

$$\begin{aligned} \dot{v}_{2i} &= \ddot{d}_i \cdot \mathbf{s}_i + 2\dot{d}_i(\boldsymbol{\omega}_i \times \mathbf{s}_i) + (d_i - e_2)\dot{\boldsymbol{\omega}}_i \times \mathbf{s}_i + (d_i - e_2)\boldsymbol{\omega}_i \times (\boldsymbol{\omega}_i \times \mathbf{s}_i) \\ &= \mathbf{J}_{v2i}\ddot{\mathbf{X}} + \dot{\mathbf{J}}_{v2i}\dot{\mathbf{X}} \end{aligned} \tag{23}$$

where $\dot{\mathbf{J}}_{v1i} = -e_1\tilde{\mathbf{s}}_i\dot{\mathbf{J}}_{wi} - e_1\mathbf{s}_i \cdot (\mathbf{J}_{wi}\dot{\mathbf{X}})^T\mathbf{J}_{wi}$, $\dot{\mathbf{J}}_{v2i} = \mathbf{s}_i \cdot \dot{\mathbf{J}}_{Oi} - 2\mathbf{J}_{Oi} \cdot \dot{\mathbf{X}} \cdot \tilde{\mathbf{s}}_i \cdot \mathbf{J}_{wi} + (e_2 - d_i)$ $[\tilde{\mathbf{s}}_i \cdot \dot{\mathbf{J}}_{wi} + [(\mathbf{J}_{wi}\dot{\mathbf{X}})\times]\tilde{\mathbf{s}} \cdot \mathbf{J}_{wi}]$

So far, the velocity and acceleration of the moving platform and sub-limbs have been solved. In the next section, we will concentrate on the derivation of dynamics of the proposed mechanism.

4 Dynamics Analysis

In this section, the dynamic model of the parallel mechanism can be deduced by resorting to the principle of virtual work and d'Alembert's formulation. Based on the previous work [13], just only the key points are presented, the following dynamic equation can be obtained

$$\mathbf{J}_0^T\boldsymbol{\tau}+\mathbf{J}_p\begin{bmatrix}\mathbf{f}_e+m_a\mathbf{g}-m_a\dot{\mathbf{v}}_p\\ \mathbf{n}_e-{}^B\mathbf{I}_p\dot{\boldsymbol{\omega}}_p-\boldsymbol{\omega}_p\times({}^B\mathbf{I}_p\boldsymbol{\omega}_p)\end{bmatrix}+\sum_i\mathbf{J}_{v\omega 1i}^T\begin{bmatrix}m_{1i}\mathbf{g}-m_{1i}\dot{\mathbf{v}}_{1i}\\ -{}^B\mathbf{I}_{1i}\dot{\boldsymbol{\omega}}_i-\boldsymbol{\omega}_i\times({}^B\mathbf{I}_{1i}\boldsymbol{\omega}_i)\end{bmatrix}$$
$$+\sum_i\mathbf{J}_{v\omega 2i}^T\begin{bmatrix}m_{2i}\mathbf{g}-m_{2i}\dot{\mathbf{v}}_{2i}\\ -{}^B\mathbf{I}_{2i}\dot{\boldsymbol{\omega}}_i-\boldsymbol{\omega}_i\times({}^B\mathbf{I}_{2i}\boldsymbol{\omega}_i)\end{bmatrix}=0 \tag{24}$$

By simplifying, the dynamics equation of the mechanism can be further deduced as

$$\boldsymbol{\tau}=-\left[\mathbf{J}_0^T\right]^+\Bigg(\mathbf{J}_p\begin{bmatrix}\mathbf{f}_e+m_a\mathbf{g}-m_a\dot{\mathbf{v}}_p\\ \mathbf{n}_e-{}^B\mathbf{I}_p\dot{\boldsymbol{\omega}}_p-\boldsymbol{\omega}_p\times({}^B\mathbf{I}_p\boldsymbol{\omega}_p)\end{bmatrix}$$
$$+\sum_i\mathbf{J}_{v\omega 1i}^T\begin{bmatrix}m_{1i}\mathbf{g}-m_{1i}\dot{\mathbf{v}}_{1i}\\ -{}^B\mathbf{I}_{1i}\dot{\boldsymbol{\omega}}_i-\boldsymbol{\omega}_i\times({}^B\mathbf{I}_{1i}\boldsymbol{\omega}_i)\end{bmatrix}$$
$$+\sum_i\mathbf{J}_{v\omega 2i}^T\begin{bmatrix}m_{2i}\mathbf{g}-m_{2i}\dot{\mathbf{v}}_{2i}\\ -{}^B\mathbf{I}_{2i}\dot{\boldsymbol{\omega}}_i-\boldsymbol{\omega}_i\times({}^B\mathbf{I}_{2i}\boldsymbol{\omega}_i)\end{bmatrix}\Bigg) \tag{25}$$

Equation (25) is a dynamics equation of the parallel mechanism, which can be simplified into the general form

$$\boldsymbol{\tau}=\mathbf{M}(\mathbf{q})\ddot{\mathbf{q}}+C(\dot{\mathbf{q}},\mathbf{q})\dot{\mathbf{q}}+\mathbf{N} \tag{26}$$

with

$$\mathbf{M}(\mathbf{q})=\left[\mathbf{J}_0^T\right]^+\left\{\mathbf{J}_p\begin{bmatrix}m_a\mathbf{J}_{vp}\\ {}^B\mathbf{I}_p\mathbf{J}_{wp}\end{bmatrix}+\sum_i^4\mathbf{J}_{v\omega 1i}^T\begin{bmatrix}m_{1i}\mathbf{J}_{v1i}\\ {}^B\mathbf{I}_{1i}\mathbf{J}_{wi}\end{bmatrix}+\sum_i^4\mathbf{J}_{v\omega 2i}^T\begin{bmatrix}m_{2i}\mathbf{J}_{v2i}\\ {}^B\mathbf{I}_{2i}\mathbf{J}_{wi}\end{bmatrix}\right\}\mathbf{J}_0^{-1}$$

$$\mathbf{C}(\dot{\mathbf{q}},\mathbf{q})=\left[\mathbf{J}_0^T\right]^+\Bigg\{\mathbf{J}_p^T\begin{bmatrix}m_a\dot{\mathbf{J}}_{vp}\\ {}^B\mathbf{I}_p\dot{\mathbf{J}}_{wp}+[(\mathbf{J}_{wp}\dot{\mathbf{X}})\times]({}^B\mathbf{I}_p\mathbf{J}_{wp})\end{bmatrix}$$
$$+\sum_i\mathbf{J}_{v\omega 1i}^T\begin{bmatrix}m_{1i}\dot{\mathbf{J}}_{v1i}\\ {}^B\mathbf{I}_{1i}\dot{\mathbf{J}}_{wi}+[(\mathbf{J}_{wi}\dot{\mathbf{X}})\times]({}^B\mathbf{I}_{1i}\mathbf{J}_{wi})\end{bmatrix}+\sum_i\mathbf{J}_{v\omega 2i}^T\begin{bmatrix}m_{2i}\dot{\mathbf{J}}_{v2i}\\ {}^B\mathbf{I}_{2i}\dot{\mathbf{J}}_{wi}+[(\mathbf{J}_{wi}\dot{\mathbf{X}})\times]({}^B\mathbf{I}_{2i}\mathbf{J}_{wi})\end{bmatrix}\Bigg\}\mathbf{J}_0^{-1}$$
$$-\left[\mathbf{J}_0^T\right]^+\left\{\mathbf{J}_p^T\begin{bmatrix}m_a\mathbf{J}_{vp}\\ {}^B\mathbf{I}_p\mathbf{J}_{wp}\end{bmatrix}+\sum_i^4\mathbf{J}_{v\omega 1i}^T\begin{bmatrix}m_{1i}\mathbf{J}_{v1i}\\ {}^B\mathbf{I}_{1i}\mathbf{J}_{wi}\end{bmatrix}+\sum_i^4\mathbf{J}_{v\omega 2i}^T\begin{bmatrix}m_{2i}\mathbf{J}_{v2i}\\ {}^B\mathbf{I}_{2i}\mathbf{J}_{wi}\end{bmatrix}\right\}\mathbf{J}_0^{-1}\dot{\mathbf{J}}_0\mathbf{J}_0^{-1}$$

$$\mathbf{N}=-\left[\mathbf{J}_0^T\right]^+\left\{\mathbf{J}_p^T\begin{bmatrix}m_a\mathbf{g}\\ \mathbf{0}\end{bmatrix}+\sum_i\mathbf{J}_{v\omega 1i}^T\begin{bmatrix}m_{1i}\mathbf{g}\\ \mathbf{0}\end{bmatrix}+\sum_i\mathbf{J}_{v\omega 2i}^T\begin{bmatrix}m_{2i}\mathbf{g}\\ \mathbf{0}\end{bmatrix}\right\}-\left[\mathbf{J}_0^T\right]^+\mathbf{J}_p\begin{bmatrix}\mathbf{f}_e\\ \mathbf{n}_e\end{bmatrix}$$

where $\mathbf{M}(\mathbf{q})$ is the inertia matrix, $\mathbf{C}(\mathbf{q})$ is the concerning velocity (Centripetal force and Coriolis force), and $\mathbf{N}$ is the gravitational force and external generalized force term.

The elements of the inertia matrix $\mathbf{M}$ is dimensionally homogeneous, so one can employ the ratio of its minimum and maximum singular eigenvalue as the dynamic performance evaluation index

$$k_M=\lambda_{\min}(\mathbf{M})/\lambda_{\max}(\mathbf{M}) \tag{27}$$

5 Simulation Examples

In order to verify the effectiveness of the kinematics and dynamics, the collaborative simulation analysis is adopted by the analysis tools, i.e. Simulink and Recurdyn. The simulation process can de described in Fig. 3.

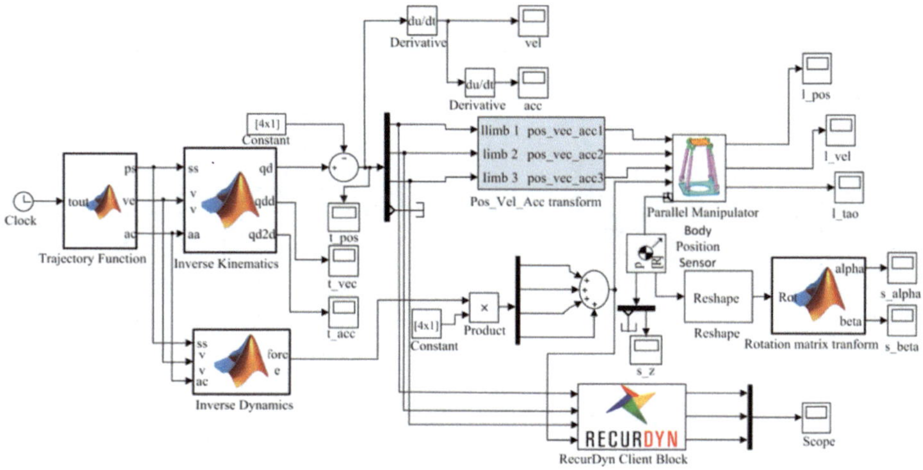

Fig. 3. Simulink and Recurdyn simulation flow block diagram

Specifically, the trajectory function of the moving platform is given as following

$$\begin{cases} \alpha = \frac{\pi}{30}\sin(\pi t) \\ \beta = \frac{\pi}{30}\cos(\pi t) \\ z = 0.7 + 0.05\sin(\pi t) \end{cases} \tag{28}$$

After the simulation, the rotation matrix of the moving platform can be obtained. Calculated by the (2), the rotation angels (α, β) of the moving platform can be required, which was compared with the desired trajectory and the illustrative results are depicted in Figs. 4 and 5 that verify the feasibility and accuracy of the parallel mechanism.

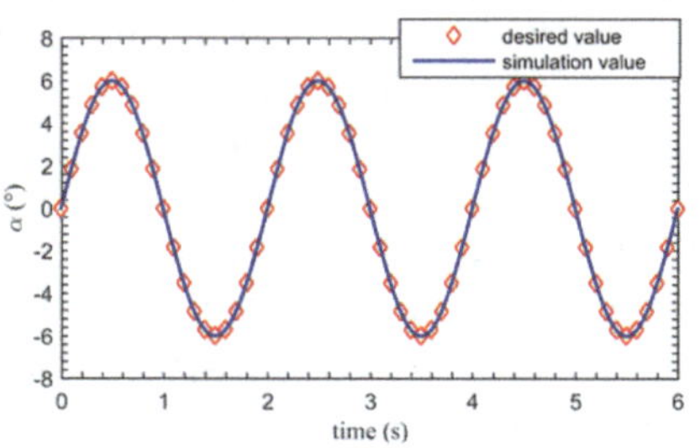

Fig. 4. Trajectory comparison with α value

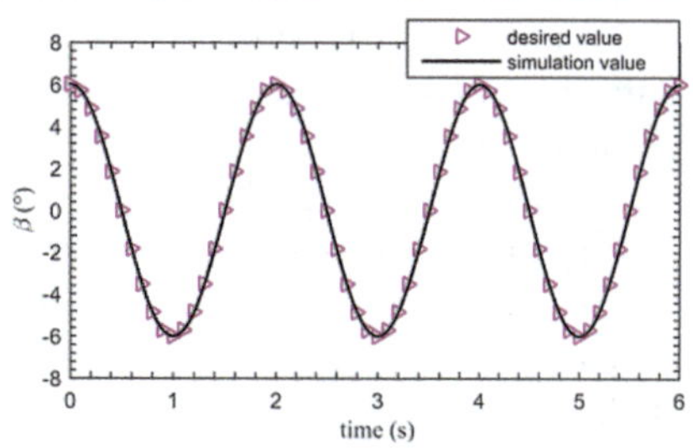

Fig. 5. Trajectory comparison with β value

Simultaneously, the positions, velocities, accelerations, and driving forces of the actuator joints can be drawn in Figs. 6, 7, 8 and 9. To facilitate analysis, just the velocity of the first and second actuators was presented, so did the acceleration of the third and fourth actuators.

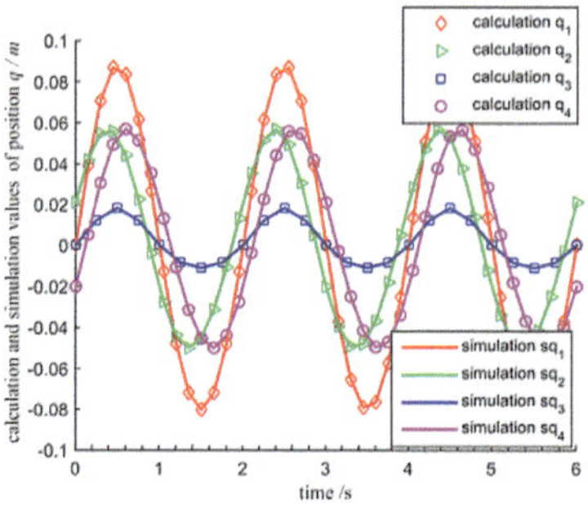

Fig. 6. Positions comparison with actuators

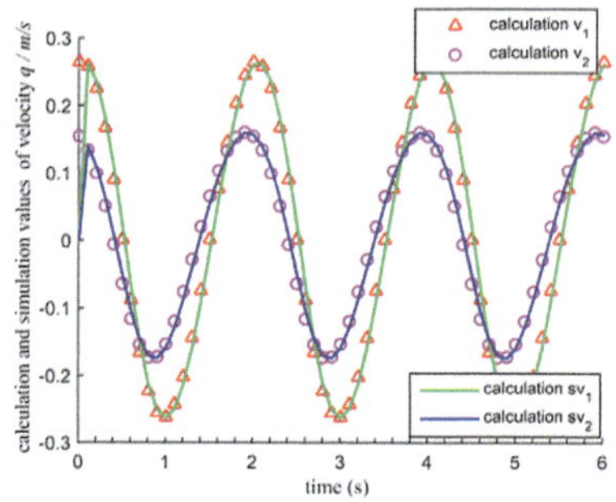

Fig. 7. Velocities comparison with actuators

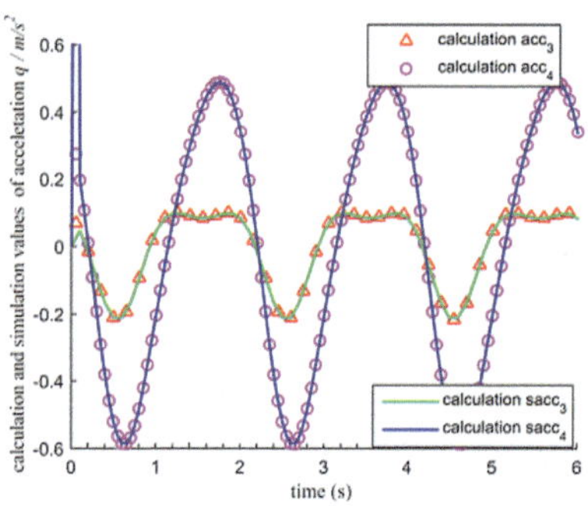

Fig. 8. Accelerations comparison with actuators

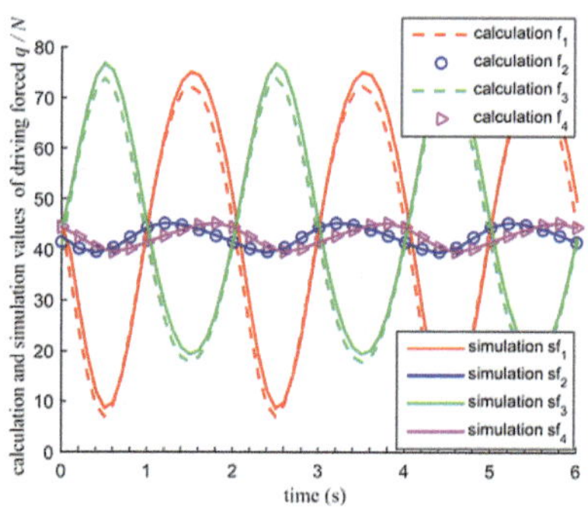

Fig. 9. Driving forces comparison with actuators

The positions, velocities, accelerations, and driving forces are varied smoothly in a certain range, and it demonstrates that the parallel mechanism has good performance characteristics of the kinematics and dynamics. Additionally, in kinematics, the analytical solutions are perfectly identical with simulation ones. But in dynamics simulation, there is a certain difference in driving force. It is due to omit the mass and velocity of the kinematic joints that affect the accuracy of calculation results.

The results simulated by the software Recurdyn were depicted in Fig. 10. On comparing the simulation and calculation results in Fig. 9, we find these results are satisfactory match, which demonstrate the analytic solutions are correct.

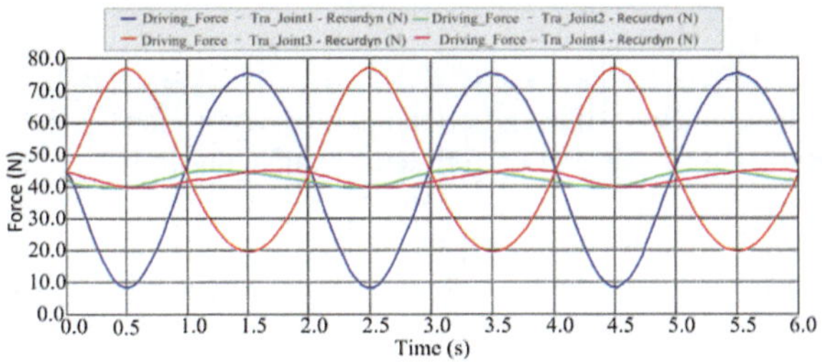

Fig. 10. Driving forces of actuators with Recurdyn simulation

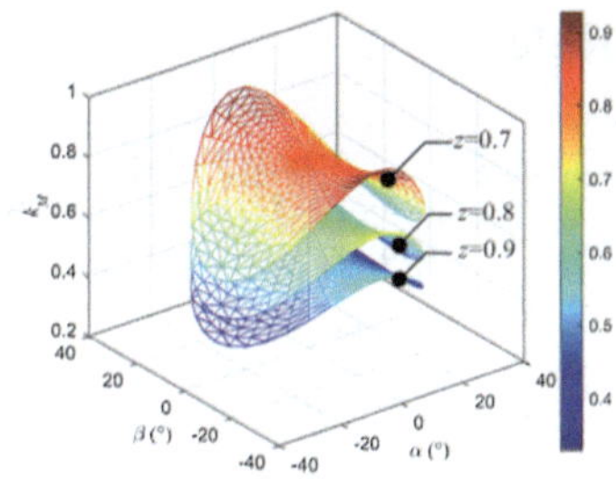

Fig. 11. Dynamic dexterity of different height

The analysis result of the dynamic dexterity is depicted in Fig. 11, where the distribution of dynamics dexterity at different heights is demonstrated. As illustrated in Fig. 11, the dynamic dexterity of the parallel mechanism is closer to 1, meaning that the parallel mechanism presents greater dynamics dexterity and higher acceleration performance within the workspace. Additionally, the distribution is uniform, the change is stable, and there is no sudden change, which illustrate that the proposed parallel manipulator has a potential application of the kinematic machine for good dynamics dexterity.

6 Conclusions

In this paper, the major contributions of this paper are as follows: the kinematics analysis of the proposed parallel manipulator has been formulated, in terms of the principle of virtual work, the dynamics formulation of the proposed parallel manipulator is established and the inertia matrix including acceleration term is especially separated and the dynamic dexterity performance index is proposed. All the derived analytical solutions of the kinematics and dynamics are verified to be correct by employing Simulink and Recurdyn collaborative simulation, and provide a theoretical foundation for the intelligent control and calibration experiment in potential applications for the kinematic machine. A further extension of this work being investigated is intelligent control design and experimental calibration.

Acknowledgement. The authors would like to acknowledge the financial support of the Fundamental Research Funds for the Central Universities under Grants No. 2018JBZ007, No. 2018YJS136, China Scholarship Council (CSC) under Grant No. 201807090079, the Natural Sciences and Engineering Research Council of Canada (NSERC), and the York Research Chairs (YRC) program. Meanwhile, the author, Haiqiang Zhang, is grateful to advanced robotics and mechatronics laboratory in York University.

References

1. Assal, S.: A novel planar parallel manipulator with high orientation capability for a hybrid machine tool: kinematics, dimensional synthesis and performance evaluation. Robotica **35**, 1031–1053 (2017)
2. Xu, P., Cheung, C., Li, B., Zhang, F.: Kinematics analysis of a hybrid manipulator for computer controlled ultra-precision freeform polishing. Robot. Comput. Integr. Manuf. **44**, 44–56 (2017)
3. Wu, Y., Yang, Z., Fu, Z.: Kinematics and dynamics analysis of a novel five-degrees-of-freedom hybrid robot. Int. J. Adv. Robot. Syst. **5**, 1–8 (2017)
4. Enferadi, J., Shahi, A.: On the position analysis of a new spherical parallel robot with orientation applications. Robot. Comput. Integr. Manuf. **37**, 151–161 (2016)
5. Zhang, H., Fang, H., Fang, Y.: Workspace analysis of a hybrid kinematic machine tool with high rotational applications. Math. Probl. Eng. **2018**, 1–12 (2018)
6. Tasi, L.W., Joshi, S.: Kinematic analysis of 3-DOF position mechanisms for use in hybrid kinematic machines. J. Mech. Des. **124**(2), 245–253 (2002)
7. Lai, Y., Liao, C., Chao, Z.: Inverse kinematics for a novel hybrid parallel–serial five-axis machine tool. Robot. Comput. Integr. Manuf. **50**, 63–79 (2018)
8. Gao, Z., Zhang, D.: Performance analysis, mapping, and multiobjective optimization of a hybrid robotic machine tool. IEEE Trans. Ind. Electron. **62**(1), 423–433 (2015)
9. Cheng, Y.M., Peng, W.X., Hsu, A.C.: Concentric hole drilling in multiple planes for experimental investigation of five-axis reconfigurable precision hybrid machine. Int. J. Adv. Manuf. Technol. **76**, 1253–1262 (2015)
10. Coppola, G., Zhang, D., Liu, K.: A 6-DOF reconfigurable hybrid parallel manipulator. Robot. Comput. Integr. Manuf. **30**(2), 99–106 (2014)
11. Xie, F.G., Liu, X.J., Zhou, Y.H.: Optimization of a redundantly actuated parallel kinematic mechanism for a 5-degree-of-freedom hybrid machine tool. Proc. Inst. Mech. Eng. Part B J. Eng. Manuf. **228**(12), 1630–1641 (2014)
12. Sun, T., Song, Y., Li, Y., Zhang, J.: Workspace decomposition based dimensional synthesis of a novel hybrid reconfigurable robot. J. Mech. Robot. **2**(3), 031009-1–8 (2010)
13. Zhang, H., Fang, H., Jiang, B.: Dynamic performance evaluation of a redundantly actuated and over-constrained parallel manipulator. Int. J. Autom. Comput. **16**(3), 274–285 (2019)

Kinematics and Singularity Analysis of a Novel Uncoupled 2-DOF Rotational Parallel Mechanism

Keming Wang, Yanbin Zhang(✉), and Xianling Jing

School of Mechatronics Engineering,
Henan University of Science and Technology, Luoyang 471003, China
yanbin_zh@163.com

Abstract. The paper presents a novel uncoupled rotational parallel mechanism with two degrees of freedom (DOFs), which consists of a moving platform connected to a fixed base by two kinematical limbs. Based on the position and orientation characteristics method, kinematic output characteristics and mobility of the mechanism are analyzed and calculated. Mathematical models of the orientation and angular velocity of the mechanism are established in terms of the actuated wrench screw theory and the closed vector approach. Singularity analysis is performed in terms of the linear dependence of the kinematic screw system and constraint screw system, respectively, and all singular configurations are given as well.

Keywords: Parallel mechanism · Kinematics analysis · Singularity · Uncoupled

1 Introduction

According to the differences of output motion characteristics of the moving platform, parallel mechanism can be divided into pure translational [1], pure rotational [2, 3] and hybrid types [4, 5]. Among them, the moving platform of the pure rotational parallel mechanism with two or three degrees of freedom can change the posture of the object in space to obtain the expected motion.

Generally, parallel mechanisms have strong motion coupling, which is beneficial to improve the rigidity and load carrying capacity of the mechanism, and which also causes some problems such as small working space, complex kinematics, and difficult control design. Therefore, decoupled parallel mechanism has become one of the hot spots in this field [6–8]. Uncoupled parallel mechanism is one special type during the decoupling parallel mechanisms. Its input-output motion has a one-to-one control relationship, which makes the kinematics and dynamic equations very simple and the control design is very easy. Xu [9] proposed two conditions for the full decoupling of two rotational DOFs of 2R1T and 2R PMs with three branches, including a just constraint branch. Jin [10] designed the 2T2R and 3T2R parallel mechanisms with large decoupled output rotational angles by using Lie group theory, and analyzed the fixed posture working space and posture space of the mechanism. Zhang [11] proposed

D. Wang et al. (Eds.): MeTrApp 2019, MMS 79, pp. 259–269, 2020.
https://doi.org/10.1007/978-981-15-0142-5_26

an uncoupled 2T1R parallel mechanism and discussed its kinematic isotropic condition. This paper proposes a novel uncoupled two-rotational parallel mechanism. Kinematic problems and singular configurations of the mechanism are studied in detail.

2 Structural Design and Output Characteristics Analysis

The novel parallel mechanism is shown in Fig. 1, which consists of a moving platform, a fixed base and two single open chains connected to two platforms. The arrangement of the first limb from the base to the platform is followed by rotational joint R_{11}, universal joint U_{12}, prismatic pair P_{13}, and universal joint U_{14}. In order to facilitate structural description of the mechanism, two universal joints U_{12} and U_{14} are replaced by rotational joint R_{12} and R_{13}, and rotational joints R_{15} and R_{16}, respectively.

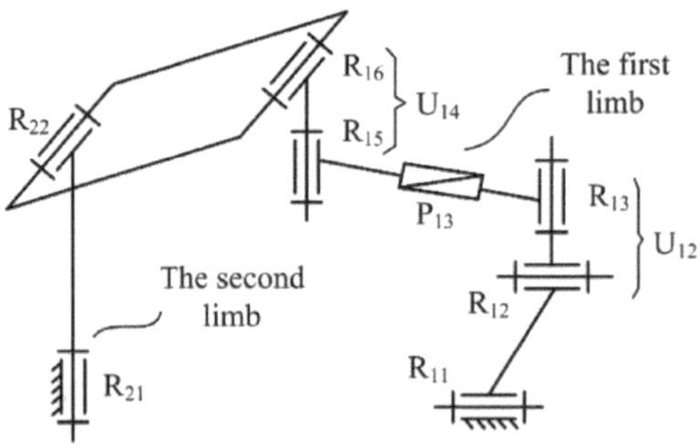

Fig. 1. Diagram of RR-RUPU parallel mechanism

In the first kinematic chain, axes of the joints R_{11} and R_{12} are parallel to each other, axes of the joints R_{13} and R_{15} are parallel to each other and perpendicular to the motion direction of the prismatic pair P_{13}. So the first chain can be recorded as $SOC\{-R_{11}//\overline{R_{12} \perp R_{13}}(\perp P_{13})//\overline{R_{15} \perp R_{16}}-\}$. The arrangement of the second limb from the base to the platform is rotational joints R_{21} and R_{22} in sequence and their axes intersect perpendicularly. This chain can also be recorded as $SOC\{-R_{21} \perp R_{22}-\}$. The assembling conditions of the two limbs are as follows: the axes of joints connected to the base are perpendicular to each other, and the axes of joints attached to the platform are parallel. The first revolute joint in each limb is selected as the actuated joint. The position and orientation characteristic sets M_{b1} and M_{b2} of the first and the second limbs can be written out, respectively, and have

$$M_{b1} = \begin{bmatrix} t^3 \\ r^2//(R_{15}, R_{16}) \cup r^1//R_{11} \end{bmatrix} \tag{1}$$

$$M_{b2} = \begin{bmatrix} t^0 \\ r^2//(R_{21}, R_{22}) \end{bmatrix} \tag{2}$$

where, the right superscript numerals denote the characteristic degree of freedom (DOF) of the limb. For example, $r^2//(R_{15}, R_{16})$ denotes that there are two rotational DOFs in the plane determined by the axes of R_{15} and R_{16}, and t^0 denotes that the number of translational DOF is 0. Therefore, the position and orientation characteristic set of the mechanism shown in Fig. 1 is as follows

$$M_b = M_{b1} \cap M_{b2} = \begin{bmatrix} t^0 \\ r^2//(R_{21}, R_{22}) \end{bmatrix} \tag{3}$$

Formula (3) shows that the platform has only two-dimensional rotational DOFs rotating around the axes of the joints R_{21} and R_{22}. It is obvious that there is without any independently translational mobility for the mechanism.

The DOF of the parallel mechanism can be calculated by Eq. (4), and

$$F = \sum_{i=1}^{m} f_i - \min \sum_{j=1}^{v} \xi_j \tag{4}$$

where, F is the DOF of the mechanism; m is the number of joints, f_i is the DOF of the ith joint, v is the number of independent loops, ξ_j is the number of independent displacement equations of the jth loop. For the mechanism shown in Fig. 1, $m = 8$, $\upsilon = 1$, $\xi_1 = 6$, so we get $F = 2$. The calculation result is consistent with the previous analysis, i.e., the mechanism only has two rotational DOFs in space.

3 Kinematic Analysis of the Mechanism

3.1 Orientation Analysis

Three-dimensional CAD model of the parallel mechanism is shown in Fig. 2. The fixed coordinate system $O - xyz$ is attached on the base. Origin point O is located at the intersection of two rotational joints of the second limb, z-axis coincides with the axis of joint R_{21}, and x-axis is parallel to the axis of joint R_{11}. The moving coordinate system $P - uvw$ is fixed on the platform, its origin point P coincides with the point O, v-axis coincides with the axis of joint R_{22} and w-axis is perpendicular to the platform. At the initial configuration, the axes of the two coordinate systems coincide with each other. Let q_{11} and q_{21} be the input angular displacement of the two actuated joints R_{11} and R_{21}, respectively. α and β are the orientation angles of the platform rotating around the v-axis and the z-axis. β is the length of the active rod in the first limb, r represents the size of the platform structure, which is measured between joints R_{16} and R_{22}.

The local coordinate system $O_1 - x_1y_1z_1$ is fixed on the base. Its origin point O_1 lies on the axis of joint R_{11} and x_1 axis coincides with it. z_1 axis is perpendicular to the platform, as shown in Fig. 2.

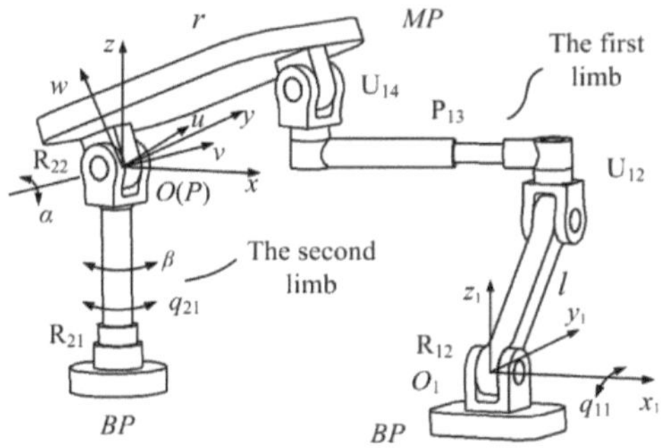

Fig. 2. CAD model of the novel parallel mechanism

The kinematic screw system of the first limb in system $O_1 - x_1y_1z_1$ is written as

$$\begin{aligned}
\$_{11} &= [1 \quad 0 \quad 0; \quad 0 \quad 0 \quad 0]^{\mathrm{T}} \\
\$_{12} &= [1 \quad 0 \quad 0; \quad 0 \quad l\sin q_{11} \quad -l\cos q_{11}]^{\mathrm{T}} \\
\$_{13} &= [0 \quad 0 \quad 1; \quad l\cos q_{11} \quad 0 \quad 0]^{\mathrm{T}} \\
\$_{14} &= [0 \quad 0 \quad 0; \quad L_{14} \quad M_{14} \quad 0]^{\mathrm{T}} \\
\$_{15} &= [0 \quad 0 \quad 1; \quad 0 \quad l\sin q_{11} \quad -l\cos q_{11}]^{\mathrm{T}} \\
\$_{16} &= [-\sin\beta \quad \cos\beta \quad 0; \quad P_{16} \quad Q_{16} \quad R_{16}]^{\mathrm{T}}
\end{aligned} \tag{5}$$

where, L_{ij} and M_{ij} are the direction vector parameters of screw $\$_{ij}$ corresponding to the jth joint in the ith chain, P_{ij}, Q_{ij} and R_{ij} are the position parameters of screw $\$_{ij}$.

According Eq. (5), we know that it is 6-system kinematic screw for the first limb, so there is no any reciprocal screw. In system $O - xyz$, the kinematic screw system of the second limb is as

$$\begin{aligned}
\$_{11} &= [0 \quad 0 \quad 1; \quad 0 \quad 0 \quad 0]^{\mathrm{T}} \\
\$_{22} &= [-\sin\beta \quad \cos\beta \quad 0; \quad 0 \quad 0 \quad 0]^{\mathrm{T}}
\end{aligned} \tag{6}$$

From formula (6), the order of kinematic screw system is 2, so there are 4 reciprocal screws for the second chain, and have

$$\begin{aligned}
\$^r_{21} &= [1 \quad 0 \quad 0; \quad 0 \quad 0 \quad 0]^{\mathrm{T}} \\
\$^r_{22} &= [0 \quad 1 \quad 0; \quad 0 \quad 0 \quad 0]^{\mathrm{T}} \\
\$^r_{23} &= [0 \quad 0 \quad 1; \quad 0 \quad 0 \quad 0]^{\mathrm{T}} \\
\$^r_{24} &= [0 \quad 0 \quad 0; \quad \cos\beta \quad \sin\beta \quad 0]^{\mathrm{T}}
\end{aligned} \tag{7}$$

If the actuated joint of the first limb is locked, the kinematic screw system $\$_1$ composed of the rest joints is

$$\$_1 = [\$_{12} \quad \$_{13} \quad \$_{14} \quad \$_{15} \quad \$_{16}]^{\mathrm{T}} \tag{8}$$

Based on the reciprocal product principle, it can be found that the reciprocal screw of $\$_1$ is obtained as

$$\$_{a1} = [\boldsymbol{S} \quad \boldsymbol{S}_0]^{\mathrm{T}} = [0 \quad 0 \quad 1; \quad P_{a1} \quad Q_{a1} \quad 0]^{\mathrm{T}} \tag{9}$$

where, $\boldsymbol{S}$ is the direction vector of screw $\$_{a1}$, $\boldsymbol{S}_0$ is the dual part of $\boldsymbol{S}$, P_{a1} and Q_{a1} are non-zero parameters related to the position of $\$_{a1}$. In terms of the definition of the actuation screw, we know that $\$_{a1}$ is the driving force applied to the moving platform by the actuated joint R_{11}, and its direction is parallel to the z-axis.

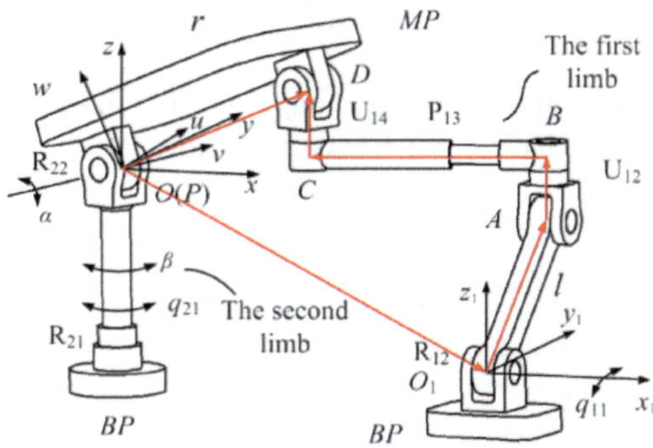

Fig. 3. Closed vector polygon diagram

For the mechanism, the forward orientation solution is to find the posture angles (α and β) of the platform if the input angular displacements (q_{11} and q_{21}) of the actuated joints are known. As shown in Fig. 3, point A is the intersection of two rotational axes in joint U_{12}, point B is the intersection of the moving direction of pair P_{13} and the vertical axis of the joint U_{12}, point C is the intersection of pair P_{13} and the vertical axis of the joint U_{14}, point D is the intersection of two axes in U_{14}. Without losing generality, vector $\boldsymbol{OD}$ can be used to describe the posture angle of the platform.

In Fig. 3, the vectors $\boldsymbol{OD}$, $\boldsymbol{OO}_1$, $\boldsymbol{O}_1\boldsymbol{A}$, $\boldsymbol{AB}$, $\boldsymbol{BC}$ and $\boldsymbol{CD}$ form a closed vector polygon. Then, have

$$\boldsymbol{OD} = \boldsymbol{OO}_1 + \boldsymbol{O}_1\boldsymbol{A} + \boldsymbol{AB} + \boldsymbol{BC} + \boldsymbol{CD} \tag{10}$$

Using the original part $\boldsymbol{S}$ of the screw $\boldsymbol{\$}_{a1}$ to multiply both sides of the Eq. (10), yields

$$\boldsymbol{S} \bullet \boldsymbol{OD} = \boldsymbol{S} \bullet \boldsymbol{OO}_1 + \boldsymbol{S} \bullet \boldsymbol{O}_1\boldsymbol{A} + \boldsymbol{S} \bullet \boldsymbol{AB} + \boldsymbol{S} \bullet \boldsymbol{BC} + \boldsymbol{S} \bullet \boldsymbol{CD} \quad (11)$$

Rearranging Eq. (11), we have

$$r \sin \alpha = O_{1z} + l \sin q_{11} + a + b \quad (12)$$

where, O_{1z} denotes the coordinate value point O_1 along z-axis in system $O - xyz$, and a and b denote the component of vector $\boldsymbol{AB}$ and $\boldsymbol{CD}$ along z axis, respectively.

Since parameters O_{1z}, a and b are only related to the structure size of the mechanism, $O_{1z} + a + b = 0$ can be meet through the reasonable design. Therefore, Eq. (12) can be simplified as follows

$$r \sin \alpha = l \sin q_{11} \quad (13)$$

Thus, the posture angle α of the platform is

$$\alpha = arc \sin(l \sin q_{11}/r) \quad (14)$$

From Fig. 3, we know that the posture angle β of the platform is only related to the angular displacement q_{21} of the actuated joint R_{21} in second limb. Then, we have

$$q_{21} = \beta + c_0 \quad (15)$$

where, c_0 is the initial angle of the platform around the z-axis.

Since both systems $O - xyz$ and $P - uvw$ are defined as coinciding to each other at the initial position, we get $c_0 = 0$. Thus formula (15) can be rewritten as follows

$$\beta = q_{21} \quad (16)$$

3.2 Angular Velocity Analysis

Differentiating Eqs. (13) and (16) with respect to time and rearranging them, the angular velocity equation is obtained

$$\boldsymbol{\omega} = \boldsymbol{J}\dot{\boldsymbol{q}} \quad (17)$$

where, $\boldsymbol{\omega} = \begin{bmatrix} \omega_v & \omega_z \end{bmatrix}^{\mathrm{T}} = \begin{bmatrix} \dot{\alpha} & \dot{\beta} \end{bmatrix}^{\mathrm{T}}$ is the output velocity vector of the platform, $\dot{\boldsymbol{q}} = \begin{bmatrix} \dot{q}_{11} & \dot{q}_{21} \end{bmatrix}^{\mathrm{T}}$ is the input velocity vector of the actuated joints, $\boldsymbol{J}$ is the velocity Jacobian matrix, and have

$$\boldsymbol{J} = \begin{bmatrix} \frac{l \cos q_{11}}{r \cos \alpha} & 0 \\ 0 & 1 \end{bmatrix} \quad (18)$$

Formula (18) shows that the Jacobian matrix $\boldsymbol{J}$ is diagonal, there is a one-to-one control mapping relationship between the inputs and the outputs, so the mechanism has uncoupled kinematics characteristics. Especially the second element on the diagonal line of the matrix is equal to 1, so the condition $\omega_z = \dot{q}_{21}$ is satisfied.

4 Singularity Analysis

Singularity is the inherent property of parallel mechanism. When the structure size of the mechanism is given, the singular configurations will be determined accordingly. Fang and Tsai [12] classified the singular types of parallel mechanisms into three types: limb singularity, actuation singularity and platform singularity.

4.1 Limb Singularity

The occurrence of this kind of singularity is related to the order of limb kinematic screw system, so it is called limb singularity. When the order of the kinematic screw system changes, the order of the constraint screw system also changes, which will lead to instantaneously gain or lose one or more degrees of freedom for the mechanism.

Formula (6) shows that the kinematic system of the second limb is always two-system, so the limb will not have any singular form. Formula (5) shows that the kinematic screw system of the first limb is six-system, and the union of the kinematic screw systems is equivalent to a square matrix $\boldsymbol{A}$. Therefore, singularity of the mechanism is determined that matrix $\boldsymbol{A}$ is full rank or not.

$$\boldsymbol{A} = \begin{bmatrix} 1 & 0 & 0 & 0 & 0 & 0 \\ 1 & 0 & 0 & 0 & l\sin q_{11} & -l\cos q_{11} \\ 0 & 0 & 1 & l\cos q_{11} & 0 & 0 \\ 0 & 0 & 0 & L_{14} & M_{14} & 0 \\ 0 & 0 & 1 & P_{15} & Q_{15} & 0 \\ -\sin\beta & \cos\beta & 0 & P_{16} & Q_{16} & R_{16} \end{bmatrix} \tag{19}$$

If matrix $\boldsymbol{A}$ is not ranked, the kinematic screw system will be linearly dependent. In a result, the determinant $\det(\boldsymbol{A})$ is equal to zero. We know that the elementary transformation of a matrix does not affect the rank of the matrix, that is to say, it does not affect whether the determinant is zero or not. According to the method of matrix block calculation, the determinant value of matrix $\boldsymbol{A}$ can be calculated as

$$\det(\boldsymbol{A}) = \det(\boldsymbol{E})\det(\boldsymbol{F}) \tag{20}$$

Where $\boldsymbol{E} = \begin{bmatrix} 1 & 0 & 0 \\ 1 & 0 & -l\cos q_{11} \\ -\sin\beta & \cos\beta & R_{16} \end{bmatrix}$, $\boldsymbol{F} = \begin{bmatrix} L_{14} & M_{14} & 0 \\ P_{15} & Q_{15} & 1 \\ l\cos q_{11} & 0 & 0 \end{bmatrix}$.

If anyone of the matrices $\boldsymbol{E}$ or $\boldsymbol{F}$ is under-rank, then the matrix $\boldsymbol{A}$ will be under-rank. Singularity of the mechanism can be analyzed as follows.

Type 1. Matrix $\boldsymbol{E}$ is under-rank

Based on matrix $\boldsymbol{E}$, we get $\det(\boldsymbol{E}) = l\cos q_{11}\cos\beta$. If $\det(\boldsymbol{E}) = 0$, then $\beta = \pm\pi/2$ or $q_{11} = \pm\pi/2$. When $\beta = \pm\pi/2$, screws $\$_{11}$, $\$_{12}$ and $\$_{16}$ are parallel to each other and also parallel to screw $\$_{22}$ of the second limb, see Fig. 4.

When the actuated joint of the first limb is locked, the rotation element of the orientation feature set of the first limb is $M_{b1}^{r} = [r^2//(R_{15}, R_{16})]$, which means that the platform still has the mobility rotating around the v axis. The singularity can be avoided by setting the distance between the two universal joints in the first limb.

When $q_{11} = \pm\pi/2$, the kinematic screw system can be computed by substituting it into formula (5). Then, the constraint screw of the limb is obtained, and have

$$\$_{11}^{r} = \begin{bmatrix} 0 & 0 & 1; & 0 & -R_{16}/\cos\beta & 0 \end{bmatrix}^{\mathrm{T}} \tag{21}$$

Formula (21) represents a zero-pitch screw parallel to z axis, which constrains the movement of the first limb along z-axis. It also implies that the platform loses the ability to translate along the same direction. At last, the rotational mobility around joint R_{22} of the platform is disappear. Due to the structural constraint, the singular configuration at $q_{11} = -\pi/2$ does not occur in the assembly condition shown in Fig. 2, and the singular configuration at $q_{11} = \pi/2$ is shown in Fig. 5.

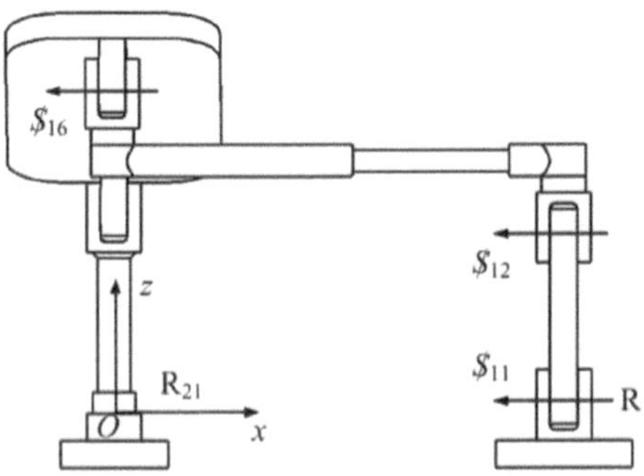

Fig. 4. Singular configuration at $\beta = \pi/2$

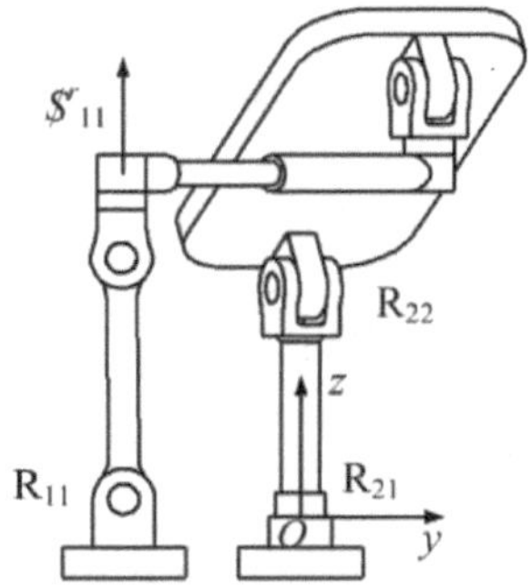

Fig. 5. Singular configuration at $q_{11} = \pi/2$

Type 2. Matrix $\boldsymbol{F}$ is under-rank

Based on matrix $\boldsymbol{F}$, we have

$$\det(\boldsymbol{F}) = M_{14} l \cos q_{11} \tag{22}$$

If $\det(\boldsymbol{F}) = 0$, then we get $q_{11} = \pm\pi/2$ or $M_{14} = 0$. When $q_{11} = \pm\pi/2$, the same results will be obtained like type 1. When $M_{14} = 0$, kinematic screw $\$_{14}$ of pair P_{13} can be rewritten as $\$_{14} = \begin{bmatrix} 0 & 0 & 0; & 1 & 0 & 0 \end{bmatrix}^{\mathrm{T}}$, which is an infinite pitch screw parallel to x axis. At this configuration, screws $\$_{12}$, $\$_{14}$ and $\$_{15}$ are coplanar and linearly dependent so that the limb kinematic singularity occurs, shown in Fig. 6.

4.2 Actuation Singularity Analysis

When the actuated joints of two limbs are locked at the same time, the constraint screws $\$_1^r$ and $\$_2^r$ of two limbs can be obtained in terms of the principle of reciprocal production, and have

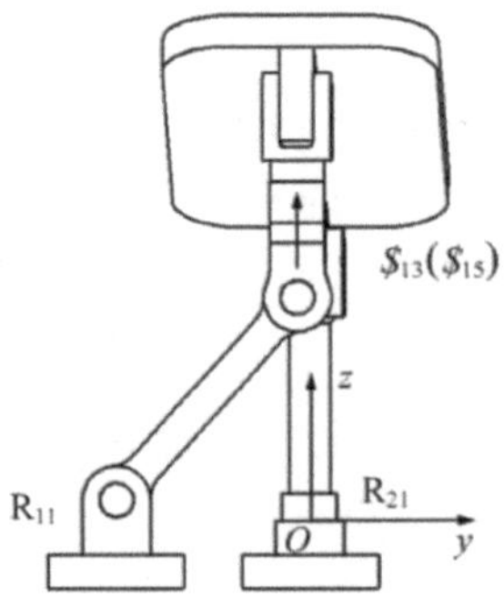

Fig. 6. Singular configuration when $M_{14} = 0$

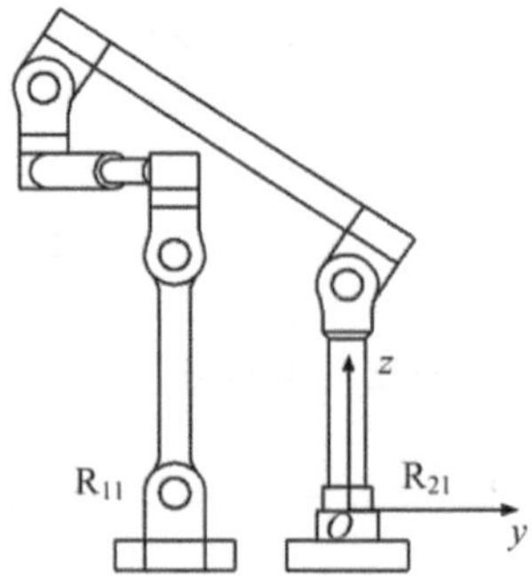

Fig. 7. Singular configuration when $\beta = -\pi/2$, and $q_{11} = \pi/2$

$$\$_1^r = [0 \quad 0 \quad 1; \quad l\cos q_{11} \quad -(\cos\beta + l\cos q_{11}\sin\beta)/R_{16} \quad 0]^{\mathrm{T}} \tag{23}$$

$$\$_2^r = \begin{cases} \$_{21}^r = [1 \quad 0 \quad 0; \quad 0 \quad 0 \quad 0]^{\mathrm{T}} \\ \$_{22}^r = [0 \quad 1 \quad 0; \quad 0 \quad 0 \quad 0]^{\mathrm{T}} \\ \$_{23}^r = [0 \quad 0 \quad 1; \quad 0 \quad 0 \quad 0]^{\mathrm{T}} \\ \$_{24}^r = [0 \quad 0 \quad 0; \quad \cos\beta \quad \sin\beta \quad 0]^{\mathrm{T}} \\ \$_{25}^r = [0 \quad 0 \quad 0; \quad 0 \quad 0 \quad 1]^{\mathrm{T}} \end{cases} \tag{24}$$

Rearranging $\$_1^r$ and $\$_2^r$ into a matrix $\boldsymbol{G}$, yields

$$\boldsymbol{G} = \left[\begin{array}{ccc|ccc} 1 & 0 & 0 & 0 & 0 & 0 \\ 0 & 1 & 0 & 0 & 0 & 0 \\ 0 & 0 & 1 & 0 & 0 & 0 \\ \hline 0 & 0 & 0 & \cos\beta & \sin\beta & 0 \\ 0 & 0 & 0 & 0 & 0 & 1 \\ 0 & 0 & 1 & l\cos q_{11} & Q & 0 \end{array}\right] \tag{25}$$

After dividing into four sub-blocks, we know if and only if the second sub-block matrix on the main diagonal is not full of rank, the matrix $\boldsymbol{G}$ is under-rank. At this condition, we have $\beta = q_{11} = \pm\pi/2$. There are total 4 sets of singular configurations, i.e.,

(a) $\beta = \pi/2,\ q_{11} = \pi/2$; (b) $\beta = -\pi/2,\ q_{11} = \pi/2$
(c) $\beta = \pi/2,\ q_{11} = -\pi/2$; (d) $\beta = -\pi/2,\ q_{11} = -\pi/2$.

Due to the structural constraints of the mechanism, the configurations (c) and (d) do not occur under the assembly condition shown in Fig. 2.

Based on the definition of the actuation screw, $\$_1^r$ represented by the Eq. (24) is the driving force spiral applied to the platform by joint R_{11} through the first limb. If the R_{11} is locked, the constraint screw $\$_1^r$ is linearly dependent to the constraint screw $\$_2^r$. So the platform loses the DOF rotating around v-axis. When $\beta = -\pi/2$, and $q_{11} = \pi/2$, singular configuration is shown in Fig. 7.

4.3 Platform Singularity Analysis

When the constraint screws of the mechanism are linear dependence, the platform will lose some ability to constraint its motions. In a result, one or more unexpected degrees of freedom will produce. From formula (7), we know that the constraint screw system applied by the limbs to the platform contains three linear independent zero-pitch screws and one infinite-pitch screw. When the assembly form of the mechanism is given as Fig. 2, there will be no platform singularity.

5 Conclusions

(1) A novel uncoupled two-rotational parallel mechanism is designed in this paper. The kinematic output characteristics of the mechanism are analyzed. Mathematical models of the kinematic problems are established.
(2) Singularity of the mechanism is discussed, including limb singularity, actuation singularity and platform singularity. The singular configurations are also given. The research provides a theoretical basis for the practical application of the mechanism.

Acknowledgments. The authors would like to thank the financial support from the Program for Science and Technology of Henan Province (192102210221), Fundamental Project of Key Scientific Research of Henan Advanced Education (18A460001), and Program for Postgraduate Innovative Foundation of Henan University of Science and Technology (CXJJ-2018-KJ03).

References

1. Zhang, Y.B., Zhao, Y.F., Jing, X.L., et al.: Type synthesis of uncoupled translational parallel manipulators based on actuation wrench screw theory. Adv. Mech. Eng. **10**(1), 1–10 (2018)
2. Di Gregorio, R.: Kinematics of a new spherical parallel manipulator with three equal legs: the 3-URC wrist[J]. J. Field Robot. **18**(5), 213–219 (2010)
3. Hou, Y.L., Duan, Y.B., Yao, J.T., et al.: Configuration optimization and static analysis of adjusting parallel mechanism for the sub-reflector of antenna. Adv. Mater. Res. **338**, 425–430 (2011)
4. Wang, C., Fang, Y., Fang, H.: Novel 2R3T and 2R2T parallel mechanisms with high rotational capability. Robotica **35**(02), 401–418 (2017)
5. Kumar, N., Piccin, O., Bayle, B., et al.: A task-based type synthesis of novel 2T2R parallel mechanisms. Mech. Mach. Theory **77**(7), 59–72 (2014)
6. Xu, Y.D., Chen, L.L., Yan, W.N., et al.: Motion decoupling analysis of a kind of 2R parallel mechanism with two continuous rotational axes. Mech. Mach. Sci. **36**, 307–313 (2016)
7. Hou, Y.L., Zeng, D.X., Duan, Y.B., et al.: The nonexistence of three degrees of freedom rotational fully decoupled parallel mechanism. Appl. Mech. Mater. **284–287**, 1951–1955 (2013)
8. Kuo, C.H., Dai, J.S.: Kinematics of a fully-decoupled remote center-of-motion parallel manipulator for minimally invasive surgery. J. Med. Devices **6**(2), 211–214 (2012)
9. Xu, Y.D., Wang, B., Wang, Z.F., et al.: Investigations on the principle of fully decoupling and type synthesis of the 2R1T and 2R parallel mechanisms. Trans. Can. Soc. Mech. Eng. **43**(2), 263–271 (2018)
10. Jin, X.D., Fang, Y.F., Qu, H.B., et al.: A class of novel 2T2R and 3T2R parallel mechanisms with large decoupled output rotational angles. Mech. Mach. Theory **114**, 156–169 (2017)
11. Zhang, Y., He, X., Wu, X., et al.: Kinematic analysis of a novel spatial uncoupled parallel manipulator. In: Proceedings of IEEE International Conference on Electronic and Mechanical Engineering and Information Technology, Harbin, China, pp. 3345–3348 (2011)
12. Fang, Y.F., Tsai, L.W.: Structure synthesis of a class of 4-DoF and 5-DoF parallel manipulators with identical limb structures. Int. J. Robot. Res. **21**(9), 799–810 (2002)

Performance Analysis and Comparison of Three Planar Parallel Manipulators

Xiaoyong Wu(✉), Ran Yan, Zhaowei Xiang, Fangyan Zheng, and Rulong Tan

School of Mechanical Engineering, Chongqing University of Technology, Chongqing 400054, China
wuxy@cqut.edu.cn

Abstract. This work presents a comprehensive performance evaluation and comparison of three planar parallel manipulators with PPR limbs. The kinematic model of the redundant 4-PPR planar parallel manipulator is established, upon which the inverse position model and singularity are analyzed. Then, the evaluations of dexterity, velocity and stiffness performance are achieved. Performance comparison between the redundant manipulator and another two non-redundant 3-PPR planar parallel manipulators, one with Δ-shape structure and the other with U-shape structure, is presented. Simulation results reveal that the U-shape manipulator owns the greatest velocity performance. Moreover, the redundant one possesses the best dexterity and stiffness performance.

Keywords: Parallel manipulator · Redundant actuation · Kinematics · Performance analysis

1 Introduction

Planar parallel manipulators (PPMs) are a special type of parallel manipulators (PMs), which can provide planar motion with one rotation around z-axis and two translations along x- and y-axes [1]. Angeles et al. [2] reported the optimum kinematic design of a 3-RRR PPM. Nonsingular assembly mode change of PPMs has been discussed in [3]. With the application of screw theory, singularity analysis of some PPMs with identical serial chains was presented in [4]. The performance of a type of PPMs was compared in [5]. The kinematics and dynamics of the known 3-PRR were analyzed in [6]. A geometrical approach for workspace analysis of PPMs was presented in [7]. Kinematics analysis of a proposed 3-PRP PPM was developed in [8]. Geometric analysis of the kinematic sensitivity of PPMs was available in [9].

For the 3-PPR PPMs, extensive researches have been reported in [10–14]. Kinematic analysis and optimal design of a 3-PPR PPM with a Δ-shape (equilateral triangle) base platform were presented in [10]. In order to maximize the workspace, a 3-PPR PPM with a U-shape (square) base platform was introduced in [11]. Dynamics and controller performance of the U-shape 3-PPR PPM was investigated in [12]. 3-PPR PPMs with Δ- and U-shape base platforms were compared with respect to kinematic sensitivity to joint clearances and workspace was addressed in [13]. Aiming at

D. Wang et al. (Eds.): MeTrApp 2019, MMS 79, pp. 270–279, 2020.
https://doi.org/10.1007/978-981-15-0142-5_27

optimum shape design of the moving platform (MP) and the base platform, a parametric model of general 3-PPR PPMs was available in [14].

As well known, redundant actuation is an effective method to optimize the kinematic and dynamic performance of PMs. In order to improve the dexterity, stiffness and dynamic performance of the 3-PPR PPM, another identical PPR limb is added to the manipulator, namely, a 4-PPR PPM is obtained. Performance comparison between this redundant one and the other two non-redundant ones are conducted.

The rest of this paper is organized as follows. Kinematics analysis of the considered redundant PPM is presented in Sect. 2. Performance indices in dexterity, velocity and stiffness are illustrated in Sect. 3. Performance evaluation and comparison between the redundant PPM and the other two non-redundant PPMs are presented in Sect. 4. Finally, conclusions are described.

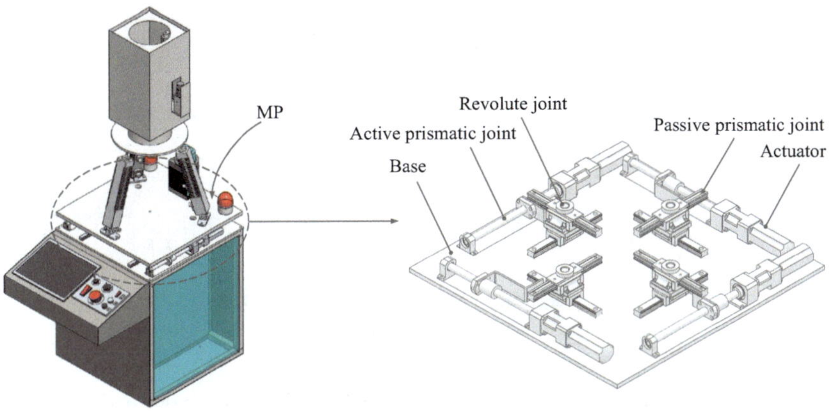

Fig. 1. 3D model of the 4-PPR redundant PPM

2 Kinematics Analysis

The proposed redundant 4-PPR PPM, aiming at providing planar motion with great dexterity, stiffness and dynamic performance for an assembly platform, is depicted in Fig. 1. In this application, the first prismatic joint in each limb is chosen as the active joint, so the proposed manipulator is a 4-PPR PPM. As depicted in Fig. 2(a), the base platform and the MP are both modeled as squares, and their side lengths are b and a, respectively. The global coordinate system $\{\boldsymbol{O}_0\}$ is fixed on the square base, with the origin coinciding with a vertex of the base. On the other hand, the local coordinate system $\{\boldsymbol{O}\}$ is attached to the center of the MP. Vector $\mathbf{p} = [x, y]^T$ provides the position of the MP with respect to the global coordinate system, and θ is the corresponding rotation angle. Moreover, $\theta = 0°$ is defined as the initial assembly mode. The considered Δ- and U-shape 3-PPR PPMs are depicted in Fig. 2(b) and (c). For the Δ-shape one, the base platform is modeled as an equilateral triangle with side length b, while the U-shape one has a square base platform with the same length.

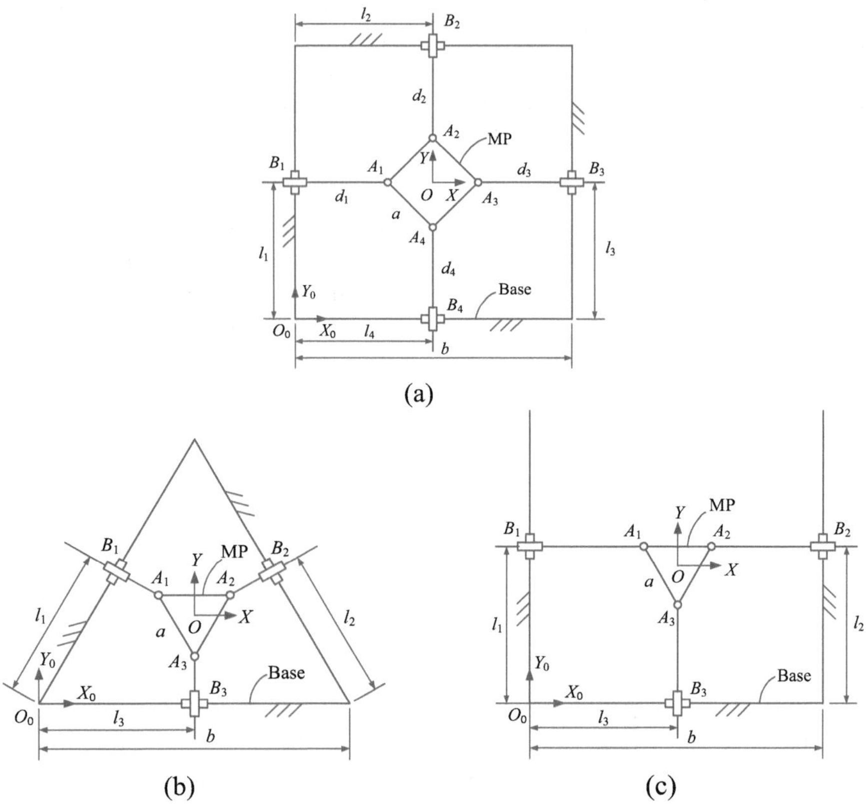

Fig. 2. Kinematic description of (a) the redundant one, (b) the Δ-shape one and (c) the U-shape one

2.1 Inverse Kinematics

The inverse kinematic problem is to determine the input of each active joint with the given posture of the MP. For the 4-PPR PPM, the posture of the MP is defined as (x, y, θ), the position vector of point A_i in the global coordinate system can be denoted by

$$\mathbf{a}_i = \mathbf{R}\mathbf{a'}_i + \mathbf{p} \tag{1}$$

where $\mathbf{R}$ represents the rotation matrix of the MP, $\mathbf{a}'_i$ denotes the position vector of point A_i in the local coordinate system, which can be expressed as

$$\mathbf{a}'_1 = \left[-\frac{\sqrt{2}}{2}a, 0\right]^T, \ \mathbf{a}'_2 = \left[0, \frac{\sqrt{2}}{2}a\right]^T, \ \mathbf{a}'_3 = \left[\frac{\sqrt{2}}{2}a, 0\right]^T, \ \mathbf{a}'_4 == \left[0, -\frac{\sqrt{2}}{2}a\right]^T \tag{2}$$

The position vector of point A_i can also be expressed in a form as follows

$$\mathbf{a}_i = \mathbf{b}_i + \mathbf{d}_i \tag{3}$$

with

$$\mathbf{b}_1 = l_1\mathbf{j},\ \mathbf{b}_2 = l_2\mathbf{i} + b\mathbf{j},\ \mathbf{b}_3 = b\mathbf{i} + l_3\mathbf{j},\ \mathbf{b}_4 = l_4\mathbf{i} \tag{4}$$

$$\mathbf{d}_1 = d_1\mathbf{i},\ \mathbf{d}_2 = -d_2\mathbf{j},\ \mathbf{d}_3 = -d_3\mathbf{i},\ \mathbf{d}_4 = d_4\mathbf{j} \tag{5}$$

where d_i represents the displacement of each passive prismatic joint, l_i denotes the input of each active prismatic joint.

Combining Eqs. (1) and (3) yields solutions to the inverse kinematic problem.

$$\begin{cases} l_1 = y - \sqrt{2}a\sin\theta/2 \\ l_2 = x - \sqrt{2}a\sin\theta/2 \\ l_3 = y + \sqrt{2}a\sin\theta/2 \\ l_4 = x + \sqrt{2}a\sin\theta/2 \end{cases} \tag{6}$$

2.2 Singularity Analysis

For general PMs, the relationship between the velocity of the active joints $\dot{\mathbf{q}}$ and the velocity of the MP $\dot{\mathbf{x}}$ can be expressed as

$$\dot{\mathbf{q}} = \mathbf{J}\dot{\mathbf{x}} \tag{7}$$

where $\mathbf{J}$ represents the Jacobian matrix.

Differentiating both sides of Eq. (6) with respect to time, we obtain

$$\dot{\mathbf{q}}_1 = \mathbf{J}_1\dot{\mathbf{x}}_1 \tag{8}$$

where $\dot{\mathbf{x}}_1$ represents the velocity vector of the MP, $\dot{\mathbf{q}}_1$ denotes the velocity vector of the active joints. $\mathbf{J}_1$ is the Jacobian matrix, which can be obtained as

$$\mathbf{J}_1 = \begin{bmatrix} 0 & 1 & -\sqrt{2}a\cos\theta/2 \\ 1 & 0 & -\sqrt{2}a\cos\theta/2 \\ 0 & 1 & \sqrt{2}a\cos\theta/2 \\ 1 & 0 & \sqrt{2}a\cos\theta/2 \end{bmatrix} \tag{9}$$

For redundant PMs, singularity will occur when the rank of the Jacobian matrix is lower than its number of rows, namely, the determinant of $\mathbf{J}^T\mathbf{J}$ is equal to zero. For the considered 4-$\underline{\text{P}}$PR PPM, we have

$$\det(\mathbf{J}_1^T\mathbf{J}_1) = 8a^2\cos^2\theta \tag{10}$$

Obviously, singularity occurs when $\theta = \pm 90°$ or $a = 0$. In order to get free from singularity, the feasible range of the workspace of the considered redundant PPM is defined as $90° < \theta < 90°$. Moreover, a is greater than 0.

3 Performance Indices

As with performance analysis, three indices are considered, namely, dexterity index, velocity index and stiffness index.

3.1 Dexterity Index

Dexterity index is an important performance index for mechanism design. The dexterity of a PM is defined as the ability to change its position and orientation arbitrarily, which can be measured in terms of condition number. For a Jacobian matrix of full rank, the corresponding condition number can be expressed as

$$k = \|\mathbf{J}\|\left\|\mathbf{J}^{-1}\right\| \tag{11}$$

where $\|\cdot\|$ is the Frobenius norm of the matrix.

As well known, the value of condition number will range from one to infinity, without an upper bound. So, the reciprocal of the condition number called as local conditioning index (LCI) is used more commonly.

$$\mathrm{LCI} = \frac{1}{k} \tag{12}$$

LCI is a local performance metric bounded between 0 and 1. The manipulator has a great dexterity performance when its LCI is large, and singular poses are characterized by LCI = 0, whereas optimal (isotropic) configurations are denoted by LCI = 1.

3.2 Velocity Index

Velocity index is another important performance index for PPMs. Referring to Eq. (7), the relationship between the velocity of the active joints and the velocity of the MP can be expressed in another form as follows

$$\dot{\mathbf{x}} = \mathbf{J}^{-1}\dot{\mathbf{q}} \tag{13}$$

Then we have

$$\dot{\mathbf{x}}^T\dot{\mathbf{x}} = \dot{\mathbf{q}}^T(\mathbf{J}^{-1})^T\mathbf{J}^{-1}\dot{\mathbf{q}} \tag{14}$$

Let the velocity vector of the active joints be a unit vector $\left(\|\dot{\mathbf{q}}\|^2 = 1\right)$, the norm of the velocity vector of the MP can be expressed as

$$\|\dot{\mathbf{x}}\|^2 = \dot{\mathbf{x}}^T\dot{\mathbf{x}} = \dot{\mathbf{q}}(\mathbf{J}^{-1})^T\mathbf{J}^{-1}\dot{\mathbf{q}} = \lambda_{vi} \tag{15}$$

where λ_{vi} is the eigenvalue of the matrix $(\mathbf{J}^{-1})^T\mathbf{J}^{-1}$.

Therefore, the extremum of the output velocity of the MP can be expressed as

$$V_{\min} = \sqrt{\min(|\lambda_{vi}|)} \tag{16}$$

$$V_{\max} = \sqrt{\max(|\lambda_{vi}|)} \tag{17}$$

For the 4-PPR PPM, the rank of matrix $(\mathbf{J}_1^+)^T\mathbf{J}_1^+ \in \mathbf{R}^{4\times4}$ is less than 4, that means there always exists a zero eigenvalue which leads to $V_{\min} = 0$, so $V_{\max}$ is chosen as the velocity index. The larger the velocity index is, the better the velocity performance is.

3.3 Stiffness Index

As reported in [15], the stiffness matrix of PPMs can be written as

$$\mathbf{K} = \mathbf{J}^T\mathbf{K}_P\mathbf{J} \tag{18}$$

where $\mathbf{K}_P$ is a matrix describing the stiffness of the platform mounted with elastic elements. For simplicity, $\mathbf{K}_P = diag(k_1, k_2, k_3)$. The deflection of the MP is thus

$$\Delta\mathbf{X} = \mathbf{K}^{-1}\mathbf{f} \tag{19}$$

where $\mathbf{f} = [f_x, f_y, M_\theta]^T$ denotes the vector of moment and forces applied to the MP, $\Delta\mathbf{X} = [\Delta x, \Delta y, \Delta\theta]^T$ represents the corresponding deflections.

It can be found that the maximum or minimum eigenvalue λ_{si} of $(\mathbf{K}^{-1})^T\mathbf{K}^{-1}$ is the limit value of $\|\Delta\mathrm{X}\|_2$. Let $k_1 = k_2 = k_3 = 1$ and $\|\mathbf{f}\| = 1$, the minimum and maximum deformations can be expressed as

$$S_{\min} = \sqrt{\min(|\lambda_{si}|)} \tag{20}$$

$$S_{\max} = \sqrt{\max(|\lambda_{si}|)} \tag{21}$$

In fact, both minimum and maximum deformations can be chosen as the stiffness indices. To keep consistent with the velocity index, the maximum deformation S_{max} is chosen as the stiffness index. The larger the deformation is, the worse the stiffness performance of the PPM is.

4 Performance Evaluation and Comparison

With the three performance indices introduced above, performance analyses of the considered PPMs are presented. As analyzed in Sect. 2.2, only θ and a have influence on the Jacobian matrix, so these two parameters are taken into consideration for performance evaluation. Note that, the three PPMs will be singular when $\theta = \pm 90°$, for this reason the feasible range of the workspace is defined as $-90° < \theta < 90°$. Moreover, the feasible range of the side length of the MP is defined as 0 mm < a < 1000 mm.

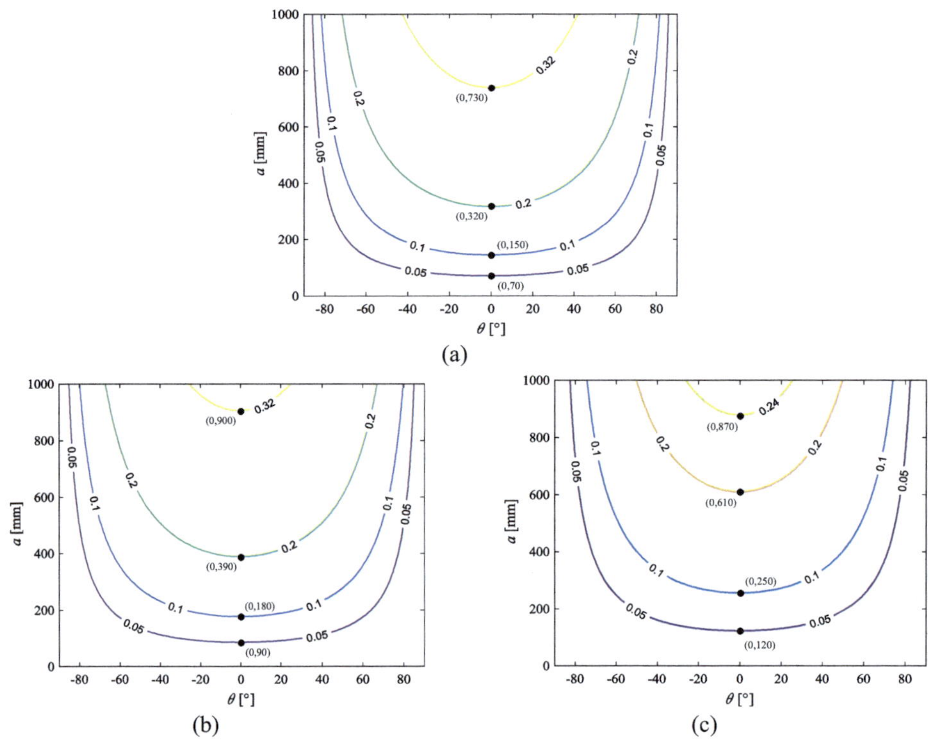

Fig. 3. Dexterity performance of (a) the redundant one, (b) the Δ-shape one and (c) the U-shape one

We look at first on the dexterity performance. As shown in Fig. 3, the distribution laws of dexterity of the three manipulators are very similar. The dexterity is symmetrical about $\theta = 0°$, and it is good near $\theta = 0°$. On the other hand, the dexterity is poor when the PPMs approach to singular configurations. Moreover, the bigger the side length a is, the better the dexterity performance is. Here, under the same situation (the same a), performance comparison between the three considered PPMs is conducted. It is found that the 4-$\underline{P}$PR redundant one owns the best dexterity performance, while that of the U-shape one is the worst.

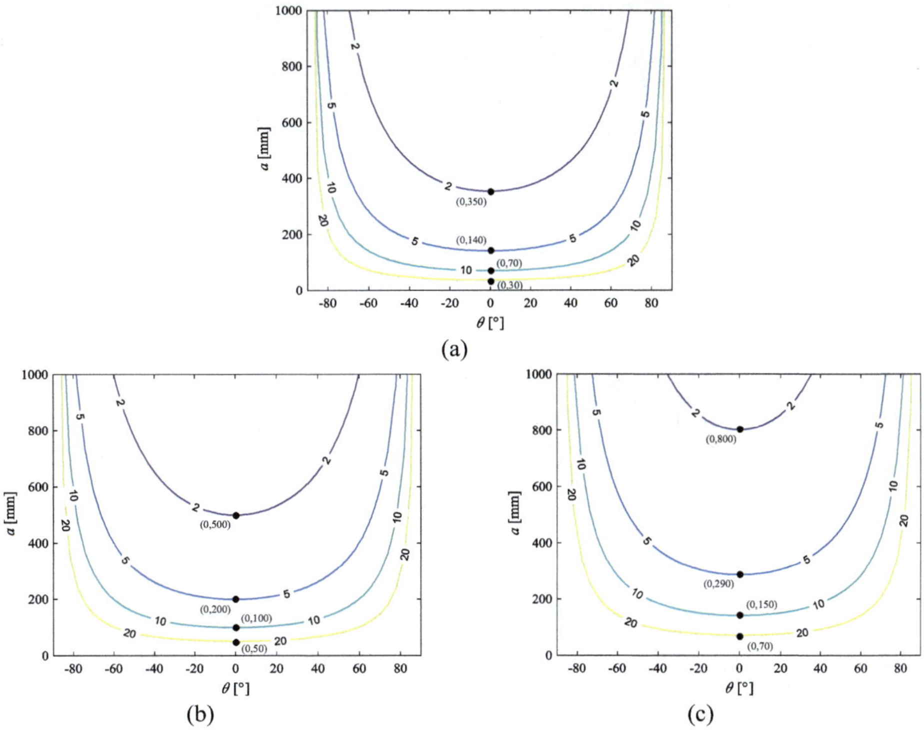

Fig. 4. Velocity performance of (a) the redundant one, (b) the Δ-shape one and (c) the U-shape one

With regard to the velocity index, the simulation results are depicted in Fig. 4. The distribution laws of velocity performance of the three manipulators are also similar, which are symmetrical about $\theta = 0°$. The velocity performance is great when the PPMs approach to singular configurations, but it is not so good near $\theta = 0°$. It is also found that the smaller the side length a is, the better the velocity performance is. Moreover, the U-shape one possesses the best velocity performance, and that of the redundant one is the worst.

At last, the stiffness index is considered, and the results are shown in Fig. 5. It is found that the distribution of stiffness performance is symmetrical about $\theta = 0°$ too. The stiffness performance will become poor when the PPMs approach to singular configurations and the zone near $\theta = 0°$ possesses good stiffness performance. The larger the side length a is, the better the stiffness performance is. Moreover, the stiffness performance of the redundant one is the best, and that of the U-shape one is the worst.

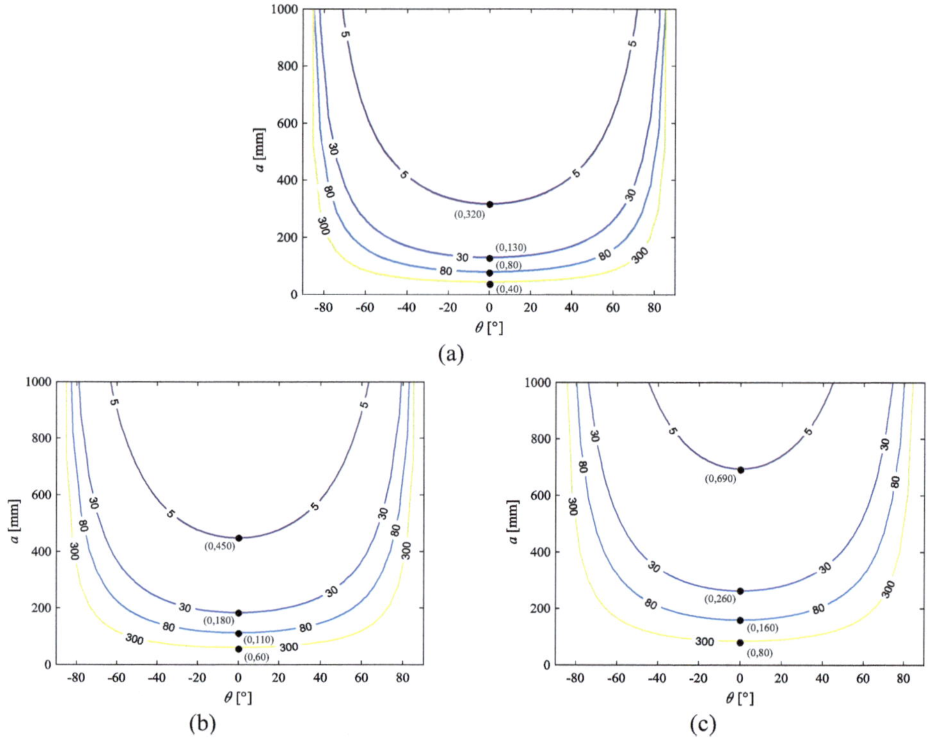

Fig. 5. Stiffness performance of (a) the redundant one, (b) the Δ-shape one and (c) the U-shape one

5 Conclusions

In this work, a novel 4-$\underline{P}$PR redundant PPM is proposed for precision assembly. The performance of the redundant manipulator and another two non-redundant ones are analyzed, while the evaluation is mainly based on their dexterity, velocity and stiffness performance. For the three considered PPMs, the distributions of all the three performances are symmetrical. Under the same situation, the velocity performance of the U-shape one is the best, and that of the redundant one is the worst. Moreover, the dexterity and stiffness performance of the redundant one is the best, while these performances of the U-shape one are the worst. In addition, all the three performances of the Δ-shape one is at the mid-level. From this opinion, it can be concluded that the dexterity and stiffness performance are improved by redundant actuation.

Funding. This work is financially supported by the Scientific Research Foundation of Chongqing University of Technology, the National Natural Science Foundation of China (No. 51805060).

References

1. Wu, J., Wang, J., You, Z.: A comparison study on the dynamics of planar 3DOF 4-RRR, 3-RRR and 2-RRR parallel manipulators. Robot. Comput. Integr. Manuf. **27**, 150–156 (2011)
2. Angeles, J., Gosselin, C.: The optimum kinematic design of a planar three-degree-of-freedom parallel manipulator. ASME J. Mech. Trans. Autom. Des. **110**, 35–41 (1988)
3. Husty, M.L.: Non-singular assembly mode change in 3-RPR parallel manipulators. In: Kecskeméthy, A., Müller, A. (eds.) Computational Kinematics: Proceedings of the 5th International Workshop on Computational Kinematics, pp. 51–60. Springer, Heidelberg (2009)
4. Bonev, I.A., Zlatanov, D., Gosselin, C.M.: Singularity analysis of 3-DOF planar parallel mechanisms via screw theory. J. Mech. Des. **125**, 573–581 (2003)
5. Zhu, Z., Dou, R.: Optimum design of 2-DOF parallel manipulators with actuation redundancy. Mechatronics **19**, 761–766 (2009)
6. Staicu, S.: Inverse dynamics of the 3-PRR planar parallel robot. Robot. Auton. Syst. **57**, 556–563 (2009)
7. Merlet, J.-P., Gosselin, C.M., Mouly, N.: Workspaces of planar parallel manipulators. Mech. Mach. Theory **33**, 7–20 (1998)
8. Zarkandi, S.: Kinematics of a star-triangle planar parallel manipulator. J. Mech. Sci. Technol. **25**, 3223–3230 (2011)
9. Saadatzi, M.H., Masouleh, M.T., Taghirad, H.D., Gosselin, C., Cardou, P.: Geometric analysis of the kinematic sensitivity of planarparallel mechanisms. Trans. Can. Soc. Mech. Eng. **35**, 477–490 (2011)
10. Choi, K.-B.: Kinematic analysis and optimal design of 3-PPR planar parallel manipulator. KSME Int. J. **17**, 528–537 (2003)
11. Bai, S., Caro, S.: Design and analysis of a 3-PPR planar robot with U-shape base. In: International Conference on Advanced Robotics, pp. 1–6 (2009)
12. Singh, Y., Vinoth, V., Kiran, Y.R., Mohanta, J.K., Mohan, S.: Inverse dynamics and control of a 3-DOF planar parallel (U-shaped 3-PPR) manipulator. Robot. Comput. Integr. Manuf. **34**, 164–179 (2015)
13. Binaud, N., Caro, S., Bai, S., Wenger, P.: Comparison of 3-PPR parallel planar manipulators based on their sensitivity to joint clearances. In: IEEE/RSJ International Conference on Intelligent Robots and Systems, pp. 2778–2783 (2010)
14. Wu, X., Xie, Z., Kepler, J.A., Bai, S.: A parametric model of 3-PPR planar parallel manipulators for optimum shape design of platforms. Mech. Mach. Theory **118**, 139–153 (2017)
15. Wu, X.: Optimal design and singularity analysis of a spatial parallel manipulator. Symmetry **11**, 551 (2019)

Singularity Loci of a New Four-Limb Parallel Schönflies-Motion Generator

Guanglei Wu[1(✉)], Huiping Shen[2], Wenkang Zhao[1], and Zirong Lin[1]

[1] School of Mechanical Engineering, Dalian University of Technology, Dalian 116024, China
gwu@dlut.edu.cn

[2] School of Mechanical Engineering, Changzhou University, Changzhou, China
shp65@126.com

Abstract. This paper studies the singularities of a four-limb parallel Schönflies-motion generator. The end-effector design adopts a planetary gear train as the amplification mechanism to realize the full-circle rotation. The singularity is analyzed by means of Jacobian-based method, which are applicable for the singularity identification graphically and numerically. Moreover, the constraint singularity loci are analyzed for avoidance. Based upon the singularity analysis, this robot admits a super-ellipsoidal workspace with the full-circle rotation of the end-effector, suitable for pick-and-place operations.

Keywords: Schönflies motion generator · Singularity · Pick-and-place · Planetary gear train

1 Introduction

Parallel Schönflies-motion generators (SMGs), are well adapted to pick-and-place (PnP) operations, thanks to their advantages of high stiffness and lightweight structure. Most of this type of robots inherit the architecture of Adept Quattro robot [1], having the symmetric structure with four identical limbs [2,6,8,10,11]. Different from the existing mobile platforms, a new parallel SMG is introduced, of which the planetary gear train (PGT) is adopted to construct the mobile platform for the full-circle rotation of the end-effector.

In the design and application of robotics, singularity analysis is a central issue due to their major effect on the robot performance. A common approach to identify the singularity is to examine the determinants of the kinematic Jacobian matrices of the manipulator [4]. Whereas, this analysis method cannot find the constraint singularity [12] that occurs in the restricted motion. Alternatively, the constraint singularity can be analyzed from the motion/force constraint performance, utilizing the approaches of Grassmann line geometry [3,7] and screw theory [5]. In this paper, the robot singularity is analyzed with the Jacobian-based method, combining the constraint singularity.

D. Wang et al. (Eds.): MeTrApp 2019, MMS 79, pp. 280–289, 2020.
https://doi.org/10.1007/978-981-15-0142-5_28

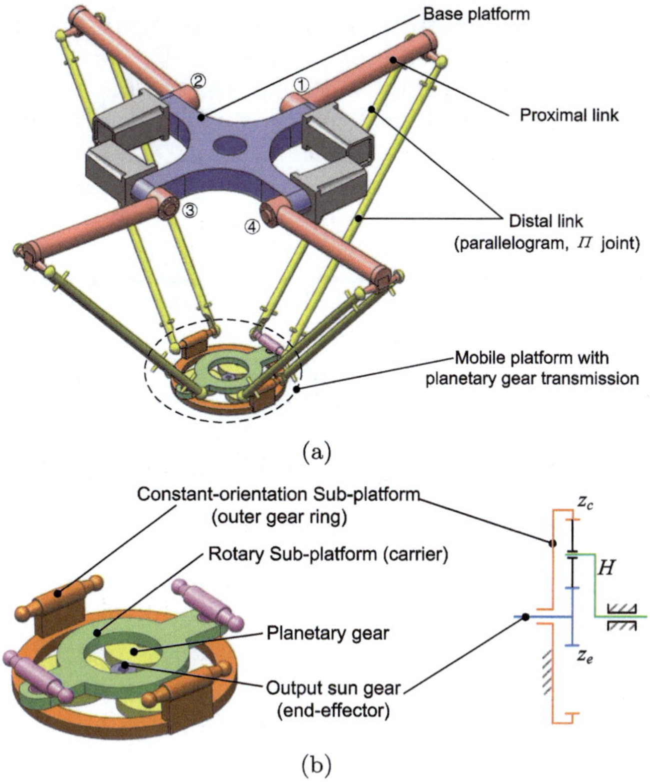

Fig. 1. The CAD model of the new SMG: (a) overall architecture; (b) mobile platform composed of planetary gear train.

In this paper, the singularity analysis is carried out, based on the Jacobian method. Two types of singularities are identified as well as the constraint singularity. Singularity loci are visualized with respect to the workspace. The rectangular workspace, namely, a super-ellipsoidal workspace with the full-circle rotation of the end-effector, is identified under a given working mode suitable for PnP operations.

2 Manipulator Architecture

Figure 1 depicts the simplified CAD model of the design of the new four-limb robot. Similar to its existing robot counterparts, each limb is composed of an actuated proximal link and a passive distal arm (parallelogram, a.k.a 'Π joint') connected to the base and mobile platforms, as shown in Fig. 1(a). The geometric centers of the four actuated joints are located at the vertices of a rectangle on the base platform to avoid the singularities when the robot is in some symmetric configurations.

Figure 1(b) shows the PGT based mobile platform, consisting of a gear ring as the constant-orientation sub-platform, a carrier as the rotational sub-platform, three symmetrical planetary gears and a sun gear as the output end-effector. The end-effector orientation is amplified by the PGT, which can be simplified as a revolute pair. The orientation of the rotational sub-platform is hence can be computed from the required end-effector orientation and speed ratio of the PGT below

$$i_{eH} = \frac{\omega_e}{\omega_H} = 1 + \frac{z_c}{z_e} \tag{1}$$

where ω_e and ω_H stand for the speeds of the end-effector speed and the rotational sub-platform, and z_c and z_e are the number of the tooth of the gear ring and the output sun gear, respectively. One example is $i_{eH} = 8$ with $z_c = 140$ and $z_e = 20$ to realize the end-effector full-circle rotation with the sub-platform rotation 45°.

3 Robot Geometry and Kinematic Jacobian Matrix

The parametrization of the robot is shown in Fig. 2. The reference coordinate frame $\mathcal{F}_b$ is built with the origin located at the geometric center of the base platform, with x-axis parallel to the segment A_2A_1 (A_3A_4). The moving coordinate frame $\mathcal{F}_p$ is attached to the rotational sub-platform with the origin located at the geometric center, and its X-axis is parallel to segment C_3C_1. In the ith limb, the axis of rotation of the actuated joint is parallel to unit vector $\mathbf{u}_i$, and unit vectors $\mathbf{v}_i$ and $\mathbf{w}_i$ are parallel to segments A_iB_i and B_iC_i, respectively.

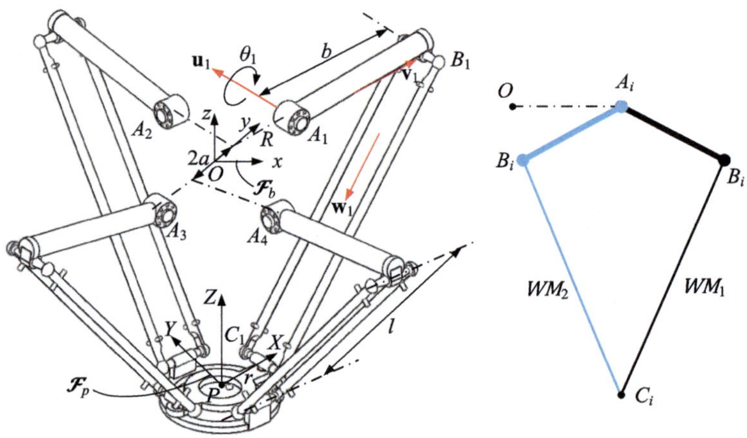

Fig. 2. The parameterization of the robot structure.

Let $\hat{\mathbf{x}}$, $\hat{\mathbf{y}}$, and $\hat{\mathbf{z}}$ denote the unit vectors along x-, y-, and z-axis, respectively, the Cartesian coordinates of points A_i and B_i in the frame $\mathcal{F}_b$ are denoted as

$$\mathbf{a}_i = \begin{cases} R\mathbf{e}_i + a\hat{\mathbf{y}},\ i = 1,\ 2 \\ R\mathbf{e}_i - a\hat{\mathbf{y}},\ i = 3,\ 4 \end{cases}; \quad \mathbf{e}_i = \begin{bmatrix}\cos\alpha_i \ \sin\alpha_i \ 0\end{bmatrix}^T,\ \alpha_i = \frac{2i-1}{4}\pi \tag{2}$$

$$\mathbf{b}_i = b\mathbf{v}_i + \mathbf{a}_i,\ \mathbf{v}_i = \mathbf{R}_z(\alpha_i)\mathbf{R}_y(\theta_i)\hat{\mathbf{x}} \tag{3}$$

and the corresponding axis of rotation of the ith actuated joint is described by $\mathbf{u}_i = \left[-\sin\alpha_i \ \cos\alpha_i \ 0\right]^T$.

Let the end-effector pose be $\chi = \left[\mathbf{p}^T \ \phi\right]^T$, $\mathbf{p} = \left[x \ y \ z\right]^T$, the Cartesian coordinates of point C_i in frame $\mathcal{F}_b$ is expressed as

$$\mathbf{c}_i = r\mathbf{Q}\hat{\mathbf{x}} + \mathbf{p},\ i = 1,\ 3;\ \mathbf{c}_i = r\mathbf{R}_z(\alpha_i)\hat{\mathbf{x}} + \mathbf{p},\ i = 2,\ 4 \tag{4}$$

where $\mathbf{Q} = \mathbf{R}_z(\phi + \phi_0 + (i-1)\pi/2)$ is the rotation matrix of the end-effector and ϕ_0 is the initial orientation angle.

By making use of the following kinematic constraints:

$$(\mathbf{c}_i - \mathbf{b}_i)^T(\mathbf{c}_i - \mathbf{b}_i) = l^2, \quad i = 1, ..., 4 \tag{5}$$

the inverse geometry problem of the robot can be solved as

$$\theta_i = 2\tan^{-1}\frac{-I_i \pm \sqrt{I_i^2 + J_i^2 - K_i^2}}{K_i - J_i} \tag{6}$$

with

$$I_i = -2b(\mathbf{c}_i - \mathbf{a}_i)^T\hat{\mathbf{z}},\ J_i = 2b(\mathbf{c}_i - \mathbf{a}_i)^T\mathbf{e}_i,\ K_i = l^2 - b^2 - \|\mathbf{c}_i - \mathbf{a}_i\|^2 \tag{7}$$

For a given pose of the robot end-effector, each limb can have two postures that are characterized by the sign "$-/+$" in Eq. (6), which means that the robot can have up to 16 working modes ($\mathcal{WM}$s). Here, the sign "+" and "−" stand for working mode 1 ($\mathcal{WM}_1$) and $\mathcal{WM}_2$ for each limb, respectively, as depicted in Fig. 2. To avoid the mechanical collision, ($\mathcal{WM}_1$) for each limb is an effective selection from the 16 working modes for this type of robot.

Differentiating the four Eq. (5) with respect to time yields

$$\mathbf{A}\dot{\mathbf{x}} = \mathbf{B}\dot{\boldsymbol{\theta}} \tag{8}$$

with

$$\mathbf{A} = \begin{bmatrix} \mathbf{w}_1^T & r\mathbf{w}_1^T\mathbf{s} \\ \mathbf{w}_2^T & 0 \\ \mathbf{w}_3^T & -r\mathbf{w}_3^T\mathbf{s} \\ \mathbf{w}_4^T & 0 \end{bmatrix},\ \dot{\mathbf{x}} = \begin{bmatrix} \dot{x} \\ \dot{y} \\ \dot{z} \\ \dot{\phi} \end{bmatrix};\ \mathbf{B} = \begin{bmatrix} h_1 & & & \\ & h_2 & & \\ & & h_3 & \\ & & & h_4 \end{bmatrix},\ \dot{\boldsymbol{\theta}} = \begin{bmatrix} \dot{\theta}_1 \\ \dot{\theta}_2 \\ \dot{\theta}_3 \\ \dot{\theta}_4 \end{bmatrix} \tag{9}$$

where

$$\mathbf{s} = \left[-\sin(\phi + \phi_0) \ \cos(\phi + \phi_0) \ 0\right]^T;\ h_i = b(\mathbf{u}_i \times \mathbf{v}_i)\cdot\mathbf{w}_i,\ \mathbf{w}_i = (\mathbf{c}_i - \mathbf{b}_i)/l \tag{10}$$

In Eq. (8), $\mathbf{A}$ and $\mathbf{B}$ are named as the forward and inverse Jacobian matrices, respectively. The kinematic Jacobian matrix is hence obtained as $\mathbf{J} = \mathbf{A}^{-1}\mathbf{B}$.

4 Singularity Analysis

It is known that the manipulator is in a singular configuration when the determinant of its forward or/and the determinant of its inverse Jacobian matrix is (are) equal to zero [4], namely,

- Type-1 singularity, also named limb singularity, occurs when $|\mathbf{B}| = 0$.
- Type-2 singularity, also named actuation singularity, occurs when $|\mathbf{A}| = 0$.

From $|\mathbf{B}| = \prod_{i=1}^{4} h_i = b^4 \prod_{i=1}^{4} (\mathbf{u}_i \times \mathbf{v}_i)^T \mathbf{w}_i$, the type-1 singularities occur when the actuated link A_iB_i and the segment of parallelogram B_iC_i are coplanar, usually existing at the reachable workspace boundaries, which can be readily identified. Therefore, the type-1 singularity will not be discussed.

When the determinant $|\mathbf{A}|$ of the forward Jacobian matrix is equal to zero, the actuation singularity occurs in the case of the general linear complex. The wrench exerting on the end-effector is generated by one of the transmission wrenches that is proportional to those that are generated by the other transmission wrenches, thus, the robot loses 1-DOF motion in this singular configuration [9].

4.1 Constraint Singularity

Besides the previous two type of singularities, the constraint singularity may exist within the workspace. For this four-limb robot, the constraint singularity usually occurs when at least two of the parallelograms are parallel [9] or the rotational sub-platform lies in the plane determined by the two parallelograms in the first and third limbs [11]:

- From $|\mathbf{A}| = r\mathbf{s}^T(\mathbf{w}_1 \otimes \mathbf{w}_3 + \mathbf{w}_3 \otimes \mathbf{w}_1)(\mathbf{w}_2 \times \mathbf{w}_4) = 0$, one constraint singularity occurs when $\mathbf{w}_2 \times \mathbf{w}_4$. It is noted that points B_2 and B_4 are always located in two parallel and noncoplanar planes, intersected with the segments C_2C_4, which implies that this singular configuration never occurs.
- When the robot is in a singular configuration of $\mathbf{w}_1 \parallel \mathbf{w}_2$, the robot has extra mobilities even though all the actuated joints are locked. Other singular configurations, for instance, $\mathbf{w}_3 \parallel \mathbf{w}_4$, $\mathbf{w}_1 \parallel \mathbf{w}_3$, and $\mathbf{w}_1 \parallel \mathbf{w}_4$ ($\mathbf{w}_2 \parallel \mathbf{w}_3$), are similar to this case.
- When the segment C_1C_3 lies in the plane determined by the vectors $\mathbf{w}_1$ and $\mathbf{w}_3$, a screw motion with the axis coincident with z-axis will be unconstrained, thus, the robot has extra mobilities.

4.2 Singularity Loci

Based on the foregoing analysis, numerical study is carried out to detect the singularity loci, where the geometric parameters in the unit of mm are set to $a = 100$, $R = 150$, $b = 300$, $l = 600$, $r = 100$.

Type-2 Singularity. As the expression of the determinant $|\mathbf{A}|$ of the forward Jacobian matrix is a complex and implicit equation, the singularity loci cannot be detected by means of the discrete searching method. The workspace is described with small increments to check $|\mathbf{A}|$ for different constant end-effector orientations. If $|\mathbf{A}| < \Delta$, $\Delta \approx 0$ being a computational tolerance, the robot can be considered in a singular configuration. Moreover, interpolation is introduced to smooth the singularity surface. Figure 3 shows the type-2 singularity loci with end-effector orientations $\phi = 0$ and $\phi = -5°$, together with the fitted regular workspace. It is observed that the singular configurations become less when the end-effector angle increases. Since the pick-and-place operations are usually confined in a long and narrow region, the lower part of a super-ellipsoid is fitted to describe the regular workspace expressed below:

$$\frac{|x|^{2.5}}{385} + \frac{|y|^{2.5}}{470} + \frac{|z+320|^{2.5}}{400} = 1 \tag{11}$$

thus, the robot admits a singularity-free rectangular workspace when the end-effector orientation is in the range of 0 and $-45°$.

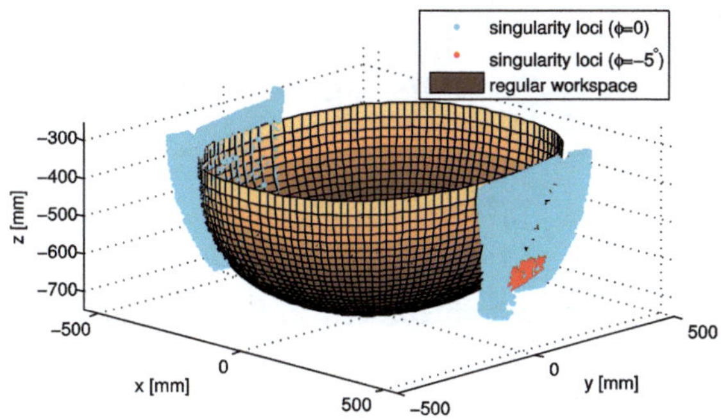

Fig. 3. Type-2 singularity loci and the fitted super-ellipsoidal workspace.

Constraint Singularity. The robot encounters constraint singularity when it is in the configuration of $\mathbf{w}_1 \parallel \mathbf{w}_2$, leading to

$$\mathbf{w}_1 - \mathbf{w}_2 = \frac{1}{2}\begin{bmatrix} \frac{1}{3}\cos(\phi_0+\phi) - \frac{\sqrt{2}}{3} - \frac{\sqrt{2}}{2}\cos\theta_1 - \frac{\sqrt{2}}{2}\cos\theta_2 \\ \frac{1}{3}\sin(\phi_0+\phi) - \frac{\sqrt{2}}{6} - \frac{\sqrt{2}}{2}\cos\theta_1 + \frac{\sqrt{2}}{2}\cos\theta_2 \\ \sin\theta_1 - \sin\theta_2 \end{bmatrix} = \mathbf{0} \tag{12}$$

with the solutions below

$$\theta_1 = \theta_2 = \pi \pm \cos^{-1}\left(\frac{1}{6}\right),\ \phi = 0;\ \theta_1 = \theta_2 = \pm\frac{2}{3}\pi,\ \phi = \frac{\pi}{2} \tag{13}$$

When $\phi = \pi/2$ or $\theta_1 = \theta_2 = \pi + \cos^{-1}(1/6)$, the mechanical collision occurs, thus, the singularity loci are located on a spherical surface expressed as

$$x^2 + \left(y - a - \frac{\sqrt{2}}{2}R - \frac{\sqrt{2}}{2}b\cos\theta\right)^2 + (z + b\sin\theta)^2 = l^2;\ \theta = \pi - \cos^{-1}\left(\frac{1}{6}\right) \quad (14)$$

for which the singularity loci are shown depicted in Fig. 4(a) together with the singularity loci when $\mathbf{w}_3 \parallel \mathbf{w}_4$. The singularity loci in these two cases are symmetrical with respect to xz plane.

Figure 4(b) shows the constraint singularity in the configuration $\mathbf{w}_1 \parallel \mathbf{w}_3$, yielding

$$\mathbf{w}_1 - \mathbf{w}_3 = \begin{bmatrix} \frac{1}{3}\cos(\phi_0 + \phi) - \frac{\sqrt{2}}{4} - \frac{\sqrt{2}}{4}\cos\theta_1 - \frac{\sqrt{2}}{4}\cos\theta_3 \\ \frac{1}{3}\sin(\phi_0 + \phi) - \frac{\sqrt{2}}{4} - \frac{\sqrt{2}}{4}\cos\theta_1 - \frac{\sqrt{2}}{4}\cos\theta_3 - \frac{1}{3} \\ \frac{1}{2}(\sin\theta_1 - \sin\theta_3) \end{bmatrix} = \mathbf{0} \quad (15)$$

with the solutions

$$\theta_1 = \theta_3 = \pm\frac{2}{3}\pi,\ \phi = \frac{\pi}{4};\ \theta_1 = \theta_3 = \pi \pm \cos^{-1}\left(\frac{2}{3} + \frac{\sqrt{2}}{2}\right),\ \phi = \frac{3}{4}\pi \quad (16)$$

To avoid the mechanical collisions, the singularity loci are located on a spherical surface expressed as

$$x^2 + y^2 + \left(z + b\sin\frac{2}{3}\pi\right)^2 = l^2 \quad (17)$$

Figure 4(c) shows the constraint singularity in the configuration $\mathbf{w}_1 \parallel \mathbf{w}_4$, resulting in

$$\mathbf{w}_1 - \mathbf{w}_4 = \frac{1}{2}\begin{bmatrix} \frac{1}{3}\cos(\phi_0 + \phi) - \frac{\sqrt{2}}{6} - \frac{\sqrt{2}}{2}\cos\theta_1 + \frac{\sqrt{2}}{2}\cos\theta_4 \\ \frac{1}{3}\sin(\phi_0 + \phi) - \frac{\sqrt{2}}{3} - \frac{\sqrt{2}}{2}\cos\theta_1 - \frac{\sqrt{2}}{2}\cos\theta_4 - \frac{1}{6} \\ \sin\theta_1 - \sin\theta_4 \end{bmatrix} = \mathbf{0} \quad (18)$$

with the solutions below

$$\theta_1 = \theta_2 = \pi \pm \cos^{-1}\left(\frac{\sqrt{2}}{3} + \frac{1}{6}\right),\ \phi = 0;\ \theta_1 = \theta_2 = \pi \pm \cos^{-1}\left(\frac{\sqrt{2}}{3} + \frac{1}{2}\right),\ \phi = -\frac{\pi}{2} \quad (19)$$

To avoid the mechanical collisions, the singularity loci are located on a spherical surface expressed as

$$\left(x - \frac{\sqrt{2}}{2}R - \frac{\sqrt{2}}{2}b\cos\theta\right)^2 + y^2 + (z + b\sin\theta)^2 = l^2;\ \theta = \pi - \cos^{-1}\left(\frac{\sqrt{2}}{3} + \frac{1}{6}\right) \quad (20)$$

The singularity loci with $\mathbf{w}_2 \parallel \mathbf{w}_3$ are symmetrical to those with $\mathbf{w}_1 \parallel \mathbf{w}_4$ with respect to yz plane.

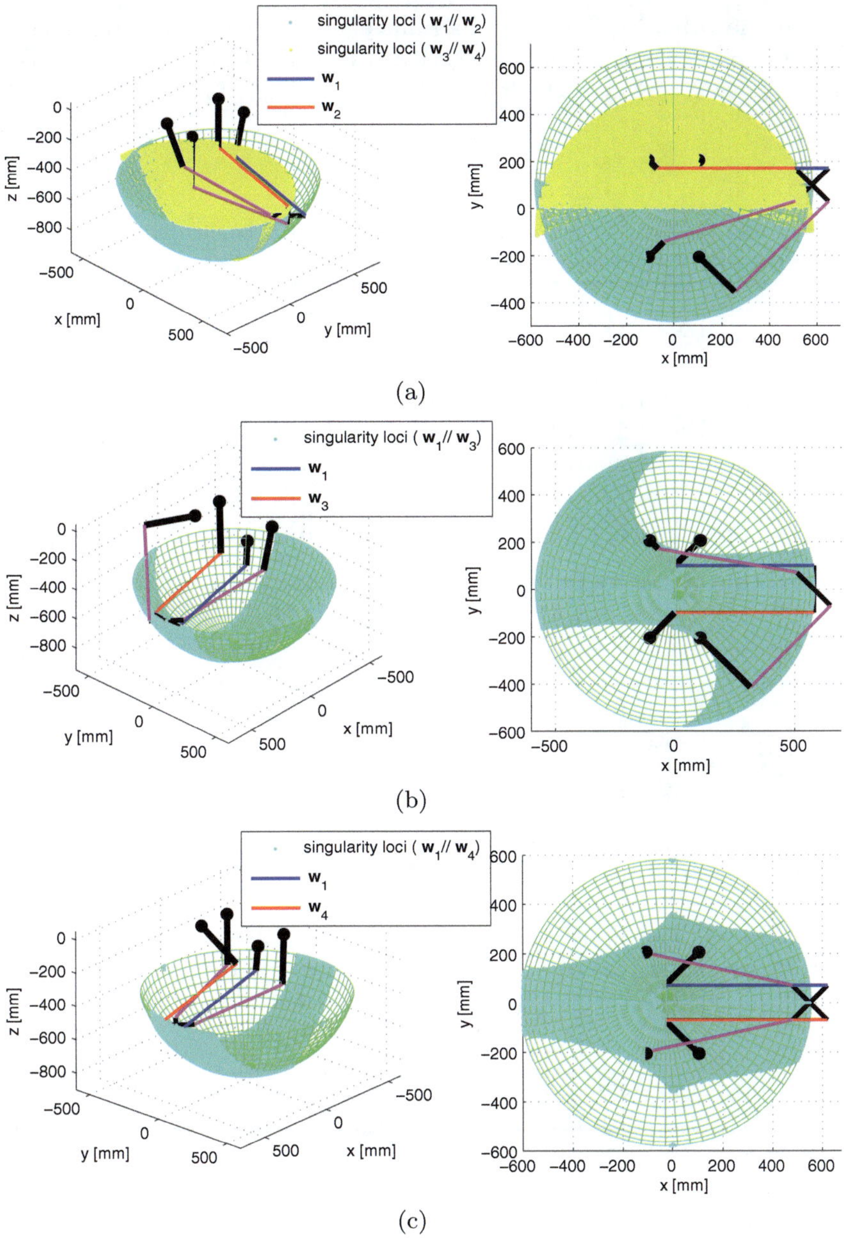

Fig. 4. The constraint singularities of the robot: (a) $\mathbf{w}_1 \parallel \mathbf{w}_2\,(\mathbf{w}_3 \parallel \mathbf{w}_4)$; (b) $\mathbf{w}_1 \parallel \mathbf{w}_3$; (c) $\mathbf{w}_1 \parallel \mathbf{w}_4$.

It is noteworthy that the robot encounters the previous constraint singularities with the corresponding limbs in $\mathcal{WM}_2$, which can be avoided by keeping the robot limb work in $\mathcal{WM}_1$ through the control design.

When the robot encounters constraint singularity with the segment C_1C_3 lying in the plane determined by the vectors $\mathbf{w}_1$ and $\mathbf{w}_3$, namely, points B_1, B_3, C_1 and C_3 being coplanar, the following geometric relationship exists

$$\phi = \tan^{-1} \frac{b_{1y} - b_{3y}}{b_{1x} - b_{3x}} - \phi_0; \ \mathbf{w}_1 \times \mathbf{w}_3 \times \mathbf{s} = \mathbf{0} \tag{21}$$

where b_{ix} and b_{iy} are the first two elements in vector $\mathbf{b}_i$, sequentially, the robot end-effector orientation is solved as

$$\phi = \tan^{-1} \frac{2\sqrt{2}a + 2R + b\cos\theta_1 + b\cos\theta_3}{2R + b\cos\theta_1 + b\cos\theta_3} - \phi_0 \tag{22}$$

then, the end-effector position can be readily solved according to the geometric configuration. The end-effector orientation with actuated joints θ_1 and θ_3 and the corresponding singularity loci are shown in Fig. 5.

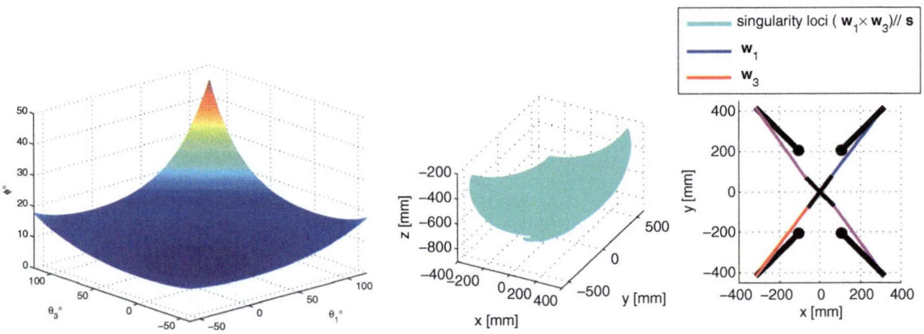

(a) (b)

Fig. 5. Constraint singularity with the segment C_1C_3 lying in the plane determined by the vectors $\mathbf{w}_1$ and $\mathbf{w}_3$: (a) end-effector orientation ϕ with the actuated joints θ_1 and θ_3; (b) singularity loci; (c) top view of one singular configuration.

Based upon the singularity analysis, this robot admits a regular workspace defined in Eq. (11), with the rotational sub-platform orientation $\phi \in [-45°, 0]$ and the full-circle rotation of the end-effector, under the working mode 1 of all limbs.

5 Conclusions

This paper introduces a new four-limb parallel Schönflies motion generator for the pick-and-place application, whose end-effector adopts a planetary gear train as the amplification mechanism to realize the full-circle rotation. The robot singularity is analyzed with the Jacobian-based method, combining the constraint singularity, where the singularity can be identified graphically and numerically.

Singularity loci are expressed analytically and visualized with respect to the workspace. The regular workspace, namely, a long and narrow super-ellipsoidal workspace with the full-circle rotation of the end-effector, is identified under a given working mode suitable for PnP operations.

Acknowledgement. The supports from the Fundamental Research Funds for the Central Universities (No. DUT19JC25), the Natural Science Foundation and the Doctoral Start up Foundation of Liaoning Province (Nos. 20180520028, 20170520134), and the Applied Basic Research Programs of Changzhou (No. CJ20180017) are gratefully acknowledged.

References

1. Adept Quattro Parallel Robots. http://www1.adept.com/main/ke/data/Archived/Quattro/sQuattro_UG.pdf
2. Altuzarra, O., Şandru, B., Pinto, C., Petuya, V.: A symmetric parallel Schönflies-motion manipulator for pick-and-place operations. Robotica **29**, 853–862 (2011)
3. Amine, S., Masouleh, M.T., Caro, S., Wenger, P., Gosselin, C.: Singularity conditions of 3T1R parallel manipulators with identical limb structures. ASME J. Mech. Robot. **4**(1), 011,011 (2012)
4. Gosselin, C., Angeles, J.: Singularity analysis of closed-loop kinematic chains. IEEE Trans. Robot. Autom. **6**(3), 281–290 (1990)
5. Kong, X., Gosselin, C.M.: Type synthesis of 3 T1R 4-DOF parallel manipulators based on screw theory. IEEE Trans. Robot. Autom. **20**(2), 181–190 (2004)
6. Krut, S., Nabat, V., Company, O., Pierrot, F.: A high-speed parallel robot for SCARA motions. IEEE Int. Conf. Robot. Autom. **4**, 4109–4115 (2004)
7. Merlet, J.P.: Singular configurations of parallel manipulators and Grassmann geometry. Int. J. Robot. Res. **8**(5), 45–56 (1989)
8. Pierrot, F., Nabat, V., Company, O., Krut, S., Poignet, P.: Optimal design of a 4-DOF parallel manipulator: from academia to industry. IEEE Trans. Robot. **25**(2), 213–224 (2009)
9. Wu, G., Bai, S., Caro, S.: A transmission quality index for a class of four-limb parallel Schönflies motion generators. ASME J. Mech. Robot. **10**(5), 051014 (2018)
10. Wu, G., Bai, S., Hjørnet, P.: Architecture optimization of a parallel Schönflies-motion robot for pick-and-place applications in a predefined workspace. Mech. Mach. Theory **106**, 148–165 (2016)
11. Xie, F., Liu, X.: Design and development of a high-speed and high-rotation robot with four identical arms and a single platform. ASME J. Mech. Robot. **7**(4), 041015 (2015)
12. Zlatanov, D., Bonev, I.A., Gosselin, C.M.: Constraint singularities of parallel mechanisms. In: IEEE International Conference on Robotics and Automation, vol. 1, pp. 496–502. IEEE (2002)

Kinematics and Multi-mode Application of a Novel Self-assembling Robot with Parallel Structure

Fuqun Zhao, Sheng Guo(✉), Xiangyang Wang, and Majun Song

School of Mechanical, Electronic and Control Engineering,
Beijing Jiaotong University, Beijing, People's Republic of China
{16116362, shguo, 17116388, 18116021}@bjtu.edu.cn

Abstract. A self-assembling robot system, which combines four deployable manipulators with an operation object, is proposed in this paper. The manipulators can be stacked in a designated area and move independently from each other before being connected to an object. After the connection with the object, the connect units which located at the end of each manipulator ensure that the manipulators can remain connection with the object for operation. The connection can be realized through a vacuum cup or an electromagnet. The resulting robot system, formed by the four deployable manipulators and the operation object, is similar to a kinematic structure of parallel mechanism. The Robot is able to achieve manipulation of objects with 2R1T by actuating exactly three joints in each mode. Each mode is formed by different three manipulators and the object. Kinematics performance and application were demonstrated in this paper. Such a manipulator–object combination possesses some advantages over existing grippers and manipulation robot. It has the repetitive and reconstruction ability with different work mode than existing grippers and manipulation robot with fixed structure parameters.

Keywords: Self-assembling robot · Parallel mechanism · Multi-mode · Workspace analysis · Path planning

1 Introduction

In areas such as space station operations and industry, the repeatability and diversity of manufacturing and operations are continuously increased. Meanwhile, the processing product quality and the production efficiency are improved. Therefore, the manipulators need to meet different tasks and ensure continuous processing. Most of the existing manipulators are designed to perform special operation tasks with fixed structured parameters. However, the use of these manipulators is limited in terms of dexterity and versatility. Usually, most of such manipulators in automated operation area need to be changed according to the work process or the task requirement, which will cause the low production efficiency and delay in production. In fact, diversified operation means that the manipulator itself has a large number of degrees of freedom, which results in a large number of links and a large volume of the mechanism.

D. Wang et al. (Eds.): MeTrApp 2019, MMS 79, pp. 290–300, 2020.
https://doi.org/10.1007/978-981-15-0142-5_29

When the multi-manipulator is connected with the operation object, it can be considered as a kind of parallel mechanism topology. In the design of this kind of mechanism, the design idea of parallel mechanism can be adopted to carry out task classification and multi-manipulation configuration design according to different operation tasks, so as to optimize the number of drives and occupation volume of a single manipulation. As mentioned above, the resulting manipulator-object system is a closed-loop structure that is similar to parallel kinematic structure. In this case, the number of actuators, required to change the pose (position and orientation) of objects, is equal to the DOF of the manipulator-object system. This will be of benefit for the control of the robot system. Therefore, motion with such a robot system is more straightforward than manipulations with grippers or existing mechanical manipulators.

In recent years, many scholars have introduced the concept of parallel mechanism into the configuration design and analysis of multi-manipulator or mechanical hand mechanism, and achieved good results. Corves et al. [1] proposed a reconfigurable multi-arm robot system through the design method of parallel mechanism, and the mechanism could achieve dynamic and other characteristics reconstruction by switching different operating arms to grasp objects. Lee et al. [2] synthesized the dexterous hand mechanism with different hand indexes according to the degree of freedom required by the task. Hunt et al. [3] proposed the manipulator of hybrid mechanism based on the screw theory, and the end-effector can gain or lose degrees of freedom under the special configuration. Gogu et al. [4, 5] applied the concept of parallel mechanism to the design of traditional dexterous hand, and proposed a class of novel grasping mechanisms. Dollar et al. [6, 7] combined the configuration design and static mechanical analysis of parallel mechanism to design and analyze the dexterous hand, and obtained a class of the dexterous hand with three fingers and two joints. At the same time, they solved the kinematics and dynamics of this kind of dexterous hand by using the method of Jacobian matrix analysis in parallel mechanism. Ebert et al. [8] summarized the similarities and differences of general grasping mechanism, parallel mechanism and mechanical arm. The advantages and disadvantages of different drive types in grasping mechanism design are analyzed by combining the features of unilateral constraint and dexterous grasping object. Bullock et al. [9] analyzed the positional manipulation workspace and finger workspaces for a single object, and studied the difference between two-fingered and three-fingered operation precision manipulation. Nefz et al. [10, 11] designed a multi-finger clamping mechanism. They compared the performance index of the integrated mechanism and obtained the ideal manipulator mechanism by taking the whole of the object and the fingers as a parallel mechanism.

Therefore, it is meaningful to the field of robot mechanism and industry to study the maneuverability and deployable characteristic of multi-manipulator mechanism, so as to reduce the carrying pressure and improve the operation performance. The rest of this paper is organized as follows. In Sect. 2, a robot system with four deployable manipulators and its working model was described in detail. Every structure of the manipulator is different. Any different three manipulators and the operation object can form a 2R1T parallel mechanism. The kinematics modeling of the parallel mechanism in different configurations was carried out, and the workspace under different modes was analyzed in Sects. 3 and 4. Based on the results and the specific path requirements, switching strategy and feasibility of the manipulator are demonstrated in Sect. 5.

2 Structural Description and Working Mode of Self-assembling Robot

2.1 Structural Description

The proposed self-assembling robot system includes four manipulators, and an operation object. As shown in Fig. 1, the manipulators can fold into the storage area. All the manipulators are able to position their contact parts independently before connecting to the object. When the robot needs to perform a task, several manipulators will unfold to contact with the object as shown in Fig. 2. In grasping phase, the manipulators can be released and combined with the operation object to change the topology of the parallel mechanism. Here, the manipulator with deployable characteristic can be treated as the deployable kinematic chain (DKC) of the formed parallel mechanism.

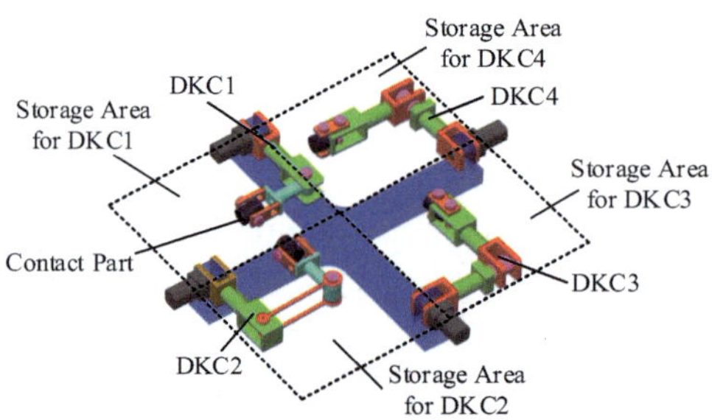

Fig. 1. The release phase of the robot

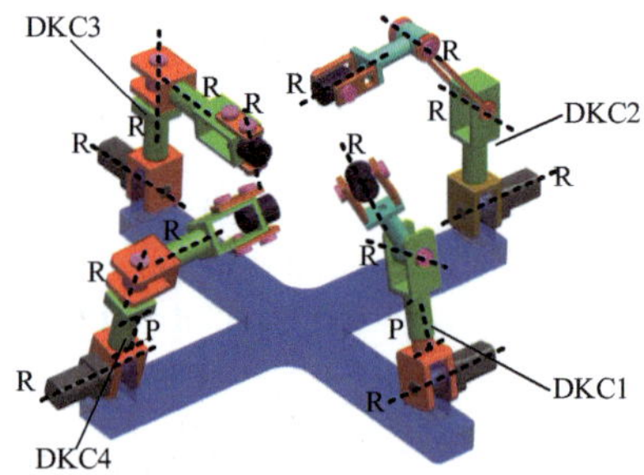

Fig. 2. The grasping phase of the robot

Figure 2 shows the four DKCs in expansion/grasping phase. The structures of DKC1-4 are RPRR, RPRRR, RPRRR and RPRR respectively, where R is rotational joint, P is prismatic joint. The functions that can be realized by the proposed robot can be summarized as follows: The manipulators possess deployable performance to improve the transportation and storage efficiency; In order to make the operation object can move in the space, the number of manipulators connected to the operation object in the grasping phase is 3. In the proposed robot, after any three manipulators are connected with the object, the working mode of the operation object is the same 2R1T motion. The different is the structure of each manipulator, so that the parallel mechanism formed by different manipulators and object has different operation performance.

2.2 Working Mode Definition

If three manipulators are selected to connect with the operation object at each time in the grasping stage, then there will be four configurations of parallel mechanism formed by the manipulators and the operation object as shown in Fig. 3.

The working modes in Fig. 3 are defined as Mode-I, Mode-II, Mode-III and Mode-IV respectively. The four working modes are realized by switching different manipulators to connect with the object. The output performance (operation workspace) of the robot system is also different under the different topology configurations of the parallel mechanism. The different workspaces under different modes are shown in Fig. 3. When the robot system operates on the operation object, the manipulator can be switched by the required output performance. These contents will be shown in Sects. 4 and 5.

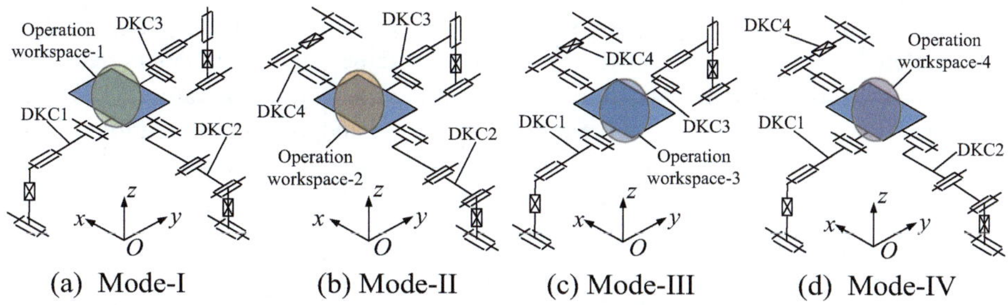

Fig. 3. The four configurations in different modes

Because the manipulators and object combination can be treated as a kinematic structure of parallel mechanism, the robot system should satisfy the characteristics of parallel mechanism. That is, the number of independent degrees of freedom (DOF) of parallel mechanism should be equal to that of independent actuators. In this section, the modified Grübler-Kutzbach [12] Eq. (1) is used to find the DOF of this mechanism [12]. The parallel mechanism in each working mode can find the DOF through Eq. (1):

$$M = d(n - g - 1) + \sum\nolimits_{i=1}^{g} f_i + v \tag{1}$$

where M is the DOF of the mechanism, d is the order of mechanism, d = 6 − λ, λ is the number of common constraints on the mechanism, n is the total number of links of the mechanism, g is the total number of the joints of the mechanism, f_i is the number of degrees of freedom of the ith joint, v is the number of redundant constraints of the mechanism.

According to the structure restraints [13] of the mechanism, there are three redundant constraints in each mode respectively. So, the total redundant constraint number leads to $v = 3$.

Taking Mode-I as an example, the DOF of the mechanism can be calculated by Eq. (1):

$$M = 6(11 - 12 - 1) + 12 + 3 = 3 \tag{2}$$

In the same way, the DOF of the other three modes can be found by Eq. (1). The calculation results of DOF in other three modes are also equal to 3. It means that only three actuators are needed when the grasping phase in each mode. In this paper, the prismatic joint in each manipulator is selected as the independent actuation.

3 Kinematics Model

3.1 Inverse Kinematics

In the robot system, the kinematics model of structure diagram that each manipulator connected to the object can be expressed as shown in Fig. 4. Since the four working modes defined in Sect. 2.2 have the same form of motion (2R1T), the four modes can be calculated at the same coordinates and given the same output parameters. As shown in Fig. 4, the fixed coordinate system *O-xyz* and the moving coordinate system *p-uvw* are respectively established at the fixed platform center and the moving platform center. In each manipulator, the axis of rotational joints A_1, A_3 and A_4 are parallel to the *x*-axis, and the axis of rotational joints A_2 is parallel to the *y*-axis.

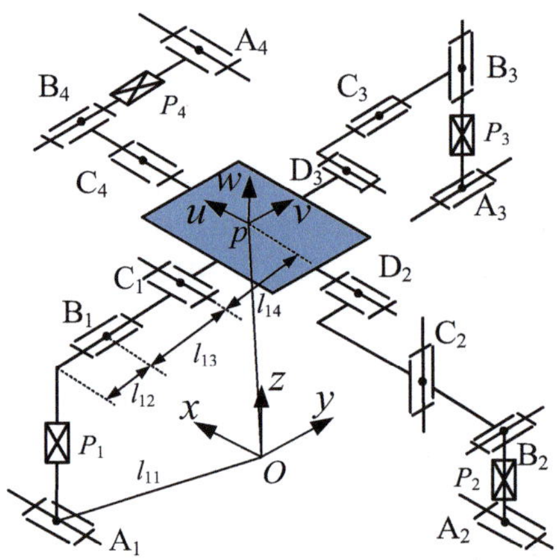

Fig. 4. Kinematics model of structure diagram in grasping phase

Let the length from the center point of rotational joint A_i to the origin of coordinates O be l_{i1}, the length from the center point of rotation joints C_1 and C_2 to the origin of coordinates p be l_{14} and l_{44}, the length from the center point of rotation pair D_2 of l_{44} to the origin of coordinates p be l_{24} and l_{34}. The length of the prismatic joint A_iB_i is P_i, The length of the link B_iC_i is l_{i3}, the length of the links C_2D_2, C_3D_3 are l_{23}, l_{33}.

When the parallel mechanism formed in each working mode, the object can rotate around x-axis and y-axis and move along z direction. The transformation matrix **T** of point p can be expressed as follows:

$$\mathbf{T} = \begin{bmatrix} {}^P_O\boldsymbol{R} & \boldsymbol{p} \\ 0 & 1 \end{bmatrix} {}^P_O\boldsymbol{R}(\alpha, \beta, 0) = \begin{bmatrix} \cos\beta & 0 & \sin\beta & x \\ \sin\alpha\sin\beta & \cos\alpha & -\sin\alpha\cos\beta & y \\ -\cos\alpha\sin\beta & \sin\alpha & \cos\alpha\cos\beta & z \\ 0 & 0 & 0 & 1 \end{bmatrix} \tag{3}$$

Let the coordinate vectors of C_i and D_i be $\boldsymbol{c}_i$, $\boldsymbol{d}_i$ in the fixed coordinate system, $\boldsymbol{c}_{Pi}$, $\boldsymbol{d}_{Pi}$ in the moving coordinate system:

$$\boldsymbol{c}_i = {}_O^P\boldsymbol{R}\boldsymbol{c}_{P_i} + \boldsymbol{p} \qquad \boldsymbol{d}_i = {}_O^P\boldsymbol{R}\boldsymbol{d}_{P_i} + \boldsymbol{p} \tag{4}$$

Assuming the coordinate vectors of A_i, B_i are $\boldsymbol{a}_i$, $\boldsymbol{b}_i$, the coordinate vectors of A_1 can be written as $\boldsymbol{a}_1 = [0, l_{11}, 0]^{\mathrm{T}}$. Since rotational joints A_1 and A_2 restrict the movement of B_1 along the x-axis and y-axis, the coordinate vector of point B_1 can be obtained: $B_1 = [0, l_{11}, p_1]^{\mathrm{T}}$. The coordinate vectors $\boldsymbol{b}_2$, $\boldsymbol{b}_3$, $\boldsymbol{b}_4$ of the center point of rotational joints in the other manipulators can be deduced in the same way.

In DKC1 and DKC4, the constraint equation can be established by using the constraint relation of the length of the B_iC_i link:

$$\begin{cases} (\boldsymbol{c}_1 - \boldsymbol{b}_1)^{\mathrm{T}}(\boldsymbol{c}_1 - \boldsymbol{b}_1) - l_{13} = 0 \\ (\boldsymbol{c}_4 - \boldsymbol{b}_4)^{\mathrm{T}}(\boldsymbol{c}_4 - \boldsymbol{b}_4) - l_{43} = 0 \end{cases} \tag{5}$$

where $\boldsymbol{c}_1 = [r(\cos\alpha\cos\beta + \sin\alpha) + r/2(\cos 2\alpha)(\cos\beta - 1), 0, -r\cos\alpha\sin\beta]^{\mathrm{T}}$, $\boldsymbol{b}_1 = [0, l_{11}, p_1]^{\mathrm{T}}$.

Equation (5) contains two unknown driving values, p_1 and p_2, which can be obtained by giving the output parameter α, β, z [14].

In DKC2 and DKC3, the constraint equation can be established by using the constraint relation of the length of the B_iC_i link:

$$\begin{cases} (\boldsymbol{c}_3 - \boldsymbol{b}_3)^{\mathrm{T}}(\boldsymbol{c}_3 - \boldsymbol{b}_3) - l_{33} = 0 \\ (\boldsymbol{c}_4 - \boldsymbol{b}_4)^{\mathrm{T}}(\boldsymbol{c}_4 - \boldsymbol{b}_4) - l_{43} = 0 \end{cases} \tag{6}$$

Meanwhile, the links C_2D_2, C_3D_3 can form the constraint equation:

$$\begin{cases} (\boldsymbol{c}_2 - \boldsymbol{d}_2)^{\mathrm{T}}(\boldsymbol{c}_2 - \boldsymbol{d}_2) - l_{23} = 0 \\ (\boldsymbol{c}_3 - \boldsymbol{d}_3)^{\mathrm{T}}(\boldsymbol{c}_3 - \boldsymbol{d}_3) - l_{33} = 0 \end{cases} \tag{7}$$

In the fixed coordinate system, the coordinates of D_2 and D_3 are known. The coordinate of points C_2 and C_3 can be expressed by the geometric parameter and input parameter through Eq. (7). Equation (6) contains two unknown driving values, p_3 and p_4, which can be obtained by giving the output parameter α, β, z.

3.2 Jacobian Matrix

In this proposed robot system, four working modes of the operation object can be completed, so there are four corresponding Jacobian matrices. In this section, we take working mode-I as an example to demonstrate the solution of the Jacobian matrix. In the configuration of mode-I, Assuming the velocity matrix of each driving value is $\boldsymbol{q} = (\dot{p}_1, \dot{p}_2, \dot{p}_3)^{\mathrm{T}}$, the moving platform velocity matrix is $\boldsymbol{x} = (\dot{\alpha}, \dot{\beta}, \dot{\gamma})^{\mathrm{T}}$. Therefore, the relation between the driving speed and the moving platform speed can be expressed as:

$$\boldsymbol{Bx} + \boldsymbol{Aq} = 0 \tag{8}$$

By differentiating the constraint equation of manipulators 1, 2 and 3 in Eqs. (5) and (7) with respect to time t, we can obtain:

$$\begin{cases} r_{11}\dot{p}_1 + a_{11}\dot{\alpha} + a_{12}\dot{\beta} + a_{13}\dot{z} = 0 \\ r_{22}\dot{p}_2 + a_{21}\dot{\alpha} + a_{22}\dot{\beta} + a_{23}\dot{z} = 0 \\ r_{33}\dot{p}_3 + a_{31}\dot{\alpha} + a_{32}\dot{\beta} + a_{33}\dot{z} = 0 \end{cases} \tag{9}$$

Sorting out driving values and moving platform parameters in Eq. (9), the coefficient matrix $\boldsymbol{A}$ and $\boldsymbol{B}$ of $\boldsymbol{q}$ and $\boldsymbol{x}$ can be expressed:

$$\boldsymbol{A} = \begin{pmatrix} r_{11} & 0 & 0 \\ 0 & r_{22} & 0 \\ 0 & 0 & r_{33} \end{pmatrix}; \quad \boldsymbol{B} = \begin{pmatrix} a_{11} & a_{12} & a_{13} \\ a_{21} & a_{22} & a_{23} \\ a_{31} & a_{32} & a_{33} \end{pmatrix}$$

where r_{ii} and a_{ij} are related to the structural and output parameters of the mechanism, i, j = 1, 2, 3, and thus Eq. (8) can be rewritten as:

$$\boldsymbol{q} = -\boldsymbol{A}^{-1}\boldsymbol{Bx} \tag{10}$$

$$\boldsymbol{J} = -\boldsymbol{A}^{-1}\boldsymbol{B} \tag{11}$$

Where $\boldsymbol{J}$ is Jacobian matrix. After the speed parameter of the moving platform of the mechanism is given, the speed of the driving value at the moment can be obtained through formula (10). The Jacobian matrix corresponding to the other three working modes can be obtained in the same manner.

4 Workspace Analysis

One of the advantages of the proposed robot system is different output performance of the mechanism can be realized by switching different manipulators. Because the structure of each manipulator is different, the topological form of the parallel mechanism is also different. Therefore, the mechanism can achieve different output performance under the same operation mode (2R1T). This allows the mechanism to perform repeatable tasks under the mechanism's performance complementarity advantage. When multiple manipulators are combined, the workspace range of the mechanism will be greatly reduced. At this time, different combinations can be selected by manipulators switching method according to the task requirements. This section will study the workspace of the mechanism in different topological forms. The workspace is mainly limited by the range of the kinematic joint angle and structural parameters of the mechanism. To find the workspaces, the selected geometric parameters in this section are shown in Table 1.

Table 1. Geometric parameters of the mechanism (all lengths in mm)

Parameters	Symbols	Values
Length O_iA_i	l_{i1}	10
Length A_iB_i	l_{i2}	5
Length B_iC_i	l_{i3}	5
Platform side pC_i, pD_i	l_{i4}	10
Length C_iD_i in DKC2,3	h_t	10
Actuator limit	p_i	[20,100]

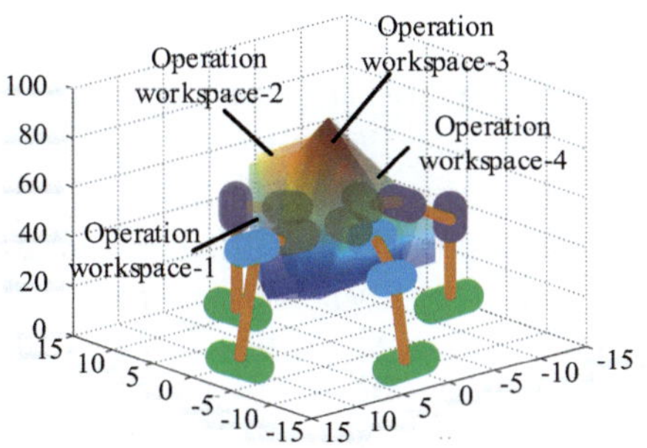

Fig. 5. Workspaces of Mode-I, II, III, IV

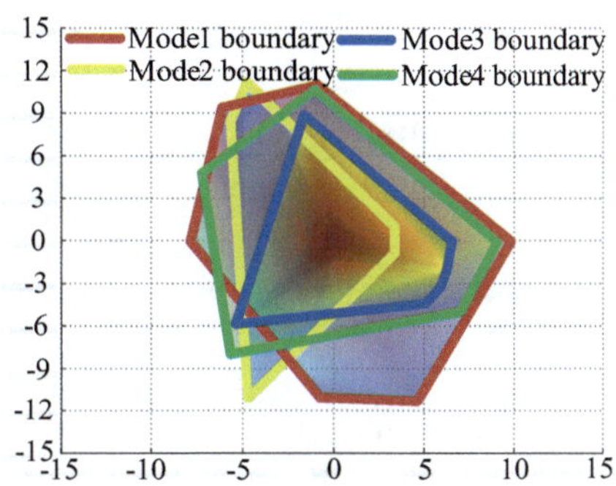

Fig. 6. Top view of the workspaces

Figure 5 shows the operation workspace under the four working modes. Based on the given geometric parameters, the workspaces under different working modes can be obtained. Therefore, the workspace range of the robot system is the superimposed region of four sub-workspaces. Figure 6 shows the top view of the workspaces, and colored solid lines represent the boundaries of each workspace. The path planning based on workspace will be expanded in next section.

5 Multi-mode Application

The robot system proposed in this paper can achieve four kinds of operation performance through four groups of different DKC combinations. Figure 7 shows the corresponding workspace diagram of the four operation modes. When the operating position needs to be changed, the switching between the manipulators can be used to achieve the repeated operation of the object in different positions under the same operation mode. When the operating position moves from point A_1 to point B_1 as shown in Fig. 7, the mechanism can move along the cross region of the corresponding workspace by switching the manipulators connected with the operation object.

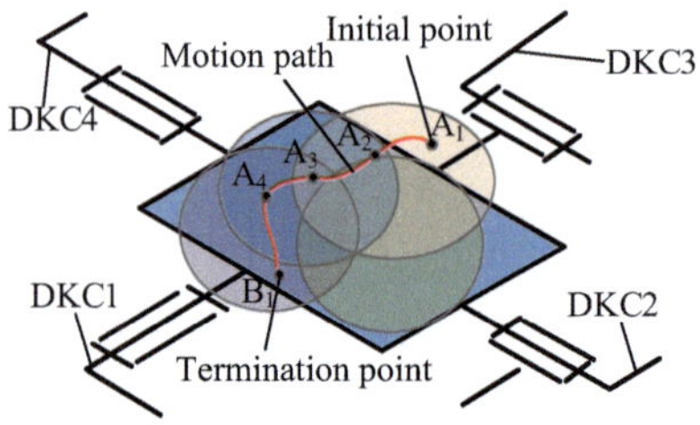

Fig. 7. Path planning

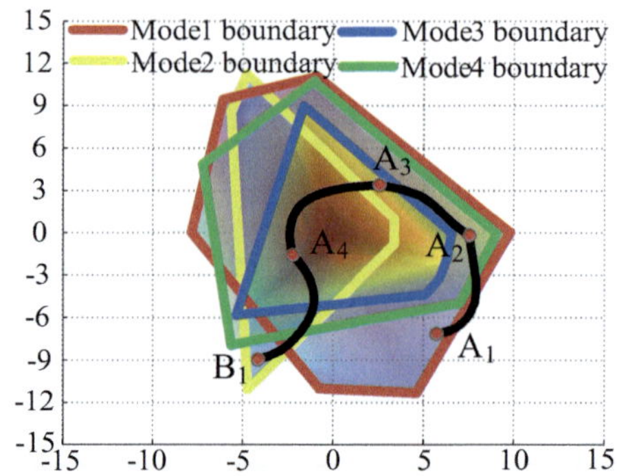

Fig. 8. Path planning within workspaces

Therefore, path planning can be described as follows: Firstly, the mechanism moves from point A_1 to point A_4 through point A_2 and A_3. Then, it moves from point A_4 to point B_1 as shown in Fig. 8. The path from A_1 to B_1 is composed of four sub-paths, and each sub-path is done in a different work mode. Based on Eqs. (5)–(7), the corresponding three actuation values of each work mode under four sub-paths can be obtained as shown in Fig. 9. Simulation time of each sub-path is set to 8 s.

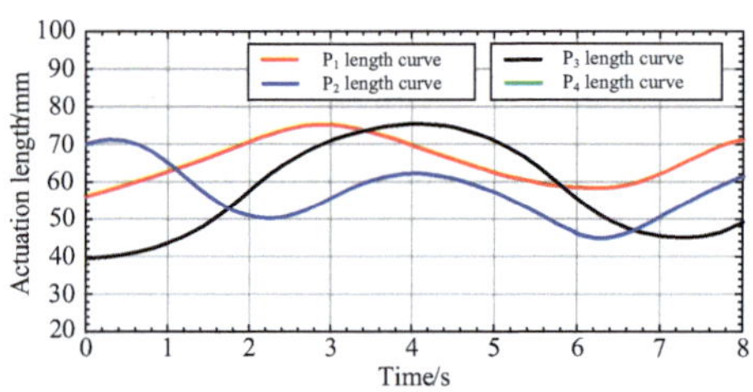

(a) Actuation length in sub-path: A_1 to A_2

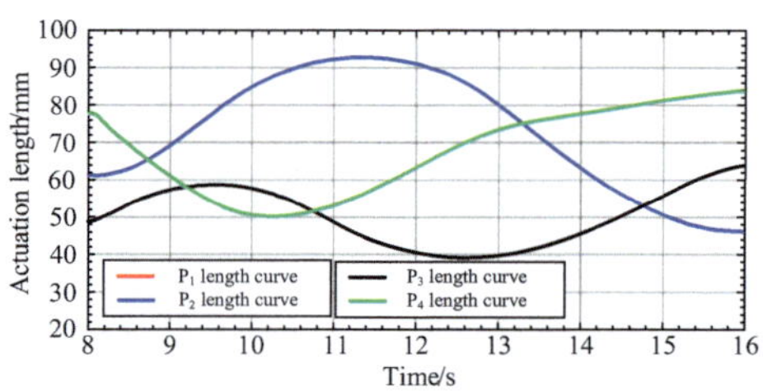

(b) Actuation length in sub-path: A_2 to A_3

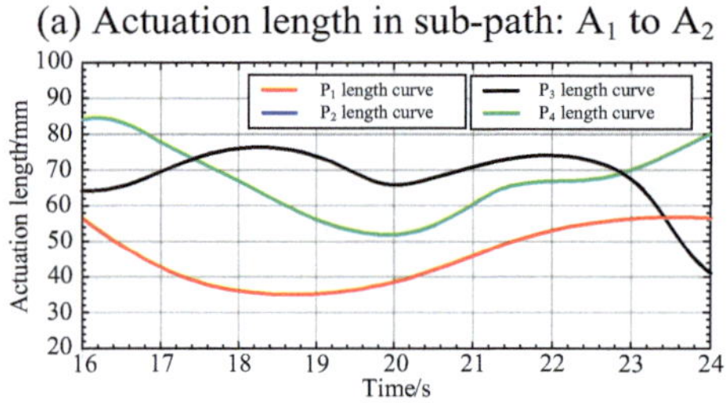

(c) Actuation length in sub-path: A_3 to A_4

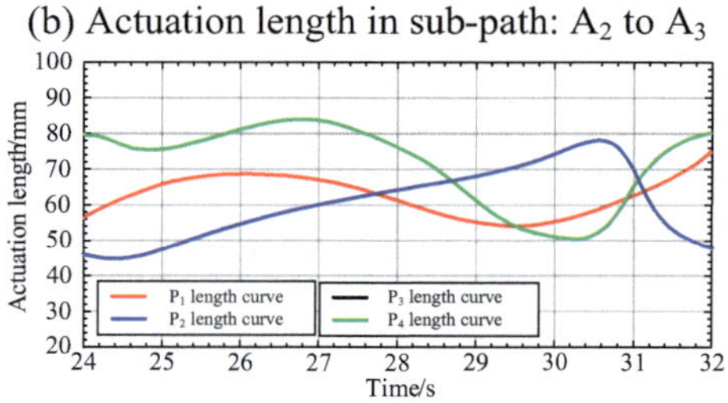

(d) Actuation length in sub-path: A_4 to B_1

Fig. 9. Actuation length curves in each sub-path

In the whole path planning simulation process, all curves are smooth without abrupt changes and sharp points, and the actuation curves are continuous, which ensures the continuity of the overall motion of the mechanism. The result shows this robot system can achieve structural reconstruction by switching the manipulators to connect with the object.

6 Conclusion

The major contributions of this paper are as follows: the robot system with four deployable manipulators was proposed, any different three manipulators and the operation object can form a 2R1T parallel mechanism. Kinematics analysis of the proposed robot system has been formulated in terms of the constraint equations, and the Jacobian matrices of the different modes are established. The operation workspaces in four modes are obtained to demonstrate the repetitive and reconstruction ability. A further extension of this work being investigated is the path planning for one of its applications. The result shows this robot can achieve performance reconstruction through switching the manipulators to connect with the object.

Acknowledgement. The authors would like to acknowledge the financial support of the National Natural Science Foundation of China under Grants No. 51875033, and the Fundamental Research Funds for the Central Universities under Grants No. 2018YJS140.

References

1. Corves, B., Mannheim, T., Riedel, M.: Re-grasping: improving capability for multi-arm robot system by dynamic reconfiguration. In: International Conference on Intelligent Robotics and Applications, pp. 132–141. Springer, Aachen (2011)
2. Lee, J., Tsai, L.: Structural synthesis of multi-fingered hands. J. Mech. Des. **124**(2), 272–276 (2002)
3. Hunt, K.H., Samuel, A.E., Mcaree, P.R.: Special configurations of multi finger multi freedom grippers-a kinematic study. Int. J. Robot. Res. **10**(10), 123–134 (1991)
4. Özgür, E., Gogu, G., Mezouar, Y.: A study on dexterous grasps via parallel manipulation analogy. J. Intell. Robot. Syst. **2017**(87), 1–12 (2017)
5. Özgür, E., Gogu, G., Mezouar, Y.: Structural synthesis of dexterous hands. In: IEEE/RSJ International Conference on Intelligent Robots and Systems, pp. 1676–1681. IEEE, Chicago (2014)
6. Borràs, J., Dollar, A.: Analyzing dexterous hands using a parallel robots framework. Aution Robot **2014**(36), 169–180 (2014)
7. Borràs, J., Dollar, A.: A parallel robots framework to study precision grasping and dexterous manipulation. In: IEEE International Conference on Robotics and Automation, pp. 1595–1601. IEEE, Karlsruhe (2013)
8. Ebertuphoff, I., Voglewede, P.: On the connections between cable-driven robots, parallel manipulators and grasping. In: IEEE International Conference on Robotics and Automation, pp. 4521–4526. IEEE, LA (2004)
9. Bullock, I.M., Feix, T., Dollar, A.: Workspace shape and characteristics for human two- and three-fingered precision manipulation. IEEE Trans. Biomed. Eng. **62**(9), 2196–2207 (2015)

10. Nefzi, M., Riedel, M., Corves, B.: Development and design of a multi-fingered gripper for dexterous manipulation. IFAC Proc. Vol. **39**(16), 133–138 (2006)
11. Riedel, M., Nefzi, M., Hüsing, M., et al.: An adjustable gripper as a reconfigurable robot with a parallel structure. In: Proceedings of the Second International Workshop on Fundamental Issues and Future Research Directions for Parallel Mechanisms and Manipulators. Rwth Aachen, Montpellier (2008)
12. Yu, J.J., Liu, X.J., Ding, X.L., et al.: Foundation of Mathematics of Robot Mechanism. China Machine Press, Beijing (2008)
13. Huang, Z., Zhao, Y.S., Zhao, T.S.: Advanced Spatial Mechanism. Higher Education Press, Beijing (2006)
14. Arun, S.R., Sandipan, B., Ashitava, G.: Analysis of the degrees-of-freedom of spatial parallel manipulators in regular and singular configurations. Mech. Mach. Theory **69**(1), 127–141 (2013)

Design and Analysis of a New 5-DOF Hybrid Serial-Parallel Manipulator

Yundou Xu[1,2], Zhenghe Xu[1], Fan Yang[1], Youen Mei[3], Yi Yue[3], Yulin Zhou[4], Jiantao Yao[1,2], and Yongsheng Zhao[1,2(✉)]

[1] Laboratory of Parallel Robot and Mechatronic System of Hebei Province, Yanshan University, Qinhuangdao 066004, China
{ydxu,yszhao}@ysu.edu.cn

[2] Key Laboratory of Advanced Forging & Stamping Technology and Science, Yanshan University, Qinhuangdao 066004, China

[3] Shanghai Aerospace Equipment Manufacture Company Limited, Shanghai 201100, China

[4] School of Mechanical Engineering, Yanshan University, Qinhuangdao 066004, China

Abstract. In this paper, a 2R1T (two rotations and one translation) over-constrained parallel mechanism R(2RPR)R/SP with two ordered continuous rotational axes is proposed, which contains fewer joints, by concatenating a 2-degree-of-freedom (DOF) RR tilting head with orientation-adjusting capability, a novel 5-DOF hybrid manipulator is constructed. On this basis, optimize the sizes with the best performance of the motion/force transmissibility as the optimization target, and the design dimensions that meet the requirements of the workspace are obtained and the structural design is carried out. The static stiffness of the manipulator are analyzed, and the analysis results are verified by simulation and compared with the products that have been successfully applied in foreign countries. An iterative searching method is used to determine the workspace of the manipulator, and evaluate whether the workspace of the hybrid manipulator meets the requirements.

Keywords: 5-DOF hybrid manipulator · Dimension optimization · Static stiffness analysis · Workspace

1 Introduction

With the rapid development of economy and the continuous improvement of science and technology in the fields of aerospace, manufacturing industry of China has broad development space [1]. The 5-DOF hybrid manipulator combines the features of large workspace and high stiffness of the series manipulator and the parallel manipulator, and there are obvious advantages in the processing of such parts.

Several of the 5-DOF hybrid manipulators that have been successfully applied abroad are designed on the basis of the 2R1T 3-DOF parallel mechanisms [2]. For example, the company of News Robotics proposed a series of manipulators named Tricept [3] and the company of Exechon proposed the hybrid machining manipulators

D. Wang et al. (Eds.): MeTrApp 2019, MMS 79, pp. 301–311, 2020.
https://doi.org/10.1007/978-981-15-0142-5_30

LINKS-EXE700 [4], which concatenated the 2-DOF RR tilting head with orientation-adjusting capability on the 2R1T parallel mechanisms 3UPS/UP and 2UPR/SPR respectively to realize the five-axis hybrid of the manipulators (R, P, U and S stand for revolute, prismatic, universal and spherical joints, respectively); Domestic researchers have also constructed other forms of 5-DOF hybrid manipulators based on some 2R1T 3-DOF parallel mechanisms, such as TriVariant [5] and TriMule [6], their parallel sections are 2UPS/UP and 2UPR/RPR/UPS, respectively, and the five-axis hybrid can be realized by concatenating the 2-DOF RR tilting head in the moving platform. Compared with foreign hybrid manipulators, these 5-DOF hybrid serial-parallel manipulators still have a certain gap in overall level, but they have high industrial application prospects, and the technology is constantly maturing.

In summary, 5-DOF hybrid serial-parallel manipulators have been successfully applied in the industry, particularly based on the spatial position-type 2R1T parallel mechanism [7] (the two rotational DOF of the mechanism are used to adjust the end's position of the manipulator, rather than the orientation). By connecting a 2-DOF RR tilting head with orientation-adjusting capability, it can achieve five-axis hybrid.

2 Work Principle and Dimensional Optimization

2.1 The Work Principle of Five-Axis Hybrid of the Manipulator

The 5-DOF manipulator proposed in this paper is based on the structure of semi-symmetric over-constrained 2R1T parallel mechanism R(2RPR)R/SP (see Fig. 1). The parallel mechanism is composed of a fixed platform, a moving platform and three driving links. Each driving link is composed of the upper and the lower link, and the upper link is connected with one side of the fixed platform, and the lower link is connected with one side of the moving platform. The links A_1a_1 and A_2a_2 are UPU structures, and the axes of the joints U which is connected with the upper and the lower link separately are parallel to each other and perpendicular to the line A_1a_1 or A_2a_2. With the two axes of joint U parallel, the two joints on the same side of the two links are connected to the moving platform or the fixed platform, these two links are also recorded as a composite link R(2RPR)R. The link A_3a_3 is an SP structure, and the line A_3a_3 is perpendicular to the plane of the moving platform.

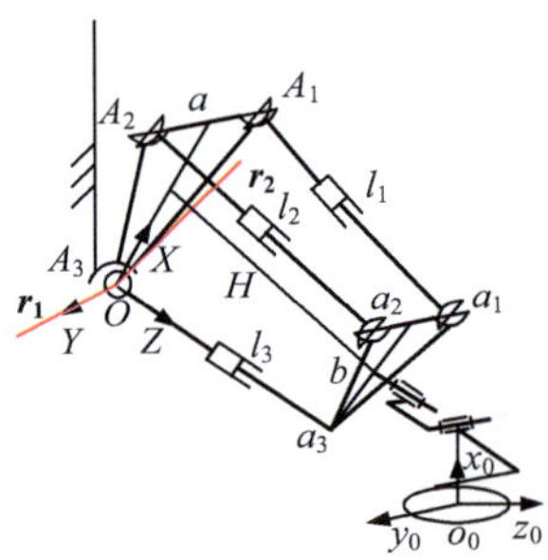

Fig. 1. The structural diagram of 5-DOF hybrid manipulator

As analyzed in literature [8], the parallel mechanism has two ordered continuous rotational axes. One rotational axes $\boldsymbol{r}_1$ is collinear with the Y axis, and the other rotating axes $\boldsymbol{r}_2$ is perpendicular to the plane defined by the links A_1a_1, A_2a_2. The two rotational axes $\boldsymbol{r}_1$ and $\boldsymbol{r}_2$ are not completely continuous, only the moving platform rotates continuously around a certain rotational axis, the other rotational axis is a continuous rotating shaft. For example, the moving platform can continuously rotate around the Y axis ($\boldsymbol{r}_1$), then, it can continuously rotate around $\boldsymbol{r}_2$ to reach any position. Since the two rotational axes are close to the fixed platform, the type of the moving platform is the spatial-position type. It can connect a 2-DOF RR tilting head with orientation-adjusting capability to achieve five-axis hybrid.

2.2 Performance Indexes and Dimensional Optimization

According to the requirements of the processing task, the workspace of the 5-DOF manipulator includes a $\phi 1300 \times 300$ mm cylinder (the workspace means the scope that the intersection of the two rotational axes of RR tilting head could touch).

In order to make the designed manipulator have excellent mechanical performance, we should comprehensively consider the performance of the motion/force transmission of the mechanism in the workspace [8], and optimize the size with the global transmission index (GTI) as the optimization target. The GTI takes the performance of the motion/force transmission of the mechanism into account and it can be described as follows:

$$\Gamma = \frac{\int_G \sum_i^n \frac{\lambda_i}{n} dw}{\int_W dw} \tag{1}$$

Where $\boldsymbol{G}$ and $\boldsymbol{W}$ represent regions and workspace with better transfer performance, respectively, n represents the number of samples, and λ is an output performance index whose expression is:

$$\lambda = \min\left(\frac{|\$_{Oi} \circ \$_{Ti}|}{|\$_{Oi} \circ \$_{Ti}|_{\max}}\right), \ (i = 1, 2, 3) \tag{2}$$

In which, the $\$_{Ti}$ represents driving force which can be regarded as a new unit transmission wrench (pure force), and the $\$_{Oi}$ represents the output twist.

The triangles formed by the hinge point of moving platform and the fixed platform of the parallel mechanism R(2RPR)R/SP are regarded as equilateral triangles, and the edge length of the fixed platform is a, the edge length of the moving platform is b, and the initial height of the moving platform is H. The three dimensions are selected to be optimized, then, set their range of variation: $a \in [750, 1000]$ mm, $b \in [350, 420]$ mm, and $H \in [1000, 1200]$ mm.

Discretize the design variables, the discrete data step size is 5 mm, calculate the workspace between the moving platform and the fixed platform for each set of the discrete data, and determine whether the workspace constraints and link length constraints are met (preset the range of link length is 600–1500 mm), and a set of design variables *a, b* and *H* that meet the requirements of the workspace are obtained. Table 1 lists the design variable dimension combinations that partially satisfy the conditions.

Table 1. Dimensions combination of design variables that satisfy workspace and link length constraints

param-eter	**Numerical value**								
	1	**2**	**3**	**4**	**5**	**6**	**7**	**8**	**9**
a	750	750	750	750	800	800	800	800	850
b	360	380	400	420	360	380	400	420	360
H	1100	1120	1120	1120	1120	1120	1140	1140	1120
param-eter	**Numerical value**								
	10	**11**	**12**	**13**	**14**	**15**	**16**	**17**	**18**
a	850	850	900	900	900	950	950	950	1000
b	400	420	360	380	400	360	380	400	360
H	1140	1160	1140	1140	1160	1140	1160	1160	1140

The λ can be regarded as a transfer index of the local performance, and the larger the value, the better the instantaneous motion/force transfer performance of the mechanism. As shown in Fig. 3, when $\lambda > 0.5$, the parallel mechanism has relatively great motion/force transfer performance. According to dimensions combinations listed in Table 1, the GTI of the motion/force transfer power global performance index in the workspace is evaluated. With formula (1), the GTI of the mechanism in the workspace is analyzed. As shown in Fig. 2, the values of GTI corresponding to each group of structural parameters are given.

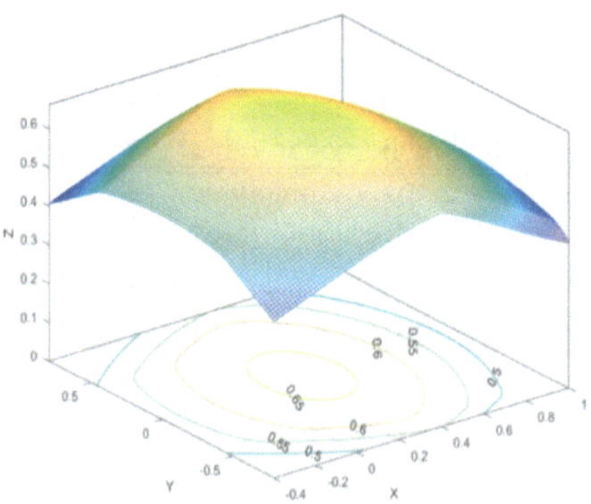

Fig. 2. The performance map of λ

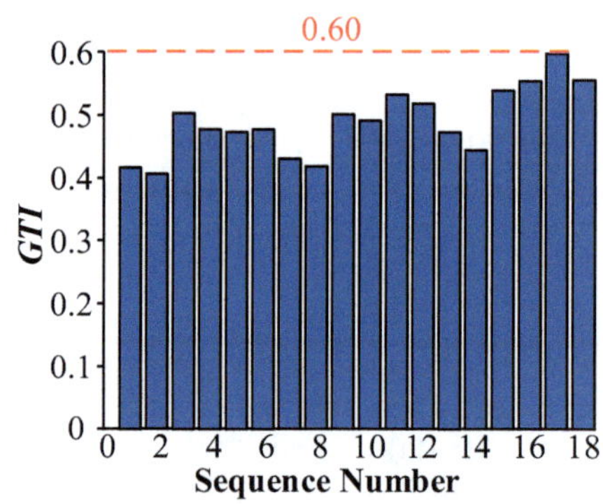

Fig. 3. The value of GTI

In the size selected by the red frame in Table 1, the GTI value of the mechanism is the highest, and thus the design variable dimensions of the mechanism is determined as shown in Table 2.

Table 2. Values of the design dimensions

Key sizes	a	b	H	The range of link length
Values (mm)	950	400	1160	600–1500

3 The Structural Design and the Static Stiffness Performance

3.1 The Structural Design

The 3D model of the hybrid manipulator is shown in Fig. 4. The hybrid manipulator is composed of a first limb, a second limb, a third limb, an upper Hooke joint, a lower Hooke joint, a moving platform and a 2-DOF RR tilting head. The first and second limbs are arranged on the top, and the third limb is arranged below. One end of the first limb and the second limb are connected to the frame through the upper Hooke joint, and the other end is connected to the moving platform through supported connecting part. One end of the third limb is connected to the frame through the lower Hooke joint, and the other end is connected with moving platform directly. The 2-DOF RR tilting head is set at the geometric center of the moving platform. The three driving limbs are all motor front.

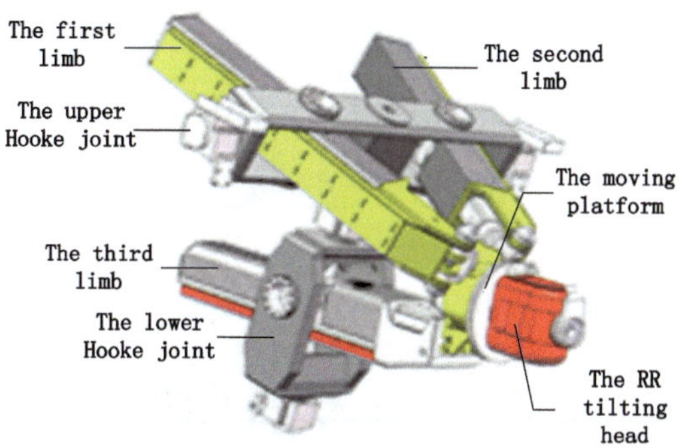

Fig. 4. The 3D model of the hybrid manipulator

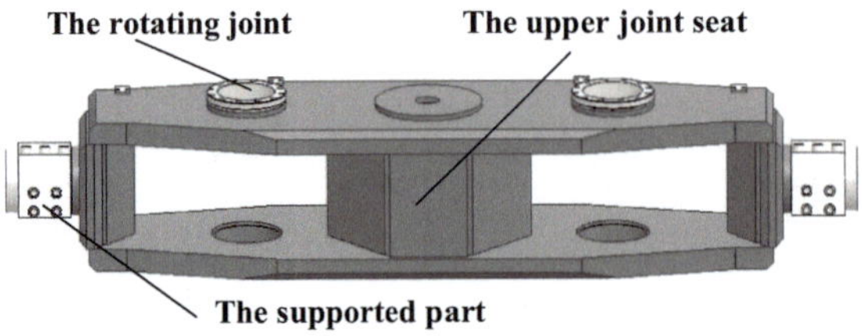

Fig. 5. The structure of the upper Hooke joint

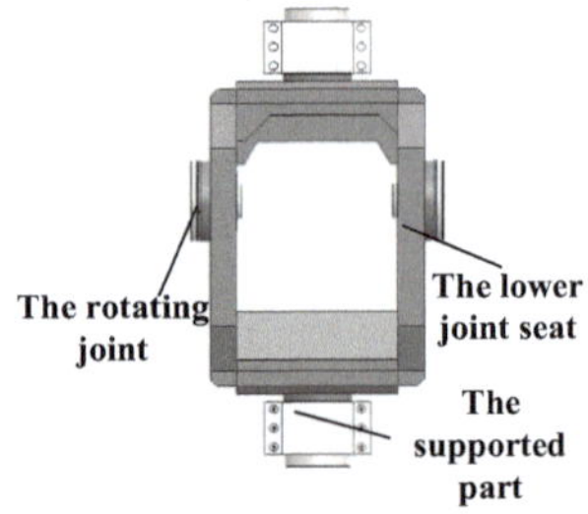

Fig. 6. The structure of the lower Hooke joint

The upper Hooke joint consists of the upper joint seat, the supported part and the rotating joint. As shown in Fig. 5, the two sides are connected to the frame through the supported part, and with connecting with the rotating joint, the first and second links respectively pass through the hollow portions.

The structure of the lower Hooke joint is shown in Fig. 6, and the principle is the same as the upper Hooke joint.

The first and second limbs are composed of the supported connecting part of moving platform, the bodies of the first and second limbs, the servo motor, the guide rail, the component of ball screw, the sliding seat, and the like, as shown in Fig. 7. The layout method putting the motor on the front improves structural stability, and the open structure facilitates heat dissipation of the motor. The middle portion is connected to the rotating joint in the upper Hooke joint by the sliding seat, and the front end is connected to the 2-DOF RR tilting head through the supported connecting part of moving platform.

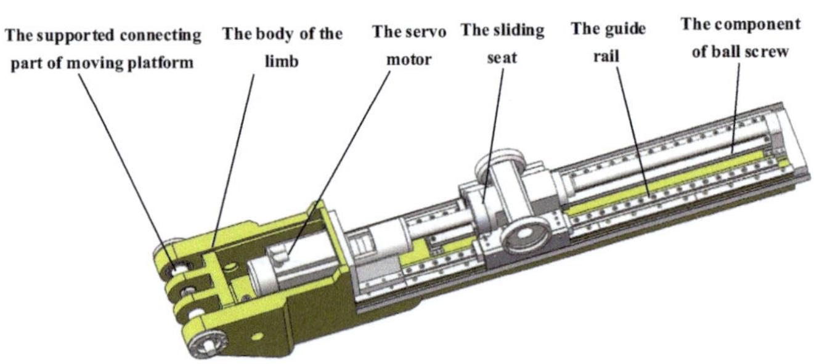

Fig. 7. The structure of the first and second limbs

The structure of the third limb is as shown in Fig. 8, and it is constituted by the body of the third limb, the servo motor, the guide rail, the component of ball screw, the rotating sliding seat, and the like. It is connected to the rotating joint in the lower Hooke joint through the rotating sliding seat, and is connected to the 2-DOF RR tilting head through the front end of the body of the third limb.

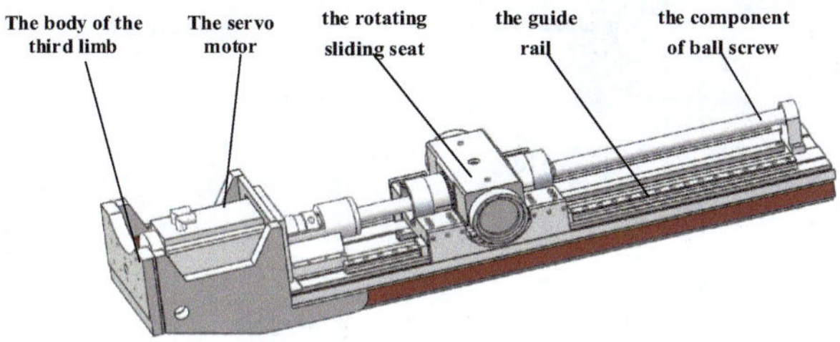

Fig. 8. The structure of the third limb

The 2-DOF RR tilting head structure is as shown in Fig. 9, which is composed of the 2-DOF head, the moving platform, the connecting part with the first and second limb, the connecting surface with the third limb, and the positioning pins.

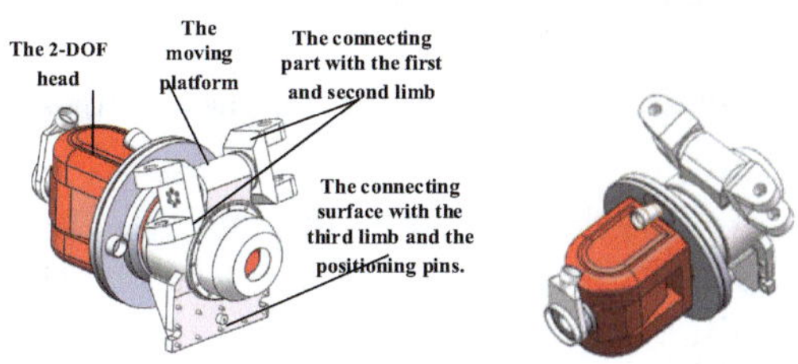

(a) View 1 **(b)** View 2

Fig. 9. The structure of the 2-DOF RR tilting head

3.2 The Static Stiffness Performance and Comparison

The static stiffness of the manipulator is analyzed with finite element software. After obtaining the finite element calculation model, since the milling and cutting force is only about 150 N during the machining process of the 5-DOF hybrid manipulator, to verify the overall rigidity of the mechanism, a load of 1000 N is added to the position that the AC swinging head mounts tool to simulate the milling and cutting load.

Milling and cutting loads are added from the axial direction, the *X*-axis direction and the *Y*-axis direction respectively, and the axial stiffness, the *X*-axis lateral stiffness and the *Y*-axis lateral stiffness are analyzed for each direction of the manipulator, and the total deformation amount of the manipulator is as shown in Fig. 10.

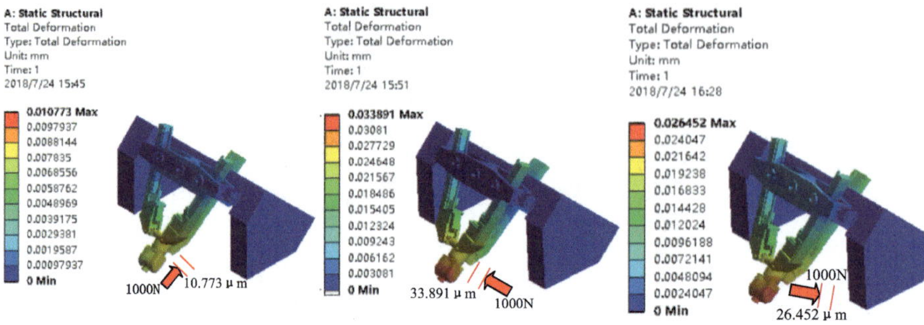

(a) Axial loading **(b)** X direction loading **(c)** Y direction loading

Fig. 10. Overall deformation of the manipulator by adding load in different directions

By arranging the results in Fig. 10, the values of the static stiffness in the respective directions as shown in Table 3 can be obtained.

Table 3. The result of static stiffness analysis

Loading method	Axial loading	X-axis loading	Y-axis loading
Maximum deformation	0.010773 mm	0.033891 mm	0.026452 mm
Static stiffness	92.8 N/um	29.5 N/um	37.8 N/um

At present, the international Exechon Company in Sweden has designed such a similar 5-DOF hybrid manipulator with relatively mature technology and has obtained certain applications. It can be seen from Fig. 11 that the lateral stiffness of the Exechon calculated by the finite element software is 30 N/um, and the finite element theoretical calculation results of the 5-DOF hybrid manipulator designed in this paper are the *X*-axis lateral stiffness and the *Y*-axis lateral stiffness, they are 29.5 N/um and 37.8 N/um, respectively. Therefore, the static stiffness of the hybrid manipulator designed in this paper is on the same order of magnitude as that of the typical similar manipulator in commercialization, which indicates that the hybrid manipulator designed in this paper has high stiffness and has a good application prospect.

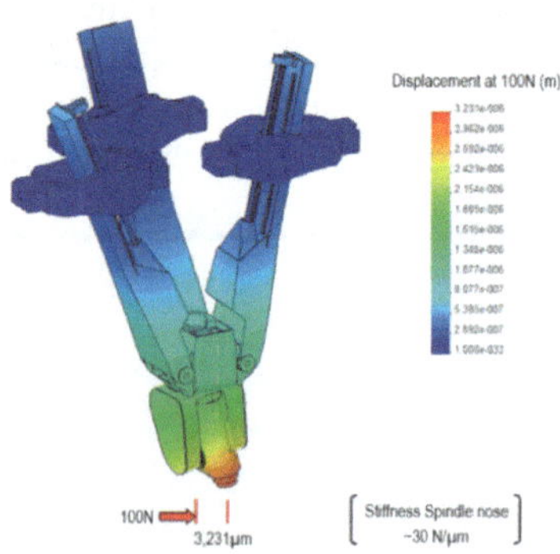

Fig. 11. Overall deformation of the manipulator Exechon

4 The Analysis of Workspace

4.1 A Iterative Searching Method of Workspace

The intersection of the two rotational axes of 2-DOF RR tilting head is selected as the reference point of the workspace. According to the actual effective running stroke of each link of the given parallel mechanism, the upper leading screw seat of each limb is taken as the initial position. In the following, based on the inverse solution of the mechanism position, an iterative searching method for judging whether the points of the end of the hybrid manipulator is in the workspace is proposed, and the workspace is obtained. The specific steps are as follows: Given a scope large enough to contain the actual workspace, the value of the length of each driving link is obtained by bringing the points in the given workspace into the inverse kinematics of the mechanism position, thereby determining whether the point is in the workspace. If yes, print the point; otherwise, judge the next point in the selected workspace until all the points in the given searching scope are searched; finally, output the points that meet the conditions into a .txt file and import them into SolidWorks software, and they will be displayed as shown in Fig. 12.

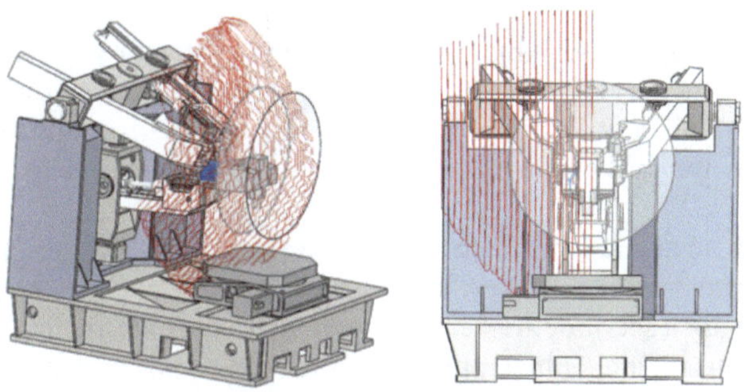

(a) View 1 **(b)** View 2

Fig. 12. Hybrid manipulator model and its workspace representing by lattice

4.2 Workspace Analysis

When the workspace height is 300 mm, the envelope diagram of the workspace envelope of the intersection of the two axes of the 2-DOF RR tilting head is shown in Fig. 13, and the 5-DOF hybrid manipulator limit workspace is a $\phi(1975 - 1345) \times 300$ truncated cone. If the mechanism is tilted down 5° on the original basis, the limit workspace can be further expanded to a maximum of $\phi(1975 - 1390) \times 300$ mm, as shown in Fig. 14.

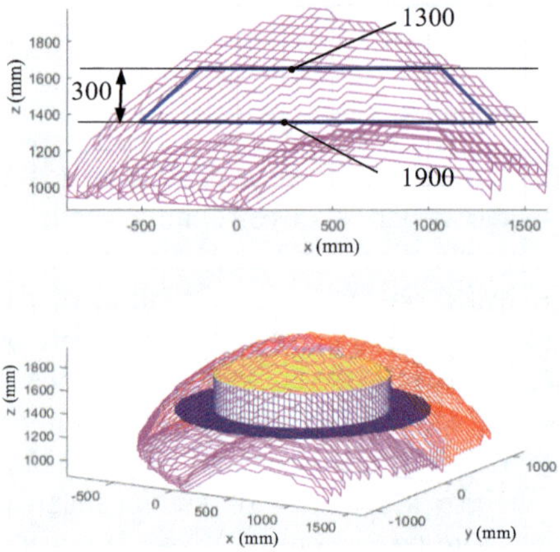

Fig. 13. The envelope diagram of workspace

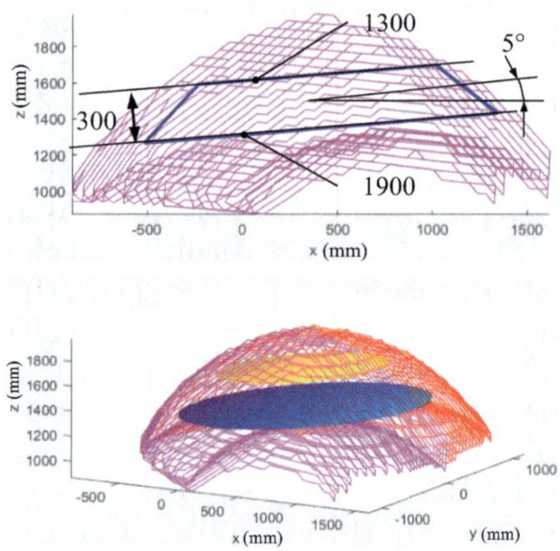

Fig. 14. The envelope diagram of workspace when the manipulator is tilted by 5°

5 Conclusion

In this paper, a novel 5-DOF hybrid manipulator is constructed based on the 2R1T parallel mechanism R(2RPR)R/SP. The parallel part has two ordered continuous rotational axes, and a 2-DOF tilting head with orientation-adjusting capability is concatenated on the moving platform of the parallel mechanism to achieve five-axis hybrid.

Considering the indexes such as workspace and the performance of the motion/force transmission, the key sizes of the parallel part of the manipulator are optimized to obtain a set of optimized sizes. On the basis, the structural design of a new 5-DOF hybrid serial-parallel manipulator is completed, and the static stiffness analysis is carried out by using the finite element analysis method.

A searching method of the workspace based on the inverse kinematics of the mechanism is proposed. The 3D model of the workspace is obtained, and the limit workspace is further analyzed to ensure that the designed 5-DOF hybrid manipulator can meet the expected workspace and leave a certain margin.

Acknowledgement. This work was supported by the National Natural Science Foundation of China under Grant 51875495 and 51405425, and the National Key R&D Program of China under Grant 2017YFB1301901.

References

1. Zhou, Z.X., Zhou, Q.Y., Ren, Y.H.: Current research and development trends of complex surface machining technology. J. Mech. Eng. **46**(17), 105–113 (2010)

2. Li, Q.C., Chai, X.X., Chen, Q.H.: Review on 2R1T 3-DOF parallel mechanisms. Sci. China **14**, 78–90 (2017)
3. Caccavale, F., Siciliano, B., Villani, L.: The tricept robot: dynamics and impedance control. IEEE/ASME Trans. Mechatronics **8**(2), 263–268 (2003)
4. Jin, Y., Bi, Z.M., Liu, H.T., et al.: Kinematic analysis and dimensional synthesis of Exechon parallel kinematic machine for large volume machining. J. Mech. Robot. **7**(4), 041004 (2015)
5. Li, M., Huang, T., Chetwynd, D.G., et al.: Forward position analysis of the 3-DOF module of the TriVariant: a 5-DOF reconfigurable hybrid robot. J. Mech. Des. **128**(1), 319–322 (2005)
6. Huang, T., Dong, C.L., Liu, H.T., et al.: Five-degree-of-freedom hybrid robot with multi-axis rotating bracket. China, 2015104012799 (2015)
7. Dong, C., Liu, H., Liu Q., et al.: An approach for type synthesis of over constrained 1T2R parallel mechanisms. In: Mechanisms and Machine Science, vol. 50. Springer, Cham (2018)
8. Zhang, D.S., Xu, Y.D., Yao, J.T., et al.: Design of a novel 5-DOF hybrid serial-parallel manipulator and theoretical analysis of its parallel part. Robot. Comput. Integr. Manuf. **53**, 228–239 (2018)
9. Liu, X.J., Wang, L.P., et al.: Design of a three-axis articulated tool head with parallel achieving desired motion/force transmission characteristics. J. Manuf. Sci. Eng. **132**(2), 021009 (2010)
10. Xu, Y., Zhou, S., Yao, J., et al.: Rotational Axes Analysis of the 2-RPU/SPR 2R1T Parallel Mechanism, 2nd edn. Springer, Cham (2014)
11. Lu, Y., Shi, Y., Hu, B.: Solving reachable workspace of some parallel manipulators by computer-aided design variation geometry. Proc. Inst. Mech. Eng. Part C **222**(9), 1773–1781 (2008)

Kinematics Research of Chewing Robot Based on Six-Axis Parallel Mechanism

Xingyu Zhou[1] and Jinghu Yu[1,2(✉)]

[1] School of Mechanical Engineering, Jiangnan University, Wuxi 214122, Jiangsu, China
1045494894@qq.com

[2] Jiangsu Province Key Laboratory of Advanced Food Manufacturing Equipment and Technology, Wuxi 214122, Jiangsu, China

Abstract. Instruments widely used in food texture detection are different from human oral environment and chewing movement parameters, which results in inconsistent detection data and taste. Experts' sensory evaluation of food also has the shortcomings of expensive, inefficient and difficult to standardize. The development and research of bionic chewing robots is regarded as a new direction to solve problems in this field. This paper presents the basic structure of a chewing robot based on six-axis parallel mechanism and establishes its kinematics model. By using the inverse kinematics method, the expressions of the position and velocity relationship between the prismatic pair driven by the motor and the moving platform were obtained through theoretical analysis. Referring to human chewing action, a set of moving platform trajectories were designed. At the same time, the displacement and velocity curves of each prismatic pair under the predetermined trajectory of the moving platform were obtained by kinematics simulation. This research will provide a basis for the motion control of chewing robots, and also contribute to the development of food texture inspection equipment.

Keywords: Chewing robot · Inverse kinematics · Kinematics simulation

1 Introduction

At present, the food industry is developing rapidly, and the research on food is more and more in-depth. Not only the safety of food needs to be analyzed, but also the objective evaluation of physical properties, especially the acceptance of texture and flavor by consumers. At present, there are two main methods for food texture assessment: sensory evaluation and instrumental determination. The former is evaluated by expert group, which has some shortcomings of subjectivity and difficulty in standardization, and a study have pointed out that people's oral processing ability also has an impact on the evaluation of food taste [1]. The latter relies on existing instruments such as texture analyzer [2]. Wee May et al. found that the eating behavior and macronutrient composition of various solid foods were correlated with texture through TPA experiments [3]. However, because the texture analyzer has only one axis motion which is different from people's actual chewing activity and the food texture has

D. Wang et al. (Eds.): MeTrApp 2019, MMS 79, pp. 312–321, 2020.
https://doi.org/10.1007/978-981-15-0142-5_31

complex characteristics, the consistency and correlation between the measured physical property data and the sensory test are not satisfactory [4]. Therefore, improving the correlation between food instrument detection and sensory evaluation is an important issue in the food industry, especially the improvement of food texture testing instruments.

The research of bionic chewing robot is considered as a new way to solve the above problems. Chewing robot can more realistically imitate the chewing activity of people, so as to obtain more objective and accurate experimental data. Raabe et al. of the University of Bristol in the United Kingdom developed a six-degree-of-freedom motion mechanism modeled on the human jaw [5]. Woda et al. of the University of Clermont-Auvergne in France developed a mechanism to imitate the human oral cavity, which simulated the masticatory cycle and amplitude of the horizontal and vertical movement of the artificial chin [6]. Researchers at Massey University simulated the activity of masticatory muscles in human chewing and designed a single-track linkage chewing robot [7]. Based on this, a flexible trajectory is realized by adding an adjustable link mechanism and a movable platform, thereby designing a new generation of chewing robots, which can mimic human chewing well [8, 9]. Cong Ming et al. of Dalian University of Technology of China established a 6-DOF chewing robot based on the biomechanics of chewing system, and obtained the displacement, speed, power and other important parameters of linear actuator through the analysis of motion characteristics [10, 11]. The above research results have important reference value for the research of this paper.

For the study of chewing robot based on six-axis parallel mechanism, kinematics analysis is the basis of other problems, especially for the mechanism control [12]. In order to control the trajectory of the moving platform correctly, the mapping relationship between the input position parameters and the output position parameters needs to be obtained by inverse kinematics solution [13]. Therefore, this paper focuses on the kinematics of the designed chewing robot. This research will provide a basis for the motion control of chewing robots, and also contribute to the development of food texture inspection equipment.

2 Establishment of Model

2.1 Description of Mechanism

The whole design structure of the chewing robot is shown in Fig. 1. The motion principle of the mechanism is based on the design of Stewart platform [14]. The design takes into account the movement of the human jaw. The upper jaw model is mounted on the static platform and remains static during the movement. The lower jaw model is mounted on the moving platform, and the simulation of the human chewing movement is completed by the synthetic movement of the six connecting rods. The force sensor can measure the load on the jaw model during the movement.

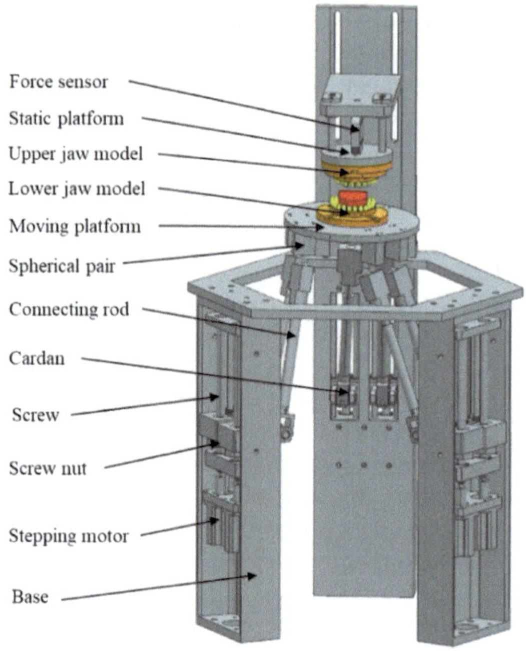

Fig. 1. The whole structure of bionic chewing robot

2.2 Kinematic Model

The schematic diagram of the mechanism and its related dynamic coordinate and fixed coordinate are shown in Fig. 2.

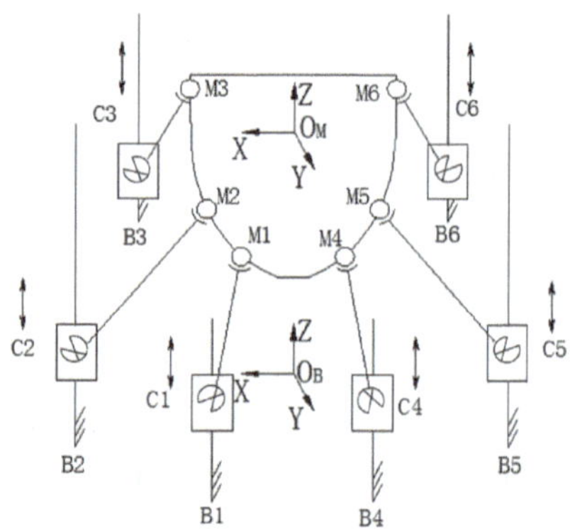

Fig. 2. Motion mechanism schematic and coordinate system

In Fig. 2, O_M is a dynamic coordinate system, which is fixed on the moving platform. O_B is a fixed coordinate system and it has a displacement vector only in the

Z-axis with the dynamic coordinate. B_i is the position where the screw is fixed on the base ($i = 1, 2\ldots 6$). C_i is a universal pair fixed on the prismatic pair ($i = 1, 2\ldots 6$). M_i is a spherical joint fixed on the moving platform ($i = 1, 2\ldots 6$).

The degree of freedom of three-dimensional spatial mechanism can be calculated according to KUTZBACH-GRUBLER formula.

$$F = 6 \times (n - g - 1) + \sum\nolimits_{i=1}^{g} f_i \tag{1}$$

In Eq. (1), F is the degree of freedom of the mechanism, n is the number of components, g is the number of motion pairs between components, and f_i is the degree of freedom of motion pair. In this mechanism:

$$F = 6 \times (14 - 18 - 1) + (6 \times 2 + 6 \times 1 + 6 \times 3) = 6 \tag{2}$$

In this coordinate system, the RPY angle is used to describe the movement of the moving platform. (α β γ) represents the rotational angle of the dynamic system O_M around the X, Y and Z axis of the fixed system O_B. The rotation matrix ${}_M^BR$ of the dynamic system O_M relative to the fixed system O_B is:

$$\begin{aligned} {}_M^BR &= R(z,\gamma)R(y,\beta)R(x,\alpha) \\ &= \begin{pmatrix} \cos\gamma & -\sin\gamma & 0 \\ \sin\gamma & \cos\gamma & 0 \\ 0 & 0 & 1 \end{pmatrix} \begin{pmatrix} \cos\beta & 0 & \sin\beta \\ 0 & 1 & 0 \\ -\sin\beta & 0 & \cos\beta \end{pmatrix} \begin{pmatrix} 1 & 0 & 0 \\ 0 & \cos\alpha & -\sin\alpha \\ 0 & \sin\alpha & \cos\alpha \end{pmatrix} \\ &= \begin{pmatrix} \cos\beta\cdot\cos\gamma & \sin\alpha\cdot\sin\beta\cdot\cos\gamma - \cos\alpha\cdot\sin\gamma & \cos\alpha\cdot\sin\beta\cdot\cos\gamma + \sin\alpha\cdot\sin\gamma \\ \cos\beta\cdot\sin\gamma & \sin\alpha\cdot\sin\beta\cdot\sin\gamma + \cos\alpha\cdot\cos\gamma & \cos\alpha\cdot\sin\beta\cdot\sin\gamma - \sin\alpha\cdot\cos\gamma \\ -\sin\beta & \sin\alpha\cdot\cos\beta & \cos\alpha\cdot\cos\beta \end{pmatrix} \end{aligned} \tag{3}$$

2.3 Inverse Solution of Position and Velocity

The position vector of the spherical pair connection point in the dynamic system O_M is $\boldsymbol{M_i}$ and in the fixed system O_B is $\boldsymbol{m_i}$. According to the composite transformation, it can be obtained that:

$$\boldsymbol{m_i} = {}_M^BR\boldsymbol{M_i}$$

$$\begin{pmatrix} m_{ix} \\ m_{iy} \\ m_{iz} \end{pmatrix} = {}_M^BR \begin{pmatrix} M_{ix} \\ M_{iy} \\ M_{iz} \end{pmatrix} \tag{4}$$

The length of the connecting rod L_i is:

$$L_i = |\boldsymbol{C_i m_i}| \tag{5}$$

$\boldsymbol{C_i}$ is the position of the universal pair (prismatic pair) in the fixed system.

$$C_i = \begin{pmatrix} C_{ix} \\ C_{iy} \\ C_{iz} \end{pmatrix} = \begin{pmatrix} B_{ix} \\ B_{iy} \\ C_{iz} \end{pmatrix} \tag{6}$$

The position of the prismatic pair in the Z-axis can be obtained as follows:

$$L_i^2 = (m_{ix} - B_{ix})^2 + (m_{iy} - B_{iy})^2 + (m_{iz} - C_{iz})^2 \tag{7}$$

$$C_{iz} = m_{iz} - \sqrt{L_i^2 - (m_{ix} - B_{ix})^2 - (m_{iy} - B_{iy})^2} \tag{8}$$

In Eq. (8), L_i^2, B_{ix} and B_{iy} are parameters that have been determined in the design of the mechanism. If the position of the moving platform is known, the position of six moving pairs can be obtained by inverse solution.

In order to obtain the expression of velocity, the derivative of Eq. (8) is obtained, as shown in Eqs. (9) and (10):

$$\begin{aligned} \frac{dC_{iz}}{dt} &= \frac{\partial C_{iz}}{\partial x} \cdot \frac{\partial x}{\partial t} + \frac{\partial C_{iz}}{\partial y} \cdot \frac{\partial y}{\partial t} + \frac{\partial C_{iz}}{\partial z} \cdot \frac{\partial z}{\partial t} + \frac{\partial C_{iz}}{\partial \alpha} \cdot \frac{\partial \alpha}{\partial t} + \frac{\partial C_{iz}}{\partial \beta} \cdot \frac{\partial \beta}{\partial t} + \frac{\partial C_{iz}}{\partial \gamma} \cdot \frac{\partial \gamma}{\partial t} \\ &= \begin{pmatrix} \frac{\partial C_{iz}}{\partial x} & \frac{\partial C_{iz}}{\partial y} & \frac{\partial C_{iz}}{\partial z} & \frac{\partial C_{iz}}{\partial \alpha} & \frac{\partial C_{iz}}{\partial \beta} & \frac{\partial C_{iz}}{\partial \gamma} \end{pmatrix} \cdot \begin{pmatrix} \dot{x} & \dot{y} & \dot{z} & \dot{\alpha} & \dot{\beta} & \dot{\gamma} \end{pmatrix}^{\mathrm{T}} \end{aligned} \tag{9}$$

$$\begin{pmatrix} \dot{C}_{1z} \\ \dot{C}_{2z} \\ \dot{C}_{3z} \\ \dot{C}_{4z} \\ \dot{C}_{5z} \\ \dot{C}_{6z} \end{pmatrix} = \begin{pmatrix} \frac{\partial C_{1z}}{\partial x} & \frac{\partial C_{1z}}{\partial y} & \frac{\partial C_{1z}}{\partial z} & \frac{\partial C_{1z}}{\partial \alpha} & \frac{\partial C_{1z}}{\partial \beta} & \frac{\partial C_{1z}}{\partial \gamma} \\ \frac{\partial C_{2z}}{\partial x} & \frac{\partial C_{2z}}{\partial y} & \frac{\partial C_{2z}}{\partial z} & \frac{\partial C_{2z}}{\partial \alpha} & \frac{\partial C_{2z}}{\partial \beta} & \frac{\partial C_{2z}}{\partial \gamma} \\ \frac{\partial C_{3z}}{\partial x} & \frac{\partial C_{3z}}{\partial y} & \frac{\partial C_{3z}}{\partial z} & \frac{\partial C_{3z}}{\partial \alpha} & \frac{\partial C_{3z}}{\partial \beta} & \frac{\partial C_{3z}}{\partial \gamma} \\ \frac{\partial C_{4z}}{\partial x} & \frac{\partial C_{4z}}{\partial y} & \frac{\partial C_{4z}}{\partial z} & \frac{\partial C_{4z}}{\partial \alpha} & \frac{\partial C_{4z}}{\partial \beta} & \frac{\partial C_{4z}}{\partial \gamma} \\ \frac{\partial C_{5z}}{\partial x} & \frac{\partial C_{5z}}{\partial y} & \frac{\partial C_{5z}}{\partial z} & \frac{\partial C_{5z}}{\partial \alpha} & \frac{\partial C_{5z}}{\partial \beta} & \frac{\partial C_{5z}}{\partial \gamma} \\ \frac{\partial C_{6z}}{\partial x} & \frac{\partial C_{6z}}{\partial y} & \frac{\partial C_{6z}}{\partial z} & \frac{\partial C_{6z}}{\partial \alpha} & \frac{\partial C_{6z}}{\partial \beta} & \frac{\partial C_{6z}}{\partial \gamma} \end{pmatrix} \cdot \begin{pmatrix} \dot{x} \\ \dot{y} \\ \dot{z} \\ \dot{\alpha} \\ \dot{\beta} \\ \dot{\gamma} \end{pmatrix} \tag{10}$$

Equation (10) can be expressed simply as follows:

$$\dot{C}_z = \boldsymbol{J} \cdot \dot{\boldsymbol{q}} \tag{11}$$

$\boldsymbol{J} = [J_{ij}]_{6\times 6}$ (i, j = 1, 2 …, 6) is the Jacobian matrix of parallel mechanism and the comprehensive influence coefficient of platform motion on six generalized coordinates. From the deduced results, it can be concluded that the establishment of inverse velocity solution model does not need complicated derivation process, and has nothing to do with the real motion speed of the mechanism, but only with the kinematic size, type and shape of the mechanism.

3 Kinematics Simulation

3.1 Model Simplification

Firstly, the mechanism model of chewing robot is simplified. Boolean sum operation is performed for those components in the model that do not move in motion, so that they can be regarded as a component. After hiding the fixed part, the moving mechanism model of the chewing robot is shown in Fig. 3. Then, the corresponding motion pairs are added for each moving part. Finally, according to a set of chewing motions of human jaw, a set of trajectory functions are designed for the moving platform in the dynamic system O_M, as shown in Eq. (12).

Fig. 3. Motion mechanism model of chewing robot

$$\begin{cases} X = \text{if}(\text{time} - 1 : 0, 0, \text{if}(\text{time} - 2.5 : 4 * \cos(2 * \text{pi}/3 * (\text{time} + 0.5)) + 4, \\ \quad 8, \text{if}(\text{time} - 3 : 8, 8, \text{if}(\text{time} - 4 : 4 * \cos(\text{pi} * (\text{time} + 3)) + 4, 0, 0)))) \\ Y = \text{if}(\text{time} - 1 : \cos(\text{pi} * \text{time}) - 1, -2, \text{if}(\text{time} - 2 : 2 * \cos(\text{pi} * \text{time}), \\ \quad 2, \text{if}(\text{time} - 4 : \cos(0.5 * \text{pi} * (\text{time} - 2)) + 1, 0, 0))) \\ Z = 2 * \cos(0.5 * \text{pi} * \text{time}) - 2 \\ \alpha = 12.5\text{d} * \cos(0.5 * \text{pi} * \text{time}) - 12.5\text{d} \\ \beta = \text{if}(\text{time} - 1 : 0, 0, \text{if}(\text{time} - 3.5 : 2\text{d} * \cos(0.4 * \text{pi} * (\text{time} + 1.5)) + 2\text{d}, \\ \quad 4\text{d}, \text{if}(\text{time} - 4 : 2\text{d} * \cos(2 * \text{pi} * (\text{time} - 3.5)) + 2\text{d}, 0, 0))) \\ \gamma = \text{if}(\text{time} - 1 : 0, 0, \text{if}(\text{time} - 2.5 : 4\text{d} * \cos(2 * \text{pi}/3 * (\text{time} - 1)) - 4\text{d}, \\ \quad -8\text{d}, \text{if}(\text{time} - 3.5 : -8\text{d}, -8\text{d}, -4\text{d} * \cos(2 * \text{pi} * (\text{time} - 3.5)) - 4\text{d}))) \end{cases} \tag{12}$$

3.2 Simulation Result

The chewing motion curve of the moving platform is shown in Fig. 4. The whole chewing cycle is 4 s. Figure 4(a) is the origin displacement curve of the dynamic coordinate, and Fig. 4(b) is the rotation curve of the dynamic coordinate relative to the fixed coordinate. During the first half cycle of chewing, the mouth opens gradually. The X-axis translation and Z-axis rotation are the preparations for the horizontal cutting posture in the right chewing movement. In the second half of the cycle, the mouth closes gradually, and the grinding action is to rotate around the X axis and move in the Z axis. Rotation around Z axis and negative translation around X axis are cutting actions. As for Y axis rotation and Y axis translation, there are both vertical grinding and horizontal cutting actions in mastication. They are auxiliary movements of vertical grinding and horizontal cutting. Their motion is the smallest in the whole mastication cycle, but cannot be ignored. The simulation curves depicted by the above functions have bionic chewing performance.

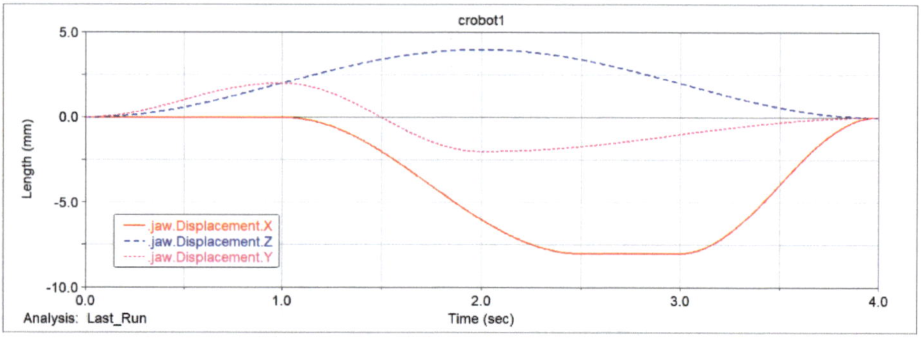

(a) Displacement curve

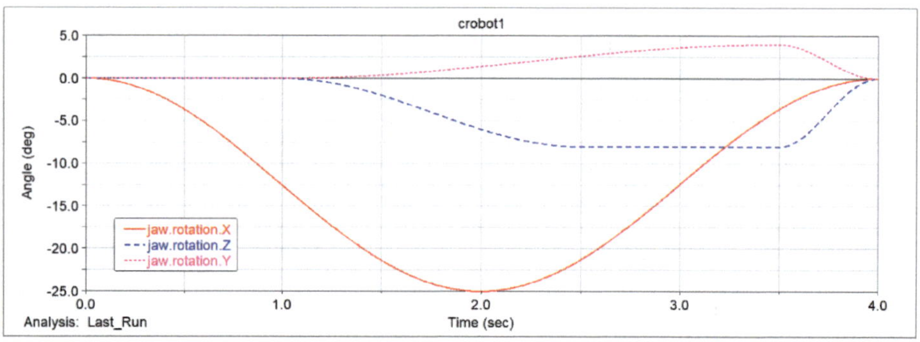

(b) Rotation curve

Fig. 4. Chewing motion curve of the moving platform

During the chewing movement, the displacement curves of the sliders are shown in Fig. 5 and the velocity curves are shown in Fig. 6.

The ultimate goal of this kinematics simulation is to obtain the displacement and velocity curves of the driving mechanism, that is, the displacement and velocity curves of the sliders. By fitting the displacement and velocity curves, the motion function of the slider can be obtained, and the motion control function of the stepping motor can also be obtained.

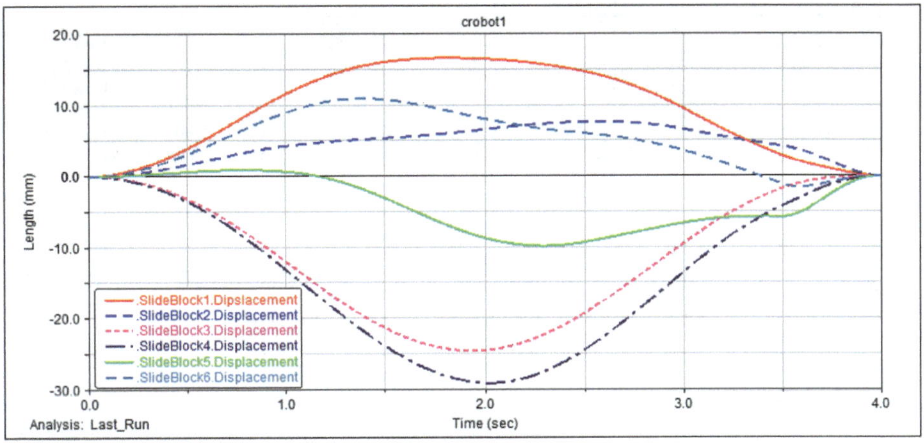

Fig. 5. Displacement curves of the sliders

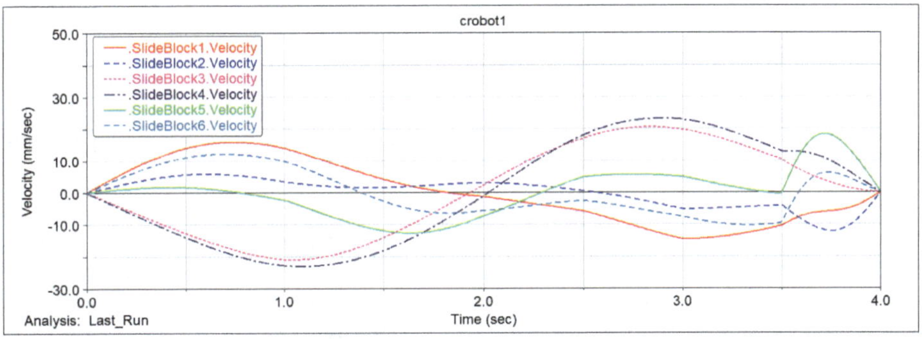

Fig. 6. Velocity curves of the sliders

The theoretical derivation of Eq. (8) is verified by Matlab programming. The coordinates of the mass centers of the prismatic pairs can be obtained by inputting the coordinates of each spherical joint obtained from the simulation results of Adams, and the displacement of the prismatic pairs in the z-axis direction can be obtained. The result obtained by MATLAB programming is shown in Fig. 7. Compared with Fig. 5, the variation trend of each curve is basically the same, but the displacement value is slightly larger. This is because the theoretical derivation is based on the simplified structural model without considering the installation size of spherical pair and cardan. Therefore, in order to improve the accuracy of theoretical calculation, it is necessary to establish a more detailed kinematics model in subsequent research.

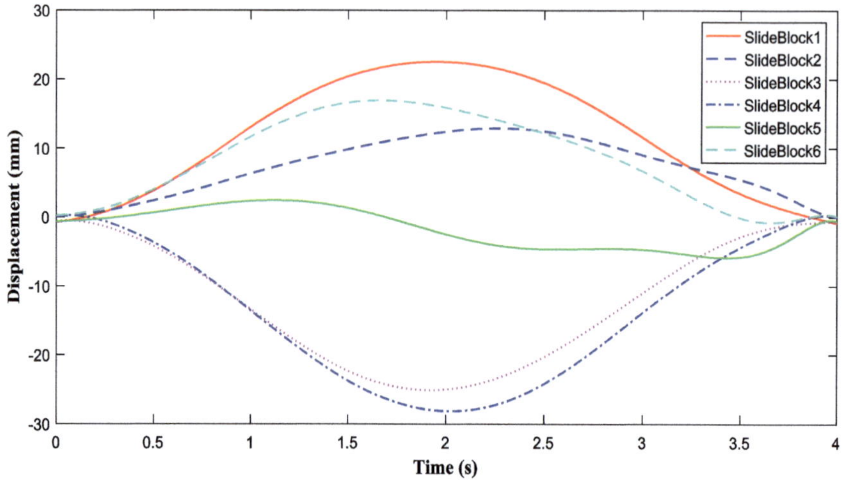

Fig. 7. Displacement curves of sliders obtained by Matlab programming

4 Conclusions

The kinematics research of chewing robot based on six-axis parallel mechanism is studied in this paper. Firstly, the coordinate system of the kinematic model of the mechanism is established and the degree of freedom of the mechanism is calculated and verified to be 6. Then, the inverse kinematics method is used to find the relationship between the displacement of each prismatic pair and the position of the moving platform. At the same time, the expression of the velocity of the prismatic pair is deduced, which shows the inverse velocity solution of prismatic pair is not related to the actual velocity of the mechanism, but only to the size and type of the mechanism. Finally, through kinematics simulation, the displacement and velocity curves of each prismatic pair under a pre-designed trajectory of moving platform are obtained. The motion control function obtained by this method is input into the stepping motor, and the moving platform of the prototype can reproduce the designed motion trajectory. However, because the simulation process does not consider the influence of the dimension accuracy and positioning accuracy of the parts on the motion, the actual motion trajectory of the prototype has errors with the designed trajectory.

Acknowledgements. This study was supported by Jiangsu Key Laboratory of Advanced Food Manufacturing Equipment and Technology (FMZ201901), National Natural Science Foundation of China (51775244) and Jiangsu Key Laboratory of Advanced Food Manufacturing Equipment and Technology (FMZ201907).

References

1. Krop, E.M.: Influence of oral processing on appetite and food intake - a systematic review and meta-analysis. Appetite **125**, 253–269 (2018)
2. Dolores Romero de ávila, M.: Rheological behaviour of commercial cooked meat products evaluated by tensile test and texture profile analysis (TPA). Meat Sci. **98**(2), 310–315 (2014)
3. Mei, W.M.S.: Correlation of mechanical properties from textural profile analysis (TPA) with eating behaviours and macronutrient composition for a wide range of solid foods. Food Funct. **48**(5), 807–811 (2018)
4. Nishinari, K.: Perception and measurement of food texture-solid foods. J. Texture Stud. **49**(2), 160–201 (2018)
5. Raabe D.: The chewing robot: a new biologically-inspired way to evaluate dental restorative materials. In: Conference Proceedings of the IEEE Engineering in Medicine and Biology Society (2009)
6. Woda, A.: Development and validation of a mastication simulator. J. Biomech. **43**(9), 1667–1673 (2010)
7. Xu, W.: Mastication Robots. Studies in Computational Intelligence, vol. 290. Springer, Heidelberg (2010)
8. Chen, B.: Opportunities and challenges in developing a next generation masticatory robot. In: 24th International Conference on Mechatronics and Machine Vision in Practice (M2VIP). IEEE (2017)
9. Wen, H.: Kinematic model and analysis of an actuation redundant parallel robot with higher kinematic pairs for jaw movement. IEEE Trans. Ind. Electron. **62**(3), 1590–1598 (2015)
10. Ming, C.: Modeling and kinematic analysis of masticatory robot based on biological structure. J. Huazhong Univ. Sci. Technol. (Nat. Sci. Ed.) **39**(S2), 13–17 (2011)
11. Wang, G.: Occlusion masticatory cycle planning and driving force optimal distribution of an actuation redundant chewing robot. Robot **39**(1), 70–80 (2017)
12. Alrashidi: In vivo mechanical assessment of human elbow kinematics using a six axis parallel mechanism developed in house. Brunel University (2011)
13. Yang, X.L.: A dual quaternion solution to the forward kinematics of a class of six-DOF parallel robots with full or reductant actuation. Mech. Mach. Theory **107**, 27–36 (2017)
14. Fichter, E.F.: A Stewart Platform Based Manipulator: General Theory and Practical Construction. Sage Publications Inc., Thousand Oaks (1986)

Dynamic Model and Simulation Analysis of a 3-DOF Parallel Mechanism

Jiaxun Li[1], Jian Liu[2(✉)], Hongjuan Hou[2], and Liyong Huang[1]

[1] School of Mechanical Engineering, Beijing Institute of Technology, Beijing 100081, China
[2] School of Mechanical and Equipment Engineering, Hebei University of Engineering, Handan 056038, China
1979267500@qq.com

Abstract. For the design requirements of heavy-duty robots, a 3-degrees of freedom 3UPS-UP parallel mechanism was used as the main feed mechanism of heavy-duty robots. The kinematics analysis of the parallel mechanism was carried out, and the complete homogeneous dimensional Jacobian matrix of the mechanism was solved. Based on the kinetic energy and potential energy of each component of the mechanism, the dynamic equation was established by used the Lagrangian method, and the dynamic example simulation was performed.

Keywords: 3UPS-UP parallel mechanism · Kinematics analysis · Lagrangian method · Dynamic equation

1 Introduction

In recent years, space less degrees of freedom parallel mechanisms have gradually become a hot topic in academic and industrial research [1, 2]. This kind of mechanism has the advantages of high rigidity, high precision and flexible movement. It has been widely used in the equal thickness processing of special-shaped structural parts in aerospace and automotive fields, and in the assembly of aviation aluminum structural parts [3, 4].

So far, many famous scholars have done a lot of research on the 1T2R parallel mechanism. Chai [5] constructed a 1T2R three-degree-of-freedom parallel mechanism of Exechon using spiral theory and group theory, and proposed two different branches of the mechanism; Bi [6] performs motion on 3-DOF 2-UPR-RPU parallel mechanism. On the basis of this, the task space of the organization is obtained through simulation; Chen [7] Comparing the force transfer performance of 3RPS and 3-PRS based on the workspace of two 1T2R parallel mechanisms.

In this paper, a 3-DOF 3UPS-UP Parallel Mechanism is used as the main feed mechanism of the heavy-duty robot. The kinematics model of the mechanism is established. The homogeneous dimensional Jacobian matrix is solved and based on the Lagrangian method. The complete dynamic model of the mechanism was established and a virtual prototype was built for simulation verification.

D. Wang et al. (Eds.): MeTrApp 2019, MMS 79, pp. 322–331, 2020.
https://doi.org/10.1007/978-981-15-0142-5_32

2 Mechanism Description

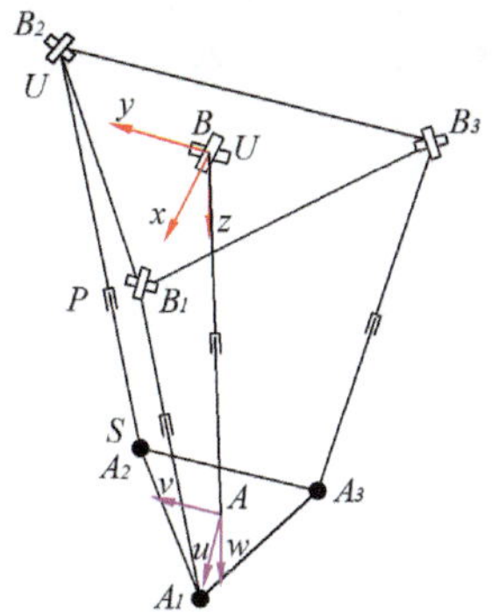

Fig. 1. 3UPS-UP mechanism diagram

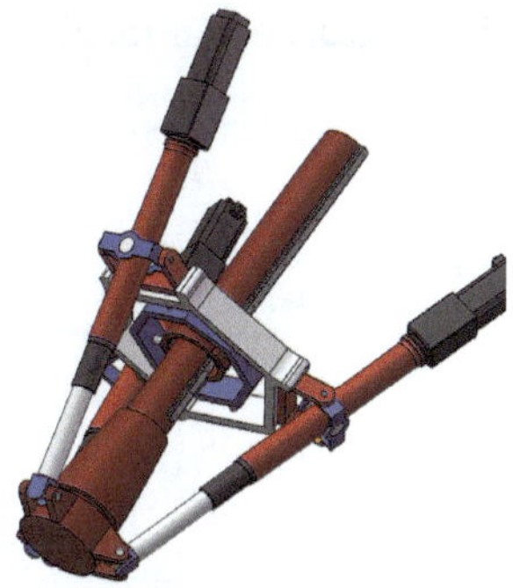

Fig. 2. 3 Three-dimensional model of UPS-UP parallel mechanism

The schematic diagram of the 3UPS-UP parallel mechanism was shown in Fig. 1. The reference coordinate system *B-xyz* fixed on the fixed platform is established. The origin *B* of the coordinate origin is the circumscribed circle center of the equilateral triangle $B_1B_2B_3$. the *x*-axis is parallel to BB_1, the *z*-axis is perpendicular to the direction of the fixed platform. The dynamic coordinate system *A-uvw* fixed on the moving platform is established. The coordinate origin *A* is the circumscribed circle center of the equilateral triangle $A_1A_2A_3$. The *u*-axis is parallel to AA_1, and the *w*-axis is perpendicular to the direction of the moving platform.

The model of the 3UPS-UP parallel mechanism is shown in Fig. 2. The three UPS branches of the mechanism do not change the degree of freedom of the UP branch. Therefore, the 3UPS-UP mechanism is a 1T2R three-degree-of-freedom parallel mechanism.

3 Kinematics Analysis of Mechanism

The position of the parallel mechanism is solved as

$$q_{\mathrm{i}} = \|\mathbf{a}_{\mathrm{i}} + \mathbf{r} - \mathbf{b}_{\mathrm{i}}\| \tag{1}$$

$$\mathbf{q}_{\mathrm{i}} = \frac{\mathbf{a}_{\mathrm{i}} + \mathbf{r} - \mathbf{b}_{\mathrm{i}}}{\|\mathbf{a}_{\mathrm{i}} + \mathbf{r} - \mathbf{b}_{\mathrm{i}}\|}. \tag{2}$$

Where, $\mathbf{a}_{\mathrm{i}} = R\mathbf{a}_{\mathrm{i0}},$

q_{i}—Length of *i*th branch
$\mathbf{q}_{\mathrm{i}}$—Unit vector of *i*th branch
$\mathbf{a}_{\mathrm{i0}}$—Vectors of Point A_i in System *A-uvw*
R—The rotation matrix of coordinate system *A-uvw* relative to coordinate system *B-xyz*,

$\mathbf{b}_i$—The vector of point B_i in system $B\text{-}xyz$,
$\mathbf{r} = [x\ y\ z]^T$—Position vector of point A in coordinate system $B\text{-}xyz$.

Then the Jacobian matrix of the parallel 3UPS drive mechanism is:

$$J_q = \begin{bmatrix} \mathbf{q}_1^T & (\mathbf{a}_1 \times \mathbf{q}_1)^T \\ \mathbf{q}_2^T & (\mathbf{a}_2 \times \mathbf{q}_2)^T \\ \mathbf{q}_3^T & (\mathbf{a}_3 \times \mathbf{q}_3)^T \end{bmatrix} \tag{3}$$

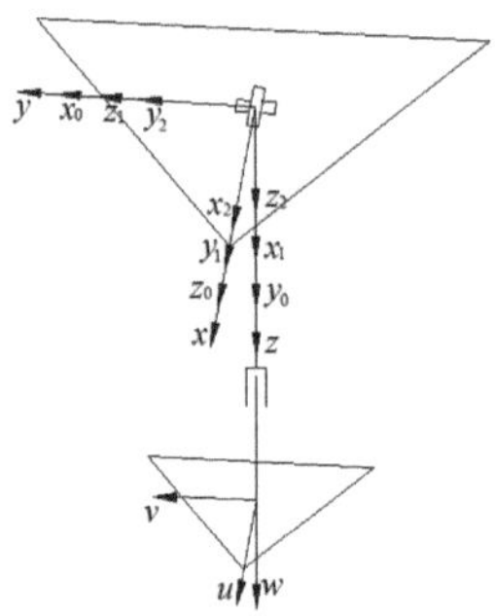

Fig. 3. Structural sketch of series branch

The series UP constrained branch can be equivalent to the series R-R-P chain [8]. The structure sketch of the constrained branch is shown in Fig. 3, and its D-H parameters are shown in Table 1.

Table 1. D-H parameters of series branch

i	$\theta_i/(°)$	d_i	a_i	$\alpha_i/(°)$
0	90	0	0	90
1	θ_1	0	0	90
2	θ_2	0	0	90
3	0	d_3	0	0

Where, θ_1, θ_2 and d_3 are variables.

The Jacobian matrix of the series branch can be solved by differential transformation method:

$$J_p = \begin{bmatrix} 0 & d_3 \sin\theta_2 & -\cos\theta_2 \\ -d_3 \sin\theta_1 \sin\theta_2 & d_3 \cos\theta_1 \cos\theta_2 & \cos\theta_1 \sin\theta_2 \\ d_3 \cos\theta_1 \sin\theta_2 & d_3 \sin\theta_1 \cos\theta_2 & \sin\theta_1 \sin\theta_2 \\ 1 & 0 & 0 \\ 0 & \sin\theta_1 & 0 \\ 0 & -\cos\theta_1 & 0 \end{bmatrix} \tag{4}$$

Which is available from Eq. (4)

$$\begin{bmatrix} \dot{\theta}_1 \\ \dot{\theta}_2 \end{bmatrix} = \begin{bmatrix} 0 & -\dfrac{\sin\theta_1}{d_3 \sin\theta_2} & \dfrac{\cos\theta_1}{d_3 \sin\theta_2} \\ \dfrac{\sin\theta_2}{d_3} & \dfrac{\cos\theta_1 \cos\theta_2}{d_3} & \dfrac{\sin\theta_1 \cos\theta_2}{d_3} \end{bmatrix} \begin{bmatrix} v_x \\ v_y \\ v_z \end{bmatrix} \tag{5}$$

$$\begin{bmatrix} w_x \\ w_y \\ w_z \end{bmatrix} = \begin{bmatrix} 1 & 0 \\ 0 & \sin\theta_1 \\ 0 & -\cos\theta_1 \end{bmatrix} \begin{bmatrix} \dot{\theta}_1 \\ \dot{\theta}_2 \end{bmatrix} \tag{6}$$

Combine Eqs. (5) and (6) to obtain:

$$\begin{bmatrix} v \\ w \end{bmatrix} = J_2[v] \tag{7}$$

Combine Eqs. (2) and (7):

$$\dot{q}_i = J_q J_2[v] \tag{8}$$

This is the homogeneous dimensional Jacobian matrix of the 3UPS-UP parallel mechanism.

4 Dynamics Analysis

The dynamic equation of 3UPS-UP parallel mechanism are solved by Lagrangian method.

4.1 System Kinetic Energy

The kinetic energy of the available platform is:

$$E_1 = \frac{1}{2} m_p v^T v + \frac{1}{2} w_u^T I_p w_u \tag{9}$$

Where,

m_p—Dynamic platform quality
$I_p = diag[I_{px}, I_{py}, I_{pz}]$—The inertia tensor of the moving platform relative to the moving coordinate system *A-uvw*.

Each UPS drive branch of the 3UPS-UP parallel mechanism is composed of a rotating seat and a push-pull rod. The kinetic energy of the driving branch is the sum of the kinetic energy of the rotating seat and the push-pull rod.

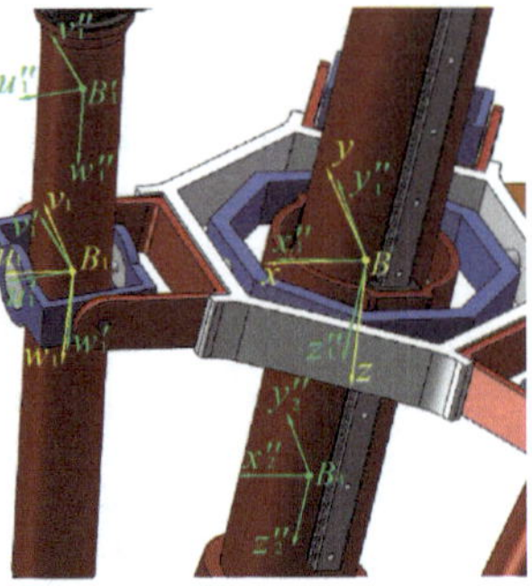

Fig. 4. Partial coordinate system

The coordinate systems of each center of mass and the center of rotation pair are established as shown in Fig. 4. B_1 is the center of U pair of the first UPS branch of the 3UPS-UP mechanism, B_i' is the center of mass of the first UPS branch, B_4 is the center of mass of the driven rod of the UPS branch, and the coordinate axis direction of B_i-$u_i v_i w_i$ is the same as the fixed coordinate system B-xyz.

The rotational angular velocity of the drive branch in the fixed coordinate system is:

$$w_i = \frac{1}{q_i}[\mathbf{q}_i \times v + \mathbf{q}_i \times (w \times \mathbf{a}_i)] \tag{10}$$

Where, $\mathbf{q}_i = [x_i y_i z_i]^T$,

According to the Eq. (10):

$$E_{1i} = \frac{1}{2} w_i^T I_{1i} w_i \tag{11}$$

Where,

$$I_{1i} = R_i \begin{bmatrix} I_{lix} + m_{l1} z_{li1}^2 & 0 & 0 \\ 0 & I_{liy} + m_{l1} z_{li1}^2 & 0 \\ 0 & 0 & I_{liz} \end{bmatrix} R_i^T,$$

R_i—The rotation transformation matrix of the coordinate system B_i-$u_i' v_i' w_i'$ relative to the coordinate system B_i-$u_i v_i w_i$ can be obtained by Euler angle transformation [9],

m_{l1}—Rotating cylinder mass,

Z_{li1}—The coordinate system B_i'-$u_i'' v_i'' w_i''$ to the coordinate system B_i-$u_i' v_i' w_i'$ translation distance,

$J_1 = diag[I_{lix}, I_{liy}, I_{liz}]$—The principal moment of inertia of the rotating cylinder.

The kinetic energy of the drawbar needs to consider the rotation of the push-pull rod around the U vice and the movement along the length of the rod. According to Eqs. (10) and (8), the kinetic energy of the push-pull rod is

$$E_{2i} = \frac{1}{2} w_i^T I_{2i} w_i + \frac{1}{2} m_{l2} \dot{q}_i^2 \tag{12}$$

Where,

$$I_{2i} = R_i \begin{bmatrix} I_{qix} + m_{l2} z_{li2}^2 & 0 & 0 \\ 0 & I_{qiy} + m_{l2} z_{li2}^2 & 0 \\ 0 & 0 & I_{qiz} \end{bmatrix} R_i^T,$$

m_{l2}—Push-pull rod quality,
z_{li2}—The center of mass of the push-pull rod is U to the center distance,
$J_2 = diag[I_{qix}, I_{qiy}, I_{qiz}]$—The main moment of inertia of the push-pull rod.

The UP branch is composed of a driven rod and a U pair, and the U pair can be decomposed into a rotating pair 1 and a rotating pair 2.

The kinetic energy of the driven rod is obtained:

$$E_{31} = \frac{1}{2} m_{u1} \dot{d}_3^2 + \frac{1}{2} w_u^T I_{31} w_u \tag{13}$$

Where,

$$I_{31} = R \begin{bmatrix} I_{ux} + m_{u1} z_u^2 & 0 & 0 \\ 0 & I_{uy} + m_{u1} z_u^2 & 0 \\ 0 & 0 & I_{uz} \end{bmatrix} R^T,$$

m_{u1}—Driven rod quality,
z_u—The coordinate system $B_4\text{-}x_2'' y_2'' z_2''$ to the coordinate system $B\text{-}x_1'' y_1'' z_1''$ translation distance,
$J_{31} = diag[I_{ux}, I_{uy}, I_{uz}]$—The main moment of inertia of the driven rod,

The kinetic energy of the rotating pair 1 is:

$$E_{32} = \frac{1}{2} w_{u1}^T I_{32} w_{u1} \tag{14}$$

Where, $w_{u1} = \begin{bmatrix} \dot{\theta}_1 & 0 & 0 \end{bmatrix}^T$,
$I_{32} = diag[I_{u1x}, I_{u1y}, I_{u1z}]$—The principal moment of inertia of rotating pair 1.

Similarly, the kinetic energy of the rotating pair 2 is:

$$E_{33} = \frac{1}{2} w_{u2}^T I_{33} w_{u2} \tag{15}$$

Where, $w_{u2} = \begin{bmatrix} 0 & \dot{\theta}_2 & 0 \end{bmatrix}^T$,
$I_{33} = diag[I_{u2x}, I_{u2y}, I_{u2z}]$—The principal moment of inertia of rotating pair 2.
Combine Eqs. (9–15), the total kinetic energy of the system is:

$$E = E_1 + E_{31} + E_{32} + E_{33} + \sum_{i=1}^{3} (E_{1i} + E_{2i}) \tag{16}$$

4.2 System Potential Energy

Set the xOy plane of the fixed coordinate system to zero potential energy surface, then the potential energy of the moving platform is

$$W_1 = -m_p g z \tag{17}$$

Rotating potential energy

$$W_{1i} = m_{l1} g z_{li1} z_i \tag{18}$$

Push-pull rod potential energy is

$$W_{2i} = -m_{l2} g z_{li2} z_i \tag{19}$$

The potential energy of the driven rod is

$$W_3 = -\frac{m_u g z_u z}{\sqrt{(x^2 + y^2 + z^2)}} \tag{20}$$

The potential energy of the rotating pairs 1, 2 is zero.
Combine Eqs. (17–20), the total potential energy of the system is

$$W = W_1 + W_3 + \sum_{i=1}^{3} (W_{1i} + W_{2i}) \tag{21}$$

4.3 3UPS-UP Parallel Mechanism Dynamics Equation

Based on the Lagrangian method, the Lagrangian dynamic equation of the 3UPS-UP parallel mechanism can be obtained by combining Eqs. (16) and (21):

$$\frac{d}{dt}\frac{\partial E}{\partial \dot{\Theta}} - \frac{\partial E}{\partial \Theta} + \frac{\partial W}{\partial \Theta} = \tau \tag{22}$$

Where, $\tau = \tau_1 + \tau_2$,

τ_1—Generalized force corresponding to driving force,
τ_2—Generalized force corresponding to external force load,

Then the driving force of each driving branch of the mechanism is:

$$f_q = \left(J^T\right)^{-1}\tau_1 \tag{23}$$

5 Driving Force Analysis and Simulation Comparison

The main parameters of the 3UPS-UP parallel mechanism model are shown in the following table:

Table 2. Mechanism main parameters

a	*b*	*m*	*h*
220	625	1040	1310

Where, a—The radius of the circumcircle of $\Delta A_1A_2A_3$,
b—The radius of the circumcircle of $\Delta B_1B_2B_3$,
m—The radius of the circumcircle of $\Delta E_1E_2E_3$,
h—BO distance,

The principal inertia moment, the inertia spindle and the center of gravity of each component of the 3UPS-UP parallel mechanism can be directly read by the mass attribute of the part in the Solidworks software, and the main inertia moment (kg*mm^2) and mass of each component read out (kg) parameters are shown in the table (Table 3) below:

Table 3. Attribute parameters of components

Moving platform		Rotating cylinder		Push rod	
I_{px}	252738	I_{lix}	10227881	I_{qix}	2739639
I_{py}	252738	I_{liy}	10214695	I_{qiy}	2740025
I_{pz}	425813	I_{liz}	202308	I_{qiz}	18677
m_p	29.4	m_{l1}	75.0	m_{l2}	29.5
Slave rod		Rotating pair 1		Rotating pair 2	
I_{ux}	29232287	I_{u1x}	480790	I_{u2x}	93558
I_{uy}	29205167	I_{u1y}	583100	I_{u2y}	69488
I_{uz}	539343	I_{u1z}	1031547	I_{u2z}	150483
m_u	106.4				

Set the running trajectory of the moving platform in the fixed coordinate system *B-xyz*:

$$\begin{cases} x = -700\tan\left(\frac{\pi}{12}\right)\cos\left(\frac{\pi}{100}t\right) \\ y = -700\tan\left(\frac{\pi}{12}\right)\sin\left(\frac{\pi}{100}t\right) \\ z = 700 \end{cases} \tag{24}$$

The external force load acting at point *A* in the moving coordinate system is:

$$\tau_2 = R[-f_1\sin(\frac{\pi t}{100}), f_1\cos(\frac{\pi t}{100}), -f_2]^T \tag{25}$$

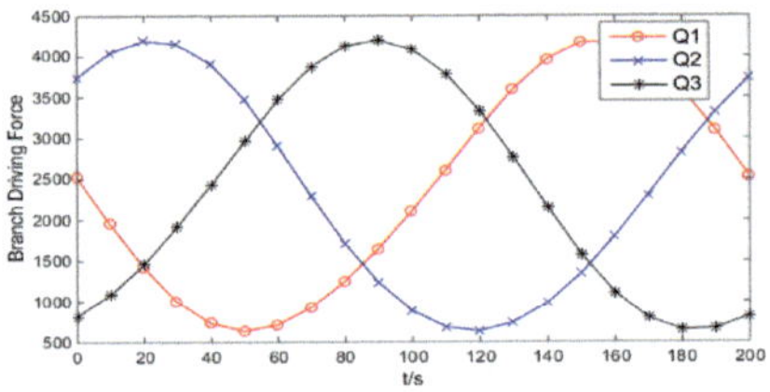

Fig. 5. Driving forces of 3UPS-UP parallel mechanism branches

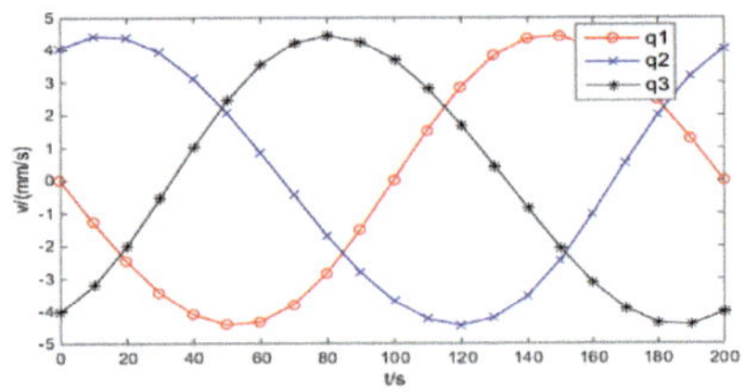

Fig. 6. Driving velocities of connecting rod

Combined Eqs. (22–25), the driving force of each branch of the 3UPS-UP parallel mechanism in the trajectory and load state is calculated by Matlab, and the result is shown in Fig. 5.

According to the Eq. (24), the 3UPS-UP parallel mechanism is used to solve the driving speed of the three UPS branches, as shown in Fig. 6.

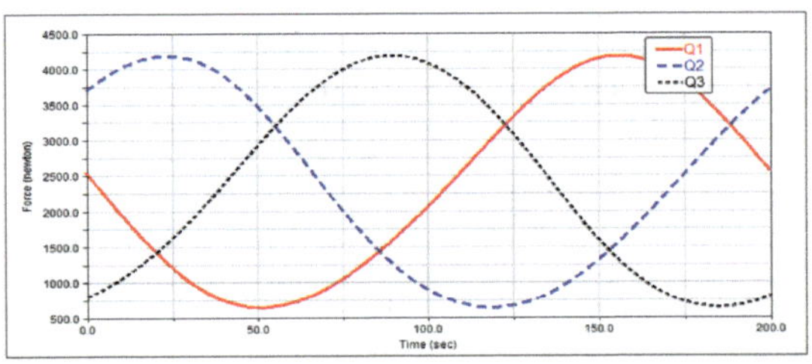

Fig. 7. Driving forces of 3UPS-UP parallel mechanism branches

According to the parameters of Table 2, the 3UPS-UP parallel mechanism is modeled in solidworks, and the model is imported into Adams. Adding the branch drive speed of the three UPS branches of the Adams prototype as shown in Fig. 6, adding external force load to the moving platform, and simulating the driving force curve of the UPS branch, as shown in Fig. 7.

It can be seen from the simulation results in Fig. 7 that the driving force curve of the parallel mechanism under the set working condition is smooth, and the driving force of the three branches is basically the same, which proves that the driving force distribution of the mechanism is relatively uniform and reasonable, suitable for heavy load. Working conditions. Comparing Fig. 7 with Fig. 5, it can be seen that the driving force simulation and calculation results are basically consistent, which proves the correctness of the dynamic model.

6 Conclusion

In this paper, a 3UPS-UP parallel mechanism is used as the main feed mechanism of heavy-duty robots. The kinematics and dynamics modeling is carried out, and the comparison simulations are carried out. The results show that the mechanism is suitable for heavy-duty conditions. Prove the correctness of the dynamic model and provide a theoretical basis for the subsequent research work.

References

1. Cui, G.H., Zhang, H.Q., Zhang, D., et al.: Analysis of the kinematic accuracy reliability of a 3-DOF parallel robot manipulator. Int. J. Adv. Robot. Syst. **12**(15) (2015). https://doi.org/10.5772/60056
2. Hu, B.: Kinematically identical manipulators for the Exechon parallel manipulator and their comparison study. Mech. Mach. Theory **103**, 17–137 (2016)
3. Huang, T.S., Li, M., Zhao, X.M., et al.: Conceptual design and dimensional synthesis for a 3 DOF module of the TriVariant a novel 5 DOF reconfigurable hybrid robot. IEEE Trans. Robot. **21**(3), 449–456 (2005)
4. Cui, G.H., Zhang, D., Zhou, H.D., et al.: Operating dexterity optimization and analysis of a 3-DOF parallel manipulator for a tunnel segment assembly system. Int. J. Mech. Mater. Des. **11**(3), 277–285 (2015)
5. Chai, X.X., Chen, Q.H., Ma, J.G., et al.: Mobility analysis of two exechon-liked parallel mechanisms. Intell. Robot. Appl. **8102**, 730–741 (2013)
6. Bi, Z.M., Jin, Y.: Kinematic modeling of Exechon parallel kinematic machine. Robot. Comput. Integr. Manuf. **27**(1), 186–193 (2011)
7. Chen, X., Liu, X.J., Xie, F.G., et al.: A comparison study on Motion/Force transmissibility of two typical 3-DOF parallel manipulators: the Sprint Z3 and A3 Tool Heads. Int. J. Adv. Robot. Syst. **11**(5), 1–10 (2014)
8. Cui, G.H., Zhang, H.Q., Xu, F., et al.: Kinematic dexterity and stiffness performance of spatial 3-PUS-UP parallel manipulator. Trans. Chin. Soc. Agric. Mach. **12**, 348–354 (2014)
9. Li, Y.Z., Wang, Y.G.: Coordinate transformation of rotating inertial tensor of robot. Robot **02**, 31–35 (1992)

U-Joint Induced Torsional Instabilities of a Family of 3-DOF Partially-Decoupled Spherical Parallel Manipulators

Guanglei Wu[1(✉)] and Stéphane Caro[2]

[1] School of Mechanical Engineering, Dalian University of Technology, Dalian 116024, China
gwu@dlut.edu.cn

[2] Laboratoire des Sciences du Numérique de Nantes, UMR CNRS 6004, Centre National de la Recherche Scientifique (CNRS), 44321 Nantes, France
stephane.caro@ls2n.fr

Abstract. This paper deals with the dynamic torsional stability problem of a family of partially-decoupled spherical parallel manipulators. The linearized equations of motion of the mechanical system are established to analyze the stability of the U-joint mechanism, resorting to the Floquet theory. Parametric stability charts of misalignment angles versus rotating speeds of the driving shaft are constructed to identify the unstable regions and critical shaft speeds, together with the effect of the parameters onto the manipulator stability. As a consequence, some criteria for the design and the operational speed of the manipulator, in terms of dynamic stability, are introduced.

Keywords: Dynamic stability · Spherical parallel manipulator · Monodromy matrix · Floquet theory · Torsional vibrations

1 Introduction

Three-degree-of-freedom (3-dof) spherical parallel manipulators (SPMs), working as robotic joints [1], orientating device [9], and devices with remote center of motion [13], have been studied in the literature for several decades. Typical symmetrical SPMs, such as 3<u>R</u>RR[1] or 3U<u>P</u>U–S ones [3, 10, 20, 21], usually have limited rotational capabilities, particularly, the torsional movement around the axis normal to the end-effector. In order to realize the unlimited rotation to extend the applications of the SPMs, an asymmetrical 2<u>R</u>RR–<u>R</u>UR–RPS SPM [24] as shown in Fig. 1(a) was proposed with a number of advantages compared to its symmetrical counterparts, such as enhanced positioning accuracy [21], infinite end-effector rotational motion [1], structural compactness and low dynamic inertia [23]. It is noted that the two actuated limbs in-parallel can be replaced with

[1] U, P, R and S stand for the universal, prismatic, revolute and spherical joints, respectively. An underlined letter indicates an actuated joint.

D. Wang et al. (Eds.): MeTrApp 2019, MMS 79, pp. 332–343, 2020.
https://doi.org/10.1007/978-981-15-0142-5_33

the UPU or PUU linkages [25] as displayed in Figs. 1(b) and (c). The common design of the three SPMs adopts a U joint to connect the driving shaft in the center to the end-effector, ensuring an infinite torsional movement in any configuration. Henceforth, it can work as an active spherical joint or a machine tool head as shown in Fig. 2. On the other hand, the U joint suffers from one major issue, namely, it transforms a constant input speed to a periodically fluctuating one [17], which may induce vibrations and wear. Accordingly, this paper deals with the parametric torsional stability issue, due to the presence of the identical the U-joint in the three SPMs in order to lead to the full rotational motion of their end-effectors.

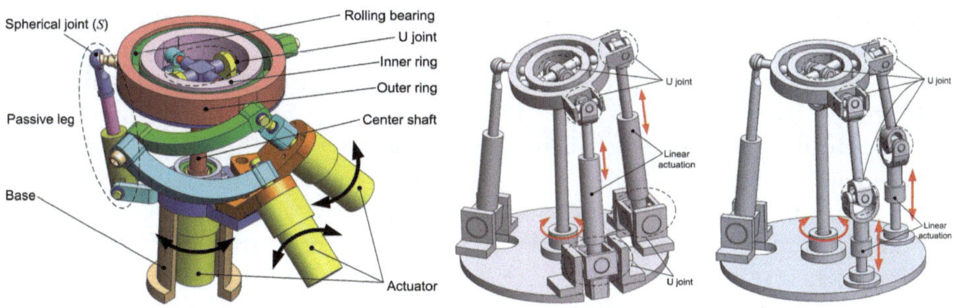

(a) (b) (c)

Fig. 1. CAD models of the asymmetrical SPMs [25]: (a) 2RRR-RUR-, (b) 2UPU-RUR-, and (c) 2PUU-RUR-RPS.

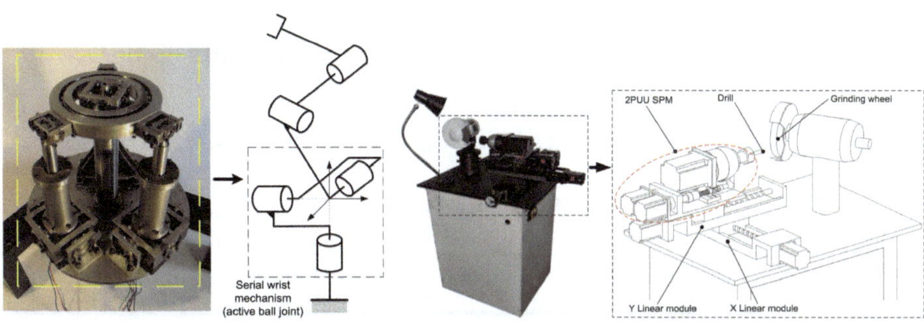

(a) (b)

Fig. 2. Applications of the asymmetrical SPMs [25]: (a) active spherical joint; (b) drill point grinder.

To the best of the authors' knowledge, Porter [15] was the first to investigate this problem. Indeed, he expressed a single degree-of-freedom linearized model to plot the stability chart by using the Floquet theory [8,12]. Porter and Gregory [16] and Éidinov *et al.* [7] revisited the same problem to build a 1-dof nonlinear dynamic model and studied the stability problem by means of

Poincaré–Lyapunov method. Later, similar modeling approaches were adopted to derive the linear or nonlinear models for the stability analysis of the U joint [2,5]. Moreover, the torsional stability of multi-shaft systems consisting of multiple shafts interconnected via Hooke's joints can also be studied using the previous various approaches [11]. According to the literature, the previous studies focus on single or multiple U-joint mechanisms. It should be noted that, a U joint working as a transmitting mechanism in parallel mechanisms has not received sufficient attention, and is the subject of this paper. As the relationship between the input and output shaft rotating speeds of the U joint is periodic [18], the Floquet theory is an effective approach to analyze the stability of the mechanisms under study, where the monodromy matrix, as the fundamental matrix of a system of ordinary differential equations (ODEs) evaluated at the period of the coefficients of a system, is used for the analysis of periodic solutions. The previous work [22] has presented some preliminary results regarding to the manipulator stability. On the other hand, the vibration behaviors, such as resonance type and emanating points, are not identified. Moreover, the analysis in the previous work is carried out in a limited motion range of the vibrating resources, where larger bands of rotating speeds should be taken into consideration to analyze the influence of parameters onto the dynamic stability problem of the robotic manipulators.

This paper deals with the dynamic torsional stability problem of a family of asymmetrical spherical parallel manipulators with a U-joint transmitting the rotational motion of the end-effector. A linear model associated to the input and output shafts interconnected via a Hooke's joint is considered. Numerical study is carried out to assess the system stability and the effects of both the geometric and dynamic parameters, resorting to the Floquet theory. Parametric stability charts of misalignment angles versus rotating speeds of the driving shaft are constructed to identify the unstable regions and critical shaft speeds.

2 SPM Architecture and the U-Joint Mechanism

The kinematic architecture of the manipulator shown in Fig. 1(a) is depicted in Fig. 3(a). The mobile platform is composed of an outer and inner rings connected to each other with a revolute joint, the revolute joint being realized with a rolling bearing. The orientation of the outer ring is controlled by two limbs in-parallel, and it is constrained by a fully passive leg that is offset from the center of the mobile platform to eliminate the rotational motion about the vertical axis. Through a universal joint, the decoupled rotation of the inner ring is generated by the center shaft, which also supports the mobile platform to improve the positioning accuracy.

Splitting the outer ring and the two parallel limbs as well as the passive one, the remaining parts of the manipulator can be equivalent to a U-joint mechanism, of which the center shaft is treated as the driving shaft and the inner ring is treated as a driven disk. The misalignment angle is denoted by β. The input and output angular velocities are denoted as Ω_{in} and Ω_{out}, respectively.

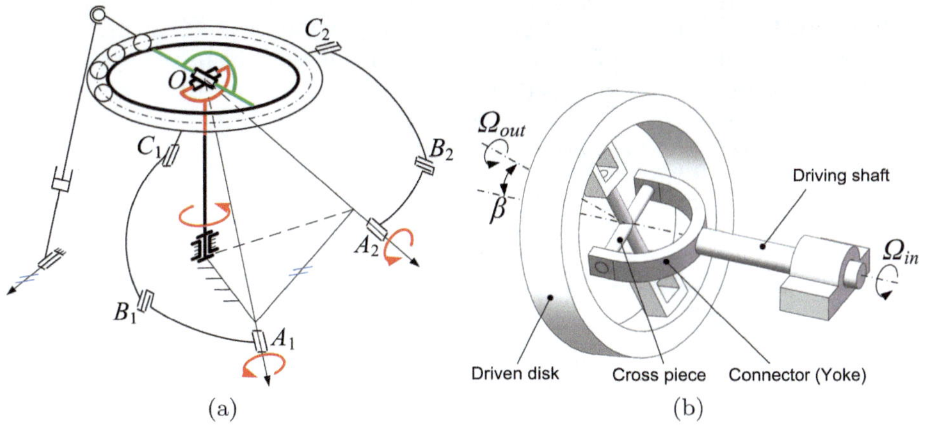

Fig. 3. Architecture of the 2$\underline{R}$RR-$\underline{R}$UR-RPS SPM (a) and its U-joint transmitting mechanism (b).

3 Equations of Motion of Torsional Vibrations

The equations of motion for the U-joint mechanism as shown in Fig. 3(b) are deduced via a synthetical approach [4, 22]. In accordance, the driving shaft and the driven disk are considered as two separate parts, as displayed in Fig. 4, where the cross piece connecting the input/output elements is supposed to be as massless. Moreover, the friction is neglected.

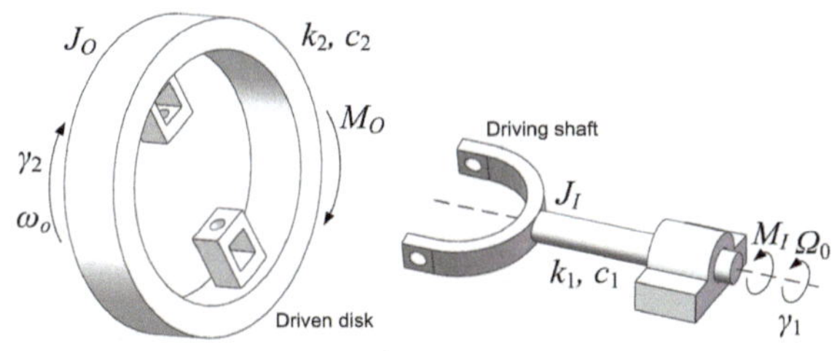

Fig. 4. The driving and driven parts of the U-joint mechanism.

Prior to dynamic modeling, the relationship between the speed of the driving part Ω_{in} and that of the driven part Ω_{out} of the Hooke's joint, as well as the relationship between the input and the output torques, can be written as [18]

$$\Omega_{out} = \Omega_{in} \cdot \eta(t), \quad M_O = \frac{M_I}{\eta(t)}; \quad \eta(t) = \frac{\cos\beta}{1 - \sin^2\beta \sin^2(\Omega_0 t + \gamma_1)} \tag{1}$$

where Ω_0 denotes the constant velocity of the driving shaft.

Skipping the detailed modeling procedure in previous work [22], the following set of equations of motion of the U-joint mechanism, with respect to the driving and driven parts, can be expressed as:

$$J_I\ddot{\gamma}_1 + c_1\dot{\gamma}_1 - \eta(t)c_2\dot{\gamma}_2 + k_1\gamma_1 - \eta(t)k_2\gamma_2 = 0 \tag{2a}$$

$$J_O\left[\ddot{\gamma}_2 + \eta(t)\ddot{\gamma}_1 + \dot{\eta}(t)(\Omega_0 + \dot{\gamma}_1)\right] + c_2\dot{\gamma}_2 + k_2\gamma_2 = 0 \tag{2b}$$

with

$$\dot{\omega}_o = \eta(t)\ddot{\gamma}_1 + \dot{\eta}(t)(\Omega_0 + \dot{\gamma}_1) \tag{3}$$

where γ_1 and γ_2 are the rotational coordinates of J_I and J_O, respectively, and M_I is the reaction torque of the input part of the Hooke's joint. Moreover, k_1, c_1 and k_2, c_2 depict the torsional stiffness and viscous damper of the driving shaft and the driven disk, respectively.

By introducing the following dimensionless parameters:

$$\Omega = \frac{\Omega_0}{\sqrt{k_1/J_I}},\ \zeta = \frac{c_1}{\sqrt{k_1 J_I}},\ \mu = \frac{c_2}{c_1},\ \nu = \frac{J_O}{J_I} \tag{4}$$

Equations (2a) and (2b) are written as [4,22]

$$\ddot{\gamma}_1 + \frac{\zeta}{\Omega}\dot{\gamma}_1 - \frac{\mu\zeta}{\Omega}\eta(\tau)\dot{\gamma}_2 + \frac{1}{\Omega^2}\gamma_1 - \frac{1}{\Omega^2\cdot\eta(\tau)}\gamma_2 = 0 \tag{5}$$

$$\ddot{\gamma}_2 + \left[\eta'(\tau) - \frac{\zeta}{\Omega}\eta(\tau)\right]\dot{\gamma}_1 + \left[\frac{\mu\zeta}{\Omega}\left(\frac{1}{\nu} + \eta^2(\tau)\right)\right]\dot{\gamma}_2 + \left[\eta''(\tau) - \frac{1}{\Omega^2}\eta(\tau)\right]\gamma_1 + \left[\frac{1}{\Omega^2}\left(\frac{1}{\nu\eta(\tau)^2} + 1\right)\right]\gamma_2 + \eta'(\tau) = 0 \tag{6}$$

where $\tau = \Omega_0 t$, and $\eta(\tau)$ is a π–periodic equation:

$$\eta(\tau) = \eta(\pi + \tau) = \frac{\cos\beta}{1 - \sin^2\beta\sin^2\tau} \tag{7}$$

Equations (5) and (6) can be cast into a matrix form, as follows

$$\begin{bmatrix}\ddot{\gamma}_1\\ \ddot{\gamma}_2\end{bmatrix} + \begin{bmatrix}\frac{\zeta}{\Omega} & -\frac{\mu\zeta}{\Omega}\eta(\tau)\\ \eta'(\tau) - \frac{\zeta}{\Omega}\eta(\tau) & \frac{\mu\zeta}{\Omega}\left(\frac{1}{\nu} + \eta^2(\tau)\right)\end{bmatrix}\begin{bmatrix}\dot{\gamma}_1\\ \dot{\gamma}_2\end{bmatrix} + \begin{bmatrix}\frac{1}{\Omega^2} & -\frac{1}{\Omega^2}\frac{1}{\eta(\tau)}\\ \eta''(\tau) - \frac{1}{\Omega^2}\eta(\tau) & \frac{1}{\Omega^2}\left(\frac{1}{\nu}\cdot\frac{1}{\eta(\tau)^2} + 1\right)\end{bmatrix}\begin{bmatrix}\gamma_1\\ \gamma_2\end{bmatrix} = \begin{bmatrix}0\\ -\eta'(\tau)\end{bmatrix} \tag{8}$$

where primes denote differentiation with respect to τ, thus, Eq. (8) consists of a set of linear differential equations with π-periodic coefficients, for the torsional vibrations of the U-joint mechanism as shown in Fig. 3(b).

4 Dynamic Stability Analysis

The right-hand term of Eq. (8) involves time-derived motions, which can lead to forced vibrations [14]. The scope of this study is about the self-excitation vibrations, namely, parametrically excited stability problem. Thus, the homogeneous parts of Eq. (8) should be considered sequentially to analyze the dynamic stability of the manipulator. Equation (8) can be expressed as in the following compact form:

$$\ddot{\boldsymbol{\gamma}} + \mathbf{U}\dot{\boldsymbol{\gamma}} + \mathbf{V}\boldsymbol{\gamma} = \mathbf{0} \tag{9}$$

with

$$\ddot{\boldsymbol{\gamma}} = \left[\ddot{\gamma}_1\ \ddot{\gamma}_2\right]^T, \dot{\boldsymbol{\gamma}} = \left[\dot{\gamma}_1\ \dot{\gamma}_2\right]^T, \boldsymbol{\gamma} = \left[\gamma_1\ \gamma_2\right]^T \tag{10a}$$

$$\mathbf{U} = \begin{bmatrix} \frac{\zeta}{\Omega} & -\frac{\mu\zeta}{\Omega}\eta(\tau) \\ \eta'(\tau) - \frac{\zeta}{\Omega}\eta(\tau) & \frac{\mu\zeta}{\Omega}\left(\frac{1}{\nu} + \eta^2(\tau)\right) \end{bmatrix} \tag{10b}$$

$$\mathbf{V} = \begin{bmatrix} \frac{1}{\Omega^2} & -\frac{1}{\Omega^2}\frac{1}{\eta(\tau)} \\ \eta''(\tau) - \frac{1}{\Omega^2}\eta(\tau) & \frac{1}{\Omega^2}\left(\frac{1}{\nu}\cdot\frac{1}{\eta(\tau)^2} + 1\right) \end{bmatrix} \tag{10c}$$

which can be represented by a state-space formulation, namely,

$$\dot{\mathbf{x}}(t) = \mathbf{A}(t)\mathbf{x}(t); \quad \mathbf{x}(t) = \begin{bmatrix} \boldsymbol{\gamma} \\ \dot{\boldsymbol{\gamma}} \end{bmatrix}, \mathbf{A}(t) = \begin{bmatrix} \mathbf{0}_2 & \mathbf{I}_2 \\ -\mathbf{V} & -\mathbf{U} \end{bmatrix} \tag{11}$$

whence $\mathbf{A}(t)$ is a 4×4 π-periodic state transition matrix, $\mathbf{I}$ being the identity matrix. According to Floquet theory, the solution to equation system (11) can be expressed as:

$$\boldsymbol{\Phi}(\tau) = \mathbf{P}(\tau)e^{\tau\mathbf{R}} \tag{12}$$

where $\mathbf{P}(\tau)$ is a π-periodic matrix and $\mathbf{R}$ is a constant matrix, which is related to another constant matrix $\mathbf{H}$, referred to as monodromy matrix, with $\mathbf{R} = \ln\mathbf{H}/\pi$. If the fundamental matrix is normalized so that $\mathbf{P}(0) = \mathbf{I}_4$, then $\mathbf{H} = \mathbf{P}(\pi)$.

The eigenvalues $\lambda_i = \kappa_i + j\sigma_i$, $i = 1, 2, 3, 4$, $j^2 = -1$, of matrix $\mathbf{H}$, referred to as Floquet multipliers, govern the stability of the system. The system is asymptotically stable if and only if the real parts κ_i of all the eigenvalues λ_i are non-positive [6]. Here, the matrix $\mathbf{H}$ is obtained numerically with the improved Runge Kutta Method [19] with a step size equal to 10^{-6}. The detailed calculation of the state transition matrix is given in Appendix A. As a result, the eigenvalues of matrix $\mathbf{H}$ can be calculated to assess stability of the system. The monodromy matrix method is a simple and reliable method to determine the stability of parametrically excited periodic systems.

5 Numerical Study and Discussions

This section is devoted to numerical stability analysis, where the stability charts are constructed on the Ω_0–β parametric planes to study the effect of parameters onto the system stability. The parameters in the analysis are given in each figure caption, and the ratio of the output and input dampers is set to $\mu = 1$ due to the same manufacturing material.

Before presentation of the stability charts, the expected features associated to the vibration behaviors will be discussed. According to Eq. (8), the U-joint mechanism is built as a 2-dof vibrating systems, henceforth, the systems has two resonance (natural) frequencies, meaning that the system will vibrate if it is excited by some periodical phenomena. Setting $\zeta = 0,\ \beta = 0,\ \Omega = 1$, these frequencies are calculated as

$$\omega_{1(2)} = \sqrt{\frac{k_1}{J_I}\left(1 - \frac{-1 \pm \sqrt{4\nu^2+1}}{2\nu}\right)} \tag{13}$$

If no damping is present, the parametric resonance occurs when the frequency $\omega_j,\ j = 1,\ 2$, is equal to the product of an integer and the rotating speed Ω_0 of the driving shaft, thus, the parametric resonances are expected to emanate from [14]

$$\Omega_{j,\,k} = \frac{\omega_j}{k};\quad j = 1,\ 2,\quad k = 1,\ 2,\ 3,\ ... \tag{14}$$

and the parametric combination resonances based on the sum or difference of the two frequencies [14] are expected to emanate from

$$\Omega_k^+ = \frac{\omega_1 + \omega_2}{2k},\quad \Omega_k^- = \frac{\omega_2 - \omega_1}{2k};\quad k = 1,\ 2,\ 3,\ ... \tag{15}$$

Sequentially, the critical rotating speeds of the driving shaft can be identified from Eqs. (14) and (15).

Figure 5 depicts the stability chart Ω_0–β to detect the instabilities of the U-joint mechanism with the prescribed parameters, where the circle-marked zones and solid markers represent the unstable parametric regions and the emanation points, respectively. It is seen that parametric resonances are located in the unstable regions. It should be noted that as long as the rotating speed Ω_0 of the driving shaft is lower than 7π rad/s, this system is stable for any misalignment angle β between 0 and 30°. On the contrary, the mechanism will be stable for any driving shaft speed smaller than 50π rad/s as long as the misalignment angle β is lower than 5°.

The effects of the parameters on the manipulator stability are depicted in Figs. 6, 7, 8, 9, where the marked areas stand for the unstable regions. Moreover, the analysis is carried out with the identical parameters but except one, for a comparative study.

Figure 6 depicts the effect of the torsional stiffness of the driving shaft on the stability, by comparing three different values of k_1. The manipulator is no longer stable if $\Omega_0 > 36\pi$ rad/s when $k_1 = 5$ Nm/rad. With lower stiffness, the critical rotating speed of the driving shaft is in a smaller threshold, according to the comparison between the cases of $k_1 = 10$ Nm/rad and $k_1 = 5$ Nm/rad. With the increasing stiffness, the stable region becomes wider. It is noteworthy that the higher the torsional stiffness of the input shaft, the more stable the manipulator.

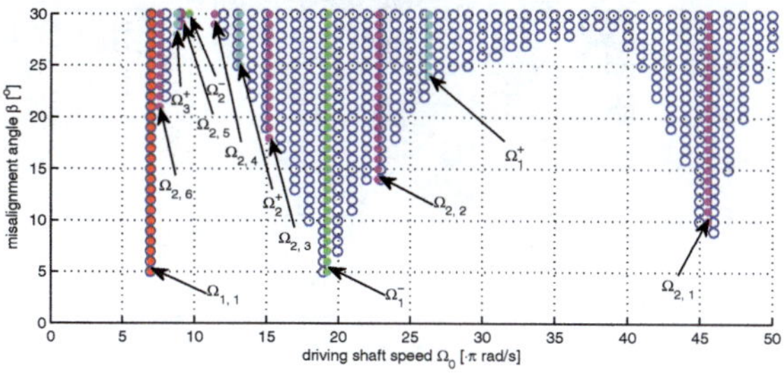

Fig. 5. Stability chart: unstable regions (circle point) with parameters $k_1 = 10\,\text{Nm/rad}$, $\nu = 10$, $J_I = 0.001\,\text{kg}\cdot\text{m}^2$, $c_1 = 0.001\,\text{Nm/(rad/s)}$.

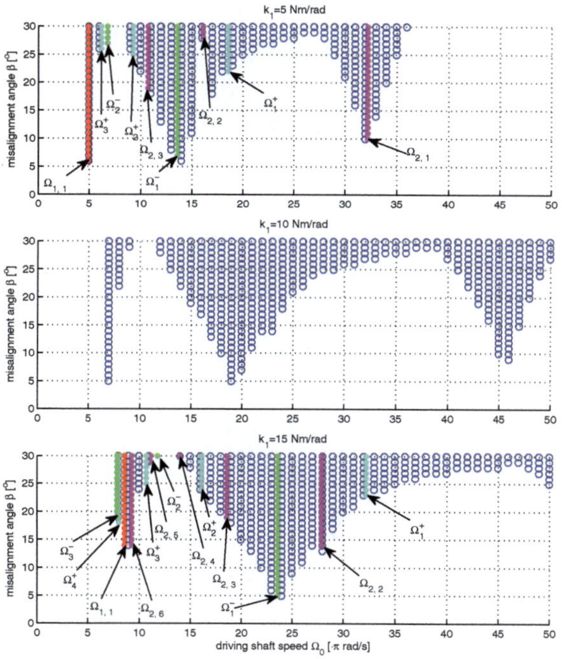

Fig. 6. Stability Chart: effects of the driving shaft stiffness k_1 with parameters $\nu = 10$, $J_I = 0.001\,\text{kg}\cdot\text{m}^2$, $c_1 = 0.001\,\text{Nm/(rad/s)}$.

Figure 7 shows the effect of the damping ratio on the system stability, from which it is seen that the unstable regions are almost coincident with the dampers that are in the range of the damping ratio of common materials, with a few exceptions when $c_1 = 0.0005\,\text{Nm/(rad/s)}$. This means that the influence of the damping ratio to the system stability is insignificant and negligible.

Similarly, Fig. 8 depicts the effect of the mass moment of inertia of the driving shaft to the parametric stability. When the inertia reduces to $J_I = 0.0005\,\text{kg}\cdot\text{m}^2$, the range of the critical misalignment angle and the shaft rotating speed become

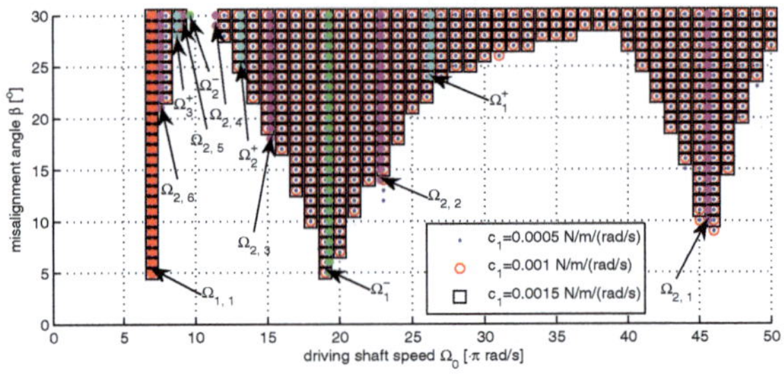

Fig. 7. Stability Chart: effects of the damping ratio c_1 with parameters $k_1 = 10\,\text{Nm/rad}$, $\nu = 10$, $J_I = 0.001\,\text{kg}\cdot\text{m}^2$.

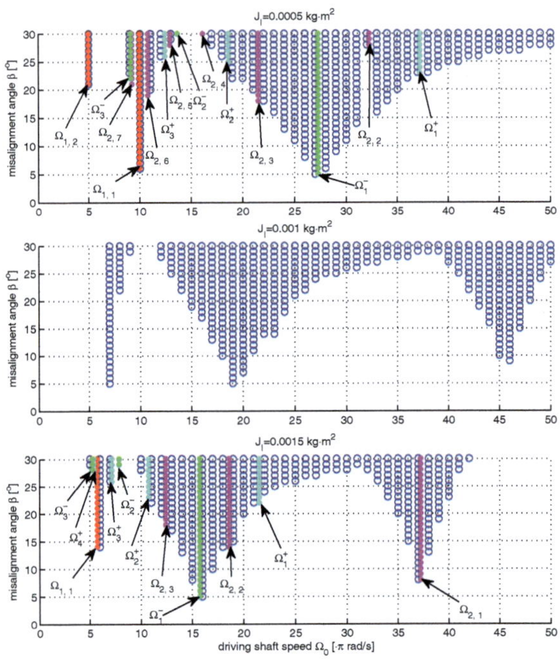

Fig. 8. Stability Chart: effects of the inertia J_I with parameters $k_1 = 10\,\text{Nm/rad}$, $\nu = 10$, $c_1 = 0.001\,\text{Nm/(rad/s)}$.

larger, together with the increasing threshold. Conversely, the increased inertia speeds up instabilities of the system.

Figure 9 displays the unstable regions of the manipulator with different ratios of the mass moment of inertia of the driven disk and the driving shaft, which shows that the influence of this parameter is not so significant. It is noteworthy that the smallest ratio can ensure the largest critical rotating speed of the driving shaft versus the misalignment angle.

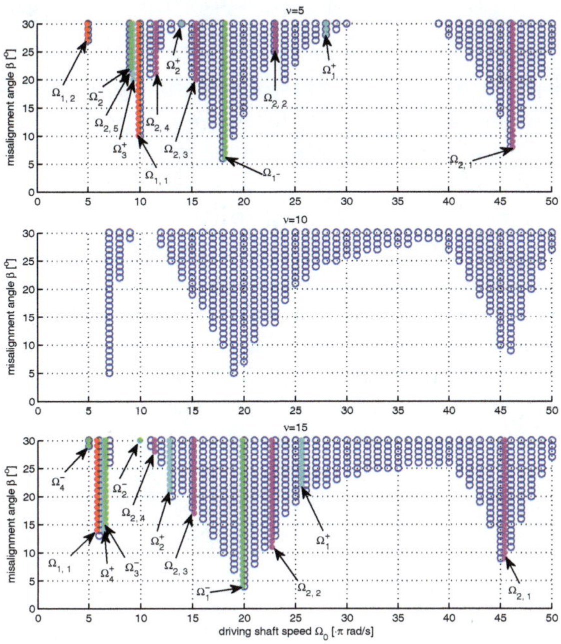

Fig. 9. Stability Chart: effects of the inertia ratio ν with parameters $k_1 = 10\,\mathrm{Nm/rad}$, $J_I = 0.001\,\mathrm{kg \cdot m^2}$, $c_1 = 0.001\,\mathrm{Nm/(rad/s)}$.

An observation from Figs. 6, 7, 8, 9 shows that all the parameters except the damping ratio heavily influence the stability of the manipulator, namely, higher actuation stiffness, lower inertia of the driving shaft and larger inertia ratio of the driven and driving parts being helpful for the dynamic stability. Moreover, an efficient approach to overcome the instability problem of the manipulator with higher rotating speed of driving shaft is to improve the actuation stiffness, for a lightweight design and high stiffness of the end-effector constrained by the driving shaft. All the stability charts show the critical rotating speeds of the driving shaft versus the misalignment angles influenced by both geometric and dynamic parameters. Accordingly, those charts can be used to find the optimum design parameters of the mechanism while guaranteeing its dynamic stability for a range of driving shaft speed.

Compared to the dynamic model in Ref. [4], the term of stiffness ratio of the output and input elements in Eq. (8) is related by the U-joint relationship, instead of a designated parameter, which can simplify the dynamic model and better characterize the manipulator stability. Moreover, the stability charts of previous numerical results [4,14] are represented by dimensionless parameters/ratios, thus, the relationship of the misalignment angles versus rotating speeds of the driving shaft in this work can be characterized intuitively.

6 Conclusions

This paper presented the dynamic torsional stability analysis of a family of 3-dof partially-decoupled spherical parallel manipulators due to the transmitting universal joint. The linearized equations of motion of the system were built to study the torsional stability problem of the manipulator due to the nonlinear input–output transmission of the U-joint, where the relationship of the stiffness between the input and output elements is involved to characterize the manipulator stability, compared to the existing model.

The approach used to analyze the torsional stability problem was numerically illustrated, wherein the unstable regions are presented graphically as well as the parametric resonances. Parametric stability charts of misalignment angles versus rotating speeds of the driving shaft are constructed to identify the unstable regions and critical shaft speeds. Results show that higher actuation stiffness, lower inertia of the driving shaft and larger inertia ratio of the driven and driving parts can ensure the manipulator stable. Moreover, an efficient approach to overcome the instability problem of the manipulator with higher rotating speed of driving shaft is to improve the actuation stiffness, for a lightweight design and high stiffness of the end-effector constrained by the driving shaft. The obtained charts can be used to find the optimum design parameters of the mechanism while guaranteeing its dynamic stability for a range of driving shaft speed.

Acknowledgement. The supports from the Fundamental Research Funds for the Central Universities (No. DUT19JC25), the Natural Science Foundation and the Doctoral Start up Foundation of Liaoning Province (Nos. 20180520028, 20170520134) are gratefully appreciated.

References

1. Asada, H., Granito, J.: Kinematic and static characterization of wrist joints and their optimal design. In: IEEE International Conference on Robotics and Automation, pp. 244–250 (1985)
2. Asokanthan, S.F., Hwang, M.C.: Torsional instabilities in a system incorporating a Hooke's joint. J. Vib. Acoust. **118**(3), 83–91 (1996)
3. Bai, S.: Optimum design of spherical parallel manipulator for a prescribed workspace. Mech. Mach. Theory **45**(2), 200–211 (2010)
4. Bulut, G., Parlar, Z.: Dynamic stability of a shaft system connected through a Hooke's joint. Mech. Mach. Theory **46**(11), 1689–1695 (2011)
5. Chang, S.I.: Torsional instabilities and non-linear oscillation of a system incorporating a Hooke's joint. J. Sound Vib. **229**(4), 993–1002 (2000)
6. Chicone, C.: Ordinary Differential Equations with Applications, chap. 2. Springer, New York (2006)
7. Éidinov, M.S., Nyrko, V.A., Éidinov, R.M., Gashukov, V.S.: Torsional vibrations of a system with Hooke's joint. Soviet Appl. Mech. **12**(3), 291–298 (1976)
8. Floquet, G.: Sur les équations différentielles linéaires à coefficients périodiques. Annales de l'École Normale Supérieure **12**, 47–88 (1883)

9. Gosselin, C., Hamel, J.: The Agile Eye: a high-performance three-degree-of-freedom camera-orienting device. In: IEEE International Conference on Robotics and Automation, pp. 781–786 (1994)
10. Kong, X., Gosselin, C.: Type synthesis of 3-DOF spherical parallel manipulators based on screw theory. ASME J. Mech. Des. **126**(1), 101–108 (2004)
11. Kotera, T.: Instability of torsional vibrations of a system with a Cardan joint. Mem. Fac. Eng. Kobe Univ. **26**, 19–30 (1980)
12. Kuchment, P.A.: Floquet Theory For Partial Differential Equations, chap. 4. Birkhauser Verlag, Boston (1993)
13. Li, T., Payandeh, S.: Design of spherical parallel mechanisms for application to laparoscopic surgery. Robotica **20**(2), 133–138 (2002)
14. Mazzei Jr., A.J., Argento, A., Scott, R.A.: Dynamic stability of a rotating shaft driven through a universal joint. J. Sound Vib. **222**(1), 19–47 (1999)
15. Porter, B.: A theoretical analysis of the torsional oscillation of a system incorporating a Hooke's joint. ARCHIVE J. Mech. Eng. Sci. **3**(4), 324–329 (1961)
16. Porter, B., Gregory, R.W.: Non-linear torsional oscillation of a system incorporating a Hooke's joint. ARCHIVE J. Mech. Eng. Sci. **5**(2), 191–209 (1963)
17. Rosenberg, R.M.: On the dynamical behavior of rotating shafts driven by universal (Hooke) couplings. ASME J. Appl. Mech. **25**(1), 47–51 (1958)
18. Seherr-Thoss, H.C., Schmelz, F., Aucktor, E.: Universal Joints and Driveshafts: Analysis, Design, Applications. Springer, Heidelberg (2006)
19. Szymkiewicz, R.: Numerical Solution of Ordinary Differential Equations, chap. 4. Academic Press, York (1971)
20. Urízar, M., Petuya, V., Altuzarra, O., Diez, M., Hernández, A.: Non-singular transitions based design methodology for parallel manipulators. Mech. Mach. Theory **91**, 168–186 (2015)
21. Wu, G., Bai, S., Kepler, J.: Mobile platform center shift in spherical parallel manipulators with flexible limbs. Mech. Mach. Theory **75**, 12–26 (2014)
22. Wu, G., Caro, S.: Torsional stability of a U-joint based parallel wrist mechanism featuring infinite torsion. In: 22nd CISM IFToMM Symposium on Robot Design. Dynamics and Control, Rennes, France, pp. 147–154 (2018)
23. Wu, G., Caro, S., Bai, S., Kepler, J.: Dynamic modeling and design optimization of a 3-DOF spherical parallel manipulator. Robot. Auto. Syst. **62**, 1377–1386 (2014)
24. Wu, G., Caro, S., Wang, J.: Design and transmission analysis of an asymmetrical spherical parallel manipulator. Mech. Mach. Theory **94**, 119–131 (2015)
25. Wu, G., Zou, P.: Comparison of 3-DOF asymmetrical spherical parallel manipulators with respect to motion/force transmission and stiffness. Mech. Mach. Theory **105**, 369–387 (2016)

Conceptual Design of a 2-DoF Planar High-Speed Industrial Parallel Manipulator

Emre Uzunoğlu[1], Merve Özkahya[1], Erkan Paksoy[1], Barış Taner[2], M. İ. Can Dede[1], and Gökhan Kiper[1(✉)]

[1] Mechanical Engineering, İzmir Institute of Technology, Izmir, Turkey
gokhankiper@iyte.edu.tr

[2] Mechanical & Aerospace Engineering, University of Texas at Arlington, Arlington, USA

Abstract. This study focuses on conceptual design alternatives for a planar high-speed/high-precision manipulator in terms of mechanism structure, control strategy, and drive system selection. These concepts are investigated specifically for planar 5-bar based parallel linkages. An over-constrained 6-bar linkage with parallelogram loops and its simply constrained version are selected for detailed design. In addition, a model-based control strategy including a stiffness model is discussed for future studies. Alternative drive systems are evaluated. Finally a prototype is presented.

Keywords: Planar parallel manipulator · Conceptual design · Control strategy

1 Introduction

Serial robots cannot provide the necessary performance during operations which require high acceleration and precise motion due to the flexibility of the manipulator; therefore manipulators with parallel kinematic structure are preferred in such cases. Although parallel manipulators can be optimally designed according to several performance criteria, the positioning accuracy is still limited in application due to factors such as manufacturing errors, flexibility of links and joints, and joint clearances. This study has emerged from problems encountered during previous studies on enhancing the positioning accuracy of a high-acceleration manipulator [1, 2]. The aim of this paper is to investigate structural and control based improvements in order to increase the positioning accuracy of planar parallel robots commonly used in industrial applications. The conceptual design alternatives are presented in terms of mechanism structure, controller strategy, and selection of drive-system components to be used for the design and prototype of a simply constrained and an over-constrained parallel manipulator.

First, a brief literature survey on calibration, balancing, and stiffness analysis are presented in this section. The highlights of these surveys are presented especially on high-speed parallel manipulators and 5-bar based mechanisms. Then, the conceptual design alternatives for the mechanism structure, control strategy and drive system are presented and discussed thoroughly. Finally, constructed prototypes are presented.

D. Wang et al. (Eds.): MeTrApp 2019, MMS 79, pp. 344–353, 2020.
https://doi.org/10.1007/978-981-15-0142-5_34

1.1 Balancing

Balancing is a common problem for high-speed machinery. If the potential energy of the system is stationary at all times when the system is under the effect of conservative forces, the system is said to be statically balanced [3]. When resultant of reaction forces and moments of the system due to inertial forces and moments are zero, the system is said to be dynamically balanced. If reaction forces are zero, the system is shaking force balanced and if reaction moments are zero, the system is shaking moment balanced [4]. A dynamically balanced system is already statically balanced.

In the literature, there are several balancing techniques, which are categorized as counter-mass balancing, spring balancing, using duplicate mechanisms, active balancing, designing the mechanism as inherently balanced, etc. In [5] and [6], some counter-mass balancing methods are investigated in detail for force and moment balancing of planar and spatial mechanisms such as full force balancing, full force and partial moment balancing, etc. Shaking forces and moments are significant problems in high-speed machinery. Van der Wijk, Herder and Demeulenaere [7] have compared and evaluated existing complete balancing principles regarding the addition of mass and the addition of inertia in order to compare the balancing performance with added mass and inertia. Complete dynamic balancing for some planar 4R (R: revolute joint), 5R, 3-RRR and spatial 3-RRS parallel mechanisms are presented in [8].

1.2 Calibration with Stiffness Model

Manipulator calibration is a cost-effective way to improve robot accuracy, and many researchers have devoted efforts to this field. Different models, measurement systems, and algorithms for identification and compensation have been developed as summarized in [9–11].

For a parallel mechanism, the global platform pose depends on all of its actuators. For our application, the global error at the end of the tool is the main concern. Some part of this error could be due to manufacturing tolerances or assembly. At high speeds, control speed and accuracy play an important role in the overall final position.

Calibration of parallel manipulators requires more complex algorithms, models and measurement methods as compared to serial manipulators. In general, the calibration methods of parallel mechanisms can be categorized as parametrical (or model-based) and numerical methods. Parametrical methods depend on the used sensors and the approach used in calibration [12]. Based on the error model, it is possible to map the geometric errors to the end-effector error of a parallel mechanism. This mapping can be used to increase the global accuracy of the structure [13].

In general, the performance of a robot manipulator is evaluated by two noteworthy points: payload capacity and positional accuracy. Metric for payload capacity is defined as payload/weight ratio in [14]. In evaluation of the positioning accuracy, usually three metrics are used as absolute accuracy, repeatability and resolution [15]. Absolute accuracy is defined as robot's ability of positioning its end-effector at a desired configuration within its workspace and it is half of the resolution.

[16] presents an off-line compensation method depending on the stiffness properties calculated under the auxiliary loads and external forces. Robot control program that

determines the trajectory is modified according to the calculated deformations. Online compensation provides best solution for the compensation of errors due to compliance of robot mechanism. The difficulty in applying an on-line compensation method stems from the complexity of the mathematical stiffness model for especially multi-degree-of-freedom (DoF) systems. Calculation of the stiffness matrix provides the relation between the forces acting on the robot and the compliant displacements of the robot mechanism, which are the reason of the compliant errors at the end-effector.

2 Conceptual Design

In order to observe the effect of over-constraint, simply and over-constrained mechanisms with similar structure are to be built and tested. In this Section, first, planar two-DoF mechanism alternatives are presented. Then, the controller strategy and drive system alternatives are discussed.

2.1 Kinematic Structure Alternatives for a Two-DoF Planar Linkage

In this Section, two-DoF planar mechanism structure alternatives are investigated. Single loop mechanisms with five links and five single-DoF (revolute (R) and prismatic (P)) joints are considered. More than two P joints cannot be used in a loop, because otherwise there will be internal mobility. In this way, although 32 different mechanisms can be enumerated (with combinations of 0P, 1P and 2P), it is possible to obtain 6 different 5-link mechanisms under the following conditions:

Actuators must be supported from the base and no actuated joints must carry the load of the other, i.e. the RPRRR is not suitable (underlined letters indicate actuated joints).

If available, the P joints must be active.

Mirror image mechanism structures are equivalent (ex. RRRPR = RPRRR).

Five-link mechanisms that satisfy these requirements are RRRRR, RRRPR, RRRRP, RPRRP, RPRPR and PRRRP [17]. Alıcı [18] has developed a method for determining the singularity positions of all these six types of mechanisms. Symmetrical structures are generally preferred because of the workspace symmetry, balancing and control simplicity [19]. The mechanisms for which the P joints are carried by the fixed revolute joints are not favorable due to high inertia of the actuated P joints. The six mechanisms mentioned just above are evaluated according to the structural symmetry, whether actuators are fixed or not and whether the actuators are of the same type (R or P) for ease of control. The evaluation is summarized in Table 1 and accordingly RRRRR (5R) and PRRRP alternatives are selected for final evaluation.

Sun, Cheung, and Lou [19] compared 5R and PRRRP mechanisms, which were optimized for the same effective square workspace, with respect to functionality, compactness and the singular values of the forward and inverse Jacobian matrices. It is concluded that a smaller size of the 5R mechanism can work in the same workspace. The 5R mechanism features lighter weight structure, better precision, resistance, and force transmission capabilities. These advantages reduce actuator power requirements

Table 1. Evaluation of five-link mechanisms considering various aspects.

Mechanism	Fixed actuators	Structural symmetry	Ease of control
RRRRR	Yes	Yes	Yes
RRRPR	No	No	No
RRRRP	Yes	No	No
RPRRP	No	No	No
RPRPR	No	Yes	No
PRRRP	Yes	Yes	Yes

and are suitable for high acceleration applications. The PRRRP mechanism, on the other hand, possesses higher velocity influence coefficient for the end-effector and is more suitable for pick-and-place applications.

5R Mechanisms: The generic 5R mechanism (Fig. 1a) is actuated at A_0 and B_0 fixed revolute joints. Input variables are θ_1 and θ_2 angles, whereas θ_3 and θ_4 are passive joint variables. Point "E" represents the end-effector point. $|A_0A|$ and $|AE|$ are respectively equal to $|B_0B|$ and $|BE|$ for a symmetric workspace. If the mechanism is evaluated in terms of construction, there is no problem about fixed revolute joint connections. However, it is difficult to assemble the tool to be used as the end-effector to point "E". Furthermore, the end-effector orientation is uncontrolled unless timing belt-pulley systems are used.

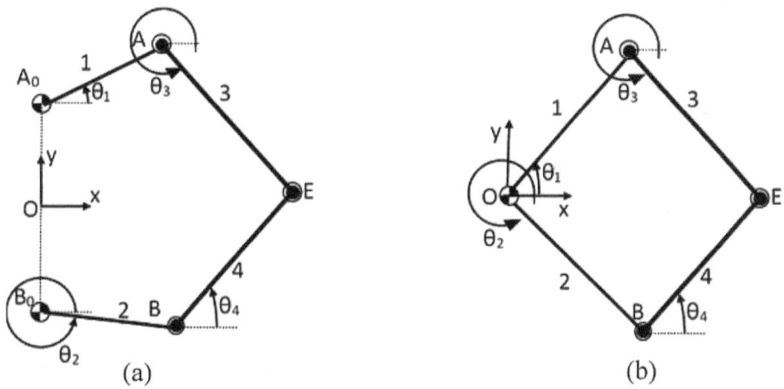

Fig. 1. (a) General 5R mechanism, (b) 5R mechanism with coincident fixed axes

Fixed revolute joints are coincident at the point "O" in Fig. 2b. All link lengths can be selected as equal to each other for ease of balancing. Although this special case has simpler kinematic structure than the general case, the motor connection construction is difficult due to the coincident fixed revolute joint axes. Coincident motor axes can be avoided by using suitable gear connections.

6R Mechanisms with Parallelogram Loops: The construction of base and end-effector for 6R mechanisms is simpler compared to 5R mechanisms. There are no coincident axes in this case. A 6R loop is obtained by introducing equal offset between coincident axes and the E joint of the 5R mechanism shown in Fig. 1b. However, parallelogram loops should be used for at least one of the arms so that the end-effector orientation is constant. The other option is to use timing belt-pulley systems within the arms. But the belt tension is a problem. If the belt is over-tensioned, excessive joint forces are observed. On the other hand, if the belt tension is not enough, there will be vibrations.

In order to obtain a simply constrained mechanism, two or three parallelogram loops must be used. The simply constrained mechanism with 2 parallelogram loops is shown in Fig. 2a. For symmetric mass distribution, parallelogram loops can be used for both arms, but in this case, the mechanism becomes an over-constrained one. A pin-in-slot (joint "L" in Fig. 2b) can be used instead of one of the passive revolute joints to have a simply constrained mechanism. The mechanism is over-constrained when only R joints are used for all four parallelogram loops (Fig. 2c).

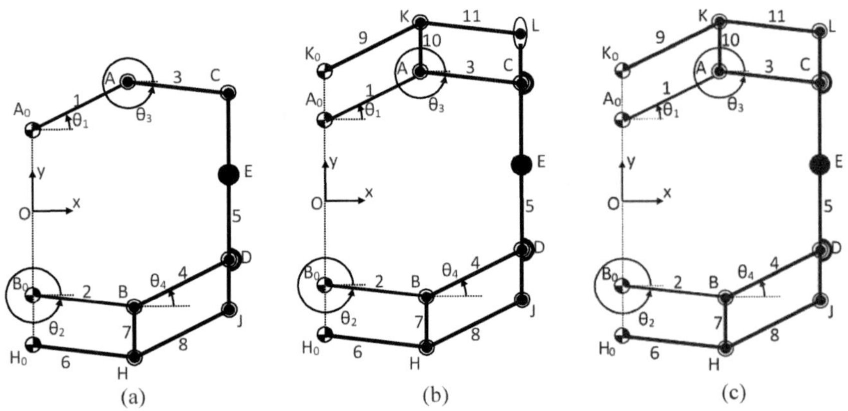

Fig. 2. (a) Simply constrained 6R mechanism with 2 parallelogram loops, (b) simply constrained 6R mechanism with pin-in-slot joint, (c) over-constrained 6R mechanism

The over-constrained 6R mechanism with four parallelogram loops is selected for investigation in this study. For comparison, another equivalent simply-constrained mechanism has to be selected and the following criteria are taken into account to select this mechanism:

1. Ease of balancing and simple mathematical model
2. To have similar construction as the over-constrained mechanism
3. Structural simplicity
4. Constant end-effector orientation
5. Structural symmetry
6. Kinematic equivalence.

The general 5R mechanism has disadvantage about criterion (1). 5R mechanism with coincident fixed axes is advantageous in terms of criteria (1), (5), (6), but it involves complexity about assembly of motors and the end-effector in terms of criteria (3) and (4). Although the 6R mechanism with two or three parallelogram loops (Fig. 2a) is disadvantageous in terms of criterion (5), it is possible to come-up with a dynamically balanced design. Moreover, it has more advantages than other alternatives in terms of criterion (2). Therefore, the simply constrained 6R mechanism with three parallelogram loops is selected for further investigation. One link of a parallelogram loop is removed from the over-constrained mechanism in order to obtain the simply constrained one.

2.2 Control Scheme Alternatives

As discussed in Sect. 1.2, a better control of a high precision parallel mechanism can be achieved by integrating the stiffness model into the controller. Two conceptual design alternatives have been devised for the control system's flowchart (Fig. 3).

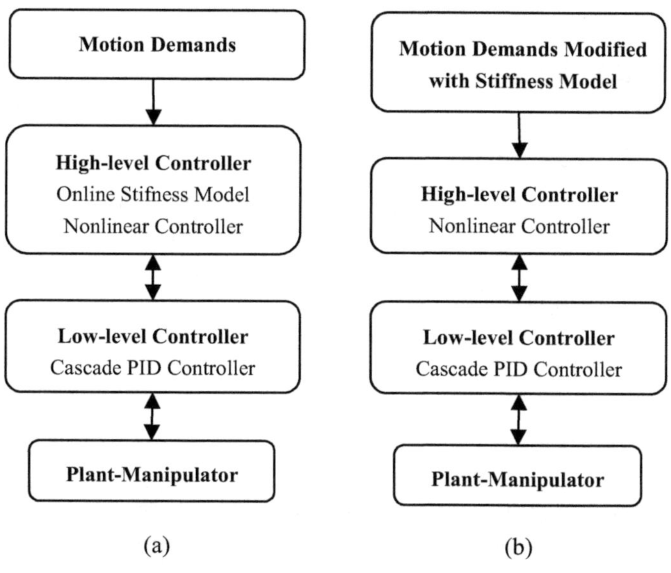

Fig. 3. Control strategy alternatives

In the first flowchart (Fig. 3a), the trajectory created for end-effector's motion demands are fed to the high-level controller. The high-level controller runs in a computer along with data acquisition device hardware, and thus, the algorithms run in a real-time operating system. The high-level controller algorithm consists of a non-linear controller and an online stiffness model. A non-linear controller such as computed – torque method or adaptive controller are tested and evaluated in simulation environment to be used in the high-level controller. However, in this scenario, the critical point is the computation time of the stiffness model, which is supposed to run in real-time at a designated frequency of 1 kHz.

In the second alternative (Fig. 3b), the end-effector trajectory is modified with a correction algorithm with stiffness model during the pre-process, which is an offline method. The working principle of the stiffness model in this scenario is to calibrate the pre-determined end-point trajectory by estimating the positioning errors due to elastic deformations of the mechanism components. In both alternatives, the low-level controller, which is embedded in motor drivers, is utilized with cascade PID controller and position feedbacks from the motors are sent back to the high-level controller.

2.3 Drive System Alternatives

In order to meet the torque and speed requirements for a high-speed industrial application, drive system selections are carried out among available alternatives of industrial servo drives and drivers. AC servo motors are favored due to the high torque per weight values, their efficiency and their frequent use in industrial systems.

Due to the high-speed and high-precision requirements in line with the other conceptual alternatives, the actuation system must be selected in such a way that it can provide high dynamic motion. Accordingly, two actuation system alternatives are evaluated: (1) a direct-drive servo motor, (2) an AC servo motor with a low-backlash (<1 arcmin) reducer that has suitable gear-ratio and efficiency.

Table 2. Drive system alternatives

Criteria	Cycloid gear + Servo motor	Direct drive servo motor
Efficiency	+	+ +
Cycle time	+	+ +
Torque-speed range	+	+ +
Clearance	+	+ +
Wight to power ratio	+	− −
Size	+	−
Ease of integration	+ +	−
Price	−	− −

+ + very good; + good; − bad; − − very bad

Direct drive systems are more efficient systems because there are no efficiency losses and backlashes due to the reducer. However, they are generally disadvantageous in terms of weight and size compared to geared motors. Also, precision equipment is required during the mounting process. In most applications, application-specific frameless motors are used and they require more complex constructional details. A geared motor is disadvantageous compared to the direct drive because of the degradation in terms of dynamic motion and efficiency. However, special low-backlash gearboxes such as harmonic drives and cycloid reducers prevail as relatively better solutions. Finally, complying with the torque and speed requirements of the mechanism, two alternatives are compared, which are: Kollmorgen DDR series direct drive motor and AKM series motor with Sumitomo fine cyclo gearbox. The evaluation table based on the technical specifications of the selected drives are provided in Table 2.

3 Conclusions

In this paper, the alternatives for mechanism structure, control strategy, and drive system are presented for a high-speed planar robotic manipulator. First, over-constrained and simply constrained mechanism alternatives are presented and evaluated. As a result, a simply-constrained mechanism with three parallelogram loops and an over-constrained 6R mechanism with 4 parallelogram loops are chosen.

Two concepts are introduced to improve the performance of controller by integrating the stiffness model. It is decided that both algorithms can be tested in future studies and these alternatives can be compared with respect to their computational cost and efficiency.

Although a direct drive system is advantageous in terms of efficiency, the system should be as light as possible and its dimensions should be within certain limits. Therefore, a motor with cycloid gear reducer is selected since it also meets the requirements of low backlash. According to alternatives presented here, a prototype for the robotic manipulator is manufactured and assembled as shown in Fig. 4. The link that is removed in Fig. 4b is replaced with two lumped masses at the associated joints in order to obtain dynamic equivalence of the two mechanisms.

(a) (b)

Fig. 4. Mechanism prototype in (a) over-constrained and (b) simply constrained configuration

Future studies include calibration, tests and design iterations. The tests will be done for both simply and over-constrained mechanisms to see the effect of over-constraint. Both mechanisms will be tested for balanced, partially force-balanced and fully force-balanced cases. The base vibrations will be monitored with accelerometers.

Control strategies including online and off-line stiffness models will be tested. The accuracy and repeatability of the end-effector will be measured using a laser cutter end-effector. In the end, the generated know-how will be shared with the academic and industrial community.

Acknowledgment. This work is financially supported by The Scientific and Technological Research Council of Turkey via grant number 116M272.

References

1. Uzunoğlu, E., Dede, M.İ.C., Kiper, G.: Trajectory planning for a planar macro-micro manipulator of a laser-cutting machine. Ind. Robot. Int. J. **43**(5), 513–523 (2016)
2. Dede, M.İ.C., Kiper, G., Uzunoğlu, E.: A macro-micro mechanism design for laser cutting process. In: Proceedings the 17th International Conference on Machine Design and Production, Bursa (2016)
3. Jean, M., Gosselin, C.M.: Static balancing of planar parallel manipulators. In: Proceedings of 1996 IEEE International Conference on Robotics and Automation, vol. 4, pp. 3732–3737 (1996)
4. Van der Wijk, V.: Methodology for analysis and synthesis of inherently force and moment-balanced mechanisms. Ph.D. thesis, University of Twente (2014)
5. Lowen, G.G., Berkof, R.S.: Survey of investigations into the balancing of linkages. J. Mech. **3**(4), 221–231 (1968)
6. Lowen, G.G., Tepper, F.R., Berkof, R.S.: Balancing of linkages-an update. Mech. Mach. Theory **18**(3), 213–220 (1983)
7. Van der Wijk, V., Herder, J.L., Demeulenaere, B.: Comparison of various dynamic balancing principles regarding additional mass and additional inertia. J. Mech. Robot. **1**, 041006 (2009)
8. Van der Wijk, V., Herder, J.L.: Synthesis of dynamically balanced mechanisms by using counter-rotary countermass balanced double pendula. J. Mech. Des. **131**, 111003 (2011)
9. Roth, Z.S., Mooring, B.W., Ravani, B.: An overview of robot calibration. In: IEEE Southcon Conference, Orlando, Florida, pp. 377–384 (1986)
10. Hollerbach, J.M.: A survey of kinematic calibration. In: Khatib, O., Craig, K.J., Lozano-Pérez, T. (eds.) The Robotics Reviews 1, pp. 207–242. MIT Press, Cambridge (1988)
11. Mooring, B.: Fundamentals of Manipulator Calibration. Wiley, New York (1991)
12. Hernández-Martínez, E., López-Cajún, C., Jáuregui-Correa, J.: Calibration of parallel manipulators and their application to machine tools. Ingeniería Investigación y Tecnología **11**, 141–154 (2010)
13. Elatta, A.Y., Li, P.G., Fan, L.Z., Yu, D., Luo, F.: Inf. Technol. J. **3**(1), 74–78 (2004)
14. Merlet, J.P.: Parallel Robots. Springer, Dordrecht (2006)
15. Klimchik, A.: Enhanced stiffness modeling of serial and parallel manipulators for robotic-based processing of high performance materials. Ph.D. thesis, Ecole Centrale de Nantes (2011)
16. Klimchik, A., Bondarenko, D., Pashkevich, A., Briot, S., Furet, B.: Compliance error compensation in robotic-based milling. In: Informatics in Control, Automation and Robotics, pp. 197–216. Springer (2014)
17. Cervantes-Sánchez, J.J., Rendón-Sánchez, J.G.: A simplified approach for obtaining the workspace of a class of 2-dof planar parallel manipulators. Mech. Mach. Theory **34**, 1057–1073 (1999)

18. Alıcı, G.: Determination of singularity contours for five-bar planar parallel manipulators. Robotica **18**, 569–575 (2000)
19. Sun, S., Cheung, J.W.F., Lou, Y.: A study on five-bar manipulators for semiconductor packaging applications. In: Proceedings 2007 IEEE International Conference on Mechatronics and Automation, pp. 1811–1816 (2007)

Compliant Mechanism

Design of a Controlled-Stiffness Flexure Mechanism Based on the FACT Method

Kaiyu Wu[1], Fan Zhang[1(✉)], Minhua Zheng[2], Hecheng Li[3], and Jing Sun[2]

[1] Shanghai University of Engineering Science, Longteng Road No. 333, Songjiang District, Shanghai 201620, People's Republic of China
pdszhangfan@sues.edu.cn

[2] Department of General Surgery, Shanghai Minimally Invasive Surgery Center, Ruijin Hospital, Shanghai Jiao Tong University School of Medicine, Shanghai 200025, People's Republic of China

[3] Department of Thoracic Surgery, Ruijin Hospital, Shanghai Jiaotong University School of Medicine, Shanghai 200025, People's Republic of China

Abstract. In complex environments such as minimally invasive surgery, mine disaster and earthquake relief, compared with the traditional rigid mechanism, variable stiffness mechanism has tremendous advantages. In this paper, a novel controlled-stiffness mechanism based on flexure element and structural constraints is proposed. The mechanism is a parallel mechanism with two branches, one of which contains a flexure element. The stiffness control of the mechanism is realized by controlling the functional direction of the flexible element. Using the traditional Grubler-Kutzbach (G-K) formula to calculate the degree of freedom (DOF) of the mechanism is difficult. In this paper, a method of freedom and constraint topology (FACT) method is introduced. This method has the advantages of visualization, simplicity, rapidity and is suitable for the analysis of DOF and motion ability of complex mechanisms. The analysis results show that the structure can realize the desired form of motion.

Keywords: Controlled-stiffness mechanism · Flexure element · Freedom and constraint topology method

1 Introduction

The traditional rigid mechanism has the advantages of high rigidity, strong bearing capacity, high precision and simple control, etc. It has been widely used in production and life at present. However, in some special scenarios, we must use variable stiffness mechanism. So variable stiffness mechanism has been widely concerned and studied. The traditional methods to achieve variable stiffness mainly include:

(a) Using magnetorheological or electrorheological materials to achieve rigid-flexible transformation [1, 2];
(b) Using phase transformation of materials to achieve rigid-flexible transformation [1, 2];

D. Wang et al. (Eds.): MeTrApp 2019, MMS 79, pp. 357–364, 2020.
https://doi.org/10.1007/978-981-15-0142-5_35

(c) Using negative pressure to make granular materials squeeze each other to achieve variable stiffness [3, 4];
(d) Increasing the tension force of wire to increase the stiffness of multi-segment structure [5].

In this paper, a structure with variable stiffness of mechanism is presented by using structural constraints to control the functional orientation of flexible body [6, 7]. The mechanism is a parallel mechanism with flexure element [8, 9]. So we will use FACT method to analyze the motion of mechanism [10]. The FACT method is unfamiliar to most people. We will give a brief introduction to this method through graphics.

2 Freedom and Constraint Topology (FACT) Method

We use the red straight line to represent the freedom of rotation and use the red straight line with arrow to represent the freedom of movement. Similarly, we use black straight line represent force constraint and black straight line with arrow to represent couple constraint. Each pair of motion of a mechanism is expressed in the form of freedom line, which form a freedom line pattern. Similarly, each constraint in a mechanism is represented as a constraint line, which forms a constraint line pattern. Any mechanical structure can be regarded as one or more intermediate in parallel, series or parallel connection. If the mechanism is in series, the freedom line pattern can be obtained by merging the freedom lines of each kinematic pair; If there are closed chains in the structure, such as parallel mechanisms, first find out the constraint line pattern of each branch chain on the platform, then merge all constraint line patterns of branch chains, finally we use Blanding rule to find out the freedom line pattern of moving platform; If it is a hybrid mechanism, we only need to disassemble it into series and parallel mechanisms and solve them separately according to the above method, so as to obtain the freedom line pattern of the target body. Finally, we use a simple parallel mechanism to demonstrate the usage of the FACT method, as shown in Fig. 1.

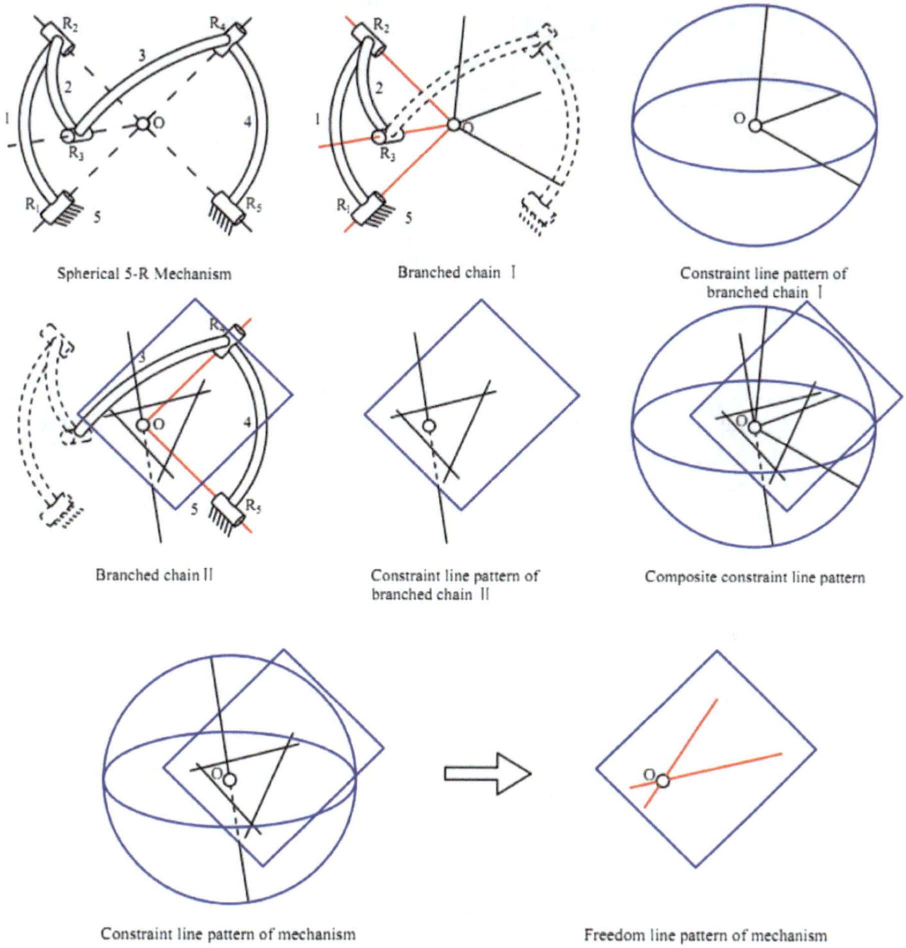

Fig. 1. Freedom analysis of Spherical 5-R Mechanisms

3 Design and Analysis of Controlled-Stiffness Mechanism Based on FACT

3.1 Design of Controlled-Stiffness Mechanism

What we want is a snake-like manipulator, which is composed of multi-segment units, each of which has a DOF of rotation, as shown in Fig. 2:

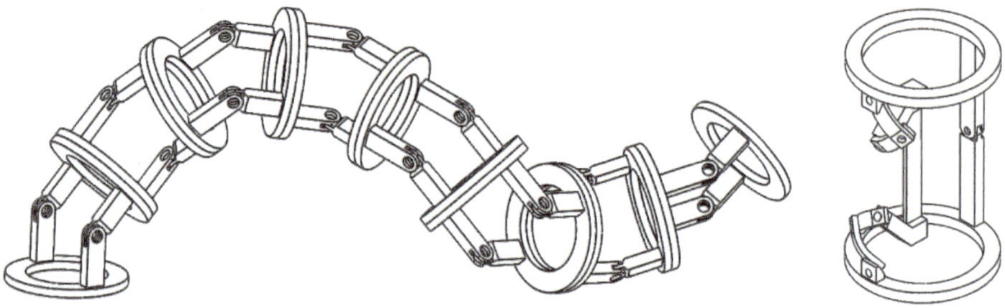

Fig. 2. Schematic diagram of snake-like manipulator and element body

According to mechanics of materials, the bending resistance of an object is determined by the moment of inertia of its cross section. The formula of moment of inertia of it's cross section is as follows:

$$I_x = \int_A y^2 dA \tag{1}$$

$$I_y = \int_A x^2 dA \tag{2}$$

For the flexible hinge shown in Fig. 3, the stiffness of the hinge is the smallest when its rotation axis is as shown in Fig. 3(a). When its rotation axis changes toward the non-functional figure, the cross-section moment of inertia I_y will increase as shown in Fig. 3(b), so the stiffness in this direction will increase.

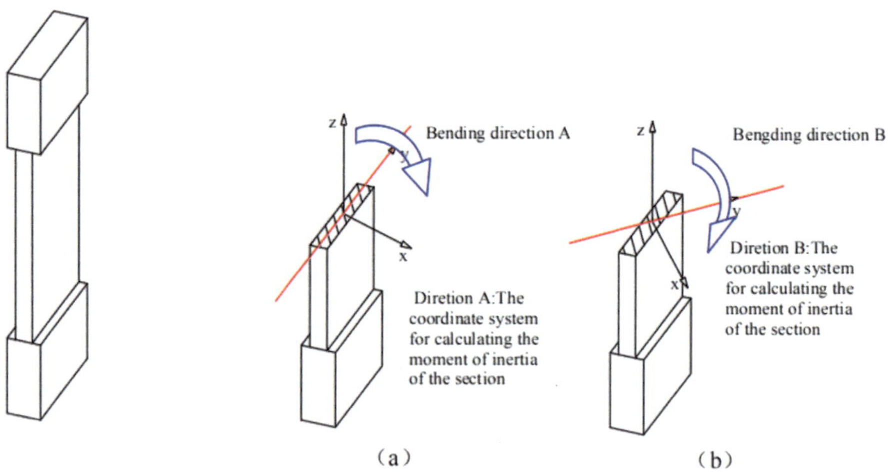

Fig. 3. Cross-sectional moments of inertia in different functional directions

By changing the functional direction of flexible body, we can change the stiffness. So we proposed a flexure mechanism with Controlled-stiffness, as shown in Fig. 4:

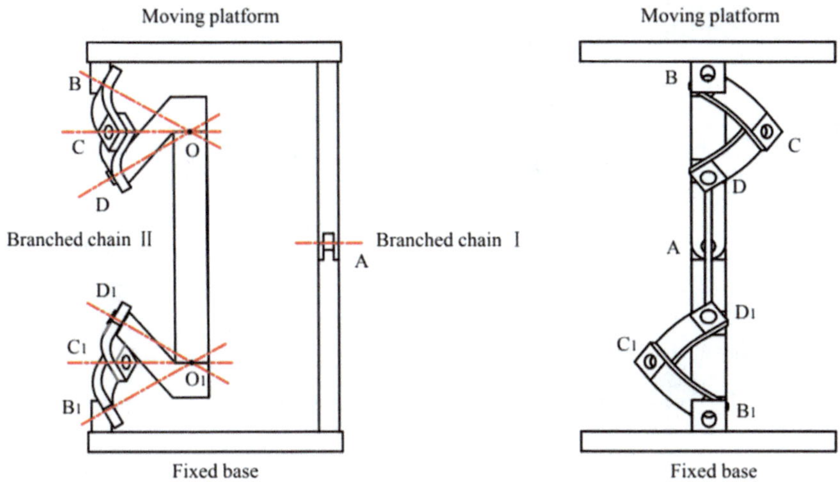

Fig. 4. Mechanical architecture with variable stiffness

3.2 Analysis of DOF of Controlled-Stiffness Mechanism

According to the FACT method, the equivalent constraint line patterns of flexible element can be expressed as follows (Fig. 5):

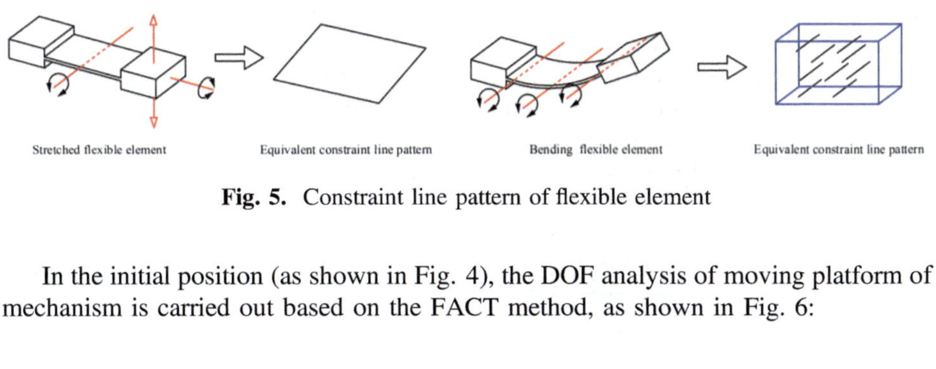

Fig. 5. Constraint line pattern of flexible element

In the initial position (as shown in Fig. 4), the DOF analysis of moving platform of mechanism is carried out based on the FACT method, as shown in Fig. 6:

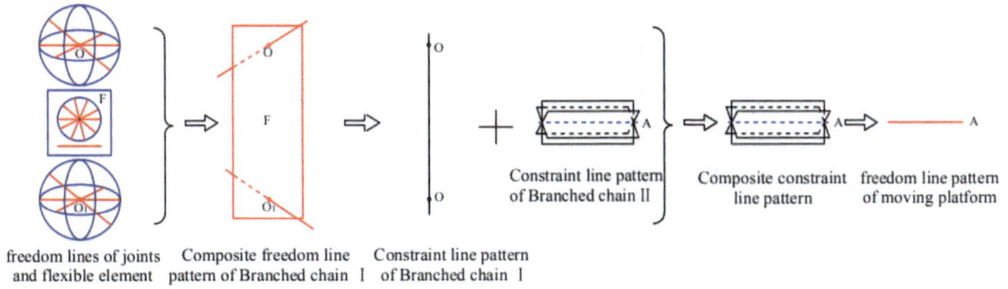

Fig. 6. Freedom line patterns of moving platform at the initial position

In the non-initial position, the line OO_1 is perpendicular but not coplanar to the axis of the rotating pair A. DOF analysis of moving platform of mechanism based on FACT method, as shown in Fig. 7:

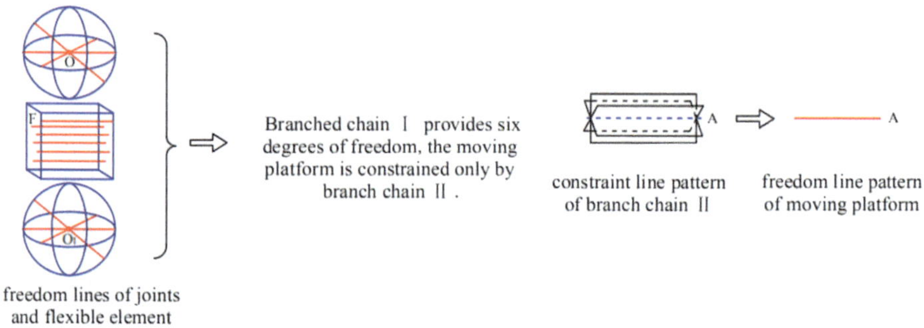

Fig. 7. Freedom line patterns of moving platform at a non-initial position

From the above analysis, we can see that the moving platform of the mechanism has a rotational DOF around the axis of the rotating pair A in both the initial and non-initial positions.

We need to control the angle of the flexible element by controlling the angle of the spherical hinge B_1, So the flexible element is regarded as the moving platform when both the moving platform and the static platform are fixed. When the branched chain consisting of spherical hinges B, C and D is regarded as the branched chain I and the branched chain consisting of spherical hinges B_1, C_1 and D_1 is regarded as the branched chain II, the DOF of the flexible element is analyzed as follows:

According to the analysis of the above figure (Fig. 8), the flexible element has a DOF of rotation around the line OO_1. At the same time, it is not difficult to find that when the relative position of the moving platform and the static platform changes, the above analysis of the flexible element will not change, so the flexible element always has a DOF of rotation around the line OO_1.

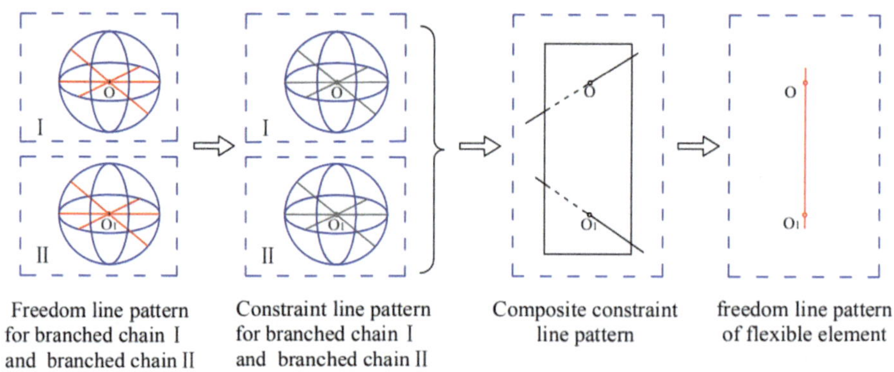

Fig. 8. Freedom line patterns of flexible element

When the fixed moving platform and spherical hinges B1, the spherical hinges B, C and D are regarded as the branched chain I, and the spherical hinges C_1 and D_1 as the branched chain II, the DOF of the flexible element is analyzed as follows (Fig. 9):

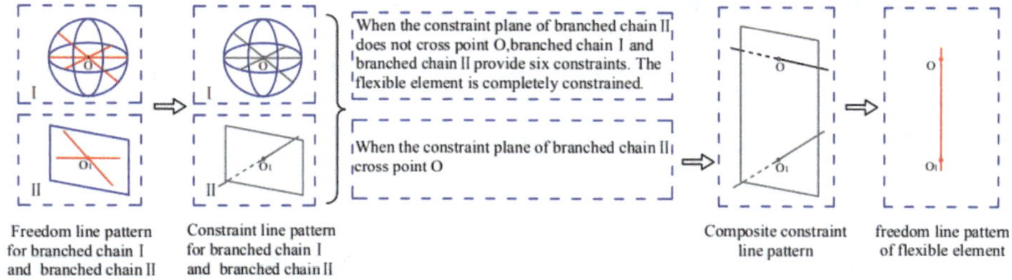

Fig. 9. Freedom line patterns of flexible element

Since the maximum rotation of flexible element is only 90 deg, the constraint plane of branched chain II will never pass through O point by choosing appropriate model parameters. As shown in Fig. 10, the points C_1, D_1, O_1 and O will not always be in the same plane during the rotation of the flexible element to 90 deg, that is to say, the flexible element will not rotate when the spherical hinge B_1 is fixed.

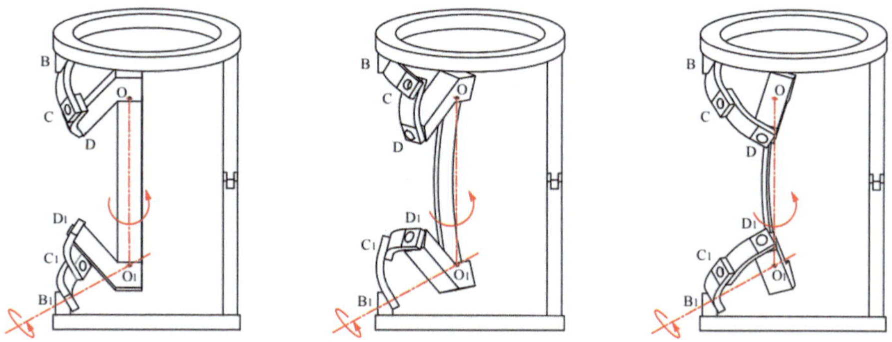

Fig. 10. Flexible body rotation angle under B_1C_1 Control

Figure 10 shows that the rotation Angle of flexible element F around line OO_1 can be controlled only by controlling the rotation Angle of spherical hinge B_1.

4 Conclusion

Through the analysis of DOF and kinematics of variable stiffness mechanism by FACT method, it shows that the analysis of complex mechanism by FACT method is intuitive, simple and fast. The feasibility of realizing controllable stiffness by changing the functional direction of flexible body is also verified by DOF analysis and modeling simulation.

References

1. Shan, W., Lu, T., Majidi, C.: Soft-matter composites with electrically tunable elastic rigidity. Smart Mater. Struct. **22**(8), 085005 (2013)
2. Shan, W., Diller, S., Tutcuoglu, A., Majidi, C.: Rigidity-tuning conductive elastomer. Smart Mater. Struct. **24**(6), 065001 (2015)
3. Lei, J., Yu, H., Wang, T.: Dynamic bending of bionic flexible body driven by pneumatic artificial muscles (PAMs) for spinning gait of quadruped robot. Chin. J. Mech. Eng. **29**(1), 11–20 (2016)
4. Polygerios, P., Wang, Z., Galloway, K.C., et al.: Soft robotic glove for combined assistance and at-home rehabilitation. Robot. Auton. Syst. **73**, 135–143 (2015)
5. Hannan, M.W., Walker, I.D.: Kinematics and the implementation of an elephant's trunk manipulator and other continuum style robots. J. Robot. Syst. **20**(2), 45–63 (2003)
6. Aukes, D.M., Heyneman, B., Ulmen, J., et al.: Design and testing of a selectively compliant underactuated hand. Int. J. Robot. Res. **33**(5), 721–735 (2014)
7. Li, X., Chen, W., Lin, W., Low, K.H.: A variable stiffness robotic gripper based on structure-controlled principle. IEEE Trans. Autom. Sci. Eng. **15**(3), 1104–1113 (2018)
8. Trease, B.P., Moon, Y.M., Kota, S.: Design of large-displacement compliant joint. J. Mech. ASME **127**(4), 788–798 (2005)
9. Timoshenko, S.P.: History of Strength of Materials, vol. 452. Courier Dover Publications, New York (1983)
10. Hopkin, J.B., Culpepper, M.L.: Synthesis of multi-degree of freedom, parallel flexure system concepts via freedom and constraint topology (FACT), Part I: principles. Precis. Eng. **34**(2), 259–270 (2010)

Compliant Mechanism Based Backbones for Continuous Manipulators and Robots

Haiyang Li[1(✉)], Qiaoling Meng[2], Guangbo Hao[3], Shujun Li[1], Yunpeng Gong[1], and Yong Zhao[1]

[1] Northeastern University, Shenyang, China
lihaiyang@me.neu.edu.cn
[2] School of Medical Instruments and Food Engineering, University of Shanghai for Science and Technology, Shanghai, China
[3] School of Engineering, University College Cork, Cork, Ireland

Abstract. The key part of a continuous backbone manipulator or robot is the backbone. A backbone is composed of several backbone modules that are assembled together in series. This paper proposes a new simple backbone module, which is a compliant mechanism. Compliant mechanisms are employed here due to their advantages over rigid body mechanisms. The compliant mechanism based backbone module is referred to as the basic compliant backbone module. An analytical model of the basic compliant backbone module is derived, showing that the stiffness of the basic compliant backbone module along and about the backbone axis is much higher than the others. The analytical model proofs the desired degrees of freedom of the basic compliant backbone module. Some other symmetric compliant backbone modules, designed by combining several copies of the basic backbone module in parallel, are also introduced in this paper. Continuous backbones, built by the proposed compliant backbone modules, enable not only to bend apart from the backbone axis but also to transmit force and torque about the backbone axis. Additionally, an application example of a proposed compliant backbone module is introduced.

Keywords: Compliant mechanism · Continuous manipulator · Continuous backbone

1 Introduction

Inspired by biological snake locomotion, snake-like robots have been created, which can provide both necessary flexibility for passing narrow slots and enough stiffness for delivering desired motions and forces [1]. Snake-like robots are classified, based on their backbones, into two groups, discrete backbone robots and continuous backbone robots. Continuous backbone robots are capable of changing their shapes to adapt to unstructured environment [2], so continuous backbone manipulators and robots have been experiencing a rapid development in recent years.

Continuous backbone robots employ flexible materials, such as elastomer, springs and bellows, to make their backbones. A classic backbone for continuous backbone robots was proposed in [3], as shown in Fig. 1. The classic backbone is consist of one

D. Wang et al. (Eds.): MeTrApp 2019, MMS 79, pp. 365–375, 2020.
https://doi.org/10.1007/978-981-15-0142-5_36

end disk, several spacer disks, one base disk, one central tube, and three stage wires. The central tube and the stage wires are flexible to be bend to any directions, but have larger stiffness along the axis directions of the central tube and the stage wires. The disks are detachable and fixed to the central tube. The stage wires are fixed to the end disk, and free to slide through the holes on the spacer disks and base disk. The spacer disks are used to prevent bucking of the central tube and to maintain equal distance between the stage wires and the central tube. The motion of the end disk can be controlled by pushing and pulling the three stage wires. In order to improve the motion characteristics of the classic backbone structure for different applications, new backbones with different geometrical structures have been proposed, such as the one shown in [4].

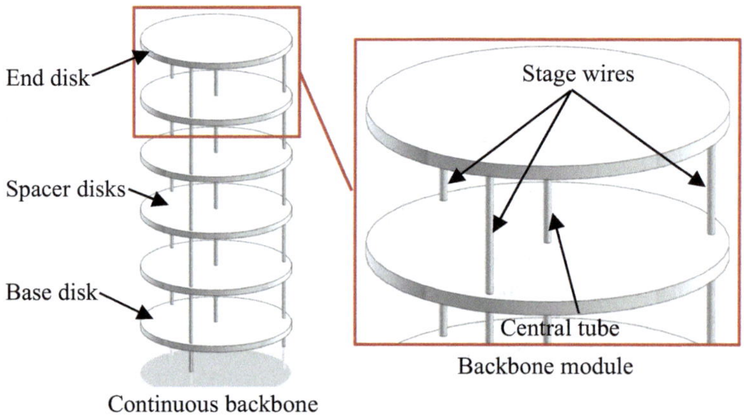

Fig. 1. A classic backbone for continuous backbone robots

In this paper, a new backbone structure is proposed especially for slender continuous backbone manipulators and robots. The new backbone is designed using compliant mechanisms, so it can be referred to as compliant backbone. Compliant mechanisms are mechanical mechanisms which transfer motions and forces by means of the deformation of their structures rather than rigid joints. Compared with rigid body mechanisms, compliant mechanisms have many advantages such as reduced number of parts, decreased assembly requirements, no friction, no need for lubrication, no backlash and ease of fabrication [5–11]. The proposed compliant backbone is non-extensible with enhanced twisting stiffness, in order to deliver forces along the backbone axis and torques about the backbone axis. The proposed compliant backbone can: (1) be manufactured in a small dimension without assembly requirements, (2) transmit forces and torques about the backbone axis, and (3) prevent too big twisting about the backbone axis. These advantages enable the compliant backbone to be suitable for many special applications. One of the applications is introduced in Sect. 4 of this paper. The compliant backbone is composed of several identical compliant modules. The compliant module is designed and modelled in Sects. 2 and 3, respectively.

2 Design of the Compliant Backbone Module

In this section, a new compliant backbone module is proposed. The new compliant backbone module is a compliant mechanism which can transmit the rotary torques and linear forces about the compliant module axis. Therefore, the compliant module is required to have four degrees of freedom (DOF). In a coordinate frame O-XYZ, the DOF of the compliant backbone module can be defined as T_x, T_z, R_x and R_z, which represent the translations and rotations about the X-and Z-axes, respectively. The linear motion T_y and rotary motion R_y along the Y-axis are constrained.

The new compliant backbone module is designed based on the compliant mechanism proposed in [12], as shown in Fig. 2. The compliant mechanism is consist of two compliant strip beams and one rigid linkage. In the O-XYZ coordinate frame, the compliant mechanism has five DOF, T_x, T_y, T_z, R_x and R_z. The compliant mechanism can be employed to transmit a single rotary motion about the Y-axis, but cannot be used to transmit the linear motion and force along the Y-axis.

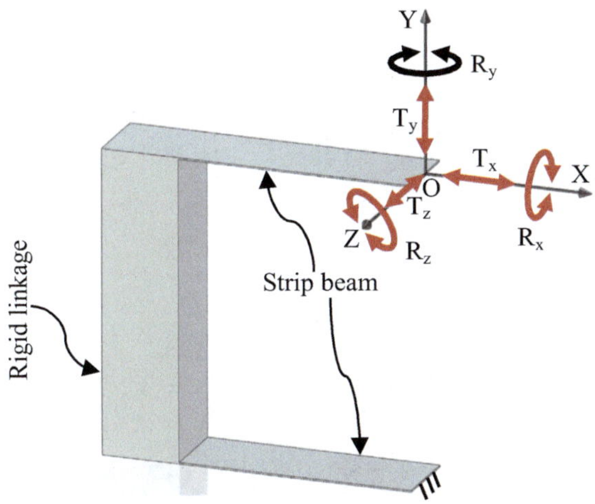

Fig. 2. A compliant mechanism with T_x-, T_y-, T_z-, R_x- and R_z-DOF

In order to enable the compliant mechanism shown in Fig. 2 to have the linear motion DOC (degree of constraint) along the Y-axis, some other compliant elements should be added to it. As well-known, a wire beam can constrain a single linear motion along the wire beam. Taking the wire beam shown in Fig. 3 as an example, the wire beam has five DOF, T_x, T_z, R_x, R_y and R_z in the O-XYZ coordinate frame. If the compliant mechanism in Fig. 2 and the wire beam in Fig. 3 are combine together in parallel, the combined mechanism should have four DOF, T_x, T_z, R_x and R_z, because the T_y-DOF of the compliant mechanism in Fig. 2 is removed by the wire beam. Therefore, the combined compliant mechanism, shown in Fig. 4, is the desired compliant backbone module. The combined compliant mechanism is defined as the basic compliant backbone module in this paper.

Based on the basic compliant backbone module, symmetric compliant backbone modules can be derived by combining several copies of the basic compliant backbone module in parallel. Figure 5 shows a symmetric compliant backbone module synthesized by four copies of the basic compliant backbone module. The symmetric design makes the compliant backbone module have better motion performance. If one hopes to minimize the manufacture difficulty of the symmetric compliant backbone module, two of the basic compliant backbone module can be removed from the symmetric compliant backbone module in Fig. 5. Actually, the amount of the copies of the basic compliant backbone module in a symmetric compliant backbone module can be modified according to different design requirements and applications. For example, a symmetric compliant backbone module is designed using three copies of the basic compliant backbone module in Sect. 4.

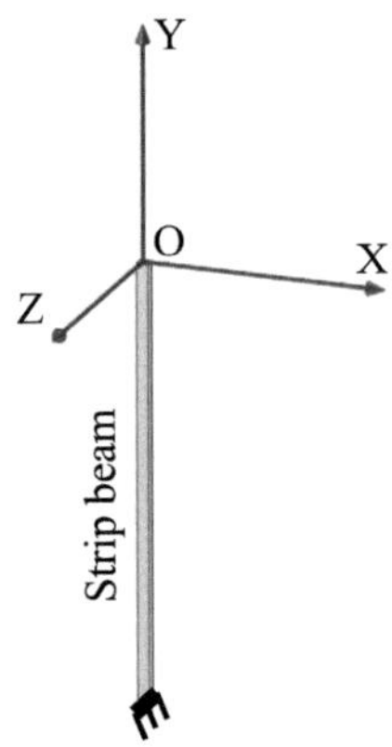

Fig. 3. A wire beam with T_X-, T_Z-, R_X-, R_Y- and R_Z-DOF

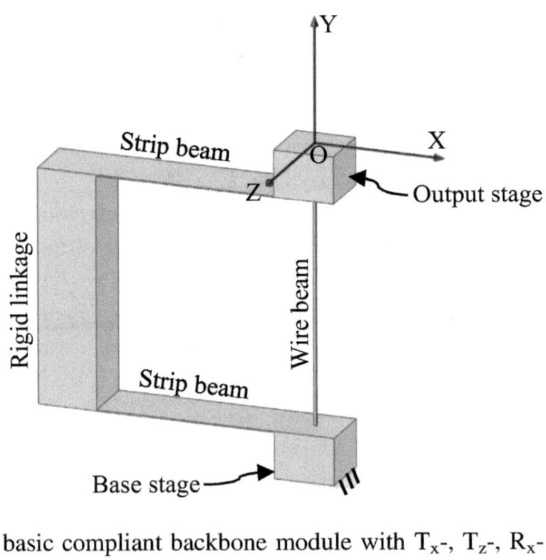

Fig. 4. The basic compliant backbone module with T_X-, T_Z-, R_X- and R_Z-DOF

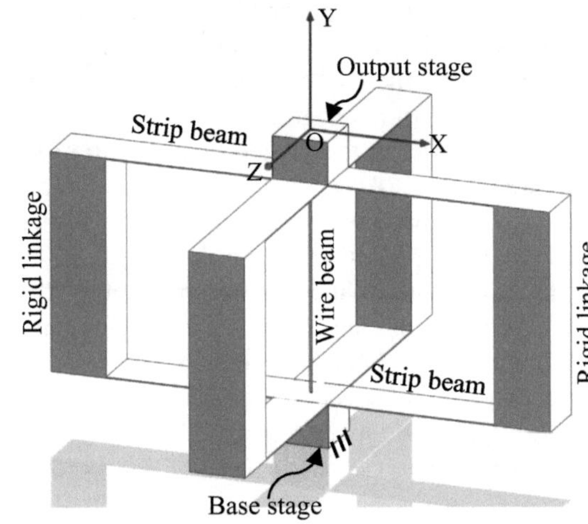

Fig. 5. A symmetric compliant backbone module with four copies of the basic compliant backbone module

3 Modelling of the Basic Compliant Backbone Module

The basic compliant backbone module, shown in Fig. 4, is modelled analytically in this section. The geometrical parameters of the basic compliant backbone module can be determined based on the analytical model to modify the stiffness along and about the three axes of the coordinate frame.

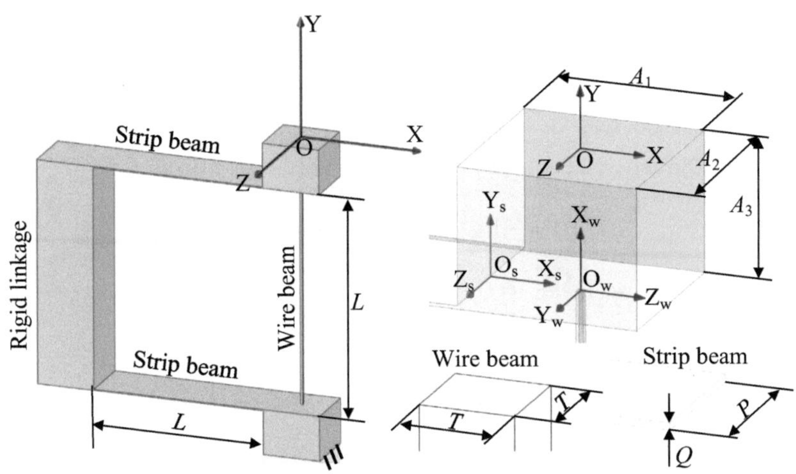

Fig. 6. Geometric dimensions of the basic compliant backbone module

For the analytical modelling in this paper, the geometrical dimension parameters are normalized by the strip/wire beam length L (in this paper, the strip beam and wire beam of the compliant backbone module are equal), force parameters normalized by $E'I/L^2$, and moment parameters normalized by $E'I/L$ [13, 14]. $E' = E/(1 - \mu^2)$, here E denotes Young's modulus and μ represents the Poisson ratio of the material. All the normalized parameters are written as the lowercase letters, compared with the capital letters of their non-normalized parameters. The geometric dimensions of the basic compliant backbone module are defined as shown in Fig. 6.

The basic compliant backbone module is a combination of the compliant mechanism shown in Fig. 2 and the wire beam shown in Fig. 3. In order to derive an analytical model of the basic compliant backbone module, the stiffness matrices of these two compliant mechanisms are needed. The stiffness matrix, $\mathbf{K}_{SS}$, of the compliant mechanism shown in Fig. 2 has been modelled in [12], and the result can be written as Eq. (1) in the O_s-$X_sY_sZ_s$ coordinate frame (the O_s-$X_sY_sZ_s$ coordinate frame is defined in Fig. 6).

$$\mathbf{K}_{SS} = \left\{ \left[\mathbf{T}_s \mathbf{K}_s (\mathbf{T}_s)^{T} \right]^{-1} + (\mathbf{K}_s)^{-1} \right\}^{-1} \tag{1}$$

where $\mathbf{K}_s$ is the stiffness matrix of the upper strip beam of the compliant mechanism in the O_s-$X_sY_sZ_s$ coordinate frame, and $\mathbf{T}_s$ is the transformation matrix used to transform the stiffness matrix of the bottom strip beam to the O_s-$X_sY_sZ_s$ coordinate frame. $\mathbf{K}_s$ and $\mathbf{T}_s$ are written in Expressions (2) and (3).

$$\mathbf{K}_s = \begin{bmatrix} 12\lambda & 0 & 0 & 0 & 0 & 0 \\ 0 & 12 & 0 & 0 & 0 & -6 \\ 0 & 0 & 12\gamma & 0 & 6\gamma & 0 \\ 0 & 0 & 0 & 2(1-\mu) & 0 & 0 \\ 0 & 0 & 6\gamma & 0 & 4\gamma & 0 \\ 0 & -6 & 0 & 0 & 0 & 4 \end{bmatrix} \tag{2}$$

Where $\lambda = (1 - \mu^2)(L/Q)^2$ and $\gamma = (1 - \mu^2)(P/Q)^2$, here P and Q are geometric dimensions of the cross section of the strip beam, as shown in Fig. 6.

$$\mathbf{T}_s = \begin{bmatrix} -1 & 0 & 0 & 0 & 0 & 0 \\ 0 & 1 & 0 & 0 & 0 & 0 \\ 0 & 0 & -1 & 0 & 0 & 0 \\ 0 & 0 & 1 & -1 & 0 & 0 \\ 0 & 0 & -1 & 0 & 1 & 0 \\ -1 & -1 & 0 & 0 & 0 & -1 \end{bmatrix} \tag{3}$$

The stiffness matrix, $\mathbf{K}_W$, of the wire beam shown in Fig. 6 can be written as Expression (4) in the O_W-$X_WY_WZ_W$ coordinate frame.

$$\mathbf{K}_{\mathrm{w}} = \begin{bmatrix} \sigma & 0 & 0 & 0 & 0 & 0 \\ 0 & 12 & 0 & 0 & 0 & -6 \\ 0 & 0 & 12 & 0 & 6 & 0 \\ 0 & 0 & 0 & \eta & 0 & 0 \\ 0 & 0 & 6 & 0 & 4 & 0 \\ 0 & -6 & 0 & 0 & 0 & 4 \end{bmatrix} \tag{4}$$

where $\sigma = 12/(T/L)^2$ and $\eta = 1/(1 + \mu)$.

The transformation matrix from the coordinate frames O_s-$X_sY_sZ_s$ and O_w-$X_wY_wZ_w$ to the coordinate frame O-XYZ can be represented by $\mathbf{T}_{s\text{-}o}$ and $\mathbf{T}_{w\text{-}o}$, so the two transformation matrices can be derived using the expressions below.

$$\mathbf{T}_{\mathrm{s-o}} = \begin{bmatrix} \mathbf{R}_{\mathrm{s}} & \mathbf{0}_{3\times3} \\ \mathbf{D}_{\mathrm{s}} \times \mathbf{R}_{\mathrm{s}} & \mathbf{R}_{\mathrm{s}} \end{bmatrix} \tag{5}$$

$$\mathbf{T}_{\mathrm{w-o}} = \begin{bmatrix} \mathbf{R}_{\mathrm{w}} & \mathbf{0}_{3\times3} \\ \mathbf{D}_{\mathrm{w}} \times \mathbf{R}_{\mathrm{w}} & \mathbf{R}_{\mathrm{w}} \end{bmatrix} \tag{6}$$

where $\mathbf{R}_s$ and $\mathbf{R}_w$ are 3 × 3 rotation matrix from the coordinate frames O_s-$X_sY_sZ_s$ and O_w-$X_wY_wZ_w$ to the coordinate frame O-XYZ, $\mathbf{0}_{3\times3}$ is a null matrix, and $\mathbf{D}_s$ and $\mathbf{D}_w$ are 3 × 3 location skew-symmetric matrix [15], as shown in Expressions (7) and (8).

$$\mathbf{D}_{\mathrm{s}} = \begin{bmatrix} 0 & -d_{\mathrm{sz}} & d_{\mathrm{sy}} \\ d_{\mathrm{sz}} & 0 & -d_{\mathrm{sx}} \\ -d_{\mathrm{sy}} & d_{\mathrm{sx}} & 0 \end{bmatrix} \tag{7}$$

$$\mathbf{D}_{\mathrm{w}} = \begin{bmatrix} 0 & -d_{\mathrm{wz}} & d_{\mathrm{wy}} \\ d_{\mathrm{wz}} & 0 & -d_{\mathrm{wx}} \\ -d_{\mathrm{wy}} & d_{\mathrm{wx}} & 0 \end{bmatrix} \tag{8}$$

where the entries d_{sx}, d_{sy} and d_{sz} are the coordinates of the origin of the coordinate frame O_s-$X_sY_sZ_s$ in the coordinate frame O-XYZ; the entries d_{wx}, d_{wy} and d_{wz} are the coordinates of the origin of the coordinate frame O_w-$X_wY_wZ_w$ in the coordinate frame O-XYZ. Therefore, the stiffness matrix, $\mathbf{K}_{basic}$, of the whole mechanism (in Fig. 4) can be obtained, as shown Eq. (9). When $A_1/L = A_2/L = P/L = \rho$, Eq. (9) can be rewritten as Eq. (10).

$$\mathbf{K}_{\mathrm{basic}} = \mathbf{T}_{\mathrm{s-o}}\mathbf{K}_{\mathrm{ss}}[\mathbf{T}_{\mathrm{s-o}}]^{\mathrm{T}} + \mathbf{T}_{\mathrm{w-o}}\mathbf{K}_{\mathrm{w}}[\mathbf{T}_{\mathrm{w-o}}]^{\mathrm{T}} \tag{9}$$

$$\mathbf{K}_{\mathrm{basic}} = \begin{bmatrix} \frac{6\lambda}{3\lambda+1} + 12 & 0 & 0 & 0 & 0 & a_1 \\ 0 & \sigma + 6 & 0 & 0 & 0 & -6\rho - 3 \\ 0 & 0 & a_2 & a_3 & a_4 & 0 \\ 0 & 0 & a_3 & a_5 & a_6 & 0 \\ 0 & 0 & a_4 & a_6 & a_7 & 0 \\ a_1 & -6\rho - 3 & 0 & 0 & 0 & a_8 \end{bmatrix} \tag{10}$$

where

$$a_1 = \frac{3(7\lambda + 2)(2\rho + 1)}{3\lambda + 1}, \quad a_2 = 12 - \frac{12\gamma(\mu - 1)}{3\gamma - 2\mu + 2},$$
$$a_3 = \frac{6(\gamma(\mu - 4) + 2(\mu - 1))(2\rho + 1)}{3\gamma - 2\mu + 2}, \quad a_4 = -\frac{6\gamma(\mu - 1)(2\rho + 1)}{3\gamma - 2\mu + 2},$$
$$a_5 = -\frac{2(-\mu^2 + 6(2\rho^2 + 2\rho + 1)\mu - 12\rho^2 - 12\rho + 3\gamma(2\mu\rho^2 - 8\rho^2 + 2\mu\rho - 8\rho + \mu - 3) - 5)}{3\gamma - 2\mu + 2},$$
$$a_6 = \frac{3\gamma(\mu - 1)(2\rho + 1)^2}{3\gamma - 2\mu + 2}, \quad a_7 = \frac{3\gamma^2 + (6\eta - 8(\mu - 1)(3\rho^2 + 3\rho + 1))\gamma - 4\eta(\mu - 1)}{6\gamma - 4\mu + 4},$$
$$a_8 = \frac{3(4(3\rho^2 + 3\rho + 1) + \lambda(40\rho^2 + 40\rho + 13))}{6\lambda + 2}.$$

The derived stiffness matrix of the basic compliant backbone module is applicable for any geometrical dimensions and materials. For a case study in this section, let L = 20 mm, T = 2 mm, P = 15 mm, Q = 0.5 mm, $A_1 = A_2$ = 15 mm, A_3 = 5 mm, E = 69 GPa and μ = 0.3. The stiffness matrix of the basic compliant backbone module is calculated using Eq. (10), as shown below.

$$\begin{bmatrix} 14.00 & 0 & 0 & 0 & 0 & 17.50 \\ 0 & \langle 1206 \rangle & 0 & 0 & 0 & -\frac{15}{2} \\ 0 & 0 & 14.80 & -18.50 & 3.50 & 0 \\ 0 & 0 & -18.50 & 24.82 & -4.37 & 0 \\ 0 & 0 & 3.50 & -4.37 & \langle \mathbf{414.64} \rangle & 0 \\ 17.50 & -\frac{15}{2} & 0 & 0 & 0 & 32.75 \end{bmatrix}$$

It can be seen from the entries of the stiffness matrix that the translational stiffness and rotational stiffness about the Y-axis are much higher than the others, which indicates the compliant backbone module has two DOC along and about the Y-axes. Using the similar modelling procedure as above, the symmetric compliant backbone modules shown in Fig. 5 can also be modelled analytically. The force-displacement relationships of the basic compliant backbone module along and about the X-, Y- and Z-axis are also obtained using COMSOL simulations, as shown in Fig. 7. The FEA simulation results shows that the stiffness along and about the Y-axis are much larger than the others, which further improves the two directions along and about the Y-axis are the DOC directions of the compliant backbone module.

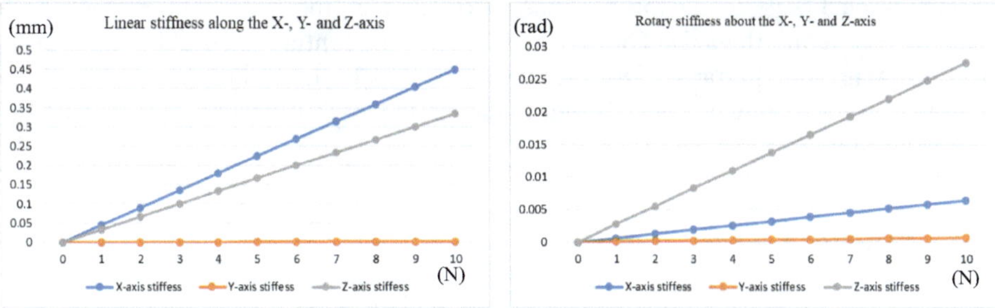

Fig. 7. Force-displacement relationships of the basic compliant backbone module

4 An Application Example of the Compliant Backbone Module

The proposed symmetric compliant backbone modules can be employed to build continuous backbone manipulators and robots, such as the one shown in Fig. 8. The slender continuous backbone shown in Fig. 8 is assembled by several symmetric compliant backbone modules in series. The symmetric compliant backbone module used here has three copies of the basic compliant backbone module. Besides, an end disk is added as an output platform of the slender continuous backbone.

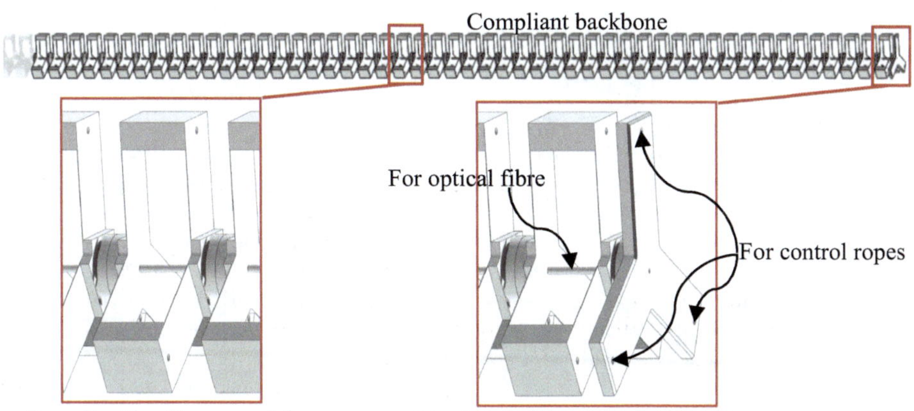

Fig. 8. An application example of the symmetric compliant backbone module, which can be used to build a fiber optic endoscope

The slender continuous backbone can be used to build a fiber optic endoscope. The fiber optic endoscope is pliable and maneuverable because of the specific flexibility of the symmetric compliant backbone module, which allows the fiber optic endoscope to access to channels in human bodies for inspection or therapy. The DOC of the slender continuous backbone along the backbone axis enables the fiber optic endoscope to transmit force along the axis. The DOC of the slender continuous backbone about the backbone axis can: (1) transmit torques about the axis to the end disk, and (2) prevent the twisting of the optical fiber, in order to reduce the loss of the signals transmitted by the optical fiber.

5 Conclusions

Continuous backbone manipulators and robots have been experiencing a rapid development because of their advantages. The key part of a continuous backbone manipulator or robot is the backbone. This paper proposes a compliant mechanism based backbone, and each unit of the backbone is defined as a compliant backbone module. In this paper, a basic compliant backbone module is designed and modelled. Symmetric compliant backbone modules are also introduced based on the basic compliant backbone module. The compliant backbone can be employed in many applications, and one application example is presented in this paper. The detailed study of the compliant backbone will be carried out in the future work.

Acknowledgement. The present study is financially supported by the Fundamental Research Funds for the Central Universities of China (No. N180304019) and the National Natural Science Foundation of China (No. 51575091). Besides, one of the authors, Haiyang Li, would also like to express his greatest gratitude to Northeastern University for awarding a Start-up Research Fund to support this work.

References

1. Liljebäck, P., Pettersen, K.Y., Stavdahl, Ø., Gravdahl, J.T.: A review on modelling, implementation, and control of snake robots. Robot. Auton. Syst. **60**(1), 29–40 (2012)
2. Walker, I.D.: Continuous backbone "continuum" robot manipulators. ISRN Robot. **2013**, 19 (2013)
3. Simaan, N., Taylor, R., Flint, P.: A dexterous system for laryngeal surgery. In: The 2004 IEEE International Conference on Robotics and Automation, pp. 351–357 (2004)
4. Dong, X., Axinte, D., Palmer, D., Cobos, S., Raffles, M., Rabani, A., Kell, J.: Development of a slender continuum robotic system for on-wing inspection/repair of gas turbine engines. Robot. Comput. Integr. Manuf. **44**, 218–229 (2017)
5. Jones, R.V.: Some uses of elasticity in instrument design. J. Sci. Instrum. **39**(5), 193 (1962)
6. Jones, R.V.: Instruments and Experiences: Papers on Measurement and Instrument Design. Wiley Series in Measurement Science and Technology, vol. 1 (1988)
7. Erdman, A.G., Sandor, G.N.: Mechanism Design (3rd edn.): Analysis and Synthesis, vol. 1. Prentice-Hall, Inc., New Jersey (1997)
8. Smith, S.T.: Flexures: Elements of Elastic Mechanisms. Gordon and Breach Science Publishers, New York (2000)
9. Howell, L.L.: Compliant Mechanisms. Wiley, New York (2001)
10. Lobontiu, N.: Compliant Mechanisms: Design of Flexure Hinges. CRC Press, Boca Raton (2002)
11. Howell, L.L.: Handbook of Compliant Mechanisms. Wiley, New York (2013)

12. Li, H., Li, S., Gong, Y., Li, Y., Hao, G., Zhao, Y.: Design of compliant mechanisms with single-rotation constraint. In: the ASME 2019 International Design Engineering Technical Conferences and Computers and Information in Engineering Conference, Anaheim, CA, USA (2019)
13. Awtar, S., Slocum, A.H., Sevincer, E.: Characteristics of beam-based flexure modules. J. Mech. Des. **129**(6), 625–639 (2006)
14. Hao, G., Kong, X.: A normalization-based approach to the mobility analysis of spatial compliant multi-beam modules. Mech. Mach. Theory **59**, 1–19 (2013)
15. Li, H., Hao, G.: A constraint and position identification (CPI) approach for the synthesis of decoupled spatial translational compliant parallel manipulators. Mech. Mach. Theory **90**, 59–83 (2015)

Modelling and Prediction of Lifetime of Metal Rubber Isolator Under Random Vibration

Hongrui Ao[1(✉)], Yong Ma[1,2], Xianbiao Wang[1], and Hongyuan Jiang[1]

[1] School of Mechatronics Engineering,
Harbin Institute of Technology, Harbin, China
hongrui_ao@hit.edu.cn
[2] The System Design Institute of Mechanical-Electrical Engineering,
CASIC, Beijing, China

Abstract. Through fatigue experiments of ring-like metal rubber (MR) isolator under random vibration load, the effects of parameters, including relative preload of metal rubber isolator, wire diameter and relative density, on isolator's lifetime are investigated. The models to describe the relationship between the lifetime and these parameters are proposed to predict the lifetime of MR isolators. An orthogonal experimental design is proceeded to evaluate the contribution of these parameters to the fatigue life of isolators, and the coefficients and functions related to the multi-parameter model are achieved experimentally. The conclusions derived from the experiments provide a new way for designer to predict the working lifetime for the MR isolators in application.

Keywords: Metal rubber isolator · Random load · Lifetime influencing factors · Lifetime prediction

1 Introduction

The entangled metallic wire material, also called Metal Rubber (MR), is a kind of porous material which has been studied extensively and used in a variety of applications such as sealing, filtering, vibration absorbers, and heat transfer equipment because of its characteristics of elasticity, adaptive structure, damping properties, and active vibration control. It exhibits good corrosion resistance, adaptation to the operational environment, especially to the harsh environment, such as low and high temperature, low vacuum, and radiative situation.

The application of MR isolator used in aerospace have drawn wide attention over the last decades. MR elements are used in the isolator as damping support [1], adsorbing vibration energy by internal hysteresis damping mechanism caused by the internal dry friction at the contact surface between adjacent metal wires [2]. Therefore, the mechanical properties of MR elements depend strongly on the metal wire material and manufacturing process. Various materials, such as titanium, have been used in the manufacturing and test of MR [3]. The optimization of materials and manufacturing parameters is of great benefit to improve the lifetime, reliability and performance of aerospace products. Up to now, the research on vibration reduction performance of MR includes the hysteresis modelling, the research on mechanical properties, the study on

D. Wang et al. (Eds.): MeTrApp 2019, MMS 79, pp. 376–384, 2020.
https://doi.org/10.1007/978-981-15-0142-5_37

energy dissipation mechanism [4] and so on. Li et al. [5] has studied the influence of low and high temperature environment on the fatigue characteristics of MR materials. The results show that the MR components have obvious work hardening after cold pressing by molds. And appropriate heat and surface treatment processes can improve the fatigue resistance of MR. The study of Wang et al. [6] indicated a cumulative damage process of MR material, and categorized its fatigue damage into three forms, that is, partial rupture, breakage and wear of the mental wires inside the MR components. This study demonstrates the performance for the bearing capacity and energy dissipation performance attenuation on the macro. Zhao et al. [7] has found that with the increase of the density of MR (material of 65 Mn), the stiffness increases while the static and dynamic loss factors decrease under small displacement, which leads to the reduction of its fatigue life. All these researches are helpful for qualitative understanding of fatigue damage factors of MR materials. However, they cannot predict the life of MR in quantify.

In this study, an isolator with ring-like MR damping elements is investigated. The factors affecting the life of MR under random vibration are studied. A life prediction model of MR isolator is proposed. These results may be beneficial to the application of MR isolators under random vibration with consideration of working lifetime.

2 Metal Rubber Specimen and Experimental System

2.1 Experimental Principle

Considering the effect of influencing factors on the lifetime of MR under random vibration, the control variable method is employed in the study. These factors include the diameter of metal wire, the relative density of MR elements, and the pre-tightening of MR damping components in vibration isolator.

Up to now, the studies on the fatigue damage of MR material are mainly focused on the condition of simple harmonic vibration. Actually, these isolators are working mostly under random vibration. The internal stress of MR component is generated under the action of vibration load when MR isolators are in operational condition.

In this study, the accelerated life test is adopted in which the internal stress level of the MR components is changed gradually. However, when MR material is loaded, it shows inherent hysteresis caused by the internal dry friction between metal coil wires in a spatial network structure. These characteristics make the stress calculation of MR more complicated than general continuous matrix materials. Therefore, factors such as self-power spectral density, vibration isolation mass and shapes of components are taken into account in the calculating.

The stress caused by pre-compression during installation depends on the relative pre-tightening and relative density of the component, expressed as [8] (the wire diameter sets as 0.1 mm):

$$\sigma_S \approx (0.613 + 1.771\varepsilon_Q - 20.4\varepsilon_Q^2 + 160\varepsilon_Q^3)\varepsilon_Q(\frac{\bar{\rho}}{0.18})^{1.7} \quad (1)$$

where $\bar{\rho} = \rho_{MR}/\rho_S$ is the relative density, ρ_{MR} is the density of the MR component and ρ_S is the density of the metal wire.

Influencing Function. Considering the influence of wire diameter d_W, the influence function $f(d_W)$ is introduced, that is,

$$\sigma_S = f(\bar{\rho}, \varepsilon_Q) \cdot f(d_w)$$

In order to obtain the specific form of the above function relationship, a series of static experiments are designed. The parameters of MR standard samples are as follows: relative density $\bar{\rho} = 0.2$, shape size is 25 mm × 25 mm × 25 mm, wire diameter d_W is 0.10 mm, 0.15 mm, 0.20 mm and 0.25 mm, respectively. The MR specimen and the static hysteresis loop are shown in Fig. 1.

a. The MR sample;

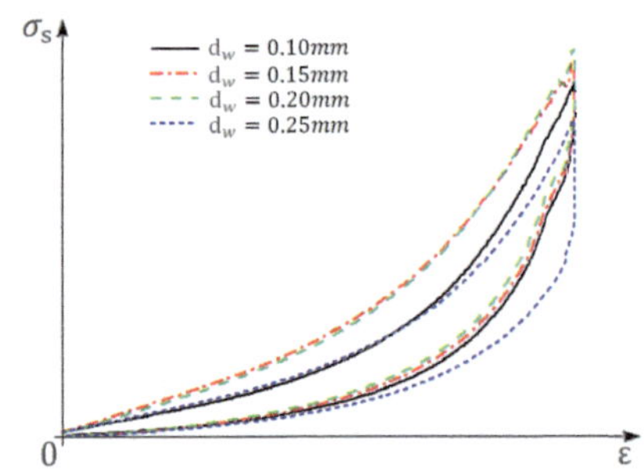

b. The hysteresis loop of different metal wires.

Fig. 1. The MR specimen and the hysteretic loops.

The loops shown in Fig. 1 are obtained when MR specimen is loaded in normal direction, which is similar to the direction of pressure forming of MR samples.

The set of hysteresis loops shows that the stiffness of each spiral coil inside the MR component increases with the increase of wire diameter. While the increase of wire diameter leads to the decrease of the wire contact points number in unit volume and the decrease of sample stiffness when the relative density of the MR component remains unchanged. According to this family of loops, an expression for the diameter influence function under this density can be obtained as:

$$f(d_w) = (-1152d_w^3 + 555.7d_w^2 - 124d_w + 18)d_w \tag{2}$$

and the calculation formula of the stress is:

$$\begin{aligned} \sigma_S \approx\ & (0.613 + 1.771\varepsilon_Q - 20.4\varepsilon_Q^2 + 160\varepsilon_Q^3)\varepsilon_Q\left(\frac{\bar{\rho}}{0.18}\right)^{1.7} \\ & \times (-1152d_w^3 + 555.7d_w^2 - 124d_w + 18)d_w \end{aligned} \tag{3}$$

Dynamic Stress. Random vibration can't be described by accurate mathematical relationship. Usually, the probability theory and statistical method are used in this kind of study [9]. If a random variable x is subject to a normal distribution of mathematical expectation μ and variance σ^2, it is denoted as $N(\mu, \sigma^2)$. The normal distribution exists the principle of 3σ, and the probability of numerical distribution in $(\mu - 3\sigma, \mu + 3\sigma)$ is 99.7%. In the fatigue experiment under random vibration, the upper limit frequency of σ in the statistical sense of random vibration is usually chosen. According to the Miles formula for solving the acceleration response by random vibration,

$$a = \sqrt{\frac{\pi A f_0 \eta_0}{2}} \tag{4}$$

where a is the resonance acceleration (g), A is the acceleration spectral density (g^2/Hz), f_0 is the natural frequency (Hz), and η_0 is the resonance transfer rate defined as $\eta_0 = a_{\text{out}}/a_{\text{in}}$ in which a_{out} and a_{in} is the output and input acceleration, respectively.

Then, the dynamic stress σ_d is

$$\sigma_d = \frac{ma}{S} = \frac{m}{S}\sqrt{\frac{\pi A f_0 \eta_0}{2}} \tag{5}$$

where S is the cross-section area (mm^2) of the circular MR elements, expressed as $S = \pi(D_1^2 - D_2^2)/4$, and m is the mass of isolation system (kg).

The internal stress σ of MR components under random vibration includes stress σ_S caused by pre-deformation and dynamic stress σ_d.

2.2 Metal Rubber Specimen

In this study, an isolator with two symmetric ring-shaped MR is tested. Its structural shapes and installation methods are shown in Fig. 2. The axis direction of the MR component is the direction of compression forming, which is the direction of applying random loads. The outer diameter of the MR component D_1 = 16 mm and the inner diameter D_2 = 6.5 mm. The effects of internal initial stress on the damping properties of MR elements are evaluated as the relative preliminary deformation ε_Q, defined as $\varepsilon_Q = (H - H')/H \times 100\%$, where H is the height of MR in free state, and H' is the height of MR components in vibration isolator after its installation.

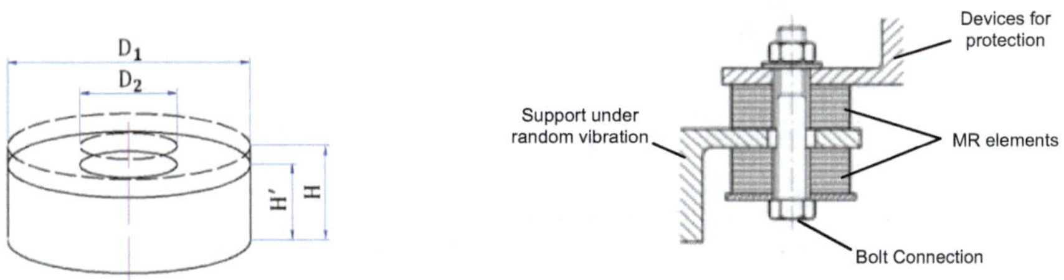

a. Pre-deformation of MR components. b. Installation of MR vibration isolator.

Fig. 2. MR isolator and its installation in the test.

Table 1 is the experimental parameters of MR components.

Table 1. Parameters of MR elements.

Diameter of metal wire d_{w}/mm	Relative density of component $\bar{\rho}$	Height of MR rings in free state H/mm
0.10	0.20	5
0.10	0.20	5
0.10	0.20	5
0.10	0.24	5
0.10	0.29	5
0.15	0.20	5
0.20	0.20	5

2.3 Experiment

The schematic diagram of the vibration test system is shown in Fig. 3. Vibration controller controls power amplifier though the driving signal to drive vibro-stand to obtain specific vibration. Sensors measure vibration of vibro-stand and feedback signal to vibration controller to form a closed-loop control system.

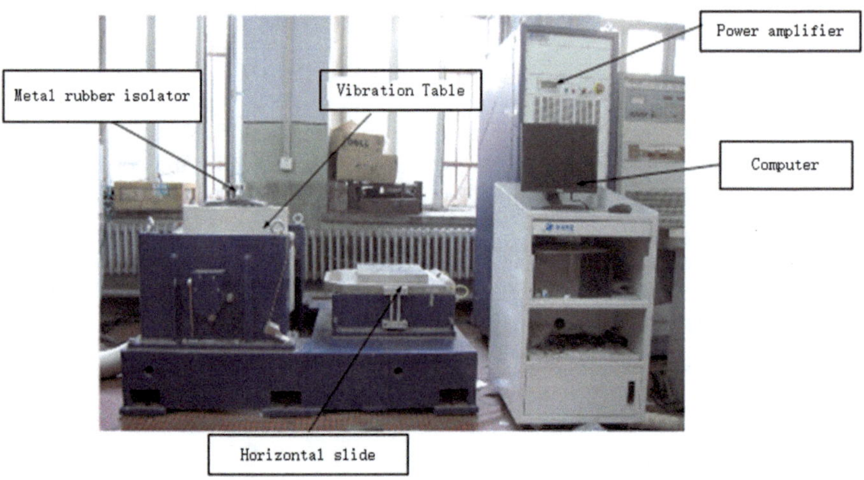

Fig. 3. The principle diagram of the experimental system.

After stabilizing the vibration of the MR isolator, the resonance frequency, f_0, and the resonance transfer rate, η_0, are obtained by a sinusoidal sweep experiment. In the study, a fatigue threshold of MR elements is set to the moment when the resonance frequency or resonance transfer rate deviates from its initial value by 20% after a period of time. That is, the relative variation of transfer rate of the isolation system is the determinate parameter for evaluating the lifetime of isolator under random vibration. The acceleration spectral density, A, can be obtained through Eq. (5) as

$$A = \frac{2\sigma_d^2 S^2}{\pi m^2 f_0 \eta_0}$$

The desired target spectrum in the test is shown in Fig. 4.

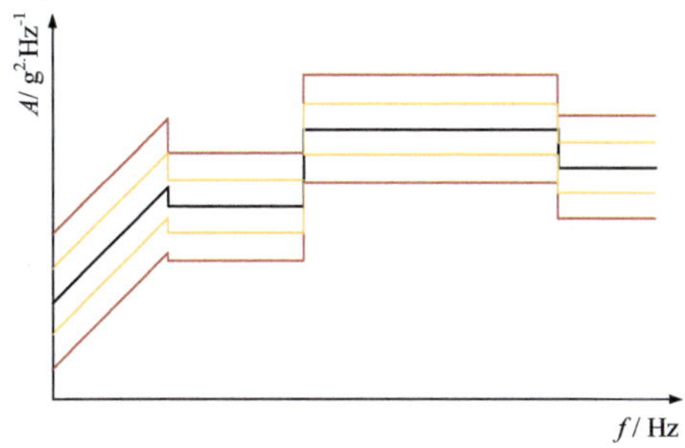

Fig. 4. The target spectrum used in the test.

In random vibration experiment, the resonance frequency and resonance transfer rate are measured by sinusoidal sweep frequency by intervals. Therefore, the invalidation of MR components is judged, and the corresponding time T (h) is recorded.

3 Experimental Results and Analysis

During the experiment, the effective working time T of MR components under random vibration load is obtained by controlling the stress level. The experimental results are shown in Fig. 5. The points in the curves represent the average values to ensure the repeatability and the credibility of the data.

As a results, the relationship between isolator lifetime and stress for a single factor show a trend which can be fitted as decaying power exponential functions based on the experimental data. Figure 6 shows a fitted curve for the description of dependency of relative deformation of MR elements on its lifetime, in which the constants A_1, t_1 and y_0 can be determined experimentally. In this study, A_1 = 0.31, t_1 = 520, y_0 = 0.22 for ε_Q = 6.4%, d_W = 0.10 mm, and $\bar{\rho}$ = 0.2. Thus, a single factor model with consideration of is obtained. In the same way, the single factor model with consideration of different wire diameters and relative densities can be obtained. The life impact coefficient under the action of influencing factors is shown in Table 2.

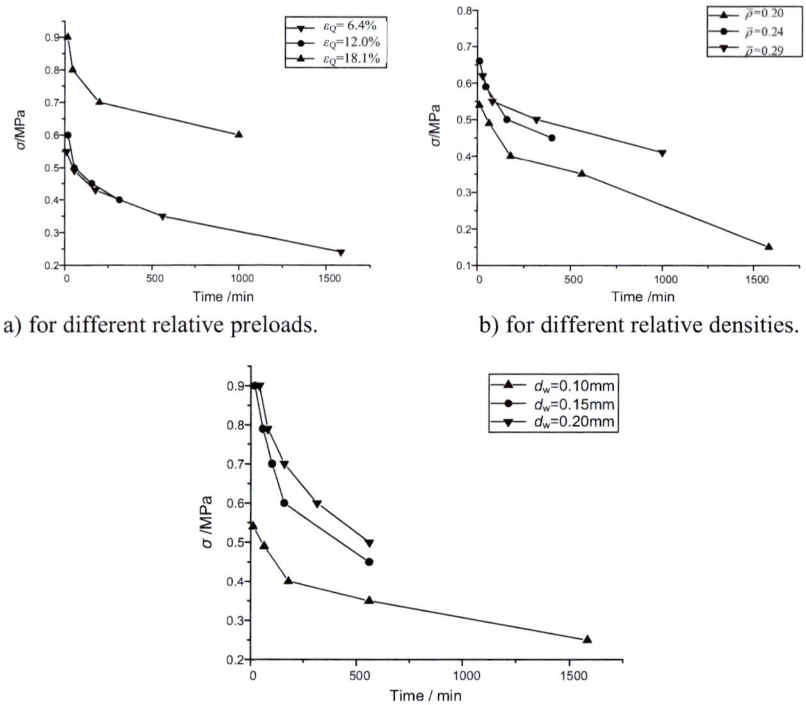

a) for different relative preloads.

b) for different relative densities.

c) for different wire diameters.

Fig. 5. Life-time curves with different wire diameter.

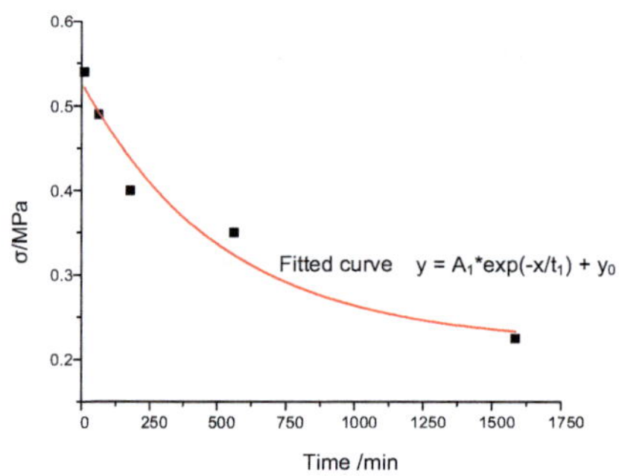

Fig. 6. Fitted curve for the description of fatigue life of MR isolator

Table 2. Coefficient values with different influence factors.

Coefficient	d_w/mm			$\bar{\rho}$			ε_Q/%		
	0.1	0.15	0.2	0.20	0.24	0.29	6.4	12.0	18.1
α_1	1	1.99	2.02	1	1.14	1.12	1	1.10	1.59
α_2	1	2.69	2.46	1	0.98	0.91	1	1.15	1.38

The above coefficients are fitted into functions according to different influencing factors. The single influencing factor of lifetime T can be obtained as: Based on the $\sigma - T$ life curve of the MR isolator with the parameter of $d_w = 0.1\,\text{mm}$, $\bar{\rho} = 0.2$, $\varepsilon_Q = 6.38\%$, the life prediction model of the MR is established. The life prediction model is set as:

$$\alpha = 0.69\alpha_1(d_W)\alpha_1(\bar{\rho})\alpha_1(\varepsilon_Q) - 0.13\alpha_2(d_W)\alpha_2(\bar{\rho})\alpha_2(\varepsilon_Q)\log T \quad (6)$$

The calculation processes are not elaborated any further for the simplification. Based on these single factor models, the multi-factors model can be derived by using the method of establishing coefficient function. According to the coefficient function for different influencing factors, the multi-factor life prediction formula is:

$$T = 10 \wedge \left[\frac{0.69\alpha_1(d_W)\alpha_1(\bar{\rho})\alpha_1(\varepsilon_Q) - \sigma}{0.13\alpha_2(d_W)\alpha_2(\bar{\rho})\alpha_2(\varepsilon_Q)}\right] \quad (7)$$

The experimental results show that the single factor life prediction formula is in good agreement with the experimental data and has high accuracy.

4 Conclusion

Metal Rubber (MR) material exhibits tunable vibration absorption properties under random dynamic vibration. Its fatigue life under random excitation is dependent on parameters such as wire diameter, relative density and pre-loading (relative deformation after installation inside the isolator). For the life prediction of MR isolator, a multi-factor formula is proposed with consideration of these three parameters listed before, and the coefficients and functions related to this model are obtained experimentally. The experimental results are supportive to the design of MR isolator and the further application under random vibration condition.

References

1. Bezborodov, S.A., Ulanov, A.M.: Calculation of vibration of pipeline bundle with damping support made of MR Material. In: Proceedings of the 3rd International Conference on Dynamics and Vibroacoustics of Machines, DVM 2016, vol. 176, pp. 169–174 (2017)
2. Li, Y.-L., Bai, H.-B., He, Z.-B., Lu, C.-H.: Experimental research on nonlinear metal rubber isolator under different incentives. Trans. Beijing Inst. Technol. **35**(03), 221–225 and 250 (2015)
3. He, G., Liu, P., Tan, Q.: Porous titanium materials with entangled wire structure for load-bearing biomedical applications. J. Mech. Behav. Biomed. Mater. **5**(1), 16–31 (2012)
4. Zou, G., Zhang, B., Chang, Z., Liu, S.: Hysteresis mechanical model and experimental study of spring metal-net rubber combination damper. Chin. J. Theor. Appl. Mech. **50**(5), 1125–1134 (2018)
5. Li, T., Bai, H., Xue, X., Wu, Y.: Fatigue properties of knitted-dapped metal rubbers under high temperature environment. China Mech. Eng. **30**(09), 1009–1017 (2019)

6. Wang, Y., Bai, H., Hou, J.: Fatigue damage performance of metal rubber material. J. Mech. Eng. **47**(02), 65–71 (2011)
7. Zhao, C., Luo, K., Liu, C.: Study on fatigue behavior of 65Mn MR made of thin wires. Noise Vibr. Control **01**, 106–111 (2010)
8. Ulanov, A.M., Bezborodov, S.A.: Calculation method of pipeline vibration with damping supports made of the MR material. In: Procedia of 2nd International Conference on Industrial Engineering, ICIE 2016, vol. 150, pp. 101–106 (2016)
9. Gonca, V., Polukoshko, S.: Buckling stability of multilayered rubber-metal vibration isolator. Vibroengineering Procedia **3**, 319–325 (2014)

A Recursive Integral-Based Dynamic Stiffness Matrix for Notch Flexure Hinges Used in Compliant Mechanisms

Mingxiang Ling(✉)

Institute of Systems Engineering, China Academy of Engineering Physics, Mianyang 621999, China
ling_mx@163.com

Abstract. Accurate dynamic analysis of flexure-hinge-based compliant mechanisms are difficult due to their complicated governing differential equations. Given that concern, this paper presents an exact dynamic stiffness matrix for free vibration analysis of notch flexure hinges based on a recursive integral without approximation. Starting from the longitudinal and transverse governing differential equations of vibration, modal functions of arbitrary notch flexure hinges are derived using a recursive integral in power series of angular frequency but not the traditional way in the well-known Frobenious method; then, the exact dynamic stiffness matrix in a general form is built. The proposed method is applied to solve the natural frequencies of flexure-hinge-based compliant mechanisms with high-precision solutions.

Keywords: Compliant mechanisms · Flexure hinges · Dynamic stiffness matrix

1 Introduction

Extensive researches have been carried out and are still receiving great attentions with regard to the dynamic analysis of the state-of-the-art compliant mechanisms, which are increasingly applied for precision positioning, micro-vibration suppression and precision gripping owing to their excellent merits of high precision and fast dynamic response without wear, friction, backlash and need no assembly [1, 2].

However, accurate dynamic analysis of these notch flexure hinge based compliant mechanisms is difficult due to their complex and even non-solvable governing equations. It appears attractive to analytically determining their vibration characteristics; meanwhile, analytical solutions are advantageous to draw general and insightful conclusions about the vibration behavior and can also be used to examine the accuracy of numerical solutions. These merits motivate the new methodology addressed in this paper: to provide a general solution for free vibration analysis of arbitrary notch flexure hinges based on the dynamic stiffness method with high accuracy.

The dynamic stiffness method for simultaneous kinetostatic and dynamic formulating of compliant mechanisms in [3–5] is based on D'Alember's principle. As a consequence, the dynamic modeling of compliant mechanisms can be simplified as a

D. Wang et al. (Eds.): MeTrApp 2019, MMS 79, pp. 385–394, 2020.
https://doi.org/10.1007/978-981-15-0142-5_38

static-similar problem without cumbersome inner-force and kinematic analyses as well as mass lumping and kinematics approximation. The dynamic stiffness method enables one accurately analyzes an infinite number of natural frequencies and vibration modes without subdividing elements [6].

The basis of the dynamic stiffness modeling method for compliant mechanisms is to build the dynamic stiffness matrix of all kinds of flexure members. Therefore, the contribution of this paper is to establish the general dynamic stiffness matrix for all kinds of notch flexure hinges in compliant mechanisms (in Sect. 2), so as to make it competent for all kinds of compliant mechanisms with notch flexure hinges. Then, a guiding flexure manipulator is employed as the case study to illustrate the high accuracy of the model (in Sect. 3). Finally, conclusions are drawn in Sect. 4.

2 Dynamic Stiffness Model for Notch Flexure Hinges

A general notch flexure hinge with arbitrary variable cross section is shown in Fig. 1 with the nodal displacement and nodal force at the two ends of the beam: axial displacements u_j and u_k; transverse deflections w_j and w_k and rotations θ_j and θ_k. A Cartesian coordinate system is defined on the beam with the origin located at node j, x-axis is aligned with the central line of the cross-section, y and z axes are coincident with the principal axes of the left-end cross-section.

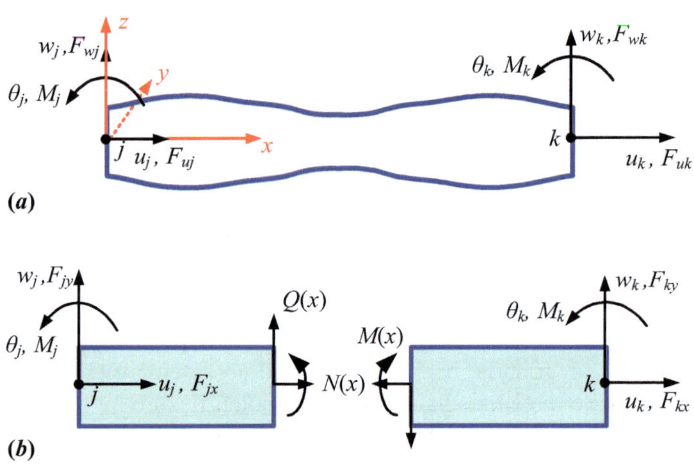

Fig. 1. Force and displacement definitions of general notch flexure hinges with arbitrary variable cross section. (a) Nodal displacement and nodal force. (b) Sign convention for positive axial, shear forces and bending moment.

The free vibration of longitudinal and transverse motions for general notch flexure hinges based on Euler-Bernoulli beam are respectively governed by the following differential equations

$$\frac{\partial}{\partial x}\left[EA(x)\frac{\partial u(x,t)}{\partial x}\right] - \rho A(x)\frac{\partial^2 u(x,t)}{\partial t^2} = 0 \tag{1}$$

$$\frac{\partial^2}{\partial x^2}\left[EI(x)\frac{\partial^2 w(x,t)}{\partial x^2}\right] + \rho A(x)\frac{\partial^2 w(x,t)}{\partial t^2} = 0 \tag{2}$$

where x is the axial coordinate, t is the time, E, ρ, $A(x)$ and $I(x)$ are the Young's modulus, density, cross-sectional area and moment of inertia varying along axial coordinate x.

Assuming a solution of the forms $u(x, t) = U_0(x)e^{i\omega t}$ and $w(x, t) = W_0(x)e^{i\omega t}$, Eqs. (1) and (2) reduce to the ordinary differential equations in the frequency domain for the modal functions $U_0(x)$ and $W_0(x)$ as

$$\frac{\mathrm{d}}{\mathrm{d}x}\left[EA(x)\frac{\mathrm{d}U_0(x)}{\mathrm{d}x}\right] + \rho A(x)\omega^2 U_0(x) = 0 \tag{3}$$

$$\frac{\mathrm{d}^2}{\mathrm{d}x^2}\left[EI(x)\frac{\mathrm{d}^2 W_0(x,t)}{\mathrm{d}x^2}\right] - \rho A(x)\omega^2 W_0(x) = 0 \tag{4}$$

where ω is the angular frequency, rad/s.

Before deriving the solutions of Eqs. (3) and (4), they are firstly transformed into non-dimensional forms by introducing the following dimensionless coefficients

$$\eta = \frac{x}{L}, U = \frac{U_0}{L}, W = \frac{W_0}{L}, \quad \alpha^2 = \frac{\rho L^2}{E}\omega^2, \beta^4 = \frac{\rho A_0 L^4}{EI_0}\omega^2,$$

$$f_1(\eta) = \frac{A(\eta L)}{A_0}, f_2(\eta) = \frac{I(\eta L)}{I_0} \tag{5}$$

where L is the length of flexure hinges. A_0 and I_0 are respectively the constant area and moment of inertia at the left end of the flexure hinges, as shown in Fig. 1.

Then, the non-dimensional governing differential equations can be expressed as

$$\frac{\mathrm{d}}{\mathrm{d}\eta}\left[f_1(\eta)\frac{\mathrm{d}U}{\mathrm{d}\eta}\right] + \alpha^2 f_1(\eta)U = 0 \tag{6}$$

$$\frac{\mathrm{d}^2}{\mathrm{d}\eta^2}\left[f_2(\eta)\frac{\mathrm{d}^2 W}{\mathrm{d}\eta^2}\right] - \beta^4 f_1(\eta)W = 0 \tag{7}$$

In the frame of Frobenious method [7], solutions of the vibration governing equations are assumed as the power series of spatial coordinate, such as x, under the conditions that the varying coefficients are polynomials or can be approximated as polynomials. In this paper, the solutions of modal functions $U(\eta)$ and $W(\eta)$ are generally assumed as the power series of angular frequency for general notch flexure hinges without the restriction of polynomial assumption in the Frobenious method, i.e.

$$U(\eta) = \sum_{n=0}^{\infty} U_n(\eta)\alpha^{2n} = U_0(\eta) + U_1(\eta)\alpha^2 + U_2(\eta)\alpha^4 + \cdots \tag{8}$$

$$W(\eta) = \sum_{n=0}^{\infty} W_n(\eta)\beta^{4n} = W_0(\eta) + W_1(\eta)\beta^4 + W_2(\eta)\beta^8 + \cdots \tag{9}$$

where wave numbers $\alpha = \omega L\sqrt{\rho/E}$, $\beta = L \cdot \sqrt{4}\omega^2 \rho A_0/EI_0$.

The logic behind Eqs. (8) and (9) is that linear vibration is the superposition of angular frequencies. And the first terms in Eqs. (8) and (9) are solutions of the static deflection. If coefficients $U_n(\eta)$ and $W_n(\eta)$ are known, vibration solutions of arbitrary non-uniform Euler-Bernoulli beams can be then confirmed. Given that concern, substituting Eqs. (8) and (9) into Eqs. (6) and (7) and collecting the coefficients of each α^{2n} and β^{4n} (n = 0, 1, 2, …) will yield the following recursive equations for $U_n(\eta)$ and $W_n(\eta)$ in a similar way of [8]

$$\begin{cases} \frac{\mathrm{d}}{\mathrm{d}\eta}\left[f_1(\eta)\frac{\mathrm{d}U_0(\eta)}{\mathrm{d}\eta}\right] = 0 \\ \frac{\mathrm{d}}{\mathrm{d}\eta}\left[f_1(\eta)\frac{\mathrm{d}U_n(\eta)}{\mathrm{d}\eta}\right] + f_1(\eta)U_{n-1}(\eta) = 0 (n = 1, 2, 3, \cdots) \end{cases} \tag{10}$$

$$\begin{cases} \frac{\mathrm{d}^2}{\mathrm{d}\eta^2}\left[f_2(\eta)\frac{\mathrm{d}^2 W_0(\eta)}{\mathrm{d}\eta^2}\right] = 0 \\ \frac{\mathrm{d}^2}{\mathrm{d}\eta^2}\left[f_2(\eta)\frac{\mathrm{d}^2 W_n(\eta)}{\mathrm{d}\eta^2}\right] = f_1(\eta)W_{n-1}(\eta) = 0\ (n = 1, 2, 3, \cdots) \end{cases} \tag{11}$$

Given the conditions that $f_1(\eta), f_2(\eta)$ are mathematically integrable, then the integral solutions of $U_n(\eta)$ and $W_n(\eta)$ in Eqs. (10) and (11) read as follows with six integration constants C_i (i = 1, …, 6)

$$\begin{cases} U_0(\eta) = C_1 + C_2\int_0^{\eta}\frac{1}{f_1(s)}ds \\ U_n(\eta) = -\int_0^{\eta}\frac{\int_0^s f_2(v)U_{n-1}(v)dv}{f_1(s)}ds (n = 1, 2, 3, \cdots) \end{cases} \tag{12}$$

$$\begin{cases} W_0(\eta) = C_3 + C_4\eta + \int_0^{\eta}(\eta - s)\frac{C_5 + C_6 s}{f_2(s)}ds \\ W_n(\eta) = \int_0^{\eta}\frac{(\eta - s)\int_0^s (s - v)f_1(v)W_{n-1}(v)dv}{f_2(s)}ds (n = 1, 2, 3, \cdots) \end{cases} \tag{13}$$

By substituting Eqs. (12) and (13) into Eqs. (8) and (9), the modal functions of the longitudinal and transverse vibrations can be expressed as

$$U(\eta) = C_1\bar{U}_1(\eta) + C_2\bar{U}_2(\eta) \tag{14}$$

$$W(\eta) = C_3\bar{W}_1(\eta) + C_4\bar{W}_2(\eta) + C_5\bar{W}_3(\eta) + C_6\bar{W}_4(\eta) \tag{15}$$

In which $\bar{U}_i(\eta)$ $(i = 1, 2)$ and $\bar{W}_i(\eta)$ $(i = 1, 2, 3, 4)$ can be analytically expressed as

$$\begin{cases} \bar{U}_i(\eta) = U_{i0}(\eta) + \sum\limits_{n=1}^{\infty} U_{in}(\eta)\ (i = 1, 2) \\ U_{in}(\eta) = -\alpha^2 \int_0^{\eta} \frac{1}{f_1(s)} \int_0^{s} f_1(v) U_{i(n-1)}(v) dv ds \ \ (n = 1, 2, 3, \cdots) \\ U_{10}(\eta) = 1, U_{20}(\eta) = \int_0^{\eta} \frac{1}{f_1(s)} ds \end{cases} \tag{16}$$

$$\begin{cases} \bar{W}_i(\eta) = W_{i0}(\eta) + \sum\limits_{n=1}^{\infty} W_{in}(\eta)\ \ (i = 1, 2, 3, 4) \\ W_{in}(\eta) = \beta^4 \int_0^{\eta} \frac{(\eta - s) \int_0^s (s-v) f_1(v) W_{i(n-1)}(v) dv}{f_2(s)} ds\ (n = 1, 2, 3, \cdots) \\ W_{10}(\eta) = 1, W_{20}(\eta) = \eta, W_{30}(\eta) = \int_0^{\eta} \frac{(\eta - s)}{f_2(s)} ds, \ \ W_{40}(\eta) = \int_0^{\eta} \frac{(\eta - s)s}{f_2(s)} ds \end{cases} \tag{17}$$

With the recursive integral in Eqs. (16) and (17), modal functions, $U(\eta)$ and $W(\eta)$, of the longitudinal and transverse vibrations can be calculated explicitly in terms of six integration constants C_i $(i = 1, \ldots, 6)$, which can be determined by imposing the boundary conditions. Equations (14) and (15) can be re-written as

$$\begin{Bmatrix} U(\eta) \\ W(\eta) \\ \Theta(\eta) \end{Bmatrix} = \begin{bmatrix} \bar{U}_1(\eta) & 0 & 0 & \bar{U}_2(\eta) & 0 & 0 \\ 0 & \bar{W}_1(\eta) & \bar{W}_2(\eta) & 0 & \bar{W}_3(\eta) & \bar{W}_4(\eta) \\ 0 & \bar{W}_1'(\eta) & \bar{W}_2'(\eta) & 0 & \bar{W}_3'(\eta) & \bar{W}_4'(\eta) \end{bmatrix} \begin{Bmatrix} C_1 \\ C_3 \\ C_4 \\ C_2 \\ C_5 \\ C_6 \end{Bmatrix} \tag{18}$$

where $\Theta(\eta)$ is the amplitude of the rotational angular due to bending, the prime denotes the derivative with respect to spatial coordinate η. $\bar{U}_i(\eta)$, $\bar{W}_i(\eta)$ and $\bar{W}_i'(\eta)$ can be calculated with the recursive integral in Eqs. (16) and (17).

Considering the boundary conditions at $\eta = 0$ and $\eta = 1$, as shown in Fig. 1, the nodal displacement vector can be calculated based on Eq. (18) as

$$\begin{Bmatrix} u_j \\ w_j \\ \theta_j \\ u_k \\ w_k \\ \theta_k \end{Bmatrix} = \begin{Bmatrix} U(0) \\ W(0) \\ \Theta(0) \\ U(1) \\ W(1) \\ \Theta(1) \end{Bmatrix} = \begin{bmatrix} \bar{U}_1(0) & 0 & 0 & \bar{U}_2(0) & 0 & 0 \\ 0 & \bar{W}_1(0) & \bar{W}_2(0) & 0 & \bar{W}_3(0) & \bar{W}_4(0) \\ 0 & \bar{W}_1'(0) & \bar{W}_2'(0) & 0 & \bar{W}_3'(0) & \bar{W}_4'(0) \\ \bar{U}_1(1) & 0 & 0 & \bar{U}_2(1) & 0 & 0 \\ 0 & \bar{W}_1(1) & \bar{W}_2(1) & 0 & \bar{W}_3(1) & \bar{W}_4(1) \\ 0 & \bar{W}_1'(1) & \bar{W}_2'(1) & 0 & \bar{W}_3'(1) & \bar{W}_4'(1) \end{bmatrix} \begin{Bmatrix} C_1 \\ C_3 \\ C_4 \\ C_2 \\ C_5 \\ C_6 \end{Bmatrix} \tag{19}$$

As defined in Fig. 1(b), the axial force, shear force and bending moment at two ends of the beam can be related to the corresponding nodal forces and moments as

$$\{F\} = \begin{Bmatrix} F_{uj} \\ F_{wj} \\ M_j \\ F_{uk} \\ F_{wk} \\ M_k \end{Bmatrix} = \begin{Bmatrix} -F(0) \\ -Q(0) \\ -M(0) \\ +F(1) \\ +Q(1) \\ +M(0) \end{Bmatrix}$$
$$= \begin{Bmatrix} -f_1(0)\dfrac{dU}{d\eta}|_{\eta=0}; \dfrac{d}{d\eta}\left(f_2(\eta)\dfrac{d^2W}{d\eta^2}\right)|_{\eta=0}; -f_2(0)\dfrac{d^2W}{d\eta^2}|_{\eta=0}; \\ f_1(1)\dfrac{dU}{d\eta}|_{\eta=1}; -\dfrac{d}{d\eta}\left(f_2(\eta)\dfrac{d^2W}{d\eta^2}\right)|_{\eta=1}; f_2(1)\dfrac{d^2W}{d\eta^2}|_{\eta=1} \end{Bmatrix} \tag{20}$$

By substituting Eq. (18) into Eq. (20), there has

$$\mathbf{F} = \begin{Bmatrix} F_{uj} \\ F_{wj} \\ M_j \\ F_{uk} \\ F_{wk} \\ M_k \end{Bmatrix} = [\mathbf{F}_1, \quad \mathbf{F}_3, \quad \mathbf{F}_4, \quad \mathbf{F}_2\,, \quad \mathbf{F}_5\,, \quad \mathbf{F}_6] \cdot \begin{Bmatrix} C_1 \\ C_2 \\ C_3 \\ C_4 \\ C_5 \\ C_6 \end{Bmatrix} \tag{21}$$

in which

$$\mathbf{F}_i = \left[-f_1(0)\tfrac{d\bar{U}_i}{d\eta}|_{\eta=0}, \quad 0, \quad 0, \quad f_1(1)\tfrac{d\bar{U}_i}{d\eta}|_{\eta=1}\,, \quad 0, \quad 0\right]^{\mathrm{T}} (i = 1, 2) \tag{22}$$

$$\mathbf{F}_i = \left[0, \frac{d}{d\eta}\left(f_2(\eta)\frac{d^2\bar{W}_{i-2}}{d\eta^2}\right)|_{\eta=0}, -f_2(0)\frac{d^2\bar{W}_{i-2}}{d\eta^2}|_{\eta=0}, 0, -\frac{d}{d\eta}\left(f_2(\eta)\frac{d^2\bar{W}_{i-2}}{d\eta^2}\right)|_{\eta=1}, f_2(1)\frac{d^2\bar{W}_{i-2}}{d\eta^2}|_{\eta=1}\right]^{\mathrm{T}}$$
$$(i = 3, 4, 5, 6) \tag{23}$$

By combining Eq. (19) with Eq. (21) and eliminating integration constants C_i (i = 1, 2, 3, 4), relations of the nodal force and nodal displacement can be expressed as

$$\begin{Bmatrix} F_{uj} \\ F_{wj} \\ M_j \\ F_{uk} \\ F_{wk} \\ M_k \end{Bmatrix} = \mathbf{D}(\omega) \cdot \begin{Bmatrix} u_j \\ w_j \\ \theta_j \\ u_k \\ w_k \\ \theta_k \end{Bmatrix}$$

$$= \left(\begin{bmatrix} \mathbf{F}_1 \\ \mathbf{F}_3 \\ \mathbf{F}_4 \\ \mathbf{F}_2 \\ \mathbf{F}_5 \\ \mathbf{F}_6 \end{bmatrix}^{\mathrm{T}} \cdot \begin{bmatrix} \bar{U}_1(0) & 0 & 0 & \bar{U}_2(0) & 0 & 0 \\ 0 & \bar{W}_1(0) & \bar{W}_2(0) & 0 & \bar{W}_3(0) & \bar{W}_4(0) \\ 0 & \bar{W}'_1(0) & \bar{W}'_2(0) & 0 & \bar{W}'_3(0) & \bar{W}'_4(0) \\ \bar{U}_1(1) & 0 & 0 & \bar{U}_2(1) & 0 & 0 \\ 0 & \bar{W}_1(1) & \bar{W}_2(1) & 0 & \bar{W}_3(1) & \bar{W}_4(1) \\ 0 & \bar{W}'_1(1) & \bar{W}'_2(1) & 0 & \bar{W}'_3(1) & \bar{W}'_4(1) \end{bmatrix}^{-1} \right) \cdot \begin{Bmatrix} u_j \\ w_j \\ \theta_j \\ u_k \\ w_k \\ \theta_k \end{Bmatrix} \quad (24)$$

Based on the dynamic stiffness matrix $\mathbf{D}(\omega)$ and applying specific boundary conditions, the problem of extracting natural frequencies is reduced to an eigenvalue problem with the following determinant of

$$\det(\boldsymbol{D}(\omega)) = 0 \quad (25)$$

3 Verification

Design, analysis and topological optimization for flexure-hinge-based compliant mechanisms have been intensively investigated in recent years, but mainly focused on the kinematic and static issues. In this section, the frequency characteristics of a guiding compliant mechanism will be analytically calculated by the proposed exact dynamic stiffness matrix. The prototype and geometric parameters are illustratively shown in Fig. 2, the exemplary compliant mechanism is a motion guiding mechanism for piezoelectric actuated precision positioning and scanning applications.

The material is selected as steel with a density of 7850 kg/m^3, Young's modulus of 200 GPa, Poisson's ratio of 0.3. The flexure hinge is an elliptic shape which is widely used in compliant mechanisms for its large deformation and small stress concentration. As shown in Fig. 2(a), the semi-minor axis and semi-major axis are respectively 5 mm and 20 mm; the depth of the whole mechanism is 10 mm. The compliant mechanism is equivalent as a non-uniform beam-based frame with four flexure hinges and a lumped mass, as shown in Fig. 2(b), the dynamic stiffness matrix of the whole mechanism can be obtained by the procedure in Ref. [3] with the proposed exactly dynamic stiffness matrix of flexure hinges in the current paper. The dimensionless cross-sectional area and moment of inertia for the elliptic flexure hinge can be expressed as

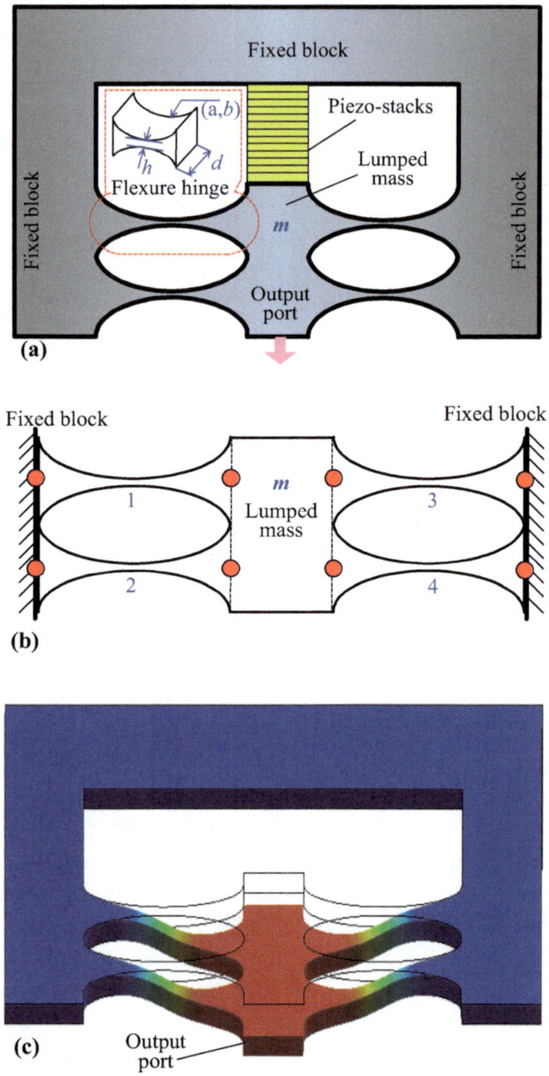

Fig. 2. An exemplary piezoelectric actuated guiding compliant mechanism. (a) The prototype and its geometric parameters. (b) Simplifying the compliant mechanism for modeling. (c) Results of the first vibration mode by ANSYS.

$$f_1(\eta) = 1 - \frac{\sqrt{1 - 4(\eta - 0.5)^2}}{h/2b + 1}, \quad f_2(\eta) = \left(1 - \frac{\sqrt{1 - 4(\eta - 0.5)^2}}{h/2b + 1}\right)^3 \tag{26}$$

For comparison, the natural frequencies of the guiding compliant mechanism were analyzed by the finite element software ANSYS. The Solid45 element was chosen to build the model and the local mesh was refined. The element size was set as 1 mm and the finite elemental results were proven to be convergent and accurate enough. The first natural frequency of one simulation set is illustratively shown in Fig. 2(c). The detailed results for three sets of geometric parameters provided by the proposed method and FEM are listed in Table 1.

Table 1. First two planar natural frequencies of the guiding compliant mechanism with different methods

Ratio: $h/2b$	ω_1			ω_2		
	FEM	Present	Error	FEM	Present	Error
0.1	1196	1147	4.10%	3084	3001	2.69%
0.2	2005	1923	4.08%	9095	8822	3.00%
0.3	2693	2605	3.26%	10470	10006	4.43%

From the calculating results, one can see that the proposed dynamic stiffness matrix matches well with the results of FEM in the whole range of the change of the geometric parameters. The maximum deviation is less than 5% for the three sets of data. The error may come from the mismatching of the geometric parameters between the proposed method and FEM. Besides, the deviation becomes larger with the increase of the ratio of parameters h to b, because the Bernoulli-Euler beam theory was used in the dynamic stiffness matrix, and the prediction accuracy will be enhanced if the shear effects are further considered. It can also be expected that the accuracy will be further improved when the number of terms of the recursive integral is increased, but the calculating time will be accordingly increased. For actual engineering applications, accuracy less than 10% is acceptable, so only the first several terms of the recursive integral could be needed in the proposed method.

4 Conclusions

Determining the vibration characteristics for design and optimization of compliant mechanisms are needed, urgent and interesting. In this paper, we propose a formulation method for the exact dynamic stiffness matrix of arbitrary notch flexure hinges based upon a recursive integral in power series of angular frequency. Comparative calculations of the natural frequencies for a flexure-hinge-based compliant mechanism with the proposed method and FEM demonstrate the feasibility of the proposed method.

References

1. Valentini, P.P., Pennestrì, E.: Compliant four-bar linkage synthesis with second-order flexure hinge approximation. Mech. Mach. Theory **128**, 225–233 (2018)
2. Lofroth, M., Avci, E.: Development of a novel modular compliant gripper for manipulation of micro objects. Micromachines **10**(5), 313 (2019)
3. Ling, M., Howell, L.L., Cao, J., Jiang, Z.: A pseudo-static model for dynamic analysis on frequency domain of distributed compliant mechanisms. J. Mech. Robot. **10**(5), 051011 (2018)
4. Ling, M.X., Cao, J.Y., Zhuang, J., et al.: Design, pseudo-static model and PVDF-based motion sensing of a piezo-actuated XYZ flexure manipulator. IEEE/ASME Trans. Mechatron. **23**(6), 2837–2848 (2018)
5. Ling, M., Cao, J., Pehrson, N.: Kinetostatic and dynamic analyses of planar compliant mechanisms via a two-port dynamic stiffness model. Precis. Eng. **57**, 149–161 (2019)
6. Tang, A.Y., Wu, J.X., Li, X.F., Lee, K.Y.: Exact frequency equations of free vibration of exponentially non-uniform functionally graded Timoshenko beams. Int. J. Mech. Sci. **89**, 1–11 (2014)
7. Attarnejad, R., Semnani, S.J., Shahba, A.: Basic displacement functions for free vibration analysis of non-prismatic Timoshenko beams. Finite Elem. Anal. Des. **46**(10), 916–929 (2010)
8. Guo, S.Q., Yang, S.P.: Transverse vibrations of arbitrary non-uniform beams. Appl. Math. Mech. **35**(5), 607–620 (2014)

Control Application

Passivity-Based Force/Position Active Disturbance Rejection Control of Dual-Arm Space Robot Clamping Capture Spacecraft

Ai Haiping, Chen Li, and Yu Xiaoyan(✉)

Fuzhou University, Fuzhou, China
cool109@163.com

Abstract. The impact effect and post-capture passivity-based force/position active disturbance rejection control of dual-arm free-floating space robot clamping capture non-cooperative spacecraft are studied. First, the dynamic models of the dual-arm space robot and the target spacecraft before capture are obtained by multi rigid body dynamics method. After that, based on the law of conservation of momentum, the constraints of kinematics and the law of force transfer, the impact effect caused by collision of capture process is analyzed. Finally, the integrated dynamic model of the closed chain composite system is derived. On the basis, a force/position active disturbance rejection control scheme based on passivity theory is proposed for post-capture closed chain composite system with consider actuator saturation, joint friction and external disturbance, which can achieve the coordinated control of the internal force and position of the target spacecraft. The method uses the extended state observer to realize the dynamic estimation of the disturbance, then combine with passivity theory to compensate it, which can improve disturbance rejection ability for the impact effect of capture effectively. The numerical simulation results verify the effectiveness of the control scheme.

Keywords: Free-floating space robot · Dual-arm clamping capture · Impact effect · Closed chain composite system · Passivity theory · Force/position active disturbance rejection control

1 Introduction

With the development of human space industry, space missions are increasing day by day. Because of the nature of space environment, it is a better choice to use space robot system to replace astronauts to complete space operation tasks [1–3]. It is noteworthy that many of the existing spacecraft in space have been abandoned for reasons such as not entering the orbit at launch or fuel exhaustion, if they are recovered, the spacecraft will still be able to work [4, 5]. Considering the preciousness of space resources and the economic effect of spacecraft recovery, the capture capability of space robot has important research value. Because the capture operation requires greater load capacity, higher load positioning accuracy, and more flexible multi-tasking collaboration capabilities, dual-arm collaborative space robot systems have become a new research trend. At present, most of the capture operation tasks of dual-arm and multi-arm space robots

D. Wang et al. (Eds.): MeTrApp 2019, MMS 79, pp. 397–408, 2020.
https://doi.org/10.1007/978-981-15-0142-5_39

take cooperative targets into consideration. Because cooperative targets have grasping hands that are convenient for grasping capture, while for non-cooperative targets, grasping capture operations are difficult to carry out, so more suitable clamping capture operations can be considered [6]. Due to the closed chain hybrid system are formed during the dual-arm grasping capture, the control should not only consider the position of the system, but also consider the coordination control of the internal force, which increases the difficulty of control.

Because of the weightless environment of space, it has strong dynamic coupling between the base and the components of the closed chain hybrid system, which makes it difficult to apply the mature ground control algorithm directly to the space robot system [7, 8]. In addition, the space robot system actuator will produce saturation phenomenon for various physical reasons. Once this saturation occurs, it will affect the output torque and affect the performance of the arm. At the same time, consider the existence of multiple complex conditions such as joint friction and external disturbance, then makes it more difficult to control the space robot. However, the existing researches mostly consider the trajectory planning of pre-capture and attitude control of post-capture, but do not consider the coordination control of position and internal force for space robot on-orbit grasping capture non-cooperative spacecraft.

Considering the influence of nonlinear disturbance term on the clamping capture control accuracy, it is necessary to design an algorithm to compensate it. In recent years, active disturbance rejection control methods have achieved good performance in compensating system nonlinear disturbances [9]. This method uses the extended state observer (ESO) to estimate the disturbance, and then improves the anti-disturbance ability of the system by adding the disturbance compensation term in the controller. Aiming at the internal force and position control of closed-chain hybrid system with actuator saturation, joint friction and disturbance, an active disturbance rejection controller for stabilize motion based on passivity theory is designed. The method divides the controller into two parts: disturbance compensation part and passive controller part. The ESO is used to estimate the total disturbance of the system, on the basis, the disturbance compensation part is designed. At the same time, the internal force and position stabilization control is designed by combining the passivity theory, which can achieve the coordination of internal force and position control [10, 12]. Finally, numerical simulations were carried out, which showed that the proposed control scheme can overcome the influence of nonlinear disturbance and achieve desired accurately.

2 Dynamic Model and Collision Impact Analysis of Space Robot Capture a Spacecraft

The structure of a dual-arm space robot consisting of a rigid base B_0, six rigid links $B_i(i = 1, 2, \ldots, 6)$ is shown in Fig. 1. The links were connected in serial and are actuated by individual actuators. Take point O as the origin, on the basis, the inertial coordinate frame XOY of system is established. At the same time, the local co-ordinate frame $x_iO_iy_i(i = 1, 2, \ldots, 6)$ of B_i is established. m_i, I_i, $l_i(i = 1, 2, \ldots, 6)$ is the mass and center inertial moment and the length of each link. m_0 is the mass of base, I_0 is the center inertial moment of base.

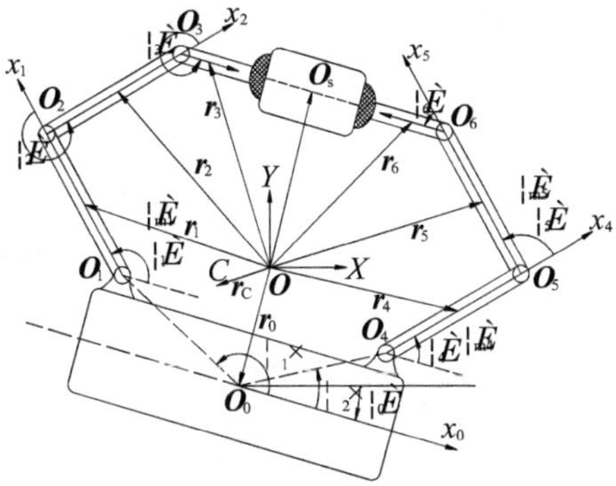

Fig. 1. Dual-arm Free-floating space robot system and target spacecraft system

Neglect the microgravity, and with Lagrange Equation, the pre-impact dynamic equations of dual-arm space robot are

$$\boldsymbol{D}(\boldsymbol{q})\ddot{\boldsymbol{q}} + H(\boldsymbol{q}, \dot{\boldsymbol{q}})\dot{\boldsymbol{q}} = [\boldsymbol{0}_{1\times 2} \quad \boldsymbol{\tau}^{\mathrm{T}}]^{\mathrm{T}} + \boldsymbol{J}^{\mathrm{T}}\boldsymbol{F} \tag{1}$$

where $\boldsymbol{q} = [x_0 \quad y_0 \quad \theta_0 \quad \boldsymbol{\theta}_L^T \quad \boldsymbol{\theta}_R^T]^T$; $\boldsymbol{D}(\boldsymbol{q}) \in \boldsymbol{R}^{9\times 9}$ is symmetric and positive inertia matrix; $\boldsymbol{H}(\boldsymbol{q}, \dot{\boldsymbol{q}})\dot{\boldsymbol{q}} \in \boldsymbol{R}^{9\times 1}$ is the Coriolis/centrifugal matrix; $\boldsymbol{\tau} = [\tau_0 \quad \boldsymbol{\tau}_L^T \quad \boldsymbol{\tau}_R^T]^T$ is the generalized control torque; $\boldsymbol{J} \in \boldsymbol{R}^{6\times 9}$ is the corresponding Jacobian matrix of the robot end-effector; $\boldsymbol{F} \in \boldsymbol{R}^{6\times 1}$ is the impact forces acting on the manipulators.

For the target spacecraft system, the dynamic equation is obtained by Newton-Euler method

$$\boldsymbol{D}_t\ddot{\boldsymbol{q}}_t = \boldsymbol{J}_t^{\mathrm{T}}\boldsymbol{F}' \tag{2}$$

where $\boldsymbol{q}_t = [x_t \quad y_t \quad \theta_t]^T$; $\boldsymbol{D}_t \in \boldsymbol{R}^{3\times 3}$ is symmetric and positive inertia matrix of the target spacecraft; $\boldsymbol{J}_t \in \boldsymbol{R}^{6\times 3}$ is the corresponding Jacobian matrix of the target spacecraft; $\boldsymbol{F}'$ is for counter-acting forces. According to Newton's third law, it satisfies $\boldsymbol{F} = -\boldsymbol{F}'$. Impact forces can be decomposed as follows

$$\boldsymbol{F}' = (\boldsymbol{J}_t^{\mathrm{T}})^+\boldsymbol{D}_t\ddot{\boldsymbol{q}}_t + \boldsymbol{F}_I \tag{3}$$

where $(\boldsymbol{J}_t^{\mathrm{T}})^+$ is the Moore-Penrose pseudo-inverse of $\boldsymbol{J}_t^{\mathrm{T}}$; $(\boldsymbol{J}_t^{\mathrm{T}})^+\boldsymbol{D}_t\ddot{\boldsymbol{q}}_t$ is the operating force of motion; $\boldsymbol{F}_I$ is internal force on the target, and satisfies $\boldsymbol{J}_t^{\mathrm{T}}\boldsymbol{F}_I = 0$.

Base on Newton's third law, and invoking (1), (2), (3), we can obtain that

$$\boldsymbol{D}(\boldsymbol{q})\ddot{\boldsymbol{q}}+\boldsymbol{H}(\boldsymbol{q},\dot{\boldsymbol{q}})\dot{\boldsymbol{q}}=[0_{1\times 2}\quad \boldsymbol{\tau}^{\mathrm{T}}]^{\mathrm{T}}-\boldsymbol{J}^{\mathrm{T}}(\boldsymbol{J}_t^{\mathrm{T}})^{+}\boldsymbol{D}_t\ddot{\boldsymbol{q}}_t-\boldsymbol{J}^{\mathrm{T}}\boldsymbol{F}_I \tag{4}$$

Integrating (4) over the momentary period of collision, according to the theorem of impulse, Eq. (4) can be simplified to:

$$\boldsymbol{D}(\boldsymbol{q})(\dot{\boldsymbol{q}}(t_0+\Delta t)-\dot{\boldsymbol{q}}(t_0))+\boldsymbol{J}^{\mathrm{T}}(\boldsymbol{J}_t^{\mathrm{T}})^{+}\boldsymbol{D}_t(\dot{\boldsymbol{q}}_t(t_0+\Delta t)-\dot{\boldsymbol{q}}_t(t_0))=0 \tag{5}$$

Consider the velocity constraint relations of post-capture closed chain composite system, we have

$$\boldsymbol{J}\dot{\boldsymbol{q}}(t_0+\Delta t)=\boldsymbol{J}_t\dot{\boldsymbol{q}}_{\mathbf{t}}(t_0+\Delta t) \tag{6}$$

Combined with (5), (6), the influence of collision impact can be solved

$$\dot{\boldsymbol{q}}(t_0+\Delta t)=\boldsymbol{N}^{-1}[\boldsymbol{D}(\boldsymbol{q})\dot{\boldsymbol{q}}(t_0)+\boldsymbol{J}^{\mathrm{T}}(\boldsymbol{J}_t^{\mathrm{T}})^{+}\boldsymbol{D}_t\dot{\boldsymbol{q}}_{\mathbf{t}}(t_0)] \tag{7}$$

where $\boldsymbol{N}=\boldsymbol{D}(\boldsymbol{q})+\boldsymbol{J}^{\mathrm{T}}(\boldsymbol{J}_t^{\mathrm{T}})^{+}\boldsymbol{D}_t\boldsymbol{J}_t^{+}\boldsymbol{J}$.

Consider the velocity constraint relations of post-capture closed chain composite system is always existed, it proves that Eq. (6) always holds. Differentiating (6), we can obtain that

$$\ddot{\boldsymbol{q}}_t=(\boldsymbol{J}_t)^{+}[\boldsymbol{J}\ddot{\boldsymbol{q}}+(\dot{\boldsymbol{J}}-\dot{\boldsymbol{J}}_t\boldsymbol{J}_t^{+}\boldsymbol{J})\dot{\boldsymbol{q}}] \tag{8}$$

Combined with (4), (8), we have

$$\boldsymbol{D}_c(\boldsymbol{q})\ddot{\boldsymbol{q}}+\boldsymbol{H}_c(\boldsymbol{q},\dot{\boldsymbol{q}})\dot{\boldsymbol{q}}=[\boldsymbol{0}_{1\times 2}\quad \boldsymbol{\tau}^{T}]^{T}-\boldsymbol{J}^{T}\boldsymbol{F}_I \tag{9}$$

where

$\boldsymbol{D}_{\mathbf{c}}(\boldsymbol{q})=\boldsymbol{D}(\boldsymbol{q})+\boldsymbol{J}^{\mathrm{T}}(\boldsymbol{J}_t^{\mathrm{T}})^{+}\boldsymbol{D}_t\boldsymbol{J}_t^{+}\boldsymbol{J};\boldsymbol{H}_{\mathbf{c}}(\boldsymbol{q},\dot{\boldsymbol{q}})=\boldsymbol{H}(\boldsymbol{q},\dot{\boldsymbol{q}})+\boldsymbol{J}^{\mathrm{T}}(\boldsymbol{J}_t^{\mathrm{T}})^{+}\boldsymbol{D}_t\boldsymbol{J}_t^{+}(\dot{\boldsymbol{J}}-\dot{\boldsymbol{J}}_t\boldsymbol{J}_t^{+}\boldsymbol{J})$.

The (10) is underactuated dynamic equation, which is not conducive to the control design. In order to convert Eq. (9) into a fully driven dynamic equation, it is written as follows

$$\begin{bmatrix}\boldsymbol{D}_{c11} & \boldsymbol{D}_{c12}\\ \boldsymbol{D}_{c21} & \boldsymbol{D}_{c22}\end{bmatrix}\begin{bmatrix}\ddot{\boldsymbol{q}}_0\\ \ddot{\boldsymbol{\theta}}_m\end{bmatrix}+\begin{bmatrix}\boldsymbol{H}_{c11} & \boldsymbol{H}_{c12}\\ \boldsymbol{H}_{c12} & \boldsymbol{H}_{c22}\end{bmatrix}\begin{bmatrix}\dot{\boldsymbol{q}}_0\\ \dot{\boldsymbol{\theta}}_m\end{bmatrix}=\begin{bmatrix}\boldsymbol{0}_{2\times 1}\\ \boldsymbol{\tau}\end{bmatrix}-\begin{bmatrix}\boldsymbol{J}_1^{\mathrm{T}}\\ \boldsymbol{J}_2^{\mathrm{T}}\end{bmatrix}\boldsymbol{F}_I \tag{10}$$

The dynamic equations of fully drive form for dual-arm g space robot with closed chain are derived by algebraic operation

$$\boldsymbol{D}_Z\ddot{\boldsymbol{\theta}}_m+\boldsymbol{H}_Z\dot{\boldsymbol{\theta}}_m=\boldsymbol{\tau}-\boldsymbol{J}_Z\boldsymbol{F}_I \tag{11}$$

Multiply both left sides $\boldsymbol{J}_t^{\mathrm{T}}\boldsymbol{J}_Z^{+}$ of Eq. (11), it can get

$$\boldsymbol{D}_a\ddot{\boldsymbol{\theta}}_m + \boldsymbol{H}_a\dot{\boldsymbol{\theta}}_m = \boldsymbol{\tau}_X \tag{12}$$

Define Lyapunov function $V_0 = \frac{1}{2}\dot{\boldsymbol{\theta}}_m\boldsymbol{D}_a\dot{\boldsymbol{\theta}}_m$. Reference [12] shows the system satisfies the traditional passive control theory, so the input $\boldsymbol{\tau}_X$ to output $\boldsymbol{\theta}_m$ is passive.

3 Passivity-Based Force/Position Active Disturbance Rejection Control

It is assumed that the joint actuators have saturated nonlinear properties. If the system control input falls outside the upper and lower limits of the system, the output signal cannot be fully executed, so there is an unexecuted part $\boldsymbol{\Delta}$, it can be expressed as

$$\boldsymbol{\Delta}_i = u_i - \tau_i = \begin{cases} u_{\max} - \tau_i,\ \tau_i \geq u_{\max}; \\ 0,\ u_{\min} < \tau_i < u_{\max}; \\ \tau_i - u_{\min},\ \tau_i \leq u_{\max}; \end{cases} \tag{13}$$

Considering the process of capturing the target spacecraft, it will lead to the liquid fuel sloshing, so there is a bounded disturbance term in the system, at the same time, there are disturbance terms such as friction torque. Therefore, the dynamic model of closed chain system with actuator saturation, a bounded disturbance and joint friction torque is considered

$$\boldsymbol{D}_Z\ddot{\boldsymbol{\theta}}_m + \boldsymbol{H}_Z\dot{\boldsymbol{\theta}}_m + \boldsymbol{d} = \boldsymbol{u} - \boldsymbol{J}_Z\boldsymbol{F}_I \tag{14}$$

where $\boldsymbol{d} = \boldsymbol{\tau}_d + [0 \quad \boldsymbol{w}_{\mathrm{f}}^{\mathrm{T}}]^{\mathrm{T}} + \boldsymbol{\Delta}$ is the total disturbance term of the closed chain system; $\boldsymbol{\tau}_d \in R^{7\times1}$ is the bounded disturbance term; $\boldsymbol{w}_{\mathrm{f}} \in \boldsymbol{R}^{6\times1}$ is the joint friction torque.

Denoting $\boldsymbol{x}_1 = \boldsymbol{\theta}_m$, $\boldsymbol{x}_2 = \dot{\boldsymbol{\theta}}_m$, $\boldsymbol{x}_3 = -\boldsymbol{D}_Z^{-1}\boldsymbol{d}$, Eq. (16) can be written to

$$\begin{cases} \dot{\boldsymbol{x}}_1 = \boldsymbol{x}_2 \\ \dot{\boldsymbol{x}}_2 = -\boldsymbol{A}\boldsymbol{x}_2 + \boldsymbol{B} \end{cases} \tag{15}$$

where $\boldsymbol{A} = \boldsymbol{D}_Z^{-1}\boldsymbol{H}_Z$; $\boldsymbol{B} = \boldsymbol{D}_Z^{-1}(\boldsymbol{u} - \boldsymbol{J}_Z\boldsymbol{F}_I) + \boldsymbol{x}_3$.

For the total disturbance term, an ESO is used to estimate it dynamically. The ESO is designed as follows

$$\begin{cases} \dot{\hat{\boldsymbol{x}}}_1 = \hat{\boldsymbol{x}}_2 + \boldsymbol{\lambda}_1\tilde{\boldsymbol{x}}_1 \\ \dot{\hat{\boldsymbol{x}}}_2 = \hat{\boldsymbol{x}}_3 + \boldsymbol{D}_Z^{-1}(\boldsymbol{u} - \boldsymbol{J}_Z\boldsymbol{F}_I - \boldsymbol{H}_Z\boldsymbol{x}_2) + \boldsymbol{\lambda}_2\tilde{\boldsymbol{x}}_1 \\ \dot{\hat{\boldsymbol{x}}}_3 = \boldsymbol{\lambda}_3\tilde{\boldsymbol{x}}_1 \end{cases} \tag{16}$$

where $\hat{\boldsymbol{x}}_1$, $\hat{\boldsymbol{x}}_2$, $\hat{\boldsymbol{x}}_3$ are estimated state of $\boldsymbol{x}_1$, $\boldsymbol{x}_2$, $\boldsymbol{x}_3$. Define the estimated error $\tilde{\boldsymbol{x}}_1 = \boldsymbol{x}_1 - \hat{\boldsymbol{x}}_1$, $\tilde{\boldsymbol{x}}_2 = \boldsymbol{x}_2 - \hat{\boldsymbol{x}}_2$, $\tilde{\boldsymbol{x}}_3 = \boldsymbol{x}_3 - \hat{\boldsymbol{x}}_3$. $\boldsymbol{\lambda}_1$, $\boldsymbol{\lambda}_2$, $\boldsymbol{\lambda}_3$ is gain of the ESO, and the selection of its value is bandwidth-based.

$$\boldsymbol{\lambda}_1 = 3\gamma \boldsymbol{I}_7,\ \boldsymbol{\lambda}_2 = 3\gamma^2 \boldsymbol{I}_7,\ \boldsymbol{\lambda}_3 = \gamma^3 \boldsymbol{I}_7 \tag{17}$$

where $\gamma > 0$ is the bandwidth of the ESO, $\boldsymbol{I}_7$ is unitary matrix.

Defining the output error as follows

$$\boldsymbol{e} = \boldsymbol{\theta}_m - \boldsymbol{\theta}_{md} \tag{18}$$

where $\boldsymbol{\theta}_{md}$ is bounded and continuous desired trajectory.

Introduce the following auxiliary signals

$$\dot{\boldsymbol{\theta}}_{mr} = \dot{\boldsymbol{\theta}}_{md} - \boldsymbol{\Lambda}_1 \boldsymbol{e},\ \dot{\boldsymbol{\theta}}_{mo} = \hat{\boldsymbol{x}}_2 - \boldsymbol{\Lambda}_2 \tilde{\boldsymbol{x}}_1 \tag{19}$$

where $\boldsymbol{\Lambda}_1 \in R^{7\times 7}$, $\boldsymbol{\Lambda}_2 \in R^{7\times 7}$ are positive definite and symmetric constant matrix.

On the basis of the above, a force/position active disturbance rejection control scheme based on passivity theory is proposed for post-capture closed chain composite system

$$\boldsymbol{u} = \boldsymbol{D}_Z \ddot{\boldsymbol{\theta}}_{mr} + \boldsymbol{H}_Z \dot{\boldsymbol{\theta}}_{mr} - \boldsymbol{K}_d(\dot{\boldsymbol{\theta}}_{mo} - \dot{\boldsymbol{\theta}}_{mr}) - \boldsymbol{D}_Z \hat{\boldsymbol{x}}_3 + \boldsymbol{J}_Z \boldsymbol{F}_{Id} \tag{20}$$

where $u_0 = -\boldsymbol{D}_Z \hat{\boldsymbol{x}}_3$ is the designed dynamic disturbance compensation term. $\boldsymbol{K}_d \in \boldsymbol{R}^{7\times 7}$ is the positive definite feedback gain matrix. $\boldsymbol{F}_{Id}$ is the expected value of the internal force $\boldsymbol{F}_I$.

Substitute Eq. (20) into Eq. (14), we have

$$\boldsymbol{D}_Z \dot{\boldsymbol{\eta}}_1 + \boldsymbol{H}_Z \boldsymbol{\eta}_1 + \boldsymbol{K}_d \boldsymbol{\eta}_1 = \boldsymbol{K}_d \boldsymbol{\eta}_2 + \boldsymbol{D}_Z \tilde{\boldsymbol{x}}_3 + \boldsymbol{J}_Z(\boldsymbol{F}_{Id} - \boldsymbol{F}_I) \tag{21}$$

where $\boldsymbol{\eta}_1 = \dot{\boldsymbol{e}} + \boldsymbol{\Lambda}_1 \boldsymbol{e}$; $\boldsymbol{\eta}_2 = \tilde{\boldsymbol{x}}_2 + \boldsymbol{\Lambda}_2 \tilde{\boldsymbol{x}}_1$.

Multiply both left sides $\boldsymbol{J}_t^T \boldsymbol{J}_Z^{+1}$ of Eq. (21), it can be expressed as

$$\begin{aligned} &\boldsymbol{J}_t^{\mathrm{T}} \boldsymbol{J}_Z^{+1} \boldsymbol{D}_Z \dot{\boldsymbol{\eta}}_1 + \boldsymbol{J}_t^{\mathrm{T}} \boldsymbol{J}_Z^{+1} \boldsymbol{H}_Z \boldsymbol{\eta}_1 + \boldsymbol{J}_t^{\mathrm{T}} \boldsymbol{J}_Z^{+1} \boldsymbol{K}_d \boldsymbol{\eta}_1 \\ &= \boldsymbol{J}_t^{\mathrm{T}} \boldsymbol{J}_Z^{+1} \boldsymbol{K}_d \boldsymbol{\eta}_2 + \boldsymbol{J}_t^{\mathrm{T}} \boldsymbol{J}_Z^{+1} \boldsymbol{D}_Z \tilde{\boldsymbol{x}}_3 + \boldsymbol{J}_t^{\mathrm{T}} (\boldsymbol{F}_{Id} - \boldsymbol{F}_I) \end{aligned} \tag{22}$$

Since, $\boldsymbol{F}_I$ and $\boldsymbol{F}_{Id}$ are internal force terms defined in the null space $\boldsymbol{J}_t^T$. Then Eq. (22) can be expressed as

$$\boldsymbol{D}_a \dot{\boldsymbol{\eta}}_1 + \boldsymbol{H}_a \boldsymbol{\eta}_1 + \boldsymbol{K}_a \boldsymbol{\eta}_1 = \boldsymbol{K}_a \boldsymbol{\eta}_2 + \boldsymbol{D}_a \tilde{\boldsymbol{x}}_3 \tag{23}$$

where $\boldsymbol{D}_a = \boldsymbol{J}_t^{\mathrm{T}} \boldsymbol{J}_Z^{+1} \boldsymbol{D}_Z$; $\boldsymbol{H}_a = \boldsymbol{J}_t^{\mathrm{T}} \boldsymbol{J}_Z^{+1} \boldsymbol{H}_Z$; $\boldsymbol{K}_a = \boldsymbol{J}_t^{\mathrm{T}} \boldsymbol{J}_Z^{+1} \boldsymbol{K}_d$.

The robot model (14) has the following structural properties [11].

Property 1: $\boldsymbol{D}_a$ is symmetric and positive definite, $\boldsymbol{D}_a$ and $\boldsymbol{D}_a^{-1}$ are bounded.

Property 2: $0 < D_{am} \le \|\boldsymbol{D}_a\| \le D_{aM}$, where $\|\boldsymbol{D}_a\| = \sqrt{\lambda_{\max}(\boldsymbol{D}_a^{\mathrm{T}} \boldsymbol{D}_a)}$. $\lambda_{\max}(\cdot)$ represent the maximum eigenvalue. $\boldsymbol{D}_{aM}$ and $\boldsymbol{D}_{am}$ are the maximum and minimum eigenvalue of $\boldsymbol{D}_a$.

Define state variables $\boldsymbol{\omega}_i = \tilde{\boldsymbol{x}}_i/\gamma^{i-1}(i=1,2,3)$, $\boldsymbol{\omega} = [\,\boldsymbol{\omega}_1^{\mathrm{T}} \quad \boldsymbol{\omega}_2^{\mathrm{T}} \quad \boldsymbol{\omega}_3^{\mathrm{T}}\,]^{\mathrm{T}}$; invoking (15), and (16), we have

$$\dot{\boldsymbol{\omega}} = r\boldsymbol{K}_v\boldsymbol{\omega} + \frac{\boldsymbol{K}_p\dot{\boldsymbol{x}}_3(t)}{\gamma^2} \tag{24}$$

Theorem 1: Assume $\dot{\boldsymbol{x}}_3$ is bounded, for any selected ESO bandwidth $\gamma > 0$, the estimated error of ESO is bounded. That satisfies the existence constant $\zeta_i > 0$, so that each element of $\tilde{\boldsymbol{x}}_i(i=1,2,3)$ is satisfied $|\tilde{\boldsymbol{x}}_{ij}| \leq \zeta_i(j=1,2,\ldots,7)$ in finite time.

Proof: Integrating (24), we have

$$\boldsymbol{\omega}(t) = e^{\gamma\boldsymbol{K}_v t}\boldsymbol{\omega}(0) + \int_0^t e^{\gamma\mathbf{K}_v(t-\tau)}\frac{\boldsymbol{K}_p\dot{\boldsymbol{x}}_3(\tau)}{\gamma^2}d\tau \tag{25}$$

Because $\dot{\boldsymbol{x}}_3(t)$ is bounded, so each element of $\dot{\boldsymbol{x}}_3(t)$ are bounded. Assuming that $|\dot{\boldsymbol{x}}_{ij}(t)| \leq \delta_j(i=1,2,3, j=1,2,\ldots,7)$, where $\boldsymbol{\delta}$ is the constant vector. Defining

$$\boldsymbol{\varepsilon}(t) = \int_0^t e^{\gamma\boldsymbol{K}_v(t-\tau)}\frac{\boldsymbol{K}_p\dot{\boldsymbol{x}}_3(\tau)}{\gamma^2}d\tau \tag{26}$$

Reference [13] related derivation shows that $|\boldsymbol{\varepsilon}_{ij}(t)|$ is bounded, and satisfy the following relationship

$$|\boldsymbol{\varepsilon}_{ij}(t)| \leq \frac{\delta\upsilon}{\gamma^{21}} + \frac{\delta r}{\gamma^{42}} \tag{27}$$

where υ and r are constants.

Defining $\omega_{\mathrm{sum}}(0) = |\omega_{\mathrm{i1}}(0)| + |\omega_{\mathrm{i2}}(0)| + \cdots |\omega_{\mathrm{ij}}(0)|$, we have

$$|[e^{\gamma\boldsymbol{K}_v t}\boldsymbol{\omega}(0)]_{ij}| \leq \frac{\omega_{\mathrm{sum}}(0)}{\gamma^{21}} \tag{28}$$

Considering Eq. (27), we can get

$$|\boldsymbol{\omega}_{ij}(t)| \leq |[e^{\gamma\boldsymbol{K}_v t}\boldsymbol{\omega}(0)]_{ij}| + |\boldsymbol{\varepsilon}_{ij}(t)| \tag{29}$$

Combining $\boldsymbol{\omega}_i = \tilde{\boldsymbol{x}}_i/r^{i-1}$, we have

$$|\tilde{\boldsymbol{x}}_{ij}(t)| \leq \frac{\tilde{x}_{\mathrm{sum}}(0)}{\gamma^{21}} + \frac{\delta\upsilon}{\gamma^{22-i}} + \frac{\delta r}{\gamma^{43-i}} \tag{30}$$

where $\tilde{x}_{\mathrm{sum}}(0) \leq |\tilde{x}_{i1}(0)| + |\tilde{x}_{i2}(0)| + \cdots + |\tilde{x}_{ij}(0)|$ $(i=1,2,3;\ j=1,2,\ldots,7)$. Therefore, the observation error of the ESO is bounded.

Construct a Lyapunov candidate as follows

$$V = \frac{1}{2}\boldsymbol{\eta}_1^{\mathrm{T}}\boldsymbol{D}_a\boldsymbol{\eta}_1 \tag{31}$$

Differentiating V with respect to time t yields

$$\begin{aligned}\dot{V} &= \boldsymbol{\eta}_1^{\mathrm{T}}\boldsymbol{D}_a\dot{\boldsymbol{\eta}}_1 \\ &\leq \|\boldsymbol{\eta}_1\|\|(\boldsymbol{K}_a\boldsymbol{\eta}_2 + \boldsymbol{D}_a\tilde{\boldsymbol{x}}_3)\| - \lambda_{\min}(\boldsymbol{K}_a)\|\boldsymbol{\eta}_1\|^2 - \boldsymbol{\eta}_1^{\mathrm{T}}\boldsymbol{H}_a\boldsymbol{\eta}_1\end{aligned} \tag{32}$$

$\boldsymbol{\eta}_2 = \tilde{\boldsymbol{x}}_2 + \boldsymbol{\Lambda}_2\tilde{\boldsymbol{x}}_1$, according to Theorem 1, the observer estimation error is bounded. Defining $\boldsymbol{K}_a\boldsymbol{\eta}_2 + \boldsymbol{D}_a\tilde{\boldsymbol{x}}_3 = \boldsymbol{w}$, assume $\|\boldsymbol{\eta}_2\| \leq K_1$, $\|\tilde{\boldsymbol{x}}_3\| \leq K_2$, we can get

$$\|\boldsymbol{w}\| \leq \lambda_{\max}(\boldsymbol{K}_a)K_1 + D_{aM}K_2 \tag{33}$$

Considering $-\boldsymbol{\eta}_1^{\mathrm{T}}\boldsymbol{H}_a\boldsymbol{\eta}_1 \leq 0$, So in order for $\dot{V} \leq 0$, it just have to satisfy that

$$\lambda_{\min}(\boldsymbol{K}_a)\|\boldsymbol{\eta}_1\| - \|\boldsymbol{w}\| \geq 0 \tag{34}$$

Whether Eq. (34) satisfies the conditions is discussed as follows

(1) $\|\boldsymbol{\eta}_1\|$ is bounded, and $\|\boldsymbol{\eta}_1\| \geq \|\boldsymbol{w}\|/\lambda_{\min}(\boldsymbol{K}_a)$. Under this condition, Eq. (38) is satisfied, $\|\boldsymbol{\eta}_1\|$ tends to be bounded, and the system is stable.
(2) $\|\boldsymbol{\eta}_1\|$ is bounded, and $\|\boldsymbol{\eta}_1\| \leq \|\boldsymbol{w}\|/\lambda_{\min}(\boldsymbol{K}_a)$. This condition does not satisfy Eq. (34), and $\dot{V} > 0$, the system energy continues to increase, it's bound to produce $\|\boldsymbol{\eta}_1\| \geq \|\boldsymbol{w}\|/\lambda_{\min}(\boldsymbol{K}_a)$; it is inconsistent with the conditions and therefore does not meet the requirements.
(3) $\|\boldsymbol{\eta}_1\|$ is unbounded. Consider $\|\boldsymbol{\eta}_1\|$ is unbounded, it must exist $\|\boldsymbol{\eta}_1\| \geq \|\boldsymbol{w}\|/\lambda_{\min}(\boldsymbol{K}_a)$. Combining Eq. (34), $\|\boldsymbol{\eta}_1\|$ tend to be bounded, it is contradicts to the conditions, so it doesn't meet the requirements.

Based on the above analysis, it can be known that $\|\boldsymbol{\eta}_1\|$ is bounded, so $\boldsymbol{e}$ is bounded, it is demonstrated that the system is global uniform ultimate boundedness.

While ensuring the stability of the position control, the boundedness of the target clamping internal force coordination control should also be considered. According to Eq. (23), we can get

$$\boldsymbol{J}_Z(\boldsymbol{F}_{Id} - \boldsymbol{F}_I) = \boldsymbol{D}_Z\dot{\boldsymbol{\eta}}_1 + \boldsymbol{H}_Z\boldsymbol{\eta}_1 + \boldsymbol{K}_d\boldsymbol{\eta}_1 - \boldsymbol{K}_d\boldsymbol{\eta}_2 - \boldsymbol{D}_Z\tilde{\boldsymbol{x}}_3 \tag{35}$$

Multiply both left sides $\boldsymbol{J}_Z^+$ of Eq. (35), it can be written as

$$(\boldsymbol{F}_{Id} - \boldsymbol{F}_I) = \boldsymbol{J}_Z^+(\boldsymbol{D}_Z\dot{\boldsymbol{\eta}}_1 + \boldsymbol{H}_Z\boldsymbol{\eta}_1 + \boldsymbol{K}_d\boldsymbol{\eta}_1 - \boldsymbol{K}_d\boldsymbol{\eta}_2 - \boldsymbol{D}_Z\tilde{\boldsymbol{x}}_3) \tag{36}$$

The norm of both ends of Eq. (36) is calculated and simplified

$$\|\boldsymbol{F}_{Id} - \boldsymbol{F}_I\| \leq \|\boldsymbol{J}_Z^+\|(\|\boldsymbol{D}_Z\|\|\dot{\boldsymbol{\eta}}_1\| + \|\boldsymbol{H}_Z\|\|\boldsymbol{\eta}_1\| + \|\boldsymbol{K}_d\|\|\boldsymbol{\eta}_1\| + \|\boldsymbol{K}_d\|\|\boldsymbol{\eta}_2\| + \|\boldsymbol{D}_Z\|\|\tilde{\boldsymbol{x}}_3\|) \quad (37)$$

Considering $\boldsymbol{\eta}_1$, $\boldsymbol{\eta}_2$ and $\tilde{\boldsymbol{x}}_3$ are bounded. $\boldsymbol{J}_Z$, $\boldsymbol{D}_Z$, $\boldsymbol{H}_Z$ satisfied property 2, so they are bounded. And $\boldsymbol{K}_d$ is bounded, so the error value of clamping internal force is also bounded, the control of clamping internal force on target is stable.

4 Simulation Examples

To show the performance of the proposed force/position active disturbance rejection controller, a simulation is carried out on a planar dual-arm space robot system shown in Fig. 1. The actual plant parameters of the system are as follows $m_0 = 200\,\text{kg}$, $I_0 = 50\,\text{kg}\cdot\text{m}^2$, $d_0 = 1.064\,\text{m}$; $m_i = 20\,\text{kg}$, $I_i = 10\,\text{kg}\cdot\text{m}^2$, $l_i = 2\,\text{m}(i = 1,2,4,5)$; $m_i = 5\,\text{kg}$, $I_i = 2\,\text{kg}\cdot\text{m}^2$, $l_i = 0.5\,\text{m}(i = 3,6)$; $m_t = 50\,\text{kg}$, $I_t = 10\,\text{kg}\cdot\text{m}^2$. $l_L = 0.5\,\text{m}$, $l_R = 0.5\,\text{m}$, $l_B = 2\,\text{m}$

The related coefficients and gains of the controller are chosen as $\boldsymbol{K}_d = \text{diag}(800, 800, 800, 800, 800, 800, 800)$; $\gamma = 60$; $u_{\max} = 20\,\text{N}\cdot\text{m}$, $u_{\min} = -20\,\text{N}\cdot\text{m}$; $C_i = 5(i = 1,2,\cdots,6)$, $\Lambda_1 = \Lambda_2 = \text{diag}(5,5,5,5,5,5,5)$. $\boldsymbol{F}_{Id} = [-10\text{N} \quad 0\text{N} \quad 0\text{N}\cdot\text{m} \quad 10\text{N} \quad 0\text{N} \quad 0\text{N}\cdot\text{m}]^{\text{T}}$.

The desired trajectory of : $\boldsymbol{\theta}_{md} = [0° \quad -120° \quad -60° \quad -60° \quad 60° \quad 60° \quad 60°]^{\text{T}}$. The initial velocity vector of load in pre-impact phase is: $\boldsymbol{v}_t = [0.1\,\text{m/s} \quad 0.1\,\text{m/s} \quad 0.35\,\text{rad/s}]^{\text{T}}$. The controller performances are shown in Figs. 2 and 3. Figure 2 shows that the system is unstable before active control is started after gripper capture. If active control is not started, the closed-chain hybrid system will roll over, which will affect the task and even damage the relevant components. However, after active stabilization control is activated, the system quickly reaches the desired position and forms a stable state.

Figure 3 shows the tracking of internal forces of clamping internal force for dual-arm space robot capture target spacecraft, the results show that the internal clamping force finally reaches the desired and stable state.

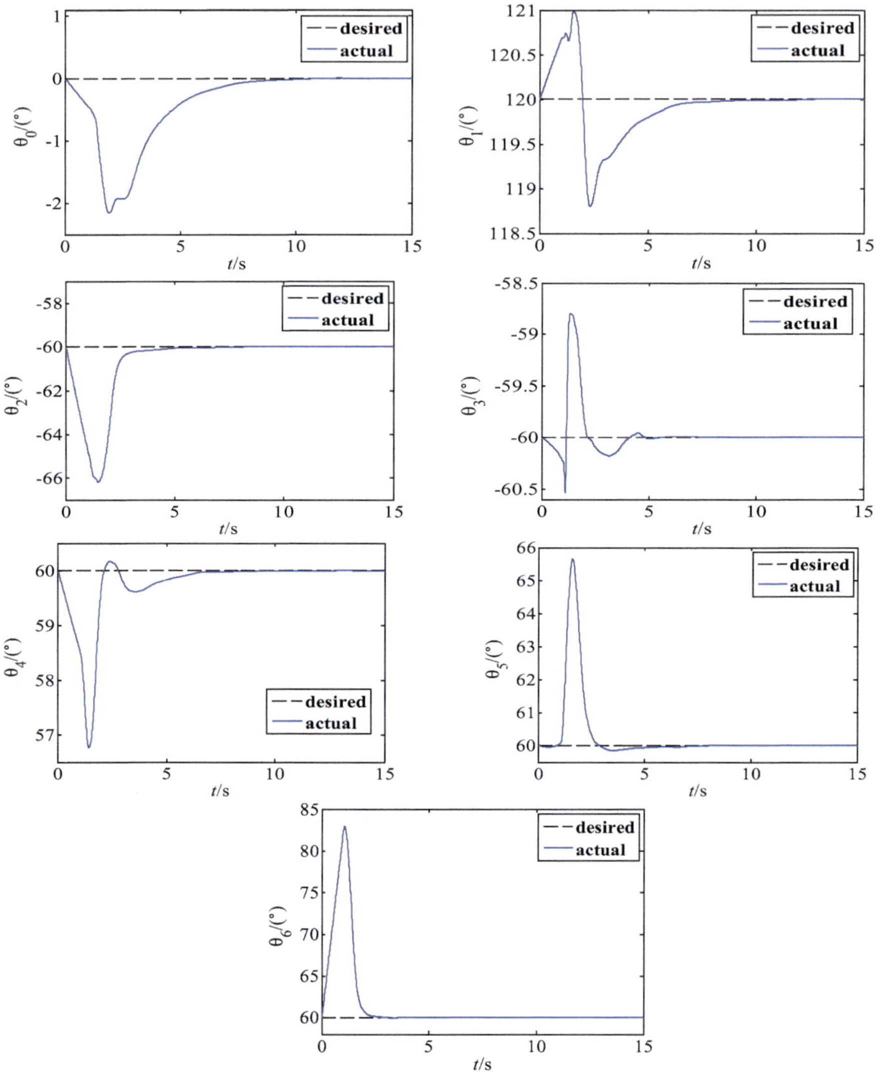

Fig. 2. The stabilization control of post-capture closed-chain dual-arm space robot system

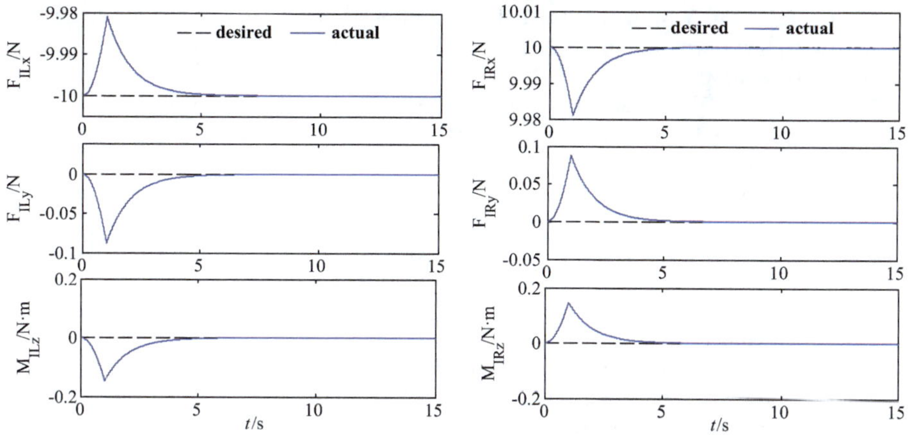

Fig. 3. Tracking of clamping internal force for dual-arm space robot capture target spacecraft

5 Conclusions

In this paper, the impact effect of the capture collision is analyzed by using the momentum theorem, the geometric relationship of motion and the law of force transfer. Then, an active force/position stabilization control algorithm based on passivity theory was proposed for the unstable close-chain system after the collision, considering the complex operating conditions of actuator saturation and joint friction. The proposed control scheme can improve the anti-disturbance ability to the impact effect after the capture operation, and has strong robustness, which can quickly realize the coordinated control of the position and internal forces of the hybrid system.

Acknowledgments. This work was supported by the National Natural Science Foundation of China (No. 51741502, No. 11372073).

References

1. Yu, X.Y., Chen, L.: Singular perturbation adaptive control and vibration suppression of free-flying flexible space manipulators. Proc. Inst. Mech. Eng., Part C: J. Mech. Eng. Sci. **229**(11), 1989–1997 (2015)
2. Yu, X.Y., Chen, L.: Modeling and observer-based augmented adaptive control of flexible-joint free-floating space manipulators. Acta Astronaut. **108**, 146–155 (2015)
3. Yu, X.Y., Chen, L.: Observer-based two-time scale robust control of free-flying flexible-joint space manipulators with external disturbances. Robotica **35**(11), 1–17 (2017)
4. Stolfi, A., Gasbarri, P., Sabatini, M.: A combined impedance-PD approach for controlling a dual-arm space manipulator in the capture of a non-cooperative target. Acta Astronaut. **139**, 243–253 (2017)
5. Abad, A., Ma, O., Pham, K., et al.: A review of space robotics technologies for on-orbit servicing. Prog. Aerosp. Sci. **68**(8), 1–26 (2014)

6. Huang, P.F., Xu, Y.S., Liang, B.: Balance control of multi-arm free-floating space robots during capture operation. In: IEEE International Conference on Robotics and Biomimetics, pp. 398–403 (2005)
7. Peng, J.Q., Xu, W.F., Liang, B., et al.: Pose measurement and motion estimation of space Non-cooperative Targets based on Laser Radar and Stereo-vision Fusion. IEEE Sens. J. **19** (8), 3008–3019 (2019)
8. Lim, J., Chung, J.: Dynamic analysis of a tethered satellite system for space debris capture. Nonlinear Dyn. **94**, 1–18 (2018)
9. Han, J.Q.: From PID to active disturbance rejection control. IEEE Trans. Ind. Electron. **56** (3), 900–906 (2009)
10. Ren, C., Ma, S.: Passivity-based model free control of an omnidirectional mobile robot. In: IEEE International Conference on Mechatronics, pp. 262–267. IEEE (2015)
11. Nicosia, S., Tomei, P.: Robot control by using only joint position measurements. IEEE Trans. Autom. Control **35**(9), 1058–1061 (1990)
12. Berghuis, H., Nijmeijer, H.: A passivity approach to controller-observer design for robots. IEEE Trans. Autom. Control **9**(6), 740–754 (1993)
13. Zheng, Q., Linda, Q., Gao, Z.: On stability analysis of active disturbance rejection control for nonlinear time-varying plants with unknown dynamics. In: Proceedings of the 46th IEEE Conference on Decision and Control, pp. 12–14 (2007)

Motion Planning of a 5-DOF Anthropomorphic Robotic Arm Under ROS Environment

Sida Zhang, Zirong Lin, and Guanglei Wu(✉)

School of Mechanical Engineering,
Dalian University of Technology, Dalian 116024, China
gwu@dlut.edu.cn

Abstract. This paper deals with motion planning of a 5-DOF anthropomorphic robotic arm. The CAD model of the robot is built and converted into a URDF file for the simulation with the ROS platform. Sequentially, the motion planning of the manipulator is completed through the MoveIt! Package. The motion process is observed in RViz, with the planned result being transmitted to the servo controller to drive the real robotic arm to reach to the target point. Moreover, the extraction and matching of key feature points based on SIFT algorithm is studied for object detection to obtain the coordinates of the targeted object, in order to realize the pick-and-place operation for the material handling.

Keywords: Motion planning · Anthropomorphic robotic arm · MoveIt! · SIFT

1 Introduction

Industrial robot plays an important role in the modern industrial production, with the application in many fields, such as material handling [1], palletizing [2], assembly [3] and surgery [4]. The use of the robotic manipulators efficiently improves productivity. With the increasing applications of the industrial robots, the motion planning becomes more and more complicated, together with the increasing demands on the control accuracy. Therefore, a system is needed to visualize the robot motion and to validate the control algorithm rapidly, with low redundancy of codes and good portability.

Among the different robot operating platforms, robot operations system (ROS) has many advantages of the other platforms. MoveIt! [5] in ROS can complete motion planning of the manipulator, and RViz [6] can be used to watch the real-time motion of the visualized model. In view of these advantages, ROS has been accepted by more and more developers in the field of robotic control. Li [7] proposed a ROS-based approach of motion planning for mobile service robots. Hernandez-Mendez [8] introduced the interface between the robot arm and the motion planner as well as the 3D viewer RViz. Lee [9] presented a fast object detection algorithm based on color histograms and LBP histograms.

In this paper, the motion planning of a 5-DOF robotic arm is presented with the MoveIt! in ROS. The planning results are applied to drive the real robotic arm, and object detection is investigated by using the key feature points based on SIFT algorithm to realize the pick-and-place operation for the material handling.

D. Wang et al. (Eds.): MeTrApp 2019, MMS 79, pp. 409–418, 2020.
https://doi.org/10.1007/978-981-15-0142-5_40

2 Robotic Arm Under Study

Figure 1 shows the model of the 5-DOF anthropomorphic robotic arm with the Servo control system. In order to verify the experiment more conveniently, the model of the manipulator can be simulated first. So it is necessary to draw the CAD model of the manipulator, and then convert the CAD model by sw2urdf plugin into the corresponding URDF file, which is in an XML language form. With the URDF file, the links, joints, kinematic and dynamic parameters, visualization model collision detection, and among others, can be defined.

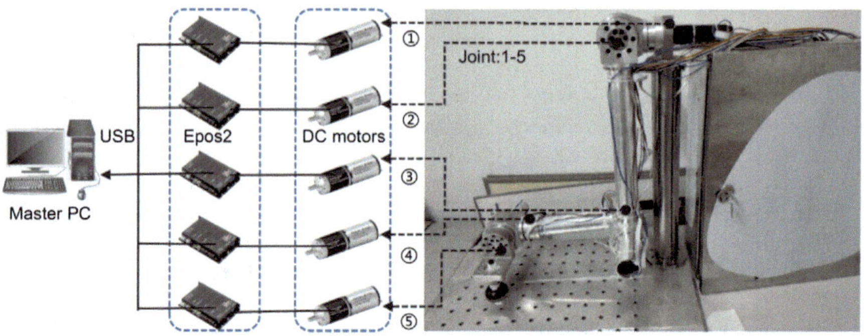

Fig. 1. The 5-DOF anthropomorphic robotic arm

Afterwards, the Setup Assistant in the MoveIt!, a graphical user interface as shown in Fig. 2(a), is adopted to configure the robot. The flow chart of the usage of the Setup Assistant is displayed in Fig. 2(b), wherein the focus is to generate the Semantic Robot Description Format (SRDF) file. In the SDRF file, the configuration information of the robot defined in the Setup Assistant is saved, such as collision detection, virtual joints, planning groups, end-effector pose, etc. To this end, all the generated files are integrated into a package that can run directly.

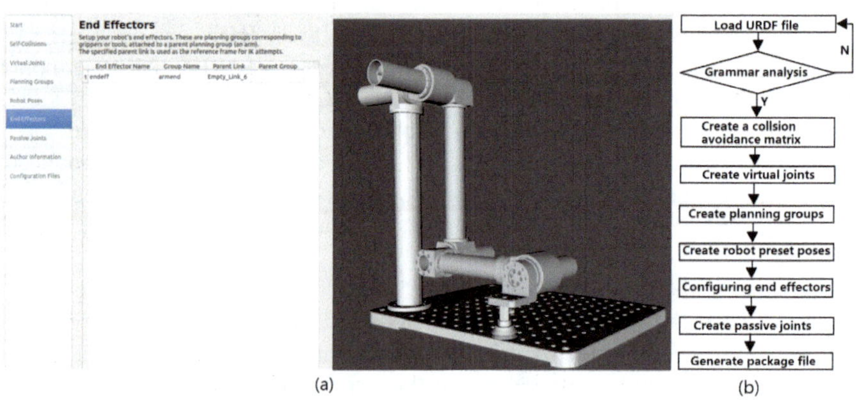

Fig. 2. (a) Setup Assistance in MoveIt! and (b) Flow chart of the configuration procedure

3 ROS System Framework for the Robotic Arm

In order to ensure the outcome of the virtual control of the robotic manipulator consistent with the real operation, a ROS based simulation system and platform needed, consisting of high-level and low-level infrastructures. Figure 3(a) depicts the high-level infrastructure, of which the Move_group node provided by MoveIt! that is treated as an integrator, to combine the independent components. The KDL library, as a plugin, accomplishes calculation of forward and inverse kinematics of the robot as well as the dynamic aspects. The OMPL provides the algorithm for the motion planning. Moreover, the URDF file mentioned in previous section is extracted from the sw_urdf_-exporter plugin in SolidWorks, with the associated parameters and models. With this framework, data packets to store the joint trajectories can be generated by pre-setting the parameters of the motion planning.

The underlying infrastructure of the simulation system is shown in Fig. 3(b), for which the inputs are from the joint trajectory data generated by the high-level one. Through the virtual controller, the algorithms such as spline interpolation is adopted for smoothing and interpolation of real-time processing. In this way, the post-processed results can be used for the control both in the virtual environment and in reality. In the virtual environment, the simulation results can be visualized with RViz.

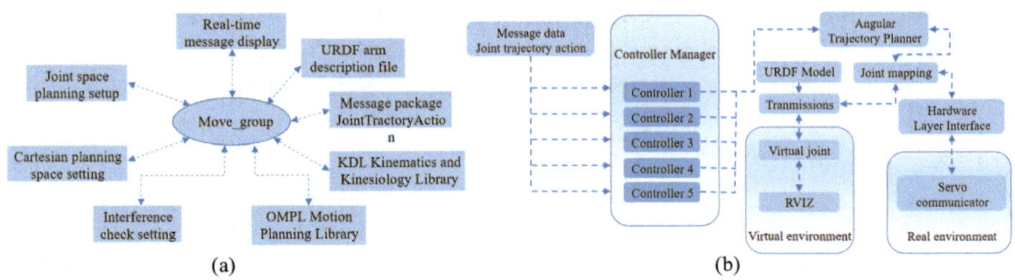

Fig. 3. The ROS System Framework for the Robotic Arm: (a) high level architecture; (b) low level architecture

4 Simulation Procedure of the Robotic Arm

To ensure the manipulator end-effector accurately and smoothly reach to the desired position, the suitable motion parameters for all the robot joints should be found to avoid the vibration and mutation during its movement. When the motion planning is conducted in the joint space, the positions of the end-effector along trajectory will be cast into the vectors of the joint angles. With the prescribed constraints on the end-effector starting position, the acceleration starting positing, the deceleration starting position, the ending position, etc., the appropriate interpolation functions are needed to fit the smooth curves of the joints, in order to achieve the interpolation computation for angle, angular velocity and angular acceleration of the joints.

For the industrial applications of the robotic arm, the trajectory planning is commonly carried out in the Cartesian space. The computation of the inverse kinematics in this procedure is extremely complex, particularly, for the multiple-DOF serial manipulators. To overcome this problem, MoveIt! is a convenient tool to simplify this procedure. It only requires the user to input the trajectory of the robot end-effector in the Cartesian space, sequentially, the planner of the MoveIt! package is employed to calculate the motion parameters of the joints through the complex numerical computation, for instance, the angular displacement, angular velocity, angular acceleration, etc.

5 Object Detection

So far, the trajectory planning of the manipulator in Cartesian space can be realized with MoveIt!. Consequently, the pick-and-place operation for material handling can be accomplished. Here, only the center position of the object can be detected to be sent to MoveIt!, for the completion of such a task. Therefore, the detection and matching of key feature points based on SIFT algorithm will be further investigated.

The key feature points extracted by the SIFT algorithm have the invariance of translation, scaling and rotation. In order to ensure correct matching under different viewing angles, brightness, distance and noise conditions, following techniques are used. Firstly, in order to solve the problem of the target distance, target size and noise, a Gaussian scale space is established to form a corresponding differential of Gaussian pyramid model. Then, in order to obtain a more stable feature point, the extracted feature points are screened to remove the edge response points and low contrast. Last according to the gradient size of the image to establish the feature vector of the corresponding feature point, thereby forming a feature descriptor.

5.1 Scale Space

The scale space of the two-dimensional image $L(x, y, \sigma)$ can be obtained by filtering the image by two-dimensional Gaussian kernel function $G(x, y, \sigma)$, expressed in the following equations:

$$G(x, y, \sigma) = \frac{1}{2\pi\sigma^2} e^{\frac{x^2+y^2}{2\sigma^2}} \tag{1}$$

$$L(x, y, \sigma) = G(x, y, \sigma) \otimes I(x, y) \tag{2}$$

where $I(x, y)$ is the original image: The constructed Gaussian pyramid is shown in Fig. 4(a). The next group of images is obtained by down-sampling the previous group. Each group is convolved with different Gaussian kernels, and then subtracted between the two layers. The difference of Gaussian pyramid is shown in Fig. 4(b), and the scale space constructed by this can detect feature points that exist at different scales. Figure 4(c) illustrates the finding of the extreme points in a DOG. $D(x, y, \sigma)$, which is represented by the differences between the adjacent scale spaces [10]:

$$D(x, y, \sigma) = L(x, y, k\sigma) - L(x, y, \sigma) \tag{3}$$

In the process of extreme point detecting, 8 pixels around the same scale of the target detection point and 18 points in the upper and lower adjacent scales are needed, thus, a total of 26 points are used to check whether the point is an extreme point or not. In this way, it is guaranteed that the point becomes an extreme point in different scale spaces.

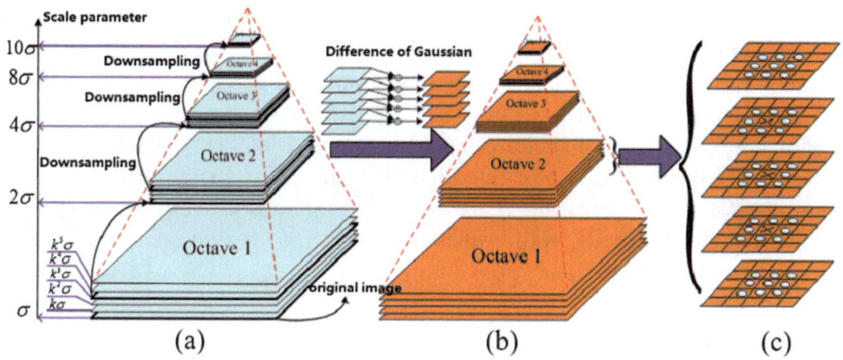

Fig. 4. Scale space model: (a) Gaussian; (b) DOG; (c) extreme value detection

5.2 Obtaining the Location and Scale of Stable Key Points

Since the extreme points detected in the DOG model are discretely distributed, and the process of differential operation is sensitive to the noise and the edge information, resulting in a strong edge response, therefore, the key points should be positioned precisely by means of Taylor expansion. Let suppose x to be candidate feature point, the contrast is equal to the absolute value of $D(x)$ as below:

$$D(x) = D + \frac{\partial D^T}{\partial x} x + \frac{1}{2} x^T \frac{\partial^2 D}{\partial x^2} x \tag{4}$$

After fitting, the orientation and scale of the key points can be determined. Removing the points with low contrast and unstable edges, the stable key points are obtained. Let the differentiated Eq. (4) be equal to zero, $\hat{x}$ can be calculated, and the substitution of the solved $\hat{x}$ into Eq. (4) yields.

$$\hat{x} = -\frac{\partial^2 D^{-1}}{\partial x^2} \frac{\partial D}{\partial x}, D(\hat{x}) = D + \frac{1}{2} \frac{\partial D^T}{\partial x} \hat{x} \tag{5}$$

If $|D(\hat{x})| \geq 0.03$, the feature point is considered stable. The neighboring pixels of each key point are represented by a histogram, which indicates the distribution of gradient values. The directional parameters corresponding to each key point ensures the rotation invariance. Here, the gradient and angle, respectively, are expressed as:

$$m(x,y) = \sqrt{(L(x+1,y) - L(x-1,y))^2 + (L(x,y+1) - L(x,y-1))^2} \quad (6)$$

$$\theta(x,y) = \arctan\frac{L(x,y+1) - L(x,y-1)}{L(x+1,y) - L(x-1,y)} \quad (7)$$

5.3 Descriptors of the Stable Feature Points

To develop the descriptors of the stable feature points, the 4 × 4 seed point from the position of the current key point are selected. Sequentially, the gradient histograms in 8 directions are drawn, according to the gradient amplitude and gradient direction of each pixel, thus, there are 8 groups of 4 × 4 seed points, which constitutes a 128-dimensional vector. When matching the feature points of the two images, it is only necessary to count their descriptor Euclidean Distance. The smaller the distance is, the more similar the two images are. This kind of neighborhood directional information joint enhances the anti-noise ability of the algorithm, to provide reasonable fault tolerance for feature matching with positioning error.

5.4 Mismatch Removal by RANSAC

In the process of feature matching of two or more pictures with the SIFT algorithm, erroneous matching may occurs, due to the large number of feature points. In order to reduce the mismatch and reduce the recognition error, the RANSAC algorithm is used for fitting. After the iterative processing, the wrong matching result is eliminated, and the matching result is verified by some constraints to ensure the unique matching relationship and improved matching.

According to the set of the feature point, a test model is built to meet the matching relationship, through which the remaining feature points can be tested by using the model. The points subject to the previous relationship is considered as the matching point. Hence, the mismatch is removed, and a mathematical model for matching the pair of feature points is derived as below.

$$\begin{bmatrix} u \\ v \end{bmatrix} = \begin{bmatrix} m_1 & m_2 \\ m_3 & m_4 \end{bmatrix} \begin{bmatrix} x \\ y \end{bmatrix} + \begin{bmatrix} t_x \\ t_y \end{bmatrix} \quad (8)$$

where (x, y) and (u, v) represent the coordinates corresponding to the feature points before and after the affine transformation respectively. Moreover, (m_1, m_2, m_3, m_4) and (t_x, t_y) are the six parameters. It can be seen from Eq. (5) that the parameters can be solved with the coordinates of 3 pairs of matching feature points. Assuming the coordinates of the three pairs of feature points are (x_1, y_1) and (u_1, v_1), (x_2, y_2) and (u_2, v_2), (x_3, y_3) and (u_3, v_3), the parameters can be solved from the following equation:

$$\begin{bmatrix} x_1 & y_1 & 0 & 0 & 1 & 0 \\ 0 & 0 & x_1 & y_1 & 0 & 1 \\ x_2 & y_2 & 0 & 0 & 1 & 0 \\ 0 & 0 & x_2 & y_2 & 0 & 1 \\ x_3 & y_3 & 0 & 0 & 1 & 0 \\ 0 & 0 & x_3 & y_3 & 0 & 1 \end{bmatrix} \begin{bmatrix} m_1 \\ m_2 \\ m_3 \\ m_4 \\ t_x \\ t_y \end{bmatrix} = \begin{bmatrix} u_1 \\ v_1 \\ u_2 \\ v_2 \\ u_3 \\ v_3 \end{bmatrix} \tag{9}$$

6 Case Study

Based upon the previous modeling and analysis, the motion planning is illustrated with the robotic arm shown in Fig. 1. Figure 5 shows the simulation of the trajectory planning of the robot in Cartesian space with MoveIt!. The interpolation algorithm of joint space planning is based on a fifth-order polynomial method. The program calculates joint angle with respect to the trajectory through continuous iteration, the trajectory being shown in the figure, and it can be seen the running process is smooth.

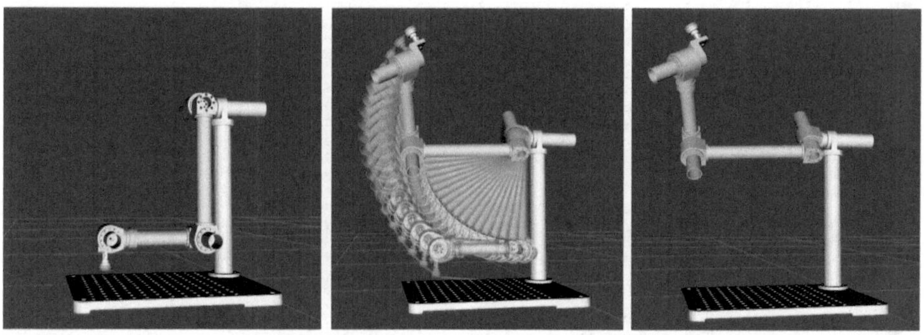

Fig. 5. Graphical simulation results of the trajectory planning in Cartesian space.

When the motion planning is implemented with MoveIt!, it is known that a set of tracking points will be generated according to the underlying architecture described in Sect. 3. In the experiments, those tracking points will be transmitted to the servo control system through the serial port communication, to drive the motors with the corresponding joint angles. The motion planning is displayed in Fig. 6.

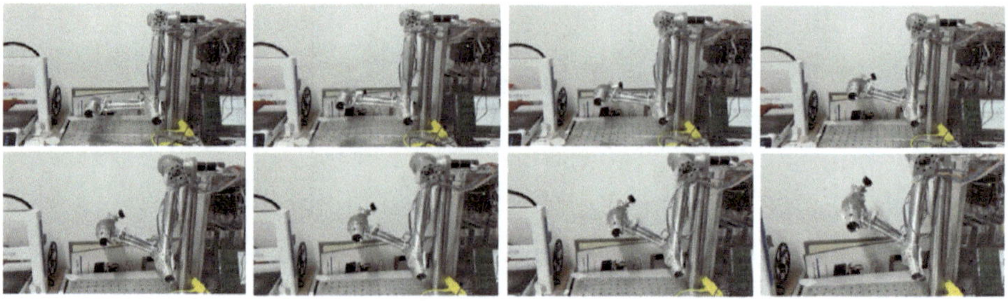

Fig. 6. Experimental motion planning of the robotic arm

Figures 7, 8 and 9 depict the object detection. As shown in Fig. 7(a), the feature point detection result of the image using the SIFT algorithm is followed by the feature matching. It can be seen that there are many matching results, but with some mismatches. Then, the RANSAC algorithm is used to match the key feature points. It can be seen that the number of RANSAC matches is reduced, with enhanced matching rate. The result in Fig. 7(b) shows high reliability, high precision and strong robustness of the RANSAC algorithm.

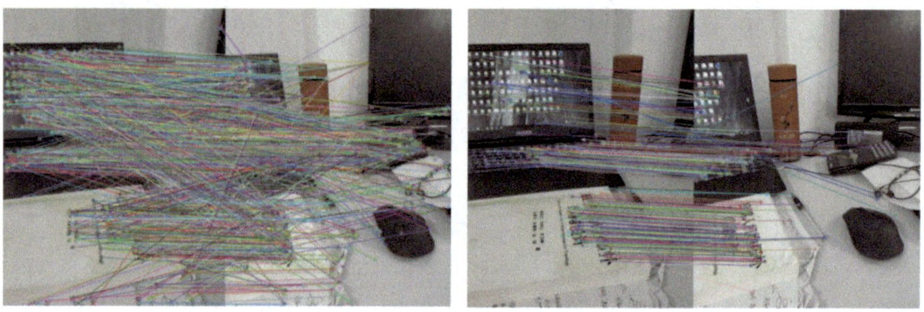

(a) (b)

Fig. 7. Key feature point matching: (a) SIFT; (b) RANSAC

In the previous experiments, it is seen that SIFT algorithm with incorporated RANSAC shows the improvement of the detection and matching of key points. Afterwards, the detection result is used to extract the positions of the object in the image, as shown in Fig. 8(a). After the template matching with the given object, the upper left corner of the figure can get its specific position in the camera vision. With the rectangle to locate the actual position of the object, the robot arm combines the results obtained by the SIFT algorithm to realize the pick-and-place operation, as shown in Fig. 8(b). Figure 9(a) shows a partial enlarged view of the object detection, from which it can be seen that the object to be picked is in the rectangular region, with the picking central position in the figure. The global coordinates of the object can be obtained by the coordinate convention between the camera coordinates and the base coordinates of

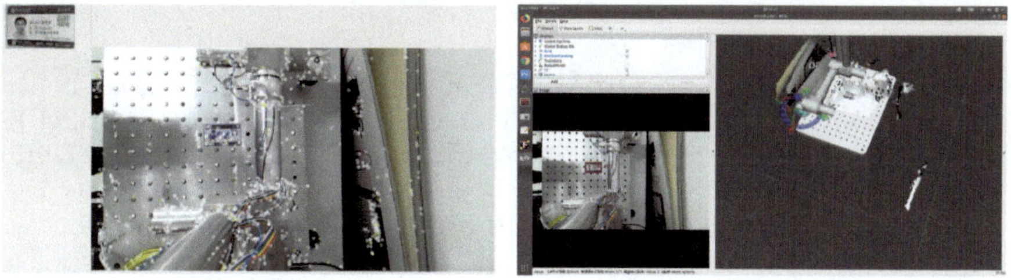

(a) (b)

Fig. 8. (a) SIFT algorithm detect a card on the platform; (b) Views in RViz

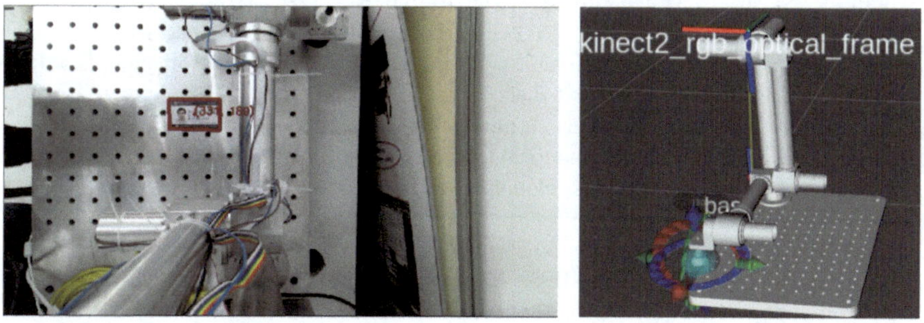

(a) (b)

Fig. 9. (a) Partial enlargement of object detection and (b) coordinate position relationship

the robot arm, as shown in Fig. 9(b). By transmitting the converted coordinates to Movelt!, the feasible path is planned to drive the end of the arm to the desired position, and the classification function of the object using the robot arm is accomplished.

7 Conclusions

In this paper, the motion simulation and control of a 5-dof manipulator under ROS is introduced. The robot operator can develop the URDF format file from the CAD model of the robot, for rapid motion simulation. Experimental results show that the use of Movelt! can not only reduce the difficulties in the robot operation, but also have advantages in terms of the algorithm validation and function exploration. Besides the motion simulation, the method proposed can be used for the real-time robot control. Moreover, the combination of the SIFT and RANSAC algorithms applied to object detection enhance the precision and robustness of the matching result. After the completion of the object matching, the coordinates of the object's center can be obtained through matrix convention, sequentially, the robot arm is driven to accomplish the pick-and-place operation.

References

1. Martinez, C., Boca, R., Zhang, B.: Automated bin picking system for randomly located industrial parts. In: IEEE International Conference on Technologies for Parctical Robot Applications, 15 Middlesex Canal Park, Woburn, MA. IEEE Computer Society (2015)
2. Zhang, G.Q., Li, L.M., Choi, S.: Robotic automated battery sorting system. In: IEE Conference on Technologies for Parctical Robot Applications, Woburn, MA, pp. 117–120. IEEE Computer Society (2012)
3. Townsend, W.: The BarrettHand grasper - programmably flexible part handling and assembly. Ind. Robot. **27**(3), 181–188 (2000)
4. Kavoussi, L.R., Moore, R.G.: Comparison of robotic versus human laparoscopic camera control. J. Urol. **158**(4), 1530 (1997)
5. Chitta, S., Sucan, I.: MoveIt! IEEE Robot. Autom. Mag. **19**(1), 18–19 (2012)
6. Hart, S., Dinh, P., Hambuchen, K.: The Affordance template ROS package for robot task programming. In: IEEE International Conference on Robotics and Automation, Seattle, WA, pp. 6227–6234. Inst. of Elec. and Elec. Eng. Inc. (2015)
7. Li, C., Hao, L., Cheng, H.: Research on motion planning system of service robot based on ROS. In: 7th IEEE Annual International Conference on CYBER Technology in Automation, Control, and Intelligent Systems, Honolulu, HI, pp. 205–209. Inst. of Elec. and Elec. Eng. Inc. (2017)
8. Hernandez-Mendez, S., Maldonado-Mendez, C.: Design and implementation of a robotic arm using ROS and MoveIt! In: IEEE International Autumn Meeting on Power, Electronics and Computing Ixtapa, Guerrero, Mexico, pp. 1–6. Inst. of Elec. and Elec. Eng. Inc. (2018)
9. Lee, K., Lee, C., Kim, S.: Fast object detection based on color histograms and local binary patterns. In: 2012 IEEE Region 10 Conference: Sustainable Development Through Humanitarian Technology, Cebu, Philippines, p. 4. Inst. of Elec. and Elec. Eng. Inc. (2012)
10. Lowe, D.G.: Distinctive image features from scale-invariant keypoints. Int. J. Comput. Vis. **60**(2), 91–110 (2004)

T-S Fuzzy Compensation Based Sliding Mode Control for a Free-Floating Space Manipulator

Zihao Chen[1], Xiaoyan Yu[1,2(✉)], and Li Chen[1]

[1] School of Mechanical Engineering and Automation, Fuzhou University, Fuzhou 350116, Fujian Province, China
cool@fzu.edu.cn

[2] Key Laboratory of Fluid Power and Intelligent Electro-Hydraulic Control, (Fuzhou University), Fujian Province University, Fuzhou 350116, Fujian Province, China

Abstract. A fuzzy sliding mode control with simple structure and fast response is proposed for the trajectory tracking in joint space of free-floating space manipulators with uncertainties such as friction, external disturbance and unmodeled dynamics. In this fuzzy sliding mode controller, the feedback linearization method is used to equivalent the known part of the dynamics, then a sliding mode control part is addressed to eliminate the uncertainties of the system, finally a T-S fuzzy compensation is proposed to simplify the controller to reduce the computational burden and speed up system response. The system stability under this fuzzy sliding mode controller is proved by Lyapunov theory. The simulation comparison with nominal model based robust sliding mode control shows the superiority of the proposed control method in convergence speed, response speed and the performance of chattering elimination.

Keywords: Free-floating space manipulator · Trajectory tracking · Fuzzy control · Sliding mode control

1 Introduction

With the development of the science and technology, human beings are constantly extending inquiry into the unknown space at a deeper domain. Against such a backdrop, space robots are playing an increasingly important role in various space activities. Since there are so many unpredictable dangers in space, the participation of space robots is very important to replace astronauts in some space activities, which can also reduce the cost of space exploration. Therefore, the research on the control theory of space robots has got more and more attention and rapid development [1–9].

Up to now, there have been a lot of control methods for free-floating space manipulators. Hyun [10] designed an adaptive controller based on T-S fuzzy model, and proposed to stabilize the control by compensating for the unknown load changing with time in the manipulator motion. Rojko [11] proposed a method to compensate for the effects of time-varying payloads by sliding mode control to control the manipulators.

However, it cannot be ignored that there are often many uncertain factors in the operation of free-floating space manipulators along with a series of uncertain factors in

D. Wang et al. (Eds.): MeTrApp 2019, MMS 79, pp. 419–429, 2020.
https://doi.org/10.1007/978-981-15-0142-5_41

space, such as friction, external interference and unmodeled dynamics [12]. Therefore, the controller of free-floating space manipulators cannot be designed only under ideal conditions, and the potential influence of these uncertain factors should be considered, otherwise the stability of the system in actual operation cannot be guaranteed.

Therefore, this paper presents a T-S fuzzy compensation based sliding mode control method, which can simply and effectively deal with the uncertain factors of free-floating space manipulators, so as to achieve ideal tracking accuracy. Different from traditional sliding mode control method, this method does well in eliminating chattering and achieving tracking trajectory more accurately with a shorter timeline.

2 The System Dynamics Model

In this paper, the two-link free-floating rigid space manipulator is deemed as the research target. Its structure can be simplified to the following figure.

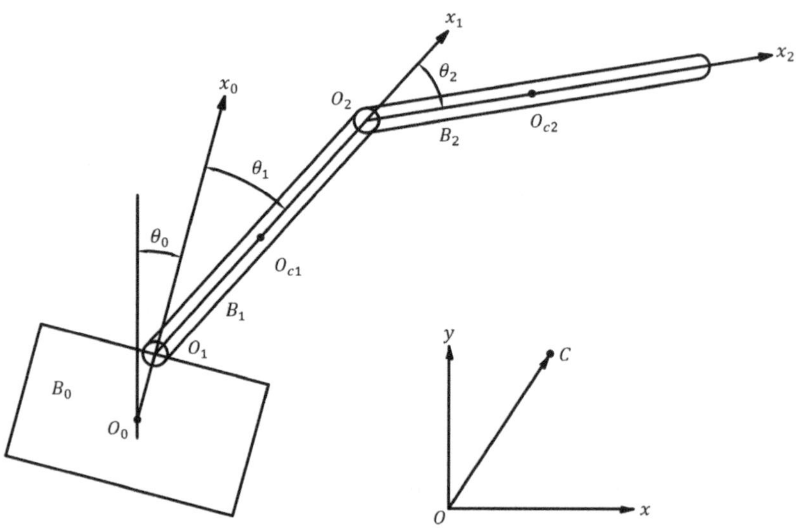

Fig. 1. Free floating space manipulator system

As shown in the figure, the main structure of the system consists of the base B_0 and the rigid arms B_1 and B_2. Where in the center of mass of the base B_0 is O_0, the centroids of the rigid arms B_1 and B_2 are respectively O_{c1} and O_{c2}; the connection point of the base B_0 and the rigid arm B_1 is O_1, and the connection point of the rigid arms B_1 and B_2 is O_2; connect O_0 and O_1 and extend it to make x_0 axis, make x_1 axis along B_1, and make x_2 axis along B_2; the centroid of the whole system is C, set the origin of coordinate to any point O in space, and establish the translational inertial coordinate system.

Since the system of free-floating space manipulators is not affected by external forces, the system as a whole will satisfy the momentum conservation relationship. Combined with the Lagrange second equation, the dynamic equation of the two-link free-floating rigid space manipulator considering the influence of external disturbances can be derived as:

$$D(q)\ddot{q} + H(q,\dot{q})\dot{q} + \tau_d = \tau(t) \tag{1}$$

where $D(q) \in R^{3\times3}$ is a symmetric positive definite mass matrix; $H(q,\dot{q}) \in R^{3\times3}$ is a matrix containing centrifugal force and Coriolis force; $q = [\,\theta_0 \quad \theta_1 \quad \theta_2\,]^T$ is a generalized coordinate column vector composed of a base attitude angle θ_0 and two joint relative rotation angles θ_1, θ_2; $\tau_d \in R^3$ is an external disturbance column vector; $\tau(t) \in R^3$ is the column vector that controls the joint torque input [13].

3 T-S Fuzzy Compensation Based Sliding Mode Control

3.1 The Feedback Linearization Method

In order to enhance the trajectory tracking performance of the free-floating space manipulator, this paper uses the feedback linearization method to obtain the known part of the equivalent dynamics [14]. The trajectory tracking of the space manipulator in joint space aims to allow its attitude angle θ_0 and joint hinge θ_1, θ_2 to track the desired trajectory $q_d \in R^3$ well.

In order to track the desired trajectory by the feedback linearization method, the input $\tau(t)$ can be written as:

$$\tau(t) = \widehat{D}(q)V(t) + \widehat{H}(q,\dot{q})\dot{q} \tag{2}$$

where $\hat{D}(q)$ is the known parts of $D(q)$, $\widehat{H}(q,\dot{q})\dot{q}$ is the known parts of $H(q,\dot{q})\dot{q}$, $V(t)$ is a new control vector.

Substitute Eq. (1) into (2), then $\mathrm{D(q)\ddot{q}}$ can be expressed as

$$\mathrm{D(q)\ddot{q}} = \widehat{D}(q)V(t) + \widehat{H}(q,\dot{q})\dot{q} - H(q,\dot{q})\dot{q} - \tau_d \tag{3}$$

Define

$$\widehat{H}(q,\dot{q}) - H(q,\dot{q}) = \Delta\mathrm{H(q,\dot{q})}$$

Then Eq. (3) is rewritten as

$$D(q)\ddot{q} = \widehat{D}(q)V(t) + \Delta H(q,\dot{q})\dot{q} - \tau_d \tag{4}$$

Then $\ddot{q}$ can be obtained as

$$\ddot{q} = D^{-1}(q)\widehat{D}(q)V(t) + D^{-1}(q)(\Delta H(q,\dot{q})\dot{q} - \tau_d) \tag{5}$$

Then Eq. (5) can be rewritten as

$$\ddot{q} = V(t) + \left(D^{-1}(q)\widehat{D}(q) - I\right)V(t) + D^{-1}(q)(\Delta H(q,\dot{q})\dot{q} - \tau_d) \tag{6}$$

Define

$$\eta = \left(D^{-1}(q)\widehat{D}(q) - I\right)V(t) + D^{-1}(q)(\Delta H(q,\dot{q})\dot{q} - \tau_d)$$

Then Eq. (6) is rewritten as

$$\ddot{q} = V(t) + \eta \tag{7}$$

It can be seen from the above equation that η contains all the uncertainties in the space manipulator system. If all the uncertainties in the system can be eliminated under ideal conditions, then $\eta = 0$.

In the above Eq. (7), The control vector V(t) is equal to

$$V(t) = \ddot{q}_d - W(t) \tag{8}$$

where $W(t)$ is a new control vector.

Utilize Eqs. (7) and (8), then $\ddot{q}$ is expressed as

$$\ddot{q} = \ddot{q}_d - W(t) + \eta \tag{9}$$

Define

$$e(t) = q - q_d$$

where $e(t)$ is the error between the desired trajectory and the actual trajectory of the attitude angle and the joint angles.

Find the second-order derivative for both sides of the above equation, and utilize it with Eq. (9), then $\ddot{e}(t)$ is expressed as

$$\ddot{e}(t) = -W(t) + \eta \tag{10}$$

Define

$$e(t) = X_1(t)$$

$$\dot{e}(t) = X_2(t)$$

Then Eq. (10) is converted to a state space expression as

$$\begin{cases} \dot{X}_1(t) = X_2(t) \\ \dot{X}_2(t) = -W(t) + \eta \end{cases} \tag{11}$$

3.2 Control Law Design and Stability Analysis

In order to design a suitable control law W(t) to make the trajectory of the attitude angle and the joint hinge better track to the desired trajectory, this paper gives a scheme based on sliding mode control. The sliding mode surface S(t) is defined as follows [15, 16]:

$$S(t) = CX_1(t) + X_2(t) \tag{12}$$

where C is a constant coefficient vector.

In this paper, the control law W(t) is split into the sum of the equivalent $W_{eq}(t)$ and the toggle item $W_s(t)$ as follows:

$$W(t) = W_{eq}(t) + W_s(t) \tag{13}$$

When the sliding mode control is in the sliding state, the sliding mode surface S (t) = 0, find the derivative for both sides of the equation and then $\dot{S}(t) = 0$ is obtained. $W_{eq}(t)$ can be used to confirm whether the system is on the sliding mode surface. $W_s(t)$ can be used to confirm whether the sliding mode exists, that is, $S(t)\dot{S}(t) < 0$.

When the system is on the sliding mode surface, find the derivative of Eq. (12), and utilize it with $\dot{S}(t) = 0$, comes out $C\dot{X}_1(t) + \dot{X}_2(t) = 0$.

Utilize the above equation with Eq. (11), comes out:

$$C\dot{X}_1(t) - W(t) + \eta = 0 \tag{14}$$

When designing $W_{eq}(t)$, it is assumed that the sliding mode surface is 0, so the existence of $W_{eq}(t)$ is to prevent the sliding mode surface from changing. According to this assumption, when designing the $W_s(t)$ part, it is assigned with 0. Then substitute Eq. (13) into Eq. (14), comes out:

$$C\dot{X}_1(t) - W_{eq}(t) + \eta = 0 \tag{15}$$

Which is also

$$W_{eq}(t) = C\dot{X}_1(t) + \eta \tag{16}$$

According to Eq. (16), it can be concluded that

$$\|W_{eq}(t)\| \leqslant C\dot{X}_1(t) + \|\eta\| \tag{17}$$

where $\|\eta\|$ is the norm of η. According to Eq. (17), $W_{eq}(t)$ can be converted to:

$$W_{eq}(t) = C\dot{X}_1(t) + \|\eta\| \tag{18}$$

Select the appropriate $W_s(t)$ to make the sliding mode surface tend to 0.

Now we use Lyapunov stability to prove the stability of the control system, the following Lyapunov function is designed as [17]:

$$V(S(t)) = \frac{1}{2}S^T S \tag{19}$$

Find the derivative for both sides of the above equation based on time t, then comes out:

$$\dot{V}(S(t)) = \dot{S}^T S \tag{20}$$

Utilize Eqs. (12) and (20), then $\dot{V}(S(t))$ can be obtained as

$$\dot{V}(S(t)) = \left(C\dot{X}_1(t) + \dot{X}_2(t)\right)^T S \tag{21}$$

Utilize Eqs. (11), (13) and (20), $V(S(t))$ can be obtained as

$$V(S(t)) = \left(CX_1(t) - \left(W_{\text{eq}}(t) + W_s(t)\right) + \eta\right)^T S \tag{22}$$

Substitute Eq. (18) into Eq. (22), comes out:

$$\dot{V}(S(t)) = \left(C\dot{X}_1(t)\right)^T S - \left(C\dot{X}_1(t)\right)^T S - \|\eta\|S - W_s^T(t)S + \eta^T S \tag{23}$$

According to the results of Eq. (23), when the following conditions are met, $\dot{V}(S(t)) < 0$, at this time:

$$W_s(t) = \begin{cases} \rho, & \text{if } S(t) > 0 \\ -\rho, & \text{if } S(t) < 0 \end{cases} \tag{24}$$

where ρ is a positive constant, according to Eqs. (13), (18) and (24), $W(t)$ can be obtained as

$$W(t) = \begin{cases} W^+(t) = W_{eq}(t) + \rho, \text{if } S(t) > 0 \\ W^-(t) = W_{eq}(t) - \rho, \text{if } S(t) < 0 \end{cases} \tag{25}$$

3.3 The Design of T-S Fuzzy System

In 1985, two scholars Takagi and Sugeno proposed a method to use a linear system to fit a nonlinear system and then use fuzzy algorithms to defuzzify and then simplify complex nonlinear models. Known as the T-S fuzzy logic system. A set of IF-THEN rules can be described as follows [18]:

R^i: If $x_1(t)$ is $\mathrm{A}_{1i}, \ldots\ldots, x_q(t)$ is A_{qi} then $\mathrm{u}_i(t) = f_i(X(t), t), i = 1, \ldots, r$.

where R^i represents the i-th fuzzy rule, x_j and $\mathrm{A}_{ij}(\mathrm{i} = 1, \ldots, \mathrm{r}, j = 1, \ldots, q)$ are variables and fuzzy sets respectively, r is the number of fuzzy rules.

Using the T-S fuzzy system, the system's input $U(t)$ can be transformed into a weighted average of the trajectory $X(t)$:

$$U(t) = \frac{\sum_{i=1}^{r} w_i(X(t))f_i(X(t), t)}{\sum_{i=1}^{r} w_i(X(t))} \tag{26}$$

where $X(t) \in R^q$.

Define

$$w_i(X(t)) = \prod_{j=1}^{q} A_{ij}(x_j(t)) \tag{27}$$

where $\sum_{i=1}^{r} w_i(X(t)) > 0$, for any $t > 0$.

The fuzzy rule base in this logical system is established as:

Rule 1: if $S(t) > 0$, then $W(t) = W^1(t) = W^+(t)$;
Rule 2: if $S(t) < 0$, then $W(t) = W^2(t) = W^-(t)$.

In the end, by using fuzzy reasoning to defuzzify, W can be obtained as:

$$W(t) = \frac{\sum_{i=1}^{2} w_i(S(t))W^i(t)}{\sum_{i=1}^{2} w_i(S(t))} \tag{28}$$

4 Simulation Analysis

Take the two-link free-floating rigid space manipulator system shown in Fig. 1 as an example, use the T-S fuzzy compensation based sliding mode control method for system simulation [19].

Select the following system model parameters: $m_0 = 40\,\text{kg}, m_1 = 2\,\text{kg}, m_2 = 1\,\text{kg}$, $m_p = 2\,\text{kg}$,, $l_0 = 1.5\,\text{m}, l_1 = 3\,\text{m}, l_2 = 3\,\text{m}$, $a_1 = 1.5\,\text{m}, a_2 = 1.5\,\text{m}$, $J_0 = 34.17\,\text{kg} \cdot \text{m}^2, J_1 = 1.5\,\text{kg} \cdot \text{m}^2, J_2 = 0.75\,\text{kg} \cdot \text{m}^2, J_p = 1.25\,\text{kg} \cdot \text{m}^2$; $C = \text{diag}([30, 30, 30]); \rho = 50$; the simulation initial value $q(0) = [0.70.51.8]^T$.

Let the pedestal attitude angle desired trajectory be: $\theta_{0d} = 0.5sin(t + \pi/2)$, it's unit is rad. The desired trajectory of the space manipulator's two joint hinges in the joint space is $\theta_{1d} = 0.5sin(2t)$, $\theta_{2d} = 1 + sin(2t)$, it's unit is rad. The external disturbance term $\tau_d(t)$ of the system is taken as $\tau_{0d} = [2.5cos(\pi t/6) - 2.5sin(\pi t/6)]$, $\tau_{1d} = 2.5cos(\pi t/6)$, $\tau_{2d} = 2.5sin(\pi t/6)$. The simulation process takes 5 s.

A nominal model based robust sliding mode control is used to make a simulation comparison with the proposed T-S fuzzy compensation based sliding mode control [20].

The simulations use the two different control methods shown in the flowing figures are performed in the same initial position, the same external disturbance term, and the same sliding mode surface parameters.

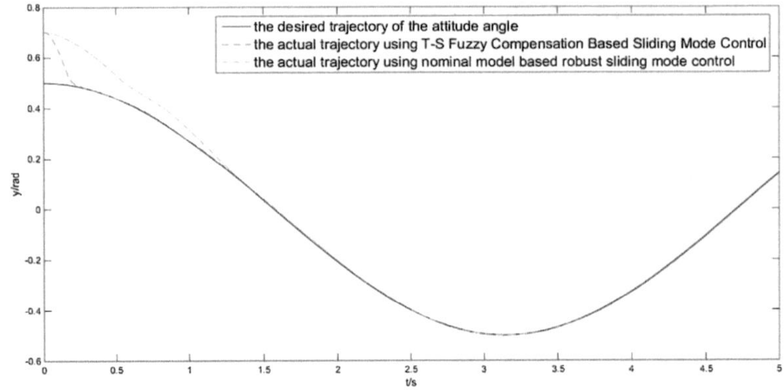

Fig. 2. The trajectory tracking of the base's attitude

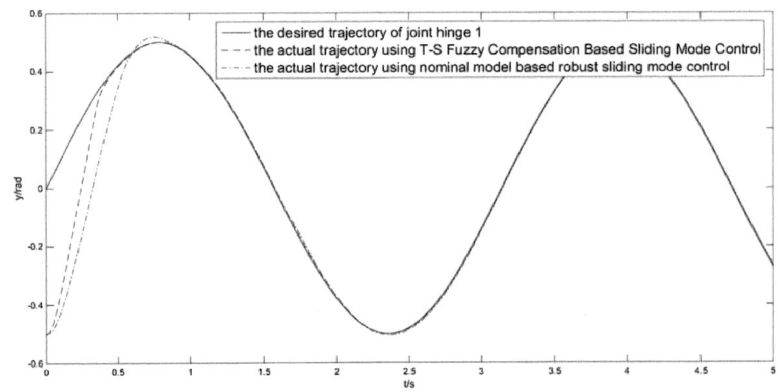

Fig. 3. The trajectory tracking of the joint hinge 1

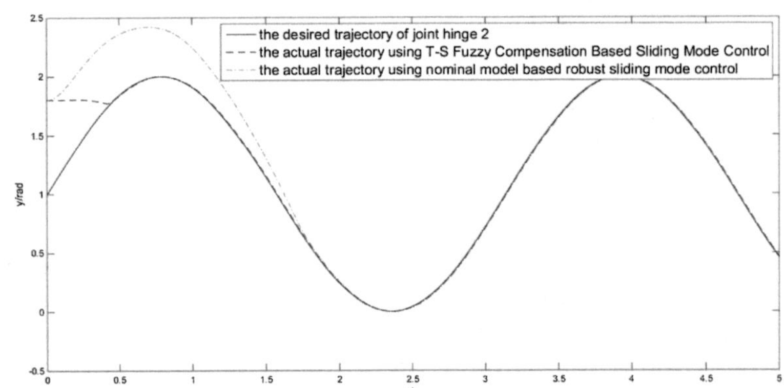

Fig. 4. The trajectory tracking of the joint hinge 2

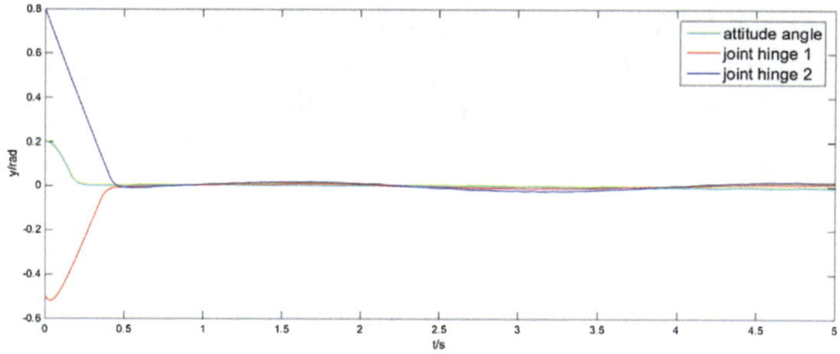

Fig. 5. Error of the T-S fuzzy compensation based sliding mode control

Figure 2 is a comparison of the attitude angle's tracking trajectory using T-S fuzzy compensation based sliding mode control and nominal model based robust sliding mode control. Figure 3 is a comparison of the joint hinge 1's tracking trajectory using T-S fuzzy compensation based sliding mode control and nominal model based robust sliding mode control. Figure 4 is a comparison of the joint hinge 2's tracking trajectory using T-S fuzzy compensation based sliding mode control and nominal model based robust sliding mode control. Figure 5 is the error of tracking trajectory of the attitude angle and two joint hinges using T-S fuzzy compensation based sliding mode control.

It can be seen from the comparison of the two control methods that the TS fuzzy compensation based sliding mode control takes about 0.5 s to make the trajectory tracking of the free-floating space manipulator base and each joint converge. However, the nominal model based robust sliding mode control takes about 2 s to make the trajectory tracking converge, and the stability after convergence is obviously not as good as the former.

The TS fuzzy compensation based sliding mode control can effectively and accurately control the free-floating space manipulator base and each joint to move according to the desired trajectory, it can eliminate the external disturbance of the system, and it can eliminate the external disturbance term τ_d of the system. The reason why the proposed method is much better than the nominal model based robust sliding mode control is that the system parameters of the TS model can be calculated offline, thus the proposed control method has small computational load, so that the response speed and convergence speed of this control is so fast.

5 Conclusions

The control of a rigid free-floating space manipulator system with uncertainties and unmodeled dynamics was discussed in this paper. First, the dynamic model of the system was derived from the Lagrange equations and momentum conservation law. And the feedback linearization method was used to obtain the known part of the equivalent dynamics. Then based on the sliding mode control theory, a suitable control law was designed. Besides, T-S fuzzy logic system was used to simplify complex nonlinear system. According to the above procedures, a sliding mode controller which combines feedback linearization and T-S fuzzy compensation is proposed. This proposed algorithm overcome the weakness of the traditional sliding mode algorithm, such as the slow response speed and convergence speed, the system's large chattering whilst it has small computational load. The feasibility and effectiveness of the proposed control method were verified by the results of numerical simulations.

Acknowledgements. This work was supported by the National Natural Science Foundation of China (No. 51741502) and the Natural Science Foundation of Fujian Province (No. 2016J01228, No. 2017J01559).

References

1. Oda, M.: Space robot experiments on NASDA's ETS-VII satellite. Adv. Robot. **13**(3), 335–336 (1998)
2. Boumans, R.: The European robotic arm for the international space station. Robot. Auton. Systems. **23**(1), 17–27 (1998)
3. Holcomb, L.B.: The NASA automation and robotics technology program. IEEE Aerosp. Electron. Syst. Mag. **2**(4), 19–26 (1987)
4. Walker, M.W.: Adaptive control of space-based robot manipulators. IEEE Trans. Robot. Autom. **7**(6), 828–835 (1991)
5. Xiaoyan, Y., Li, C.: Singular perturbation adaptive control and vibration suppression of free-flying flexible space manipulators. Proc. IMechE Part C J. Mech. Eng. Sci. **229**(11), 1989–1997 (2015)
6. Xiaoyan, Y., Li, C.: Augmented robust control of a free-floating flexible space robot. Proc. IMechE Part G J. Aerosp. Eng. **229**(5), 947–957 (2015)
7. Wang, M.: A non-linear model predictive controller with obstacle avoidance for a space robot. Adv. Space Res. **57**(8), 1737–1746 (2016)
8. Xiaoyan, Y., Li, C.: Modeling and observer-based augmented adaptive control of flexible-joint free-floating space manipulators. Acta Austronautica **108**, 146–155 (2015)
9. Xiaoyan, Y., Li, C.: Observer-based two-time scale robust control of free-flying flexible-joint space manipulators with external disturbances. Robotica **35**(11), 2201–2217 (2017)
10. Hyun, C.: Takagi-Sugeno fuzzy model based indirect adaptive fuzzy observer and controller design. Inf. Sci. **180**(11), 2314–23270 (2010)
11. Rojko, A.: Sliding-mode motion controller with adaptive fuzzy disturbance estimation. IEEE Trans. Ind. Electron. **51**(5), 963–971 (2004)
12. Rybus, T.: Control system for free-floating space manipulator based on nonlinear model predictive control. J. Intell. Robot. Syst. **85**(3–4), 491–509 (2017)
13. Green, A.: Robot manipulator control for rigid and assumed mode flexible dynamics models. In: AIAA Guidance, Navigation, & Control Conference & Exhibit (2015)
14. Cambera, J.C.: Input-state feedback linearization control of a single-link flexible robot arm moving under gravity and joint friction. Robot. Auton. Syst. **88**, 24–36 (2016)
15. Baek, J.: A new adaptive sliding-mode control scheme for application to robot manipulators. IEEE Trans. Ind. Electron. **63**(6), 3628–3637 (2016)
16. Hashemzadeh, S.M.: Optimal sliding mode control of a robot manipulator under uncertainty using PSO. Nonlinear Dyn. **84**(4), 2227–2239 (2016)
17. Jing, Z.: Adaptive dynamic sliding mode control for space manipulator with external disturbance. J. Control. Decis., 1–16 (2018)

18. Wai, R.J.: Adaptive fuzzy neural network control design via a t–s fuzzy model for a robot manipulator including actuator dynamics. IEEE Trans. Syst. Man Cybern. Part B **38**(5), 1326–1346 (2008)
19. Jinkun, L.: Robot Control System Design and Simulation of MATLAB. Tsinghua University Press, Beijing, Beijing (2008)
20. Liu, X.: An intelligent sliding mode controller based on nominal model of robotic manipulator. In: International Conference on Natural Computation IEEE (2008)

Controlling Strategy of Rail Grinding Mechanism for the Constant Working Power

Jin Xie[1(✉)], Zhongqing Yi[2], and Zhaohui Liu[1]

[1] Southwest Jiaotong University, Chengdu 610031, People's Republic of China
xj_6302@263.net
[2] Dongfeng Nissan Passenger Vehicle Company,
Guangzhou 510800, People's Republic of China

Abstract. The rail grinding mechanism comprises grinding subsystem and pressing subsystem. The grinding subsystem, in which the grinding wheel is directly driven by a motor, is mounted to a platform whose posture is adjusted by three pressing cylinders. The platform and pressing cylinders compose the pressing subsystem. The two subsystems are dynamically coupled but actuated separately. This paper focuses on design of controlling system for rail grinding mechanism to realizing constant grinding power. To this end, two controllers are designed to control the subsystems respectively. For the grinding subsystem, PI controller is employed to keep the grinding angular velocity of the motor to be constant. And for the pressing subsystem, two Active Disturbance Rejection Controllers (ADRC) are used to control the pressure in the rear end and the displacement of proportional decompressing valve element respectively, it insures that the pressing force, furthermore, the grinding torque is constant. The effectiveness of the controller is demonstrated by the results of simulation.

Keywords: Rail grinding mechanism · Mechatronics · Controlling · PI controller · ADRC

1 Introduction

Mechatronic systems are more and more used in both daily life and industry as well. The popular examples are computer numeric controlled machine (CNC) [1], robot [2], micro-electro-mechanism system (MEMS) [3].

The design and control of mechatronic system have drawn much attention of researchers. The greatest majority of research literature in this field, as a whole, focuses on motion control, such as tracking trajectory [4–7], or controlling the motion of a body in machine or mechanism [8, 9]. Next majority focuses on compensating the errors due to dynamics [10] and the thermal deformations [11]. The controlling methods employed in mechatronic system involve almost all controlling methods both conventional and advanced as well. A comparison of different controlling methods utilized in CNC is presented in Ref. [12].

Nevertheless, for some kinds of mechatronic system, the goal of designing is to achieve a constant working power under certain circumstances. One of such system is rail grinding mechanism. It has been proved by experiments that only when the power

D. Wang et al. (Eds.): MeTrApp 2019, MMS 79, pp. 430–439, 2020.
https://doi.org/10.1007/978-981-15-0142-5_42

used to grind rail is constant, is it possible to archive higher quality of grinding. By now, the controlling methods for rail grinding mechanism are mainly classical Proportional-Integral- Derivative (PID) control [13]. But the efficiency of controlling is unsatisfied. The reasons for that are likely that the system is a coupled nonlinear dynamic system, on the other hand, and the anti-interference ability of PID method is not strong enough. In general practice, more than 48 rail grinding mechanisms are mounted on a vehicle moving along the railway rail at high speed. All the rail grinding mechanisms work simultaneously to grand the rail from different direction, and the working condition of each mechanism is different from others. The vibration of the vehicle cannot be prevented. Such vibration will in turn, as interference, impact the performance of each rail grand mechanism. Based on this observation, it is necessary to propose a new strategy for controlling rail grinding mechanism.

In this paper, we divide the rail grinding mechanism into two subsystems, grinding subsystem and pressing subsystem, and apply PID and Active Disturbance Rejection Control (ADRC) [14] to the subsystems respectively. This controlling strategy makes use of structure characteristic of the rail grinding mechanism and high anti-interference ability of ADRC.

The remainder of this paper is organized as follows. In Sect. 2, the structure composition of rail grinding mechanism is presented. Owing to its composition manner, the hardware of grinding mechanism is divided into grinding subsystem and pressing subsystem. And the controlling strategy for achieve constant grinding power is proposed. In Sect. 3, the details of designing of controller for grinding subsystem and pressing subsystem are presented. Section 4 offers conclusions.

2 Rail Grinding Mechanism and Controlling Plan

2.1 The Hardware Composition of Rail Grinding Mechanism

The mechanical system of a rail grinding mechanism is depicted in Fig. 1. It can be divided into two subsystems. The grinding motor is connected directly with the grinding wheel. They form a so-called grinding subsystem. The grinding subsystem is mounted on a platform whose posture is determined by three pressing cylinders. With adjusting the lengths of three pressing cylinders, the platform moves toward to the rail and the grinding wheel grinds the rail in different direction. The platform and pressing cylinder compose a so-called pressing subsystem.

2.2 The Planning of Controlling System

From the points of view of structure composition of mechanism, rail grinding mechanism possesses two degree of freedom: one for the grinding subsystem which is actuated by the motor, and the other one for pressing subsystem which is actuated by the pressing cylinders. Two subsystems are coupled with the variable M_L, which denotes the grinding torque.

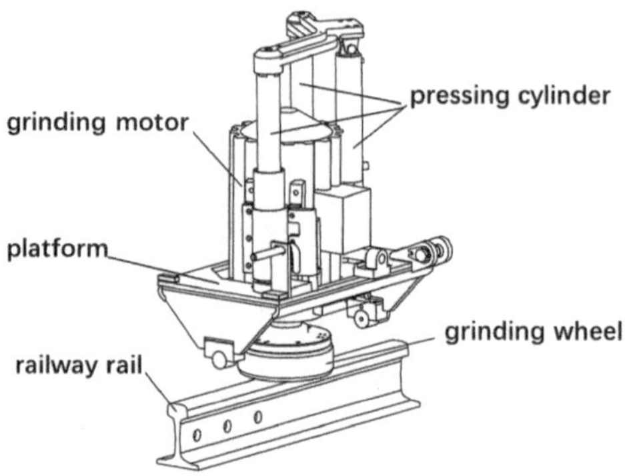

Fig. 1. The sketch of rail grinding mechanism

The grinding power P can be calculated by

$$P = \omega_r M_L \tag{1}$$

where ω_r represent the angular velocity of the grinding wheel. Since the grinding wheel is directly connected with the grinding motor, ω_r is the output angular velocity of the motor.

The grinding torque M_L is the coupled variable between the two subsystems. For the grinding subsystem it is the input torque, or load, and for the pressing subsystem it is the output torque. The magnitude of M_L dependents on pressing force F_n provided by the pressing cylinder, as shown in Fig. 2a. To determine the relationship between M_L and F_n, we simplify the grinding rail into cutting amount grain with cone shape [16]. The specific cutting force is denoted as F_p, half cone angle as θ_z, and the depth as a_g. Then relationship between the tangential grinding force F_T and the normal grinding force F_N can be represented as

$$F_T = \frac{\pi}{4} F_N \cot \theta_z \tag{2}$$

And from Fig. 2b, we have

$$M_L = 2\sum_{R_1}^{R_2} dF_T(r)r = \frac{\pi}{4} F_n \cot \theta_z (R_1 + R_2) \tag{3}$$

Equation (3) means that the grinding torque M_L is proportional to the pressing force F_n.

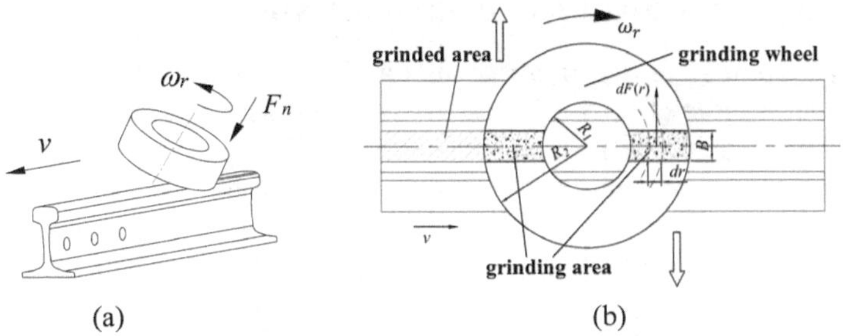

Fig. 2. The model for calculation of the grinding torque

As mentioned above, the grinding subsystem and the pressing subsystem are actuated separately. it allows to design two controllers for the two subsystems respectively, one is for the grinding subsystem to keep the ω_r to be a constant, and the other one to keep M_L, or F_n to be a constant, further to gain a constant grinding power.

To realize this controlling plan, we employ PI controller to control the output angular velocity of the motor ω_r. In order to qualify the controlling system with the ability of anti-interference, we employ ADRC to control the grinding torque M_L, actually to control the pressing force F_n.

The block diagram of overall controlling system is depicted in Fig. 3. In the figure, the superscripts "*" denote the desired values of controlling. As shown in Fig. 3, we utilize two ADRCs in controlling of pressing subsystem to insure that the grinding power is constant more precisely.

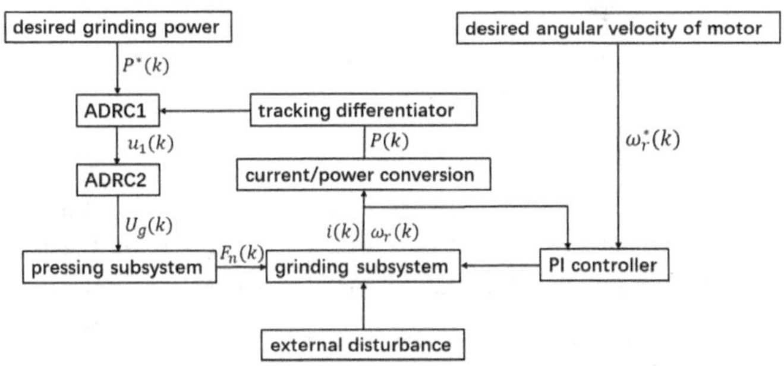

Fig. 3. The controlling scheme for rail grinding mechanism

3 Design of Controlling System

In this section, we will discuss the parameters design of the controller for grinding subsystem and pressing subsystem respectively.

3.1 Designing of Controller for Grinding Subsystem

The grinding motor is of squirrel-cage motor. To analyze of the motor motion, the principle of rotor indirect field oriented vector control is applied. This method, comparing with direct field oriented vector control, doesn't need to directly measure the rotor magnetic field, and achieve the directional control through controlling the rotate difference frequency. In addition to this, this method is easier to be implemented.

In order to control the output angular velocity to be a constant, PI controller is used. The concept of the controller for the grinding subsystem is shown in Fig. 4. In Fig. 4, the superscripts "*" denotes the desired value of variables and ψ_r denotes the flux of the motor, L, i denote the inductance and current respectively, and the subscripts m, t denote the orthogonal coordinate of rotor indirect field.

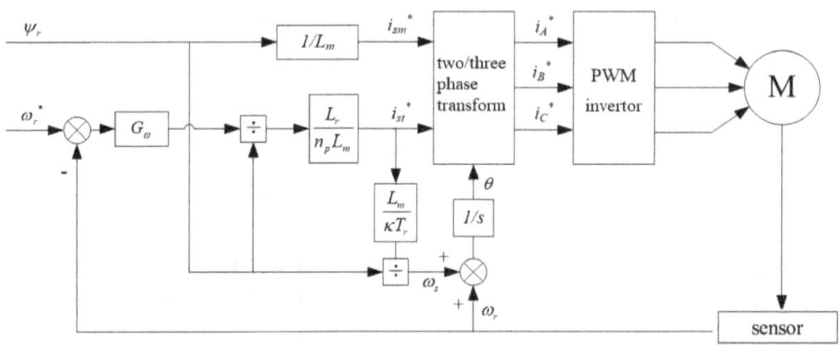

Fig. 4. The concept of controller for grinding subsystem

Proportional-Integral (PI) controller is represented as

$$T_e^* = k_p(\omega_r^* - \omega_r) + k_i \int (\omega_r^* - \omega_r)dt \tag{4}$$

where ω^* is the desired angular velocity of the motor, and ω_r is the real angular velocity of the motor, k_p is the proportional gain, and k_i is the integral gain. Then the key to design a PI controller is to find out appropriate values of k_p and k_i.

Based on Fig. 4, we get the dynamics equations of grinding subsystem with controller as following

$$\begin{cases} \frac{d\psi_{rm}}{dt} = -\eta\psi_{rm} + \frac{1}{\kappa T_r i_{sm}^*}\psi_{rt} i_{st} + L_m \eta i_{sm}^* \\ \frac{d\psi_{rt}}{dt} = -\eta\psi_{rt} - \frac{1}{\kappa T_r i_{sm}^*}\psi_{rm} i_{st} + L_m \eta i_{st}^* \\ \frac{d\omega_r}{dt} = \delta(\psi_{rm} i_{st} - i_{sm}^* \psi_{rt}) - \lambda M_L \\ \frac{di_{st}}{dt} = \varepsilon\{k_p[\omega_r^* - \delta(i_{st}\psi_{rm} - i_{sm}^*\psi_{rt}) - \lambda M_L] + k_i(\omega_r^* - \omega_r)\} \end{cases} \tag{5}$$

where $i_{sm}^* = \frac{\psi_r^*}{\hat{L}_m}$, $i_{st}^* = \frac{\hat{L}_r T_e^*}{n_p \hat{L}_m \psi_{rm}}$, $\omega_s^* = \frac{\hat{L}_m}{\hat{T}_r \psi_r^*} i_{st}^*$, and η, λ, δ are the parameters of the motor.

For simplifying the analysis, we take $k_i = 10k_p$, and plot the bifurcation diagrams of flux of the motor versus the integral gain, as shown in Fig. 5. It can be found that the output angular velocity of motor finally is stabilized at the desired value ω^* as $k_i \geq 1.7$. Therefore, we take the proportional gain $k_p = 1$ and the integral gain $k_i = 10$.

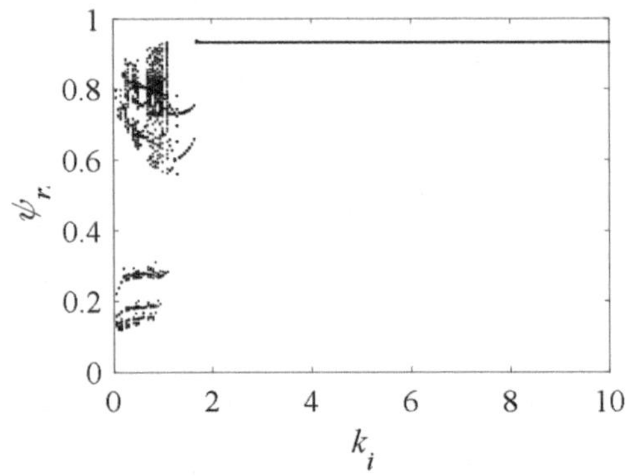

Fig. 5. The bifurcation diagram of the grinding subsystem versus integral gain k_i

3.2 Designing of Controller for Pressing Subsystem

The principle of the pressing subsystem is shown in Fig. 6.

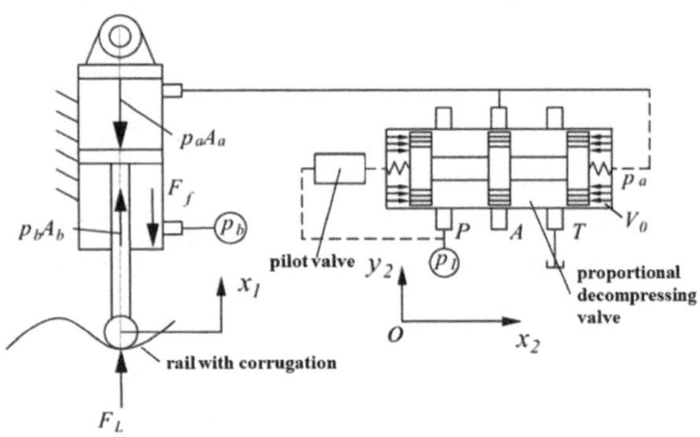

Fig. 6. The principle of the pressing subsystem

It can be seen from Fig. 6 that since the grinding wheel contacts with the rail, and the rail corrugation can be described with sinusoid [17], the motion of the rod in pressing cylinder can be written as

$$\begin{cases} x_1 = A_m \sin(\omega_C t) \\ \dot{x}_1 = A_m \omega_C \cos(\omega_C t) \\ \ddot{x}_1 = -A_m \omega_C^2 sin(\omega_C t) \end{cases} \tag{6}$$

Such motion of rod will cause the platform oscillating, and the pressure in the rear end, p_a, varying.

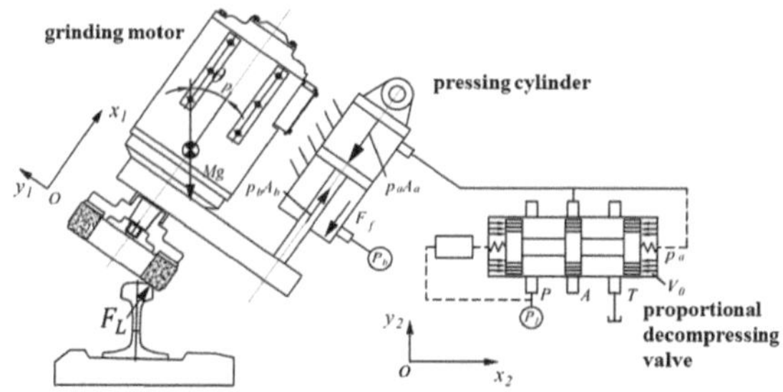

Fig. 7. The model for calculation of the pressure p_a

Figure 7 describes the conversion of force from the pressing cylinder to the grinding wheel. From Fig. 7, we get

$$F_L = F_n - Mg\cos\theta_p - Mg\mu\sin\theta_p \tag{7}$$

and

$$\begin{aligned} p_a =& [F_n - Mg\cos\theta_p - Mg\mu\sin\theta_p + p_b A_b \\ &+ MA_m\omega_C^2\sin(\omega_c t) - B_{m1}A_m\omega_C\cos(\omega_c t)]/A_a \end{aligned} \tag{8}$$

From Eq. (8), it can be seen that only when p_a is a constant, is it possible to keep the pressing force, F_n, to be a constant. So the goal of controlling pressing subsystem can be stated as keeping the pressure in the rear end, p_a, to be a constant.

Also based Fig. 7, and Eq. (8), the dynamics equations of the pressing subsystem can be derived, they can be written as

$$\begin{cases} \frac{dp_a}{dt} = \frac{\beta_e}{V_{E1}+V_{E2}}[f_3 + f_6 + A_0\frac{dx_2}{dt} - C_{ic}p_a - A_0A_m\frac{2\pi v}{\lambda_p}\cos\left(\frac{2\pi v}{\lambda_p}t\right) - C_{ec}(p_a - p_b)] \\ \frac{dx_2}{dt} = v_2 \\ \frac{dv_2}{dt} = \frac{u - kx_2 - B_{m2}\frac{dx_2}{dt} - A_0p_a}{m} \end{cases} \tag{9}$$

where β_e represents the elastic modulus of the hydraulic oil; V_{E1} and V_{E2} represent the volumes of rear end and right chamber of the master valve respectively; f_3 and f_6 represent the flow rates at the P port and T port of the proportional decompressing valve respectively; A_0 represents the end area of valve core of master valve; C_{ic} and C_{ec} represent the coefficients of external and internal leakage respectively; u represents the control variable proportional decompressing valve; B_{m2} represent the adhesion coefficients of valve core of master valve; k and m represent the stiffness of the spring and the mass of valve core of master valve respectively.

To control p_a, we utilize ADRC controller. ADRC inherits from Proportional-Integral-Derivative (PID) controlling [14]. Over PID controlling, ADRC has the advantages of high precision and high anti-interference ability. Roughly speaking, three components involved in ADRC play important roles. The first component is the tracking differentiator (TD) used to arrange the ideal transient process of the system; the second component is the extended state observer (ESO) used to estimate all the external disturbances; the third component is the nonlinear state error feedback (NLSEF) used to obtain the control input.

The control scheme of pressing subsystem proposed here is that the displacement of proportional decompressing valve element, x_2, is controlled by electromagnetic force, and it in turn control the pressure in the rear end, p_a. So we should design two ADRCs, one for controlling of x_2, the other one for controlling of p_a.

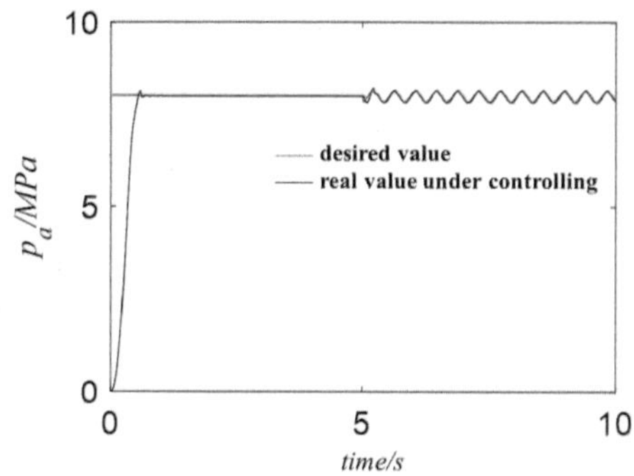

Fig. 8. The pressure tracking response of the controller for pressing subsystem

In order to examine the efficiency of the controller, the simulations are carried out with Eq. (9). Figure 8 presents the results of simulations. At the beginning of simulation, the grinding wheel is separated from the rail, and the pressure in the rear end $p_a = 0$; at the first second, p_a rises to 7.21 MPa; and at the fifth second, an interference of rail corrugation, $2\sin(15t)$, is applied to the pressure p_a. Thereafter, the real value of pressure p_a under controlling matches desired value very well.

4 Conclusion

We proposed a controlling system for rail grinding mechanism to address the problem of realizing the constant working power. The rail grinding mechanism is a typical mechatronic system, and can be divided into grinding subsystem and pressing subsystem, and the subsystems are actuated by motor and cylinder respectively. Such hardware configuration of rail grinding mechanism allows to design the controller for each subsystem with different aim. For the grinding subsystem, the controller is designed based on principle of rotor indirect field oriented vector control and PI controller with the aim at keeping a constant angular velocity of the motor. While for the pressing subsystem, the controller is designed utilizing Active Disturbance Rejection Controlling (ADRC) with the aim at keeping a constant pressing force and, furthermore, a constant grinding torque. The effectiveness of the controlling system is demonstrated by the results of simulation. And in near future, some experiments should be conducted to verify the effectiveness.

The problem of controlling design for constant working power not only concerns with the motion of the system but the dynamics of the system. In order to arrive this goal, the design of hardware and software both are equal important. Well-designed hardware facilities the design of the software.

The methods presented in this paper are valuable help to the design of other mechatronic systems.

Acknowledgements. The authors would like to acknowledge the financial support of NSFC (National Natural Science Foundation of China) under the grant No. 51575457.

References

1. Altintas, Y.: Manufacturing Automation: Metal Cutting Mechanics, Machine Tool Vibrations, and CNC Design. Cambridge University Press, Cambridge (2012)
2. Craig, J.J.: Introduction to Robotics: Mechanics and Control, 4th edn. Pearson Education Inc., London (2018)
3. Liu, C.: Foundations of MEMS, 2nd edn. Pearson Education Inc., London (2012)
4. Chen, C., Chang, Y., Tsao, T.: Dynamic trajectory tracking by synergistic dual-stage actuation and control. IEEE/ASME Trans. Mechatron. **22**(6), 2600–2610 (2017)
5. Qiu, L., Shi, Y., Pan, J.: Positioning-tracking controller design of a linear motion control system based on vectorization technique. IEEE/ASME Trans. Mechatron. **23**(4), 1512–1520 (2018)
6. Rezayi, S., Arbabtafti, M.: A new model-based control structure for position tracking in an electro-hydraulic servo system with acceleration constraint. J. Dyn. Syst. Meas. Control. **139**(12), 121006-1-21006-121011 (2017)
7. Wu, X., Shen, W., Lou, P., et al.: An automated guided mechatronic tractor for path tracking of heavy-duty robotic vehicles. Mechatronics **35**, 23–31 (2016)
8. Dymov, I.S., Kotin, D.A.: Synthesis of automatic control system of the rotor axial displacement of the mechatronic motion module. In: 11th International Forum on Strategic Technology (IFOST), pp. 48–51 (2016)
9. Flores-Hernández, D.A., Palomino-Resendiz, S., Lozada-Castillo, N., et al.: Mechatronic design and implementation of a two axes sun tracking photovoltaic system driven by a robotic sensor. Mechatronics **47**, 148–159 (2017)
10. Keck, A., Sawodny, O., Gronle, M., et al.: Model-based compensation of dynamic errors in measuring machines and machine tools. IEEE/ASME Trans. Mechatron. **23**(5), 2252–2262 (2018)
11. Assaad, B., Benkara, K.E., Friedrich, G., et al.: An advanced simulation tool for considering the electric machines thermal performance in mechatronic systems. In: 11th France-Japan & 9th Europe-Asia Congress on Mechatronics (MECATRONICS)/17th International Conference on Research and Education in Mechatronics (REM), pp. 197–202 (2016)

12. Pushpkant, Jha, S.K.: Comparative study of different classical and modern control techniques for the position control of sophisticated mechatronic system. Procedia Comput. Sci. **93**, 1038–1045 (2016)
13. Zeng, H., Yu, G., et al.: Design of rail grinding control system. Electr. Drive Locomot. **4**, 79–82 (2018). (in Chinese)
14. Han, J.: From PID to active disturbance rejection control. IEEE Trans. Ind. Electron. **56**(3), 900–906 (2009)
15. Duindam, V., Macchelli, A., Stramigioli, I.S., Bruyninckx, H. (eds.): Modeling and Control of Complex Physical System. Springer, Berlin (2009)
16. King, R.I., Hahn, R.S.: Handbook of Modern Grinding Technology. Springer, New York (1986)
17. Jin, X.-S., Wang, K.-Y., et al.: Effect of rail corrugation on vertical dynamics of railway vehicle coupled with a track. Acta. Mech. Sin. **21**(1), 95–102 (2005)
18. Xia, Y., Fu, M.: Overview of ADRC. In: Compound Control Methodology for Flight Vehicles, Lecture Notes in Control and Information Science, vol. 438. Springer (2013)

Terminal Sliding Mode Adaptive Fuzzy Controller of a Free-Floating Space Manipulator Based on Time Delay Estimation

Jianyu Zhang[1], Xiaoyan Yu[1,2(✉)], and Li Chen[1]

[1] School of Mechanical Engineering and Automation, Fuzhou University, Fuzhou 350116, Fujian Province, China
cool@fzu.edu.cn

[2] Key Laboratory of Fluid Power and Intelligent Electro-Hydraulic Control, (Fuzhou University), Fujian Province University, Fuzhou 350116, Fujian Province, China

Abstract. A time delay estimation based terminal sliding mode adaptive fuzzy control for trajectory tracking control of free-floating robot manipulators is presented. This controller consisted of a time delay estimation element, a terminal sliding mode and an adaptive fuzzy system. Based on time delay estimation technology, which can estimate most unknowns of the system for reducing the negative impact of uncertainty on stability caused by system inertia parameter, a terminal sliding mode adaptive fuzzy controller is designed. A nonlinear error dynamic model is established by terminal sliding mode, which can also make actual trajectory track the desired trajectory well in a short time. The adaptive fuzzy system would compensate for the error caused by the delay estimation technique and suppresses the chattering problem caused by the terminal sliding mode. It is proved that the tracking errors converge into zero asymptotically by using Lyapunov stability theory. Last numerical simulation shows that the controller has good tracking performance.

Keywords: Free-Floating space robot · Time delay estimation · Terminal sliding mode control · Adaptive fuzzy control

1 Introduction

In recent years, with the advancement of research work, the exploration of space has broken through the limitations of technology, and human space activities in space are increasing [1, 2]. Because of the characteristics of microgravity, high vacuum, and strong radiation in the space environment, the space robot that can replace the astronauts in the extravehicular operation has been well developed, especially for the dynamic control research of space robots, which has received more attention [3, 4]. In normal, there are some parameters in the actual space arm system which are uncertain, such as the quality of the clamped load. In the case of some parameters being uncertain, the design of the control system is very difficult [5]. Shin [6], Paraktuna [7], Gu [8] proposed an improved adaptive control method for space robots with uncontrolled position and uncontrolled attitude. A robust control and adaptive control scheme was

D. Wang et al. (Eds.): MeTrApp 2019, MMS 79, pp. 440–449, 2020.
https://doi.org/10.1007/978-981-15-0142-5_43

designed by Li Chen [9] to solved the control problem with uncertain parameters but the error range is known. Yu [10] proposed a terminal sliding mode control scheme for a floating-based flexible space robot with uncontrolled carrier position and attitude.

The delay estimation technique (TDE) was proposed around 1990 and has received wide attention as its low dependence on the dynamics model of the system [11]. This feature makes it more suitable for applications because it does not require the full knowledge of the controlled object and has good robustness even under large disturbances and parameter variations [12, 13]. Due to the nonlinear error dynamic model, the terminal sliding mode (TSM) can achieve faster convergence. But, without taking the compensation for the TDE error into account, TSM also has chattering problems [14].

In this paper, the trajectory tracking and control problem of free-floating space robot system is analyzed. Firstly, based on the TDE technique, which can estimate most of the unknowns in the kinetic model, a TSM adaptive fuzzy controller is designed. TSM is used as the dynamic model of expected error to pursue fast convergence of error. The fuzzy logic system is exploited to suppress the effect of the TDE error and the chattering problem caused by the TSM.

This paper is organized as follows. The next section builds the dynamics of free-floating space manipulator. The third section proposes a TSM adaptive fuzzy controller of a free-floating space manipulator based on TDE and the controller is proved by using Lyapunov stability theory. Finally, conclusions are given.

2 Dynamics of Free-Floating Space Manipulator

Without any loss of generality, a free-floating space manipulator with the position uncontrolled for saving the jet fuel is considered, as shown in Fig. 1. The system consists of the floating base B_0 and connected with two rigid arms B_1, B_2, where P is the target satellite at the end of the system. The following Table 1 gives other symbols.

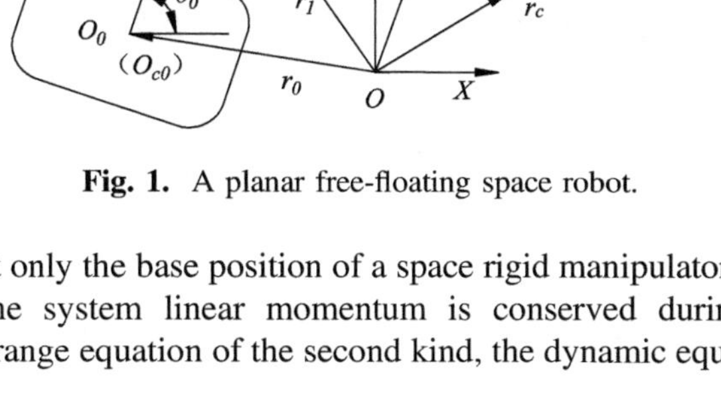

Fig. 1. A planar free-floating space robot.

Assuming that only the base position of a space rigid manipulator is not controlled, we know that the system linear momentum is conserved during the operation. According to Lagrange equation of the second kind, the dynamic equation is expressed as follows:

Table 1. Symbols in dynamic model.

Symbols	Definition
$(O - XY)$	Inertial coordinate
$(O_i - x_i y_i)$	Local coordinate local coordinate frame of $B_i(\mathrm{i} = 0, 1, 2)$
r_i	Position vector of $B_i(\mathrm{i} = 0, 1, 2)$
r_c	Position vector of the mass center C of the entire system
d_i	Distance from joint $O_i$to the mass center $O_{Ci}(\mathrm{i} = 1, 2)$;
m_i	Mass of $B_i(\mathrm{i} = 0, 1, 2)$
J_i	Inertial moment of $B_i$with respect to its mass center $O_{Ci}(\mathrm{i} = 0, 1, 2)$
θ_0	The angle between the X axis and the $x_0$axis
θ_i	The angle between the x_{i-1}axis and the $x_0$axis$(\mathrm{i} = 1, 2)$

$$D(\theta)\ddot{\theta} + H(\theta, \dot{\theta})\dot{\theta} = \tau \tag{1}$$

where $\mathrm{D}(\theta) \in R^{3\times3}$ is the symmetric positive definite inertial matrix, $\mathrm{H}(\theta, \dot{\theta}) \in R^{3\times1}$ is the vector of the Coriolis and centrifugal force, $\tau = [\tau_0\, \tau_1\, \tau_2]^T \in R^{3\times1}$ is the control input, $\theta = (q_0\, q_1\, q_2)^T$ is the generalized coordinates vector of space manipulator, with $\dot{\theta}$, $\ddot{\theta}$ the velocity and acceleration of the joints.

For a free-floating space manipulator, it is very difficult to establish a model that can tracks the desired trajectory well. Therefore, a model that does not require stricter parameters is needed. Equation (1) can be rearranged as:

$$\tau = \overline{D}\ddot{\theta} + \overline{H} \tag{2}$$

where $\overline{D}$ is a positive diagonal matrix to be designed, and

$$\overline{H} = (D - \overline{D})\ddot{\theta} + H\left(\theta, \dot{\theta}, \ddot{\theta}\right) \tag{3}$$

which would be estimated by TDE.

3 Combined Control Law Design

3.1 Terminal Sliding Mode Control and Sliding Surface

The nonlinear error estimate is used to compensate the error, the convergence speed will increase, and it can converge in a limited time. In order to make the free-floating space manipulator to track the actual trajectory quickly and accurately, a TSM sliding surface is defined as

$$s = e + K\dot{e}^{\frac{p}{q}} \tag{4}$$

where e denotes the error between the desired trajectory and actual trajectory, $e = \theta_d - \theta$, $\dot{e} = \dot{\theta}_d - \dot{\theta}$, and $\ddot{e} = \ddot{\theta}_d - \ddot{\theta}$; where K is a constant diagonal matrix; p > 0, q > 0, 1 < p/q < 2 and are both odd.

3.2 Controller Structure

The control method based on TDE of TSM adaptive fuzzy can be designed as

$$\tau = \overline{D}\gamma + \widehat{H}\left(\theta, \dot{\theta}, \ddot{\theta}\right) \tag{5}$$

$$\gamma = \ddot{\theta}_d + \tfrac{q}{p} K^{-1} \dot{e}^{2-\frac{p}{q}} + k + \rho s \tag{6}$$

where $\widehat{H}(\theta, \dot{\theta}, \ddot{\theta})$ is the estimate of $\overline{H}$. And $k = [k_1, \cdots k_i \cdots, k_n]$ denotes the fuzzy logic system, with the corresponding compensation item as ρs. Where $\rho = \mathrm{diag}[a_1 + \sigma_1, \cdots, a_i + \sigma_i, \cdots, a_n + \sigma_n]$ is a positive diagonal coefficient matrix, $a_i > 0$ and $\sigma_i > 0$.

Here $\widehat{H}(\theta, \dot{\theta}, \ddot{\theta})$ can be estimated by the TDE [10], as

$$\overline{H} \approx \overline{H}_{t-L} \tag{7}$$

where $\cdot_{t-L}$ is the previous status of $\cdot$, t is current control time, L is the estimated time delay which can be taken directly as the sampling frequency, and when L is infinitely small, it can be considered that (7) is more accurate.

Therefore, TDE improves system robustness in the following form

$$\widehat{H}\left(\theta, \dot{\theta}, \ddot{\theta}\right) = \overline{H}_{t-L} = u_{t-L} - \overline{M}\ddot{\theta}_{t-L} \tag{8}$$

Thus, with the combination of (5), (6), (8) the control law is expressed by

$$\tau = \overline{D}\left(\ddot{\theta}_d + \tfrac{q}{p} K^{-1} \dot{e}^{2-\frac{p}{q}} + k + \rho s\right) + \tau_{t-L} - \overline{D}\ddot{\theta}_{t-L} \tag{9}$$

Incidentally, the past acceleration $\ddot{\theta}_{t-L}$ is given by numerical differentiation as

$$\ddot{\theta}_{t-L} = \frac{\theta_t - 2\theta_{t-L} + \theta_{t-2L}}{L^2} \tag{10}$$

According to the study of Chang and Lee [15] and Lee and Chang [16], in order to prove boundedness, the conditions should be:

$$\left|\phi_i\left(D_k^{-1}\overline{D} - I_n\right)\right| < 1 \tag{11}$$

where $\left|\phi_i\left(D_k^{-1}\overline{D} - I_n\right)\right|$ represents the i th eigenvalue of $D_k^{-1}\overline{D} - I_n$. The determination of the value of $\overline{D}$ needs to satisfy the above conditions.

The core idea of TDE technique is to estimate the total set of system state quantities at the current time from the system state quantity at the previous moment.

$$\overline{H} - \widehat{H} = \overline{H}_t - \overline{H}_{t-L} = \overline{D}\left(\gamma - \ddot{\theta}\right) \tag{12}$$

with the TDE error δ defined as

$$\delta = \overline{D}^{-1}\left[\overline{H}_t - \widehat{H}_t\right] = \gamma - \ddot{\theta} \tag{13}$$

3.3 Fuzzy Logic System

The typical system of fuzzy controller consists of four parts: fuzzy interface, knowledge base, inference engine and defuzzification interface.

According to the analysis of the designed sliding mode sliding surface and the fuzzy module, when s_i is large, k_i is also large; when s_i is small, k_i is also small; when s_i is zero, k_i is also zero. Then the fuzzy rules take the following form:

$$\text{If } s_i \text{ is } A_i^m, \text{ then } k_i \text{ is } B_i^m$$

where A_i^m and B_i^m are the fuzzy sets. The membership function of s_i and k_i is: NB, NS, ZE, PS and PB. Define it as Gaussian membership function, expressed as:

$$\mu_A(x_i) = \exp\left[-\left(\tfrac{x_i - \alpha}{\sigma}\right)^2\right] \tag{14}$$

where A represents fuzzy sets, x_i represents k_i and s_i, α is the center of A, and σ is the width of A.

The definition of fuzzy rules is:

If s_i is NB, then k_i is NB.
If s_i is NS, then $k_i$is NS.
If s_i is ZE, then k_i is ZE.
If s_i is PS, then k_i is PS.
If s_i is PB, then k_i is PB.

Take the singleton fuzzifer, the product inference engine, and the center-average defuzzifer; then, the output variable k_i can be described as

$$k_i = \frac{\sum_{m=1}^{M} \theta_{k_i}^m \mu_A^m(s_i)}{\sum_{m=1}^{M} \mu_A^m(s_i)} = \theta_{k_i}^T \psi_{k_i}(s_i) \tag{15}$$

where M is the number of fuzzy rules, $\theta_{k_i} = [\theta_{k_i}^1, \cdots, \theta_{k_i}^m, \cdots, \theta_{k_i}^M]^T$ with $\theta_{k_i}^m$ being an adjustable value, and $\psi_{k_i}(s_i) = [\psi_{k_i}^1(s_i), \cdots, \psi_{k_i}^m(s_i), \cdots, \psi_{k_i}^M(s_i)]^T$ is a fuzzy basis vector, whose element is of the form

$$\psi_{k_i}^m(s_i) = \frac{\mu_A^m(s_i)}{\sum_{m=1}^M \mu_A^m(s_i)} \tag{16}$$

3.4 Adaptive Module

Substituting (9) into (2), we obtain the following error:

$$\delta = \ddot{e} + \tfrac{q}{p} K^{-1} \dot{e}^{2-\frac{p}{q}} + k + \rho s \tag{17}$$

Let the estimated adjustable value be $\theta_{k_i}^*$, the optimal estimate error can be defined as follows [17]:

$$k_i = \theta_{k_i}^{*T} \psi_{k_i}(s_i) \tag{18}$$

When there is $w_i > 0$, and satisfied:

$$\left| \delta_i - \theta_{k_i}^{*T} \psi_{k_i}(s_i) \right| \le w_i \tag{19}$$

The k_i can be defined as follows:

$$k_i = \widetilde{\theta}_{k_i}^T \psi_{k_i}(s_i) + \theta_{k_i}^{*T} \psi_{k_i}(s_i) \tag{20}$$

where $\widetilde{\theta}_{k_i} = \theta_{k_i} - \theta_{k_i}^*$.

The boundary conditions of the compensation item should meet the following requirements:

$$w_i \le \sigma_i^* |s_i| \tag{21}$$

Then we can get:

$$\rho_i s_i = (a_i + \sigma_i) s_i = a_i s_i + \sigma_i s_i \tag{22}$$

where $\widetilde{\sigma}_i = \sigma_i - \sigma_i^*$.

Design the adaptive law as follows:

$$\dot{\widetilde{\theta}}_{k_i} = \tfrac{p_i}{q_i} K_i \dot{e}^{p_i/q_i - 1} \eta_{\theta_{k_i}} s_i \psi_{k_i}(s_i) \tag{23}$$

$$\dot{\widetilde{\sigma}}_i = \tfrac{p_i}{q_i} K_i \dot{e}^{p_i/q_i - 1} \eta_{\sigma_i} s_i^2 \tag{24}$$

where $\eta_{\theta_{k_i}}$ and η_{σ_i} are Positive fixed parameters.

Proof. Construct a Lyapunov function as follows:

$$V = \frac{1}{2}s^T s + \frac{1}{2}\sum_{i=1}^{n}\left(\frac{1}{\eta_{\theta_{k_i}}}\tilde{\theta}_{k_i}^T\tilde{\theta}_{k_i}\right) + \frac{1}{2}\sum_{i=1}^{n}\left(\frac{1}{\eta_{\sigma_i}}\tilde{\sigma}_i^T\tilde{\sigma}_i\right) \tag{25}$$

Differentiating V with respect to time yields:

$$\begin{aligned}\dot{V} &= s^T\dot{s} + \sum_{i=1}^{n}\frac{1}{\eta_{\theta_{k_i}}}\tilde{\theta}_{k_i}^T\dot{\tilde{\theta}}_{k_i} + \sum_{i=1}^{n}\frac{1}{\eta_{\sigma_i}}\tilde{\sigma}_i^T\dot{\tilde{\sigma}}_i \\ &= s^T\left(\dot{e} + \frac{p}{q}K_i\dot{e}^{p/q-1}\ddot{e}\right) + \sum_{i=1}^{n}\frac{1}{\eta_{\theta_{k_i}}}\tilde{\theta}_{k_i}^T\dot{\tilde{\theta}}_{k_i} + \sum_{i=1}^{n}\frac{1}{\eta_{\sigma_i}}\tilde{\sigma}_i^T\dot{\tilde{\sigma}}_i\end{aligned} \tag{26}$$

Substituting $\delta = \ddot{e} + \frac{q}{p}K^{-1}\dot{e}^{2-\frac{p}{q}} + k + \rho s$ into Eq. (26):

$$\begin{aligned}\dot{V} &= s^T\left[\dot{e} + \frac{p}{q}K_i\dot{e}^{p/q-1}\left(\delta - \frac{q}{p}K^{-1}\dot{e}^{2-\frac{p}{q}} - k - \rho s\right)\right] + \sum_{i=1}^{n}\frac{1}{\eta_{\theta_{k_i}}}\tilde{\theta}_{k_i}^T\dot{\tilde{\theta}}_{k_i} \\ &\quad + \sum_{i=1}^{n}\frac{1}{\eta_{\sigma_i}}\tilde{\sigma}_i^T\dot{\tilde{\sigma}}_i \\ &= s^T\frac{p}{q}K_i\dot{e}^{p/q-1}(\delta - k - \rho s) + \sum_{i=1}^{n}\frac{1}{\eta_{\theta_{k_i}}}\tilde{\theta}_{k_i}^T\dot{\tilde{\theta}}_{k_i} + \sum_{i=1}^{n}\frac{1}{\eta_{\sigma_i}}\tilde{\sigma}_i^T\dot{\tilde{\sigma}}_i \\ &= \sum_{i=1}^{n}s_i\frac{p_i}{q_i}K_i\dot{e}^{p_i/q_i-1}(\delta_i - k_i - \rho_i s_i) + \sum_{i=1}^{n}\frac{1}{\eta_{\theta_{k_i}}}\tilde{\theta}_{k_i}^T\dot{\tilde{\theta}}_{k_i} + \sum_{i=1}^{n}\frac{1}{\eta_{\sigma_i}}\tilde{\sigma}_i^T\dot{\tilde{\sigma}}_i\end{aligned} \tag{27}$$

Substituting $k_i = \tilde{\theta}_{k_i}^T\psi_{k_i}(s_i) + \theta_{k_i}^{*T}\psi_{k_i}(s_i)$ and $\rho_i s_i = a_i s_i + \sigma_i s_i$ into Eq. (27):

$$\dot{V} = \sum_{i=1}^{n}s_i\frac{p_i}{q_i}K_i\dot{e}^{p_i/q_i-1}\left[\delta_i - \tilde{\theta}_{k_i}^T\psi_{k_i}(s_i) - \theta_{k_i}^{*T}\psi_{k_i}(s_i) - a_i s_i - \sigma_i s_i\right] \tag{28}$$

Substituting the adaptive law into Eq. (28):

$$\dot{V} = -\sum_{i=1}^{n}\frac{p_i}{q_i}K_i\dot{e}^{p_i/q_i-1}s_i a_i s_i + \sum_{i=1}^{n}\frac{p_i}{q_i}K_i\dot{e}^{p_i/q_i-1}s_i\left[\delta_i - \theta_{k_i}^{*T}\psi_{k_i}(s_i) - \sigma_i^{*T}s_i\right] \tag{29}$$

Since $\left|\delta_i - \theta_{k_i}^{*T}\psi_{k_i}(s_i)\right| \le w_i$ and $w_i \le \sigma_i^*|s_i|$, it can be obtained that:

$$\begin{aligned}\frac{p_i}{q_i}K_i\dot{e}^{p_i/q_i-1}s_i\left[\delta_i - \theta_{k_i}^{*T}\psi_{k_i}(s_i)\right] &\le \frac{p_i}{q_i}K_i\dot{e}^{p_i/q_i-1}|s_i|\left|\delta_i - \theta_{k_i}^{*T}\psi_{k_i}(s_i)\right| \\ &\le \frac{p_i}{q_i}K_i\dot{e}^{p_i/q_i-1}|s_i|w_i \le \frac{p_i}{q_i}K_i\dot{e}^{p_i/q_i-1}\sigma_i^* s_i^2\end{aligned} \tag{30}$$

From Eq. (29) we can obtain that:

$$\begin{aligned} \dot{V} \leq & -\sum_{i=1}^{n} \frac{p_i}{q_i} K_i \dot{e}^{p_i/q_i - 1} s_i a_i s_i + \sum_{i=1}^{n} \frac{p_i}{q_i} K_i \dot{e}^{p_i/q_i - 1} \left(|s_i| w_i - \sigma_i^* s_i^2 \right) \\ & \leq -\sum_{i=1}^{n} \frac{p_i}{q_i} K_i \dot{e}^{p_i/q_i - 1} s_i a_i s_i \end{aligned} \tag{31}$$

Since $a_i > 0$, when $\dot{e} \neq 0$ and $s \neq 0$, we can obtain $\dot{V} < 0$. Substituting $\dot{e} = 0$ into Eqs. (2), (5), (8):

$$\ddot{e} = -k - \rho s + \delta \neq 0 \tag{32}$$

Therefore, $\dot{e} = 0$ does not exist. Only when $s = 0$, $\dot{V} = 0$ does exist.

Therefore, Eq. (31) denotes that the Lyapunov function is negative, and all state variables of the closed-loop control system (9) are bounded. The system trajectory will converge in a finite amount of time.

4 Numerical Simulation

The tracking performance of the controller applied to a planar free-floating space manipulator system shown in Fig. 1 was verified with the desired trajectories $\theta_{1d} = -1.5\sin(\pi t/5) + \pi/8(\text{rad})$, $\theta_{2d} = 1.5\sin(\pi t/5) + \pi/4(\text{rad})$. The inertial parameters of composite system are given as: m0 = 40 kg, m1 = 2 kg, m2p = 3 kg, J1 = 37.4 kg · m2, J1 = 1.5 kg · m2, J2p = 2.5 kg · m2, l0 = 1.5 m, l1 = 3 m, d1 = 1.5 m, d2 = 1.5 m, L = 0.001 s, t = 10 s. The parameters of the controller are $\overline{D} = diag(0.5, 0.5, 0.1)$, $\eta_{\theta_{k_i}} = diag(50, 50, 10)$, $\eta_{\sigma_i} = diag(50, 50, 10)$, $a = diag(80, 80, 20)$, $p = 5I_{3\times3}$, $q = 3I_{3\times3}$, $K = 5I_{3\times3}$. The initial state of the space manipulator is $\theta_0(0) = 0.5\,rad$, $\theta_1(0) = 0.5\,rad$, $\theta_2(0) = 0.7\,rad$.

The controller performances are shown in Figs. 2 and 3.

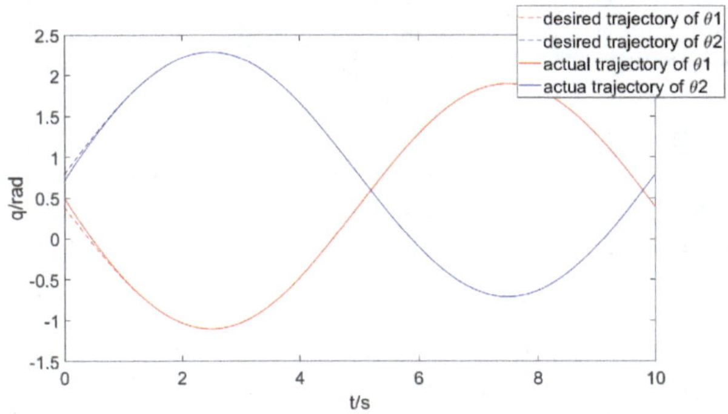

Fig. 2. The comparison between the desired trajectory and the actual one of link joints.

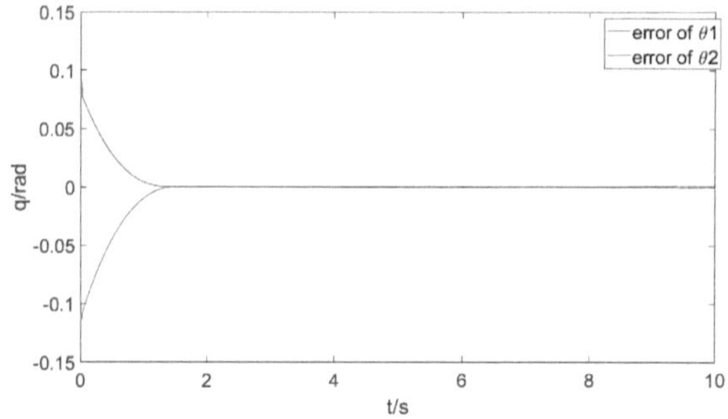

Fig. 3. The observing errors of joints.

5 Conclusions

The control problem of a free- floating space robot system with unknown payload parameters and nonlinearities was discussed. The dynamic model of the system was derived from the Lagrange equations and momentum conservation law. A sliding mode adaptive fuzzy control method based on TDE for the free-floating space manipulator is proposed. The stability was verified using the Lyapunov function. The method does not require any system inertia parameters, which greatly simplifies the construction of the space manipulator system, and can eliminate the negative effects caused by some system parameter uncertainties, and has relatively strong robustness. It can be seen from the simulation results that the actual trajectory tracks the desired trajectory well in a short time.

Acknowledgements. This work was supported by the National Natural Science Foundation of China (No. 51741502) and the Natural Science Foundation of Fujian Province (No. 2016J01228, No. 2017J01559).

References

1. Flores-Abad, A., Ma, O., Pham, K., Ulrich, S.: A review of space robotics technologies for on-orbit servicing. Prog. Aerosp. Sci. **68**, 1–26 (2014)
2. Yu, X.Y., Chen, L.: Singular perturbation adaptive control and vibration suppression of free-flying flexible space manipulators. Proc. IMechE Part C Mech. Eng. Sci. **229**(11), 1989–1997 (2015)
3. Chen, L., Liu, Y.Z.: Adaptive and robust control schemes of coordinated motion of space-based robot system. Chin. J. Mech. Eng. **37**(08), 18–22 (2001)
4. Yu, X.Y., Chen, L.: Augmented robust control of a free-floating flexible space robot. Proc. IMechE Part G. Aerosp. Eng. **229**(5), 947–957 (2015)
5. Yu, X.Y., Chen, L.: Observer-based two-time scale robust control of free-flying flexible-joint space manipulators with external disturbances. Robotica **35**(11) (2017)
6. Shin, J.H., Lee, J.J.: Dynamic control with adaptive identification for free-flying space robots in joint space. Robotica **12**(6), 541–551 (1994)

7. Parlaktuna, O., Ozkan, M.: Adaptive control of free-floating space manipulators using dynamically equivalent manipulator model. Robot. Auton. Syst. **46**(3), 185–193 (2004)
8. Gu, Y.L., Xu, Y.: A normal form augmentation approach to adaptive control of space robot systems 1993, pp. 731–737. IEEE Computer Society Press (1993)
9. Chen, L., Liu, Y.Z., Wu, W.L.: Adaptive and robust control of space manipulator system. Control Theory Appl. **16**(1), 71–75 (1999)
10. Yu, X.Y., Chen, L.: Terminal Sliding Mode Control of a Free-floating Flexible Space Robot. China Mech. Eng. **26**(07), 936–942 (2015)
11. Wang, Y., Chen, J., Gu, L., Li, X.: Time delay control of hydraulic manipulators with continuous nonsingular terminal sliding mode. J. Cent. South Univ. **22**(12), 4616–4624 (2015)
12. Chang, P.H., Park, S.H.: On improving time-delay control under certain hard nonlinearities. Mechatronics **13**(4), 393–412 (2003)
13. Wang, Y., Luo, G., Gu, L., Li, X.: Fractional-order nonsingular terminal sliding mode control of hydraulic manipulators using time delay estimation. J. Vib. Control **22**(19), 849454738 (2015)
14. Liang, X.L.X., Wan, Y.W.Y., Zhang, C.Z.C., Kou, Y.K.Y., Xin, Q.X.Q., Yi, W.Y.W.: Robust position control of hydraulic manipulators using time delay estimation and nonsingular fast terminal sliding mode. Proc. Inst. Mech. Eng. Part I J. Syst. Control. Eng. **232**(1), 50–61 (2018)
15. Chang, P.H., Lee, S.: A straight-line motion tracking control of hydraulic excavator system. Mechatronics **12**(1), 119–138 (2002)
16. Lee, S., Chang, P.H.: Control of a heavy-duty robotic excavator using time delay control with integral sliding surface. Control Eng. Pract. **10**(7), 697–711 (2002)
17. Guo, Y., Woo, P.Y.: An adaptive fuzzy sliding mode controller for robotic manipulators. IEEE Trans. Syst. Man Cybern. Part Syst. Hum. **33**(2), 149–159 (2003)

Robots and Mechatronics

Dimensional Synthesis for the Chang'e-Type Legged Mobile Lander Based on Performance Atlas

Youcheng Han and Weizhong Guo(✉)

State Key Laboratory of Mechanical Systems and Vibration,
School of Mechanical Engineering, Shanghai Jiao Tong University,
Shanghai, China
{youcheng, wzguo}@sjtu.edu.cn

Abstract. Legged mobile lander (LML) possesses both landing and walking functions with the promising application prospect in extraterrestrial exploration. There are three challenges on the dimensional synthesis method for the LML, including the derivation of the performance criteria for both landing and walking behaviors, the establishment of the dimensional optimization design model, and the demonstration and expression method for dimensional synthesis. To solve these problems, the paper will firstly introduce the mechanism topology of the proposed Chang'e-Type LML; then, derive the optimization parameters and the optimization criteria; thereafter, implement the non-dimensional optimization process for both single leg and overall robot respectively by the performance atlas approach; finally, obtain the dimensional global optimum solution. The study can help one understand the relationship between different dimensions and the corresponding performances, and obtain the global optimum dimensions for the Chang'e-Type LML.

Keywords: Legged mobile lander · Dimensional synthesis · Performance atlas · Optimization design · Chang'e

1 Introduction

Recently, extraterrestrial exploration has become a hot topic, and many schemes for the exploration system are proposed by academic institutes, commercial organizations, and aerospace agencies. The lander carrying with the rover is recognized as a conventional method applied among different nations, such as American's Apollo Project (1969–1972) [1], Soviet/Russian's LK Lunar Craft Program (1970–1971) [2], Chinese Chang'e 3 & 4 Program (2013–2019) [3] and so on. On the other hand, the mobile landing exploration robot is proposed with both landing and walking functions, which is also designed with complicated mechanism and structure, such as the robotic and crewed centaur lander by Lockheed Martin [4], the Mars cargo lander by Boeing Company [5], the wheel-legged mobile manned lunar lander by China Academy of Space Technology [6] and so on. In this paper, we propose a new legged mobile lander (LML) with the similar mechanism topology of Chang'e 3 & 4 lunar lander, and

D. Wang et al. (Eds.): MeTrApp 2019, MMS 79, pp. 453–463, 2020.
https://doi.org/10.1007/978-981-15-0142-5_44

commit to implement the dimensional synthesis of the Chang'e-Type LML with the best landing and walking performances.

As shown in the previous researches [7–9], the structural design method is widely used for the past legged landers, which takes the lander as truss structure with buffering characteristics, then investigates the structural strength, reliability, finite element analysis, etc. All that matters is there is not a cogent approach to determine the preliminary dimensions of the truss structure, which depends on the experience and intuition of designers. Moreover, the evaluation indices for both landing and walking behaviors are not sufficient and reasonable at present. Most criteria are suited to the analysis or test process rather than the design process, such as the four statistical landing performance criteria used in Apollo Project [10] and the five comprehensive evaluation indices used in Chang'e 3 Program [7] are all applied in landing tests. Furthermore, the demonstration and expression method for dimensional synthesis is of vital importance as well. For a long time, the cost function approach is widely used by the multi-object numerical algorithm, for instance, Wu et al. [11] utilize the second generation of non-inferior sorting genetic algorithm to optimize the legged lander with multiple objections. However, this approach is easy to get a local optimum solution, or else it requires more computing time and effort to get the global optimum solution. Furthermore, it's still not perceptual to compare different solutions by visualization, and the designers don't know the optimal degree of optimization results.

In conclusion, there is almost no directly feasible method for the dimensional synthesis of the LML at present. Hence, the paper is structured as follows: Step 1, introduces the mechanism topology of the proposed Chang'e-Type LML; Step 2, derives the optimization parameters and the optimization criteria; Step 3, demonstrates the non-dimensional optimization process of both single leg and overall robot respectively with the performance-chart based design methodology (PCbDM); Step 4, obtains the dimensional global optimum solution of the Chang'e-Type LML.

2 Structure and Optimization Criteria

2.1 Structure and Mechanism

Figure 1 shows the structure and mechanism of the LML among different stages. The LML contains one main body and four legs with both landing and walking functions. Each limb in the leg includes one actuation sleeve, one damper sleeve, and one damper piston. Their coupling relationships are illustrated in the figure detailedly, where U denotes the universal hinge, S denotes the spherical hinge, $\underline{\mathrm{P}}$ denotes the prismatic hinge between the actuation sleeve and the damper sleeve, while $\overline{\mathrm{P}}$ denotes the equivalent prismatic hinge for the damper when it's shortened or stretched by landing impact. In pre-landing stage, the LML can be seen as a truss structure with zero mobility. In landing stage, the damper in main limb is shortened while the damper in secondary limb is both possible to be shortened or stretched under different landing conditions, so the single leg is recognized as the hybrid mechanism $\left(2\mathrm{U}\overline{\mathrm{P}}\mathrm{S}\&U\right) - \overline{\mathrm{P}}$

with 2R1T motion characteristics. In walking stage, all dampers are fixed with constant lengths. The damper sleeve is actuated by the actuation sleeve, so the single leg is recognized as the parallel mechanism 2UP̲S&*U*P̲ with 2R1T motion characteristics.

In this figure, $l_1 + l_3$ represents the length of main limb, l_2 and l_4 represent two secondary limbs. For the single leg, the static platform isosceles triangle $\Delta U_1U_2U_3$ is similar to the moving platform isosceles triangle ΔAS_2S_3, their bottom edges are $|U_2U_3| = 2a_1$ and $|S_2S_3| = 2a_2$, heights are b_1 and b_2. Moreover, the main body is a $2d \times 2d \times d$ cuboid, θ_{in} is the installation angle between single leg and main body.

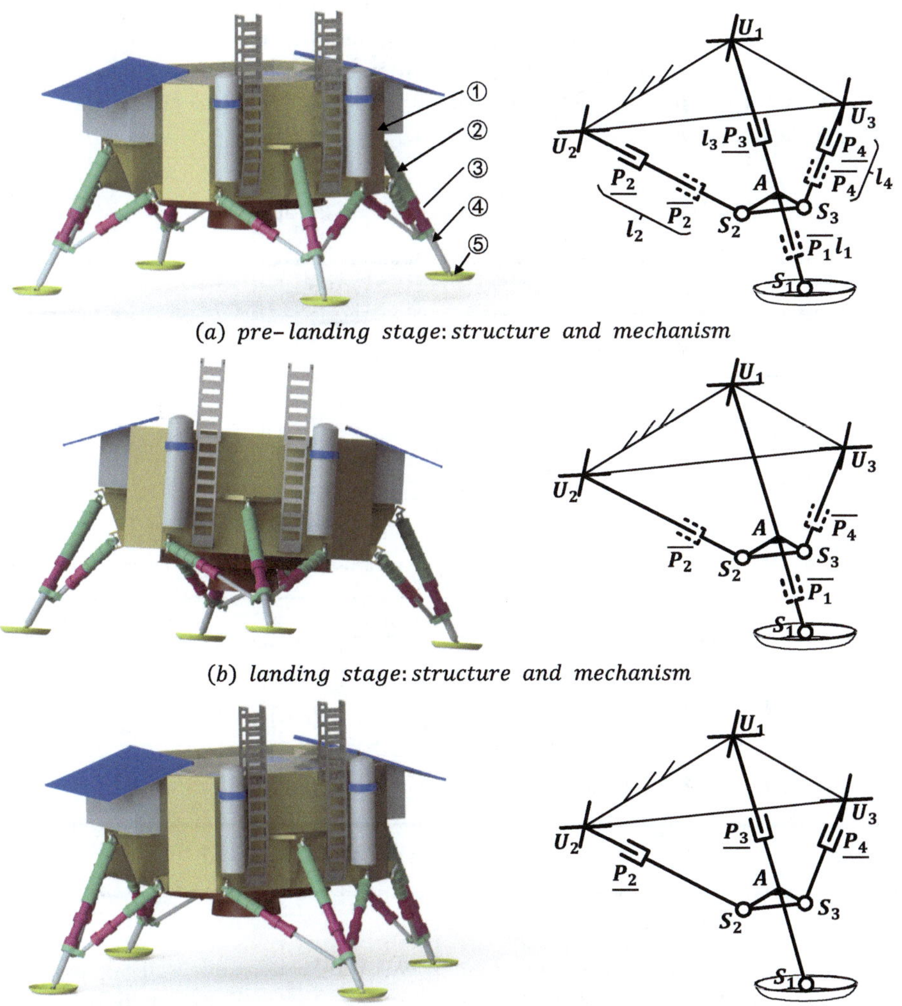

(a) pre-landing stage: structure and mechanism

(b) landing stage: structure and mechanism

(c) walking stage: structure and mechanism

① main body; ② actuation sleeve; ③ damper sleeve; ④ damper piston; ⑤ footpad

Fig. 1. Structure and mechanism of the LML among different stages

There are five optimization parameters for the single leg, including a_1, b_1, a_2, b_2, l_3, and three optimization parameters for the overall robot, including l_3, R_{ms}, R_{sp}, where R_{ms} represents the distance between the main body axis and the main strut universal hinge, which can be considered as the radius of the moving platform of the LML, $R_{ms} = d + b_1 sin\theta_{in}$; R_{sp} represents the radius of the circle formed by four footpads, which can also be considered as the radius of the static platform of the LML.

2.2 Optimization Criteria

There are five performance indices for evaluating the landing and walking behaviors of the LML shown in Table 1, where "★" denotes the index is appropriate for the single leg or overall robot. Among them, the first four indices are based on the works by Gosselin, and Liu, etc. [12, 13], while the fifth index is first put forward based on the work by Papadopoulos [14].

Table 1. Performances indices for single leg and overall robot

Types	Leg	LML	Expression	
WVI	★	★	$\eta_W = \int_0^{2\pi} d\theta \int_0^{\rho_{max}} \rho d\rho \int_{z_{min}}^{z_{max}} dz$	(1)
GCI	★	★	$\eta_{Jv} = \frac{\int_V \mu_{Jv} dW}{\int_V dW}$	(2)
		★	$\eta_{J\omega} = \frac{\int_V \mu_{J\omega} dW}{\int_V dW}$	(3)
GPI	★	★	$\eta_{Fmax} = \frac{\int_V \sqrt{max(\lvert\lambda_{Fi}\rvert)} dW}{\int_V dW}, \eta_{Fmin} = \frac{\int_V \sqrt{min(\lvert\lambda_{Fi}\rvert)} dW}{\int_V dW}$	(4)
		★	$\eta_{Mmax} = \frac{\int_V \sqrt{max(\lvert\lambda_{Mi}\rvert)} dW}{\int_V dW}, \eta_{Mmin} = \frac{\int_V \sqrt{min(\lvert\lambda_{Mi}\rvert)} dW}{\int_V dW}$	(5)
GSI	★	★	$\eta_{Pmax} = \frac{\int_V \sqrt{max(\lvert\lambda_{Pi}\rvert)} dW}{\int_V dW}, \eta_{Pmin} = \frac{\int_V \sqrt{min(\lvert\lambda_{Pi}\rvert)} dW}{\int_V dW}$	(6)
		★	$\eta_{Omax} = \frac{\int_V \sqrt{max(\lvert\lambda_{Oi}\rvert)} dW}{\int_V dW}, \eta_{Omin} = \frac{\int_V \sqrt{min(\lvert\lambda_{Oi}\rvert)} dW}{\int_V dW}$	(7)
GLSI		★	$\eta_S = \frac{\int_V S_k dW}{\int_V dW}, S_k = min(\theta_i \cdot \lVert \boldsymbol{d}_i \rVert \cdot \lVert \boldsymbol{F}_i^* \rVert)$	(8)

In the table, the workspace volume index (WVI) represents the buffering extent and ability and is the basis to establish all other performances; the global conditioning index (GCI) evaluates the dexterity, isotropy and the ability away from singularity; the global payload index (GPI) evaluates the extremum of the force or moment in landing and walking stages; the global stiffness index (GSI) evaluates the linear and angular displacement deformation; the global landing stability index (GLSI) evaluates both horizontal and vertical stabilities to make the landing or walking stage secure. Furthermore, we hope a bigger WVI, bigger GCIs, bigger GPIs, smaller GSIs and bigger GLSI for the LML as the best multi-objective performances. In the following, to balance all these performance criteria, we'll transform the multi-objective optimization problem into the single-objective optimization problem to implement the dimensional synthesis process.

3 Non-dimensional Optimization of the Single Leg

For the single leg, the dimensional synthesis problem can be abstracted into the dimensional optimization design model as follows:

$$\begin{cases} \text{Parameters: } X_{LEG} = \{a_1, b_1, a_2, b_2, l_3\} \\ \text{Maximum: } F(X_{LEG}) = m_1\tilde{\eta}_W + m_2\tilde{\eta}_{Jv} + m_3\tilde{\eta}_{Fmax} + m_4\tilde{\eta}_{Fmin} \\ \qquad\qquad + m_5(1-\tilde{\eta}_{Pmax}) + m_6(1-\tilde{\eta}_{Pmin}) \\ \text{Subject to: } \begin{cases} l_{imin} \le l_{ji} \le l_{imax} \\ \beta_{min} \le \beta_{ji} \le \beta_{max} \\ \gamma_{min} \le \gamma_{ji} \le \gamma_{max} \\ \xi_{min} \le \xi_{ji} \le \xi_{max} \end{cases} \end{cases} \tag{9}$$

where $\tilde{\eta}_i$ $(i \in [1,6])$ is the normalization form of η_i, and $\tilde{\eta}_i \in [0,1]$; m_i is the effective weighting factor of $\tilde{\eta}_i$ to construct the single-objective optimization problem, and $\sum m_i = 1$. Moreover, the constraint conditions are the angle ranges of $\beta_{ji}, \gamma_{ji}, \xi_{ji}$ for universal and spherical hinges, and the lengths ranges of l_{ji} for dampers.

Based on the PCbDM method, the normalized equation can be obtained by:

$$r_1 + r_2 + r_3 + r_4 + r_5 = 5 \tag{10}$$

where $r_1 = l_3/R$, $r_2 = b_1/R$, $r_3 = b_2/R$, $r_4 = a_1/R$, $r_5 = a_2/R$, and R is the normalization factor. Considering that the low DOF parallel mechanism usually owns the geometrically similar static platform and moving platform to get better performances [15], hereby we let $r_4/r_2 = r_5/r_3 = \chi$, and assign $\chi = 1.5$ on account of the supporting stiffness and landing stability. Equation (11) indicates the parameter design space, which is illustrated in Fig. 2 by the rectangular pyramid $ABCDE$.

$$r_1 + (1+\chi)(r_2 + r_3) = 5 \tag{11}$$

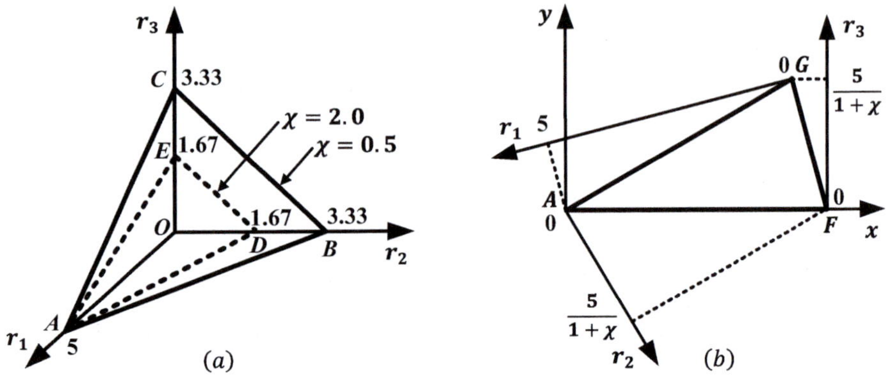

Fig. 2. Parameter design space of the single leg

Figure 3 shows the comprehensive performance atlas of the single leg, where $m_1, m_2, m_4, m_5 = 0.2$, and $m_3, m_6 = 0.1$. One can see that: (1) the comprehensive performance in the design space is almost inversely proportional to r_3 when r_1 is specified, which means the dimensions of the moving platform of the single leg should be smaller than that of the static platform; (2) for the single leg with the best comprehensive performance, the optimum solution region is $\Omega_{LEG} = \{(r_1, r_2, r_3)|2.0 < r_1 < 3.5, 0.6 < r_2 < 1.2, 0 < r_3 < 0.4\}$.

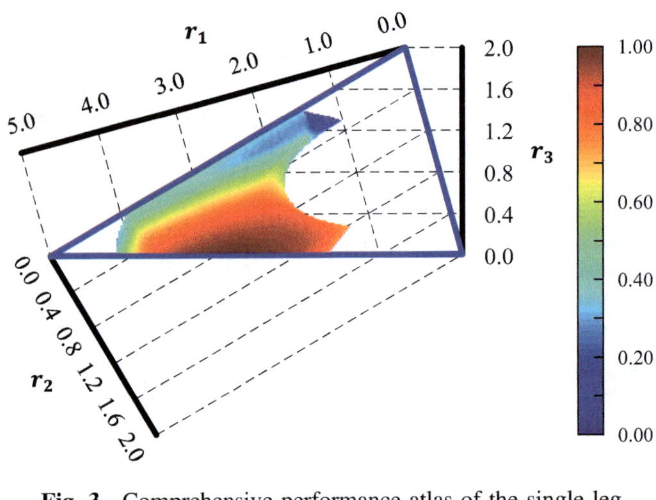

Fig. 3. Comprehensive performance atlas of the single leg

Table 2 gives the numerical analysis for the multi-objective performances of the single leg, and lists ten different dimension combinations and the corresponding performances in Ω_{LEG}. For the reason that there is not an unique optimum solution for the multi-objective optimization problem, and it relies on the designer to decide the importance degree of different performance indices. By contrast, we determine to choose No. 8 as the optimum non-dimensional parameters for the following investigation, to get a better landing and walking performances.

Table 2. Numerical analysis for the multi-objective performance of the single leg

No.	(r_1, r_2, r_3)	$(\eta_W, \eta_{Jv}, \eta_{Fmax}, \eta_{Fmin}, \eta_{Pmax}, \eta_{Pmin})$
1	(2.00,0.80,0.40)	(0.40,0.61,1.10,0.67,2.42,0.88)
2	(2.00,0.90,0.30)	(1.13,0.62,1.23,0.74,2.03,0.78)
3	(2.00,1.00,0.20)	(2.65,0.60,1.42,0.79,1.81,0.68)
4	(2.00,1.10,0.10)	(5.28,0.57,1.59,0.80,1.76,0.61)
5	(2.50,0.60,0.40)	(15.55,0.50,1.48,0.66,3.21,0.77)
6	(2.50,0.70,0.30)	(14.13,0.53,1.58,0.72,2.44,0.71)
7	(2.50,0.80,0.20)	(15.03,0.53,1.70,0.76,2.08,0.61)
8	(2.50,0.90,0.10)	(17.06,0.52,1.83,0.79,1.88,0.57)
9	(3.00,0.60,0.20)	(37.29,0.46,1.70,0.66,3.75,0.69)
10	(3.00,0.70,0.10)	(37.10,0.47,1.84,0.72,2.83,0.62)

4 Non-dimensional Optimization of the Overall Robot

For the overall robot, the dimensional synthesis problem can be abstracted into the dimensional optimization design model as follows:

$$\begin{cases} \text{Parameters: } X_{LML} = \{l_3, R_{ms}, R_{sp}\} \\ \text{Maximum: } F(X_{LML}) = n_1\tilde{\eta}_W + n_2\tilde{\eta}_{Jv} + n_3\tilde{\eta}_{J\omega} + n_4\tilde{\eta}_{Fmax} + n_5\tilde{\eta}_{Fmin} \\ \qquad + n_6\tilde{\eta}_{Mmax} + n_7\tilde{\eta}_{Mmin} + n_8(1-\tilde{\eta}_{Pmax}) + n_9(1-\tilde{\eta}_{Pmin}) \\ \qquad + n_{10}(1-\tilde{\eta}_{Omax}) + n_{11}(1-\tilde{\eta}_{Omin}) + n_{12}\tilde{\eta}_S \\ \text{Subject to: } \begin{cases} L_v - [L_v] \geq 0 \\ L_h - [L_h] \geq 0 \\ L_{imin} \leq L_{ji} \leq L_{imax} \\ \beta_{min} \leq \beta_{ji} \leq \beta_{max} \\ \gamma_{min} \leq \gamma_{ji} \leq \gamma_{max} \\ \xi_{min} \leq \xi_{ji} \leq \xi_{max} \end{cases} \end{cases} \tag{12}$$

where $\tilde{\eta}_i$ ($i \in [1, 12]$) is the normalization form of η_i, and $\tilde{\eta}_i \in [0, 1]$; n_i is the effective weighting factor of $\tilde{\eta}_i$ to construct the single-objective optimization problem, and $\sum n_i = 1$. Moreover, the constraint conditions are the landing stability constraints in both horizontal and vertical directions, the angle ranges of $\beta_{ji}, \gamma_{ji}, \xi_{ji}$ for universal and spherical hinges, the lengths ranges of L_{ji} for actuators, where L_v is the distance between the engine nozzle skirt and the extraterrestrial surface, L_h is the distance between the main body axis and the connecting line of two adjacent footpads, $[L_v]$ and $[L_h]$ are the minimum value of L_v and L_h that ensure the LML secure.

Based on the PCbDM method, the normalized equation can be obtained by:

$$t_1 + t_2 + t_3 = 3 \tag{13}$$

where $t_1 = l_3/T$, $t_2 = R_{ms}/T$, $t_3 = R_{sp}/T$, and T is the normalization factor. Hence, Eq. (13) indicates the parameter design space, which is illustrated in Fig. 4 as well.

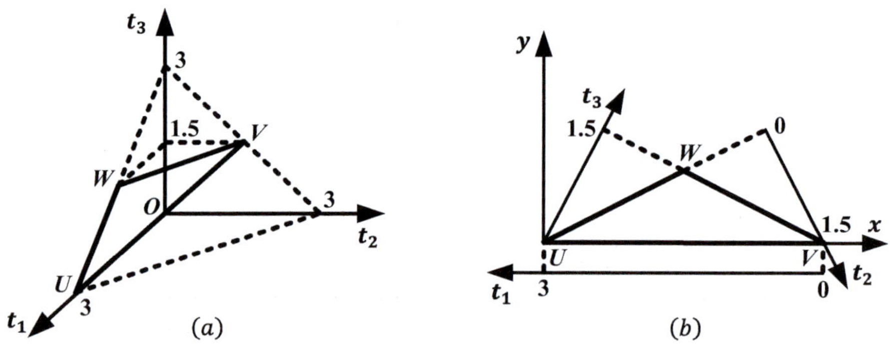

Fig. 4. Parameter design space of the overall robot

Figure 5 shows the comprehensive performance atlas of the overall robot, where $n_1, n_2, n_3, n_5, n_7, n_8, n_{10}, n_{12} = 0.1$, and $n_4, n_6, n_9, n_{11} = 0.05$. One can see that: (1) the comprehensive performance in the design space is almost proportional to t_3 when t_1 or t_2 is specified, which means R_{sp} should be bigger than R_{ms}. (2) for the overall robot with the best comprehensive performance, the optimum solution region is $\Omega_{LML} = \{(t_1, t_2, t_3) | 0.4 < t_1 < 1.6, 0 < t_2 < 1.1, 1.3 < t_3 < 1.5\}$.

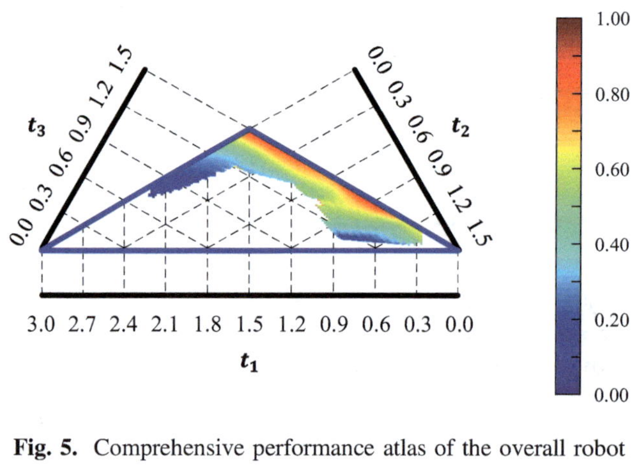

Fig. 5. Comprehensive performance atlas of the overall robot

Table 3 gives the numerical analysis for the multi-objective performances of the overall robot, and lists thirteen different dimension combinations and the corresponding performances in Ω_{LML}. By contrast, we determine to choose No. 2 as the optimum non-dimensional parameters, to get a better landing and walking performance.

Table 3. Numerical analysis for the multi-objective performance of the overall robot

No.	(t_1, t_2, t_3)	$(\eta_W, \eta_{Jv}, \eta_{J\omega}, \eta_{Fmax}, \eta_{Fmin}, \eta_{Mmax}, \eta_{Mmin}, \eta_{Pmax}, \eta_{Pmin}, \eta_{Omax}, \eta_{Omin}, \eta_S)$
1	(0.4,1.1,1.5)	(0.26,0.25,0.23,2.09,0.49,0.60,0.12,6.18,0.24,131.39,5.17,0.60)
2	(0.6,0.9,1.5)	(0.57,0.51,0.26,1.74,0.83,1.03,0.19,1.96,0.35,35.80,4.90,0.66)
3	(0.6,1,1.4)	(0.19,0.31,0.28,2.00,0.59,0.34,0.08,3.93,0.25,213.60,17.13,0.41)
4	(0.8,0.7,1.5)	(0.87,0.61,0.18,1.64,0.97,1.39,0.21,1.16,0.39,24.13,1.69,0.69)
5	(0.8,0.8,1.4)	(0.22,0.45,0.22,1.89,0.83,0.90,0.12,1.78,0.28,82.72,9.24,0.45)
6	(0.8,0.9,1.3)	(0.10,0.40,0.27,1.95,0.77,0.21,0.05,1.91,0.26,386.36,30.28,0.28)
7	(1,0.5,1.5)	(0.97,0.58,0.14,1.67,0.94,1.51,0.21,1.21,0.38,24.29,0.56,0.74)
8	(1,0.6,1.4)	(0.46,0.29,0.11,2.01,0.57,1.43,0.16,3.99,0.25,41.93,0.70,0.54)
9	(1.2,0.3,1.5)	(0.89,0.49,0.14,1.75,0.84,1.48,0.21,1.51,0.34,24.03,0.48,0.85)
10	(1.2,0.4,1.4)	(0.67,0.24,0.11,2.08,0.49,1.39,0.16,4.66,0.23,40.98,0.53,0.54)
11	(1.4,0.1,1.5)	(0.53,0.39,0.14,1.90,0.72,1.41,0.20,2.06,0.29,26.37,0.55,1.08)
12	(1.4,0.2,1.4)	(0.51,0.2,0.13,2.14,0.49,1.17,0.15,4.34,0.22,47.47,0.75,0.74)
13	(1.4,0.3,1.3)	(0.46,0.14,0.11,2.26,0.32,0.94,0.11,10.32,0.20,87.59,1.15,0.45)

5 Dimensional Global Optimum Solution

Given in the stowed stage, the LML should meet the envelope demand of the carrying vehicle, that is to say, the diameter of the LML should be less than the envelope diameter ϕ_e of the carrying vehicle, given by:

$$max\left(2\sqrt{2}d, 2R_{ms}\right) < \phi_e \tag{14}$$

According to the equation $d = R_{ms} - b_1 sin\theta_{in}$ and the non-dimensional parameters solved in Sects. 3 and 4, we can deduce $2\sqrt{2}d < \phi_e$. Here we take $\phi_e = 3650$ mm in reference to the envelope demand of Chang'e 3 [16], and thus obtain $d < 1290$ mm, so we let $d = 1200$ mm for the following steps. Then, the normalization factor of the overall robot can be calculated by: $T = R_{ms}/t_2 = 1363.64/0.9 = 1515.15$ mm. So the optimum dimensional parameters of the overall robot can be obtained by: $l_3 = t_1 \cdot T = 909.09$mm, $R_{sp} = t_3 \cdot T = 2272.73$ mm. Next, the normalization factor of the single leg can be ensured uniquely by: $R = l_3/r_1 = 909.09/2.5 = 363.64$ mm, so the optimum dimensional parameters of the single leg can be obtained by: $b_1 = r_2 \cdot R = 327.27\,\text{mm}$, $b_2 = r_3 \cdot R = 36.36\,\text{mm}$, $a_1 = b_1 \cdot \chi = 490.91\,\text{mm}$, $a_2 = b_2 \cdot \chi = 54.54\,\text{mm}$. As a consequence, all dimensional parameters of the LML are optimized with the best comprehensive performance.

6 Conclusion

The paper proposes the optimum dimension design method for the Chang'e-Type Legged Mobile Lander, establishes a reliable and comprehensive optimization design model with proper mechanism topological parameters, performance criteria and constraint conditions. The study develops a systematic strategy to optimize both single leg and overall robot respectively. Finally, the relationship between different possible dimensions and the corresponding performances is analyzed, and the global optimum dimensions for the LML are obtained. The results shown by the performance atlases are perceptual and credible.

Acknowledgments. The author thanks the financial support under the projects from the National Natural Science Foundation of China (Grant No. 51735009) and the Research Fund of State Key Lab of MSV, China (Grant No. MSV-ZD-2016-08).

References

1. Orloff, R.W.: Apollo by the Numbers: A Statistical Reference for the Manned Phase of Project Apollo (2000)
2. Harvey, B.: Soviet and Russian Lunar Exploration. Praxis, New York (2007). https://doi.org/10.1007/978-0-387-73976-2
3. Li, C., Liu, J., Ren, X., et al.: The chang'e 3 mission overview. Space Sci. Rev. **190**(1–4), 85–101 (2015). https://doi.org/10.1007/s11214-014-0134-7
4. Birckenstaedt, B., Kutter, B.F., Zegler, F.: Centaur application to robotic and crewed lunar lander evolution. In: AIP Conference Proceedings, vol. 880, no. 1, p. 779 (2007). https://doi.org/10.1063/1.2437517
5. Mark, G.B.: A conceptual mars exploration vehicle architecture with chemical propulsion, near-term technology, and high modularity to enable near-term human missions to mars. In: AIAA SPACE 2015 Conference and Exposition, California, US (2015). https://doi.org/10.2514/6.2015-4611
6. Zhang, Z.X., Liang, L., Guo, L.L.: Concept of a wheeled-legged mobile manned lunar lander. Manned Space, 202–209 (2016). https://doi.org/10.3969/j.issn.1674-5825.2016.02.010
7. Yang, J.: Spacecraft Landing Buffer Mechanism. China Aerospace Publishing House, Beijing (2015)
8. NASA: Monte Carlo Approach to Touchdown Dynamics for Soft Lunar Landing, United States (1965)
9. NASA: A Mathematical Procedure for Predicting the Touchdown Dynamics of a Soft-Landing Vehicle, United States (1971)
10. NASA: Apollo Experience Report: Lunar Module Landing Gear Subsystem, United States (1972)

11. Wu, H., Wang, C., Ding, J., et al.: Soft Landing performance optimization of the new type lander based on multiple operating conditions. J. Beijing Univ. Aeronaut. Astronaut., 1–7 (2016). https://doi.org/10.13700/j.bh.1001-5965.2016.0296
12. Gosselin, C., Angeles, J.: A global performance index for the kinematic optimization of robotic manipulators. Trans. ASME. J. Mech. Des. **113**(3), 220–226 (1991). https://doi.org/10.1115/1.2912772
13. Liu, X.J., Wang, J.S., Pritschow, G.: Performance atlases and optimum design of planar 5R symmetrical parallel mechanisms. Mech. Mach. Theory **41**(2), 119–144 (2006). https://doi.org/10.1016/j.mechmachtheory.2005.05.003
14. Papadopoulos, E., Rey, D.A.: The force-angle measure of tipover stability margin for mobile manipulators. Vehicle Syst. Dyn. **33**(1), 29–48 (2000). https://doi.org/10.1076/0042-3114(200001)33:1;1-5;FT029
15. Liu, X.J., Wang, J.: A new methodology for optimal kinematic design of parallel mechanisms. Mech. Mach. Theory (2007). https://doi.org/10.1016/j.mechmachtheory.2006.08.002
16. Sun, Z., Zhang, T., Zhang, H., et al.: Technical design and achievements of Chang'e-3 probe. China Sci. Technol. Sci. (04), 331–343 (2014). https://doi.org/10.1360/092014-37

Kinematic Modeling and Simulation of a Leg-Wheel Robot for Unexplored Rough Terrain Environment

Xu Cai, Jun He(✉), and Feng Gao

School of Mechanical Engineering,
Shanghai Jiao Tong University, Shanghai 200240, China
jhe@sjtu.edu.cn

Abstract. A quadruped leg-wheel robot over rough terrain is proposed for planetary exploration. The closed-form solution for the inverse displacement model is developed, and then the forward displacement equations are also obtained. The Jacobian matrix is derived from the displacement model. Using kinematic equations, the co-simulation model is established, and the motion planning and dynamic simulation are performed in MATLAB and UG software, respectively. After that, a typical climbing simulation has been carried out based on the co-simulation model. Accordingly, the driving torques of all joints are obtained, which can be used for the design of prototype.

Keywords: Leg-wheel robot · Kinematic · Co-simulation · Planetary exploration

1 Introduction

In the future, the planetary exploration robot should accomplish various missions in unexplored environments, which requires over-obstacle ability, driving placidity, traction controllability and power-saving characteristics. The wheeled robot, owing to its advantages of simple structure, low weight, high speed, high efficiency, etc., has been the main research choice of space-exploring around the world. Carrying out the research work of space exploration robots is conducive to making some progress in the innovative design of high-speed mobile robots. This can provide technical support and high-performance mobile platform for space exploration in China [1].

Space exploration robot has been developed for many years in the world, especially in Russia and the United States. Both of them have very mature experience in space exploration technology [2]. In the motion structure of robots, most of them adopt special independent suspension or change the traditional single motion structure. Accordingly, the robot has multiple structural functions such as wheel-track, wheel-leg and so on. It also enhances the adaptability for terrain and the ability to overcome obstacles.

It is especially important that robots can overcome complex terrain environment [7]. Studies have shown that the research on space exploration robots mainly focuses on the following two aspects. They are the integration of wheel and leg, and the

D. Wang et al. (Eds.): MeTrApp 2019, MMS 79, pp. 464–473, 2020.
https://doi.org/10.1007/978-981-15-0142-5_45

problems of suspension. The most representatives are the American parallelogram suspension [3], the Russian series combined suspension [4], and the parallel structure suspension of Beijing Institute of Technology [5].

Generally speaking, the structure of the robot is relatively complicated, and the existing structure of the robot is not fully utilized. For example, the robot body no longer has any effect when it is retracted, which causes a waste of energy [6]. Active suspension detectors are difficult to climb over 0.5 m obstacles. The simple wheel-leg method improves maneuverability, but it is not easy to satisfy the landing buffer function. The wheel-foot self-deformation scheme is still in the stage of concept, and it is difficult to control.

Based on this, this paper proposes the design concept of a wheel-legged robot. The leg-wheel robot has the function of landing buffer. It works on planes, slopes and obstacles, which has a simple structure and is easy to control.

2 Structural Design

According to the design requirements, the overall design was carried out by referring to the advantages and disadvantages of various mobile methods of Reference [7]. This quadruped wheel-legged robot can be divided into mechanical structural components and electrical control systems. The mechanical structure consists of body frame, wheel-leg parts, and cushion leg. The electrical control system mainly includes the motor, reducer, brake, driver, etc. The envelope of the robot's collapsed space is 1400 × 1030 × 900 mm, and the general working space is 2000 × 1030 × 1400 mm.

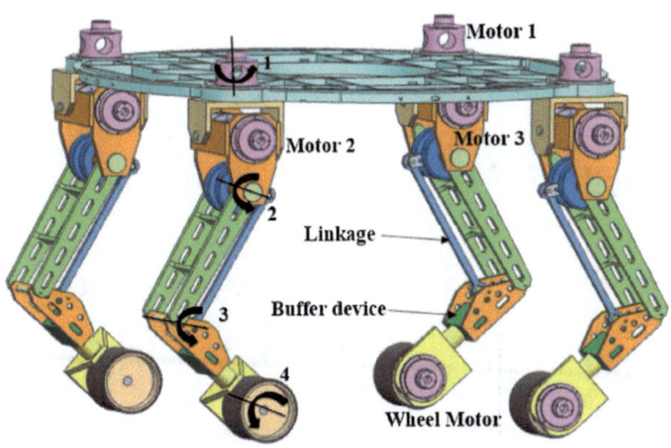

Fig. 1. Quadruped wheel-legged robot model

In order to reduce the inertia of the moving parts, three drives of each leg are mounted on the robotic body. The azimuth angles 1 around the hip joint controls the orientation of forward. The pitch angles 2 around the hip joint controls the height of the leg from the ground. The angles 3 around the knee joint controls the height of the leg from the ground. Finally, the wheel drive 4 is mainly used on wheels. The specific location is shown in Fig. 1.

The position of each drive is depicted in Fig. 1, and the position of the parallelogram linkage and the buffer device is indicated. The linkage is mainly used to drive the knee joint, and the driving torque can be transmitted through a parallelogram. The transmission efficiency is high and the structure is simple. The buffer device is mainly used in the landing process. It can convert the most of the impact potential energy into the elastic potential energy of the spring in the buffer device, so as to absorbing energy in the buffer.

3 Kinematic Modeling

3.1 Forward Kinematics

The structural parameters can be known after determining the structure of the robot. The structural parameters are different as long as the coordinate system is different, even for the same type of robot [8]. In this paper, the one-leg coordinate system of the robot is established by using the D-H method in Reference [9].

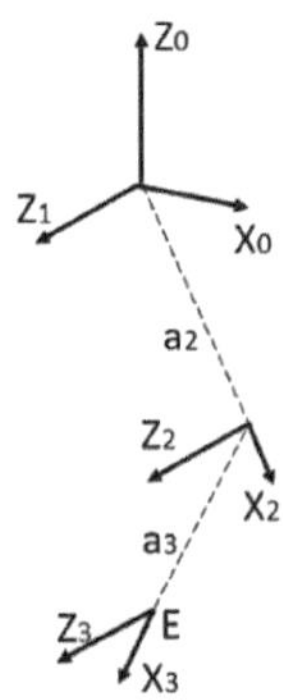

Fig. 2. 3-DOF coordinate system of the robot

According to the definition of the D-H matrix, the homogeneous coordinate transformation matrix of two adjacent coordinate systems can be described as

$$ {}^{i-1}_{i}T = \begin{bmatrix} c\theta_i & -s\theta_i c\alpha_i & s\theta_i s\alpha_i & a_i c\theta_i \\ s\theta_i & c\theta_i c\alpha_i & -c\theta_i s\alpha_i & a_i s\theta_i \\ 0 & s\alpha_i & c\alpha_i & d_i \\ 0 & 0 & 0 & 1 \end{bmatrix} \tag{1} $$

where, $c\theta_i$, $s\theta_i$, $c\alpha_i$ and $s\alpha_i$ are respectively expressed as $cos\theta_i$, $sin\theta_i$, $cos\alpha_i$ and $sin\alpha_i$.

In this paper, the model only considers the single mechanical legs of angle 1, angle 2 and angle 3. The coordinate system is shown in Fig. 2. The origin of the base coordinate system 0 is at the intersection of z_0 and z_1. Simultaneously, z_1 and z_2 are parallel, and x_1 and x_2 are respectively aligned with the corresponding link directions. The D-H matrix parameters are shown in Table 1. The parameters of robot are $a_2 = 350\,\text{mm}$, $a_3 = 324.05\,\text{mm}$.

Table 1. D-H matrix parameters

Link	a_i	d_i	α_i	θ_i
1	0	0	$\pi/2$	θ_1
2	a_2	0	0	θ_2
3	a_3	0	0	θ_3

The coordinate rotation matrix is obtained by substituting the parameters of Table 1 into Eq. (1).

$$
{}_1^0T = \begin{bmatrix} c_1 & 0 & s_1 & 0 \\ s_1 & 0 & -c_1 & 0 \\ 0 & 1 & 0 & 0 \\ 0 & 0 & 0 & 1 \end{bmatrix};\ {}_2^1T = \begin{bmatrix} c_2 & -s_2 & 0 & a_2c_2 \\ s_2 & c_2 & 0 & a_2s_2 \\ 0 & 0 & 1 & 0 \\ 0 & 0 & 0 & 1 \end{bmatrix};\ {}_3^2T = \begin{bmatrix} c_3 & -s_3 & 0 & a_3c_3 \\ s_3 & c_3 & 0 & a_3s_3 \\ 0 & 0 & 1 & 0 \\ 0 & 0 & 0 & 1 \end{bmatrix}
$$

$$
{}_3^0T = {}_1^0T\,{}_2^1T\,{}_3^2T = \begin{bmatrix} c_1c_{23} & -c_1s_{23} & s_1 & c_1(a_2c_2 + a_3c_{23}) \\ s_1c_{23} & -s_1s_{23} & -c_1 & s_1(a_2c_2 + a_3c_{23}) \\ s_{23} & c_{23} & 1 & a_2s_2 + a_3s_{23} \\ 0 & 0 & 0 & 1 \end{bmatrix} \tag{2}
$$

where, $c_1 \cdots c_3$, $s_1 \cdots s_3$, c_{23} and s_{23} are respectively expressed as $cos\theta_1 \cdots cos\theta_3$, $sin\theta_1 \cdots sin\theta_3$, $cos(\theta_2 + \theta_3)$ and $sin(\theta_2 + \theta_3)$. In the base coordinate system, the coordinates of the end point E can be described as

$$
\begin{cases} E_x = c_1(a_2c_2 + a_3c_{23}) \\ E_y = s_1(a_2c_2 + a_3c_{23}) \\ E_z = a_2s_2 + a_3s_{23} \end{cases} \tag{3}
$$

3.2 Inverse Kinematics

Solving the inverse kinematics of a robot is equivalent to solving three unknown angles which can be solved by the end coordinates of the foot $E(E_x, E_y, E_z)$. The forward kinematics equation of E-point is given by Eq. (2), and the coordinates of the E-point in the base coordinate system are given by Eq. (3).

The Eq. (3) can be described as

$$
E_x^2 + E_y^2 + E_z^2 = a_2^2 + a_3^2 + 2a_2a_3c_3 \tag{4}
$$

The formula above can be written as

$$
c_3 = \frac{E_x^2 + E_y^2 + E_z^2 - a_2^2 - a_3^2}{2a_2a_3} \tag{5}
$$

Among them, the range of the defined joint angle is [−π, π], and the range of the solution is [−1, 1]. Therefore, the formula can be written as

$$s_3 = \pm\sqrt{1 - c_3^2} \tag{6}$$

$$\theta_3 = Atan2(s_3, c_3) \tag{7}$$

θ_2 can be calculated after getting θ_3. The Eq. (3) can be described as

$$E_x^2 + E_y^2 = (a_2 c_2 + a_3 c_{23})^2 \tag{8}$$

The formula above can be written as

$$a_2 c_2 + a_3 c_{23} = \pm\sqrt{E_x^2 + E_y^2} \tag{9}$$

According to Eqs. (3) and (9), the formula above can be written as

$$c_2 = \frac{\pm\sqrt{E_x^2 + E_y^2}(a_2 + a_3 c_3) + E_z a_3 s_3}{a_2^2 + a_3^2 + 2a_2 a_3 c_3} \tag{10}$$

$$s_2 = \frac{E_z(a_2 + a_3 c_3) \pm \sqrt{E_x^2 + E_y^2} a_3 s_3}{a_2^2 + a_3^2 + 2a_2 a_3 c_3} \tag{11}$$

According to Eqs. (10) and (11), the formula above can be written as

$$\theta_2 = Atan2(s_2, c_2) \tag{12}$$

Finally, θ_1 can be solved, and using Eq. (9) to rewrite the first two lines of Eq. (3) as

$$E_x = \pm c_1\sqrt{E_x^2 + E_y^2} \tag{13}$$

$$E_y = \pm s_1\sqrt{E_x^2 + E_y^2} \tag{14}$$

The formula can be described as

$$\theta_{1,I} = Atan2(E_y, E_x) \tag{15}$$

$$\theta_{1,II} = Atan2(-E_y, -E_x) \tag{16}$$

Hereto, the three unknown joint angles θ_1, θ_2, θ_3 can be solved. Four sets of corresponding solutions can be obtained according to the range of the defined joint angle.

3.3 Kinematic Jacobian Matrix

The Jacobian matrix can be described as

$$\begin{bmatrix} v \\ w \end{bmatrix} = J(q)\dot{q} \tag{17}$$

where, v, w—linear velocity and angular velocity; $\dot{q}$—joint velocity vector [10].

In the robot coordinate system, the angular velocity of the I axis can be written as

$$w_i = z_{i-1}\dot{q}_i \tag{18}$$

The linear velocity can be written as

$$v_i = [z_{i-1} \times (p_3 - p_{i-1})]\dot{q}_i \tag{19}$$

$$z_{i-1} = R^0_{i-1} z_0, p_{i-1} = {}^0_{i-1}Tp_0, z_0 = \begin{bmatrix} 0 & 0 & 1 \end{bmatrix}^T, p_0 = \begin{bmatrix} 0 & 0 & 0 & 1 \end{bmatrix}^T \tag{20}$$

where, R^0_{i-1} is a rotation matrix. The form of the Jacobian matrix can be written as

$$J(q) = \begin{bmatrix} z_0 \times (p_3 - p_0) & z_1 \times (p_3 - p_1) & z_0 \times (p_3 - p_2) \\ z_0 & z_1 & z_2 \end{bmatrix} \tag{21}$$

$$p_1 = {}^0_1Tp_0 = \begin{bmatrix} c_1 & 0 & s_1 & 0 \\ s_1 & 0 & -c_1 & 0 \\ 0 & 1 & 0 & 0 \\ 0 & 0 & 0 & 1 \end{bmatrix} \begin{bmatrix} 0 \\ 0 \\ 0 \\ 1 \end{bmatrix} = \begin{bmatrix} 0 \\ 0 \\ 0 \\ 1 \end{bmatrix} = p_0 \tag{22}$$

$$p_2 = {}^0_2Tp_0 = \begin{bmatrix} c_1c_2 & -c_1s_2 & s_1 & c_1a_2c_2 \\ s_1c_2 & -s_1s_2 & -c_1 & s_1a_2c_2 \\ s_2 & c_2 & 0 & a_2s_2 \\ 0 & 0 & 0 & 1 \end{bmatrix} \begin{bmatrix} 0 \\ 0 \\ 0 \\ 1 \end{bmatrix} = \begin{bmatrix} c_1a_2c_2 \\ s_1a_2c_2 \\ a_2s_2 \\ 1 \end{bmatrix} \tag{23}$$

$$p_3 = \begin{bmatrix} c_1c_{23} & -c_1s_{23} & s_1 & c_1(a_2c_2 + a_3c_{23}) \\ s_1c_{23} & -s_1s_{23} & -c_1 & s_1(a_2c_2 + a_3c_{23}) \\ s_{23} & c_{23} & 1 & a_2s_2 + a_3s_{23} \\ 0 & 0 & 0 & 1 \end{bmatrix} \begin{bmatrix} 0 \\ 0 \\ 0 \\ 1 \end{bmatrix} = \begin{bmatrix} c_1(a_2c_2 + a_3c_{23}) \\ s_1(a_2c_2 + a_3c_{23}) \\ a_2s_2 + a_3s_{23} \\ 1 \end{bmatrix} \tag{24}$$

The unit vector of the rotating joint axis can be written as

$$z_0 = \begin{bmatrix} 0 & 0 & 1 \end{bmatrix}^T; z_1 = z_2 = \begin{bmatrix} s_1 & -c_1 & 0 \end{bmatrix}^T \tag{25}$$

Therefore, the Jacobian matrix can be described as

$$J = \begin{bmatrix} -s_1(a_2c_2 + a_3c_{23}) & -c_1(a_2s_2 + a_3s_{23}) & -a_3c_1s_{23} \\ c_1(a_2c_2 + a_3c_{23}) & -s_1(a_2s_2 + a_3s_{23}) & -a_3s_1s_{23} \\ 0 & a_2c_2 + a_3c_{23} & a_3c_{23} \\ 0 & s_1 & s_1 \\ 0 & -c_1 & -c_1 \\ 1 & 0 & 0 \end{bmatrix} \tag{26}$$

4 Numerical Simulation

The numerical simulation include verification of kinematic equations and typical climbing simulation with Simulink and UG software.

4.1 Verification of Kinematic Equation

In order to verify the correctness of kinematics equation, it is necessary to know the programming idea of MATLAB. First, this paper puts forward angles of three rotating joints. In the second step, θ_1 and θ_3 are constant value, and θ_2 is rotated at a constant speed. Accordingly, the motion trajectory of the end point E can be solved. According to the motion trajectory of the end point E, the corresponding angles can be solved by kinematic inverse solution. In order to verify the correctness of the equation [11], we compared the value of the joint angle with the solution result.

The corresponding inverse solution values $\theta_2^{'}$ $\theta_3^{'}$, etc. can be solved by

$$\theta_1 = 0, \theta_2 = \frac{2}{3}\pi t (0 \le \mathrm{t} \le 0.5), \theta_3 = 0 \tag{27}$$

The curves of θ_2 and $\theta_2^{'}$ are shown in Fig. 3. The analysis shows that the maximum error is less than 2.54×10^{-14}, which verifies the correctness of the forward and inverse kinematics model.

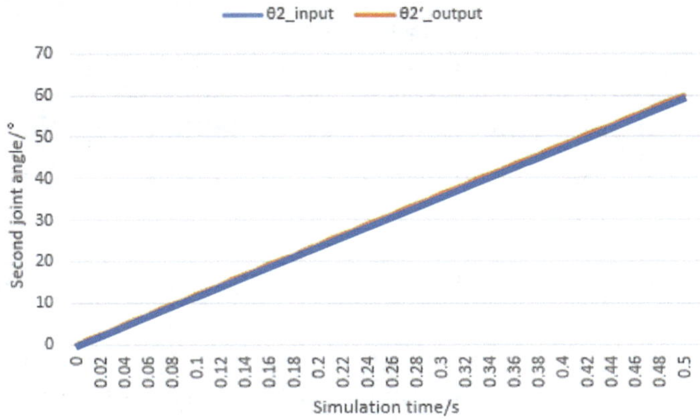

Fig. 3. Verification of kinematics equation

4.2 Typical Climbing Simulation

In order to verify the feasibility of the climbing, this section will simulate at 25° slope. The co-simulation is carried out in MATLAB and UG. The physical model, sensor parameters and contact parameters are establish in UG. The control output in MATLAB/Simulink is solved by the virtual controller. The 3D contact collision stiffness is 10000 N/mm and the damping is 50 N/(mm/s). The friction force is coulomb friction, the friction coefficient is 0.6, and the simulation time is 4 s. The co-simulation principle of MATLAB and UG is shown in Fig. 4.

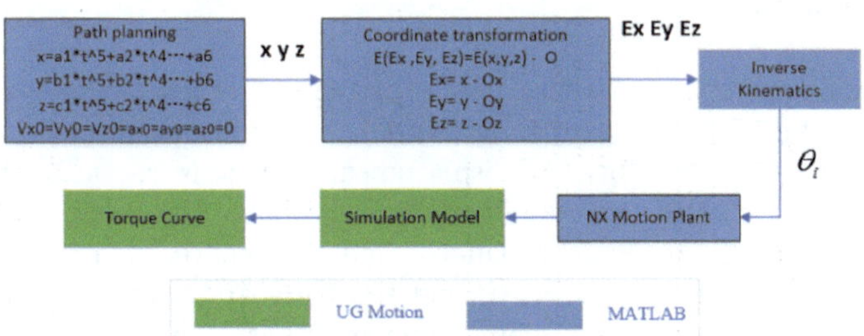

Fig. 4. Simulink/UG co-simulation

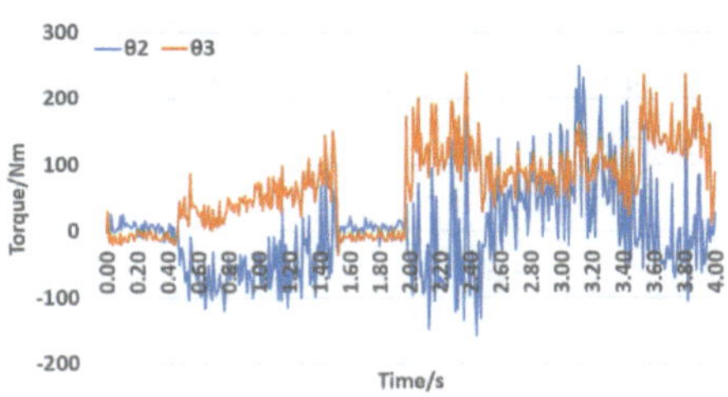

Fig. 5. Driving torque of leg A

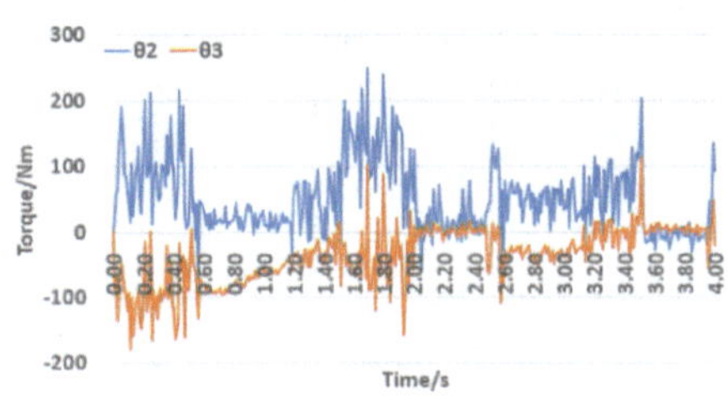

Fig. 6. Driving torque of leg B

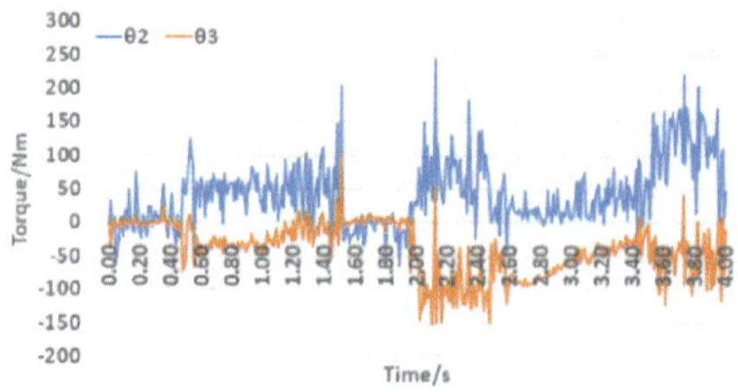

Fig. 7. Driving torque of leg C

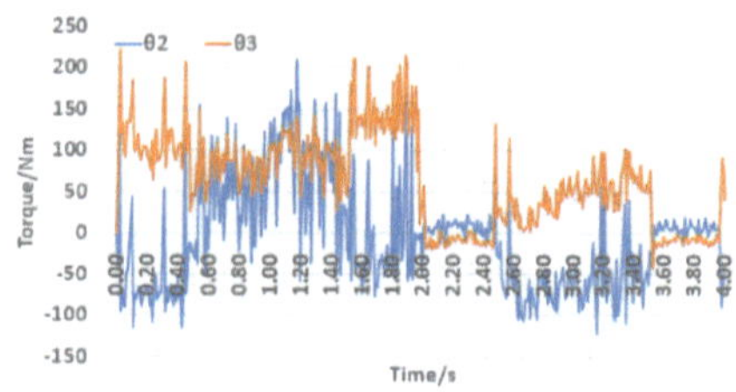

Fig. 8. Driving torque of leg D

The simulation diagram of the driving torque in four legs is shown in Figs. 5, 6, 7 and 8. It can be found that the sudden change of the driving torque is mainly concentrated on the start and stop timing of each joint motion. The joint driving torque can be controlled by trajectory planning. The maximum torque results of each joint are shown in Table 2. The maximum joint torque occurs when the joint motion is discontinuous, which can be reduced by trajectory planning. In the table, the drive torque does not exceed 260 Nm, which meets the design requirements.

Table 2. Maximum Torque of Joints

Name	Leg A		Leg B		Leg C		Leg D	
	θ_2	θ_3	θ_2	θ_3	θ_2	θ_3	θ_2	θ_3
Maximum torque (Nm)	248	237	250	113	244	102	210	221

5 Conclusions

In this paper, a quadruped leg-wheel robot is proposed for planetary exploration. Based on the proposed model, the closed-form solution for the inverse displacement model is established, and then the forward displacement equations are also obtained. The Jacobian matrix is derived from the displacement model. The co-simulation model is established, by using kinematic equations, and the motion planning and dynamic simulation are performed in MATLAB and UG software, respectively. Based on the co-simulation model, a typical climbing simulation has been carried out. Finally, the driving torques of all joints are obtained, which can be used for the design of prototype.

Acknowledgments. This project is supported by National Natural Science Foundation of China (No. 51575337), Equipment Pre-research Aerospace Joint Fund (No. 6141B06220407), and Key Laboratory Fund of Science and Technology on Space Intelligent Control (No. HTKJ2019KL502011).

References

1. Shang, J.Z.: Research on the Locomotion Mechanism and System of Space Exploring Robot, pp. 1–13. China Academic Journal Electronic Publishing House (2006)
2. Tong, A.L.: The world's space nations once again target the moon. J. Aerosp. China **6**, 41–45 (1995)
3. Deng, Z.Q., Li, S.J., Gao, H.B.: Research and development of passive rocker suspension of planetary exploration rover. J. Astronaut. **29**(6), 1695–1700+1722 (2008)
4. Tao, J.G.: Research on Six-Wheeled Rover Mobile System with Series Multi-Articulated Suspension and Its Key Technology, pp. 3–11. China Academic Journal Electronic Publishing House (2009)
5. Liu, D.C., Wang, J.Z., Wang, S.K., Hui, P.: Omnidirectional gait switching method of electric parallel wheel-foot robot based on velocity vector. J. Mech. Eng. (2019)
6. Guan, G.S.: Research on the System of Small Morphing Leg-Wheel Quadruped Robot, pp. 2–11. China Academic Journal Electronic Publishing House (2017)
7. Zhao, D.P.: Research on the Design and Gait Planning of Wheel-legged Mobile Robot on Unstructured Environment, pp. 37–40. China Academic Journal Electronic Publishing House (2010)
8. Jiao, E.Z., Chen, M.H.: Study on velocity simulation and Jacobian matrix solution of 6R serial robots. Mach. Tool Hydraul. (2010)
9. Siciliano, B., Sciavicco, L., Villani, L., et al.: Robotics (2009)
10. Peng-Cheng, Z., Tie, Z.: Analysis on solution of 6D of robot Jacobian matrix and singularity configuration based on vector product method. Mach. Des. Manuf. (2011)
11. Sun, Z.C.: Design of a three−finger dexterous hand driven by artificial torsional fibers for space debris capturing. Mach. Des. Res. (2019)
12. Zheng, H.C.: A Hardware-in-the-loop simulation strategy of space fine manipulation based on admittance control. Aerosp. Control. Appl. (2018)

Mechanism Design and Leg Kinematics Analysis of Four-Legged Robot

G. M. Wang, G. Y. Ma(✉), Y. L. Yao, and R. C. Liu

School of Mechanical, Electrical & Information Engineering, Shandong University, Weihai 264209, China
gyma@sdu.edu.cn

Abstract. Aiming at the shortcomings of serial quadruped robot, such as small load/dead weight ratio and large moment of inertia of the legs, a new quadruped robot with serial-parallel coupling mechanism is designed. The hybrid leg mechanism can realize the change in rotating pair to moving pair through redundant drive, and has the characteristics of stable movement, large work space range, strong bearing capacity and large stiffness. Firstly, the degree of freedom of the mechanism is analyzed, and its position and velocity kinematics are analyzed. The foot position of the leg was analyzed by positive and negative kinematics solutions. MATLAB and SOILDWORKS software are used for theory and simulation verification of the foot movement space; the kinematics simulation of the leg mechanism is carried out by ADAMS software, and the displacement, velocity and acceleration curves of the foot end are obtained. The results show that the motion speed of the mechanism is stable and there is no obvious mutation. This study has important guiding significance for the dynamics and gait planning of the follow-up leg mechanism.

Keywords: Quadruped robot · Hybrid mechanism · Redundant drive · Kinematics analysis · ADAMS simulation

1 Introduction

With the constant improvement of the national industrialization level, the dependence on high-tech products like robot is deepening, and the research of the robot is deepening, from the initial series arm to, the parallel mechanical platform such as Stewart [1], and then to today's string of compound, the robot is developing to the direction of diversification and high reliability. Series robots and parallel robots are mainly the existing foot robots. BigDog [2], SpotMini, hydraulic-driven quadruped bionic robot [3, 4] developed by Shandong University, and Hy-Q [5] developed by Italy University of technology are all series robots, which are simple in structure and easy to control, but have large inertia and low bearing capacity [6]. At present, Delta mechanism, Tricept mechanism and Z3 head are the most successful parallel robots [7]. Professor Huang Zhen of Yanshan University developed the first 6 DOFs parallel robot in 1991, and Professor Huang Tian of Tianjin University improved the Tricept mechanism and obtained the Trivariant mechanism [8]. The parallel robot have a compact structure, high bearing capacity, small sports load, but small working space [9]. In order to

D. Wang et al. (Eds.): MeTrApp 2019, MMS 79, pp. 474–483, 2020.
https://doi.org/10.1007/978-981-15-0142-5_46

overcome the small carrying capacity of serial robots and the limitation of the small working space of parallel robots, a new kind of four series and hybrid robot is put forward. By means of analytical method, the freedom of the leg mechanism of the robot and the positive and negative solution of the position of the space mechanism are analyzed. The foot motion space was calculated and simulated by MATLAB and SOLIDWORKS software. Finally, the mechanism kinematics simulation is carried out in ADAMS software, which provides the basis for subsequent dynamics and gait planning.

2 Structure Description of Leg Mechanism of Quadruped Robot

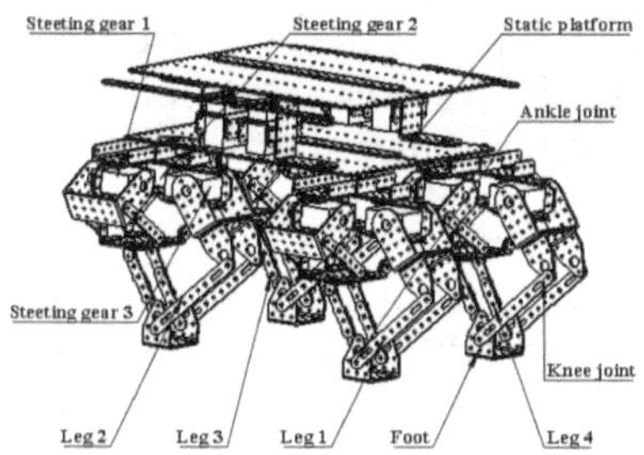

Fig. 1. The mechanism of Quadruped Robot

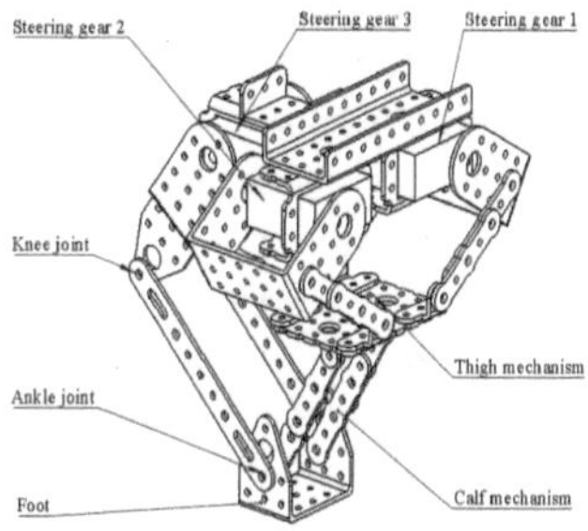

Fig. 2. The leg mechanism of Quadruped Robot

Figure 1 shows the overall structure of the quadruped robot, and Fig. 2 shows the structural con-figuration of the legs. The leg mechanism is fixed on the static platform of the body. The thigh and the static platform are connected by rotating pairs, which are composed of two r-r-r branch chains in parallel. The leg lifting movement towards the thigh can be completed through the control of two steering gear 1 and 2 fixed on the static platform. The lower leg is attached to the thigh and the lower leg to the ankle joint. Because the ankle joint is a rotating pair, the foot can rotate around the ankle flexibly to adapt to the road conditions. The motion chains on which the knee joint is located are driven by a steering gear 3 fixed on the frame. The knee joint, located between the thigh and calf, is responsible for the knee bend, as shown in Fig. 2. When the robot walks, Leg1 and Leg3 lift (the other two legs touch the ground), drive 3 drives the branched chain of the knee forward at an angle, and then both legs fall to the ground. In addition, Leg2 and Leg4 lift and rotate their legs in the same way to realize the walking of the robot. When designing the fuselage structure, the size and bearing capacity of the fuselage should be taken into account. If the space between legs is too small, the robot structure is too compact and the movement space of legs is reduced; if the space is too large, the structure is too loose and the bearing capacity will decline.

3 Position Analysis of the Leg Mechanism of Quadruped Robot

At present, there are many formulas for calculating the degree of freedom of the mechanism. The quadruped robot mechanism proposed to this paper can be equivalent to a planar mechanism. Taking into account the redundant constraints and local degrees of freedom of the mechanism, the revised Grubler-Kutzbach formula [10] is selected to calculate:

$$F_1 = (6 - \lambda)(n_1 - p - 1) + \sum_{i=1}^{p} f_i + p' - F' \tag{1}$$

F_1 - the number of degrees of freedom of the leg mechanism, F' - the number of local degrees of freedom in the leg mechanism, λ - the number of common constraints in the leg mechanism, n_1 - the number of members in the leg mechanism (including the frame), p - the leg mechanism The number of middle motion pairs, p' - the number of redundant constraints in the leg mechanism; f_i - the number of relative degrees of freedom of the i motion pair in the leg mechanism.

By analyzing the leg mechanism of the quadruped robot and drawing a schematic diagram of the leg mechanism as shown in Figs. 3 and 4. it can be concluded that the number of common points λ of the mechanism is 0, including 9 members and 11 rotating pairs, n_1 is 9, p is 10, the redundancy of the mechanism. The number of constraints p' is 4, the number of local degrees of freedom of the mechanism F' is 0, and the relative degrees of freedom of each motion pair f_i are 1. Bring the above data into formula (1) to get:

$$F_1 = 6 \times (9 - 10 - 1) + 10 + 4 = 2$$

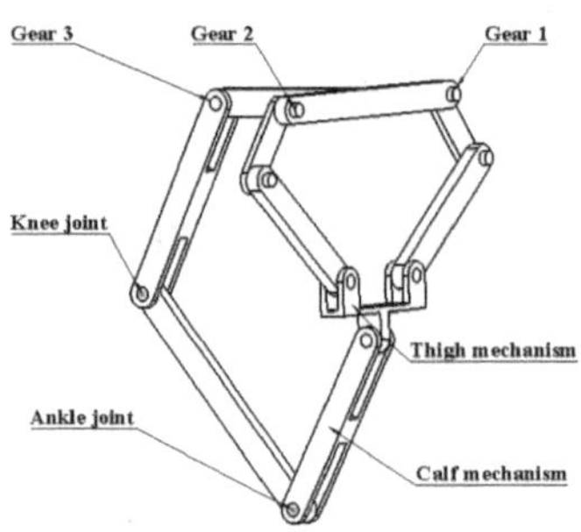

Fig. 3. The leg mechanism diagram of Robot

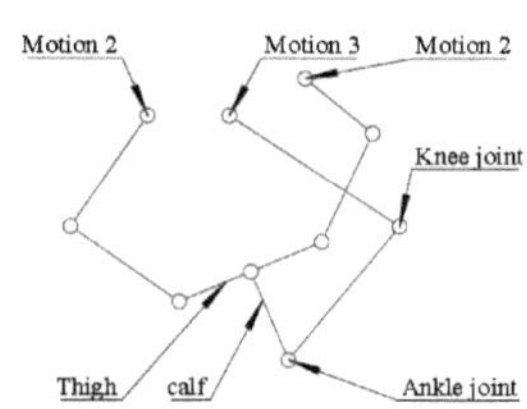

Fig. 4. The leg structure diagram of Robot

3.1 Positive Position of the Leg Mechanism

Since the designed mechanism can be equivalent to a plane mechanism and the static platform where the thigh and steering engine are located is equivalent to a hydraulic lever, the basic coordinate system O-X-Y can be established in the schematic diagram of the mechanism, as shown in Fig. 5.

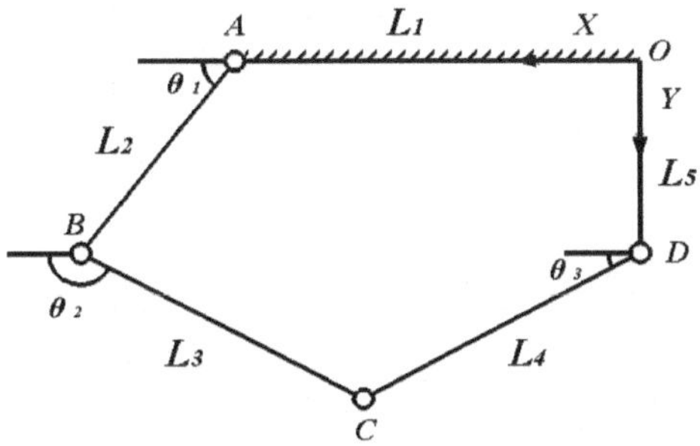

Fig. 5. The leg structure diagram of Four-legged robot

The positive position solution is known to input the original corner *A* and the length of the hydraulic rod L_5, establishing the relationship between the input parameters and the coordinates of the end effector. The equations can be listed from Fig. 5:

$$\begin{cases} x_c = L_1 + L_2 \cos \theta_1 + L_3 \cos \theta_2 = L_4 \cos \theta_3 \\ y_c = L_2 \sin \theta_1 + L_3 \sin \theta_2 = L_5 + L_4 \sin \theta_3 \end{cases} \tag{2}$$

After the formula (2) is removed, it can be obtained:

$$(A - B) \tan^2 \frac{\theta_2}{2} + 2C \tan \frac{\theta_2}{2} + A + B = 0 \tag{3}$$

among them

$$\begin{aligned} A &= L_3^2 - L_4^2 + (L_1 + L_2 \cos \theta_1)^2 + (L_2 \sin \theta_1 - L_5)^2 \\ B &= 2L_3(L_1 + L_2 \cos \theta_1) \\ C &= 2L_3(L_2 \sin \theta_1 - L_5) \end{aligned} \tag{4}$$

Solve the Eq. (3) to get the relationship:

$$\theta_2 = 2 \arctan\left(\frac{-C \pm \sqrt{C^2 + B^2 - A^2}}{A - B}\right) \tag{5}$$

Then the coordinates of point C can be obtained.

$$\begin{cases} x_c = L_1 + L_2 \cos\theta_1 + L_3 \cos\theta_2 \\ y_c = L_2 \sin\theta_1 + L_3 \sin\theta_2 \end{cases} \tag{6}$$

Therefore, in the MATLAB software, the RAND function is used to randomly take eight sets of corner and length data in the maximum motion range of the mechanism, and the coordinates of the foot end C are obtained by the positional positive solution formula. The calculation results are shown in Table 1.

Table 1. Positive location data

(θ_1, L_5)	x_c	y_c
(89.62°, 61.7284)	−19.206	131.1195
(13.97°, 85.7150)	71.7834	91.2950
(100.47°, 71.5613)	−28.3982	137.7243
(69.56°, 81.0084)	26.1153	148.1053
(10.73°, 61.2566)	68.6695	82.9015
(30.64°, 69.6528)	60.1673	109.1989
(60.15°, 84.4721)	39.7933	144.4762
(105.33°, 80.7662)	−23.2086	148.9231

3.2 Reversal Position of the Leg Mechanism

The position inverse solution C is known as the foot end position, and the input active member parameters A, that is, the rotation angle of the rotating pair θ_1 and the length L_5 are solved.

$$\begin{cases} x_c = L_1 + L_2 \cos\theta_1 + L_3 \cos\theta_2 \\ y_c = L_2 \sin\theta_1 + L_3 \sin\theta_2 \end{cases} \tag{7}$$

θ_2 Can be eliminated:

$$K \tan^2\left(\frac{\theta_1}{2}\right) + M \tan\left(\frac{\theta_1}{2}\right) + N = 0 \tag{8}$$

among them

$$\begin{aligned} K &= L_2^2 - L_3^2 + (x_c - L_1)^2 + y_c^2 + 2L_2(x_c - L_1) \\ M &= -4y_c L_2 \\ N &= L_2^2 - L_3^2 + (x_c - L_1)^2 + y_c^2 + 2L_2(x_c - L_1) \end{aligned} \tag{9}$$

Solve the Eq. (8) to obtain the servo motor R1 rotation angle

$$\theta_1 = 2\arctan(\frac{-M \pm \sqrt{M^2 - 4MN}}{2K}) \tag{10}$$

Further finishing by geometric relationship can obtain linear drive length L_5

$$L_5 = y_c - \sqrt{L_4^2 - x_c^2} \tag{11}$$

As the design of robot legs is similar to that of planar four-bar mechanism, due to the limitation of Angle, its value range is 0–110°. Only when the position of end-actuator is determined, the inverse solution of planar four-bar mechanism are unique. Corresponding to driving component parameters, only unique solution, Angle and data are shown in Table 2.

Table 2. Inverse location data

(x_c, y_c)	θ_1	L_5
(−19.2060, 131.1195)	88.54°	61.3487
(71.7834, 91.2950)	12.26°	87.6482
(−28.3982, 137.7243)	103.44°	72.3461
(26.1153, 148.1053)	68.87°	77.9590
(68.6695, 82.9015)	11.21°	63.1655
(60.1673, 109.1989)	28.54°	70.0684
(39.7933, 144.4762)	61.08°	82.4157
(−23.2086, 148.9231)	106.29°	79.5114

It can be seen from the comparison of data Table 1 and data Table 2 that the position of inverse analysis obtained by Angle and length is not obviously different from the preset value of the positive solution, so the mathematical model of reversed position is the correct solution.

4 Speed Analysis of the Foot of Quadruped Robot

Equation (9) can be organized into:

$$\begin{cases} L_1 + L_2\cos\theta_1 + L_3\cos\theta_2 - L_4\cos\theta_3 = 0 \\ L_2\sin\theta_1 + L_3\sin\theta_2 - L_5 - L_4\cos\theta_3 = 0 \end{cases} \tag{12}$$

Derivation of Eq. (12) can be obtained:

$$\begin{cases} \omega_2 L_2 \sin\theta_1 + \omega_{32} L_3 \sin\theta_2 - \omega_{45} L_4 \sin\theta_3 = 0 \\ \omega_2 L_2 \cos\theta_1 + \omega_{32} L_3 \cos\theta_2 - v_5 - \omega_{45} L_4 \cos\theta_3 = 0 \end{cases} \tag{13}$$

Where is the angular velocity of the *i* rod relative to the *j* rod, and then the two ends are eliminated ω_{45} in Eq. (13)

$$\begin{cases} v_{cx} = -\omega_2 L_2 \sin\theta_1 - \omega_{32} \sin\theta_2 \\ v_{cy} = \omega_2 L_2 \cos\theta_1 + \omega_{32} \cos\theta_2 \end{cases} \tag{14}$$

It is the positive solution of speed. From Eq. (14):

$$\begin{cases} v_{cx} \cos\theta_2 = -\omega_2 l_2 \sin\theta_1 \cos\theta_2 - \omega_{32} \sin\theta_2 \cos\theta_2 \\ v_{cy} \sin\theta_2 = \omega_2 l_2 \cos\theta_1 \sin\theta_2 + \omega_{32} \cos\theta_2 \sin\theta_2 \end{cases} \tag{15}$$

$$\begin{cases} v_{cx} \cos\theta_1 = -\omega_2 l_2 \sin\theta_1 \cos\theta_1 - \omega_{32} \sin\theta_2 \cos\theta_1 \\ v_{cy} \sin\theta_1 = \omega_2 l_2 \cos\theta_1 \sin\theta_2 + \omega_{32} \cos\theta_2 \sin\theta_1 \end{cases} \tag{16}$$

Combine (15) and (16) to simplify

$$\begin{cases} \omega_2 = \dfrac{v_{cx} \cos\theta_2 + v_{cy} \sin\theta_2}{l_2 \sin(\theta_2 - \theta_1)} \\ \omega_{32} = \dfrac{v_{cx} \cos\theta_1 + v_{cy} \sin\theta_1}{\sin(\theta_1 - \theta_2)} \end{cases} \tag{17}$$

Substituting Eq. (16) into Eq. (13) can get the velocity v_5 of the moving sub-movement L_5.

The semicircle track was selected for the foot end movement towards Fig. 6. The radius of the track was 25 mm, the foot end step distance reached 50 mm, and the highest point was 25 mm from the ground. At the same time, the foot end step distance and the highest point height could be guaranteed to meet the design requirements.

In the simulation movement as shown in Fig. 7, the foot end completes the leg lift from point 1 to point 2, the leg kick from point 2 to point 3, the split kick from point 3 to point 4 to point 5, the leg retraction from point 5 to point 6, and the leg drop from point 6 to point 7.

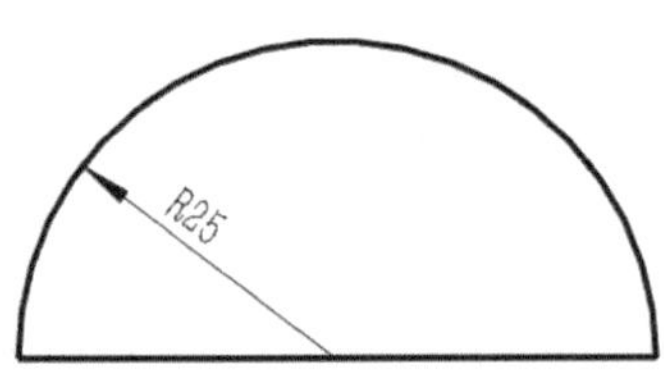

Fig. 6. Foot end trajectory of Robot

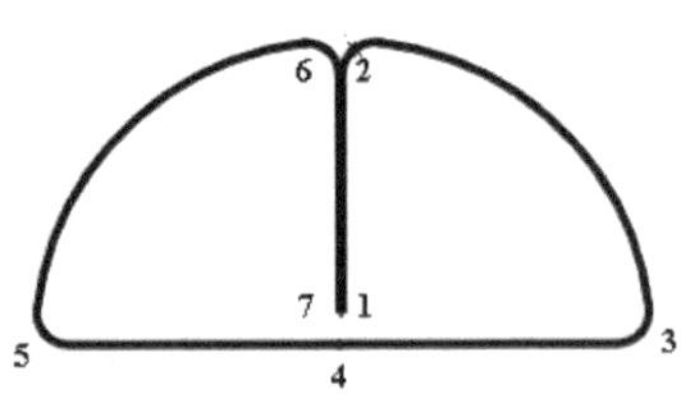

Fig. 7. Simulates the motion trajectory

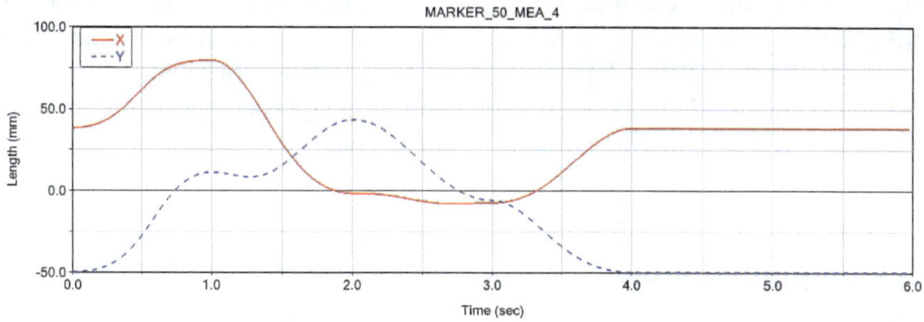

Fig. 8. The displacement amplitude curve of the robot foot

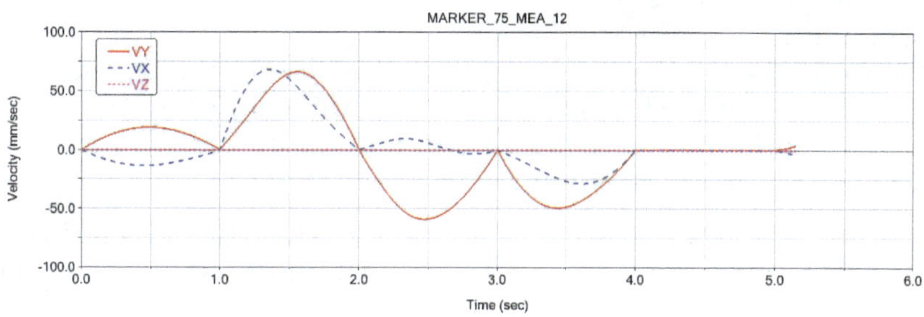

Fig. 9. The speed curve of the robot foot

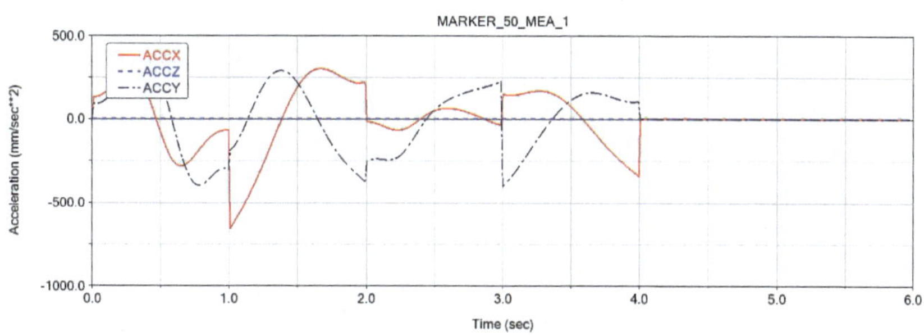

Fig. 10. The acceleration curve of the robot foot

According to this trajectory, the simulation of foot displacement, velocity and acceleration in ADAMS is shown in the Figs. 8, 9 and 10. According to the position analysis of foot space, the X-direction range and Y-direction range of reachable workspace are approximately 140 mm and 80 mm respectively. it can be seen that in the actual movement, in order to avoid the locked state, the X-direction motion range is smaller than the theoretical value. Among them, the speed in the X and Y directions changes greatly during the swinging process. As shown in Fig. 9, the velocity amplitude is greater than the amplitude during the ascent due to gravity, but the motion is basically stable, which can meet the needs of the quadruped robot to complete simple movements such as slow walking and turning. It can be seen from the combination of Figs. 8 and 9 that the designed quadruped robot leg mechanism runs smoothly, has no sudden change in speed, and has no major impact during exercise, and can complete the basic motion requirements of the quadruped robot.

4.1 Simulation of Foot Motion Space

The specific simulation process of foot end motion space is divided into the following four steps:

(1) At the beginning, the Angle between rod 5 and the frame is at 150°, while the Angle between rod 2 and the frame is rotated from 90° to 165°;
(2) Keep the Angle between rod 2 and the frame at 165° and the Angle between rod 5 and the frame at 120° from 150°;
(3) The Angle between the holder rod 5 and the frame is at 120°, and the Angle between the holder rod 2 and the frame is rotated from 165° to 90°;
(4) Keep the Angle between rod 2 and the frame at 90°, and the Angle between rod 5 and the frame at 150° from 120°;

Complete a periodic motion in this order. Through the motion function of SOILDWORKS, the motion space of the foot end can be simulated as shown in Fig. 11.

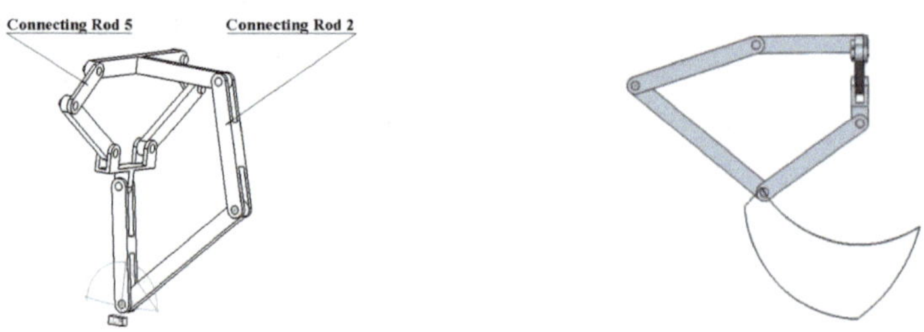

Fig. 11. Simulated foot end motion space

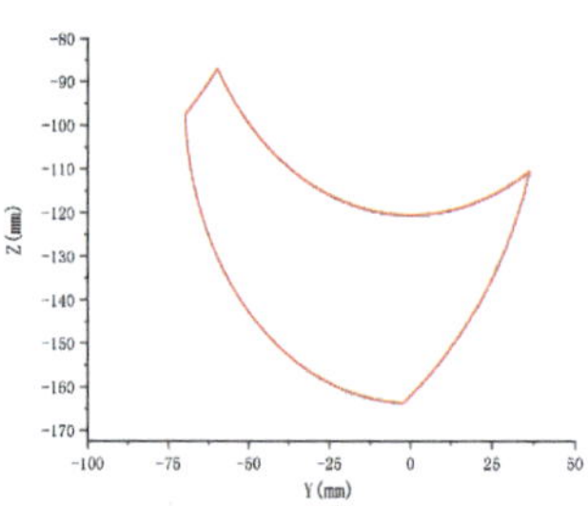

Fig. 12. Fitting the motion space of the foot

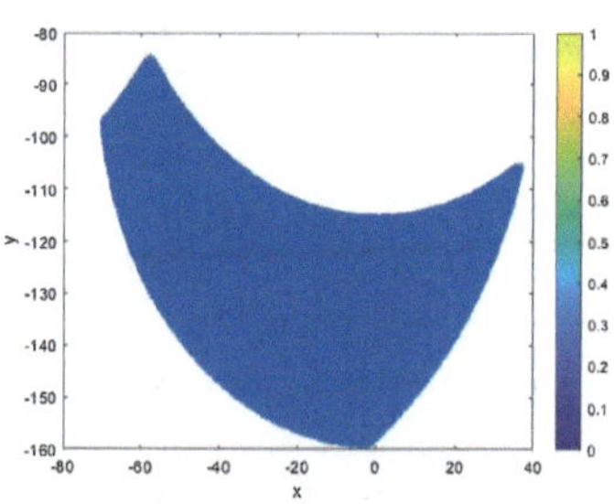

Fig. 13. The space simulation of motion of the foot

Figure 12 is the foot end motion space fitted according to the trajectory, and Fig. 13 is the result calculated according to the positive solution of the leg mechanism. It can be seen that the spatial range of motion in the two figures is basically the same, and the correctness of the spatial position calculation results can be determined.

5 Conclusion

A hybrid quadruped robot is proposed. The degree of freedom of the mechanism is analyzed. The positional forward and reverse solution calculations of the leg mechanism of the robot are carried out by analytical method. Through the analysis and comparison of the foot movement space by MATLAB and SOLIDWORKS software, it is proved that the mechanism has a large leg movement space. Finally, the kinematics simulation of the designed model is carried out in the ADAMS software, which verifies the stability and feasibility of the mechanism motion, and lays a theoretical foundation for the next step of dynamic analysis and gait planning.

Acknowledgements. We gratefully acknowledge the financial support provided by the Key Research Project of Shan-dong Province with Grant No. 2017GGX30112.

References

1. Stewart, D.: A Platform with Six Degrees of Freedom. The Institution of Mechanical Engineers, London (1965)
2. Rraibert, M., Blankespoor, K., et al.: BigDog, the rough-terrain quadruped robot. In: Proceedings of the 17th IFAC World Congress, Oxfora, pp. 10822–10825 (2018)
3. Li, M., Jiang, Z., Wei, G., et al.: One-legged system of four-legged bionic robot. Robot **36**(1), 21–28 (2014)
4. Zhang, X., Qin, X., Feng, H., et al.: Design and experimental study of one-legged rigid-flexible four-legged robot for running sports. Robots **35**(5), 582–588 (2013)
5. Semini, C., Tsagarakis, Ng., Guglielmino, E., et al.: Design of HyQ-a hydraulically and electrically actuated quadruped robot. Proc. Inst. Mech. Eng. Part I J. Syst. Control. Eng. **255**(6), 831–849 (2011)
6. Ren, G., Chen, W., Chen, B., et al.: Design and analysis of Roach-robot based on double four-bar mechanism. J. Mech. Eng. **47**(11), 14–22 (2011)
7. Zeng, C.: Study on design method of parallel and serial-parallel hybrid mechanism. School of Mechanical and Electrical Engineering, Harbin Institute of Technology, Harbin (2017)
8. Hong, Z., Mei, J., Zhao, X., et al.: Error modeling and sensitivity analysis of re-configurable hybrid manipulator-Trivariant. J. Mech. Eng. **42**(12), 6569 (2006)
9. Gao, J., Li, M., Hou, B., Wang, B.: Kinematics analysis of a series of four-legged walking robots with a hybrid leg. Opt. Precis. Eng. **23**(11), 3147–3160 (2015)
10. Zhou, J., Wang, W., Ou, Y., Huang, S., He, L., Yan, Y.: Calculating the degree of freedom of rod structure. Mech. Eng. **41**(01), 95–98 (2019)

Optimization Design and Analysis of an Adaptive Variable Magnetic Adsorption Climbing Robot

Denghui Guo and Yuan Chen(✉)

School of Mechanical, Electrical & Information Engineering,
Shandong University at Weihai, Weihai, China
cyzghysy@sdu.edu.cn

Abstract. In order to improve the efficiency of magnetic energy utilization, this paper proposes a non-contact variable magnetic adsorption device that can automatically adapt to different diameter changes. Firstly, the kinematics model of the mechanism is established by D-H parameter method and the motion variation characteristics of the magnetic adsorption unit are obtained. The feasibility of the approximate circular arc mechanism motion and the correctness of the mathematical model are verified by ADAMS simulation analysis. Then through MAXWELL, the magnetic field distribution characteristics of the magnetic adsorption unit and the variation characteristics of the magnetic adsorption force are simulated and analyzed, which proves that the mechanism can effectively improve the magnetic adsorption force and provides an effective method for further optimizing the efficiency of magnetic energy utilization.

Keywords: Non-contact · Variable magnetic adsorption · Variable diameter · Kinematics model

1 Introduction

In recent years, with the increase of marine resources development, the development of underwater robot technology has matured. As an important branch of underwater robots, magnetic adsorption climbing robots play a role in pipeline cleaning, submarine operations, pipeline inspection etc. Magnetic adsorption is generally classified into permanent magnet adsorption and electromagnetic adsorption. Due to the long working time of the underwater pipeline and the complex task environment, it is necessary to provide reliable and stable adsorption force. Therefore, the permanent magnet adsorption method is generally adopted as the main adsorption mode. In order to overcome the contradiction between adsorption capacity and motion flexibility, the current magnetic adsorption pipeline climbing robots adopt the variable magnetic gap adsorption method.

However, for cylindrical pipes, due to the rectangular magnets used in the magnets, the magnetic gap between the magnet and the pipes is constantly changing. When the farther away from the center of the pipe, the vertical magnetic gap is also increasing. Although for the larger diameter cylinder, the precise control of the magnetic gap can

D. Wang et al. (Eds.): MeTrApp 2019, MMS 79, pp. 484–493, 2020.
https://doi.org/10.1007/978-981-15-0142-5_47

still meet the requirements of stable magnetic adsorption force, the size of the magnet is limited to reduce the maximum load capacity and reduce the utilization efficiency of magnetic energy.

In order to improve the magnetic energy utilization efficiency and adapt to the pipe requirements of different diameters, this paper designs an adaptive non-contact variable magnetic pipe adsorption device that can automatically adapt to different diameter pipes based on the original variable magnetic gap control adsorption force. The kinematics model of the mechanism is established by D-H parameter method and obtain the motion state of the approximate circular arc mechanism. The correctness of the kinematics model is verified by ADAMS simulation. The comparative analysis of ANSOFT MAXWELL proves that approximate arc type magnet mechanism can effectively improve the magnetic energy utilization efficiency.

2 Introduction to Structure and Function

2.1 Structure Introduction

The variable magnetic adsorption mechanism is the basis of the whole magnetic adsorption robot. In order to adapt to different diameter changes and working environment, this paper designs a new non-contact variable magnetic adsorption mechanism that can automatically adapt to different diameter changes. This mechanism not only overcomes the contradiction between adsorption capacity and motion flexibility in the mode, but the utilization efficiency of the magnetic energy is also improved. It provides a more flexible, safe and stable working environment for pipe climbing robots. As shown in Fig. 1, the mechanism is composed of frame, diameter detecting mechanism, variable magnetic gap mechanism, variable diameter approximate circular arc mechanism, and a moving traveling mechanism. The variable diameter mechanism can detect a change in diameter to generate a certain displacement signal. The variable magnetic gap mechanism can control the distance between the entire variable diameter mechanism and the pipe surface of the pipe through the lead screw. The variable diameter mechanism is composed of three rectangular magnets of the same shape and connected by rotating pair. The displacement signals generated by the two movable members of the diameter detecting mechanism control the three magnets of the variable diameter mechanism to form an approximate arc to change the magnetic gap with the pipe. Due to the strong adaptability of the mechanism to the pipe surface, it can not only provide safe and reliable adsorption operation requirements on the outer cylindrical surface, but also can be satisfied for the plane and the inner cylindrical surface. Due to the need for all-round cleaning of the pipeline, the use of a universal wheel is used.

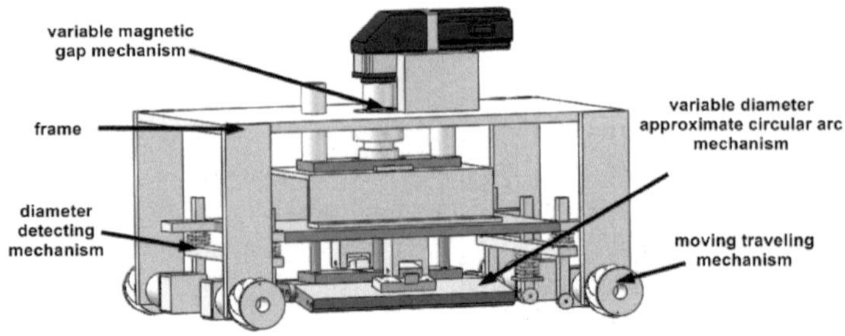

Fig. 1. Robot structure

2.2 Introduction to Adaptive Variable Magnetic Mechanism

As the basis of the force of the whole organization, the extreme value of the magnetic adsorption force and the adaptability determine the quality of the entire mechanism. Due to the sensitivity of the magnetic adsorption force to the magnetic gap and the fixed shape of the magnet, when working on the cylindrical pipe. The magnetic gap of the pipe surface is unevenly distributed. When the farther away from the central axis, the larger the magnetic gap is, which severely limits the utilization efficiency of magnetic energy and reduces the extreme value of magnetic adsorption force.

In order to overcome this problem, as shown in Fig. 2; the variable magnetic gap adjusting mechanism first designs a diameter detecting device for detecting the change of the pipe diameter, and the fixed support and the movable member are in contact with each other when moving through the pipe surface. The relative displacement of the piece can be used to obtain variable diameter mechanism displacement. The variable diameter mechanism is an approximate circular arc mechanism composed of three rectangular magnets with rotating pair. By controlling the up and down movement of the intermediate movable magnet, the rotating magnets on the left and the right sides can be rotated correspondingly. Thereby the corresponding approximate arc is the same as the arc of the pipe. At the initial position when the mechanism is working, the variable magnetic gap mechanism changes the magnetic gap between the approximate circular arc mechanism and the cylindrical pipe through the lead screw, and the sensor detects the relative displacement value of the movable member to control the moving distance of the moving magnet to form an approximate circular arc mechanism. In addition to working on the outer wall of the pipeline, when the moving amount of the movable member of the detecting mechanism is zero, the three magnet blocks form a plane to form a plane adsorption; when the mechanism moves on the inner cylindrical surface, the displacement amount of the movable member of the detecting mechanism is the opposite of the outer cylindrical surface which allows the three magnets to form an outer arc to form an approximately circular arc surface with the inner surface of the pipe.

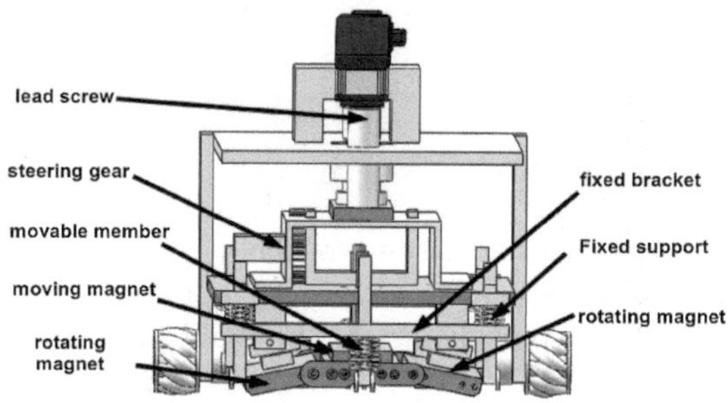

Fig. 2. Adaptive variable magnetic gap optimization mechanism

3 Kinematics Analysis of Variable Magnetic Mechanism

3.1 Analysis of Variable Diameter Detection Mechanism

In order to automatically adapt to different pipe diameters, the displacement of the moving magnet of the approximate circular arc mechanism needs to be accurately changed according to the change of the diameter of the climbing pipe. The displacement amount of the movable member with respect to the fixed support is the same as the displacement amount of the moving magnet. The displacement amount y is transmitted to the control unit through the sensor and the movement of the magnet is controlled by the control unit to complete the change of the approximate circular arc mechanism. Therefore, for the entire adaptive variable magnetic gap mechanism, the automatic detection mechanism is the basis of the entire automatic magnetic control mechanism. As shown in Fig. 3, the fixed support and the movable member are in contact with the cylinder, and the formula (1) can be obtained according to the geometrical characteristics of the mechanism.

$$y = r - \sqrt{r^2 - (b/2)^2} \tag{1}$$

Where r is the radius of the pipe being tested;

b is the distance between two fixed supports;

y is the amount of displacement of the movable member relative to the fixed support.

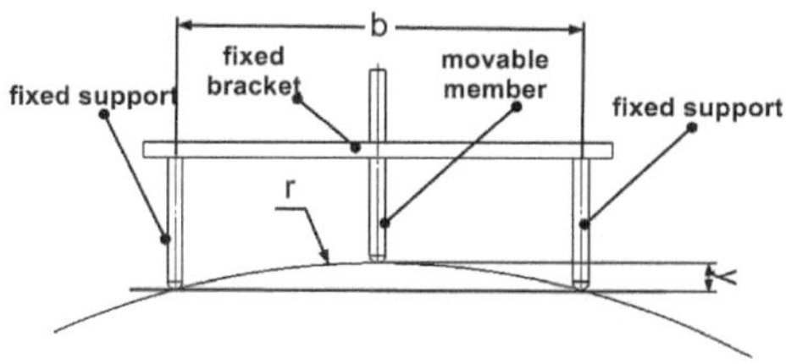

Fig. 3. Schematic diagram of variable diameter inspection mechanism

By simplifying:

$$r = \frac{4y^2 + b^2}{8y} \tag{2}$$

Since b is a known amount, when the y of the movable member is measured, the diameter of the mechanism can be obtained. In addition, in order to obtain the change of the mechanism for different pipe diameters, the relative displacement of the movable member at a diameter of 1000 mm to 2000 mm was measured. The motion of the movable member as a function of the diameter is obtained, as shown in Fig. 4. It can be seen from the figure that as the diameter becomes larger, the displacement amount y is reduced.

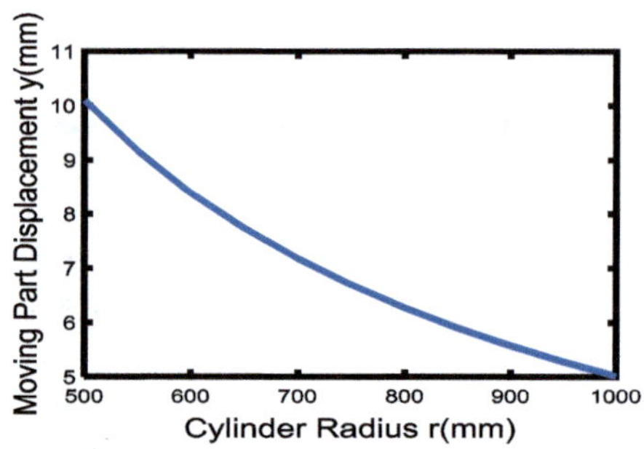

Fig. 4. Moving parts displacement

3.2 Kinematic Modeling of Variable Diameter Mechanism

In order to observe the variation characteristics of the three magnet plates during the movement of the variable diameter approximate arc mechanism, the D-H parameter method was used to analyze the kinematics of the mechanism. Since the structure has symmetry and the movable characteristics of the rotating magnets are the same, it is only necessary to analyze the motion characteristics of the left rotating magnet by the displacement amount y of the movable magnet. We establish $O_1 - X_1Y_1$, $O_2 - X_2Y_2$, $O_3 - X_3Y_3$ the coordinate system respectively on the movable magnet, the rotating magnet and the rotary slider, as shown in Fig. 5.

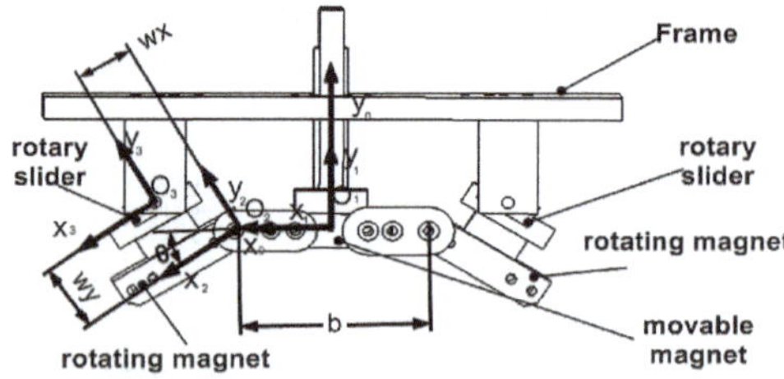

Fig. 5. Approximate circular arc machine D-H parameter coordinate system

As shown in Fig. 5, the $O_0 - X_0Y_0$ is a fixed coordinate system. Movable coordinates $O_1 - X_1Y_1$ move along the Y_0 axis of the static coordinates $O_0 - X_0Y_0$, Its vertical displacement is y. The rotation angle of coordinate $O_2 - X_2Y_2$ relative to coordinate $O_1 - X_1Y_1$ is θ_2. The displacement of coordinate $O_3 - X_3Y_3$ relative to coordinate $O_2 - X_2Y_2$ is a_3^*, The expression of its transformation matrix is:

$$T_3^0 = Trans(x_1, y)\, Rot(z_2, \theta_2)\, Trans(x_3, a_3^*) \tag{3}$$

So the transformation matrix:

$$T_3^0 = \begin{bmatrix} c\theta_2 & -s\theta_2 & 0 & y + \frac{b}{2} + w_x c\theta_2 - w_y s\theta_2 \\ s\theta_2 & c\theta_2 & 0 & w_x s\theta_2 - w_y c\theta_2 \\ 0 & 0 & 1 & 0 \\ 0 & 0 & 0 & 1 \end{bmatrix} \tag{4}$$

a_3^* is the distance between two origins of coordinates $O_2 - X_2Y_2$ and coordinates $O_3 - X_3Y_3$.

w_x is the horizontal coordinate value of o_3 in coordinate $O_2 - X_2Y_2$.

w_y is the ordinate coordinate value of o_3 in coordinate $O_2 - X_2Y_2$.

$c\theta_2^*$ and $s\theta_2^*$ represent the trigonometric functions $\cos\theta_2^*$ and $\sin\theta_2^*$ respectively.

Table 1. Corresponding parameter list

Link	a_i	α_i	d_i	θ_i
1	y	0	0	0
2	b/2	0	0	θ_2^*
3	a_3^*	0	0	0

Since the center of rotation o_3 of the rotary slider is fixed relative to the fixed coordinate system $O_0 - X_0Y_0$, It can be obtained with respect to position q^2 and q^0 of coordinate system $O_3 - X_3Y_3$. The parameter list is shown in Table 1. w_{x0} is the initial value and l is the displacement of o_3 on the abscissa in coordinates $O_2 - X_2Y_2$. So build system of equations:

$$\begin{cases} q^2 = [0 \quad 0 \quad 0 \quad 1] \\ q^0 = [w_{x0} + l + b/2 \quad w_y \quad 0 \quad 1] \\ q^0 = T_3^0 q^2 \end{cases} \tag{5}$$

Simplified equations: two left and right rotating magnet displacement expressions l and moving parts relative rotation expressions θ_2:

$$l = w_x - \sqrt{w_x^2 + (w_y - y)^2 - w_y^2} \tag{6}$$

$$\theta_2 = \pi/2 - \arccos(w_y/\sqrt{w_x^2 + (w_y - y)^2}) - \arccos(w_x/\sqrt{w_x^2/4 + (w_y - y)^2}) \quad (7)$$

The expression shows the motion characteristics of the left and right rotating magnets when the amount of displacement of the intermediate magnet is y.

3.3 Simulation Analysis of Variable Magnetic Mechanism

It can be seen from the above section that the motion expressions of the angle and displacement changes of the rotating magnets have been obtained by D-H parameter analysis. In order to verify the correctness of the results, it is first analyzed by MATLAB to obtain the results. The curve of change is shown in Fig. 6. Then by analyzing the motion state of the mechanism in ADAMS, the motion characteristics of the rotating magnet are obtained, as shown in Fig. 7. It can be seen from the above figures that when the displacement of the displacement magnet from the variable diameter mechanism is from 0 mm to 10 mm, the rotation angle of the rotating magnets on the left and right sides and the relative displacement with respect to the slider are gradually increased. the approximate arc mechanism formed is formed reduce size of the magnetic gap greatly. The results obtained by motion analysis and mathematical modeling are the same, which verifies the correctness of the mathematical modeling of the mechanism.

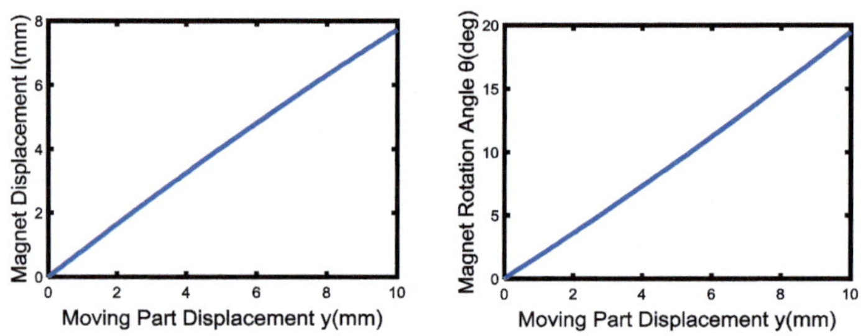

(a) (b)

Fig. 6. Rotational magnet motion characteristics

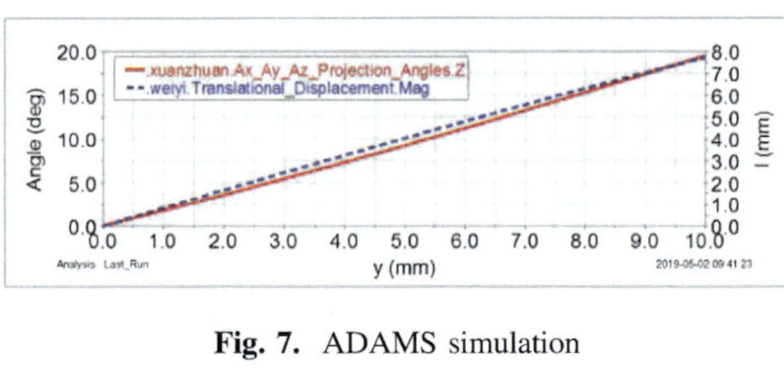

Fig. 7. ADAMS simulation

4 Variable Magnetic Analysis

4.1 Analysis of Magnetic Force and Magnetic Gap

The surface of the pipe is non-uniform with respect to the perpendicular magnetic gap of the surface of the rectangular magnet. The farther away from the central axis, the greater the value of the vertical magnetic gap. In this paper, the relationship between magnetic gap and magnetic adsorption force in plane contact is analyzed firstly. A 150 $\times$ 200 mm rectangular magnet is used to form a magnetic circuit with a low-carbon steel of the same area. We analyze the trend of change of the adsorption force of a rectangular magnet on the planar magnetic gap with the magnetic gap by using finite element analysis method. As shown in Fig. 8(a), the magnetic adsorption force is very sensitive with the change of the magnetic gap. With the increase of the magnetic gap, the magnetic adsorption force decreases rapidly. As shown in the Fig. 8(b), the magnetic field strength vector obtained by MAXWELL analysis. In this paper, parameter λ is introduced and the setting parameter is $\lambda = F/S$, where F is the adsorption force, S is the area of the working magnet area. The magnitude of the adsorption force generated by the magnet per unit area is known by λ. It can be seen from the figure that at a magnetic position of 5 mm, F = 1775 N; therefore, λ = 0.059 N/mm^2. It is also specified that λ is used as a standard value when the magnetic gap is 5 mm.

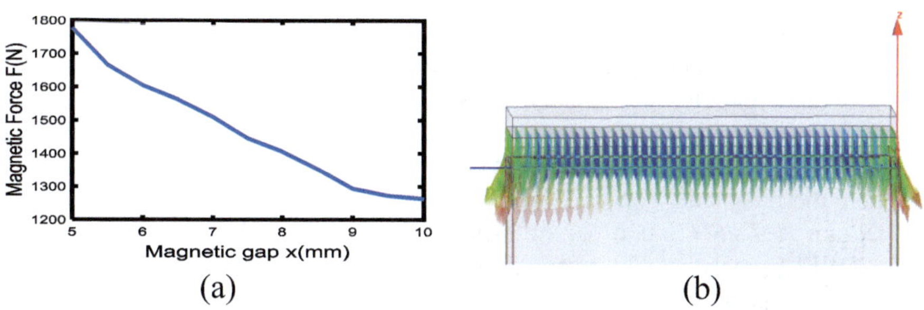

Fig. 8. Variation of magnetic force and magnetic field distribution in a planar magnetic gap

In order to analyze the change of the magnetic attraction force of the rectangular magnet at different distances from the central axis of the pipe, a rectangular magnet of 15 $\times$ 200 mm is placed at a distance of 5 mm from the highest point of the cylindrical surface on a cylindrical iron pipe with a diameter of 1000 mm. The horizontal direction gradually increases from the central axis. As shown in Fig. 9(a), it can be seen that as the horizontal distance increases, the adsorption force of the rectangular magnet to the pipe decreases rapidly. When the horizontal distance is 150 mm, the adsorption force has dropped to about 10 N, which is far from meeting the adsorption requirement. As shown in the Fig. 9(b), the magnetic field strength vector is obtained by MAXWELL analysis.

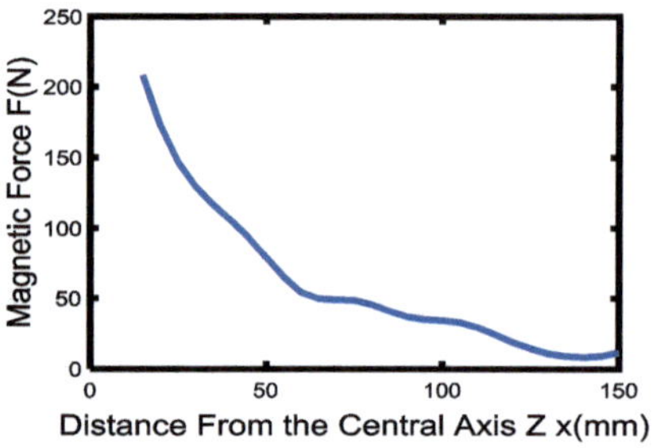

(a)

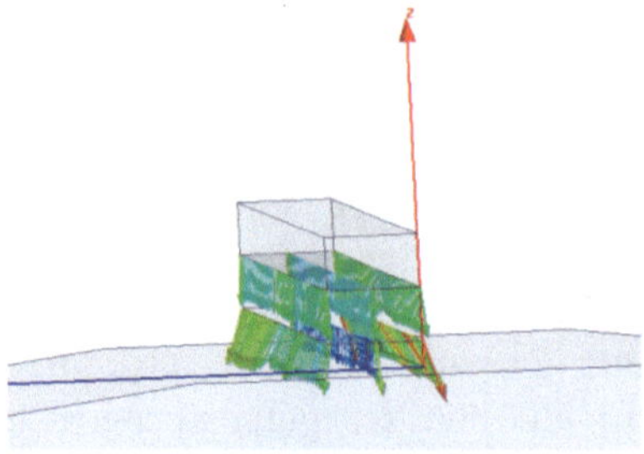

(b)

Fig. 9. Magnetic force changes with distance from the central axis

4.2 Comparative Analysis of Optimized Design

It can be obtained from the analysis in the previous section, which is necessary for the optimal design of rectangular magnets. In order to analyze the optimization results of the variable diameter mechanism for magnetic field strength and magnetic adsorption force, this paper analyzes the magnetic force variation trend of rectangular magnets with approximate magnet area of 150 × 200 mm and approximate arc mechanism under different magnetic gaps when the pipe diameter is 1000 mm. Through the method of finite element analysis, ANSOFT MAXWELL was used to analyze the change of magnetic attraction force under the two conditions, as shown in Fig. 10(a). It can be seen from the figure that as the magnetic gap increases, the adsorption force of the approximate circular arc mechanism is higher than that of the planar rectangular magnet. λ_1 and λ_2 respectively represent the magnet adsorption force per unit area of the variable-diameter magnet mechanism and the planar rectangular magnet.

When the magnetic gap is 5 mm, As showed in Fig. 10(a), $\lambda > \lambda_1 = F_1/S = 0.047\ \text{N/mm}^2 > F_2/S = 0.036\ \text{N/mm}^2$. And the magnetic field distribution is obtained by MAXWELL analysis. The magnetic field distribution of the approximate circular arc mechanism is shown in Fig. 10(b) when the magnetic gap is 5 mm. Because the magnetic gap distribution is relatively uniform, the influence on the magnetic field is small.

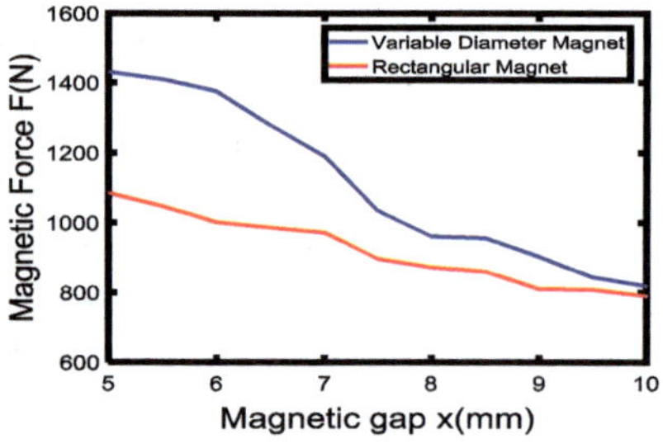

(a)

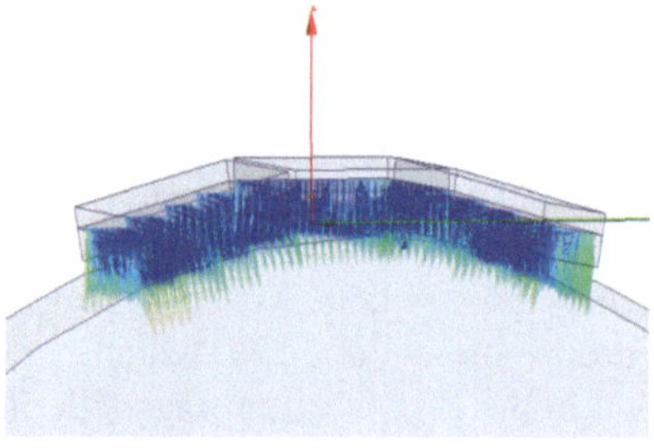

(b)

Fig. 10. Comparative analysis of magnetic forces generated by two kinds of magnets

5 Summary

In order to improve the utilization efficiency of the magnetic energy of the variable magnetic gap mechanism, a new adaptive variable magnetic adsorption mechanism is proposed in this paper. The motion model of the variable magnetic mechanism is constructed according to the mechanism characteristics of the variable magnetic mechanism and the motion model is derived. And it prove the feasibility of a variable magnetic mechanism on the outer wall of the pipeline. According to the method of finite element analysis, the comparative analysis of ANSOFT MAXWELL proves that the variable diameter mechanism can effectively improve the efficiency of magnetic energy utilization and provides a highly efficient magnetic adsorption mechanism for the variable magnetic climbing robot.

Acknowledgements. Grateful acknowledgement is given to the National Natural Science Foundation of China with Grant N0. 51375246, Major Scientific and Technological Innovation Project of Shandong Province with Grant No. 2017CXGC0923, Key Research and Development Program of Shandong Province with Grant No. 2018GGX103025, the Foundation Research Funds for the Central Universities with Grant No. 2019ZRJC006, and Natural Science Foundation of Shandong Province with Grant No. ZR2019MEE019.

References

1. Wen, J., Dun, X.M., Shan, D.L.: Structure design and weld seam surmounting characteristic of a wall-climbing robot with variable magnetic adsorption force device. Robot **33**(4), 405–410 (2011)
2. Lv, Q.Y., Wang, X.B., Jiao, H.K.: Design of small magnetic wall-climbing car for the wind power maintenance. J. Chang. Univ. Sci. Technol. **36**(4), 89–91 (2013)
3. Yuan, F.C., Wang, L.Z.: Development of a transmission system for the adsorption and movement robot on underwater hull. Manufacture **9**, 88–91 (2010)
4. Xu, Z.L, Ma, P.S.: Analysis of permanent magnetic tracked wall-climbing robot's turning agile capability. J. Shanghai Jiaotong Univ. **37**(11), 6–11 (2003)
5. Ding, W.S., Wang, X.C., Tang, C.T.: Structural design of permanent-magnet adsorption tracked ship wall-climbing robot. Hydromechatronics Eng. **44**(6), 88–91 (2003)
6. Wang, L., Lu, X.L., Zhao, S.P., Cao, Z.H.: Design and analysis of linear rod units for electricity pylons climbing robot. Manufacture **3**, 52–54 (2012)
7. Yuan, F.C., Lu, N.L., Qu, X.Q.: Design and study on magnetic absorbing of underwater robot for cleaning hull. Chin. J. Mech. Eng. **19**(4), 388–391 (2008)
8. Meng, X.Y., Dong, H.L.: Research on the structure design of climbing robot and the key technology of surface magnetic force absorption. Manuf. Autom. **40**(6), 19–22 (2018)

The Design of a Kind of Two-Joint Mechanical Fish

Hui-hui You[1], Chang-tang You[1], Xiao-yan Yu[1,2(✉)], Chang-ling Zhang[1], and Kai-long Wang[1]

[1] School of Mechanical Engineering and Automation, Fuzhou University, Fuzhou 350116, Fujian Province, People's Republic of China
cool@fzu.edu.cn

[2] Key Laboratory of Fluid Power and Intelligent Electro-Hydraulic Control, (Fuzhou University), Fujian Province University, Fuzhou 350116 Fujian Province, China

Abstract. Bionic mechanical fishes mainly involve single-joint mechanical fishes and multi-joint mechanical fishes. Although the motions of multi-joint mechanical fishes are more similar with those of fishes in nature, micro servos, whose torques are limited, are chosen as the driving mechanism in multi-joint mechanical fishes, and the movement speed of such mechanical fishes are not high. In contrast, DC motors with greater power and torques can be used as the driving mechanism in single-joint mechanical fishes, which is simple in structure but has higher oscillation frequency and can achieve faster forward speed. Therefore, a new type of two-joint caudal fin propulsion bionic robotic fish is designed. It has a novel tail structure that uses one DC motor to drive two tail joints so that it can achieve a higher tail fin swing frequency than most of the traditional multi-joints mechanical fish. 3D modeling is carried out for the mechanical fish, and then dynamics analysis and strength analysis are done by ADAMS and ANSYS Workbench. The result shows that the design is feasible and reasonable.

Keywords: Bionic robotic fish · Two-joint caudal fin propulsion · 3D modeling · Dynamic analysis · Strength analysis

1 Introduction

The area of the oceans is more than twice that of the land. The oceans are not only rich in resources, but also of great significance to national politics and security. Bionic robotic fish, as the name suggests, is an underwater mobile robot that imitates biological fish and has a wide range of applications [1–4]. The specific performance is:

(1) Display the swimming mechanism of fishes in various popular science exhibitions about fishes;
(2) Monitoring water quality parameters by installing various sensors on the mechanical fish;
(3) Equipped with a variety of different military sensors, using its flexible swimming characteristics to complete a variety of special military tasks;

D. Wang et al. (Eds.): MeTrApp 2019, MMS 79, pp. 494–502, 2020.
https://doi.org/10.1007/978-981-15-0142-5_48

(4) Provide theoretical basis and technical support for the design of high-performance AUV (autonomous underwater vehicle) [5].

Foreign research in the field of bionic mechanical fish is generally leading in China. Valdivia y Alvarado and Youcef-Toumi designed a mechanical fish FILOSE that can swim for a long time and control its depth and plane trajectory by swimming in three-dimensional space [6]. Foreign experts also verified that SMC based on dynamic model can significantly improve the speed tracking performance of multi-link robot. Good research and development results have also been achieved in China. The bionic mechanical fish SPC-II developed by Beihang University has the characteristics of fast movement, small power and high appearance simulation. The bionic machine developed by Harbin Engineering University can complete a variety of underwater movement, and has been successfully verified in lab and sea [7]. Some experts use IPMC to make robotic fish [8–10]. At present, bionic mechanical fishes involve single-joint mechanical fishes and multi-joint mechanical fish. Although the motions of multi-joint mechanical fishes are more similar with those of fishes in nature, micro servos, whose torques are limited, are chosen as the driving mechanism in multi-joint mechanical fishes, and the movement speed of such mechanical fishes are not high. In contrast, DC motors with greater power and torques can be used as the driving mechanism in single-joint mechanical fishes, which is simple in structure but has higher oscillation frequency and can achieve faster forward speed.

Therefore, this paper presents a new type of two-joint caudal fin propulsion bionic robotic fish. It has a novel tail structure that uses one DC motor to drive two tail joints so that it can achieve a higher tail fin swing frequency than most of the traditional multi-joints mechanical fish. 3D modeling is carried out for the mechanical fish, and then dynamics analysis and strength analysis are done by ADAMS and ANSYS Workbench.

2 Mechanical Fish Design

The movement of the mechanical fish in the water involves advancing forward, diving upward-downward and turning. The turning of the mechanical fish can be realized through the propulsion mechanism. Therefore, the scheme design of mechanical fish mainly includes propulsion mechanism design and dive mechanism design.

2.1 Propulsion Mechanism Design

In this motion mode, the caudal fin actually makes a compound motion of oscillation-translation. The oscillations and motions are not completely synchronous, and there is a phase difference between them. This phase difference is beneficial to the fish's propulsion. Studies show that the swing - translation phase difference of fish in nature is roughly PI/4.

For this purpose, a two-joint propulsion mechanism of the caudal fin is designed, whose structure is shown in Fig. 1. The DC motor fixed on the tail frame drives the front eccentric and the rear eccentric through the planetary gear head, and the rotating front eccentric and rear eccentric respectively drives the front follow-up frame and rear follow-up frame in contact with them to shift left and right on the tail frame. The front follow-up frame converts its translational motion into the swing of the tail frame through the front output rod, forming the first joint; the second joint of the mechanical fish is formed by the translation of the rear follow-up frame into the swing of the caudal fin through the rear output bar. There is PI/4phase difference between the front and rear eccentric, so two joints movement also exist phase difference, more in line with the movement of biological fish.

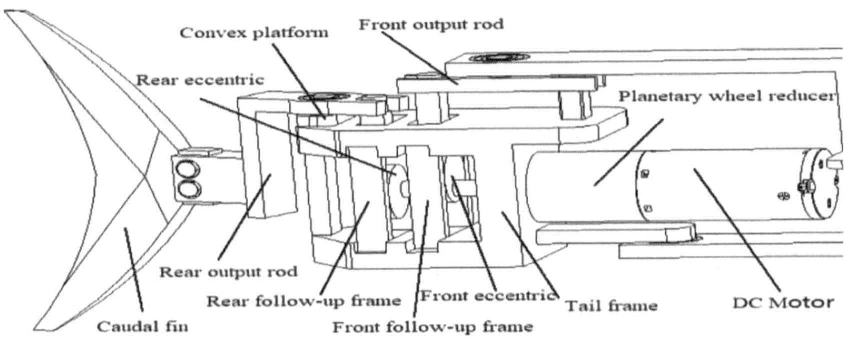

Fig. 1. The caudal fin propulsion mechanism

The selected driver is Maxon a-max26 110940 DC motor with graphite brush and Maxon 406764 planetary gear head. In this design, whenever the dc motor rotates once, the mechanical fishtail will complete one cycle of movement correspondingly. Therefore, the number of turns of the output shaft of the reducer is equal to the swing frequency of the mechanical fishtail in value. As mentioned above, the rated speed of the DC motor n_1 used in this design is 6910r/min; the deceleration ratio i of planetary wheel reduceris 27:1. Therefore, the output speed can be got as Eq. (1):

$$\mathrm{n} = \frac{\mathrm{n}_1}{i} = \frac{6910}{27} = 255.9\mathrm{r/min} \tag{1}$$

Taking $\mathrm{n} = 256\mathrm{r/min}$, the oscillation frequency can be got as Eq. (2):

$$f = \frac{\mathrm{n}}{60} = 4.27\,\mathrm{Hz} \tag{2}$$

Considering the influence of practical factors, $f = 4\,\mathrm{Hz}$ is taken in the subsequent software simulation.

2.2 Propulsion Mechanism Design

As shown in Fig. 2, this paper uses a servo to drive two pectoral fins at the same time, therefore, connected by a connection plate and frame of the u-shaped frame to connect two pectoral fins axis, and the cylindrical convex platform above the cylinder as a positioning convex surface to match the connection plate positioning hole, to ensure that the two pectoral fins shaft at the center of the shaft has a better alignment. For the pectoral fin movement is relatively simple, so only need to use the EMAX es08A 9g servo is enough.

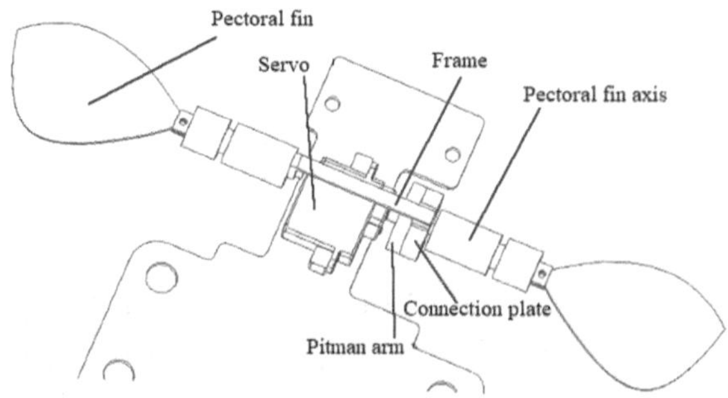

Fig. 2. The dive mechanism diagram

After the propulsion mechanism is selected, the mechanical fish is designed according to the size of selected parts and the 3D modelling is completed. The 3d model of the mechanical fish is shown in Fig. 3.

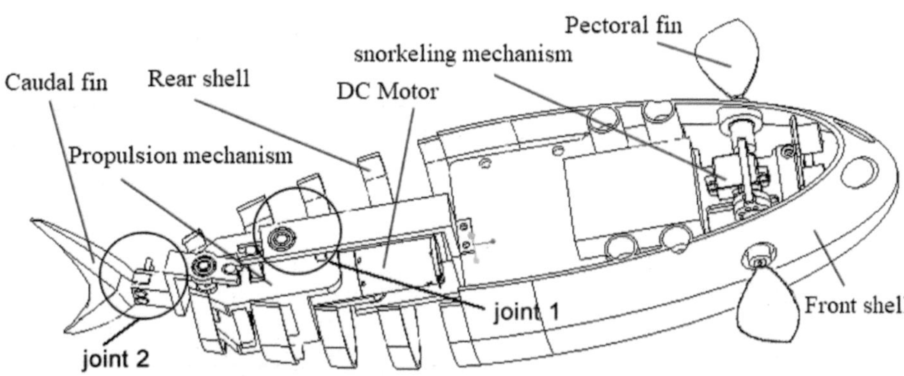

Fig. 3. The 3D model diagram of mechanical fish

3 Mechanical Fish Tail Fin Dynamic Simulation

The dynamic analysis of the 3D model of mechanical fish in ADAMS obtained the angular displacement, angular velocity and angular acceleration curves of the two joints as shown in Figs. 4, 5, 6, 7, 8 and 9. Butterworth low-pass filtering was performed on the angular acceleration curve in order to show the change law more clearly.

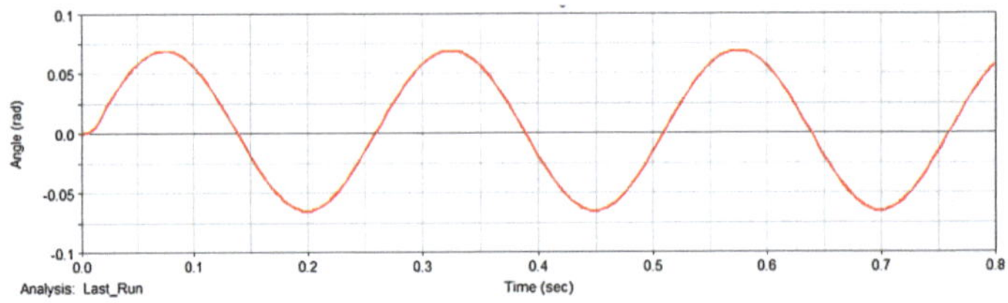

Fig. 4. The angular displacement of joint 1

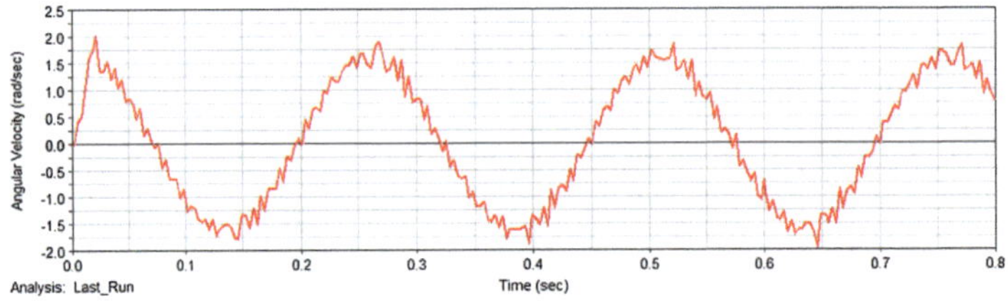

Fig. 5. The angular velocity of joint 1

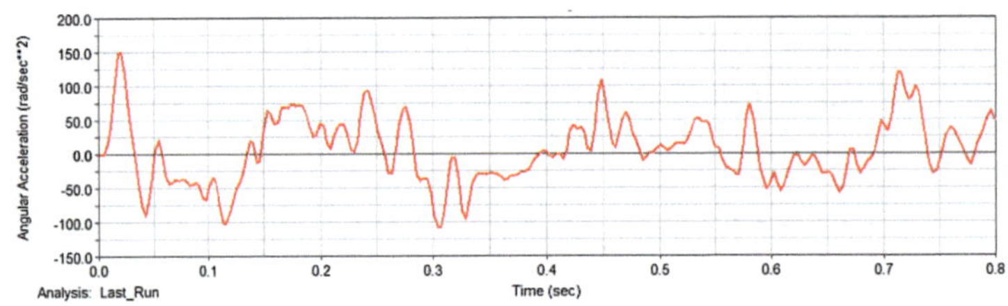

Fig. 6. The angular acceleration of joint 1 (after filtering)

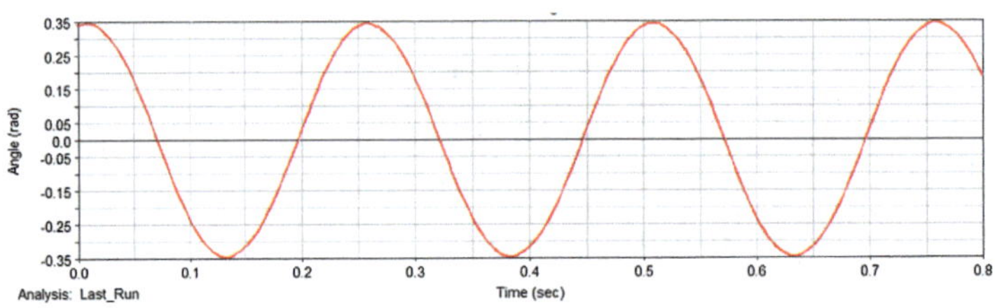

Fig. 7. The angular displacement of joint 2 relative to joint 1

4 Mechanical Fish Shell Strength Analysis

This design adopts ANSYS Workbench for shell strength check. Compared with the traditional interface, this design is more simple and easy to use. The stress analysis can be directly conducted through its static structural analysis. After the dynamic analysis of ADAMS, set up the working environment of underwater 1 m two dangerous points

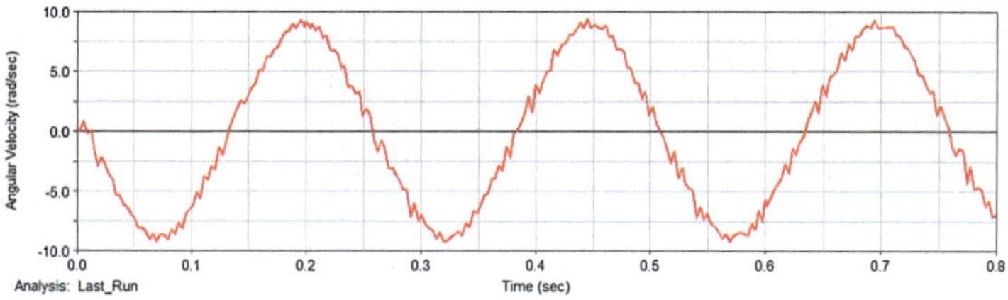

Fig. 8. The angular velocity joint 2 relative to joint 1

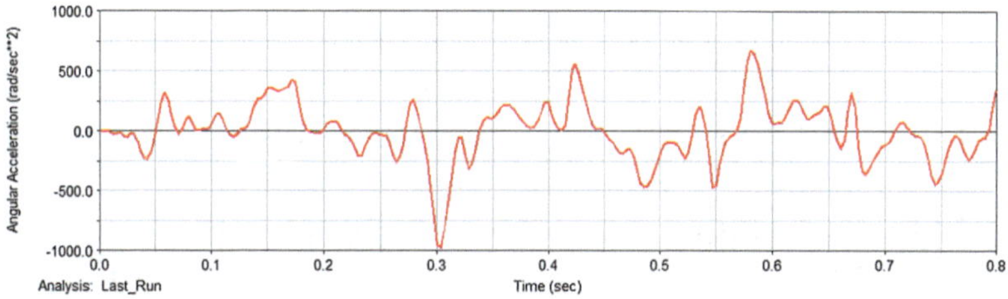

Fig. 9. The angular acceleration of joint 2 relative to joint 1 (after filtering)

were obtained and the stress was analyzed respectively. The grid nodes of the shell are 25,993, and the number of elements is 5641.

When the mechanical fish moves in the water, there are mainly three kinds of loads applied to the shell, which are the pressure of the water, the resistance of water and the pressure at the connection position. In this paper, it is assumed that the pressure of the water distribute on average, diving depth of mechanical fish h is 1 m and the speed u is 0.5 m/s, so we can easily get the pressure of the water P_1, the resistance of water P_2 according to the as Eq. (3) and as Eq. (4).

$$P_1 = \rho g h \tag{3}$$

$$P_2 = \frac{1}{2} C_d \cdot \rho \cdot u^2 \tag{4}$$

Where ρ in the equation is the density of water, g is acceleration of gravity, C_d is the drag coefficient. ADAMS is used to get the pressure at the connection position. Firstly, 4 markers are set at the connection position as shown in Fig. 10, and get the force direction of the markers in Fig. 11.

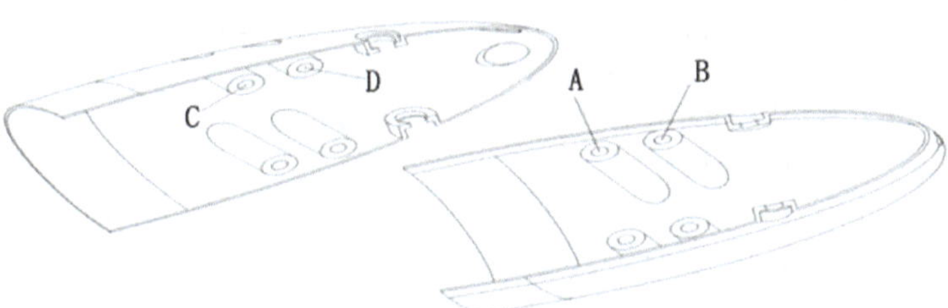

Fig. 10. The markers in the shell

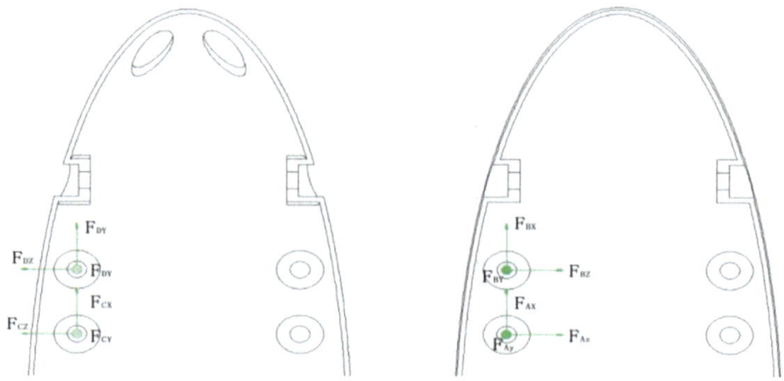

Fig. 11. The force direction of the markers

Secondly, dynamic analysis used to get the varying graph of the marker, which is show in the Fig. 12.

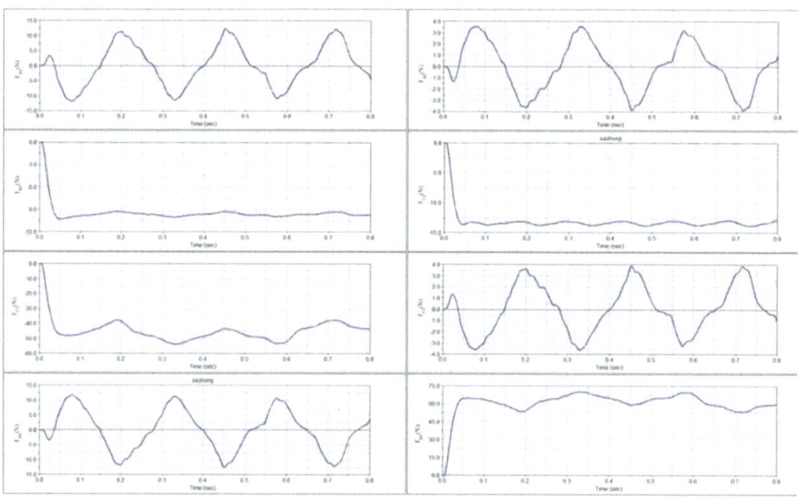

Fig. 12. The varying graph of Force

It is easily to find that there are 2 danger moment in the shell. One is about 0.2 s, the other is 0.325 s. By analyzing the loads applied to the shell in Workbench, we get the equivalent stress distribution diagrams which are shown in Figs. 13 and 14. Since the material used is 2A12 aluminum plate, which is a plastic material, the third strength theory and the fourth strength are used for checking.

① For 0.325 s, the maximum stress occurs at node 6480. The stress condition is as follows: $\sigma 1 = 19.14$ Mpa; $\sigma 2 = 15.23$ Mpa; $\sigma 3 = -0.9$ Mpa;Stress obtained according to the third strength theory: SINT = 20.04 Mpa; Stress obtained according to the fourth strength theory: SEQU = 18.4 Mpa.

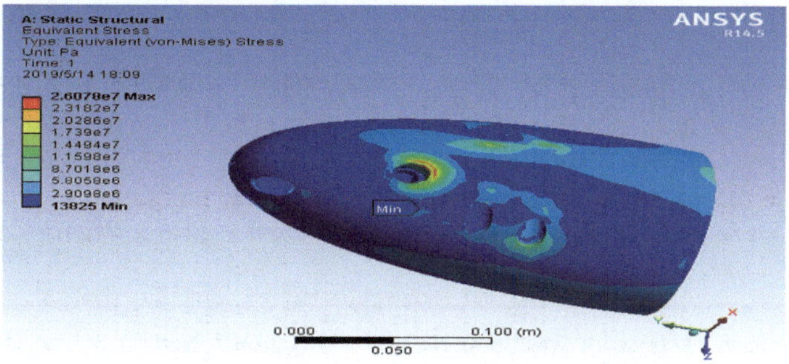

Fig. 13. The equivalent stress distribution of 0.2 s

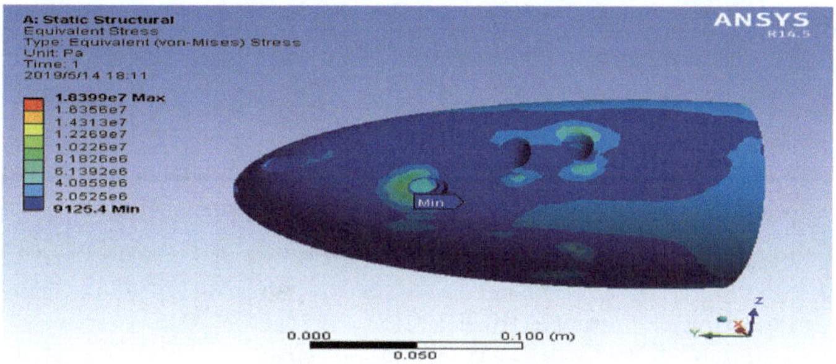

Fig. 14. The equivalent stress distribution of 0.325 s

② For 0.2 s, the maximum stress occurs at node 6480. The stress condition is as follows: $\sigma 1 = 26.97$ Mpa; $\sigma 2 = 21.84$ Mpa; $\sigma 3 = -1.29$ Mpa; Stress obtained according to the third strength theory: SINT = 28.26 Mpa; Stress obtained according to the fourth strength theory: SEQU = 26.08 Mpa.

Since the force on the mechanical fish shell in water is variable stress, the allowable safety coefficient can be 1.3–1.7, and the maximum value in this paper is 1.7. The yield strength σ_s of 2A12 material is 325 MPa, so the allowable stress as Eq. (5).

$$[\sigma] = \frac{\sigma_s}{n} = 191.2\,\text{MPa} \tag{5}$$

Obviously, the stresses obtained at both moments according to the third strength theory and the fourth strength theory are far less than the allowable stress, so the shell is strong enough.

5 Conclusion

A new type of two-joint caudal fin propulsion bionic robotic fish is designed. The bionic robotic fish has a novel tail which can get a 4 Hz tail fin oscillation frequency,so it is faster than using servos as drive, and at the same time the turning motion is also driven by the tail. A servo is used to drive two pectoral fins to make mechanical fish dive. The analysis shows that the strength of the shell is also enough to complete the design requirements.

Acknowledgment. This work was supported by the National Natural Science Foundation of China (No. 51741502) and the Natural Science Foundation of Fujian Province (No. 2016J01228, No. 2017J01559).

References

1. Yen, W.-K., Guo, J.: Phase controller for a robotic fish to follow an oscillating source. Ocean Eng. **161**, 77–87 (2018)
2. Zhong, Y., Song, J., Yu, H.: Towards a transform method from lighthill fish swimming model to biomimetic robot fish. IEEE Robot. Autom. Lett., 1 (2018)
3. Li, T., Li, G., Liang, Y.: Fast-moving soft electronic fish. Sci. Adv. **3**(4), e1602045 (2017)
4. Elege, N., Solapurkar, S,, Joordens, M.: Eye sensor for swarm robotic fish. In: System of Systems Engineering Conference. IEEE (2017)
5. Naik, M.-S., Singh, S.-N., Mittal, R.: Indirect adaptive output feedback control of a biorobotic AUV using pectoral-like mechanical fins. Bioinspiration Biomim. **4**(2), 026001 (2009)
6. Kruusmaa, M., Fiorini, P., Megill, W.: FILOSE for Svenning: a flow sensing bioinspired robot. IEEE Robot. Autom. Mag. **21**(3), 51–62 (2014)
7. Katzschmann, R.K., Marchese, A.D., Rus, D.: Hydraulic autonomous soft robotic fish for 3D swimming. In: Experimental Robotics. Springer (2016)
8. Chen, Z.: A review on robotic fish enabled by ionic polymer–metal composite artificial muscles. Robot. Biomim. **4**(1), 24 (2017)
9. Yang, T., Chen, Z.: Development of 2D maneuverable robotic fish propelled by multiple ionic polymer-metal composite artificial fins. In: IEEE International Conference on Robotics & Biomimetics IEEE (2017)
10. Palmre, V., Hubbard, J-J., Fleming, M.: An IPMC-enabled bio-inspired bending/twisting fin for underwater applications. Smart Mater. Struct. **22**(1), 014003 (2016)

Research on Knowledge Template Technology for Body Design of Transfer Robot

Wei Wang[1,2(✉)], Wenlin Gao[1], and Hongmin Yu[1]

[1] School of Mechanical Engineering and Automation, Fuzhou University, Fuzhou 350116, Fujian Province, China
mkwang@fzu.edu.cn
[2] Fujian High-end Equipment Manufacturing Co-innovation Center, Fuzhou 350116, Fujian Province, China

Abstract. In order to investigate the problems of low knowledge reuse rate and long product design cycle in the body design process of transfer robot, knowledge template system architecture and main knowledge template structures are established based on knowledge engineering theory and template technology. The knowledge template system is hierarchical, which are divided into integral transfer robot product family template of top level, the diverse transfer robot product template of middle level, and the functional block templates of lower level. The transfer robot product family template, the wrist template and the arm template structures are given. An intelligent design system framework for transfer robot body design based on the design templates is presented, which is composed of interactive interface, application design module, inference engine module, three-dimensional parametric design module, knowledge base and management module. An intelligent design prototype system is developed by using object-oriented technology and the knowledge template technology, and the mechanism configuration design, structural design of functional blocks are implemented in the system. The research results provide the basis for establishing a more integrated design system and the template technology is promising for speeding the design process.

Keywords: Transfer robot · Body · Modular design · Knowledge template

1 Introduction

The application of industrial robots in China has been expanding, gradually from the automobile manufacturing industry to food, medical, service and national defense and other fields [1]. Transfer robot is a kind of industrial robot used for carrying operations. Robot body design is the foundation of realizing robot function, including the body structure and transmission mechanism [2]. The rationality of its design will directly affect the motion, power and structural strength of the robot system.

The traditional body design process of transfer robot relies on the personal experience of designers. The design burden is heavy, product development cycle is long, and it is difficulty to improve the design quality. The rapid progress of Knowledge Based Engineering (KBE) technology, template technology, three-dimensional parametric

D. Wang et al. (Eds.): MeTrApp 2019, MMS 79, pp. 503–511, 2020.
https://doi.org/10.1007/978-981-15-0142-5_49

design technology and computer technology lay the foundation for the reuse of body design knowledge of transfer robot [3].

KBE is a computer integrated processing technology for product and engineering life cycle and design process. It combines CAD/CAE/CAM and AI technology, takes knowledge as the core, and solves engineering practical problems through dynamic expansion, acquisition, inheritance, reproduction, integration, management and application of the whole process knowledge [4]. The final objective is to reduce time and cost of product development [5].

Modular design can solve the contradiction between the variety of robots, the small batch size and the long design and manufacture cycle and the high cost. Modular partitioning is the first step in modular design [6]. Whether module partitioning is reasonable directly affects the function, performance and cost of modular system.

In this paper, the body intelligent design method of transfer robot based on knowledge template technology is studied. Knowledge engineering and template technology are used to encapsulate design knowledge structurally, drive the design process based on knowledge template, and realize the effective reuse of design knowledge.

2 Knowledge Template Technology for Body Design of Transfer Robot

2.1 Knowledge Template System of Transfer Robot

For modular design, the analysis of product structure and function is the main component of the design, but also the basis of the design [7]. The total function material transportation of transfer robot are divided into support function, rotation function, lifting function, wrist holding function, grasping function etc.

The body structure modules of transfer robot are shown in Fig. 1, which include base module, waist module, arm module, wrist module, end-effector module, joint module and other modules. The arm module is further divided into upper arm module and lower arm module, and the joint module is further divided into four axle joint modules.

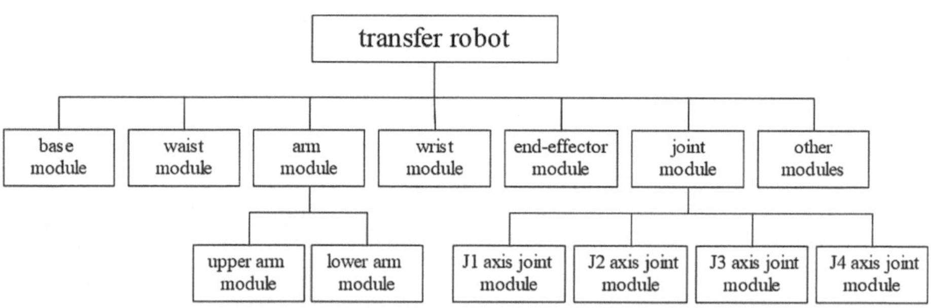

Fig. 1. Composition of the body module of the transfer robot.

In the process of modular design of transfer robot body, object-oriented technology is used to encapsulate the design knowledge of body and its modules into classes, which constitute the knowledge templates of transfer robot. The hierarchical knowledge template system are shown in Fig. 2.

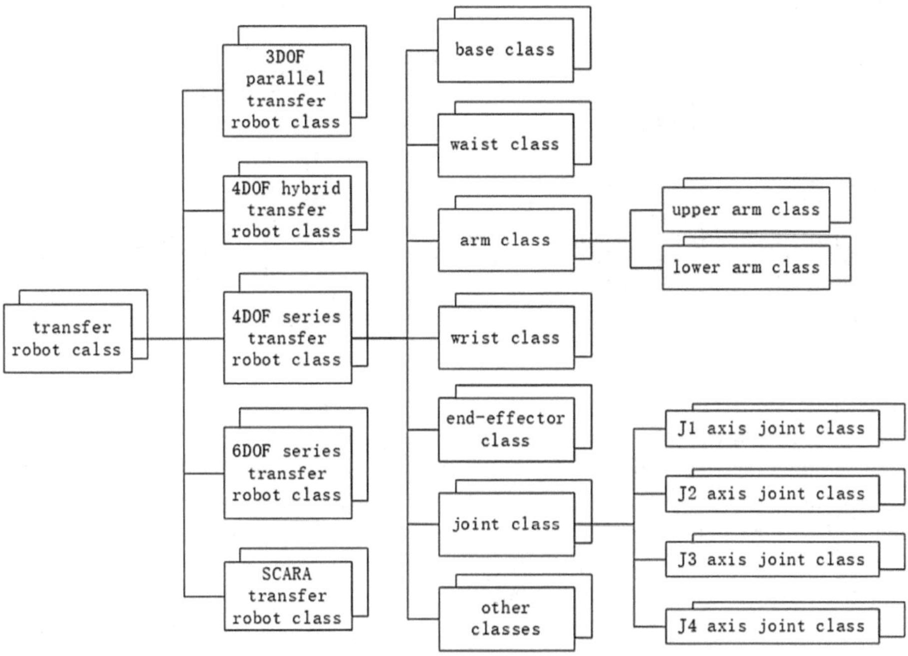

Fig. 2. The hierarchical knowledge template system.

2.2 Knowledge Template Construction of Major Modules

Different modules can be represented by corresponding knowledge templates. Template is a characteristic set, which presenting a group of similar things. The knowledge template is expressed in the form of object-oriented classes, which include attributes and methods. The attributes represent the design parameters used in the design process, and methods represent the design methods and design criteria used in the design process.

2.2.1 The Transfer Robot Product Family Template

The transfer robot product family template is expressed as {attribute | [load level, repetitive positioning accuracy, mechanism size, flexibility requirement, maximum speed, end effector attitude keeping requirements, mechanism configuration], method | [configuration design]}. According to the template attributes, the design decision tree for determining the configuration of the mechanism is shown in Fig. 3. The knowledge of the design decision tree is expressed in the form of rules, and the configuration design is realized in the way of rule-based reasoning.

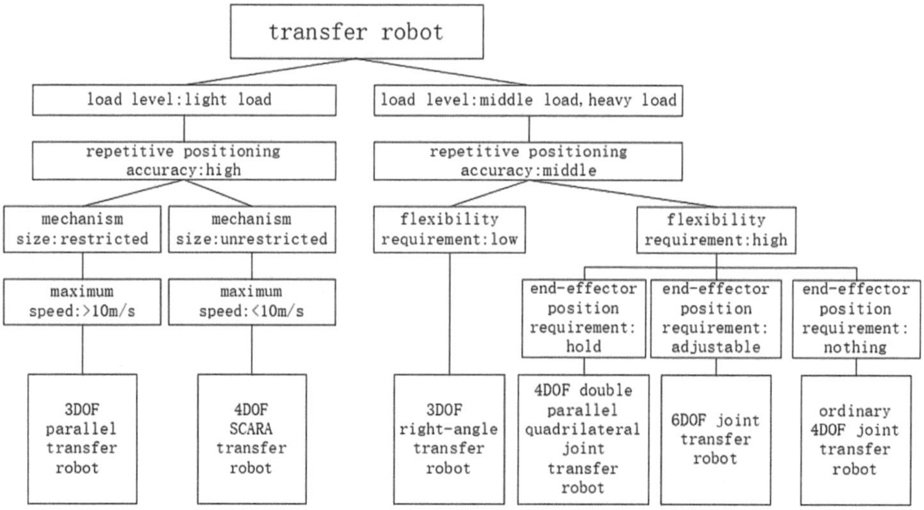

Fig. 3. Decision tree for configuration design of transfer robot.

2.2.2 Wrist Template

Wrist template is expressed as {attribute | [material name, material property, structural dimensions, maximum working radius, pay load, maximum rotating speed of axis], method | [material selection, structural dimensions design]}. Because the relationship between wrist structural parameters is complex and it is difficult to determine parameters, the wrist design adopt the case based reasoning method, and the design results are retrieved by matching the attributes similarity between the design tasks and the design cases of the case base.

For example, a case base C, which is composed of m case instances $\{c_1, c_2, \cdots c_m\}$. The attribute set of case instance i are composed of n case attributes $\{c_i^1, c_i^2, \cdots c_i^n\}$. The attribute set of design task c_t is $\{c_t^1, c_t^2 \cdots c_t^n\}$. The overall similarity between the design task and case instance i can be expressed as follows:

$$sim(c_i, c_t) = \sum_{r=1}^{n} w_r \frac{\min(c_i^r, c_t^r)}{\max(c_i^r, c_t^r)} \tag{1}$$

Where,

$sim(c_i, c_t)$—overall similarity

w_r—Relative weight of attributes, $w_r \in [0, 1]$, and $\sum_{r=1}^{n} w_r = 1$

A given design task is {Load 60 kg, maximum working radius 1950 mm, maximum rotating speed of J1 axis 90°/s, maximum rotating speed of J2 axis 90°/s, maximum rotating speed of J3 axis 90°/s and maximum rotating speed of J4 axis 180°/s}. The weights of all attributes are taken to 1/6. From the cases of the case base, the case with the maximum overall similarity is selected for the task. The selected case attributes follow as {Load 60 kg, maximum working radius 2100 mm, maximum rotating speed of J1 axis 90°/s, maximum rotating speed of J2 axis 85°/s, maximum rotating speed of J3 axis 85°/s and maximum rotating speed of J4 axis 240°/s}, and the overall similarity is 0.95. The structural parameters of the selected case are used as the basis of wrist structure design for the given task.

2.2.3 Arm Template

The arm template is represented as {Attribute | [Material Name, Material Attribute, Kinematic Size, Section Shape, Maximum Working Radius, Pay load, Load Type], Method | [Material Selection, Kinematic Dimension Determination, Section Design]}, According to the maximum working radius, pay load and load type, the kinematic dimension, material type and cross-section shape of the arm can be determined.

The kinematics dimension of the joint manipulator can be based on the three moving links theory [8]. The relationship equation of three individual link length and maximum working radius can be established by fitting the data from the upper limb dimension parameter database of national standard handbook (GB 10001998).

Circular hollow section and square hollow section are two commonly used cross-section types, square one is used for heavy payload.

Other templates including axle joint template and end effector template are also constructed.

3 Intelligent Prototype System for Body Design of Transfer Robot Based on Knowledge Template

3.1 Framework of Body Intelligent Design System

The body design process of the transfer robot mainly includes mechanism configuration selection, overall design, kinematic dimension determination, structure design of key components, driving system design, and three-dimensional design modelling.

Similar to the general body design process, the framework of the intelligent body design system of the transfer robot is shown in Fig. 4. The system consists of five modules, which are human-computer interaction interface, application design module, inference engine module, knowledge base and management module, and three-dimensional parametric design module. The human-computer interaction interface provides convenient and intuitive interaction and data input and output between users and the system. The application design module mainly includes five sub-modules: configuration selection, overall design, kinematics dimension design, key component structure design and driving system design. Knowledge base and management module mainly include knowledge template base, case base, rule base, database and CAD template base. Inference engine module includes rule-based reasoning and case-based reasoning. Three-dimensional parametric design module realizes rapid construction and modification of three-dimensional product model.

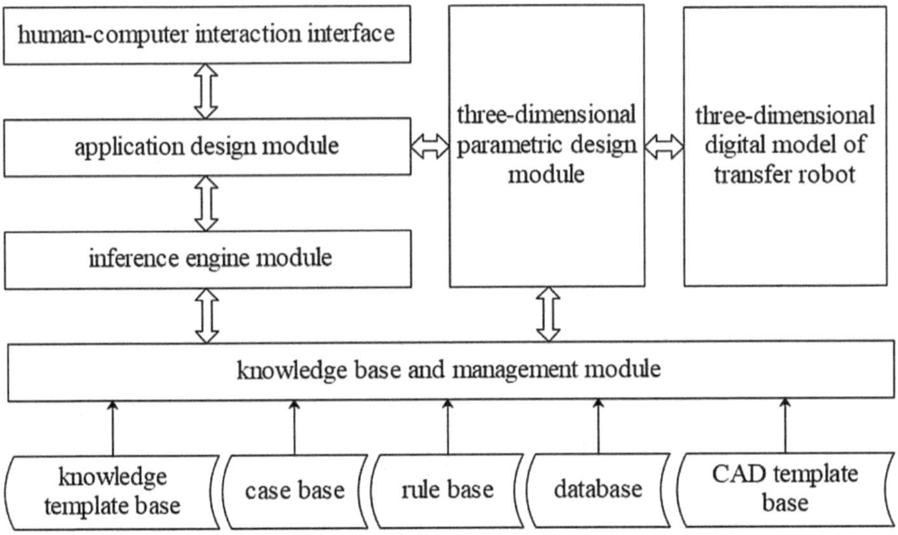

Fig. 4. The framework of the intelligent body design system of the transfer robot.

3.2 Prototype System Development and Functional Verification

With Microsoft Visual Studio as the system integration development environment, the human-computer interaction interface and functional modules of the prototype system are developed by using object-oriented Visual C++. Based on SolidWorks 2012 three-dimensional parametric design platform, the three-dimensional parametric model library of parts and assembly model driven by parameters are developed [9, 10]. Figure 5 are some design interfaces of robot body intelligent design prototype system.

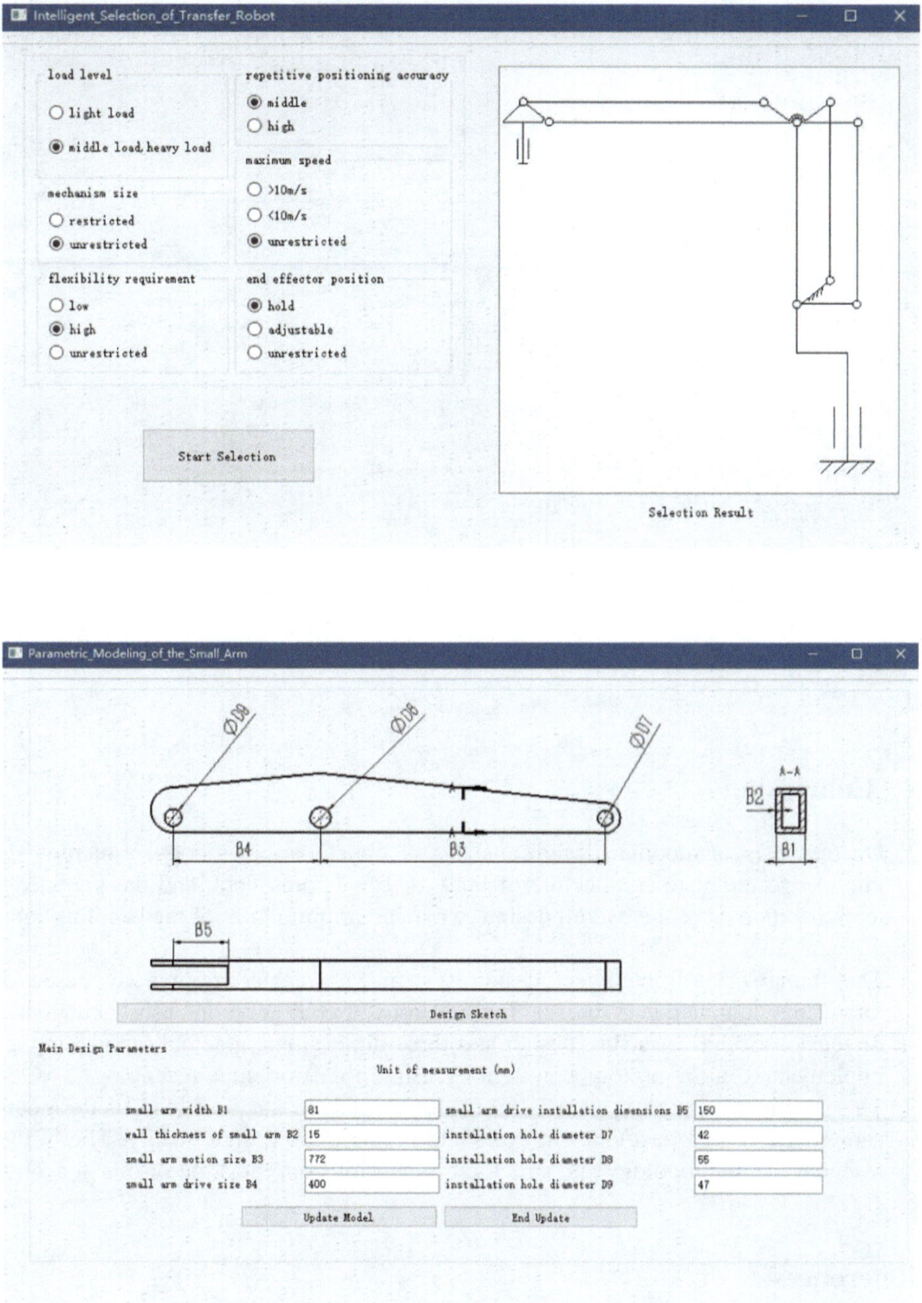

Fig. 5. Some design interfaces of robot body intelligent design prototype system.

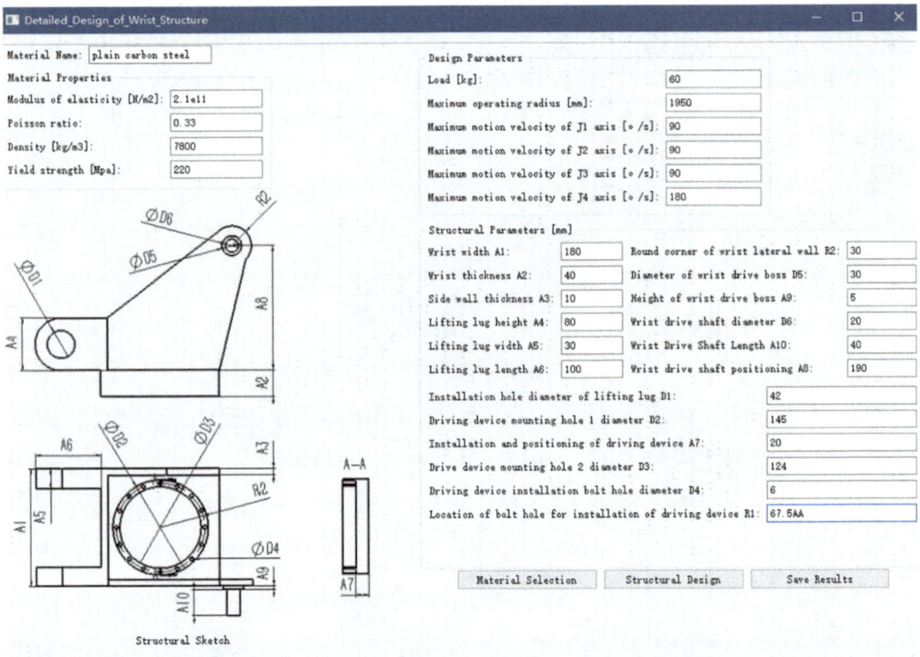

Fig. 5. (*continued*)

4 Conclusion

1. On the basis of modular design analysis of transfer robot s body, a hierarchical knowledge template architecture of body design is presented, and the knowledge template models of the overall design, wrist and arm modules of the handling robot are analyzed.
2. The framework of intelligent design system for transfer robot body based on knowledge template is given, and the functions of configuration design, kinematics dimension design, structure design and three-dimensional assembly modelling are implemented in the prototype system by using object-oriented technology.
3. The intelligent design system is promising to reduce the workload of designers, shorten the design development cycle and improve the design quality. The further research include topology and structural dimensions optimization of transfer robot.

References

1. Gao, F., Guo, W.Z.: Thinking of the development strategy of robots in China. J. Mech. Eng. **52**(7), 1–5 (2016)
2. Jiang, H., Xu, Y.S., Xie, Y.: Research on module partition method of mechanical products. Manuf. Technol. Mach. Tools **3**, 7–9 (1999)

3. Tor, S.B., Lee, S.G., Britton, G.A., Zhang, W.Y.: Knowledge-based functional design of industrial robots. Int. J. Prod. Res. **46**(16), 4501–4519 (2008)
4. Rusu, C., Tiuca, T.L., Noveanu, S., Mândru, D.: A KBE tool for solving the mechanisms kinematics. In: IOP Conference Series: Materials Science and Engineering, vol. 147, no. 1, pp. 12080–12085 (2016)
5. Yuan, Z.P.: Knowledge-based Functional Design of Industrial Robots. Master, Hunan University (2015)
6. Xiao, C., Zhou, Y.L., Sheng, H.Y., et al.: Modular design of mechanical noumenon for industrial robots. China Mech. Eng. **27**(8), 1018–1025 (2016)
7. Chen, X., Lan, M.F., Liu, Q., et al.: Module division and product structure model establishment of automobile chassis. J. Chongqing Inst. Technol. **29**(10), 19–23 (2015)
8. Yang, J.L., Gao, F.: The design of biomimetic robot based on the theory of three moving links. In: 15th International Conference on Mechanism and Machine Science, Yinchuan, pp. 242–245 (2006)
9. Li, W.L., Wang, Y.X.: Parametric design of hydraulic cylinder with secondary development based on solidworks. Mach. Build. Autom. **46**(1), 74–77 (2017)
10. Zhang, J.W., Wang, S.J., Zhang, X.L.: The 3D modeling of ball screw based on secondary development of solidworks dimension driven. Modul. Mach. Tool Autom. Manuf. Tech. **11**, 22–24 (2015)

Kinematics Analysis and Reconfiguration Planning of the Reconfigurable Robot

Bin Ye, Feng Gao(✉), and Weixing Chen

School of Mechanical Engineering, Shanghai Jiao Tong University, Shanghai 200240, China
gaofengsjtu@gmail.com

Abstract. Though the unknown environment can be dangerous, people are always interested in it. However, the problem is that current robots cannot adapt to work in an unknown environment. This paper designs a reconfigurable robot that can perform tasks in unknown environments. The robot has four different configurations. And they contain a six legs configuration, a two wheels configuration, a three wheels configuration, and a star shape configuration. Four different configurations can be transformed to each other. This paper mainly introduces the kinematics of the reconfigurable robot and the schemes of the transformations according to the kinematics model. Finally, in order to verify the feasibility of the simulation reconfiguration schemes, experiments are carried out.

Keywords: Reconfigurable · Kinematics · Transform · Simulation

1 Introduction

Human beings always have strong interests in exploring the unknown environment from ancient time to today. With the continuous development of science and technology, more and more robots are designed to explore the unknown environment. Because the unknown environment can be complex and dangerous, in order to explore more it effectively, new types of robots adapting to complex work environments need to be developed.

Wheeled robots have advantages in high stability, simple control, high speed as well as high energy efficiency [1]. Wheeled robots are one of the first categories of mobile robots to be used in unknown environments. As early as 1970, the Soviet Union had launched Lunakhod series of lunar vehicles for exploring the lunar surface. Since then, the National Aeronautics and Space Administration (NASA) has developed the Travellers, the Courage and the Curious for Mars exploration in 1997, 2004 and 2011 respectively and in the year of 2008, the ATHLETE robot was developed for the lunar surface exploration [2]. Subsequently, the China Aerospace Fifth Institute also developed the robot, named Yutu for the moon exploration in 2013. However, wheeled mobile robots have overt shortcomings: the unknown environmental conditions can be complex and most of them are uneven. These surfaces cannot be suitable for wheeled mobile robots.

In order to solve this problem, the researchers introduced legged mobile robots with more terrain adaptability [3]. Legged mobile robots are flexible and can pass over

D. Wang et al. (Eds.): MeTrApp 2019, MMS 79, pp. 512–523, 2020.
https://doi.org/10.1007/978-981-15-0142-5_50

ditches and obstacles [4]. But legged mobile robots also have their shortcomings: the leg-type mobile robots' control system is complex, and the stability of legged robots is not as good as the wheeled mobile robot.

Based on the above background, in order to complement the advantages of the wheeled mobile robot and the legged mobile robot, the researchers are turning the attention to the reconfigurable robot, which can switch between the wheeled robot and the legged robot [5].

At present, there are many designs of reconfigurable robots. Luo constructed a reconfigurable hybrid wheel-track mobile robot based on a Watt II six-bar linkage [6]. Guan used the D-H method to analyze the kinematics of the robot's single-legged mechanism, and based on this, he designed a small robot suitable for field walking [7]; Li established a model of chain wheel-legged robots in ADAMS, and obtained theoretical optimal parameters of chain wheeled-legged robots based on optimization analysis of parameters [8]. According to biomimetic bionics, Chi proposed a reconfigurable robot that can switch between wheeled and legged configurations [9]. Wang proposed a reconfigurable mobile robot with eight modes witch had five configurations [10].

Based on this, this paper introduces a new kind of reconfigurable robot which has four kinds of configurations. This robot integrates the advantages of wheeled robots and legged robots and is capable of working in complex environment.

2 The Prototype of the Robot

The prototype of the robot is showed in Fig. 1.

Fig. 1. The prototype of the reconfigurable robot

In general, this configurable robot has six legs with one arm, which are all driven by motors. The size of the robot is 1100 × 900 × 600 mm. The legs of the robot have the same mechanical structure. In order not to cause interference, the three legs on each side are not set on the line. So, when the robot is transformed into two wheels configuration or three wheels configuration, the wheels of the robot are not a complete circle. And in order to make the robot walk conveniently, the robot is powered by lithium batteries. Besides, TWINCAT, which is a kind of motor controlling software, is used to control the motors. When the configurable robot gets certain signals sent by sensors attached in the robot, TWINCAT will control the motors to execute it, then the robot can be transformed into different configurations.

Unknown environments may have complex surfaces. In order to enable robots to work effectively in it, the configurable robots designed in this paper have four different configurations that can be switched to suit multiple working environments. The configurations are six legs configuration, two wheels configuration (with one arm), three wheels configuration and star shape configuration. The applications of each configuration are shown below.

(1) The six legs configuration is suitable for bumpy ground walking;
(2) Two wheels configuration (with one arm) is suitable for performing tasks by controlling the arm;
(3) The three wheels configuration is suitable for fast walking on the ground;
(4) The star shape configuration is suitable for overturning large obstacles or deep gullies.

3 Kinematics Analysis

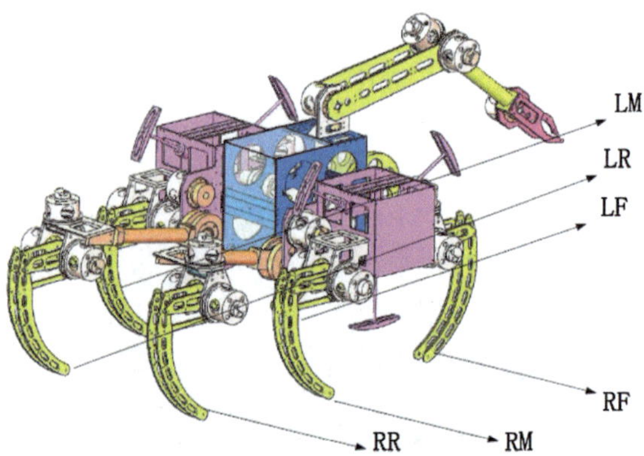

Fig. 2. The modeling of configurable robot

The model of the configurable robot is shown in Fig. 2. The legs are named as LF, RF, LM, RM, LR, RR respectively. The body's coordinate system is named *B*. Besides, the *i*-th leg's coordinate system is named L_i.

3.1 Forward Kinematics of One Leg

Taking LF leg as the research object, the schematic diagram of mechanism is showed in Fig. 3.

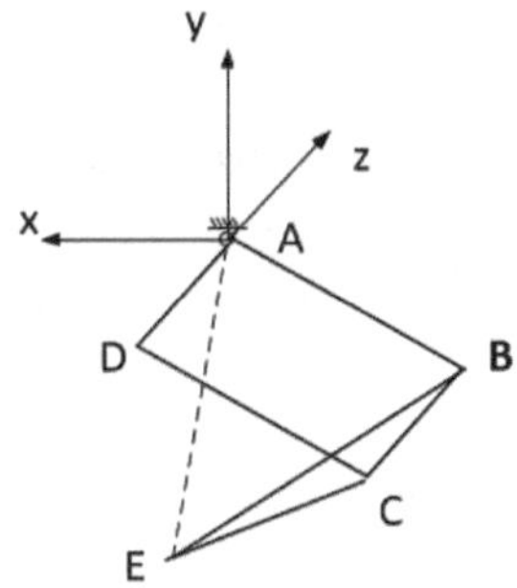

Fig. 3. The schematic diagram of mechanism of LF leg

Establish a three-dimensional coordinate system L_1 with the point A of the LF leg as the coordinate origin. The input angles are θ_{11}, θ_{12}, θ_{13}. Among them, θ_{11} is the angle between rod AB and the x-axis in the LF leg; θ_{12} is the angle between rod AD and the x-axis in the LF leg; θ_{13} is the angle between the line AE, which is projected in the plane x-A-z, and the x-axis. θ is the angle between rod BE and rod CB and is a fixed value. l_{AB} and l_{BE} are the length of rod AB and rod BE respectively. According to $\theta_{11}, \theta_{12}, \theta_{13}$, the equations of finding the coordinates of the point E in the L_1 coordinate system, $[L_{1x} \quad L_{1y} \quad L_{1z}]$, are shown below:

$$L_{1x} = (l_{AB}cos\theta_{11} + l_{BE}cos(\theta_{12} + \theta))cos\theta_{13} \tag{1}$$

$$L_{1y} = l_{AB}sin\theta_{11} + l_{BE}sin(\theta_{12} + \theta) \tag{2}$$

$$L_{1z} = (l_{AB}cos\theta_{11} + l_{BE}\cos(\theta_{12} + \theta))sin\theta_{13} \tag{3}$$

3.2 Inverse Kinematics of One Leg

In the LF leg, the coordinates of the point E in the coordinate system L_1, $[L_{1x} \quad L_{1y} \quad L_{1z}]$, are known and according to them, the angles, $[\theta_{11} \quad \theta_{12} \quad \theta_{13}]$, can be obtained.

$$\theta_{13} = \sin^{-1}\left(\frac{L_{1z}}{\sqrt{L_{1x}^2 + L_{1z}^2}}\right) + \pi \tag{7}$$

$$x_1 = \frac{L_{1x}}{\cos(\theta_{13} - \pi)} \tag{8}$$

$$l_{AE} = \sqrt{L_{1x}^2 + L_{1y}^2 + L_{1z}^2} \tag{9}$$

l_{AE} is the length of the rod AE.

If $atan2(L_{1y}, x_1) > 0$, then:

$$\theta_E = atan2(L_{1y}, x_1) \tag{10}$$

Else:

$$\theta_E = atan2(L_{1y}, x_1) + 2\pi \tag{11}$$

θ_E is the angle between plane x-A-z and rod AE.
According to Cosine theorem, in ΔBEA, we can get $\angle BEA$ as:

$$\angle BEA = \cos^{-1}\left(\frac{l_{BE}^2 + l_{AE}^2 - l_{AB}^2}{2l_{AE}l_{BE}}\right) \tag{12}$$

So:

$$\theta_{12} = \angle BEA + \theta_E - \theta \tag{13}$$

According to Cosine theorem, in ΔBEA, we can get $\angle EBA$ as:

$$\angle EBA = \cos^{-1}\left(\frac{l_{BE}^2 + l_{AB}^2 - l_{AE}^2}{2l_{BE}l_{AB}}\right) \tag{14}$$

So:

$$\theta_{11} = \theta_{12} - (\pi - \theta - \angle EBA) \tag{15}$$

In summary:

$$\begin{cases} \theta_{11} = \theta_{12} - (\pi - \theta - \angle EBA) \\ \theta_{12} = \angle BEA + \theta_E - \theta \\ \theta_{13} = \sin^{-1}\left(L_{1z}/\sqrt{L_{1x}^2 + L_{1z}^2}\right) + \pi \end{cases} \tag{16}$$

3.3 Forward Kinematics of Full Body

When the robot body coordinate system B rotates β angle, γ angle, α angle around z-axis, y-axis, x-axis respectively, the rotation matrices are shown below.

$$R_z = \begin{bmatrix} \cos(\beta) & -\sin(\beta) & 0 \\ \sin(\beta) & \cos(\beta) & 0 \\ 0 & 0 & 1 \end{bmatrix} \tag{17}$$

$$R_y = \begin{bmatrix} \cos(\gamma) & 0 & \sin(\gamma) \\ 0 & 1 & 0 \\ -\sin(\gamma) & 0 & \cos(\gamma) \end{bmatrix} \tag{18}$$

$$R_x = \begin{bmatrix} 1 & 0 & 0 \\ 0 & \cos(\alpha) & -\sin(\alpha) \\ 0 & \sin(\alpha) & \cos(\alpha) \end{bmatrix} \tag{19}$$

Then, the robot body coordinate system B is rotated by β angle around the z-axis, then rotated by γ angle around the y-axis, and finally rotated by α angle around the x-axis to obtain the B_1 coordinate system. The rotation matrix of B_1 relative to B is shown below.

$$\begin{aligned} {}^{B}_{B_1}R &= R_z R_y R_x \\ &= \begin{bmatrix} \cos(\beta) & -\sin(\beta) & 0 \\ \sin(\beta) & \cos(\beta) & 0 \\ 0 & 0 & 1 \end{bmatrix} \begin{bmatrix} \cos(\gamma) & 0 & \sin(\gamma) \\ 0 & 1 & 0 \\ -\sin(\gamma) & 0 & \cos(\gamma) \end{bmatrix} \begin{bmatrix} 1 & 0 & 0 \\ 0 & \cos(\alpha) & -\sin(\alpha) \\ 0 & \sin(\alpha) & \cos(\alpha) \end{bmatrix} \\ &= \begin{bmatrix} c\beta c\gamma & s\gamma s\alpha c\beta - s\beta c\alpha & s\beta s\alpha + c\beta c\alpha \\ s\beta c\gamma & c\beta c\alpha + s\alpha s\beta s\gamma & s\beta s\gamma c\alpha - s\alpha c\beta \\ -s\gamma & s\alpha c\gamma & c\alpha c\gamma \end{bmatrix} \end{aligned} \tag{20}$$

In Eq. (20), $s\alpha, c\alpha, s\beta, c\beta, s\gamma, c\gamma$ represent $\sin(\alpha), \cos(\alpha), \sin(\beta), \cos(\beta), \sin(\gamma)$, $\cos(\gamma)$ respectively.

Let L_i be the leg coordinate system of the i-th leg (i = 1, 2, 3, 4, 5, 6), and $P_{L_i} = [L_{i\_x}\, L_{i\_y}\, L_{i\_z}]^T$ represents the coordinates of the coordinate origin of the i-th leg in the coordinate system B. Then the attitude transformation matrix of the leg coordinate system L_i relative to the body coordinate system B is shown below.

$${}^{B}_{L_i}T = \begin{bmatrix} {}^{B}_{B_1}R & P_{L_i} \\ 0_{1\times 3} & 1_{1\times 1} \end{bmatrix} \tag{21}$$

From the forward kinematics analysis of the single leg, we can get that when the angles $[\theta_{i1}\, \theta_{i2}\, \theta_{i3}]$ (i = 1, 2, 3, 4, 5, 6) are known, according to (1), (2) and (3) equations, the coordinates of point E in the coordinate system L_i, $P_E^{L_i} = [L_{ix}\, L_{iy}\, L_{iz}]^T$, can be obtained. Then the coordinates P_E^B of the point E of each leg in the coordinate system B are shown below.

$$P_E^B = {}^{B}_{L_i}T P_E^{L_i} \tag{22}$$

3.4 Inverse Kinematics of Full Body

${}^{B}_{L_i}T$ is the attitude transformation matrix of the leg coordinate system L_i relative to the body coordinate system B; then the attitude transformation matrix of the body coordinate system B relative to the leg coordinate system L_i is shown below.

$$ {}^{L_i}_{B}T = {}^{B}_{L_i}T^{-1} \tag{23} $$

Knowing the coordinates P_E^B of each E point in the body coordinate system B, the coordinates $P_E^{L_i}$ of each point E in the respective leg coordinate system L_i are shown below.

$$ P_E^{L_i} = {}^{L_i}_{B}TP_E^B \tag{24} $$

Then, according to Eq. (16), $[\theta_{i1}\ \theta_{i2}\ \theta_{i3}]$ can be obtained.

4 Reconfiguration Planning: Simulations and Experiments

The robot has four configurations, which can be derived from the six legs configuration and transformed into two wheels configuration (with one arm), three wheels configuration and star shape configuration.

In this paper, the simulations are done by a software named SOLIDWORKS. The simulations show the whole process of transforming. In order to verify the feasibility of the reconfigurations scheme, experiments are also carried out. In the experiments, we chose TWINCAT to control all the motors.

TWINCAT is a controller developed by German company BECHHOFF. The biggest feature of the controller is that the software is separated from both the controller hardware and IO hardware, which is convenient for changing the IO modules. The whole process is shown by Fig. 4.

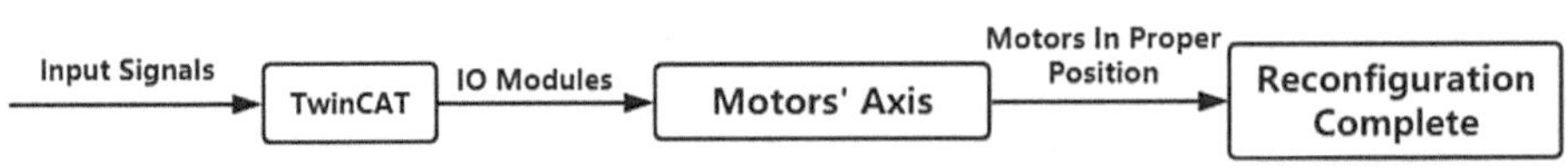

Fig. 4. The process of TWINCAT control motors

4.1 From Six Legs Configuration to Two Wheels Configuration

The robot in two wheels configuration can perform tasks. So, when the robot receive the signals that it need to execute certain tasks, it can transformed into two wheels configuration to do it. And the simulation is showed below.

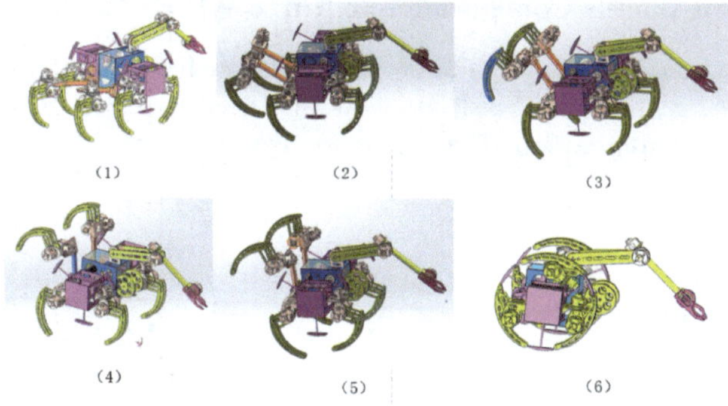

Fig. 5. From six legs configuration to two wheels configuration (simulation)

As shown in Fig. 5, (1) the original shape of the robot; (2) the LR and RR legs of the robot are turned up; (3) in order to keep balance, the LR and RR legs continue to rotate upward, and the center of gravity of the robot is offset to the middle of the body; (4) LF, RF, LM, RM legs are gathered directly below the final position; (5) LF, RF, LM, RM four legs do not move, the robot body moves down; (6) the robot completes the deformation.

In the same time, experiment is carried out to verify the feasibility. The experiment is shown in Fig. 6.

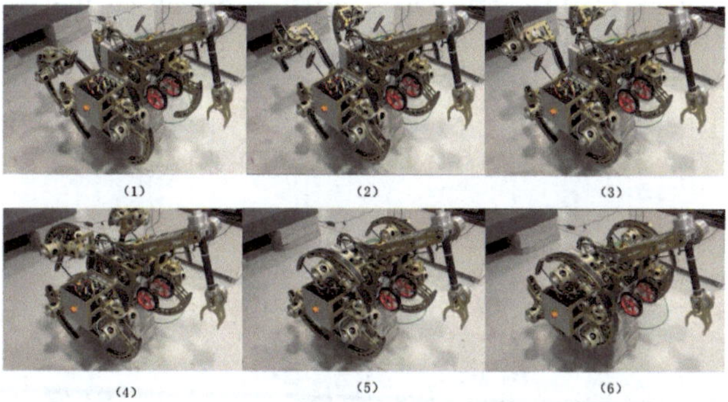

Fig. 6. From six legs configuration to two wheels configuration (experiment)

The experiment is basically consistent with the simulation expectations. This indicates that the scheme, which can transform the robot from six legs configuration to two wheels configuration, is feasible.

4.2 From Two Wheels Configuration to Three Wheels Configuration

The robot in three wheels configuration can walk much faster than it in two wheels configuration. After the robot finish the work and need to leave fast, it can transform into three wheels configuration. The simulation is shown below.

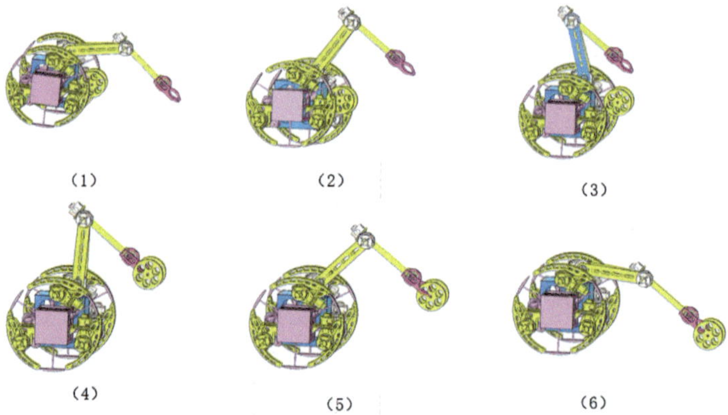

Fig. 7. From two wheels configuration to three wheels configuration (simulation)

As shown in Fig. 7, (1) the robot is in two wheels configuration; (2) the robot rotates the arm upward until it is in the proper position; (3) the arm reaches the position just above the wheel; (4) the arm grabs the wheel placed on the robot; (5) the arm moves forward with the captured wheel; (6) the arm places the captured wheel on the ground to complete the deformation.

The experiment of transforming from two wheels configuration into three wheels configuration had also been carried out. And the process is shown in Fig. 8.

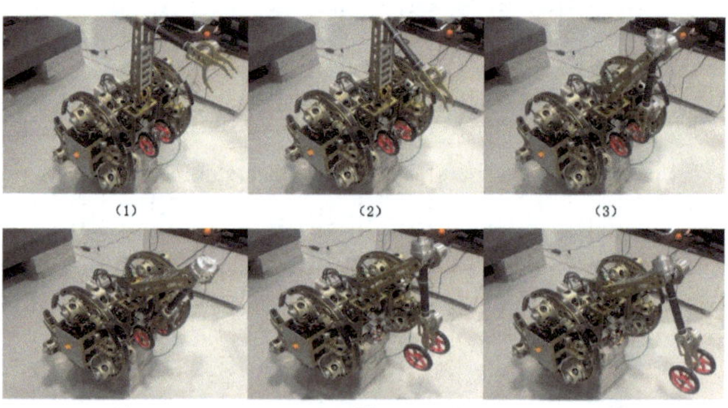

Fig. 8. From two wheels configuration to three wheels configuration (experiment)

Compared with Figs. 7 and 8, the experimental results are basically consistent with the simulation expectations. It indicates that the arm can reach the position where the small wheel is placed and the scheme is feasible.

4.3 From Six Legs Configuration to Star Shape Configuration

When the robot needs to pass large obstacles or deep gullies, which six legs configuration cannot complete to pass, the star shape configuration is needed. Star shape configuration has stretched structure, which make it easy to pass complex terrain. And the simulation is shown below (Fig. 9).

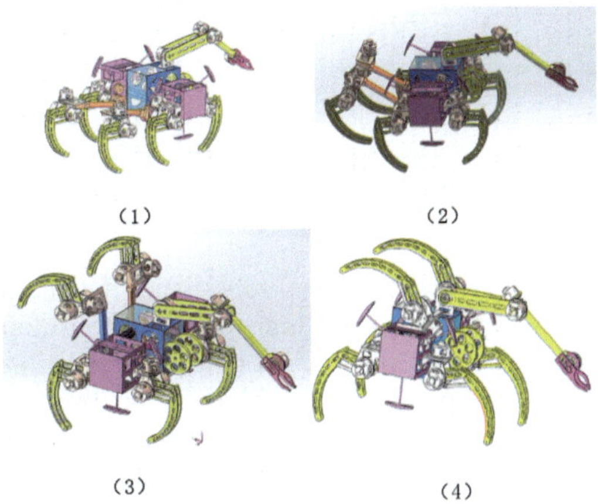

Fig. 9. From six legs configuration to star shape configuration (simulation)

As shown in Fig. 5, (1) the six legs configuration of the robot; (2) the LR and RR legs are rotated up; (3) when the LR and RR legs reach a certain position, the LF, RF, LM and RM legs are respectively extend outward; (4) when six legs are stretched to a certain position, the reconfiguration is completed.

In order to verify the feasibility of the scheme, the experiment of transforming from six legs configuration into star shape configuration was done. And the process is shown in Fig. 10.

Compared with the results of the simulation with the results of the experiment, they are basically consistent with each other. It indicates that the reconfigurable robot can stretch to the star shape configuration and the scheme is feasible.

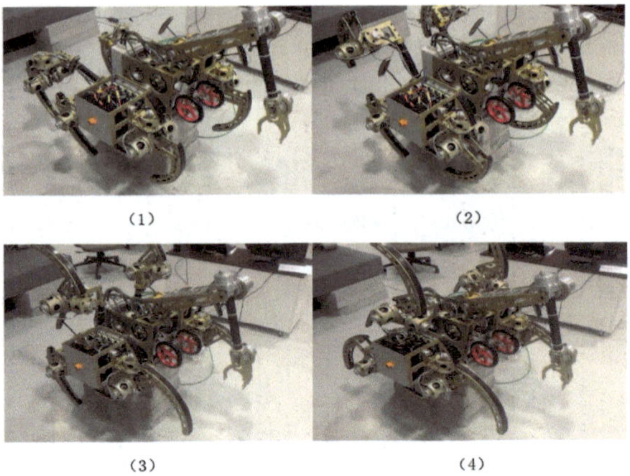

Fig. 10. From six legs configuration to star shape configuration (experiment)

5 Conclusions

Based on the kinematics analysis of the robot and the usage of the controller, TWINCAT, the reconfigurable robot can transform into four different configurations, and the feasibility of the schemes are verified both by simulations and experiments.

In the follow-up work, people can place vision sensors in the robot. With the vision sensor, the robot can explore the environment by itself and send signals to the controller, TWINCAT, automatically. So, the robot can adapt to complex environments.

This reconfigurable robot also has the disadvantage of high weight and this will cause high power consumption. To solve the problems, materials with lower density will be used in the robot.

References

1. Kassaeiyan, P., Tarvirdizadeh, B., Alipour, K.: Control of tractor-trailer wheeled robots considering self-collision effect and actuator saturation limitations. Mech. Syst. Signal Process. **127**, 388–411 (2019)
2. SunSpiral, V., Wheeler, D.W., Chavez-Clemente, D., Mittman, D.: Development and field testing of the FootFall planning system for the ATHLETE robots. J. Field Robot. **29**(3), 483–505 (2012)
3. Liu, J., Zhao, X., Tan, M.: A review of the research on legged robots. Robot, **28**(1), 81–88
4. Gultekin, G.K., Saranli, A.: Feature detection performance based benchmarking of motion deblurring methods: applications to vision for legged robots. Image Vis. Comput. **82**, 26–38 (2019)
5. Saab, W., Racioppo, P., Ben-Tzvi, P.: A review of coupling mechanism designs for modular reconfigurable robots. Robotica **37**(2), 378–403 (2018)
6. Luo, Z., Shang, J., Wei, G., Ren, L.: A reconfigurable hybrid wheel-track mobile robot based on Watt II six-bar linkage. Mech. Mach. Theory **128**, 16–32 (2018)

7. Guan, G.: Research on Small Variable-Form Wheel-Legged Robot System. Harbin Institute of Technology (2017)
8. Li, C., Han, L., Yuan, S., Zhang, Y.: Optimization design of six-chain wheel-legged lunar robot. Manned Space Flight **25**(01), 37–41 (2019)
9. Chi, Z., Song, X., Yang, L.: Design of a wheel-leg composite bionic robot. J. Tianjin Agric. Coll. **25**(03), 87–91+104 (2018)
10. Wang, J., Yao, Y., Kong, X.: A reconfigurable tri-prism mobile robot with eight modes. Robotica **36**(10), 1454–1476 (2018)

Development of Robotic Wiping System for Piezo Components

Lulu Chu[1], Yi Luo[1,2(✉)], and Xiaodong Wang[1,2]

[1] Key Laboratory for Micro/Nano Technology and System of Liaoning Province, Dalian University of Technology, Dalian 116024, China
luoy@dlut.edu.cn

[2] Key Laboratory for Precision and Non-traditional Machining of Ministry of Education, Dalian University of Technology, Dalian 116024, China

Abstract. In order to clean the oxide layer and contaminations on the surface of piezoelectric ceramic sheets and solder rings, an automatic wiping system based on the cooperative robot UR3 is designed. The wiping system takes UR3 robot as the moving unit, and employs the micro force sensor to detect the wiping force. The end effector of the robot is designed according to the size and shape of the parts. Vacuum adsorption stages for fixed piezoelectric components were designed, in which silicon wafer etched by MEMS process was served as the adsorption head for fixing solder rings. The vacuum generator is selected according to the stress of piezo components. Finally, the feasibility of the adsorption device was proved by experiments, and the actual vacuum adsorption force under different exhaust pressures was measured. The device completes the surface cleaning of the piezoelectric ceramic sheets and the solder rings, thereby improving the reliability of subsequent soldering.

Keywords: Automatic cleaning equipment · Piezoelectric ceramic sheets · Solder rings

1 Introduction

Piezo components can be served as precision actuators and is widely used in precision systems [1]. The piezo components are composed of piezoelectric ceramic sheets and solder rings. Piezoelectric ceramic sheets have inverse piezoelectric effect. When the voltage is applied at both ends, piezoelectric ceramics will deform slightly along the polarization direction [2]. Piezoelectric ceramic micro-displacement actuators can achieve micrometer-level displacement compensation to adjust the expansion or contraction of the laser cavity length to achieve laser stabilization [3]. The composition of the solder ring is tin and lead. The tin-lead solder ring is widely used in microelectronics due to its good wettability, low cost, mature process, low reflow temperature, etc. [4]. The surface of the piezoelectric ceramic sheet is plated with an electrode layer, and the oxidized slag generated by the air oxidation of the solder ring may be brought into the welding point or the welding surface, resulting in false welding, affecting the thermal, electrical and mechanical properties, and the product quality is not guaranteed [5]. Therefore, the removal of oxide layer and contaminations on the surface of piezo

D. Wang et al. (Eds.): MeTrApp 2019, MMS 79, pp. 524–532, 2020.
https://doi.org/10.1007/978-981-15-0142-5_51

components is very important to ensure the reliability of welding. Many enterprises and universities at home and abroad have begun the research and development of the corresponding technology, and have achieved some results. Alshaer [6] used 100 ns of Nd: YAG laser to clean and weld the oxide layer on the surface of 4043 and AC-170PX aluminum alloy sheets. It was found that laser cleaning can effectively reduce the porosity in welded joints. Messaoudi [7] used picosecond laser technology to study the removal of Al and Si coating on the surface of hot rolled 22MnB5 steel, and obtained the influence rule of laser wavelength, pulse energy and scanning parameters on the removal depth and removal rate. Yan [8] used single-side rough grinding and lapping processing of fused silica glass with free abrasive. The experiments show that the material removal rate is the highest when the mass fraction of abrasive, the rotational speed of grinding disc and the grinding flow rate are at a fixed value. In 2019, Xie [9] discussed the influence of pickling solution types and pickling conditions on the pickling effect of oxidation layer of steel skeleton and the bonding performance between steel and NR. The experiment showed that the pickling effect of cx-206 was the best, and the bonding performance of the product after pickling was improved.

At present, workers wipe piezoelectric ceramic sheets with hand-held abrasive rubber and scrape solder rings with a knife. The fixing of the piezoelectric ceramic sheets and the solder rings becomes a problem, and the worker cannot accurately control the wiping force during the wiping process, which is liable to cause damage of the parts and high rejection rate. This paper develops an automatic device for removing the oxide layer or electrode layer on the surface of piezo components, which can solve the problems of low efficiency and easy to damage parts by manual wiping.

2 Design of Robot Wiping System

Robot wiping system consists of UR3 robot, abrasive rubber head, fixing stage for piezoelectric ceramic sheet, solder ring adsorption stage and vacuum adsorption gas path (see Fig. 1). Piezoelectric ceramic sheet is circular and has a thickness of 0.4 mm. Solder ring is circular and has a thickness of 0.03 mm. The specific process of the equipment is that the robot wipes the piezoelectric ceramic sheet with the abrasive rubber, while for the piezoelectric ceramic sheet, it is placed on the fixing table, and the solder ring is fixed by the vacuum adsorption stage and manually cleaned.

2.1 UR3 Robot

The UR 3 robot is a six-degree-of-freedom articulated robot whose main features are light weight and safety. Its workload can reach 3 kg, and the repetitive positioning accuracy is 0.05 mm. Therefore, it can meet the requirements of work. Comparison to the Cartesian coordinates robot, the UR3, which is a serial robot, has larger working space. Since the UR3 robot has light weight and low inertia. Meanwhile, it has a current sensor in the joint which will immediately stop the robot when it hits the object, so it can be work with people friendly [10]. Since this kind of cobot will be a trend in the future robotic application, the UR3 robot is selected in this design as the motion unit.

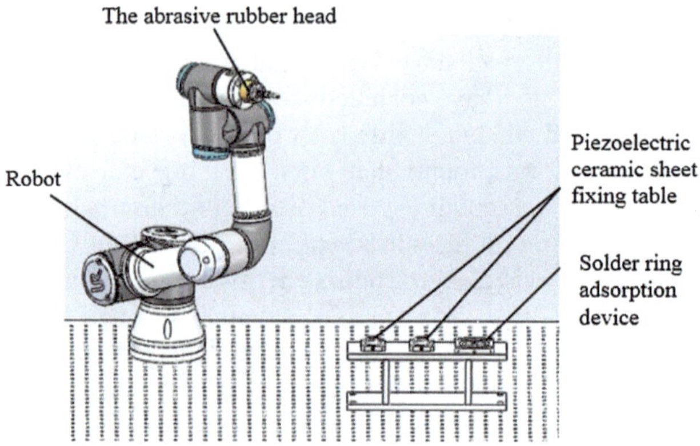

Fig. 1. The layout of the robot wiping system

2.2 Structural Design of the Abrasive Rubber Head

The abrasive rubber assembly consists of a sensor, two adapter plates, an aluminum shaft, and an abrasive rubber (see Fig. 2). The abrasive rubber is installed in the stepped shaft of the aluminum shaft by interference fit and is connected to the adapter plate through the chuck. Alumina abrasive is doped in the abrasive rubber as the abradant. The abrasive force of the rubber to polishing piezoelectric ceramic sheet is selected as 1 N based on the previous experiments. The force sensor which can withstand tension and pressure is selected to detect the contact force in real time, with a range of 0–30 N.

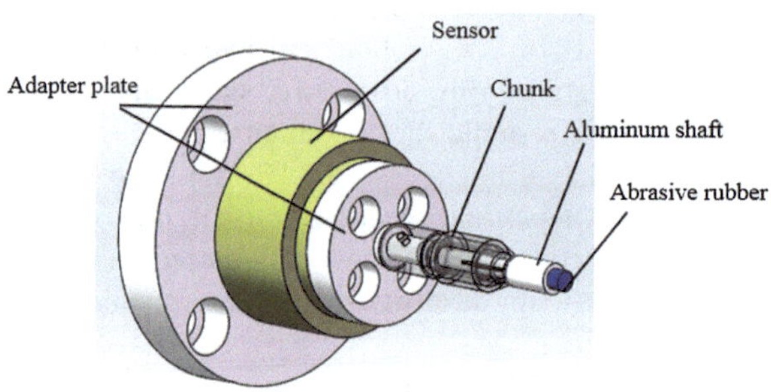

Fig. 2. The diagram of the abrasive rubber head

2.3 Design of Piezoelectric Ceramic Placing Table

The piezoelectric ceramic sheets are brittle so in the abrasive process, adsorption fixing is better than clamping. In order to avoid the ceramic sheet slipping from the fixing table during the abrasion, the edge of the placing table has a protruding shape, and the openings on both sides are for manual placement and picking up. The ceramic sheets will rotate during the scrubbing eraser process. To prevent this from happening, the holes is opened in the placing table and connected to the vacuum suction device through the rotary joint, and the piezoelectric ceramic sheet is fixed on the fixing stage by vacuum suction (see Fig. 3).

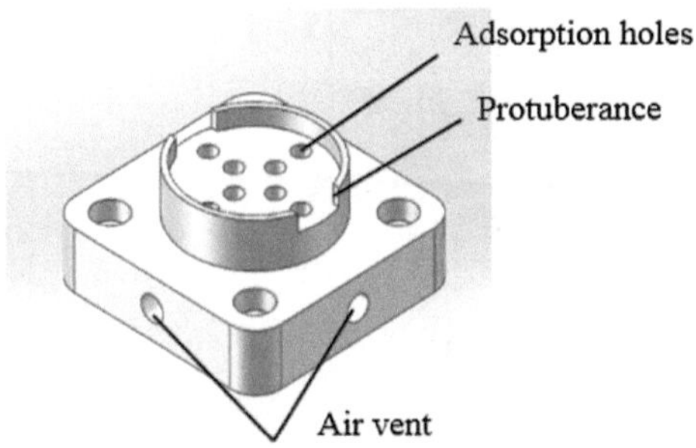

Fig. 3. Fixing stage for piezoelectric ceramic sheet

2.4 Design of Solder Ring Adsorption Device

The outer diameter of solder ring is 20 mm, the inner diameter is 6 mm, and the thickness is 0.03 mm. It is adsorption to the fixing stage while abrasion. However, if it contacted the fixing stage directly, solder ring is easy to wrinkle and damage. Therefore, micro scale adsorption holes are formed on the silicon wafer by the MEMS process for adsorption. The mass of the solder ring is 0.078 g, thus the gravity is negligible. Because the solder rings are easily deformed and damaged, the force used to manually scrape the solder rings is small. Therefore, it is only necessary to firmly fix the solder rings on the silicon wafer adsorption head.

The material used to fabricate the auxiliary part was a 2-inch silicon wafer (N type 100 orientation) with a thickness of 250 μm. Double-sided wet etching method were selected to fabricate the bars for adsorption. The top width of the adsorption bars was 250 μm, and according to the orientation of silicon wafer, the narrowest width of the adsorption bars was 73 μm. In order to avoid air leakage during the adsorption process, all the long holes are between 10 mm and 18 mm in diameter. The wafer is cleaned, oxidized, lithography, wet etched, etc. [11], and the etched silicon wafer is shown (see Fig. 4).

The solder ring adsorption stage is shown (see Fig. 4). Stepped surface is used to place auxiliary wafers. At the center of the fixing stage, an extruding platform with the diameter of 10 mm was design to support the auxiliary wafer to avoid crushing of silicon chips in the process of scratching of the solder ring.

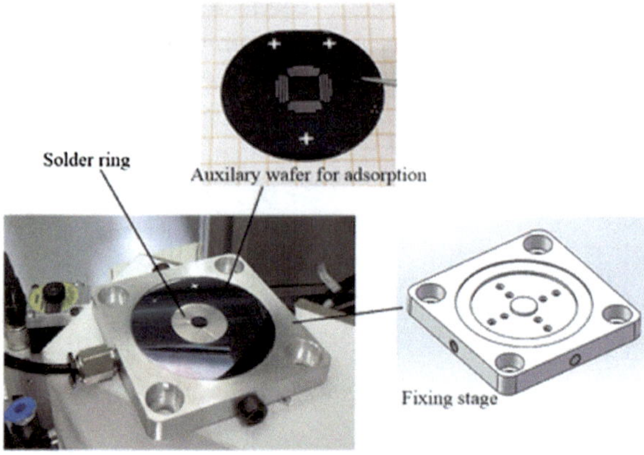

Fig. 4. The adsorption stage for solder ring

2.5 Gas Path Design

The fixation of piezoelectric ceramic sheets and solder rings are completed by vacuum adsorption. Vacuum adsorption needs negative pressure as power source, and negative pressure can be generated by vacuum generator. The two gas paths are connected in the same way. When the solenoid valve is connected, the gas path is connected, and the vacuum generator works. When the solenoid valve is not connected, the vacuum generator stops working (see Fig. 5).

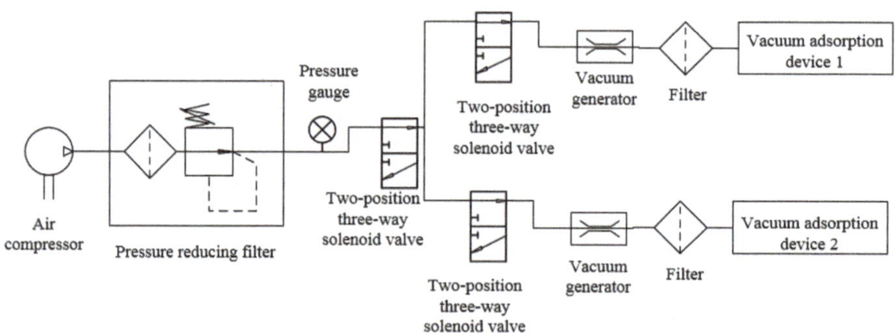

Fig. 5. Vacuum adsorption gas path

3 Experiments

3.1 The Friction Coefficient Measurements

The static friction force generated by the suction and pressure on the piezoelectric ceramic sheet is greater than the sliding friction force generated by the sanding eraser during the polishing process. The force required to move the piezoelectric ceramic sheet on the aluminum block or the sanding rubber should be measured by experiment, thereby calculating the friction coefficient between the contact faces. The piezoelectric ceramic sheet is placed on the aluminum block, and the micro force sensor is installed on the linear motor. The clamp is installed in front of the micro force sensor, and one end of the hairspring is connected by the clamp, and the other end of the hairspring is connected by the piezoelectric ceramic sheet. The motion of linear motor drives the motion of piezoelectric ceramic sheet through the hairspring, so that the force needed to pull the piezoelectric ceramic sheet can be measured (see Fig. 6). To calculate the friction coefficient between rubber and the ceramic sheet, using the same testing unit, but change the aluminum block to rubber.

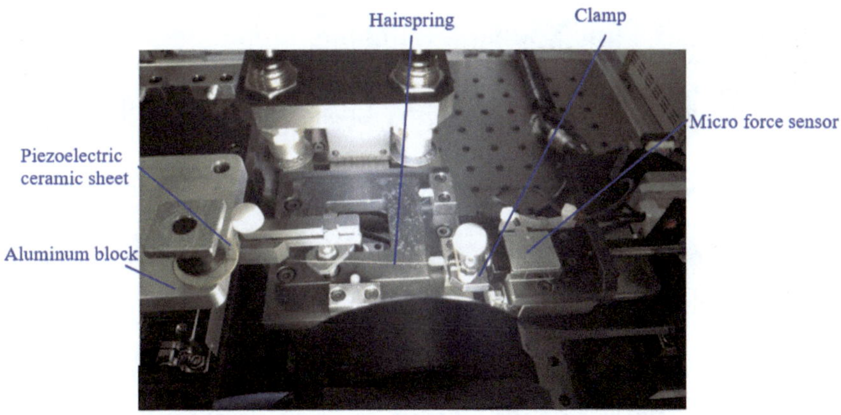

Fig. 6. Measuring the force of piezoelectric ceramic sheet moving in aluminum block

The force to pull the ceramic sheet justly is 0.023 N and the load of piezoelectric ceramic sheet is 0.084 N. The tension of the piezoelectric ceramic sheet in the process of motion is 0.05 N, and the load of piezoelectric ceramic sheet is 0.084 N. The coefficient of friction μ should meet the requirement of Eq. (1):

$$F = \mu \times N \tag{1}$$

μ - the coefficient of friction;
N - the load of piezoelectric ceramic sheet, N;

It is found that the maximum static friction coefficient between the aluminum block and the piezoelectric ceramic sheet is 0.28, and the sliding friction coefficient between the surface of the frosted rubber and the piezoelectric ceramic sheet is 0.6. If the piezoelectric ceramic sheet does not produce displacement or rotation in the abrasion, the adsorption force F_s should meet the requirement of Eq. (2), and get the minimum F_s is 1.14 N.

$$0.6 \times F_e \leq 0.28 \times (F_s + F_e) \tag{2}$$

F_e - the pressure of the rubber, N;
F_s - adsorption force, N;

There are eight adsorption holes of the piezoelectric ceramic sheet placement table have a diameter of 3 mm (see Fig. 3). The degree of vacuum p should meet the requirement of Eq. (3):

$$p = Gt/(Sn) \tag{3}$$

p - the required vacuum, MPa;
G - theoretical adsorption, N;
t - safety factor, horizontal hanging $t \geq 4$, vertical hanging $t \geq 8$;
S - adsorption head bottom area, mm^2;
n - the number of adsorption heads;

Where $G = 1.14$ N, $t = 4$ [12], $S = 7.1$ mm^2, $n = 8$, brought into Eq. (3) can be calculated $p = 0.081$ MPa. The vacuum generator is VML-05-04. The main specifications are shown in Table 1:

Table 1. Main parameters for vacuum generator

Ultimate vacuum degree (kPa)	Air consumption flow (l/min)	Working pressure range (MPa)	Inhalation flow (l/min)	Nozzle diameter (mm)
−91.8	11.5	0–0.9	7	0.5

3.2 Solder Ring Adsorption Design

The solder ring adsorption experiment was carried out according to the built vacuum gas path. The silicon wafer was placed on the solder ring adsorption stage, and the solder ring was placed on the adsorption channel of the silicon wafer. One of the gas holes is connected with the air passage through the adapter, and the other three are sealed with screw to prevent air leakage. When the vacuum generator works, the solder ring is tightly adsorbed on the silicon wafer (see Fig. 7).

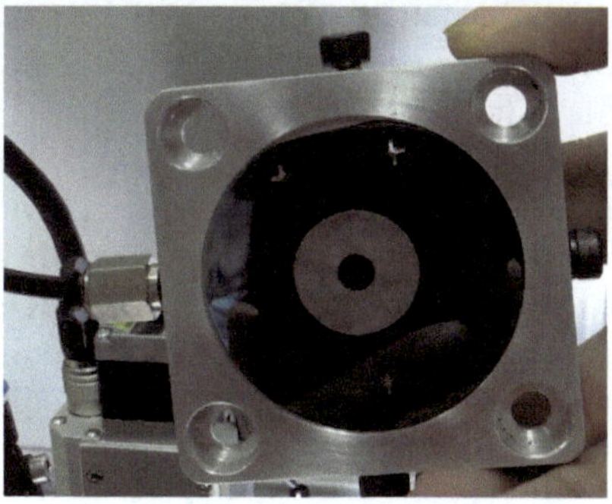

Fig. 7. Solder ring adsorption test

The piezoelectric ceramic sheet was used instead of the solder ring to investigate the adsorption force generated by the solder ring adsorption stage under different exhaust pressures. A spring dynamometer with a range of 250 N and an accuracy of 0.05 N is used to measure the adsorption force. The rated exhaust pressure of the air compressor used in the work is 0.6 MPa. The pressure of the filter pressure reducing valve is set to different values, as shown in Table 2:

Table 2. Adsorption force under different exhaust pressures

Pressures (MPa)	0.55	0.5	0.45	0.4	0.35	0.3	0.25	0.2
Adsorption force (N)	1.90	1.60	1.10	0.80	0.60	0.50	0.45	0.25

4 Results and Discussion

The adsorption channel on the silicon wafer is 8.825 mm^2, and when the pressure of the pressure reducing valve is 0.55 MPa, the theoretical adsorption force is 1.2 N according to the Eq. (3). According to Table 2, the actual adsorption force is 1.9 N. The reason maybe the actual adsorption force is affected by multiple factors, the uniformity of the vent hole, the pressure loss in change the direction, and the stability of the compressed air flow. The thickness of solder ring is very thin, and the adsorption force should be optimized. If the adsorptive force is greater, the solder ring will be damaged, and if the adsorptive force is smaller, the solder ring will move during wiping. In the solder ring adsorption experiment, the optimum pressure is 0.4 MPa. The solder ring can tightly adsorbed on the silicon wafer without wrinkled, and can sustain the scratching force.

5 Conclusion

In this paper, a robotic wiping system for cleaning the piezo components is designed and fabricated. Using UR3 robot as the moving unit, the piezoelectric ceramic sheets and the solder rings are fixed by vacuum adsorption device. The frosted rubber assembly is mounted on the robot, and the removal of the surface electrode layer of the piezoelectric ceramic sheet is completed by the circular motion of the robot combined with the micro force sensor. The etched silicon wafer is used to adsorb the solder rings to avoid wrinkling and damaging of the solder rings during manual scratching. Experiments have proved the feasibility of the system. The piezoelectric ceramic sheets and the solder rings can be fixed on the vacuum adsorption tables reliably. The robot can carry out the wiping movement of the wiping. The cycle time is 35 s.

Future work will be focus on optimized the movement of the end effector. The contact pressure is set in the case where the surface removal rate is the largest, thereby shortening the wiping time.

Acknowledgment. This research work was supported by Major Project of Basic Scientific Research of Chinese Ministry (Grant No. JCYK 2016 205 A003).

References

1. Liu, G., Li, L.: Study of key technique on ultraprecision feeding mechanism with piezoelectric actuator. Piezoelectric and Acoustooptic **04**, 411–413 (2006)
2. Qian, W., Zeng, Z., Yang, X.: Tracking of frequency of ultrasonic motor drive. J. Tianjin Univ. **03**, 375–377 (2000)
3. Zhao, L., Zhuang, L., Zhang, C., et al.: The piezoelectric ceramic controlling system and it's application to laser. Piezoelectric and Acoustooptic **05**, 550–552 (2007)
4. Tian, F., Tian, W., Zhou, M.: Reliability study of eutectic tin-lead solder and thin gold solder joints. Solid State Electron. Res. Prog. **39**(01), 72–76 (2019)
5. Huang, Z., Zhang, Z., Zhang, Y., et al.: Tin-lead solder oxidation resistance method and K-O type solder. Sichuan Metall. **02**, 81–85 (1984)
6. Alshaer, A.W., Li, L., Mistry, A.: The effects of short pulse laser surface cleaning on porosity formation and reduction in laser welding of aluminium alloy for automotive component manufacture. Opt. Laser Technol. **64**, 162–171 (2014)
7. Messaoudi, H., Mehrafsun, S., Tromenschlager, W.: Picosecond laser cleaning of hot stamped 22 MnB5 steel. J. Mater. Sci. Surf. Eng. **5**(1), 504–508 (2017)
8. Yan, Q., Li, J., Pan, J.: Experimental research on plane lapping of fused silica glass substrate. Diam. Abras. Eng. **39**(01), 60–66 (2019)
9. Xie, Y., Wang, L., Li, B., et al.: Research on technology of acid cleaning of steel skeleton oxide layer. Spec. Rubber Prod. **40**(02), 53–56 (2019)
10. Xu, X.: Research on Welding Trajectory Planning and Control Based on UR3 Robot. Wuhan University of Science and Technology (2018)
11. Tang, B.: Research on Deep Silicon Etching Process for MEMS Packaging. Tianjin University (2014)
12. Yang, J.: Development of Micro-Displacement Automatic Measuring Equipment for Piezoelectric Ceramic Sheets. Dalian University of Technology (2016)

Position and Pose Optimization of Work Piece Based on Image Processing Data

Feng Qian, Sichen Cai, and Huimin Dong(✉)

School of Mechanical Engineering, Dalian University of Technology, Dalian 116024, China
donghm@dlut.edu.cn

Abstract. This paper presents a position and pose optimization method of work piece based on image processing data, in which the minimum distribution error of the processing allowance is as the optimization objective, the position and pose of the work piece relative to the center hole as optimal variables. According to the work piece data obtained by image processing of the rough casting, the center position of circular is firstly determined from coordinate transformation. An optimization model of the work piece position and pose is then established based on the requirement of the processing process and the work piece margin data. An example is conducted by the method presented in this paper. The result shows that the method can deal very well with casting defects existing on the blank work piece contour, and replace the traditional process effectively, which will promote the development of processing automation.

Keywords: Position and pose · Optimization · Image processing data · Work piece contour

1 Introduction

With the development of the intelligent manufacturing technology the automatic level in the manufacturing industry, companies has increased gradually replacing manual operations. It not only reduces the production cost of the enterprise, but also improves the production efficiency [1, 2]. Therefore, how to automate the manufacturing process is a key issue to improve production value.

During the production of the part, the blank will retain some margin. It will result in the contact surface not being the design benchmark when the blank is placed on the scribing plate. And the casting defects of the blank will make the surface uneven, which will result in the contact surfaces not being able to fit the scribing plate. The blank will be placed at an angle, and this condition causes an angle between the blank and the horizontal plane. Therefore, the blank cannot use the actual contact surface as the original reference. Otherwise, the processing results will not conform to the design requirements, and even the workpiece will probably be scrapped. The traditional way to solve this problem is artificial scribing before processing. This work requires the fitter to measure the dimensions of each machined surface, and then allocates margins to each surface based on experience and requirements. Finally, the appropriate processing reference for subsequent processing is marked. This process is time-consuming and

D. Wang et al. (Eds.): MeTrApp 2019, MMS 79, pp. 533–541, 2020.
https://doi.org/10.1007/978-981-15-0142-5_52

laborious [3]. And according to the actual situation of the blank, the time spent on the work is also different, resulting in unstable production beat.

With the rapid development of technologies such as computer technology and image processing, machine vision technology has become more and more widely used in the detection and measurement of industrial products. Machine vision detection has the advantages of high precision, low cost and good flexibility. In the process of production, the use of machine vision can greatly improve the degree of automation, and increase production efficiency [4–6].

In order to get over the shortcomings of artificial scribing, in this paper, we propose a method that can replace the traditional process. In this way, this process can be integrated into an intelligent manufacturing processing system. First, we establish a theoretical model of the target blank, i.e. write the optimization equation according to the design requirements. Then, a two-dimensional image with image processing data is analyzed, and the edge single pixel point set of the key surface is obtained. Using the point set data and the optimization equation, the processing position and pose of blank are determined. In this way, the tilt of the workpiece can be corrected, the symmetry can be ensured, and the margin of each surface is as uniform as possible.

2 Workpiece Characteristics

Figure 1 shows a face of a typical part with protruding fixations on both sides, and with a circular hole in the center. The projected contour of the part on this surface is constituted of basic primitives, such as line and circles. By milling the plane where the contour is located, we can process finished parts.

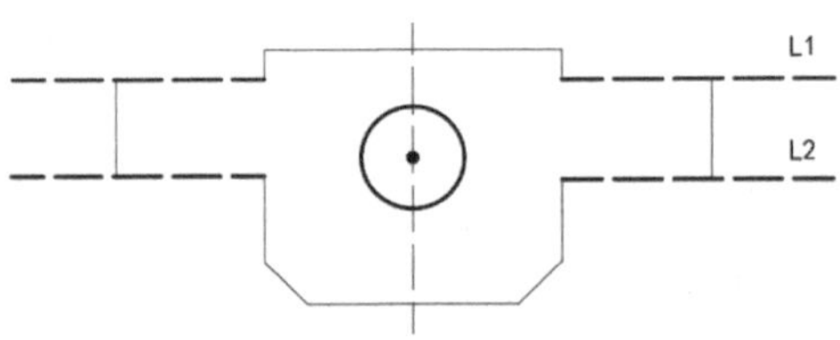

Fig. 1. A view of the workpiece

According to the design basis of the workpiece, the lower surface of the fixed table on both sides is completely contacted with the positioning block, and the contact surface is used as a positioning benchmark. The projection of the contact surface shown in Fig. 1 belongs to a horizontal line, which is called L_1. At this time, the line at the upper plane of the workpiece fixing table is called L_2.

Since the blank of the workpiece is a cast blank, each surface has a certain margin. Therefore, the part cannot be directly processed according to the design benchmark. And the uneven surface of the blank will affect the benchmark positioning of the workpiece. If the blank surface margins and the bulged portion are not taken into account, placing the workpiece directly on the fixture for positioning will result in a processing error. The processing benchmark will be displaced and tilted from the design basis. Therefore, the benchmark used in the processing process of the workpiece will occur errors, and the symmetry of the workpiece is destroyed, the processing margin is uneven.

In the processing process designed before, L_2 is first placing on the positioning block. Based on this benchmark, L_1 is viewed as a rough benchmark, and this rough benchmark is used for the next processing. we also need to consider the location and radius of the circular hole. Therefore, for this workpiece, the upper plane of the fixed table, the lower plane of the fixed table, and the circular hole are the most important elements that need serious consideration. Since the size and position characteristics of these three parts need to be reflected in the two-dimensional image, the most convenient view is selected for analyzing this problem.

Therefore, in the traditional process, if fitter does not carry out artificial scribing, we cannot carry on the next processing. we desire to design a new method for pre-treating the workpiece blank before processing. This method can directly determine the benchmark without artificial scribing. Finally, we correct the workpiece processing posture by calculation result, making the processing result can fulfill the precision requirements.

3 Optimization Model

In order to be able to calculate the processing benchmarks that fulfill the design requirements, we establish an optimization model of the workpiece based on the actual processing conditions.

3.1 Margin Calculation

As shown in Fig. 2, there are margins on the workpiece blank, and these margins are not uniform. The left side of the blank has more margins, causing the workpiece tilts clockwise. In this circumstance, the center axis of the workpiece deviates from the vertical direction at an angle. Therefore, the processing benchmark is not easy to determine, resulting in the upper plane cannot be directly processed.

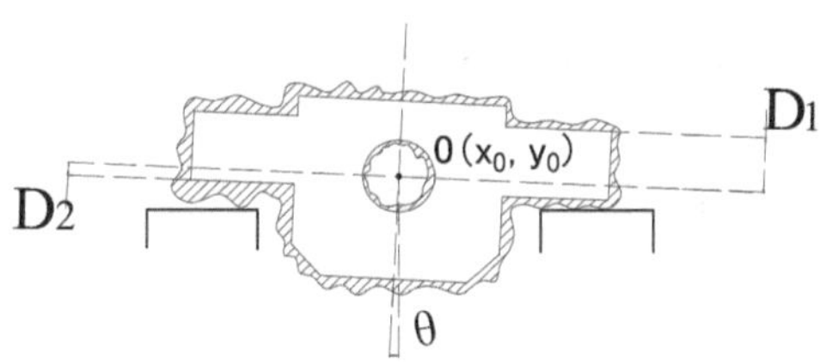

Fig. 2. Uneven margins lead to tilt

At this time, the actual processing plane of the workpiece does not parallel to the horizontal plane, we can find there is a certain angle assumed to be θ. In this circumstance, the center position of the hole on the workpiece also be changed, and we assume that the center of the hole in the actual processing is $O(x_0, y_0)$.

Thus, in the actual profile, the upper plane of the fixed table can be located, and the equation of the projection line L_1 in the Fig. 2 is:

$$\tan\theta x + y - [\mathrm{y}_0 + \tan\theta\, \mathrm{x}_0 + D_1(\tan\theta\ \sin\ \theta + \ \cos\theta)] = 0 \tag{1}$$

In Eq. (1), D_1 is the distance from the design center of the hole to the line L_1.

Similarly, the lower plane of the fixed table in the Fig. 2 can also be located, the projection line L_2 is:

$$\tan\theta x + y - [\mathrm{y}_0 + \tan\theta\, \mathrm{x}_0 - D_2(\tan\theta\ \sin\ \theta + \ \cos\theta)] = 0 \tag{2}$$

In Eq. (2), D_2 is the distance from the design center of the hole to the line L_2.

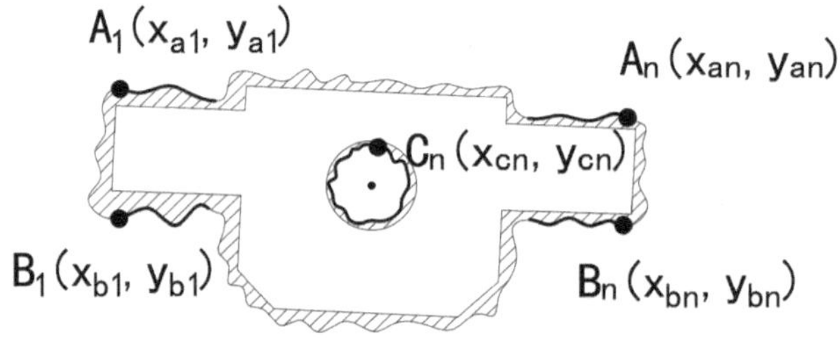

Fig. 3. Contour point in different processing surfaces

For the actual blank, as shown in Fig. 3, according to reasonable measurement means, we obtain the position of the plane contour points, which located on the left and right fixed table of the blank. These contour points of upper planes are recorded as point set A, including points $A_1(x_{a1}, y_{a1})$ to $A_n(x_{an}, y_{an})$. Similarly, contour points of the lower plane are recorded as point set B, including $B_1(x_{b1}, y_{b1})$ to $B_n(x_{bn}, y_{bn})$. The center hole contour points are recorded as point set C, including $C_1(x_{c1}, y_{c1})$ to $C_n(x_{cn}, y_{cn})$.

Therefore, the size of the plane processing margins, which can be evaluated by the accumulated value of the distance from each point in the point set A to the line L_1 [7]. For a single point, the distance formula from the nth point in point set A to the line L_1 is:

$$d_{An} = \frac{\mathrm{x}_{an}\tan\theta + y_{an} - [\mathrm{y}_0 + \tan\theta\, \mathrm{x}_0\, D_1(\tan\theta\ \sin\theta + \ \cos\theta)]}{\sqrt{\tan\theta^2 + 1}} \tag{3}$$

Similarly, the distance from the nth point in point set B to the line L_2 is:

$$d_{Bn} = \frac{\mathrm{x}_{bn}\tan\theta + y_{bn} - [\mathrm{y}_0 + \tan\theta\, \mathrm{x}_0 - D_2(\tan\theta\ \sin\theta + \ \cos\theta)]}{\sqrt{\tan\theta^2 + 1}} \tag{4}$$

For the circular hole, the distance from the nth point in the point set C to the actually processed hole contour is:

$$d_{Cn} = D_3 - D_{cn} \tag{5}$$

In Eq. (5), D_3 is the design radius of the center hole. D_c is the distance from the nth point in point set C to the center position $O(x_0, y_0)$. The calculation formula is as Eq. (6):

$$D_{cn} = \sqrt{(x_0 - x_{cn})^2 + (y_0 - y_{cn})^2} \tag{6}$$

Adding all the values of d_A, d_B and d_C, and we can get the total margin value.

3.2 Coordinate Transformation

In actual processing, firstly, we need to obtain a two-dimensional image of the blank by means of machine vision technology. Then, we use the image processing technology to acquire single pixels which distributed in the outline of the processing surface. Finally, we substitute single pixel contour data into the theoretical model and the optimization equation for solving.

Since this paper only discusses pose optimization based on the blank contour image, we won't illustrate how to acquire contour single pixels by machine vision. Therefore, we will use a hypothetical workpiece blank binary map for analyzing an example.

When we construct a theoretical model, we select the center of the hole as the origin is appropriate. The relative position of the hole center and other surfaces is always constant. Therefore, when we establish the theoretical model coordinate system, as shown in Fig. 4, the hole center is used as the origin of the coordinate system. Further, the angle θ with the horizontal direction is the X-axis, and the angle θ with the vertical direction is the Y-axis. However, when we use Matlab for image processing, the default coordinate system is different. As shown in Fig. 4, the origin O' is the up-left corner of the image, the vertical direction is the X'-axis, and the horizontal direction is the Y'-axis. Simultaneously, the measurement units of the two coordinate systems are also different. The unit of measure in the theoretical model coordinate system is: mm, while the unit of measure in the Matlab coordinate system is: pixels. Therefore, we also require to transform size.

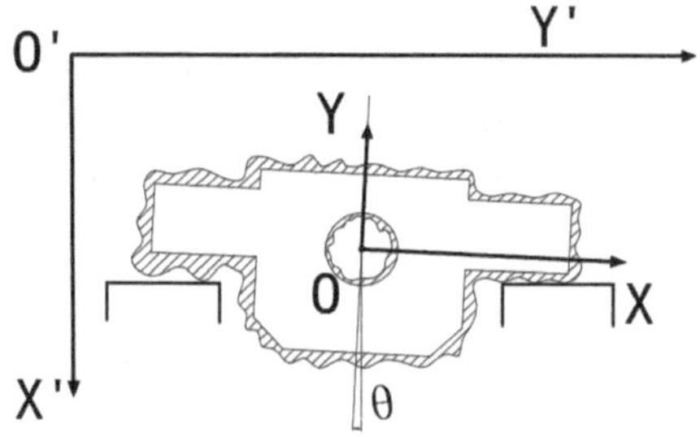

Fig. 4. Theoretical model and Matlab coordinate system

If you want to substitute the data obtained by image processing into the optimization equation, you must perform a coordinate transformation firstly. The transformation relationship is as follows:

$$M_a \begin{bmatrix} \sin\theta & \cos\theta & 0 \\ -\cos\theta & \sin\theta & 0 \\ 0 & 0 & 1 \end{bmatrix} \begin{bmatrix} 1 & 0 & 0 \\ 0 & 1 & 0 \\ Y_m & X_m & 1 \end{bmatrix} \begin{bmatrix} f_0 & 0 & 0 \\ 0 & f_0 & 0 \\ 0 & 0 & 1 \end{bmatrix} = M_b \tag{7}$$

In Eq. (7), M_a refers to the point or point set in the Matlab image coordinate system. M_b refers to the points or point set in the theoretical model coordinate system. Y_m refers to the Y-axis shift. X_m refers to the X-axis shift. f_0 refers to the size scaling factor.

3.3 Optimization Equation

In order to determine a suitable processing method, we necessary to find an optimal processing pose. The original processing benchmark can be included in the blank workpiece edge after rotation and translation, and the margin of each processing surface can be as uniform as possible. According to the least squares optimization criterion, when the square-sums of all points processing distances is smallest, it can be considered that the margins allocation is most uniform. At this situation, the processing contour can be considered to be the best. Since the processing contour cannot exceed the region of the blank actual contour, the distance-related parameters cannot be less than zero [8, 9].

Therefore, according to the above formula and conclusion, we can establish the objective function and constraints, the optimization equation of the workpiece pose should be:

$$\min f(x_m, y_m, \theta) = \sum_{i=1,j=1}^{n} d_A^2(x_i, y_j) + \sum_{p=1,q=1}^{n} d_B^2(x_p, y_q) + \sum_{s=1,t=1}^{n} d_C^2(x_s, y_t)$$
$$s.t. \begin{cases} d_A(x_i, y_j) \geq 0 \\ d_B(x_p, y_q) \geq 0 \\ d_C(x_s, y_t) \geq 0 \end{cases} \tag{8}$$

Finally, we can substitute the actual data to solve the optimization equation. By solving the optimization equation, we can obtain the optimal values of X_m, Y_m and θ. According to the optimization results, we can change the coordinate system of the machining center, and process blank according to the required posture.

4 Case Study

To verify the method correct and feasible, a case study is conducted, in which the part of design drawing is shown in Fig. 5. we will solve the optimization Eq. (8) by actual data.

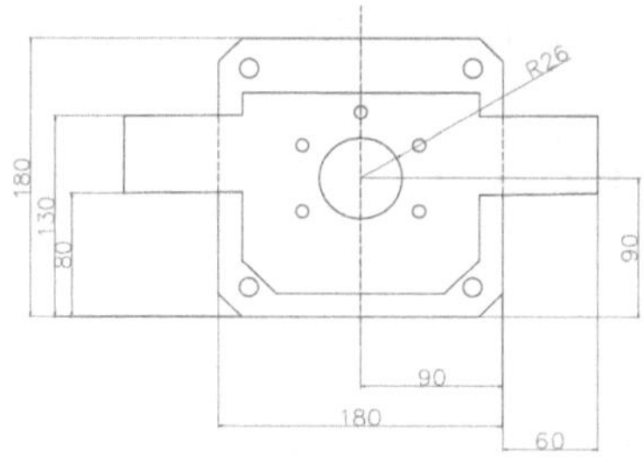

Fig. 5. A part of design drawing

In addition to the point set data, as shown in Fig. 5, $D_1 = 40$, $D_2 = 10$, and $D_3 = 26$ are also substituted according to the design size. f_0 can be obtained by calculating the corresponding pixel value of the known actual size. In this case, the size scaling factor $f_0 = \frac{120}{433}$.

Because the center of the circle is regarded as the origin, so $X_0 = 0$, $Y_0 = 0$.

Since the center of the circle should not exceed the range of the center hole, set the initial value to: $X_m = 250$, $Y_m = 250$, $\theta = 0$.

The value range is set to: $X_m \in (1, 1352)$, $Y_m \in (1, 788)$, $\theta \in \left(-\frac{\pi}{2}, \frac{\pi}{2}\right)$.

In the end, the solution results are: $\begin{bmatrix} X_m \\ Y_m \\ \theta \end{bmatrix} = \begin{bmatrix} 258 \\ 670 \\ 0.038 \end{bmatrix}$.

According to the optimization result, the distance of every points (including the upper plane, the lower plane and the center hole) is drawn in the figure, as shown in Fig. 6:

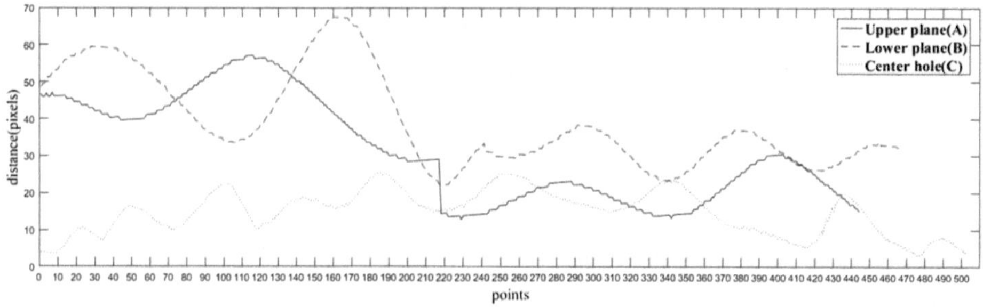

Fig. 6. The distance of every points

For the data related to the points involved in the calculation, as shown in Table 1:

Table 1. Points involved in the calculation

	Upper plane(A)	Lower plane(B)	Center hole(C)
Number of points	444	466	503
Average distance(pixels)	31.34	39.13	14.75
Minimum distance(pixels)	12.77	22.08	2.90
Maximum distance(pixels)	57.15	67.40	25.4

It can be seen from the data in Table 1 that the margins allocated to each processing surfaces are greater than zero. The constraints are fulfilled and the margins are evenly distributed, fulfilling the processing requirements.

5 Conclusions

This paper proposes a method based on image processing data to optimize the workpiece processing trajectory. Firstly, we analyzed that the casting defects existing on the surface of the blank will affect the processing benchmark, according to this situation, we establish the theoretical model of the workpiece. An optimization model of the work piece position and pose is established based on the requirement of the processing process and the work piece margin data. A case study shows that the method can deal very well with casting defects existing on the blank work piece contour, and replace the traditional process effectively, which verify that this is method is feasible, can replace the traditional scribing process in some range. On the other hand, this technology can be applied in various intelligent manufacturing systems.

Acknowledgements. This work was supported by the Liaoning Provincial Development and Reform Commission Northeast Revitalization New Kinetic Energy Cultivation Platform and Facilities Construction Special (No. 2017.812).

References

1. Meng, F., Zhao, G.: Research on the factors affecting the development of traditional manufacturing to intelligent manufacturing. J. Sci. Technol. Prog. Countermeasures **35**(01), 66–72 (2018)
2. Zhao, S., Jia, X.: Research progress and trends of intelligent manufacturing and its core information equipment. J. Mech. Sci. Technol. **36**(01), 1–16 (2017)
3. Huang, Y.: Difficulties in the scribing and processing techniques of arc embossed plate inserts. J. China High-tech Zone **10**, 170–171 (2018)
4. Qin, Z., Chen, J., Hong, R.: Rolling bearing detection of rolling bearings based on machine vision. J. Combined Mach. Tools Autom. Process. Technol. **01**, 118–121 (2019)
5. Chen, G., Guan, G., Wei, X.: On-line detection of surface defects of stamping parts based on machine vision. J. Prog. Laser Optoelectron. **55**(01), 341–347 (2018)

6. Li, C., Sun, J.: Weld defect detection and classification algorithm based on machine vision method. J. Comput. Eng. Appl. **54**(06), 264–270 (2018)
7. Zeng, Y.: Research on Online Measurement and Margin Calculation of Large-scale Propeller NC Processing. Huazhong University of Science and Technology (2005)
8. Yan, S., Zhou, Y., Peng, F., Lai, X.: Research on uniform optimization of processing margin for large complex surface parts. J. Huazhong Univ. Sci. Technol. **10**, 35–37 (2002)
9. Zhang, Y., Liu, M., Zhang, D., Wu, B.: Adaptive optimization method for blade processing allowance based on online detection. J. Comput. Technol. Dev. **24**(11), 226–229 + 233 (2014)

Development of Conductive Hydrogel Driven Dielectric Elastomer

Xiaoru Niu, Jianzhong Shang(✉), Zirong Luo, and Tao Jiang

Department of Intellectual Mechanisms and Devices,
National University of Defense Technology, Changsha, Hunan 410073, China
shangjianzhong@nudt.edu.cn

Abstract. Conductive hydrogel-based dielectric elastomer drives can perform flexible motion under high voltages (in kV) stimulus. Such a drive is typically comprised of a dielectric elastomer layer sandwiched by two conductive hydrogel layers as electrodes. The high voltage applied between the electrodes leads to an attraction force within the two due to Maxwell effects, which squeezes the elastomer in the middle and causes remarkable lateral expansions. Contrast to the conventional rigid metal electrodes, the flexible properties of hydrogel electrodes make them suitable for dielectric elastomers that can sustain large deformations. Researches on hydrogel based dielectric elastomer drives in the recent decade have achieved intriguing outcomes and sparked numerous applications in various fields such as material science, electrical engineering, biomedical and bioengineering. This review covers the most recent developments of hydrogel based dielectric elastomers, with emphases on the materials compositions, mechano-electrical properties, adhesive capabilities and structural stabilities. Comparisons of the key features and functions of different drives are listed, and typical applications of these drives are given. Finally, the review provides a perspective of the possible practical usages of this promising device in the not-so-distant future.

Keywords: Dielectric elastomer drives · Conductive hydrogel · Elastomer actuators · Elastomer sensors

1 Introduction

Dielectric elastomer actuators have received extensive attention due to their excellent tensile properties for studying flexible devices. In 1998, Pelrine firstly applied dielectric elastomer to artificial muscles with a sandwich shape which a thin layer of elastomer in the middle and the conductive electrode on both of side [1]. When applying a voltage to the electrode the free-moving ionic would accumulate at the interfaces between the elastomer and the electrode with opposite polarities that generates Maxwell force to move the two electrodes toward each other, thereby thinning the thickness of the elastomer and expanding the area, achieving over 100% electrical strain as shown in Fig. 1. A variety of devices have been developed using the large deformation properties, such as flexible sensors [2], artificial muscles [3], soft robots etc [4].

D. Wang et al. (Eds.): MeTrApp 2019, MMS 79, pp. 542–555, 2020.
https://doi.org/10.1007/978-981-15-0142-5_53

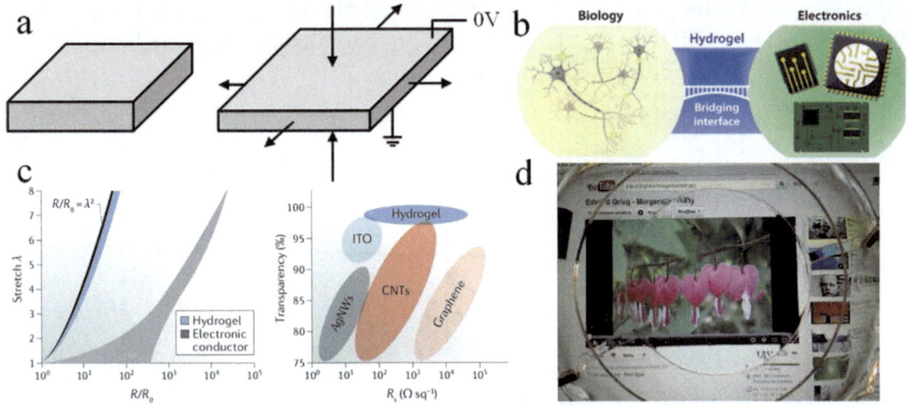

Fig. 1. Dielectric elastomer working diagram and characteristics of hydrogels. (a) Dielectric elastomer working principle. (b) Hydrogel as a bridge between biology and electronics [5]. (c) Conductor stretchability and transparency [6]. (d) The transparent loudspeaker [7].

The dielectric elastomer film model built by Pelrine can be described $P = \varepsilon\varepsilon_o E^2$, where P ε ε_oE is the electrostatic pressure, the relative permittivity of the elastomer, the vacuum permittivity and the electrostatic field respective [1]. It can be seen when increasing the electric field strength or the relative permittivity the pressure could be increased either.

The dielectric elastomer with a high dielectric constant, which can change shape or volume under external electrical stimulation. When the stimulation is cancelled it can recover the original shape. The elastomer materials currently used are VHB produced by 3M Company, polydimethylsiloxane (PDMS) [8], Ecoflex (Smooth-On). The VHB has very high stretchability (Stretched to 9 times the original length using standard tensile specimens) and high transmission (at a thickness of 1 mm the transmittance >90%) The stretchability of PDMS is not as good as VHB (Stretched to 2 times the original length using standard tensile specimens), and the transmission is good (at a thickness of 1 mm the transmittance >90%). Ecoflex has high stretchability (Stretched to 7 times the original length using standard tensile specimens), but poor transmission (at a thickness of 1 mm becomes opaque) [6]. Among the three materials, the VHB has been cross-linked, so the plasticity is poor. The PDMS and Ecoflex are all liquids, which can be easily cross-linked to different shapes according to needs.

The flexible electrode has the ability to deform the shapes under external force. There are three main types of electrode materials: (1) Metal electrodes (such as metal film electrodes, liquid metals, metal nanowires) [9–11]. (2) Carbon-containing electrodes (such as carbon-based electrodes, graphene, carbon nanotubes) [12, 13]. (3) conductive polymer electrodes (such as flexible hydrogel conductive electrode) [14, 15]. These electrodes are extensible, depending on different materials the advantages are different [16]. The graphene flexible electrode manufactured by Zang can achieve 500% stretching and the transmittance can reach 60% [17]. However, the price of carbon nanotubes and graphene are highly. Most of the conventional graphite electrodes have poor transmittance, which limits in optical applications. The flexible hydrogel electrode has good stretchability and transmission, which can be stretched to 2000% [18]. The flexible hydrogel electrodes can be biocompatible, such as biological tissue culture sodium alginate hydrogels [19] (Fig. 2).

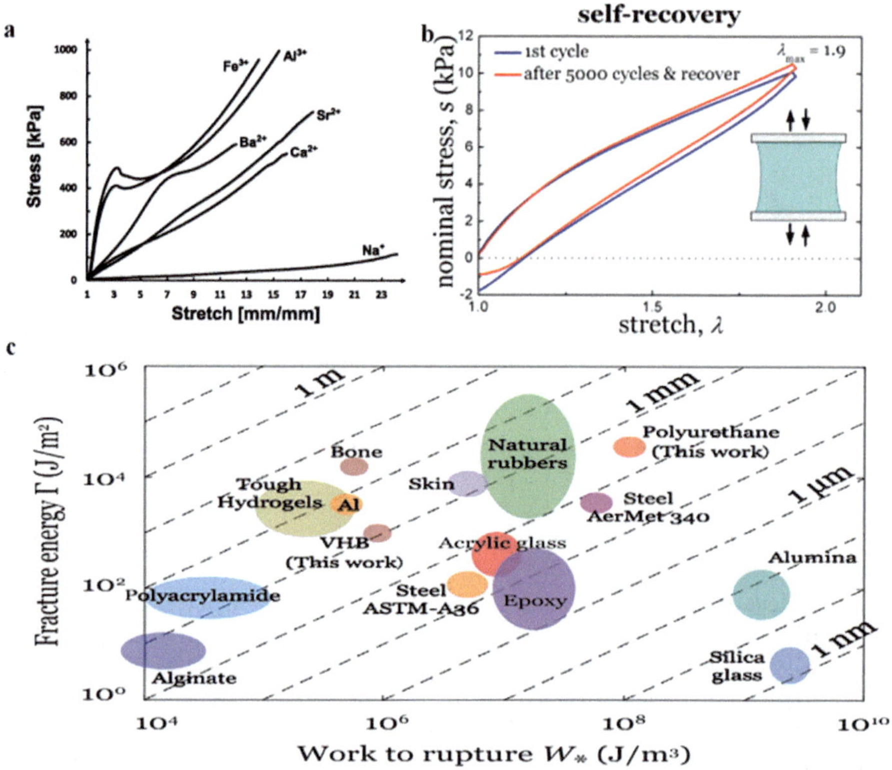

Fig. 2. Hydrogel tensile properties with different ions. (a) Stress-stretch curve with different ion hydrogels [20]. (b) Self-recovery hydrogel under cyclic loading [21]. (c) Fracture energy of different materials [22].

The flexible hydrogel electrodes can be divided into the ionic conductor hydrogel and the conductive polymer hydrogel. The ionic conductor hydrogel is a solid polymer network containing a large amount of free moving ions and when applying voltage to the electrodes, the moving ions generate current, the conductivity of the hydrogel depends on the ion concentration. The conductive polymer hydrogel is already conductive and does not require the addition of conductive particles [23].

This review introduces the key technologies of dielectric elastomer actuators which includes the mechanical and physicochemical properties of conductive hydrogels, the fabrication methods of hydrogels, the applications in current, the trend of development of dielectric elastomer actuators. This article is divided into the following chapters: 1 Introduction, 2 Conductive hydrogels, 3 Dielectric elastomer devices, 4 Summary and perspective.

2 Conductive Hydrogels

2.1 Mechanical Properties

The hydrogel consists of water and polymer networks. The polymer networks are much larger than a single water molecule so the water molecular properties are maintained. The hydrogel is solid which contains water >90% but is not soluble in water. Parts of hydrogels have good stretchability, transmission capacity and conductivity. They can transmit electrical signals in long distance at high frequency [24]. Researchers work hard to advance hydrogels into variety of devices, such as artificial muscles [25], artificial skin [26], touch screens [27], liquid crystal devices [28], software robots [29], etc. Hydrogels with high stretchability and flexibility have been produced, such as dual-network hydrogels [30, 31], topologies hydrogel [32], nanocomposite hydrogel [33] etc. The dual-network hybrid hydrogel designed by Sun can be stretched to 20 times its original length with the fracture energy up to 9000 Jm^{-1} which is comparable to natural rubber (breaking energy is 10000 Jm^{-1}), opening up the future application markets (Fig. 3).

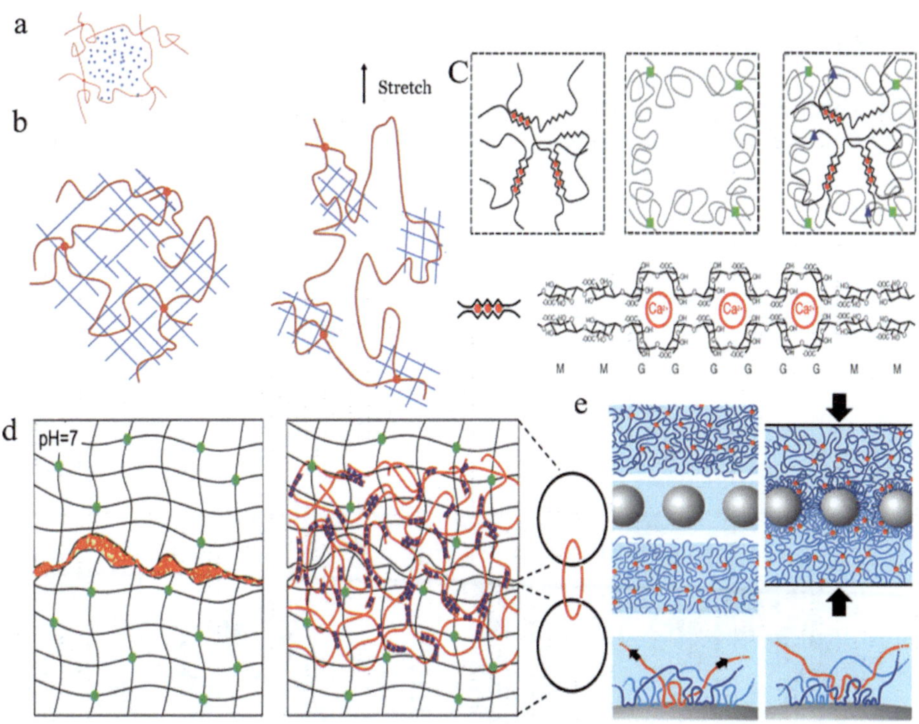

Fig. 3. The Relationship between mechanical properties and hydrogel connection. (a) (b) The single network and dual networks [31]. (c) Ionic polymer networks [18]. (d) Topology connection networks [32]. (e) Nanoparticle connection networks [33].

Suo and his team members soak the cured hydrogels in different ions solution to explore the mechanical strength bout stretchability and flexibility of hydrogels [20].

Meanwhile they explore the different ion effect on mechanical properties of hydrogels, finding hydrogels which contain divalent or trivalent cations have better mechanical properties.

The researchers also explore the anti-fatigue properties of hydrogels. Bai makes two polyacrylamide hydrogels which are bonded by covalently and non-covalently but one is intact the other is cut into a small mouth to test the fatigue resistance under cyclic stress [21]. Under cyclic stress the incision will increase with the increasing cycle numbers until fractured but the completed one which has constant performance as the number of cycles increase is called self-healing hydrogel or anti-fatigue hydrogel [34]. Such hydrogels typically work together by covalent polymer networks and non-covalent polymer networks.

2.2 Conductivities

Generally, conductive hydrogels can be divided into three groups:

(1) Non-conducting hydrogels, which are formed by adding ion to the gel matrix to form conductive gels.
(2) Self-conductive polymer, which can conjugate polymer solution to conductive gels.
(3) Conductive hydrogels are made by linking the polymer to the hydrogel network structure. Except the conductive polymer gels, conductivity is gained from conductive networks formed by conductive ion within gels.

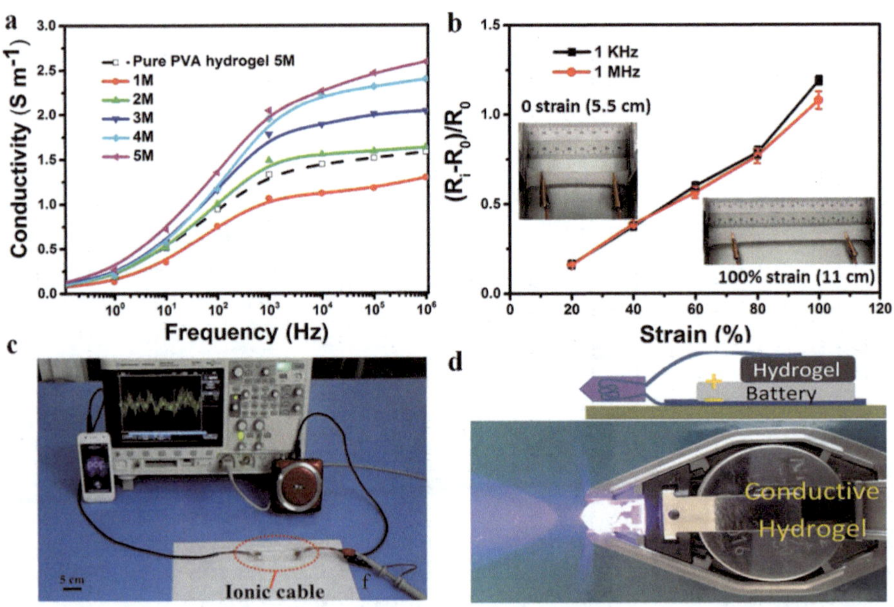

Fig. 4. Conductive characteristic curves of different kinds of hydrogels and the applications. (a) Conductivity of hydrogels with different concentrations as a function of frequency. (b) The relationship between hydrogel conductivity and strain [35]. (c) the ionic cable [24]. (d) The conductive hydrogel lights bulb [36].

The physical and biochemical properties of hydrogels are similar to many tissues and organs of animals and plants [19]. For example, the nerves in living organisms are ionic conductors. The hydrogels consist of polymer networks containing water molecules which incorporate ions in polymer networks to provide moving-ions, the conductivity is related to the type and concentration ions. Hydrogels have a resistivity about 10^{-1} Ωm [37], which is much higher than indium tin oxide ($\sim 10^{-5}$ Ωm) [38]. After stretching a hydrogel to its λ times of original length, the resistance value becomes λ^2 times of original value. As the thickness increases, the resistance of hydrogels is decrease. A part of hydrogels is high transparent and stretchable [39] (Fig. 4).

A new type of axon transparent ion wire designed by Yang mimics the function of human axons, transmitting on signals to sensing systems and decision systems. They design a sandwich structure [24]. Artificial ion wires can transmit signals at a frequency of 100 MHz in the distance of 10 cm or more. In this experiment when the concentration of lithium chloride is 8.0 mol/L, the resistivity is $\rho = 1.6 \times 10^{-2}$ Ωm, when the concentration of lithium is 0.01 mol/L, $\rho = 2.0$ Ωm. Conductive polymer hydrogels have electronic and ionic conductivity unlike traditional hydrogels Naficy in 2012 made (3,4ethylenedioxythiophene) polystyrene sulfonate (PEDOT:PSS) hydrogels was 4.3 S cm^{-1} [36].

2.3 Stabilities

The stability of ion conductor hydrogels is a necessary condition to ensure devices performe well. Ions in hydrogels provide conductive particles, water acts as solvent. The water in hydrogel affects the performance, such as the conductivity, the stretchability and the hardness etc. Therefore, the retention of water in hydrogels is basic to effect other properties. Bai explores water retention in four different hydrogels at room temperature and humidity is 20% [37]. She dissolves LiCl, KAc, MgC12 and NaCl four types of ions respectively in acrylamide monomer at a concentration of 4 mol/L, it is observed that water retention in gel is smallest with NaCl.

Dissolving the hygroscopic ions in water can reduce hydrogel vapor pressure and maintain water. Small molecules such as water molecules and oxygen molecules diffuse in the hydrogel as in polymeric liquid, it is easily to dehydrate in the air. Bai has studied and found that water retention in hydrogels can be improved by changing the type and concentration of dissolving ions. Researchers attempt to make hydrogels not dry out in the air by adding hygroscopic ions which can swell in the environment depend on the humidity, but the study is lack.

Other methods to maintain water in hydrogels is encapsulation [40]. The surface of hydrogels is covered by thin layer of elastomers, which will slow down but not completely prevent dehydration. The elastomers allow molecules diffuse which are smaller than the size of elastomer mesh at molecular scale and diffuse out through elastic membrane in a lower humidity environment. Increasing the thickness of the elastomer can increase the diffusion time until achieving the new balance in the set environment. In addition, the elastomer can also increase the temperature endurance of hydrogels. Liu throws a hydrogel which coated a thin elastomer film to the boiling water but the hydrogel remains intact without blistering [41].

At present, the bond between hydrogels and elastomers mainly through surface modification. These methods make iontronic hydrogels to be designed wearable [42], washable and conductive. Although different methods have been developed, it is still a challenge to gain strong adhesion between any material hydrogels and elastomers (Fig. 5).

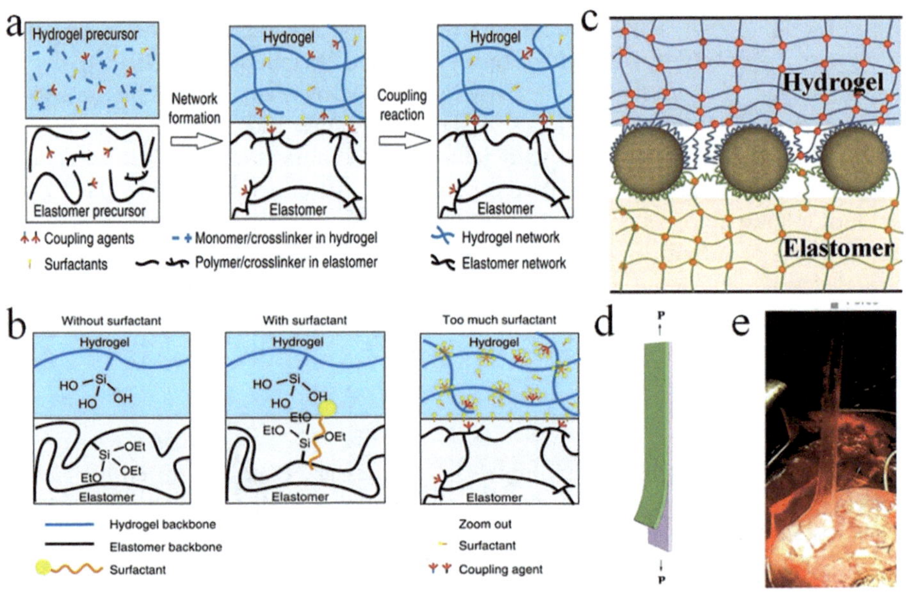

Fig. 5. Different bonding methods between hydrogel and elastomer result in different bond strengths. (a) (b) Hydrogel and elastomer connection after surface modification [41]. (c) Polymer chains from both the elastomer and the hydrogel adsorb on the surface of the nanoparticles [43]. (d) In a deformed state, elastomer and hydrogel stretching schematic [43]. (e) In vivo test on a beating porcine heart with blood exposure [44].

2.4 Making Methods

At present, the manufacturing methods of conductive hydrogel can be divided into in-situ curing [45] and 3D printing [46, 47]. The in-situ curing method transfers the hydrogel solution into a pre-made mold and is cured, then it is machined into the desired shape by a laser cutter. This method is more efficient for regular shapes but difficult for irregular shapes. Researchers try to make complex shapes by using the fast-growing 3D printing technology. The prepared hydrogel solution is squeezed out through the nozzle to print different shapes. For photocurable materials, adding UV light to the nozzle can make it cured quickly [48].

Using the in-situ solidification method, Kellaris made a hydraulically amplified self-recovering muscle [49], which firstly solidified the electrode material on a flexible substrate and then machined it into desired shape using laser cutter.

Recently, Yang and his group members studied a method to print hydrogels and elastomer structures in any shapes. The hydrogel fracture energy was 10000 Jm^{-2}, the elastomer fracture energy was 6000 Jm^{-2}, and the adhesion energy was 5000 Jm^{-2} [50]. As a precursor of hydrogels and elastomers, each ink is elasto-viscoplastic non-Newtonian fluids. They flow through the nozzles under pressure and maintain their shapes under gravity and capillary action. In particular, one of the inks carries an initiator to induce covalent interconnection between the two different networks (Fig. 6).

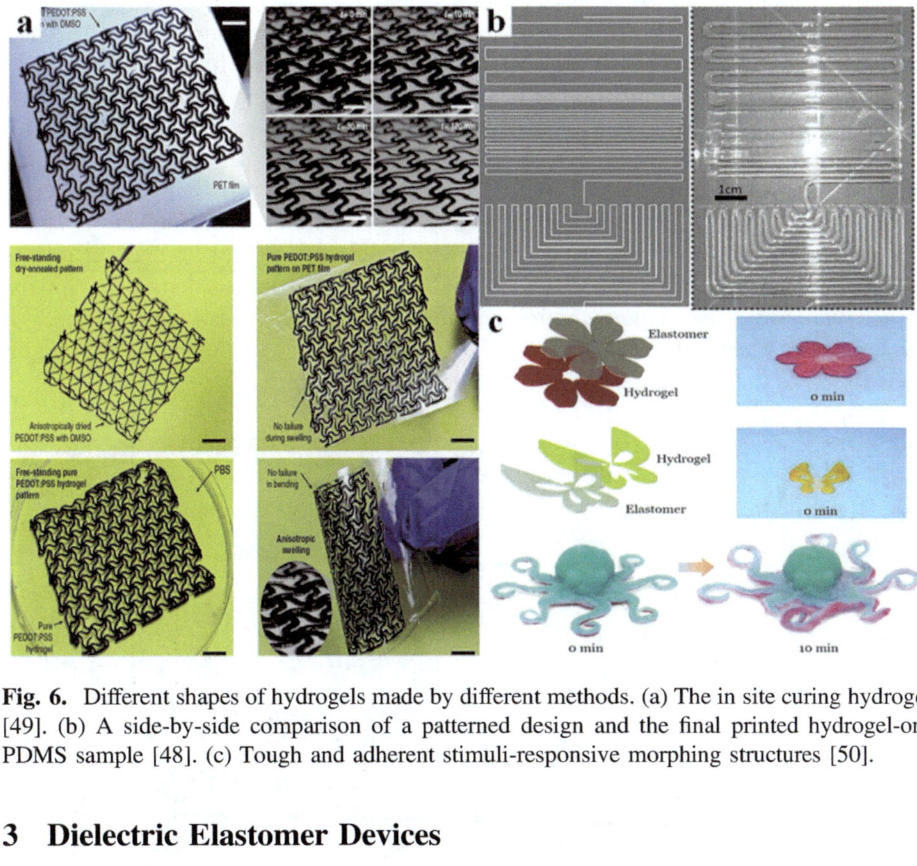

Fig. 6. Different shapes of hydrogels made by different methods. (a) The in site curing hydrogel [49]. (b) A side-by-side comparison of a patterned design and the final printed hydrogel-on-PDMS sample [48]. (c) Tough and adherent stimuli-responsive morphing structures [50].

3 Dielectric Elastomer Devices

Most of the existing drives are connected by rigid components but most of the natural creatures have flexible structures and the power directly output by muscle which eliminate the intermediate transmission, more efficient, and more environmentally adaptable. People have utilized the design variables offered by the dielectric elastomers and developed many devices. The devices can be classified according to the application mode of dielectric elastomer, into actuators, sensors [51, 52] etc.

3.1 Dielectric Elastomer Actuators

The dielectric elastomer actuators are composed of a flexible medium and a flexible electrode. The artificial muscle has large strain and self-sensing capability base on the dielectric elastomers. Oscillator developed by Benjamin [25] is driven by a dielectric elastomeric artificial muscle. He applied carbon grease to the surface of the elastomer as the drive electrode, and the plasticity of the carbon grease provided the rheological functionality for the device. The VHB were pre-stretched by 250%. The electrodes are tightly bonded to the surface of the elastomer and maintain stability to support a ball rolling in guide. When applying the voltage and controlling the magnitude of the voltage the muscle active that the oscillator produces different deformation under Maxwell pressure which can tilt the track rolling the ball under its gravity.

In terms of bionics, Professor Li designed a bionic fish [29] which does not have any motor and is fully driven by dielectric elastomer actuator, achieving the highly agility with a maximum speed of 13.5 cm/s, which is comparable to the speed of a real fish, the minimum turning radius is 8.5 cm. The bionic fish can swim at an average speed of 6.4 cm/s, which is equivalent to 0.69 times its own length, twice as fast as other bionic fish. For example, the SMAs-driven bionic fish speed is 0.35 times its own length, and the IPMCs-driven bionic fish speed is 0.25 times its own length. Bionic fish equips with power and remote-control system to remote control can work normally in the temperature range from 0.4 °C to 74.2 °C, and because of the material is highly transparent, to a certain extent, it has a stealth effect (Fig. 7).

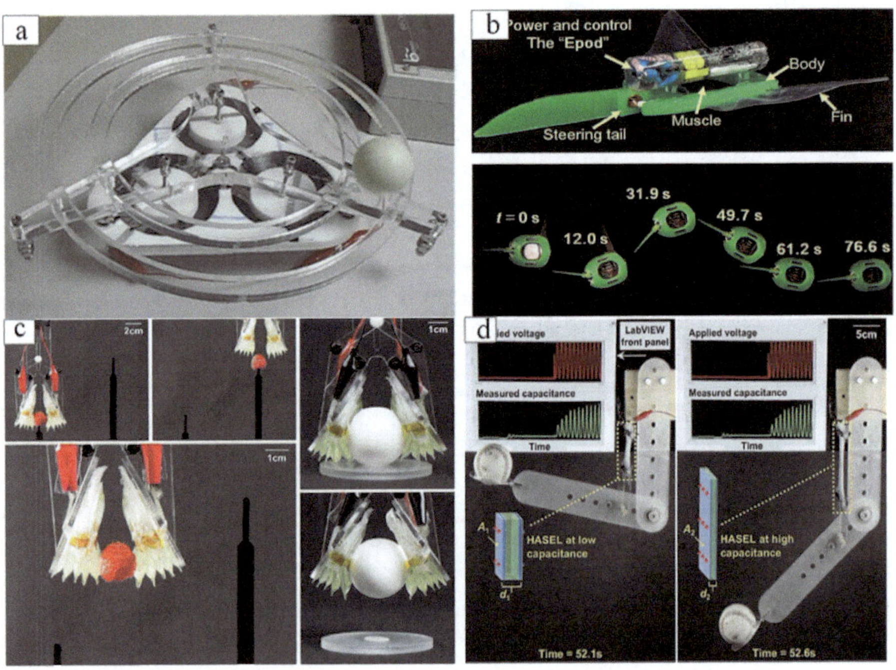

Fig. 7. The applications of dielectric elastomer drivers. (a) Ball rolling oscillator [25]. (b) Performance of the untethered fish [29]. (c) (d) Different shapes of dielectric elastomer artificial muscle [53].

Existing software robots mainly rely on fluid to drive. However fluid transport must be carried out through pipes and pipeline systems which limit the drive speed and efficiency. Acome designed a dielectric elastomer actuator-hydraulically amplified self-recovering artificial muscles to grip fragile objects flexibly [53], and according to different shapes of the actuator it can obtain multiple function and more efficient. A single-cell HASEL actuator was used to test the cycle life at high mechanical output power. The actuator provides an average (peak 586 W/kg) specific power of 358 W/kg when contracted, and mechanical breakage occurred at 158061 cycles. At one full test the energy conversion efficiency of a single unit during the cycle was 21%. Dielectric elastomer drivers are an important direction for the development in future artificial muscles.

3.2 Dielectric Elastomer Sensors

Sun designed transparent ionic skin which can be used as a capacitive sensor and pressure sensor and this device is similar to a capacitive device [26]. When the shape of the device changed, the capacitance value changes. When pressure is applied to the device, the deformation causes a change in capacitance value, and the corresponding force can be measured according to the deformation relationship. He uses the transparent ion conductor hydrogel containing sodium chloride as electrode. A layer of VHB as dielectric elastomer which sandwiched by the hydrogels, and the outer layer of the hydrogel is encapsulated by two pieces of elastomer films to isolate them from the external environment to reduce evaporation. The hydrogels are externally connected to two metal electrodes. When the sensor is mounted to a finger, the ionic skin stretches and follows the joint moving. As the finger repeatedly bends the sensor records the change of capacitance value, and the ionic skin can be compatible with the finger when the finger undergoes a large bending deformation (Fig. 8).

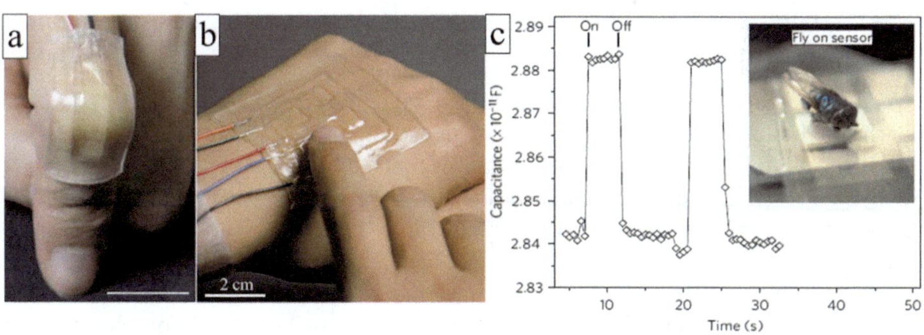

Fig. 8. Dielectric elastomer sensors. (a) (b) Ionic skin sensor [26]. (c) The microstructure films are able to sense the application of very small pressures [54].

4 Summary and Perspectives

This paper mainly describes the key technologies of hydrogels and dielectric elastomers. Firstly, introduces the dielectric elastomer actuator mechanism. Secondly, discusses hydrogel electrodes in details such as the mechanical properties, the electrical conductivity, stability and the fabrication methods. Thirdly, introduces the use of dielectric elastomer actuators. The research indicates that the dielectric elastomers have a series of significant advantages, and have broaden the prospective in software robots, flexible electronic devices, and human interaction. The results indicates in order to optimize hydrogel elastomers some challenges need to be overcome:

(1) One goal is to improve the bonding between hydrogel and elastomer interface The development of manufacturing processes can often lead to advance in technology. The bonding between hydrogels and elastomer interfaces can maintain the stability of devices [55]. Most of the existing hydrogel-elastomer devices are bonded by a flat lamination method, but the adhesion is poor. Researching new adhesion methods and manufacturing progresses are important to improve the existing bonding methods.
(2) Another goal is to obtain a breakthrough of new hydrogel conductors and dielectric elastomer materials.

The development of new materials can promote technological innovation. Hydrogel dielectric elastomer drivers are required to operate at high voltages at present [56], requiring high voltage chassis to supply power, which limits the use of the devices. High drive voltage is highly dangerous which can result in electrical breakdown and affect the equipment life. It is necessary to explore hydrogels with high conductivity, high draw ratio, high mechanical strength, and new elastomer materials with high dielectric constant and high transparency is necessary to design a safe, efficient, long-life, wide-area dielectric elastomeric equipment.

(3) The third goal is to minimize the mismatch between biological tissue electrode and mechanics interface

The biocompatible properties of some hydrogels have attracted widespread attention in the biological field. In the past few decades, some hydrogels have become one of the most studied engineering materials in many fields such as tissue engineering, biomedicine, because of its biocompatibility [57]. As researchers continue to delve into the complexity of tissue-electrode interactions, and recent advances in neuroscience and engineering have led to physical and mechanical mismatch in tissue-electrode interfaces, which is one of the major sources of neuroinflammation and corresponding failures in bioelectronic applications. Future research in hydrogel bioelectronics will build on understanding the fundamental mechanisms of tissue-electrode interactions, addressing the mismatch between biology and electronics, and achieving a seamless integration of biology and electronics.

(4) Future goals

The emergence of dielectric elastomer drivers is a global trend. Nature is full of optimally designed devices. There for it is useful to use Biomimetics to develop innovative new devices. As an emerging technology, dielectric elastomer flexible actuators are still in the preliminary research, and some key technical problems need to be solved urgently. Innovative global research precedes construction of new devices. It is expected that new technological breakthrough is obtained and a new generation of innovative devices are realized in the near future.

References

1. Pelrine, R.: High-speed electrically actuated elastomers with strain greater than 100%. Science **287**(5454), 836–839 (2000)
2. Boutry, C.M., et al.: A stretchable and biodegradable strain and pressure sensor for orthopaedic application. Nat. Electron. **1**(5), 314–321 (2018)
3. Kellaris, N., et al.: Peano-HASEL actuators: muscle-mimetic, electrohydraulic transducers that linearly contract on activation. Sci. Rob. **3**(14), eaar3276 (2018)
4. Lu, N., Kim, D.-H.: Flexible and stretchable electronics paving the way for soft robotics. Soft Rob. **1**(1), 53–62 (2014)
5. Yuk, H., Lu, B., Zhao, X.: Hydrogel bioelectronics. Chem. Soc. Rev. **48**(6), 1642–1667 (2019)
6. Yang, C., Suo, Z.: Hydrogel ionotronics. Nat. Rev. Mater. **3**(6), 125–142 (2018)
7. Keplinger, C., et al.: Stretchable, transparent, ionic conductors. Science **341**, 984–987 (2013)
8. Madsen, F.B., et al.: Dipolar cross-linkers for PDMS networks with enhanced dielectric permittivity and low dielectric loss. Smart Mater. Struct. **22**(10), 104002 (2013)
9. De, S., et al.: Silver nanowire networks as flexible, transparent, conducting films: extremely high DC to optical conductivity ratios. ACS Nano **3**(7), 1767–1774 (2009)
10. Yu, Z., et al.: Highly flexible silver nanowire electrodes for shape-memory polymer light-emitting diodes. Adv. Mater. **23**(5), 664–668 (2011)
11. Hu, W., et al.: Intrinsically stretchable transparent electrodes based on silver-nanowire-crosslinked-polyacrylate composites. Nanotechnology **23**(34), 344002 (2012)
12. Kim, K.S., et al.: Large-scale pattern growth of graphene films for stretchable transparent electrodes. Nature **457**(7230), 706–710 (2009)
13. Scardaci, V., Coull, R., Coleman, J.N.: Very thin transparent, conductive carbon nanotube films on flexible substrates. Appl. Phys. Lett. **97**(2), 023114 (2010)
14. Shian, S., Diebold, R.M., Clarke, D.R.: Tunable lenses using transparent dielectric elastomer actuators. Opt. Express **21**(7), 8669–8676 (2013)
15. Zhang, S., et al.: Flexible highly sensitive pressure sensor based on ionic liquid gel film. ACS Omega **3**(3), 3014–3021 (2018)
16. Rosset, S., Shea, H.R.: Flexible and stretchable electrodes for dielectric elastomer actuators. Appl. Phys. A **110**(2), 281–307 (2012)
17. Zang, J., et al.: Multifunctionality and control of the crumpling and unfolding of large-area graphene. Nat. Mater. **12**(4), 321–325 (2013)
18. Sun, J.Y., et al.: Highly stretchable and tough hydrogels. Nature **489**(7414), 133–136 (2012)
19. Horvath, M.A., et al.: Towards alternative approaches for coupling of a soft robotic sleeve to the heart. Ann. Biomed. Eng. **46**(10), 1534–1547 (2018)

20. Yang, C.H., et al.: Strengthening alginate/polyacrylamide hydrogels using various multivalent cations. ACS Appl. Mater. Interfaces. **5**(21), 10418–10422 (2013)
21. Bai, R., et al.: Fatigue fracture of self-recovery hydrogels. ACS Macro Lett. **7**(3), 312–317 (2018)
22. Chen, C., Wang, Z., Suo, Z.: Flaw sensitivity of highly stretchable materials. Extreme Mech. Lett. **10**, 50–57 (2017)
23. Zhang, Y.S., Khademhosseini, A.: Advances in engineering hydrogels. Science **356**(6337), eaaf3627 (2017)
24. Yang, C.H., et al.: Ionic cable. Extreme Mech. Lett. **3**, 59–65 (2015)
25. O'Brien, B.M., et al.: Dielectric elastomer switches for smart artificial muscles. Appl. Phys. A **100**(2), 385–389 (2010)
26. Sun, J.Y., et al.: Ionic skin. Adv. Mater. **26**(45), 7608–7614 (2014)
27. Kim, C.C., et al.: Highly stretchable, transparent ionic touch panel. Science **353**, 682–687 (2016)
28. Yang, C.H., et al.: Organic liquid-crystal devices based on ionic conductors. Mater. Horiz. **4** (6), 1102–1109 (2017)
29. Li, T., et al.: Fast-moving soft electronic fish. Sci. Adv. **3**, e1602045 (2017)
30. Gong, J.P., et al.: Double-network hydrogels with extremely high mechanical strength. Adv. Mater. **15**(14), 1155–1158 (2003)
31. Zhang, W., et al.: Fatigue of double-network hydrogels. Eng. Fract. Mech. **187**, 74–93 (2018)
32. Yang, J., Bai, R., Suo, Z.: Topological adhesion of wet materials. Adv. Mater. **30**(25), e1800671 (2018)
33. Rose, S., et al.: Nanoparticle solutions as adhesives for gels and biological tissues. Nature **505**, 382 (2013)
34. Lin, S., et al.: Anti-fatigue-fracture hydrogels (2018)
35. Zhou, Y., et al.: Highly stretchable, elastic, and ionic conductive hydrogel for artificial soft electronics. Adv. Funct. Mater. **29**(1), 1806220 (2019)
36. Naficy, S., et al.: Electrically conductive, tough hydrogels with pH sensitivity. Chem. Mater. **24**(17), 3425–3433 (2012)
37. Bai, Y., et al.: Transparent hydrogel with enhanced water retention capacity by introducing highly hydratable salt. Appl. Phys. Lett. **105**(15), 151903 (2014)
38. Yang, S., Ng, E., Lu, N.: Indium Tin Oxide (ITO) serpentine ribbons on soft substrates stretched beyond 100%. Extreme Mech. Lett. **2**, 37–45 (2015)
39. Chen, B., et al.: Stretchable and transparent hydrogels as soft conductors for dielectric elastomer actuators. J. Polym. Sci., Part B: Polym. Phys. **52**(16), 1055–1060 (2014)
40. Wang, Z., et al.: Elastic leak for a better seal. J. Appl. Mech. **82**, 081010 (2015)
41. Liu, Q., et al.: Bonding dissimilar polymer networks in various manufacturing processes. Nat. Commun. **9**(1), 846 (2018)
42. Someya, T., Bao, Z., Malliaras, G.G.: The rise of plastic bioelectronics. Nature **540**(7633), 379–385 (2016)
43. Tang, J., et al.: Adhesion between highly stretchable materials. Soft Matter **12**(4), 1093–1099 (2016)
44. Li, J., et al.: Tough adhesives for diverse wet surfaces. Science **357**, 378–381 (2017)
45. Lin, S., et al.: Stretchable hydrogel electronics and devices. Adv. Mater. **28**(22), 4497–4505 (2016)
46. Haghiashtiani, G., et al.: 3D printed electrically-driven soft actuators. Extreme Mech. Lett. **21**, 1–8 (2018)
47. Peng, X., et al.: Surface patterning of hydrogels for programmable and complex shape deformations by Ion Inkjet Printing. Adv. Func. Mater. **27**(33), 1701962 (2017)

48. Tian, K., et al.: 3D printing of transparent and conductive heterogeneous hydrogel-elastomer systems. Adv. Mater. **29**(10), 1604827 (2017)
49. Lu, B., et al.: Pure PEDOT:PSS hydrogels. Nat. Commun. **10**(1), 1043 (2019)
50. Yang, H., et al.: Printing hydrogels and elastomers in arbitrary sequence with strong adhesion. Adv. Funct. Mater. 1901721 (2019)
51. Larson, C., et al.: Highly stretchable electroluminescent skin for optical signaling and tactile sensing. Science **351**(6277), 1071–1074 (2016)
52. Saquib Sarwar, M., et al.: Bend, stretch, and touch: Locating a finger on an actively deformed transparent sensor array. Sci. Adv. **3**, 1602200 (2017)
53. Acome, E., et al.: Hydraulically amplified self-healing electrostatic actuators with muscle-like performance. Science **359**, 61–65 (2018)
54. Mannsfeld, S.C., et al.: Highly sensitive flexible pressure sensors with microstructured rubber dielectric layers. Nat. Mater. **9**(10), 859–864 (2010)
55. Madsen, F.B., et al.: Silicone elastomers with high dielectric permittivity and high dielectric breakdown strength based on dipolar copolymers. Polymer **55**(24), 6212–6219 (2014)
56. Suo, Z.: Theory of dielectric elastomers. Acta Mech. Solida Sin. **23**(6), 549–578 (2010)
57. Darnell, M.C., et al.: Performance and biocompatibility of extremely tough alginate/polyacrylamide hydrogels. Biomaterials **34**(33), 8042–8048 (2013)

Kangaroo-Inspired Jumping Robot with Controllable Takeoff

Yonghong Zhang, Sizhe Liu, Lei Zheng, Xiaojuan Mo, and Wenjie Ge(✉)

Department of Mechanical Engineering and Automation, School of Mechanical Engineering, Northwestern Polytechnical University, 127 West Youyi Road, Xi'an, Shaanxi, PR China
gwj@nwpu.edu.cn

Abstract. Kangaroo-inspired continuous jumping robot "Zbot", with two degrees of freedom (DOFs) and three joints, is proposed. The robotic knee and ankle joints of "Zbot" are coupled using a geared five-bar (GFB) mechanism, in which the parameters are optimized in order to mimic the locomotion of kangaroo hind legs. The hip and knee joints are actuated by two hydraulic cylinders individually. The robot can adjust the takeoff velocity and angle, which is called controllable takeoff. The kinematic and dynamic simulation show that the robot has a good jumping performance and bionic characteristic during takeoff, and the experiments demonstrate that "Zbot" can make controllable jumping, including varied takeoff angles and velocities.

Keywords: Jumping robot · Kinematic analysis · Controllable takeoff · Geared five-bar mechanism

1 Introduction

In order to overcome obstacles in unstructured environments, avoid predators and improve hunting efficiency, many animals adopt jumping as their main locomotion modes. Animals' brilliant performance in unmanned environments, inspired researchers to develop bio-inspired jumping robots [1]. Jumping robots can be applied to military reconnaissance, field search and life rescue.

Based on the jumping strategies adopted by jumping robots, they are classified into two categories: (1) intermittent jumping robot; Normally after one jump, it needs a long time to reset itself after landing, accumulate energy and adjust the direction for the next jump, such as MSU jumper [2], "Grillo" series [3]. (2) Continuous jumping robot; The time interval between two continuous jumps is very short. The balancing and resetting of body posture, the adjustment of jumping angle and the accumulation of jumping energy, are completed during jumping. A lot of continuous jumping robots are designed mimicking kangaroos, galagoes and cheetahs [4].

In our work, kangaroos are adopted as bionic objects. Zeglin developed a kangaroo inspired jumping robot, "Uniroo" [5]. This legged robot used four hydraulic cylinders to actuate four joints individually, which increases the mass of leg. Barkan et al. developed a jumping robot "Ken-2" [6], which has 3 DOFs (degree of freedom)

D. Wang et al. (Eds.): MeTrApp 2019, MMS 79, pp. 556–565, 2020.
https://doi.org/10.1007/978-981-15-0142-5_54

including ankle, knee, and hip joints. Three motors were exploited to actuate these joints respectively. In 2013, the Festo Bionic Learning Network developed a hopping robot "BionicKangaroo" [7], with a four-bar mechanism leg driven by pneumatic actuators. Elastic tendons are connected with actuators in the leg and used to store the energy and to provide the power for hopping. In addition, the hip and tail joints are equipped with electrical drivers to balance and stabilize during whole jumping process.

From the research of current continuous jumping robots, we can conclude that most of which have a single jumping mode. They cannot change takeoff velocity and or takeoff angle at the same time. In addition, their jumping performance decreases sharply with increased load. In this paper, we propose a jumping robot named "Zbot", which can adjust takeoff angles and velocities with fewer DOFs. The ankle and knee joints of the "Zbot" are coupled using a gear five-bar mechanism (GFB), which decreases the DOFs in leg mechanism, reduce the leg mass, simplify the control strategy.

The rest of this paper organized as follow: Sect. 2 presents the mechanism design and mechanism optimization of the "Zbot"; In Sect. 3, the kinematics are carried out and controllable takeoff strategy is explained; In Sect. 4, experimental results are shown to demonstrate the performance of the "Zbot"; In Sect. 5, conclusion and future works are provided.

2 Mechanism Design and Optimization

2.1 Bioinspired Mechanism Design

Comparing to the intermittent jumping strategy adopted by miniature jumping animals like locusts, fleas and click beetles, kangaroos adopts continuous jumping strategy. The jumping of a kangaroo is simplified as a mass-spring model, as shown in Fig. 1(a). The weight of the robot is focused on the trunk, and the leg works as a spring connected with an actuator. The spring is modeled as achilles tendon and used to store or release the energy. Researchers found that the kangaroo can absorb 35% energy during landing, and reuse the stored energy for next jump.

The open-chain three-joint mechanism with three actuators was adopted in our previous work, as illustrated in Fig. 1(b). While the open-chain model is driven by three hydraulic cylinders, which makes it difficult to control of three DOFs, and the efficiency decreases because the excessive mass of three actuators. To solve existing problems, gear five-bar mechanism (GFB) is adopted, as shown in Fig. 2, to combine the movements of ankle joint and knee joint, it helps to reduce the leg mass and simplify the control process.

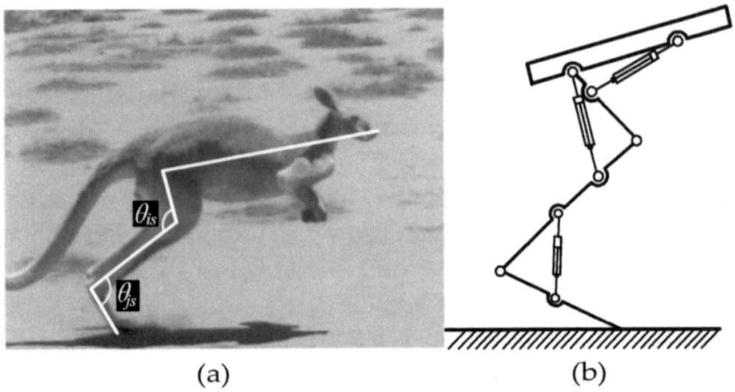

Fig. 1. (a) Jumping kangaroo; (b) Jumping robot model.

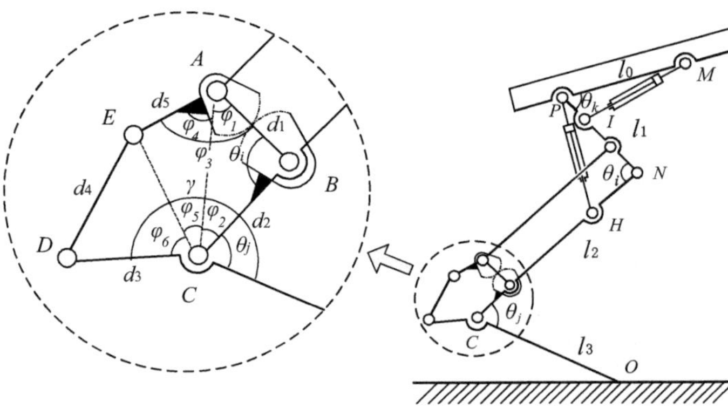

Fig. 2. The geared five-bar mechanism

It is well known that GFB mechanism is more diverse than four-bar mechanism, both in trajectory and transmission [8]. The adopted GFB mechanism is designed, as shown in Fig. 2. The four-bar parallelogram mechanism is used to transmit the motion of the knee joint to the GFB mechanism. The GFB mechanism works as a movement conversion device, which coordinate the knee-ankle joint movement. The freedom of the hip is designed to help the robot leg land smoothly. The knee and ankle are coupled by a GFB mechanism, which helps to reduce one actuator but also improves the mechanical advantage (MA) [9].

2.2 Mechanism Optimization

The GFB mechanism is the essential part of the robot. It constricts the relationship between the ankle angle (θ_j) and knee angle (θ_i). Bar *AB*, *BC*, *CD*, *DE*, *EA* are hinged, as illustrated in Fig. 2. Two gears located at point *A* and *B*, and they are fixed with bar *AE* and *BC* respectively. The characteristics of the mechanism can be expressed by nine variables $(d_1, d_2, d_3, d_4, d_5, i, \theta_{i0}, \varphi_{30}, \gamma)$, d_1, d_2, d_3, d_4, d_5 represent the lengths of each bar, i is the gear transmission ratio. (θ_{i0}, φ_{30}) are the initial angle of θ_i and θ_j, since bar *BC* and *EA* are fixed with the gear, the following equations are got:

$$\varphi_3 = \varphi_{30} + i(\theta_i - \theta_{i0}) \tag{1}$$

If d_{AC} is defined as the distance between point A and point C (The same below). The following equation is derived.

$$d_{AC}^2 = d_1^2 + d_2^2 - 2d_1 d_2 \cos\theta_i \tag{2}$$

$$\frac{\sin\varphi_1}{d_2} = \frac{\sin\theta_i}{d_{AC}} \tag{3}$$

Equation (3) can be written in following form.

$$\varphi_1 = \arcsin(\frac{d_2}{d_{AC}}\sin\theta_i) \tag{4}$$

$$\varphi_2 = 180^\circ - \theta_i - \varphi_1 \tag{5}$$

In ΔACE, the following equation can be derived.

$$\varphi_4 = \varphi_3 - \varphi_1 \tag{6}$$

$$d_{CE}^2 = d_5^2 + d_{AC}^2 - 2d_5 d_{AC}\cos\varphi_4 \tag{7}$$

$$\frac{\sin\varphi_4}{d_{CE}} = \frac{\sin\varphi_5}{d_5} \tag{8}$$

Equation (8) can be written in following form.

$$\varphi_5 = \arcsin(\frac{d_5}{d_{CE}}\sin\varphi_4) \tag{9}$$

And in $\triangle ECD$, the following equation can be drawn.

$$d_4^2 = d_3^2 + d_{\mathrm{CE}}^2 - 2d_3 d_{\mathrm{CE}}\cos\varphi_6$$

$$\varphi_6 = \arccos(\frac{d_3^2 + d_{CE}^2 - d_4^2}{2d_3 d_{CE}}) \tag{10}$$

According to Fig. 2 in the relationship between the angle calculated.

$$\theta_j = \gamma - (\varphi_2 + \varphi_5 + \varphi_6) \tag{11}$$

From Eq. (3), the relationship between θ_i and θ_j can be derived.

$$\theta_j = f(\theta_i) \tag{12}$$

Equation (12) is the function of θ_j, which is related with parameters ($d_1, d_2, d_3, d_4, d_5, i, \theta_{i0}, \varphi_{30}, \gamma$). A reasonable set of parameters can make the robot knee ankle joint motion characteristics to achieve the best state, these parameters have been optimized to make the robotic knee and ankle angle change mimic the change of knee-ankle joint angle of kangaroo jumping during takeoff as completely as possible. The optimized results are shown in Table 1. By substituting the results of optimization into Eq. (12), we can obtain ankle angle, and comparing optimization curve with the measured discrete point of knee-ankle joint.

Table 1. Optimized results.

Optimized variables	d_1 (mm)	d_2 (mm)	d_3 (mm)	d_4 (mm)	d_5 (mm)	i	θ_{i0} (°)	φ_{30} (°)	γ (°)
Optimized results	54	56	60	60	32	1	80	95	198

3 Kinematics Analysis and Controllable Takeoff

3.1 Kinematics Analysis

According to the D-H method of kinematics [10], the center of mass (COM) of the robotic body can be described by all the joint angles and bar length. In order to get the relationship between body COM and active joints, the kinematics cquation needs to be established. As shown in Fig. 3, point O is coordinate origin, and the coordinate systems for each bar at the joint of the robot are set up. λ is the angle of the body relative to x-axis. The value of λ has little fluctuation (near 0°) when kangaroo jumping.

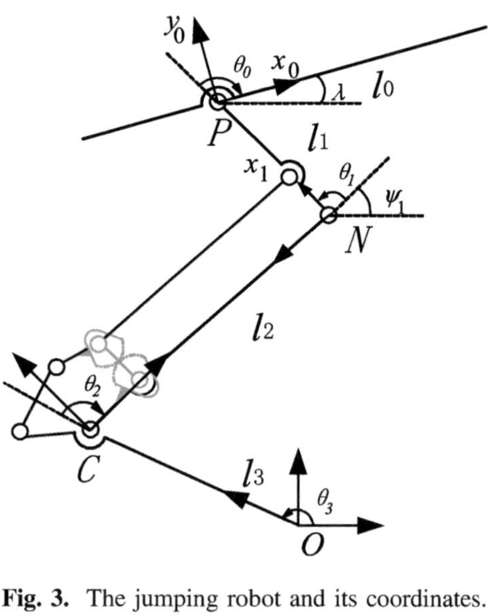

Fig. 3. The jumping robot and its coordinates.

The rotation matrix between two coordinates can be expressed as follows,

$$A_i^j = \begin{bmatrix} \cos\theta_i & -\sin\theta_i & 0 & a_i\cos\theta_i \\ \sin\theta_i & \cos\theta_i & 0 & a_i\sin\theta_i \\ 0 & 0 & 1 & 0 \\ 0 & 0 & 0 & 1 \end{bmatrix} \quad (13)$$

In Eq. (13), the A_i^j is rotation matrix of coordinate {i} in {j}. Because the COM of the body is point $S(a, b)$ in local coordinate $\{0\}$. According to the principle of D-H method, by substituting the parameters into Eq. (2). The COM position of body in global coordinate $\{O\}$ can be obtained as follows,

$$S = A_3^O A_2^3 A_1^2 A_0^1 \begin{bmatrix} a \\ b \\ 0 \\ 1 \end{bmatrix} \quad (14)$$

The Eq. (14) has four unknown variables, and the robot has two DOFs. So the other two constraints are necessary to determine the position of body COM. As mentioned above, assume that λ keeps level, another limitation is Eq. (12) that GFB mechanism couples the relative movement of knee and ankle joints.

The ranges of hip and knee angles could reach (36°–87°) and (78°–129°) respectively. We choose the hip joint for research, the operational space of the robot is shown in Fig. 4, from which we can observe that the functional area can satisfy any jumping directions from 0°–90°.

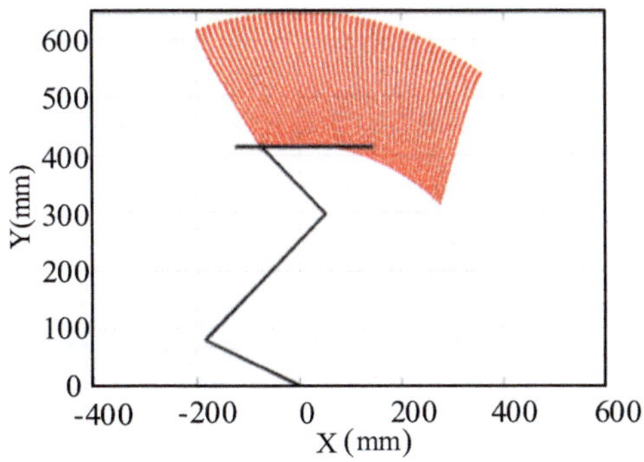

Fig. 4. Operational space of the robot.

3.2 Controllable Takeoff

3.2.1 The Strategy of Controllable Takeoff

Controllable takeoff means that the jumping speed and direction of the robot can be adjusted according to obstacles. The key to the robotic takeoff is that the direction of the reaction force from the field passes the body COM. Under the action of energy, the body moves in a straight line. The 2 DOFs jumping robot can realize this technology by controlling actuators. The end of the robot (body) maybe moves in a linear function, a quadratic function or a high order function of time. To avoid the impact to the leg by the actuators, we choose the quadratic function as the strategy of the end of the robot. The steps of controllable takeoff are shown in Fig. 5.

3.2.2 Inverse Kinematic Solving

The hip joint was chosen as reference point of the robot. v and α represent the speed and direction of the body when it leaves the ground. Here, the trajectory of the hip joint of the robot is planned as a uniformly accelerated rectilinear motion, as Fig. 6.

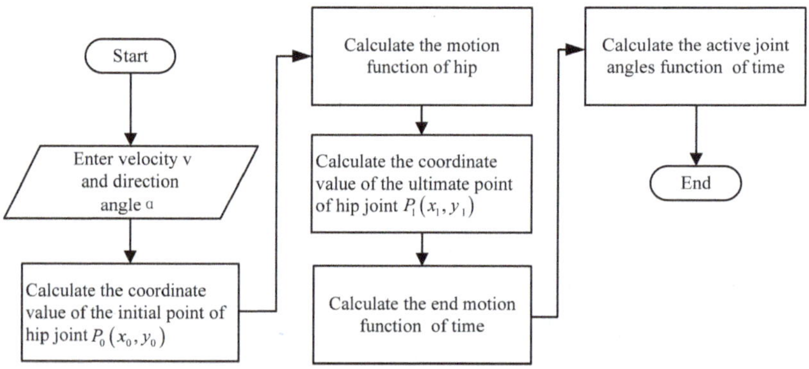

Fig. 5. The steps of controllable takeoff.

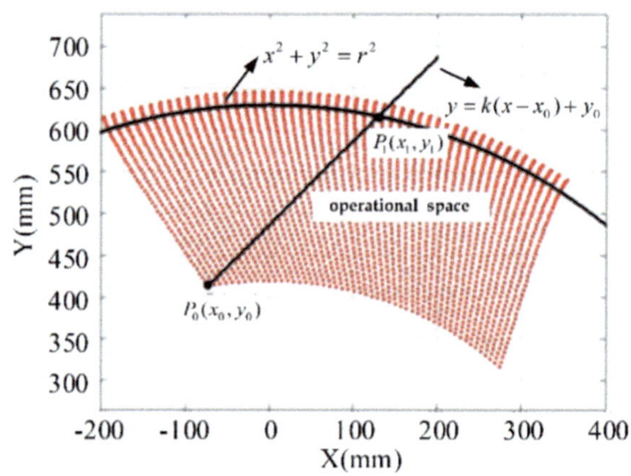

Fig. 6. The trajectory of the hip.

Combined the straight trajectory and arc function,$P_1(X_1, Y_1)$ can be calculated. If the point (p_x, p_y) represents any point in the straight line P_0P_1. According to the law of evenly accelerated movement, we can get the trajectory function of time

$$\begin{cases} p_x = \frac{v_{1x}^2}{4(x_1 - x_0)} \cdot t^2 + x_0 \\ p_y = \frac{v_{1y}^2}{4(y_1 - y_0)} \cdot t^2 + y_0 \end{cases} \tag{15}$$

where the v_{1x} and v_{1y} is the projection of v on the horizontal and vertical directions, so

$$v_{1x} = v\cos\alpha, v_{1y} = v\sin\alpha \tag{16}$$

When we enter the v and α, we could solve the angles of the joints by inverse kinematics, Finally, the jumping process of the robot is shown in Fig. 7.

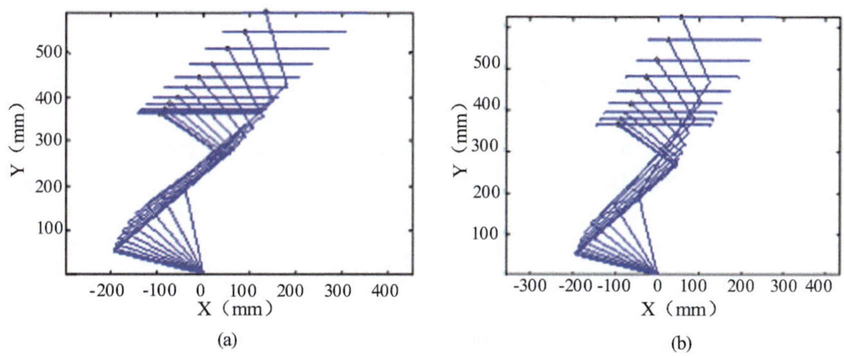

Fig. 7. The process of the robotic taking off: (a) jump with takeoff of 45°, (b) jump with takeoff angle of 60° (all of these are the demonstrates at $v = 3$ m/s)

4 Experiments

According to the robot's mathematical model, as shown in Fig. 2, and the optimization results of the GFB mechanism, the robotic prototype "Zbot" is fabricated as shown in Fig. 8. The overall size is 375 × 380 mm, and the major material is aluminum. "Zbot" weighs 4.5 kg and is driven by two hydraulic cylinders installed at hip and knee joints. The hydraulic system produces the maximum pressure of 9 Mpa, and "Zbot" joints are controlled by the mathematical model of the hydraulic system. In the experiment, in order to reduce the effect of tubing dragging the robot, the 0.8 kg weights were fixed on the body to increase the inertia of the body. The elastic foot and actuator are in series connection.

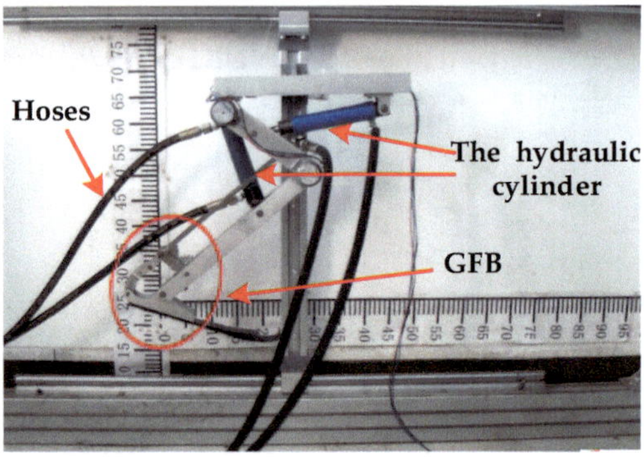

Fig. 8. The prototype "Zbot".

In order to evaluate the jumping performance of "Zbot", we will test two jumping modes, one is in the jumping directions of 45° and the other is in 60° at the velocity of 3 m/s respectively.

Figure 9 is the image sequence of "Zbot" jumping video with takeoff angle at 45°. It takes 0.26 s to take off. The hip joint is marked with black dot and the takeoff trajectory of the robot is recorded. The recorded trajectory is close to our predicted result (red line), when "Zbot" jumping with takeoff angle at 60°, Similarly, the hip joint is marked with black dot and the takeoff trajectory of the robot is recorded. The recorded trajectory is also close to our predicted result. but there are some minor deviations. The possible reasons for the deviations is analyzed and listed as follows: (1) the slippage between the feet and the ground; (2) external disturbance caused by the pull of hoses. The vertical displacement of hip joint is measured during takeoff, the height reaches 520 mm, which is very close to our simulation results. In addition, the robot could maintain initial jumping posture during the whole takeoff process, which can reduce the burden of the balance control system.

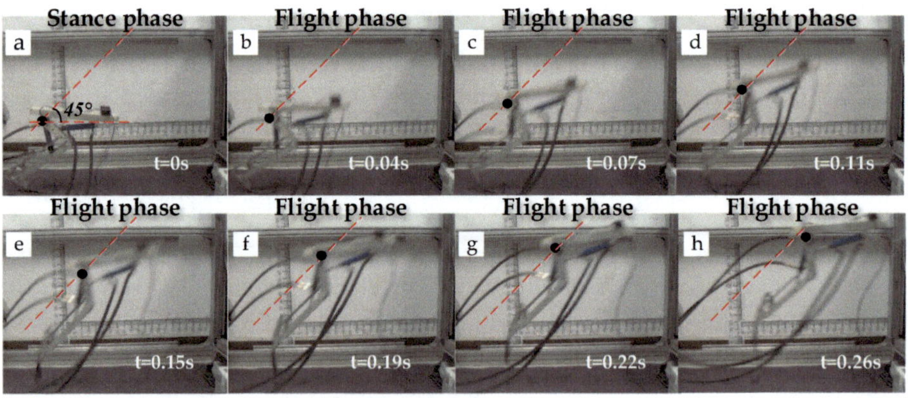

Fig. 9. "Zbot" jumps with takeoff angle at 45°. The red dotted line is reference of 45°

5 Conclusion

To increase the jumping efficiency, this paper presents a 2 DOFs and three-joint kangaroo inspired jumping robot “Zbot”. The knee and ankle joints are coupled using a GFB mechanism. The parameters of mechanism are optimized to make the robot to move as kangaroo. Different from existing continuous jumping robots, the takeoff velocity and angle of “Zbot” can be controlled at the same time. Moreover, the robot has excellent jumping performance under loading. Experimental results show that kangaroo inspired jumping robot “Zbot” can adjust takeoff angles using similar strategy as its bionic object, kangaroo. In future work, we will study postural righting of the robot in flight phase and tail-assisted balancing and the stability of the robot in landing phase. Thanks to the brilliant jumping performance under loading, “Zbot” has a broad application in the future.

References

1. Bai, L., Ge, W.-J., Chen, X.-H., Meng, X.-Y.: Hopping capabilities of a bio-inspired and mininally actuated hopping robot. In: 2011 International Conference on Electronics, Communications and Control (ICECC), pp. 1485–1489. IEEE (2011)
2. Zhao, J., Xi, N., Cintrón, F.J., Mutka, M.W., Xiao, L.: A single motor actuated miniature steerable jumping robot. In: 2012 IEEE/RSJ International Conference on Intelligent Robots and Systems, pp. 4274–4275. IEEE (2012)
3. Scarfogliero, U., Stefanini, C., Dario, P.: Design and development of the long-jumping “grillo” mini robot. In: Proceedings 2007 IEEE International Conference on Robotics and Automation, pp. 467–472. IEEE (2007)
4. Armour, R., Paskins, K., Bowyer, A., Vincent, J., Megill, W.: Jumping robots: a biomimetic solution to locomotion across rough terrain. Bioinspiration Biomimetics **2**, S65 (2007)
5. Zeglin, G.J.: Uniroo–a one legged dynamic hopping robot. Massachusetts Institute of Technology (1991)
6. Ugurlu, B., Kawamura, A.: ZMP-based online jumping pattern generation for a one-legged robot. IEEE Trans. Ind. Electron. **57**, 1701–1709 (2010)
7. Graichen, K., Hentzelt, S., Hildebrandt, A., Kärcher, N., Gaißert, N., Knubben, E.: Control design for a bionic kangaroo. Control Eng. Pract. **42**, 106–117 (2015)
8. Sun, Y., Ge, W., Zheng, J., Dong, D.: Design and evaluation of a prosthetic knee joint using the geared five-bar mechanism. IEEE Trans. Neural Syst. Rehabil. Eng. **23**, 1031–1038 (2015)
9. Haldane, D.W., Plecnik, M.M., Yim, J.K., Fearing, R.S.: Robotic vertical jumping agility via series-elastic power modulation. Sci. Rob. **1**, (2016)
10. Young, K.-Y., Chen, J.-J., Wang, C.-C.: An automated robot calibration system based on a variable DH parameter model. In: Proceedings of 35th IEEE Conference on Decision and Control, pp. 881–886. IEEE (1996)

Hydrodynamic Coefficients Calculation of Complex-Shaped AUV Based on Lattice Boltzmann Method

Yinpo Yan and Yuan Chen(✉)

School of Mechanical, Electrical and Information Engineering, Shandong University at Weihai, Weihai, China
cyzghysy@sdu.edu.cn

Abstract. The lattice Boltzmann method is a new computational fluid dynamics method, which can simulate several kinds of fluid problems. In this paper, a part of hydrodynamic coefficients of an AUV with a complex shape were obtained by a numerical simulation in CFD software based on lattice Boltzmann method. The pure pitch and pure heave motion of PMM experiment were conducted in CFD software Xflow and corresponding resistance and resistance moment were measured. The Wall-Adapting Local Eddy (WALE) viscosity model was employed to stabilize the simulation. In order to obtain the corresponding hydrodynamic coefficients, Response spectrum analysis was performed on the results derived from CFD simulation. The complex work of meshing is avoided and the workload of numerical simulation is significantly reduced.

Keywords: AUV · Hydrodynamic coefficients · CFD · Lattice boltzmann method

1 Introduction

At the end of 1980s, with the progress of artificial intelligence technology, micro-electronics technology, control hardware and computer technology, the intelligent underwater robot technology had developed rapidly. Many coastal countries, especially developed countries, are committed to the research and development of intelligent underwater robot technology product research and development [1]. In general, Unmanned Underwater Vehicles (UUVs) can be classified as Remotely Operated Vehicles (ROVs) and Autonomous Underwater Vehicles (AUVs). Compared with ROV, AUV without the restrictions of a rope has a wider range of activities, better maneuverability and intelligence. In the civil field, AUV can be used for laying off-shore oil and gas pipelines, hydrological environment survey, search and rescue, off-shore platform pipeline cleaning and maintenance and repair of underwater communication equipment, etc. In the military field, AUV is increasingly used for reconnaissance, mine clearance and intelligence collection due to the advantages of superior concealment and the wider range of activities. In essence, the use of under-water robotic system will greatly benefit all kinds of ocean activities [2].

D. Wang et al. (Eds.): MeTrApp 2019, MMS 79, pp. 566–575, 2020.
https://doi.org/10.1007/978-981-15-0142-5_55

A dynamics model is essential in the design of guidance, navigation and control (GNC) systems of an underwater vehicle and the need for performance improvements in GNC has motivated deeper investigation into hydrodynamics modeling and the identification of the open frame ROV dynamics [3]. However, accurate hydrodynamic coefficients are very challenging to obtain. At present, there are mainly three methods for calculating hydrodynamic coefficient: confined or free self-navigation model test, computer numerical simulation, and semi-theoretical and semi-empirical estimation methods [4]. The model experiment method, including the constrained model test and the self-propelled model test of the vehicle, is the most important method to determine the hydrodynamic coefficient so far. The Planar Motion Mechanism (PMM) system proposed by Goodman and Gertler can easily solve the linear acceleration derivative and linear velocity derivative of various forces and torques. However, building a planar motion mechanism is expensive and testing is time-consuming. Secondly, although the methods of theoretical calculation and empirical formula estimation are more economical than the model experimental method, the hydrodynamic accuracy obtained cannot meet the engineering requirements, and is often used as the object of comparative analysis with other methods.

It is a huge project to obtain most of the hydrodynamic coefficients of submarine by mode test. The rapid development of CFD technology and the continuous improvement of computer performance provide technical support for the solution of this problem. More and more scholars try to solve this problem by numerical simulation. The numerical method can not only provide an ideal environment for submarine hydrodynamic research free from external factors, but also make up for the defect that various working conditions cannot be tested due to the limitation of test facilities [5]. In order to obtain the hydrodynamic coefficient of the rotary movement, Xu et al. [6] calculated the damping force/torque of ROV with complex shape in the rotary motion with overlapping grid method based on CFD calculation software STAR-CCM + and moving reference frame and overset methods are combined to calculate the rotational hydrodynamic force to accelerate the convergence rate and improve reliability of CFD simulations. Zhang [7] numerically simulated the oblique motion experiment and PMM experiment in constraint mode test of SUBOFF by solving the RANS equation based on CFD technology and the analysis shows that the viscous hydrodynamic coefficients obtained by the oblique motion experiment are more consistent with the test value, and inertial hydrodynamic coefficients obtained by the PMM experiment are more accurate. Li et al. [8] calculate the hydrodynamic coefficients including inertial hydrodynamic coefficients, viscous hydrodynamic coefficients by simulating virtually a steady-state motion simulation test and an unsteady-state motion simulation test in CFD and the accuracy and reliability of hydrodynamic coefficients was proved by comparing to the prototype test results. Gwangsoo and Hyung [9] put forward a new methodology to determine all components of linear and nonlinear damping coefficients of a tow-fish underwater vehicle CFD simulations without any pre-elimination.

In a word, existing numerical fluid simulations by solving Reynolds average Navier-Stokes equation (RANS) based on CFD software Fluent, Star CCM +, CFX and so on show that the precision of the calculation result is very dependent on the meshing quality and the setting of numerical model. Meshing will also cost a lot of time and brought certain limits to CFD simulation.

In this paper, pure heave motion and pure pitch motion of the PMM experiment of an AUV with a complex shape were simulated in CFD software Xflow which is based on the Lattice Boltzmann Method (LBM) and corresponding resistance and resistance moment were measured, and then the corresponding hydrodynamic coefficients were obtained.

2 Basic Theory

2.1 Description of the Model

The research object of this paper is an underwater robot used for pipeline cleaning, which is equipped with a high-pressure spray gun and can replace people to clean the attached Marine organisms on the supporting pipeline under the ocean platform. The underwater robot has a symmetrical surface, and its 3D model is shown in Fig. 1. The AUV consists mainly of a main cabin, two small cabins, six propellers, a high-pressure spray gun, and magnetic adsorption devices. The size of the body is $1.23 \times 1.31 \times 0.60$ m (i.e. length $\times$ width $\times$ height), the mass and the drainage volume was respectively 55.582 kg and 0.065 m^3. The maximum forward thrust and reverse thrust of a single propeller are respectively 150 N and 60 N, and the maximum adsorption force and power of a single electromagnet are respectively 350 N and 8 W.

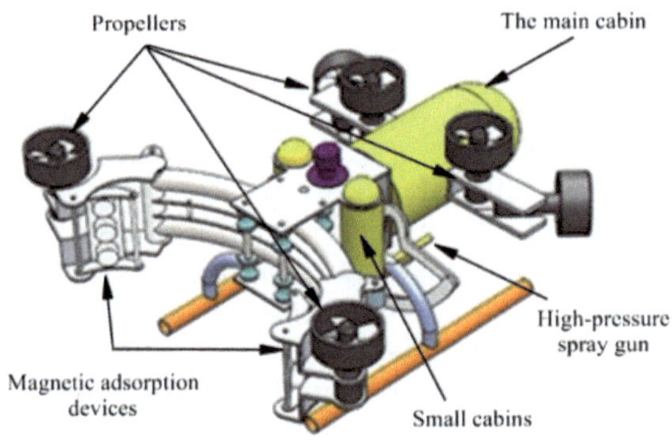

Fig. 1. 3D model of the research object.

2.2 Governing Equation

There exist two methods to describe fluid motion: continuum method based on macroscopic level, from which Navier-Stokes equation is derived; Statistical mechanics method based on microscopic level, from which Boltzmann equation is derived [10]. The basic idea is to construct a simplified dynamic model, which satisfies the basic physical laws of mass conservation, momentum conservation and energy conservation on the macro level, but dynamic calculation for a large number of molecules on the micro level is not performed. Compared with other traditional CFD methods, the advantages of this method is that fluid interaction is simpler to describe, complex boundary is easier to set, parallel calculation is easier to perform and program is easier to implement.

Evolution Equation. The lattice Boltzmann transport equation is defined as follows:

$$\frac{\partial f_\alpha}{\partial t} + e_\partial \cdot \nabla f_\alpha = \Omega_\alpha, \alpha = 1, \ldots, b, \tag{1}$$

where f_α is the particle distribution function in the direction α, e_α the corresponding discrete velocity and Ω_α the collision operator which is used to replace the collision term in Boltzmann equation.

Collision Operators. Asingle-relaxation time based on the Bhatnagar-Gross-Krook (BKG) approximation is a common approach because of its simplicity. However it has a poor performance even numerical instabilities in high Mach number applications. In order to over the limitations of the BGK, the collision operator based on a multiple relaxation time scheme is applied in the simulation in this work where the collision process is carried out in moment space instead of the usual velocity space

$$\Omega_\alpha^{MRT} = M_{\alpha\beta}^{-1} \hat{S}_{\alpha\beta} \left(m_\alpha^{eq} - m_\alpha \right) \tag{2}$$

Where $M_{\alpha\beta}$ is the transformation matrix, the collision matrix $\hat{S}_{\alpha\beta}$ is diagonal and m_α^{eq} is the equilibrium value of the moment m_i [11].

Raw moments and the central moments can be respectively defined as

$$\mu x^k y^l z^m = \sum_\alpha^N f_i e_{\alpha x}^k e_{\alpha y}^l e_{\alpha z}^m \tag{3}$$

$$\tilde{\mu} x^k y^l z^m = \sum_\alpha^N f_\alpha (e_{\alpha x} - u_x)^k \left(e_{\alpha y} - u_y\right)^i (e_{\alpha z} - u_z)^m \tag{4}$$

Turbulence Model. The Large Eddy Simulation (LES) applied to turbulence modeling is the wall-adapting local eddy viscosity model which provides a consistent local eddy-viscosity and near wall behavior.

2.3 Valid Condition

Simulation is performed in CFD software Xflow which is a new generation of fluid dynamics simulation software from Next Limit Technology Company based on particles and the complete Lagrange function [12]. In calculation, Large-Eddy simulation (LES) is adopted for the turbulence model. The scale of solution domain is $3 \times 3 \times 6$ m; the viscosity model is Newton viscosity model and the dynamic viscosity is 1.14×10^{-3} Pa·s; resolved scale is 0.03 m and refinement algorithm was adopted; target resolved scale is 0.015 m and wake resolution is 0.015 m. The 3D model and simulation domain is shown in Fig. 2.

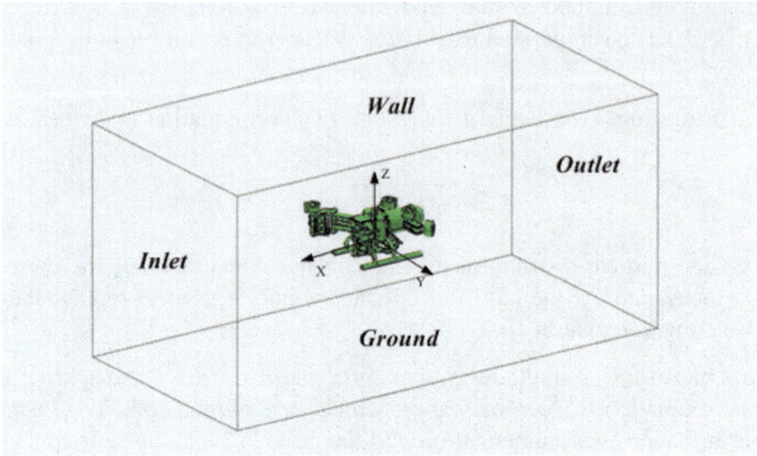

Fig. 2. 3D model and simulation domain

3 Hydrodynamic Simulation Test

3.1 Pure Heave Motion and Pure Pitch Motion Simulation Test

In this paper, the pure pitch and pure heave motion of PMM experiment were conducted as an example and the corresponding inertial hydrodynamic coefficients and viscous hydrodynamic coefficients were obtained.

The pure heave motion of the model was expressed as:

$$\begin{cases} \theta = \dot{\theta} = 0 \\ \xi = a \sin \omega t \\ w = \dot{\xi} = a\omega \cos \omega t \\ \dot{w} = -a\omega^2 \sin \omega t \end{cases} \tag{5}$$

where θ, $\dot{\theta}$-angle and angular velocity rotated along y axis of AUV mode;

ξ- vertical displacement of AUV mode;
a- amplitude of the heave motion of AUV model;
ω- circular frequency of heave motion of AUV model;
$w, \dot{w}$- vertical velocity and acceleration of the heave motion of AUV model.

The pure pitch motion of the model was expressed as:

$$\begin{cases} w = \dot{w} = 0 \\ \theta = \theta_0 \sin \omega t \\ q = \dot{\theta} = \theta_0 \omega \cos \omega t \\ \dot{q} = -\theta_0 \omega^2 \sin \omega t \end{cases} \tag{6}$$

where θ, $\dot{\theta}$- angle and angular velocity rotated along y axis of AUV mode;

$q, \dot{q}$- angle and angular velocity in pitch motion of AUV mode;
ω- circular frequency of pitch motion of AUV model;
$w, \dot{w}$-vertical velocity and acceleration along the z direction of the local coordinate.

In the simulation, different frequency values ω were adopted for the pitch motion and heave motion (0.2 Hz, 0.25 Hz, 0.4 Hz, 0.5 Hz, 0.6 Hz, and 0.8 Hz); the amplitude of heave motion was set as a = 0.04 m; each simulation runs for 8 s. Time step model and Courant were respectively set as Fixed automatic and 1. The motion of AUV can be regarded as a combined motion along the horizontal plane and the vertical plane. The relative velocity of AUV in the horizontal direction along x-axis was -1 m/s which was represented by the inlet velocity of the simulation domain.

3.2 Data Processing

The mean value of the last three cycles stable data obtained in different simulation frequency is used for post-processing. Figure 3(a) shows the contour of fluid velocity at the end of the pure heave motion simulation with circular frequency 0.5 Hz. Figure 3(b) shows the contour of fluid velocity at the end of the pure pitch motion simulation with circular frequency 0.5 Hz. Figure 3(c) shows that the value of stability parameter of the simulation is stable between 0 and 1, which proves that the simulation result is reliable.

Results derived from the simulation calculation are a series of discrete points of force and moment with regard to *t*. Dealing with the result data as Fourier fitting using Matlab software yields

$$f(t) = K_a \cdot \sin \omega t + K_b \cdot \cos \omega t + K_0. \tag{7}$$

The corresponding hydrodynamic coefficients can be obtained by respectively comparing Eq. (4) with the force and moment functions of the pure heave motion and the pitch motion:

The pure heave motion

$$Z = Z_{\dot{w}} \sin \omega t + Z_w \cos \omega t + Z_0 \tag{8}$$

$$M = M_{\dot{w}} \sin \omega t + M_w \cos \omega t + M_0 \tag{9}$$

where $Z_{\dot{w}} = \frac{K_a}{a\omega^2}$; $Z_w = -\frac{K_b}{a\omega}$; $M_{\dot{w}} = \frac{K_a}{a\omega^2}$; $M_w = -\frac{K_b}{a\omega}$,

and the pure pitch motion

$$Z = Z_{\dot{q}} \sin \omega t + Z_q \cos \omega t + Z_0 \tag{10}$$

$$M = M_{\dot{q}} \sin \omega t + M_q \cos \omega t + M_0 \tag{11}$$

where $Z_{\dot{q}} = \frac{K_a}{\theta_0\omega^2}$; $Z_q = -\frac{K_b}{\theta_0\omega}$; $M_{\dot{q}} = \frac{K_a}{\theta_0\omega^2}$; $M_q = -\frac{K_b}{\theta_0\omega}$.

Figure 4(a) and (b) show respectively the Fourier expansion fitting diagram of the heave motion force and torque; (c) and (d) show respectively the Fourier expansion fitting diagram of the pitch motion force and torque.

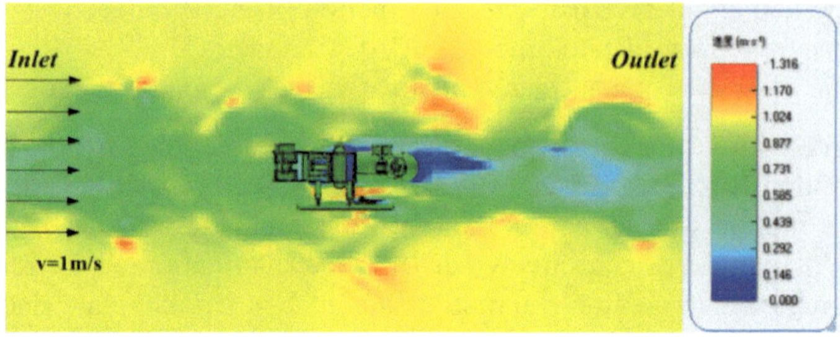

(a)

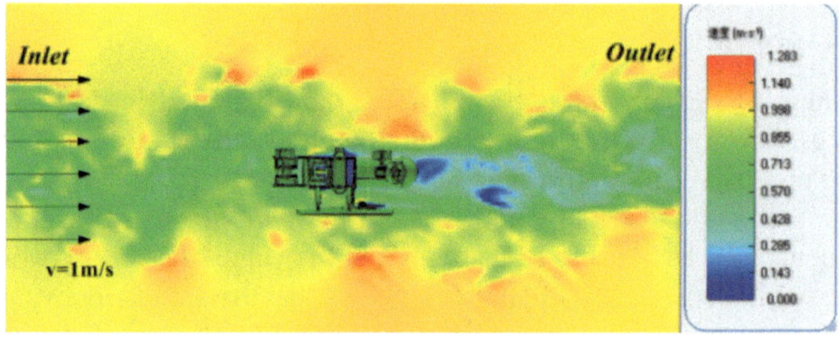

(b)

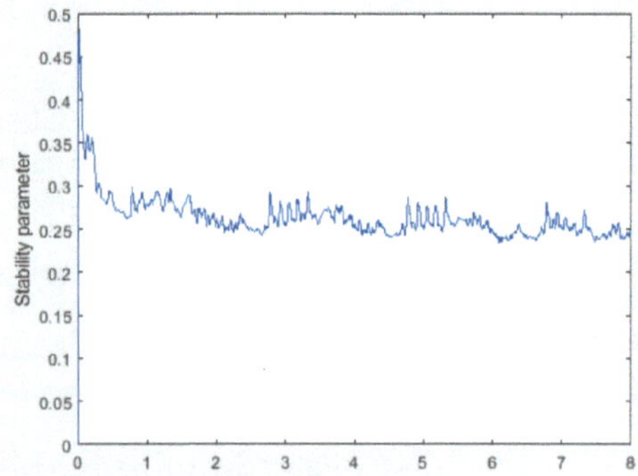

(c)

Fig. 3. (a) contour of fluid velocity of the pure heave motion at $\omega = 0.5$ Hz and $t = 8$ s; (b) contour of fluid velocity of the pure heave motion at $\omega = 0.5$ Hz and $t = 8$ s; (c) stability parameter curve

The following 8 hydrodynamic coefficients can be obtained by Fourier fitting results: $Z_{\dot{w}}$, Z_w, $M_{\dot{w}}$, M_w, $Z_{\dot{q}}$, Z_q $M_{\dot{q}}$, M_q (see Table 1).

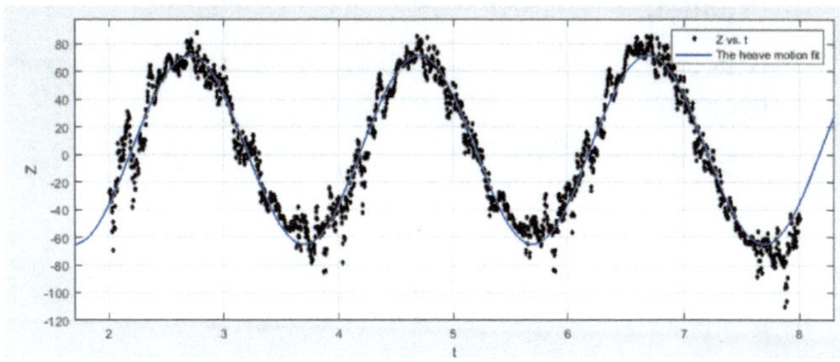

(a)

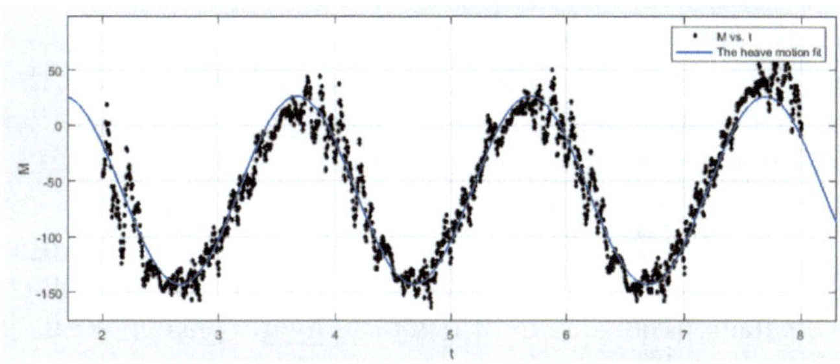

(b)

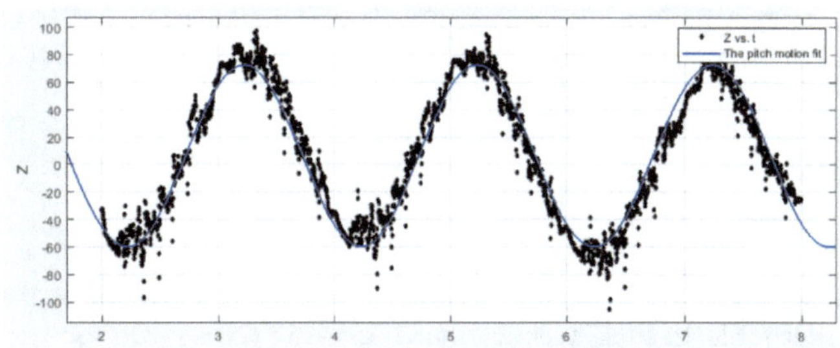

(c)

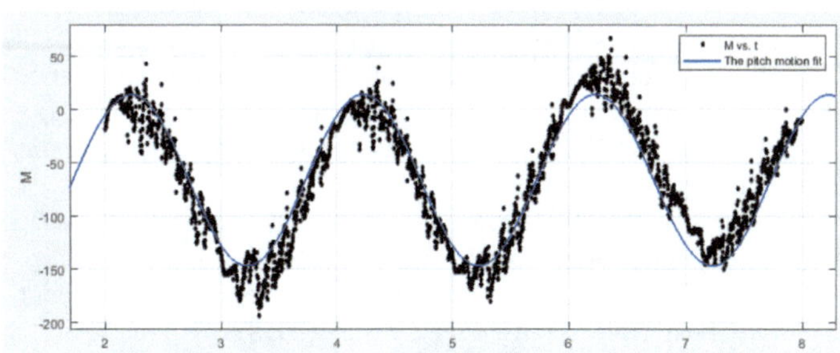

(d)

Fig. 4. (a) the heave motion force Z; (b) the heave motion moment M; (c) the pitch motion force Z; (d) the pitch motion moment M

Table 1. Calculated hydrodynamic coefficients

Hydrodynamic coefficients	Value	Hydrodynamic coefficients	Value
$Z_{\dot{w}}$	−102.82	$Z_{\dot{q}}$	−32.5
Z_w	−437.02	Z_q	−133.07
$M_{\dot{w}}$	99.27	$M_{\dot{q}}$	42.21
M_w	597.37	M_q	155.71

4 Conclusions

In this paper, the pure heave motion and pure pitch motion of PMM test were simulated in CFD software Xflow which is based on lattice Boltzmann method, and part of hydrodynamic coefficients of an AUV with a complex shape were obtained. The accuracy of simulation results is no more depended on the quality of meshing and parameters setting, and the workload of calculating hydrodynamic coefficients of underwater robot by CFD simulation is significantly reduced. It is just an attempt to acquire the hydrodynamic coefficients of underwater robot based on lattice Boltzmann method and the simulation calculation of complete hydrodynamic coefficients and the verification of accuracy will be carried out in the following work.

Acknowledgements. Grateful acknowledgements are given to the Major Scientific and Technological Innovation Project of Shandong Province with Grant No. 2017CXGC0923, and Key Research and Development Program of Shandong Province Grant No. 2017GGX30112 and 2018GGX103025.

References

1. Yu, M., Zhang, X., Chen, Z.: Summary of Autonomous Underwater Vehicle (AUV) technology. Mech. Electr. Eng. Technol. **46**(8), 155–157 (2017)
2. Eng, Y.H., Lau, W.S., Low, E.: Estimation of the hydrodynamics coefficients of an ROV using free decay pendulum motion. Eng. Lett. **16**(3), 326–331 (2008)
3. Caccia, M., Indiveri, G., Veruggio, G.: Modeling and identification of open-frame variable configuration unmanned underwater vehicles. IEEE J. Oceanic Eng. **25**(2), 227–240 (2000)
4. Dong, M., Zhang, X., Yu, C., Gui, H.: Research development on calculation of underwater vehicle's hydrodynamic coefficients. Ship Sci. Technol. **41**(1), 1–6 (2019)
5. Zhou, G., Ou, Y., Gao, X., Liu, R.: Progress and prospect of maneuverability prediction for submarine. Hipbuilding China **59**(3), 203–214 (2018)
6. Xu, M., Feng, Z.: Rotational hydrodynamic calculation of complex-shaped underwater vehicle. J. ShangHai Jiao Tong Univ. **52**(7), 764–769 (2018)
7. Zhang, F.: Research on prediction method of hydrodynamic coefficient of submarine maneuverability. China Water Trans. **18**(1), 13–15 (2018)
8. Li, Z., Tao, J., Luo, Y.: Hydrodynamic coefficients calculation of a complex-shaped underwater robot y simulation and prototype testing. In: 2017 2nd International Conference on Advanced Robotics and Mechatronics, pp. 474–478. Institute of Electrical and Electronics Engineers Inc, Washington (2017)

9. Gwangsoo, G., Hyung, T.A.: Hydrodynamic derivative determination based on CFD and motion simulation for a tow-fish. Appl. Ocean Res. **82**, 191–209 (2019)
10. Zhou, L.: Applied study of lattice Boltzmann method for the computational fluid dynamics. Dalian University of Technology (2013)
11. Holman, D.M., Brionnaud, R.M., Abiza, Z.: Solution to industry benchmark problems with the lattice-Boltzmann code XFlow. In: 6th European Congress on Computational Methods in Applied Sciences and Engineering, pp. 6809–6824. DAAAM International Vienna, Vienna (2012)
12. Wu, P., Chen, C., Zheng, X.: Comparative analysis of preliminary estimation methods for planning boat resistance. Shipbuilding Technol. Res., pp. 63–69 (2018)

Study on the Development of Toner Xerography Digital Printing Technology

Gaosheng He[1] and Yan Li[2,3,4(✉)]

[1] Beijing People's Electric Plant Co., Ltd., Beijing 102600, China
591526804@qq.com
[2] Beijing Key Laboratory of Digitization Printing Equipment, Beijing 102600, China
137330424@qq.com
[3] Printing Equipment Engineering Research Center of Beijing Higher Education Institute, Beijing 102600, China
[4] Beijing Printing Electronic Engineering Technology Research Center, Beijing 102600, China

Abstract. Xerography digital printing technology is one of the main technologies of digital printing. Among them, toner xerography technology plays an important role in the invention and development of xerography digital printing technology. In order to study the origin and development of toner xerography technology, from the social needs Departing, on behalf of patented technology and representative products, by search of US patents related to toner xerography technology from 1938 to 2019 on the "State Intellectual Property Office" patent search platform and the research on technology and product development history of the manufacture of toner xerography digital printing equipment manufacturers, on the basis of analyzing the technology of patented and representative products of toner xerography, summarized the development history of toner xerography digital printing technology.

Keywords: Digital printing · Toner xerography · Patents · History of development

1 Introduction

Beginning with the invention of typography printing in the 11th century (Chinese Northern Song Dynasty), printing technology entered the digital stage after 10 centuries of development and change. Digital printing is gradually replacing traditional printing and becoming the core productivity of the current printing field. At present, the digital printing field is dominated by two core technologies of xerography and inkjet imaging. Xerography technology is the main force of production digital printing at this stage because of its close proximity to offset printing. The application of toner xerography technology is more for a wide range [1, 2]. The development of toner xerography digital printing technology can be divided into four stages, namely, germination period, invention period, development period and maturity period.

D. Wang et al. (Eds.): MeTrApp 2019, MMS 79, pp. 576–585, 2020.
https://doi.org/10.1007/978-981-15-0142-5_56

In order to solve the social contradiction of obtaining multiple copies of the same documents in life production with Time-consuming and labor-intensive, people sought a technique and equipment capable of rewriting and invented carbon paper and type-writers. The common use of carbon paper and typewriters realized early Non-manually rewritten, in order to further improve the scope of replication copying speed and copy quality, Chester Carlson invented the electrostatic copying technology, "Haloid" company based on Carlson's continuous development and laid The basis of modern toner xerography digital printing technology [3]. The realization of technology takes products as the carrier, and the product is the milestone of technology development. The development of technology must be accompanied by the application of patents. Therefore, patents are the tools for the technology development research; to research development of patents in the field of toner xerography technology is a feasible solution for its technology development research.

2 Principle of Toner Xerography Digital Printing Technology

The core of toner xerography digital printing technology is the light guiding property of photoconductor, The photoconductor is charged under the dark environment, and then the photo-exposure material is scanned by the light-emitting material carrying the image information to form an electrostatic latent image, the exposure and non-exposure of the photoconductor imaging surface will generate a potential difference, and the electrostatic latent image has a certain potential, The electrostatic attraction between the poor toner and the electrostatic latent image transfers the toner to the photoconductive drum, and corona treatment of the paper causes a certain potential difference with the developed toner on the photoconductive drum, thereby forming an electrostatic attraction, Then, by means of heat and pressure, the toner is transferred to the substrate to form a stable image. The principle of toner xerography digital printing technology is shown in Fig. 1 [2–4].

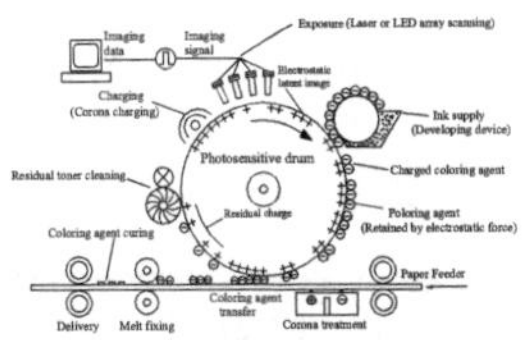

Fig. 1. Principle of digital electrostatic printing technology for toner xerography [3]

3 The Budding Period of Digital Printing Technology (1714–1938)

The invention of printing solves the contradiction between human demand for civilization communication and lack of convenient recording technology, but it cannot solve the need of human beings to present their ideas on the page immediately, nor can it meet the needs of convenient copying of office documents. The emergence of typewriters and carbon papers make this contradiction was alleviated and eventually solved almost perfectly due to the emergence of digital printing technology. In 1714, the British engineer Henry Mill applied for a patent on a writing machine. As far as the current verification is concerned, it can be considered as the earliest technical record of the text output device. In 1806, British Wedgwood invented the carbon paper and obtained the patent on the "copy letter file device" [6]. In 1808, the world's first typewriter was invented by the Italian Pellelinito, and then the Gengens Machinery Company in Copenhagen, Denmark, and the American Christopher Latham Sholes also developed corresponding Typewriters, but they have not officially entered the market and sold in large quantities, but the invention of C. Sholes played a huge role in the development of typewriters, known as the father of printers. The first to enter the market and succeeded was the 26-key typewriter invented by the Jost, original Shoals of C. Sholes. It officially entered the market in 1874 and brought the era of typewriters. After that, the typewriter was combined with carbon paper, to generate multiple copies of a document. However, the problem of low efficiency and the inability to print and copy slightly complex graphics has led to digital printing technology being pursued and developed [7]. The earliest digital printing technology was the Siphon recording device based on the siphon principle invented by William Thomson in 1858. This is also the origin of modern inkjet digital imaging technology. Compared with the invention of inkjet imaging digital printing technology, the xerography digital printing technology's invention was about 81 years late, the xerography digital printing technique was invented in 1938 by American physicist and lawyer Chester Carlson [8].

4 The Invention Period of Toner Xerography Technology (1938–1948)

In 1785, the French physicist Charles-Augustin de Coulomb discovered Coulomb's law, one of the core principles of xerography. In 1937, Bulgarian physicist and academician of the former Soviet Union Academy of Sciences Georgi Nadjakov discovered the light guiding effect, another core principles of xerography. In the 1930s, most of the document reproduction was manual copying using carbon paper, typewriters with carbon paper, and silver salt photography. At the time, Chester Carlson, an American worked at the patent office, often needed to prepare for a lot of patent literature. Manual copying using carbon paper and typewriters with carbon paper are inefficient and labor intensive, and the silver salt photocopying operation is inconvenient and costly. In order to solve the above text copying problem, after years of research, On October 22, 1938, Chester Carlson invented xerography technology based on Coulomb's law and light guiding effect, and copied the short text "ARTORIAl0-22-38", as shown in Fig. 2. On April 4, 1939, Chester Carlson submitted a patent application with the publication number US2297691A, representing the initial invention of xerography technology. The patent information is shown in Table 1. The patent has a total of 27 claims. In the patent, Carlson detailed description of charging, exposure, development, transfer, etc. In xerography, the patent drawing is shown in Fig. 3 [8–11].

Table 1. Patent Information of US2297691A

Invention name	Application number	Application date	Patent number	Public day	IPC	Assignee	Inventors	Priority number	Priority date
Electrophotography	US19390265925	1939.04.04	US2297691A	1942.10.06	G03G15/26; G03G5/082; G03G9/08	CARLSON CHESTER F	CARLSON CHESTER F	US26592539	1939.04.04 14

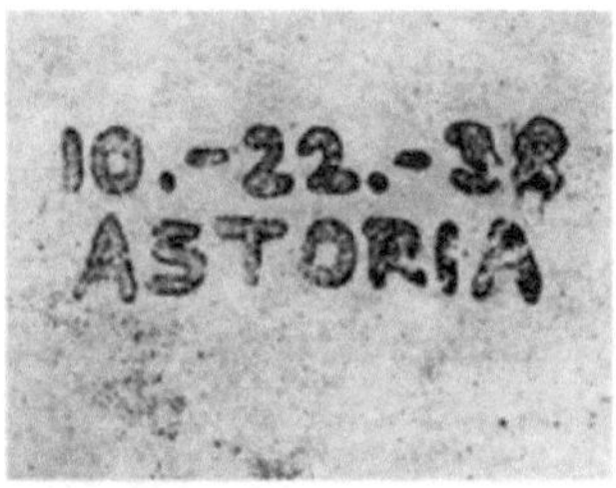

Fig. 2. The first copy of xerography

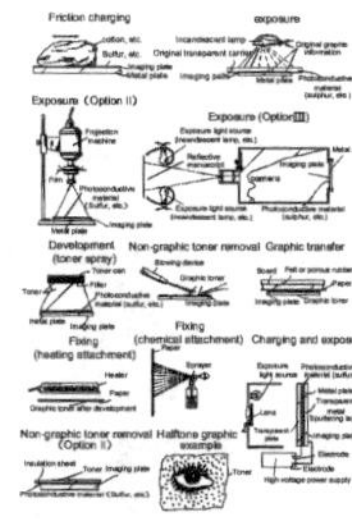

Fig. 3. The drawing of US2297691A [11]

At the end of 1940, "Haloid" purchased Carlson's patented technology and started commercial development [7]. "Haloid" developed a new xerography process in 1948 after Carlson invented xerography technology for 10 years and applied for a patent number US2588675A, marking the invention of modern xerography technology [8–10, 12].

5 Development Period of Toner Xerography Technology (1948–2002)

After 1948, "Haloid" continued to invest in xerography technology, especially residual charge elimination, photoconductive imaging plate and copiers. In 1949, "Haloid" invented a photoconductive imaging plate and manufacturing method, US Pat. No. 2663636A. In 1950, "Haloid" invented a residual static elimination device, US Pat. No. 2777957. In 1950, "Haloid" invented a method for producing a photoconductive imaging plate, US Pat. No. 2657152A. In 1950, based on the above patented technology, "Haloid" invented the world's first copier "Model A", US Pat. No. 2600580A, the patent drawing is shown in Fig. 4, "Model A" spent 2–3 min copying a page and used a special copy paper [8–11].

Although "Model A" realized the convenient copying of text, its slow copying speed and the process of using special copy paper could not meet the growth of office copying requirements at that time. For this reason, "Haloid" continued to invest in research and development, In 1959, "914" ordinary paper copier was invented, US Pat. No. 2945434A, patent drawing is shown in Fig. 5, "914" created the technology of ordinary paper copying, and the copying speed had also been improved to 6 pages per minute, "Haloid" company gradually changed its name to Xerox according to the analysis of the patent applicant name since 1959 [3, 5, 8–10, 12].

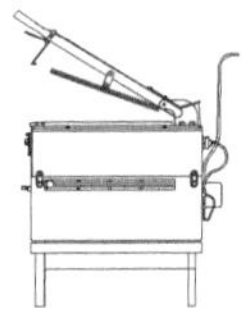

Fig. 4. The drawing of US2600580A

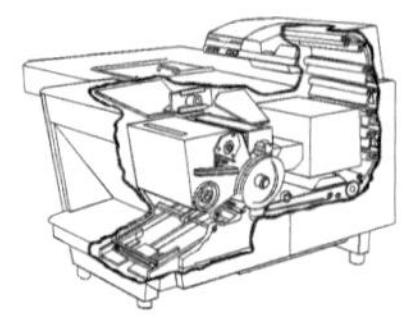

Fig. 5. The drawing of US2945 434A

Since 1948, Xerox has applied for a large number of patents to protect its electrostatic copying technology. Patent protection had made Xerox the dominant player in the field of copiers. It has also limited the progress of xerography technology to some extent. It was not until 1970 that xerography technology core patents expired, giving other companies an opportunity in 1971, Canon launched the "Canon NP-1100" black and white copier based on the improved photoconductive imaging plate imaging version, US Pat. No.3784297A, the patent drawing is shown in Fig. 6 [3, 5, 8–10, 14, 15].

The addition of competitors promoted Xerox's technology research and development. In 1971, Xerox invented an automatic development control method, US Pat. No. 3754821A. In 1973, Xerox invented a color standard and method for calibrating a color electrostatic copying machine, US Pat. No. 3799668A. In 1973, Xerox invented an image density control device, US Pat. No. 3815988A. Based on the above three core inventions. IN 1973, Xerox invented a color electrostatic copying machine "6500", US Pat. No.3869203A, and the patent drawing is shown in Fig. 7 [3, 5, 8–10, 14, 15].

Fig. 6. The drawing of US3784 297A

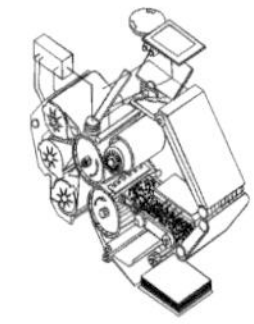

Fig. 7. The drawing of US3869203A

Since 1970, the emergence of large-scale and very large-scale integrated circuits has promoted the development of computers, especially Personal computers. Printers have become another widely used graphic output device after copiers, benefiting from the familiarity of the computer market. IN 1974, IBM invented the world's first laser printer "IBM3800", mainly for continuous printing, US Pat. No. 4031519A, and the patent drawing is shown in Fig. 8 [3, 5, 8–10, 14, 15].

Lasers have been used for expose light sources in the early stage of xerography technology. However, with the advancement of xerography technology, traditional solid or gas lasers have large volume, complicated structure, low efficiency and high price, which restrict the development of xerography technology. The high efficiency, small size and low cost characteristics just make up for a series of problems in the xerography technology of traditional lasers. In 1987, a patent on Electrophotography printer comprising an exposure energy correcting means for the optical character generator was applied for by Siemens, US Pat. No. 4780731A, The LED exposure light source technology is fully disclosed for the first time in the patent, and the patent drawing is shown in Fig. 9 [3, 5, 8–10, 14, 15].

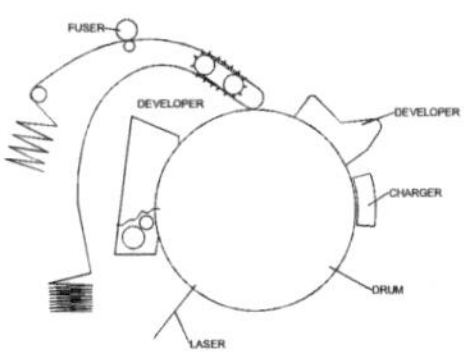

Fig. 8. The drawing of US4031519A

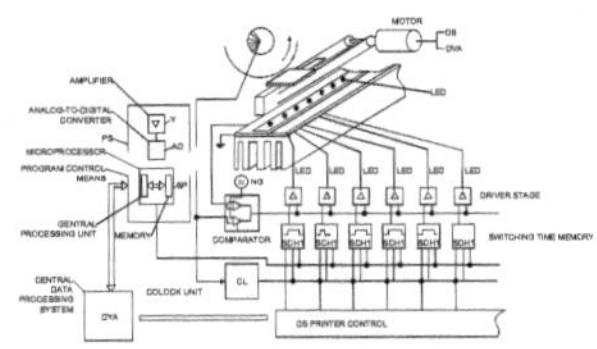

Fig. 9. The drawing of US4837613A

In November 1982, Adobe was founded by Warnock and the leader during of work at Xerox, Chuck Geschke. Adobe finally decided to design a postscript print controller. In 1988, "Postscript" page description language was applied for a patent for invention, US Pat. No. 4837613A, patent drawings shown in Fig. 10. Postscript promoted the development of copying and printing technology [9, 10, 16].

The emergence of photoconductive imaging plate, computer and page description languages has solved the core problem of xerography technology. The rise of personalized printing and direct mail printing has made digital printing a new demand. At the end of the 20th century, xerography technology basically met the needs of ordinary commercial printing in terms of printing quality and process convenience, except printing speed. In 1990, Xerox invented a high-speed electrostatic digital printing machine for paper feeder, US Pat. No. 5081595A. In order to improve the printing speed of xerography digital printing technology, the patent drawing is shown in Fig. 11. In the same year, Xerox launched the world's first black and white digital printing machine "DocuTech135" based on high speed paper feeding technology. The printing speed was 135 pages per minute. At this point, it means toner electrostatic imaging digital printing technology has entered the commercial printing field [3, 5, 8–10, 14, 15].

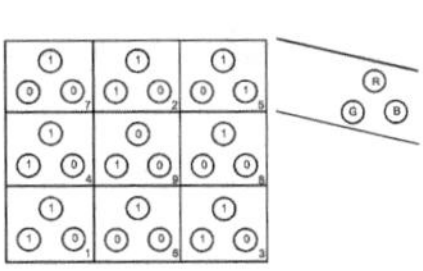

Fig. 10. The drawing of US4780 731A

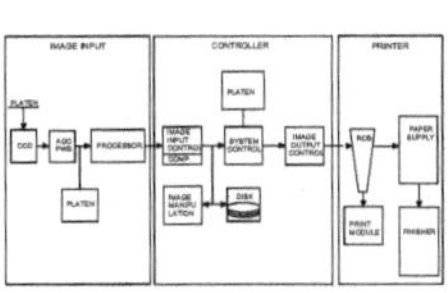

Fig. 11. The drawing of US5081595A

With the development of society and economy, color digital printing presses have become a new demand. In 1992, Xerox invented color printer calibration architecture, US Pat. No. 5,305,119A. In 1993, Xerox invented a structured image format for describing complex color raster images, US Pat. No. 5485568A. In 1994, Xerox invented a hybrid quantification method for color document reproduction, US Pat. No. 5,394,252 A. In 1995, Fuji Xerox invented a print job scheduling system, US Pat. No. 6213652B1. Based on the above technology, Xerox invented the world's fastest color digital press "Docucolor 4040" in 1995. From the core patent analysis integrated by Xerox "Docucolor4040", the main research and development direction of Xerox had shifted from basic mechanism, principle and function to color reproduction control, printing precision improvement, speed improvement and printing management. As the leader of xerography technology, Xerox's research direction represents the key technology of toner xerography digital printing begins to mature [3, 5, 8–10, 14, 15].

After 1995, xerography digital printing technology was developed in the direction of improving the quality of printed products and optimizing system operation. To this end, Xerox has invented a number of corresponding technologies. For example, in 1998, Xerox invented a system and method for directional sharpness segmentation resolution compression and decompression, US Pat. No. 6771827B1; in 1999, Xerox invention a printing processing job scheduling system, US Pat. No. 7630092B1; in 1999, Xerox invented a printing processing job scheduling system, US Pat. No. 6606163B1; in 1999, Xerox invented a printing engine scheduling method and Device, US Pat. No. 6850336B1; in 1999, Xerox invented a melting unit cleaning device, US Pat. No. 6215975B1; in 1999, Xerox invented a method for dynamically setting air system pressure based on real-time form acquisition time data System and method, US Pat. No. 6,279,896 B1; in 1999, Xerox Corporation invented a feedback control system for toner concentration in an imaging system, US Pat. No. 6167213A. Based on the above patents, in 2000, Xerox released the "Docucolor 2000" series of color toner xerography digital presses containing "Docucolor 2045" and "Docucolor 2060" [3, 5, 8–10, 14, 15].

Since 2000, xerography digital printing technology has been developing towards intelligent, high-efficiency and high-quality printing. In 2001, Xerox invented an object-optimized printing system and method, US Pat. No. 6327043B1; in 2001, Xerox invented a method and system for cleaning and preparing a printer, US Pat. No.6517066B2. Based on the above objectives and invention patents, Xerox introduced "Smartpress" technology and DocuColor iGen3 digital printing machine based on this technology in 2002 [3, 5, 8–10, 14, 15].

6 Maturity of Toner Xerography Technology (2002 – Present)

Since 2002, the development of toner xerography technology has begun to mature. Many companies began to enter the toner xerography digital printing industry and invested in research and development to improve the display size, imaging quality, printing speed, reduce energy consumption, equipment miniaturization, preparation time, and cut costs. A large number of advanced technologies and products have been launched based on the above objectives, as shown in Table 2 [3, 5, 8–10, 14, 15, 17].

Since 2010, the development of toner xerography technology has slowed down. There is no representative technology has been introduced. Instead, some companies

Table 2. Development of toner xerography technology after 2002

Applicant	Time	Invention name	Patent number	Purpose of the invention	Characteristics
Minolta	2002	Toner for developing an electrostatic image	US6475690B2	Increase display size	2400dpi × 600dpi
Océ	2003	Low temperature toner bonding and fixing technology	US7245866B2	Improve image quality	The quality of printed products is close to the offset printing
Delphax	2003	Charge deposition print head and printing method	US5886723A	Increase printing speed	Electron beam exposure
Océ	2004	Gemini print technology	US7266335B2	Increase printing speed	Finishing double-sided printing in one pass
Eastman Kodak	2004	Tandem printing method "NexPresss"	US7502582B2	Improve image quality	Having at least five color printing stations, applying a clear toner to the image after fixing
Ricoh	2006	Toner technology (PXP EQ toner)	US7374851B2	Improve image quality; Reduce energy consumption	Delivers brilliant image chromatic aberration with a minimum toner
Fuji Xerox	2006	LED array head and image recording device	US2007109395A1	Improve image quality; miniaturization	LEDs integrated on one drive circuit
Fuji Xerox	2006	Polymerized toner technology	US7858285B2	Improve image quality; Reduce energy consumption	Smaller toner particles and lower fixing temperature
Ricoh	2007	Laser print head	US8089498B2	Increase the display size	2,400 × 4,800(dpi)
Fuji Xerox	2007	LED white light scanning technology	US7864381B2	Reduce energy consumption Increase speed	It's 1.5 times the brightness of the xenon lamp, and 30% of the power consumption
Océ	2007	Variable speed printing technology	US8231287B2	Increase speed	Fast start and shorter processing times
Fuji Xerox	2008	Tandem printing method	US7894739B2	Increase speed	The printing speed is 220 pages per minute, 30% faster than iGen3
Fuji Xerox	2009	Heating device, fixing device and imaging device	US8041278B2	Reduce energy consumption and preparation time	IH fixing technology
Xerox	2010	Transparent toner	US8859175B2	Improve image quality;	Improve print quality

have turned to inkjet imaging technology even though they originally engaged in toner xerography technology, including Xerox. Through the above research on the patents and representative equipment of toner xerography technology from 1938 to 2010, the development history of toner xerography technology is summarized in Fig. 12 [10, 15].

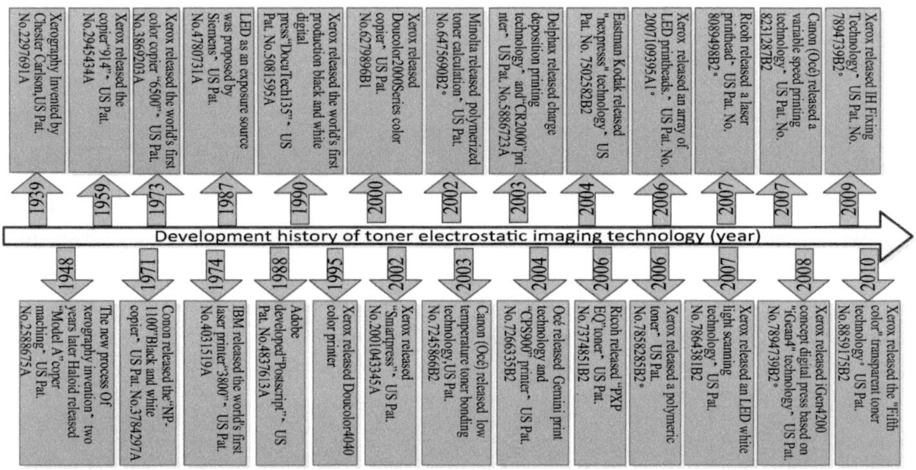

Fig. 12. Development history of toner xerography technology

7 In Conclusion

Through the research on the development history of toner xerography digital printing technology, digital printing technology originated from people's demand for file copying. The initial technical invention of xerography was completed by Carlson. "Haloid" company improved and invented modern electrostatic imaging technology based on Carlson. In the development of toner xerography technology, Xerox (formerly "Haloid", later Fuji Xerox) brought the xerography technology from the invention to maturity, at the same time, ADOBE and IBM made great contributions to the invention of printing software and printers. From 2000 to 2002, the toner xerography technology matured. After that, many companies in this field have invented various new technologies and equipment for improving the display size, image quality, the printing speed, reducing energy consumption, miniaturization, the preparation time, and the cost. In the research of patent search and representative enterprises, core enterprises in the field of toner xerography technology have begun to develop inkjet imaging technology, the development of Landa Nano inkjet technology and the aging of print head MEMS manufacturing technology mean that next-generation inkjet imaging technology will become the future of digital printing technology.

References

1. Wang, C.C.: Development status of electrostatic imaging digital printing. Screen Printing **3**, 44–47 (2011)
2. Qi, F.B.: Electrostatic imaging digital printing technology and its printing machine (Series 1). Print World **1**, 43–46 (2013)
3. Yao, H.G.: Digital Printing. China Light Industry Press, Beijing (2010)
4. Jin, L., Zhou, K.P., et al.: Development and analysis of key technologies of electrostatic imaging digital printing based on technology roadmap. Beijing Inst. Graphic Commun. **3**, 7–10 (2018)
5. Song, H.H.: The history of digital printing growth. Printing Technol. **5**, 36–44 (2007)
6. Yang, Q.Y.: Discussion on carbon paper. In: Proceedings of the 11th Annual Conference of Shandong Papermaking Society, pp. 166–167 (2006)
7. Pingsheng: The development of typewriters. Foreign Sci. Technol. News **06**, 29–30 (1998)
8. Yao, H.G.: The origin and development of digital printing. China Printed Packag. Res. **5**, 1–12 (2010)
9. https://www.fujixerox.com.cn/zh-CN/Company/brand-gene
10. SIPO. http://www.cnipa.gov.cn/
11. Carlson Chester, F.: Electrophotography: US, US2297691A [P]. 1942.10.06
12. Walkup Lewis, E., Copley Harold, E. et al.: Electrocopy apparatus: US, US2588675A [P]. 1952.03.11
13. Zhang, J.: Development and Evaluation of POSTSCRIPT Printing Control Strip. Wuhan University, Wuhan (2005)
14. Yao, H.G.: Review and prospect of high speed electro photographic digital printing. Printing Eng. **4**, 28–31 (2008)
15. Luo, T.Y.: The Discriminating Method of Core Patents and the Application Analysis in the Industrial Field. China Institute of Science and Technology Information, Beijing (2012)
16. Wang, S.Z.: Research on Color Management of Digital Printing Machine Based on Electrostatic Imaging Technology, p. 2. Wuhan University, Wuhan (2005)
17. Lux, R., Yuh, H.-J.: International Conference on Digital Printing Technologies, Final Program and Proceedings. Society for Imaging Science and Technology, United States, pp. 323–327 (2004)

Education on Theory of Mechanisms and Machines

Project-Oriented Learning/Teaching Applied to Theory of Mechanisms and Machines

Guanglei Wu(✉), Delun Wang, Huimin Dong, Yuan Gao, and Feng Qian

School of Mechanical Engineering, Dalian University of Technology, 116024 Dalian, China
gwu@dlut.edu.cn

Abstract. This paper deals with the experience of the learning/teaching on Theory of Mechanisms and Machines carried out in Dalian University of Technology (DUT, China) for the undergraduates in Mechanical Engineering. The project-based learning methodology is adopted, and the learning/teaching approach of DUT is briefly presented. The education aspects, for instance, the textbook, the projects, the learning/teaching mode, multimedia, evaluation, are depicted. Results acquired from the learning/teaching on Theory of Mechanisms and Machines are discussed to show the effectiveness.

Keywords: Theory of Mechanisms and Machines · Problem-based learning · Creative design of mechanisms and machines · Type synthesis · Dimensional synthesis · Performance design

1 Introduction

As a fundamental course for the undergraduates in Mechanical Engineering [5], Theory of Mechanisms and Machines (TMM) plays an important role. It is the bridge between the General Foundation Requisites and Subject Foundation Requisites, which covers the contents of Mathematics, Physics, Theoretical Mechanics, Electrotechnics, Electronics, Engineering Graphics, Computing Essentials, etc. Based upon ten years' curriculum construction and experience of engineering projects, Dalian University of Technology (DUT) has been the leading pioneer in China, regarding the field of the educational activity on TMM teaching, by incorporating the multidisciplinary knowledge. In this paper, the learning and teaching on TMM of DUT is introduced.

The education for the undergraduates in Mechanical Engineering in China lasts four years, where the TMM is arranged in the third year. In order to merge the General Foundation Requisites into the TMM, DUT adopts the methodology of Project-Based Learning (PBL) [1–3] in the TMM teaching, through which the TMM scopes can be covered, for instance, the principle of the mechanism composition, type synthesis, dimensional synthesis, kinematics/statics/dynamics

D. Wang et al. (Eds.): MeTrApp 2019, MMS 79, pp. 589–597, 2020.
https://doi.org/10.1007/978-981-15-0142-5_57

of mechanisms and machines. Besides, the curriculum design and experimental teaching of TMM are synchronously implemented along with its teaching to enhance the learning ability of the students. Henceforth, DUT's education on TMM, including the framework, contents and procedure of TMM curriculum, behaves the innovation and practice of the teaching reform of in mechanical speciality. Here, we will briefly present the learning and teaching of TMM in DUT.

This remaining of this paper is organized as follow. Section 2 introduces the learning/teaching mode of the TMM in DUT. The results of learning/teaching is presented in Sect. 3. The last section concludes the paper.

2 TMM Learning/Teaching in DUT

This section briefly introduces DUT's TMM learning/teaching, namely, textbook, learning/teaching procedure, subjects of the case studies, evaluation, etc.

2.1 Textbook

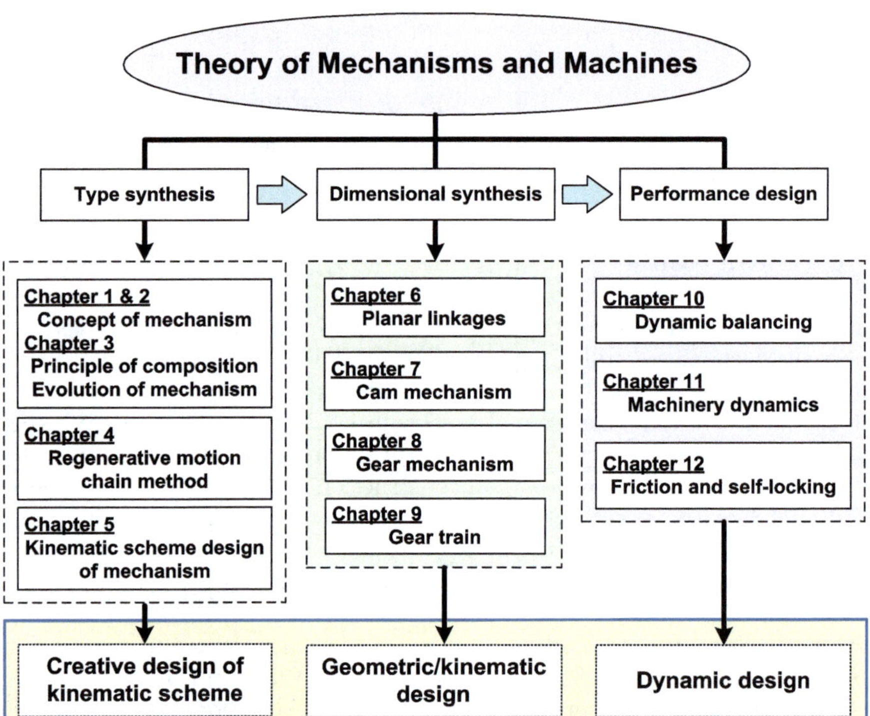

Fig. 1. The contents of the textbook used in DUT's TMM teaching.

Figure 1 shows the contents of the TMM textbook for the DUT undergraduates [7], which is edited by the DUT lecturers. The TMM contents are grouped into three modules, i.e., type synthesis, dimensional synthesis and performance design, of which covers the TMM fundamental knowledge. By virtue of the project-based learning on TMM, this textbook builds the connections among the knowledge in different chapters, instead of the traditional textbooks with the independent knowledge.

2.2 Project Topics

As the key roles of the PBL, the projects adopted in TMM teaching should cover all the contents of the textbook as possible and should be acceptable for the third-year undergraduates. The considerations in the selection of the projects are listed below:

- The mechanism can be found in the daily life, for which the literature review and survey can be readily carried out.
- The mechanism includes the six-bar linkage, applicable for the use of regenerative motion chain method.
- The mechanism is composed of the linkage, cam, gear, and among other mechanisms as much as possible.
- The mechanism consists of the actuation (electronic/hydraulic/pneumatic actuator), transmission and execution mechanisms.

Sequentially, some selected project topics are displayed in Fig. 2. It is noteworthy that it is difficult to ensure all the projects meet the teaching requirements, thus, the missed contents displayed in Fig. 1 in the project will be substituted by the exercises in the textbook or designated by the lecturers.

2.3 Learning/Teaching Mode

The learning/teaching on TMM have been developed around a project in order to apply the so-called PBL methodology. The project implementation is consisted of a theoretical part and associated practical sessions, experimental study and curriculum design that are oriented to the project of the course. The design project involved in the practical sessions and curriculum design along with the textbook teaching and experimental study.

The Gantt chart of the learning/teaching mode is illustrated with Fig. 3. The textbook teaching lasts 14 weeks with four class hours per week. The exercises and curriculum design are assigned along with the teaching progress.

To carry out the project, the students need to apply the knowledge of different fields [6] acquired from the General Foundation Requisites, such as Mathematics, Physics, Theoretical Mechanics, Electrotechnics, Electronics, Engineering Graphics, Computing Essentials, etc. They need to conduct the literature review and grasp the use of the engineering softwares (MATLAB, 3D modeling and FEA softwares, etc.). Finally, each student has to document a technical report following the table of contents of the textbook and show the work-done through a

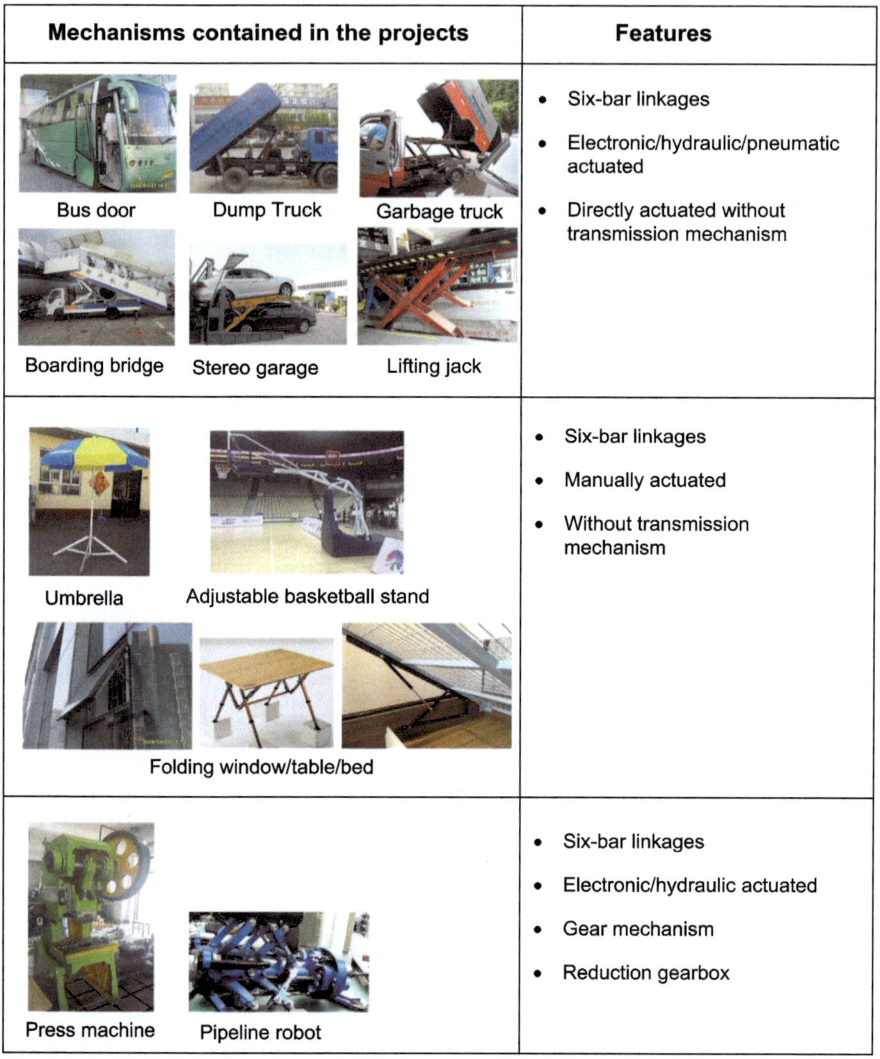

Fig. 2. Some projects of the projects used in DUT's TMM teaching.

public presentation (5-min oral presentation plus 3-min question session by the lecturers, teaching assistants (TAs) and students).

The students have to work in groups of five or six people (depending on the number of the students). Each group selects one project, but with different designs (working principle, mechanism composition, dimensions, application background, design specifications, etc.) for any group member. The group discussion among the members are made to talk about the problems encountered in the exercises and curriculum design together with the design procedure for their projects and technical reports.

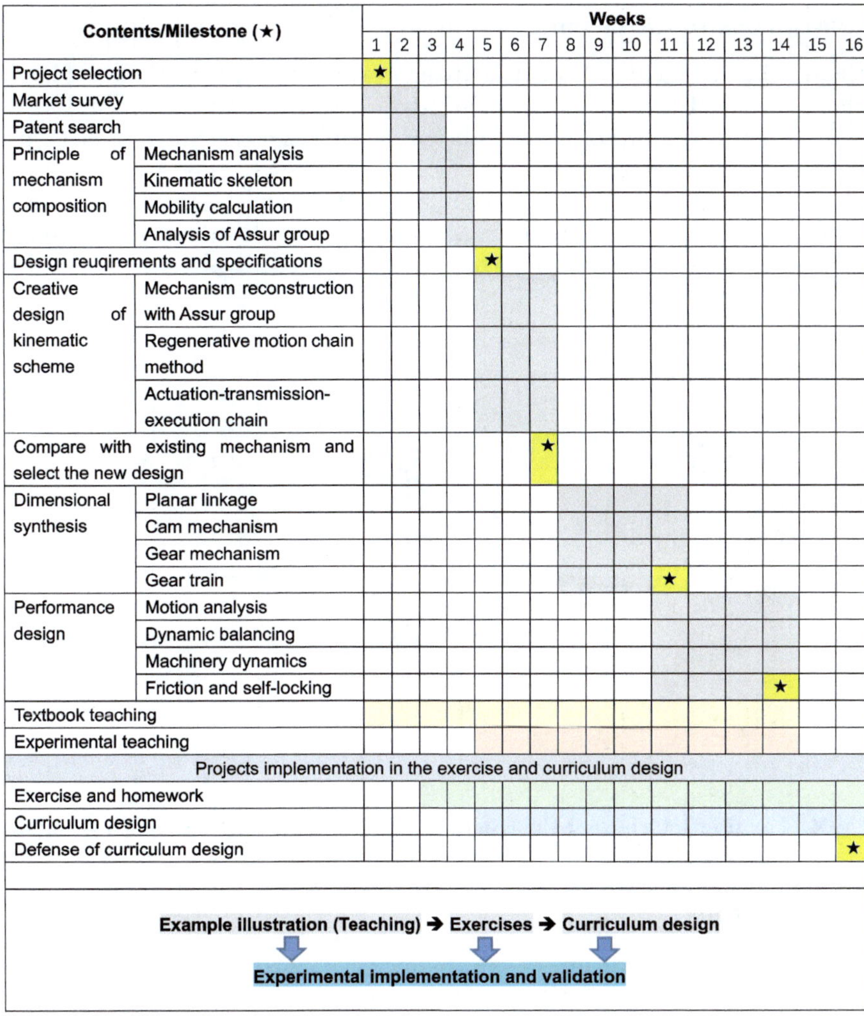

Fig. 3. Gantt Chart of DUT's TMM learning/teaching.

The groups are responsible for preparing their own work plan and dividing the work among the group members. The plan should agree with the teaching schedule and meet the requirements imposed by the lecturers. The lecturers and TAs supervise the students to carry out the project work and organize to check the progress of each group every two weeks.

After the completion of the textbook teaching, the students take two weeks to document the final technical report and presentation slides, and all of them participate a formal defense of their report.

2.4 Multimedia of Learning/Teaching

In the textbook teaching, the APP teaching software on the smart phone as displayed in Fig. 4 is used, besides the traditional lecture with the slides. With the APP, the lecturers post the questions during the teaching to check if the students have learned the knowledge or not. Moreover, this APP can also be used to measure the individual performance in the classroom teaching/exercises, e.g., the response (answers and time consumption) to the questions posted. The students can raise their questions through the APP.

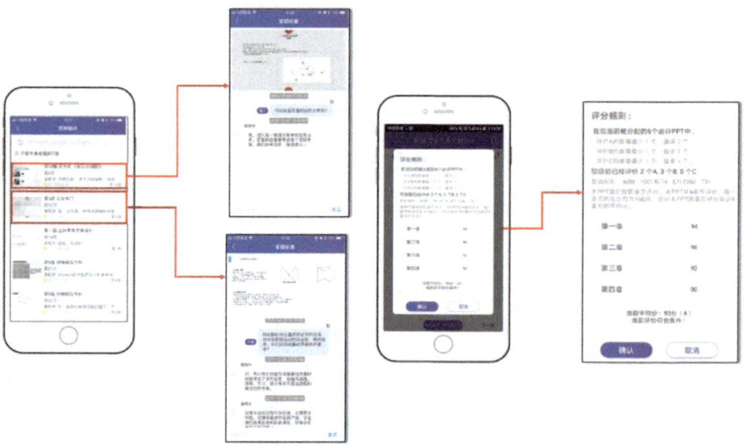

Fig. 4. The user interface of the teaching APP on the smart phone (in Chinese).

2.5 Evaluation

To measure the TMM learning of the students, the total scores (hundred-grading system) Q of each student are calculated by means of the following equation:

$$Q = 40\%Q_e + 40\%Q_c + 20\%Q_t \tag{1}$$

where Q_e, Q_c and Q_t stands for the scores of the final examination, curriculum design and APP evaluations, respectively. The scores of the final examination completely come from the test, and the scores of the curriculum design are calculated by the mean scores from the lecturers, TAs and the remaining students except himself, depending on the quality of the report, the presentation and the group discussion. The scores of the APP evaluation are assessed by the individual performance in the classroom teaching/exercises.

3 Results

Through the project-oriented learning/teaching methodology, the students recall the knowledge acquired from the General Foundation Requisites. The projects attract the students' interest to TMM learning. The procedure of the curriculum design impress the students, to enhance the capability to use the knowledge they have learned. The students are no longer limited to analyze the existing mechanisms or machines, but they can use the methodology acquired in the TMM teaching to create the new designs, for instance, the creative design of the press machine based on the original one as illustrated in Fig. 5.

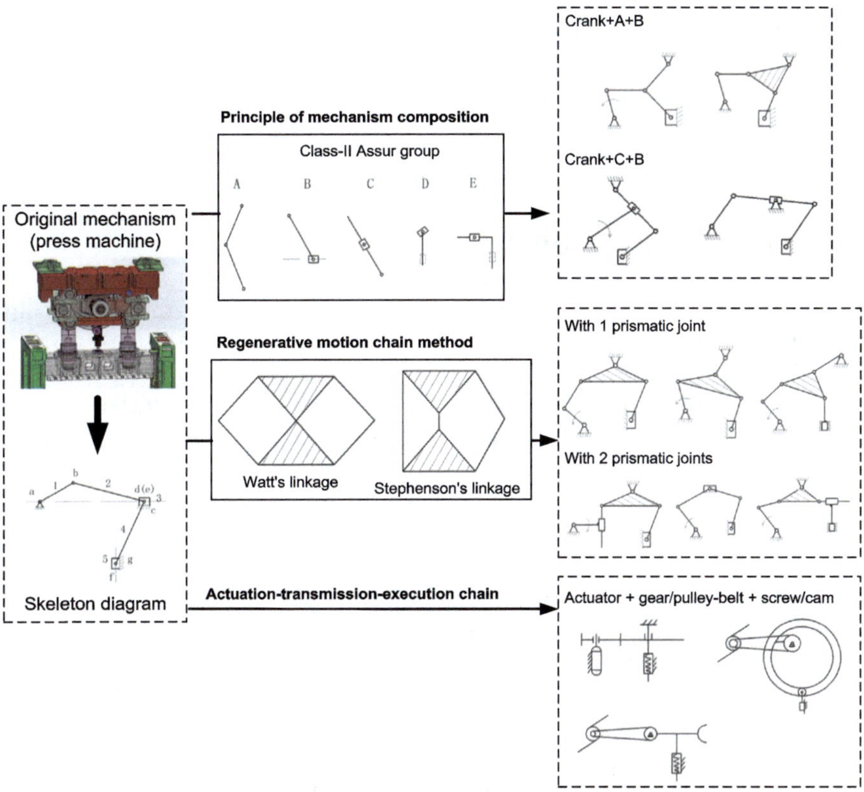

Fig. 5. Some creative designs of the press machine using the methodology in DUT's TMM teaching (more designs are not shown due to the page limit).

The results of the such a learning/teaching methodology shows the effectiveness below:

- The contents of TMM teaching has been innovated, compared to the Chinese conventional textbooks. The projects are oriented to the industrial application, which helps to build the training system of the engineering talents. Creative design of machines has been introduced in the TMM learning.
- The content innovation and teaching methodology can attract the students attentions, which interests the students to learn.

- The practical survey allows the students to acknowledge the market development of the mechanical products, and improves the students' ability of communications.
- The evaluation of the students is composed of the final exams, curriculum design and exercises in the classroom/homework, which can better reflect the study progress of the students. This changes the situation that some students do not study in the daily life until the coming exams.
- With different design tasks, the students has to accomplish their tasks by themselves through the discussion with the lecturers, TAs, and group members, avoiding the duplication in the exercise or homework.
- Most of the students had improved their skills to make technical documentation and oral presentations, compared to those skills obtained in their previous course studies.

4 Discussion and Conclusions

The experience of the learning/teaching on Theory of Mechanisms and Machines carried out in Dalian University of Technology for the undergraduates in Mechanical Engineering has been successful and well evaluated by both students and lecturers. The experience is helpful for other lecturers involved in the undergraduate courses who are now planning to modify their lectures for the upcoming semester.

The effectiveness of such a learning/teaching methodology has been achieved, including the content innovation, methodology of creative design, improvement of the students' study interests, enhanced technical and social abilities, etc. It is also helpful to the students when they encounter the interviews of the Master's entrance examination and job seeking.

However, there are some problems to solve in the future. The current project work is limited to theoretical part, thus, prototyping work can better trains the students, wherein the cost is a major obstacle.

Some unexpected accidents may occurs when the students use the APP in the classroom. The network is sometimes offline and their answers to the questions raised during the lectures cannot be submitted on time, which will influence the evaluation on their exercises.

Another problem is that the duplication may still exist. The creative design for the same project has the similar even same procedures, thus, these parts can be shared among the members of the same group.

Finally, there is another unexpected problem in the same semester: students have dedicated many more hours than initially planned to this subject [4]. Thus, the time they spend on other subjects become less.

Acknowledgement. The authors would like to acknowledge the support from 2018 Central Education Reform Project.

References

1. Barron, B.J., Schwartz, D.L., Vye, N.J., Moore, A., Petrosino, A., Zech, L., Bransford, J.D.: Doing with understanding: lessons from research on problem-and project-based learning. J. Learn. Sci. **7**(3–4), 271–311 (1998)
2. Bell, S.: Project-based learning for the 21st century: skills for the future. Clearing House **83**(2), 39–43 (2010)
3. Blumenfeld, P.C., Soloway, E., Marx, R.W., Krajcik, J.S., Guzdial, M., Palincsar, A.: Motivating project-based learning: sustaining the doing, supporting the learning. Educ. Psychol. **26**(3–4), 369–398 (1991)
4. Noriega, A., Sierra, J., Cortizo, J., Prieto, M., Linera, F., Martín, J.: Project-based learning applied to mechatronics teaching. In: New Trends in Educational Activity in the Field of Mechanism and Machine Theory, pp. 49–56. Springer (2019)
5. Norton, R.L.: Design of Machinery. McGraw-Hill, New York (2004)
6. Soares, F.O., Sepúlveda, M.J., Monteiro, S., Lima, R.M., Dinis-Carvalho, J.: An integrated project of entrepreneurship and innovation in engineering education. Mechatronics **23**(8), 987–996 (2013)
7. Wang, D., Gao, Y.: Theory of Mechanisms and Machines. China Machine Press, Beijing (2014). (in Chinese)

APP Based Teaching/Learning Mode and Quality Evaluation of MMT Course

Huimin Dong(✉), Delun Wang, Yuan Gao, Feng Qian, and Guanglei Wu

School of Mechanical Engineering, Dalian University of Technology, Dalian 116024, China
donghm@dlut.edu.cn

Abstract. With the development and popularization of internet and cellphone, Internet + education has become a trend for the teaching/learning mode reform. To improve teaching/learning quality and efficiency of mechanism and machine theory (MMT) course, an APP teaching/learning mode and quality evaluation approach are presented in this paper, aiming at cultivating excellent engineers in the new era. Based on APP network platform, the function of the APP teaching/learning, data tracking and analyzing are designed, by means of which an interaction and discussion teaching/learning environment for teacher-student and student-student is provided. Thus, the APP intelligent interactive teaching/learning mode with teaching strategy adapting, student-centered autonomous learning and cooperative learning, and the quality evaluation approach of learning process assessment are established. The teaching/learning of MMT course of ME major in Dalian University of Technology shows that stimulating students' initiative in learning and improving teaching quality embodies the feasibility and practicability of application-based teaching mode and evaluation method, which facilitate teachers to master students' learning situation and adjust teaching strategies.

Keywords: Mechanism and machine theory course · Teaching/learning mode · APP network platform · Quality evaluation · CLC number: TH212 · TH213.3 document code: A

With the development of era and the advancement of science and technology, the new generation of information technology based on the Internet has had a wide and profound impact on the production, work and learning of human society [1]. Under the background of "Internet +" as well as the coexistence of opportunities and challenges, it is necessary to combine its own characteristics, recognize new requirements, explore development and innovation, and promote the deep integration and development between the new generation of Internet-based information technology and higher education [2, 3].

The 2018 National Undergraduate Education Work Conference opened a new era of undergraduate education of colleges and universities in China. The Ministry of Education requires [4] to comprehensively improve the ability of university personnel training, pay close attention to the problems of teachers' energy input and students'

D. Wang et al. (Eds.): MeTrApp 2019, MMS 79, pp. 598–606, 2020.
https://doi.org/10.1007/978-981-15-0142-5_58

energy input in teaching/learning, and proposes to strengthen classroom teaching management, establish classroom teaching rules, and strengthen learning process management. Quality is put together to eliminate the low level course and create a high level course to effectively improve the quality of course teaching/learning. How to make use of the advantages of the Internet, it is imperative to reform the traditional teaching mode and explore the new teaching mode that adapts to the educational needs of the new era in light of the problem of higher education personnel training and teaching quality [5, 6].

At present, Internet + education has become a new trend of teaching reform, such as MOOC teaching flip classroom [6], smart classroom [7], APP teaching [8] and so on. Based on the content of the "Type synthesis-dimension synthesis-performance synthesis" of the mechanism and machine theory (MMT) course of Dalian University of Technology (DLUT) [9], and teaching method of the engineering case-based "project-leading and problem-driving" [10], we found that some problems of the current teaching/learning mode of MMT need to be improved: (1) teacher-student/student-student interaction, learning atmosphere; (2) teaching/learning situation tracing; (3) making the adapting teaching strategy with students learning status; (4) the process data comprehensive evaluation; (5) the student's potential release. This paper, in order to solve these problems, the exploration and practice of the new mode and quality evaluation of MMT course based on APP network platform, is implemented to adapt to the training of talents in the new era and to effectively improve the quality of teaching.

According to the characteristics of the internet, this paper, using APP platform, designs a new APP teaching/learning mode for MMT course, which will realize teacher-student interaction, student exchange discussion and teaching/learning data tracing and analyzing. The APP MMT teaching activity and data analysis and process evaluation method are also established based on APP platform. The APP teaching/learning mode has been practiced in the two-term students of the ME majors in DLUT. The practice shows that this new mode and evaluation approach stimulates the enthusiasm of students to learn, and it is convenient for teachers to grasp the students' learning status in time, and then adjust teaching strategies to improve teaching effectiveness and quality.

1 APP Teaching/Learning Mode of MMT Course

Based on the APP network platform, the APP teaching/learning function is designed by giving full play to the role of network + smart phone in teaching, taking students as the center, and breaking through the time/space/equipment limitation of teacher-student/student exchange in teaching.

1.1 Teaching/Learning Function of APP System

Using the APP "Learning to Learn" software, the APP MMT teaching/learning function is built, in this system, the server and data warehouse mode and various mobile terminal device cross-platform technologies are adopted. Besides, the system is also compatible with smartphone Android android and Apple IOS (Internet-working Operating System) dedicated system as is shown in Fig. 1.

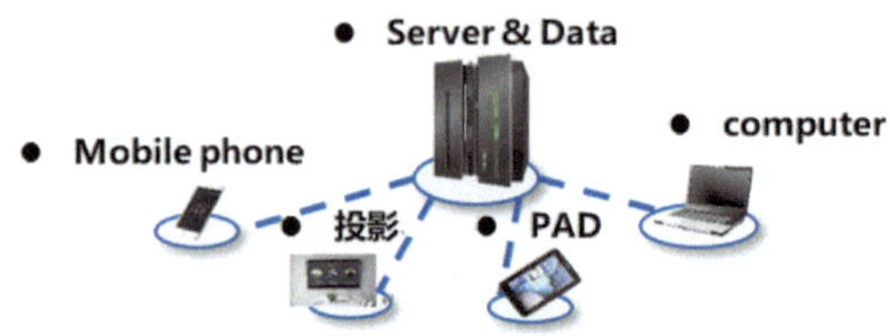

Fig. 1. System framework of the APP network platform

The platform provides three kinds of clients, PPT plug-in, mobile APP and Web, and its teaching function is shown in Fig. 2.

The PPT plug-in provides teachers with three main functions: lesson preparing, teaching and test. In the preparation stage, the teacher can make PPT courseware and set quiz questions. In the teaching stage, the PPT explanation, question, and test modules can be used by teachers, and the teaching status can be also displayed in real time on the client and the relevant data will be upload to a remote database at the same time.

The mobile APP provides students with the functions of synchronous class learning, quiz/query, self-learning management, etc. Students can follow the teacher's lectures in the classroom, receive the questions/tests pushed by the instructors, and submit answers, as well as ask questions. After class, they can choose homework, self-study to participate in the discussion, The mobile APP can record students' learning behavior, view teacher/teaching assistant/same group ratings and evaluations.

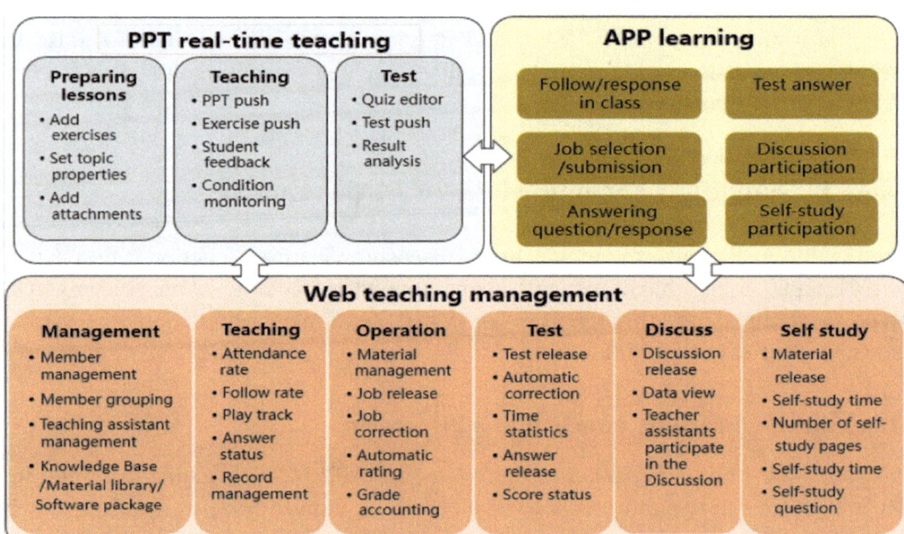

Fig. 2. Teaching/learning function of mobile APP platform system

The web provides teachers with database management, including grouping, random assignment of students' interactive reading, teaching analysis, and adjustment/editing functions. Teachers can view the course's teaching behavior on the web, as well as score and evaluate tests, assignments, discussions, and innovative design activities.

Based on the background of this APP teaching software, teachers can flexibly formulate curriculum teaching modes according to their respective teaching characteristics, organize/manage teaching activities, control teaching effects, and ensure teaching quality with the support of teaching resources.

1.2 APP Teaching/Learning Mode of MMT Course

The content of the MMT course emphasizes the abstraction of the actual machine as the mechanism, the motion scheme, and dimension and performance innovation design on the abstract configuration. "Type synthesis-dimension synthesis-performance synthesis" content system is the feature of the MMT course of Dalian University of Technology, combined with the APP network platform. The paper proposes a new mode of " Engineering project leading - problem driving - capability orientation " MMT APP teaching where this mode focus on learning process = classroom + homework/online learning + case application; highlight communication discussion = homework evaluation + mobile APP exchange/discussion; strengthen flexible application = case analysis in each chapter + modeling calculation + innovation design; focus on ability orientation = ability evaluation + knowledge evaluation.

In teaching/learning, it is required that: in class: understanding concepts + positive thinking + flexible applying; after class: practice/master engineering methods + observation of life + principle application. Each chapter must examine the analysis and design of application cases, the in-class and post-class teacher-student interaction, student interaction/discussion peer response, and the learning atmosphere of the large class teaching group learning, as shown in Figs. 3 and 4.

The specific implementation of the teaching mode:

(1) Before the start of the course (1 week/1 day), students install the mobile APP of MMT course, register and set up learning group.
(2) The first lecture: introduce the teaching content, APP based teaching/learning mode and evaluation methods of the course, and publish/explain the different levels of projects titles and requirements throughout the overall course duration, and provide some learning materials to the students when they log in to the APP.
(3) For each lecture, the students log in to the APP, and the PPT page of the lecture is displayed synchronously on the screen of their mobile phone, and the attendance rate is automatically displayed on the screen of the professor's computer, also the questions sent from students' mobile phone are displayed which is convenient for adjustment of teaching.
(4) Lectures in each chapter in the classroom teaching: take an engineering case as leading, solve problem as driving, to teach the fundamentals and methods of MMT in analysis and design. By asking questions (1 min small questions) about key knowledge points, setting classroom quizzes, the professor can properly adjust teaching strategies according to the student's feedback.

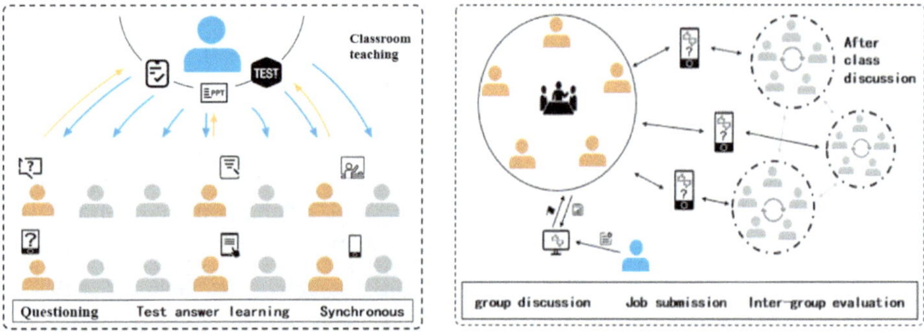

Fig. 3. Interactive relationship of the APP MMT teaching/learning mode

(5) Students learn synchronously, promptly ask questions on the mobile APP, feedback for the professor's questions by means of the mobile APP and teaching computer. Thus the interactive teaching/learning of teacher-student in the classroom is achieved, which is convenient for the teacher to adjust the speed of the explanation. Take the communication in class as an example to realize the interaction between teachers and students in the student-centered classroom teaching.
(6) After-school exercises and small project design: students self-study, group discussion, and team members have different emphasis on analyzing and solving engineering case problems and flexible use of course knowledge and methods to achieve "Engineering project leading - problem driving - capability orientation".
(7) Teach students in accordance with their aptitude through teacher evaluation, group/group peer evaluation mechanism, and analysis of application learning data curve.

The classroom teaching mode is based on response data mainly includes classroom questions and answers, PPT courseware explanation and response following, classroom questioning and feedback, classroom test and summary, etc. Question/answer: the instructor sets inspirational or assessment questions in the classroom, which help the teacher to guide the students understanding knowledge points and quickly grasping the points, and then, Each student responds and answers questions quickly in the class. This real-time interactive response allows students to focus on the content during the class, and the response to the generated data can also be the standard for evaluating their classroom performance. Explanation/following: the feedback of the students' on the PPT courseware (several pages) sent to the instructor in real time, and thus the teacher adjusts the teaching rhythm, and guarantees the quality of the teaching. Questioning: students can ask questions in real time through the mobile phone in the classroom, promptly feedback the questions and difficulties of the course content, and answer questions in the teacher's class or after class. Quiz/summary: using the APP to evaluate the mastery of the content of the students in the classroom, the rapid data feedback and finishing mechanism allows the teacher to easily perform classification and hierarchical teaching in the class.

This APP teaching mode is that with process quality assurance through teaching tracking and process evaluation, timely adjustment/supervision.

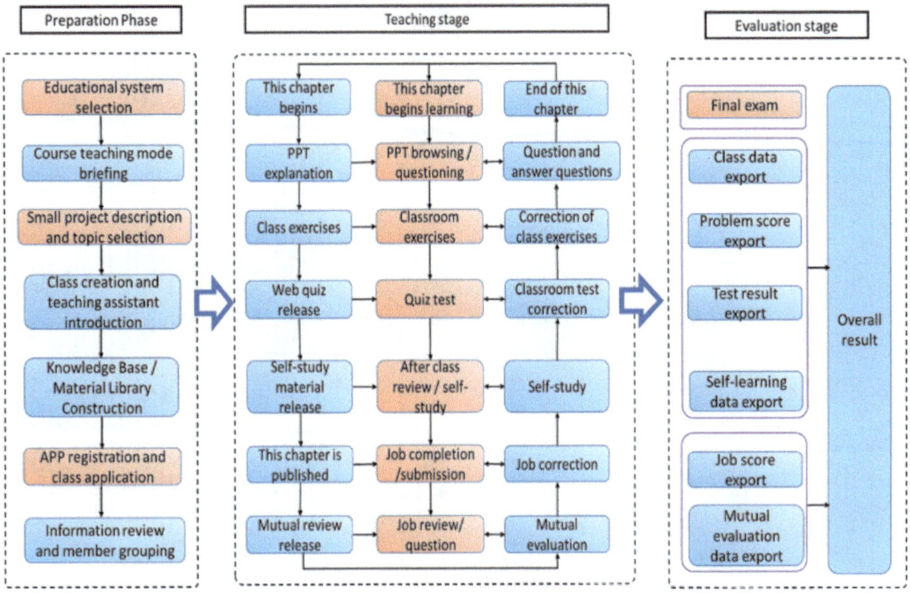

Fig. 4. The specific implementation of the teaching mode

2 The Teaching/Learning Tracking and Quality Evaluation

With the help of the network-assisted teaching/learning system on the APP, teaching/learning data acquisition and tracking the data of teaching/learning, the establishment of a process-oriented and multi-section method of evaluation can be achieved. By using multiple statistical distribution model for evaluation, students can be guided to pay more attention to the entire learning process, reduce the phenomenon of Examination-oriented education and improve the quality of teaching/learning.

2.1 Analyzing and Tracking of Teaching/Learning Circumstance

According to data during the teaching procedure based on the APP, teaching/learning procedure can be analyzed easily. To take class attendance for an example, it was an average of 85.1% in 2017 and 92.9% in 2018. For Another Example, teacher-student interaction of teachers' questions/students' questions: it was 161/47 in 2017 and 21/58 in 2018; the graph of quiz results, shown in Fig. 5, reflects the response of synchronous learning and classroom teaching interaction.

It's convenient for teacher to determine student's learning condition and how much knowledge they have gained through fixed answers and the feedback from students. For example, According to the statistical analysis of quiz results, the scores of the first and seventh chapters are low, indicating that the mastery of these chapters is weak and the teachers need to strengthen them (Fig. 6).

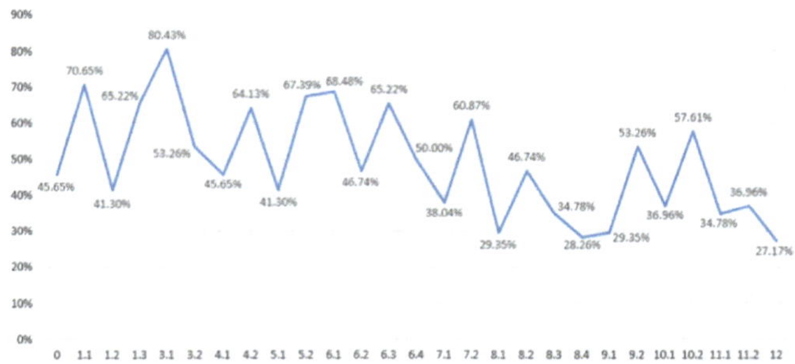

Fig. 5. Class attendance and the rate of following teaching

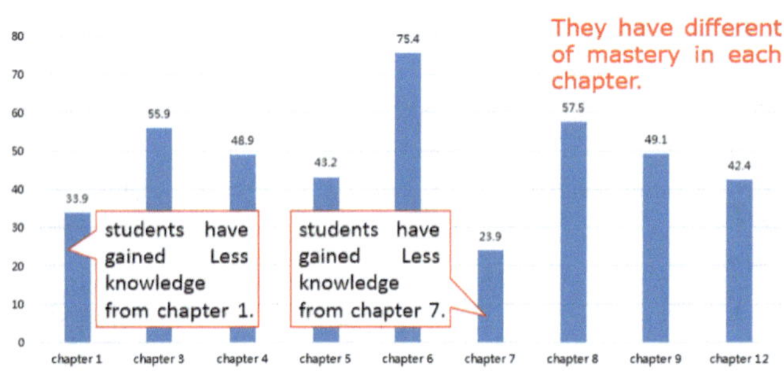

Fig. 6. The average scores of quizzes

And through the length of self-study, determine the students' learning behavior. According to the analysis of various data curves, teachers develop corresponding teaching improvements to continuously improve teaching effects and teaching/learning quality.

2.2 Performance Evaluation Method

We established a process-oriented performance for evaluation method, the score consists of three parts, $S_1 + S_2 + S_3$: class performance S_1 = quizzes + Q&A/ask questions/call in question, 20%; job performance S_2 = job completion + the evaluation of homework, 40%; test scores S_3 = final examination (ability testing), 40%.

The evaluation indicators are based on the "engineering case leading-problem-driving-capability-oriented" teaching/learning model to grasp the basic concepts/principles/methods to assess knowledge/capabilities, determine the results level of ABCD: not understood knowledge is D; understanding the knowledge is C (60–75); the correct application of knowledge + ability is B (76–89); flexible use of knowledge + ability is A-(90–95); reasoning to develop knowledge + ability is A (96–100).

The final grade is evaluated by the normal distribution of the percentages $S_{\sigma1}/S_{\sigma2}/S_{\sigma3}$, then $S_{\sigma1}$ 20% + $S_{\sigma2}$ 40% + $S_{\sigma3}$ 40%. In 2018, the pass rate of 92 students was 91.30%, with an average score of 73.14 points and a maximum score of 98 points of theory of machines and mechanisms (Fig. 7).

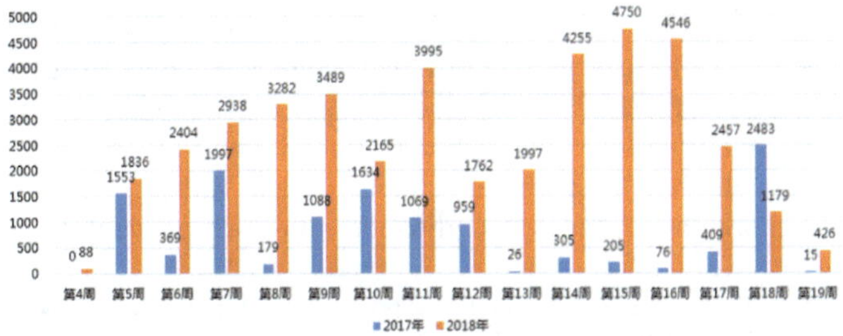

Fig. 7. The average time taken in self-study in each chapter

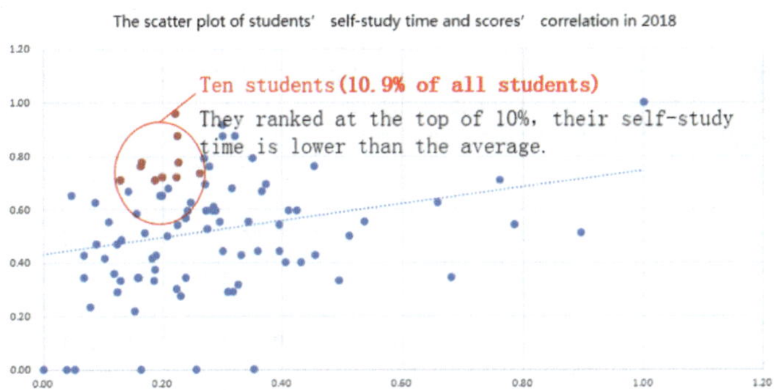

Fig. 8. The scatter plot of students' self-study time and scores' correlation

The analysis of the factors related to the distribution of students' achievement and self-study time, the scatter plot of students' self-study time and performance correlation, as shown in Fig. 8, found that the top 10% of students ranked, the self-study time is lower than the average. This shows that the teaching/learning mode of this paper directly or indirectly reflects the role of ability-oriented, effectively avoiding the phenomenon of bookish study and reading books mechanically.

3 Conclusion

Internet + education is a new opportunity and challenge for teaching reform. The MMT course APP teaching/learning approach and evaluation method constructed by DLUT show that the new mode and evaluation system have stimulated the students' enthusiasm for learning, and promoted the interaction between teachers and students.

And it also provide convenient and quik way to communication for students which obviously improved teaching efficiency. Students' performance evaluation runs through the whole semester which can effectively avoided the phenomenon of late-stage surprise examination. This evaluation method improves the authenticity and objectivity of students' performance. And according to the students' periodical performance, teachers can adjust teaching strategies, which obtained achievements. The evaluation method, which use the ability-oriented multiple statistical distribution model based on the analysis of teaching/learning data, guides students to focus on learning process and ability training, alleviating the test-taking phenomenon and ensuring the quality of personnel training. This mode and method have promoted the efficient improvement of teachers' teaching methods, and improved the quality of teaching/learning, which can be extended to the follow-up curriculum teaching.

Acknowledgment. This work is financially supported by the first-class MMT course construction project of Dalian University of Technology of China.

References

1. McFarlane, A.E.: Devices and desires: competing visions of a good education in the digital age. Br. J. Edu. Technol. **50**(3), 1125–1136 (2019)
2. Gikas, J., Grant, M.M.: Mobile computing devices in higher education: student perspectives on learning with cellphones, smartphones & social media. Internet High. Educ. **19**, 18–26 (2013)
3. Xu, X., Zhan, D., Zhang, C., et al.: The MOOC/SPOC based "1+M+N" multi-university collaborative teaching and learning mode:practice and experience. **288**(12), 5–10 (2018)
4. Jiao Gaohan No. 8 (2018)
5. Wang, X.H., Wang, J.P., Wen, F.J., et al.: Exploration and practice of blended teaching model based flipped classroom and spoc in Higher University. J. Educ. Pract. **7**, 99–104 (2016)
6. Zhan, D., Sun, C., Xu, X.: An exploration of MOOC + SPOCs teaching mode. Software Eng. Educ. Going Agile (2016)
7. Alelaiwi, A., Alghamdi, A., Shorfuzzaman, M., et al.: Enhanced engineering education using smart class environment. Comput. Hum. Behav. **51**, 852–856 (2015)
8. Dou, J., Wen, S.: Research on the reform of college english turning classroom teaching based on APP. High. Educ. Res. Heilongjiang (5), 162–167 (2015)
9. Wang, D., Gao, Y.: Mechanism and Machine Theory, p. 9. Mechanical Industry Press, Beijing (2012)
10. Dong, H., Wang, D., Gao, Y.: The new model of institutional innovation design for the case teaching of mechanism and machine theory course. In: Proceedings of the 12th National MMT Teaching Experience Exchange Conference: Connotation Development and Quality Improvement of MMT Course, 2017, August 6th Guiyang (2017)

Experimental Study on Mechanical System Dynamics Performance

Yuan Gao[1,2(✉)], Hong Zhang[1], Delun Wang[1], Huimin Dong[1], and Cong Chen[1]

[1] School of Mechanical Engineering, Dalian University of Technology, 116024 Dalian, China
gaoyuan@dlut.edu.cn
[2] Mechanical Engineering Experimental Teaching Center, Dalian University of Technology, 116024 Dalian, China

Abstract. This paper deals with the teaching-oriented experiments of Theory of Mechanisms and Machines carried out in Dalian University of Technology (DUT, China) for the undergraduates in Mechanical Engineering. Complying with the teaching concept on Theory of Mechanisms and Machines, i.e., taking engineering case as traction and using comprehensive performance test as the mean to cultivate students' comprehensive design ability, a series of comprehensive experimental platforms for mechanical system dynamics have been developed to set up experiments, including mechanical system dynamics performance experiment, flywheel speed regulation and energy saving experiment, mechanism balance design experiment. Research-based, design-based and comprehensive experiments are implemented to cultivate students' ability of solving complex issues in modern engineering design.

Keywords: Theory of Mechanisms and Machines · Teaching-oriented experiment · Mechanical dynamics · Mechanism balance · Flywheel speed regulation

1 Introduction

In higher education, engineering practice is a significant way to cultivate college students' ability of innovation and engineering practice [1], as well as an important link to enhance college students' innovation consciousness, engineering consciousness and teamwork spirit. Therefore, the research and exploration of engineering practice teaching is urgent to be strengthened to cultivate new engineering talents, who can adapt to new technologies, new products, new environment and new situations.

With the goal of cultivating students' ability for comprehensive design and complying with the teaching concept on Theory of Mechanisms and Machines, i.e., taking engineering cases as traction and using comprehensive performance tests as the mean to cultivate students' comprehensive design ability, emphasis is laid on the application of experimental technology to study the typical engineering issues [2]. Dalian University of Technology (DUT, China) developed experimental platform based on industrial presses, set up research-based mechanical system dynamics experiments to

D. Wang et al. (Eds.): MeTrApp 2019, MMS 79, pp. 607–616, 2020.
https://doi.org/10.1007/978-981-15-0142-5_59

strengthen the engineering experimental ability of students to enable them to solve complex engineering design issues.

2 Teaching-Oriented Experimental Platform Design

2.1 Mechanical Device

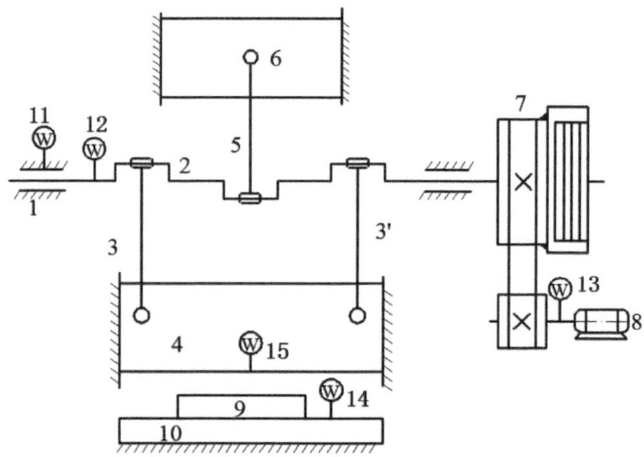

1. Frame 2. Crank 3. Link 4. Slider 5. Link 6. Balance slider 7. Flywheel assembly 8. Motor 9. Stamping workpiece 10. Workbench 11. Vibration sensor 12. Angular sensor 13. Torque sensor 14. Pressure sensor 15. Displacement sensor

Fig. 1. Composition diagram of the experimental platform

The research object of mechanical system dynamics experiments is the double-point crank slider press for engineering application. The mechanical device of the experimental platform is composed of the drive system, the transmission system, the execution system and the auxiliary mechanism, including the balance mechanism and the flywheel assembly. The platforms are respectively driven by three-phase asynchronous motor with variable frequency speed regulation and servo motor. The execution system are the crank slider mechanism and the double-slider six-bar mechanism respectively. The type of the transmission system is the belt drive, whose large pulley fixed with flywheel, to realize the deceleration. The balancing mechanism is a multi-bar mechanism arranged in reverse symmetry. Figure 1 is a schematic diagram of the mechanical system dynamics test platform, and Fig. 2 is a picture of the mechanical system dynamics test platform.

Fig. 2. Mechanical system dynamics test platform

The series of press platform, which is based on engineering practical applications, can realize a variety of kinematic programs through various actuators and a combination of various execution systems. In the mechanical system dynamics performance experiment, the stamping load's change of the test platform can be realized by selecting different stamping workpieces to study the influence of load variations on the mechanical performance. In the flywheel speed regulation and energy saving experiment, the change of the flywheel moment of inertia can be realized by removing and installing flywheel assembly, which is convenient for studying the influence of flywheel inertia on speed regulation and energy saving function of mechanical system dynamics. In the mechanism balance design experiment, the mass of the balance slider is variable, thereby various balance mechanism design programs can be realized. In this way, researching the balance design theory of the mechanism is implemented more effectively. In short, the modularization of the experimental platform and the design with variable parameters ensure the realization of a variety of experimental design programs, which promotes the in-depth development of the design-oriented and research-oriented experiments in Theory of Mechanisms and Machines.

2.2 Test System

The hardware components of the platform's test system include torque sensors, pressure sensors, angular sensors, displacement sensor, vibration sensors, data acquisition cards and a computer. The torque sensor is mounted on the motor drive shaft to measure the motor driving torque. The pressure sensor is set on the workbench to measure the pressing force when different workpieces are being machined. The vibration sensor is installed on frame to measure the vibration in the three directions. The angular sensor is mounted on the crank to measure the angular displacement and angular velocity of the crank. The displacement sensor is set on the slider to measure the dynamic parameters of the slider. The data of experimental platform are acquired and processed by the NI Compact DAQ system, and displayed by the computer finally (Fig. 3).

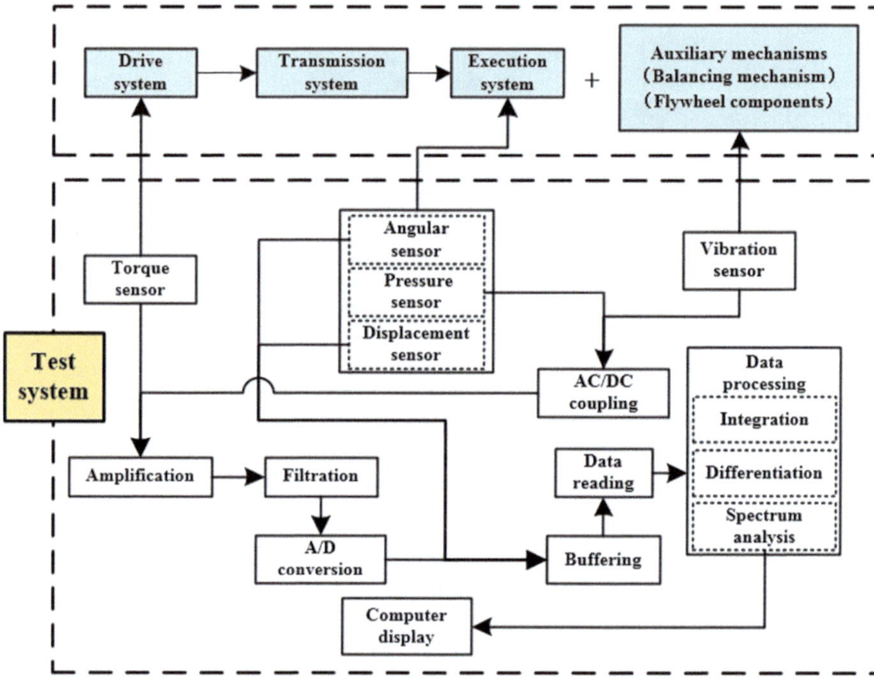

Fig. 3. Experimental sensor application

To process the sensor-generated data, the signal processing software, which is based on Lab View virtual environment and NI Compact DAQ system, is developed to realize continuous reading of data, buffer storage, time-frequency conversion, parameter reading and waveform display. Its main function is to perform Fourier transform on the acquired time domain vibration signal and obtain the frequency domain signal, then selects the data according to the preset sampling frequency and time parameter, finally displays it in the form of waveform by computer.

3 Experimental Projects and Theories

3.1 Mechanical System Dynamics Performance Experiment

Mechanical system dynamics is to study the force of the mechanical process, the relationship between the mass of the mechanical components and the mechanical motion [3]. The mechanical performance is measured by mechanical dynamics analysis, which is important to guide mechanical kinematics and dynamics design.

The experimental platform machines specifications of stamping workpieces to achieve different stamping loads. By measuring the motor driving power and the actual dynamic parameters of the slider, the influence of load and moment of inertia on mechanical system dynamics as well as the test methods of measuring the dynamics of the machine can be mastered.

A model of mechanical system dynamics is established to analysis experimental equipment's dynamics performance.

$$d\left[\frac{1}{2}J_e(\varphi)\omega^2\right] = [M_{ed}(\omega) - M_{er}(\varphi)]d\varphi \tag{1}$$

Where φ, ω, J_e, M_{ed} and M_{er} stand for the equivalent component angular displacement, the equivalent component angular velocity, equivalent moment of inertia for mechanical systems, driving torque and impedance torque, respectively.

Using the difference method to solve the nonlinear Eq. (1), you can get

$$\omega_{i+1} = \frac{[M_{ed}(\varphi,\omega) - M_{er}(\varphi,\omega)]\Delta\varphi}{J_i\omega_i} + \frac{3J_i - J_{i+1}}{2J_i}\omega_i \tag{2}$$

The driving torque of mechanical system can be solved according to the real angular velocity of the equivalent component. Figure 4 shows the results of theoretical analysis.

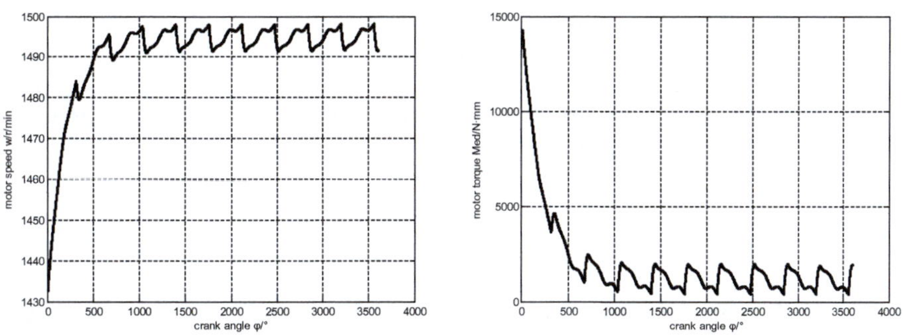

Fig. 4. The results of theoretical analysis.

The simulation software ADAMS is used to simulate the mechanical system dynamics. The machine simulation model is established as shown in Fig. 5 and results are shown in Fig. 6.

Fig. 5. Simulation model

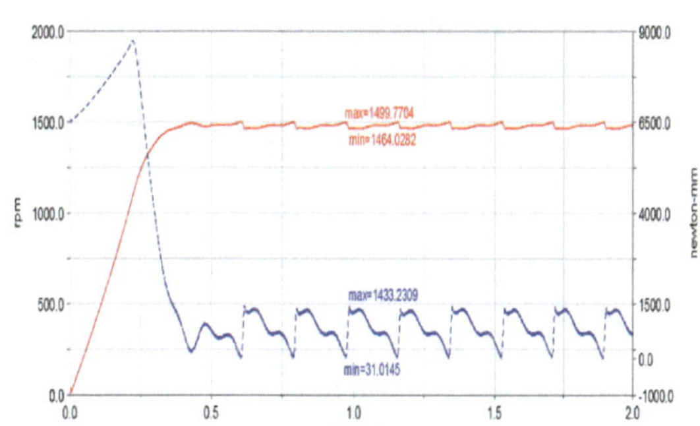

Fig. 6. Machine dynamics simulation results

Through the above mechanical system dynamics performance test, theoretical calculation and simulation analysis, students are cultivated to master the design method of mechanical system dynamics.

3.2 Flywheel Speed Regulation and Energy Saving Experiment

The flywheel assembly stores energy in the non-working section and releases energy in the working section, which can regulate the periodic fluctuation of the machine. For the mechanical system subjected to instantaneous impact load, the flywheel has energy-saving effect [4]. The machine is equipped with a flywheel assembly to change the equivalent moment of inertia of the mechanical system.

$$J_e = J + J_f \tag{3}$$

Where J_e is mechanical system equivalent moment of inertia, J is the inertia of the system without the flywheel assembly, J_f is the moment of inertia of the flywheel assembly.

The optional flywheel assembly with different inertia is equipped. By testing the motor drive power and the dynamic performance parameters of the slider, the relationship between the machine's moment of inertia and the machine dynamics performance is investigated, and the principle of flywheel speed regulation and energy saving is studied [5]. The relevant formula is as follows.

$$J_f \geq \frac{[W]}{\omega_m^2[\delta]} \tag{4}$$

Where $[W]$ is maximum surplus and deficit work, $[\delta]$ is allowable speed fluctuation coefficient of the system, ω_m is average angular velocity of the equivalent component.

The speed fluctuation coefficient is the evaluation index of the flywheel speed regulation, and the formula is as follows.

$$\delta = \frac{\omega_{max} - \omega_{min}}{\omega_m} \tag{5}$$

Where $\omega_{\max}$ and $\omega_{\min}$ is the maximum and minimum angular velocity of the equivalent component.

The motor output energy when platform is subjected to the impact load is index of flywheel saving energy, and the formula is as follows.

$$\mathrm{E} = \int_{\varphi_1}^{\varphi_2} M_{ed} d\varphi \tag{6}$$

Where: M_{ed} is the drive torque,φ_1 and φ_2 stand for the motor angular positions when platform is subjected to the impact load.

The theoretical and simulation analysis of platform with the different moment of inertia of the flywheel assembly are carried out to achieve the system speed fluctuation coefficient and motor torque, and to evaluate the effect of flywheel in speed regulation and energy saving. Figures 7 and 8 show the theoretical calculation results.

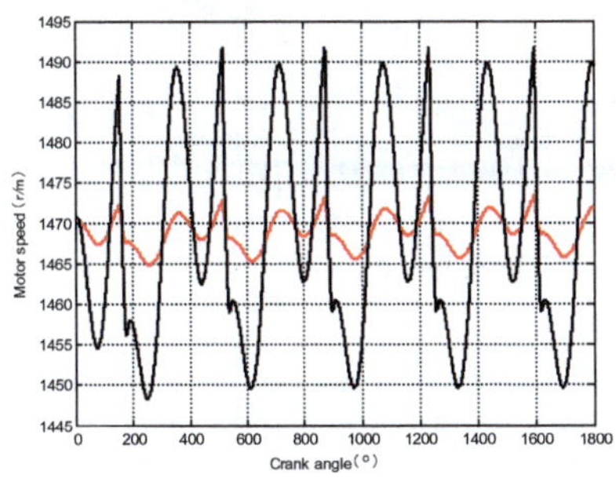

Fig. 7. Motor speed of different flywheel

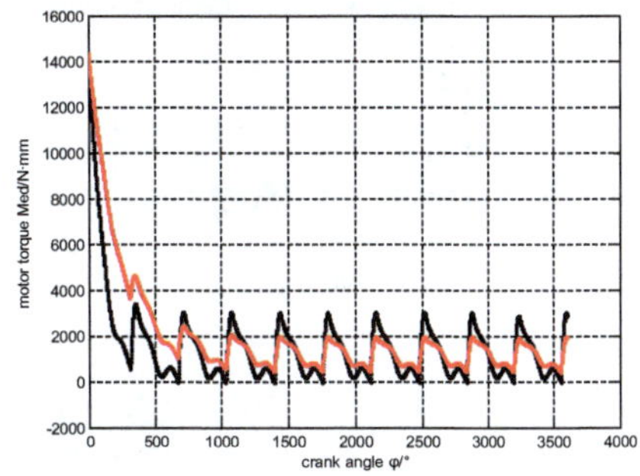

Fig. 8. Driving torque of different flywheel

Through the above test, theoretical calculation and simulation analysis, students are fostered to master the design method for flywheel speed regulation and energy saving.

3.3 Design Experiment of Mechanism Balance

The moving components of the mechanical system will generate inertial loads that fluctuate periodically, which is the fundamental cause of the imbalance of the mechanism [6]. The methods of balance design include partial balance, complete balance and the others. The Fig. 9 is the mechanism balance program based on the design of the execution system. The execution system is a double-point crank slider mechanism, and the balancer is reverse crank slider mechanism.

The completely balance mechanism needs to meet the requirement that the sum of inertial force and moment are both zero.

$$\begin{cases} \sum F_c = 0 \\ \sum M_c = 0 \end{cases} \tag{7}$$

The mass (moment of inertia) and kinematic parameters of each component determine its force and moment of inertia, which effects the mechanism balance. Compared with other components, the mass of balance slider has the most significant influence on the mechanism balance [6]. A balance program based on the theoretical calculation is carried out according to Eq. 7, and the theoretical analysis result is shown in Fig. 10.

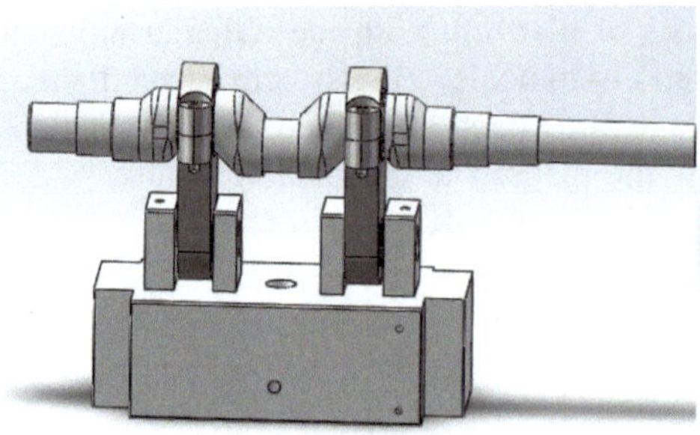

a) Execution system

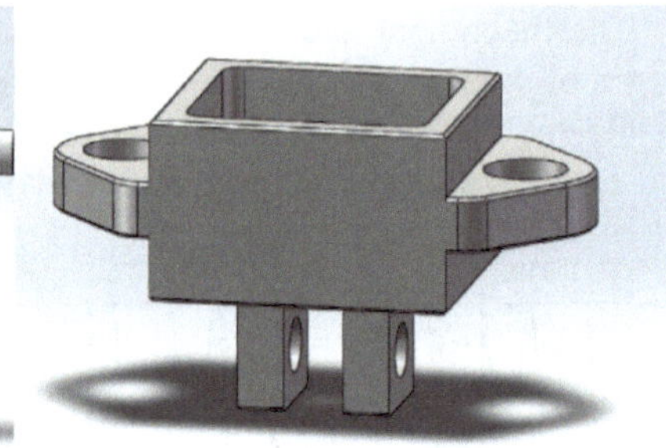

b) The balance mechanism

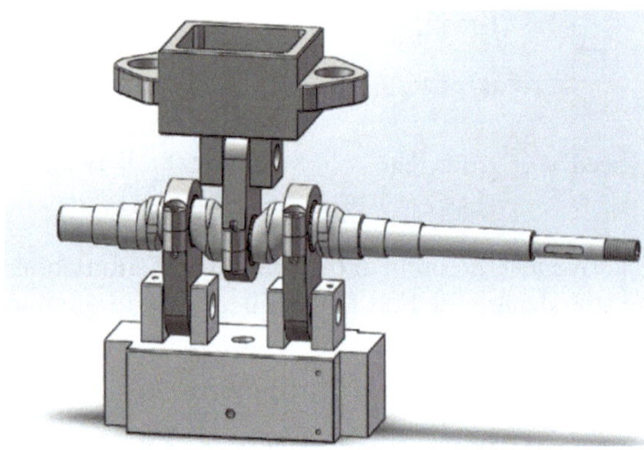

c) Execution system with balance mechanism

Fig. 9. The mechanism balance design program

As can be seen from the figure, the maximum fluctuation of the horizontal inertial force before balance is 101 N, after the balance is 50 N, which is reduced by about 50%, and the maximum value of the horizontal inertial force is also reduced by about 50%. The maximum fluctuation of the vertical inertial force before balance is 644 N, after the balance is 62 N, which is reduced by about 90%, and the maximum value of the vertical inertial force is reduced by about 80%.

Assuming that other conditions remain unchanged, the balance mechanism of the experimental system is equipped with sliders of different mass. The balance program of the mechanism is evaluated by testing the vibration response of the frame. The experimental results are shown in Fig. 11.

It can be seen that, when the system is unbalanced, rotary frequency is in the great peak, which is 0.02 g, and higher component is small; multiple frequencies amplitude difference can't be in the same order of magnitude, which is unbalancing typical characteristic [4]. After the balance, the peak appears at the quadruple frequency, which is 0.0015 g, and rotary frequency is much smaller than this value. The vibration amplitude is greatly reduced, and the main cause of vibration is no longer the imbalance of the mechanism. Thus the vibration caused by unbalancing has removed and the balance effect is obvious.

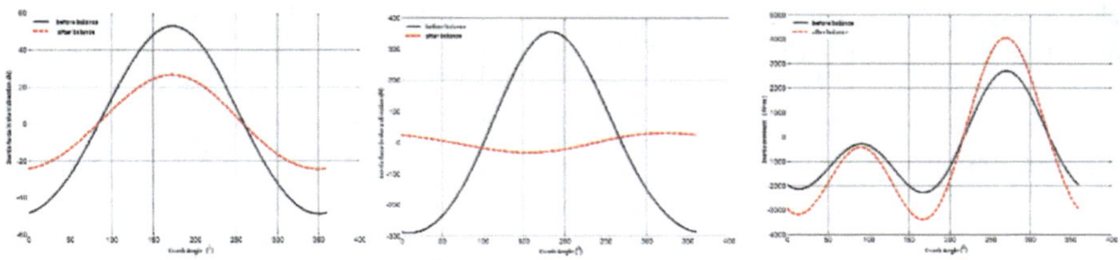

a) Horizontal inertial force b) Vertical inertial force c) Moment of inertia

Fig. 10. Effect of balance design

Through the theoretical calculation and experimental test above, students will master the method of mechanism balance design.

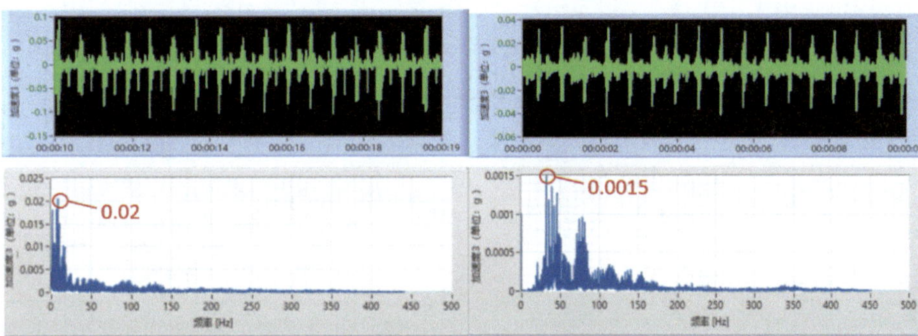

a) Unbalanced mechanical system b) Balanced mechanical system

Fig. 11. Results of experiment

4 Conclusion

The course of Theory of Mechanisms and Machines focuses on the implementation of theoretical teaching content system, but the teaching-oriented experiment is fragmentary and lack of combination with theoretical teaching content. It is difficult for students to systematically apply theoretical knowledge combined with experimental technology to analyze and solve engineering issues, leading students to lack of engineering thinking and practical skills. Under the background of new engineering, the construction of experimental teaching of Theory of Mechanisms and Machines is implemented and the integration of theoretical teaching and curriculum experiment is advocated, a new teaching-oriented experimental platform was developed to strengthen the integration of experimental teaching and modern engineering. Implement a series of comprehensive, research-based and design-based experiments of mechanical system dynamics to cultivate students' engineering consciousness, practical ability, innovation consciousness and realize the goal of training new engineering talents.

Acknowledgment. This work is financially supported by the first-class MMT course construction project of Dalian University of Technology of China.

References

1. Gao, Y., Wang, D.L., et al.: Research and practice of experimental teaching system in mechanical design series. In: Proceedings of the 12th National MMT Teaching Experience Exchange Conference: Connotation Development and Quality Improvement of MMT Course, Guiyang, China (2017)
2. Dong, H.M., Wang, D.L., Gao, Y.: The new model of institutional innovation design for the case teaching of mechanism and machine theory course. In: Proceedings of the 12th National MMT Teaching Experience Exchange Conference: Connotation Development and Quality Improvement of MMT Course, Guiyang, China (2017)
3. Hibbeler, R.C.: Engineering Mechanics: Dynamics. Pearson Prentice Hall, New Jersey (2010)
4. Wang, D.L., Gao, Y.: Theory of Mechanisms and Machines. China Machine Press, Beijing (2012)
5. Dong, C.Y.: Design of Comprehensive Performance Test Platform of Crank Press. Dalian University of Technology, China (2018)
6. Sun, Y.B.: Optimum Design and Dynamic Performance Research of High-speed Press. Dalian University of Technology, China (2019)

Construction of Golden Course Group for the Innovative Talent Cultivation in New Engineering of Machinery

Kedong Bi, Yajing Kan, Ruiming Qian(✉), Guodong Yin, and Zhonghua Ni

School of Mechanical Engineering, Southeast University, Nanjing 211189, China
rmqian@seu.edu.cn

Abstract. The main task of the new engineering construction is to cultivate innovative talents for the new economic development characterized by new technologies, new industries, new formats and new models, and the course is the core element of talent cultivation. Considering that design courses play an important role in the training of mechanical undergraduate talents and industrial needs, the comprehensive ability and innovation ability of students can be improved to solve complex mechanical design problems through the construction of "golden course group of "Design Principles and Methods" project for the innovative talent cultivation in new engineering of machinery".

Keywords: New engineering · Golden course · Mechanical principle · Mechanical design · Talent cultivation

In response to the national implementation of innovation-driven development, "Made in China 2025", "Belt and Road", "Internet+" and other major strategies, the new engineering construction, "Fudan Consensus", was reached at the seminar on the development strategy of higher engineering education in February 2017 after the concept of new engineering was put forward by the Higher Education Department of the Ministry of Education in 2016 [1]. The first proposal of "New Engineering" provided a new perspective and "China Program" for the reform and exploration of the higher engineering education in the world [2]. For colleges and universities, talent cultivation was the most important core task for new engineering construction to meet the needs of current and future industry development [3]. In June 2018, the Ministry of Education proposed that the course was the core element of personnel training at the Undergraduate Education Work Conference of Chinese Higher Education in the New Era. All colleges and universities should comprehensively sort out the teaching content of each course, eliminate the "water course" and create the "golden course" with high-level, innovation and challenges [4].

The "Design Principles and Methods" course group of Southeast University was established by a national teaching team and it was selected as the first batch of "golden course" construction platform promotion plan for colleges and universities in Jiangsu Province in 2019. Considering that design courses play an important role in the training

D. Wang et al. (Eds.): MeTrApp 2019, MMS 79, pp. 617–626, 2020.
https://doi.org/10.1007/978-981-15-0142-5_60

of mechanical undergraduate talents and industrial needs, the construction of "golden course group of "Design Principles and Methods" project for the innovative talent cultivation in new engineering of machinery" was proposed and then carried out.

1 Construction Goal

1.1 Positioning Goal of the Project

Taking the current "New Engineering" national innovative talent cultivation strategy as the opportunity, we are aiming at the direction of networked and intelligent development of design technology based on decades of construction experience accumulated in our school and other colleges and universities, achievements and domestic influence. We will reorganize the content system of the main courses of the existing design majors of mechanical engineering, optimize the system and integrate them systematically, and create the golden course group of "Design Principles and Methods" for the innovative talent cultivation in new engineering of machinery. In addition, we will explore the new model of "project-oriented", "interest-driven", "online and offline" diversified courses, construct a diversified "study and fusion" practice carrier and create an educational environment for inquiry thinking and individualized development, thus comprehensively improving the comprehensive ability and innovative ability of students to solve complex mechanical design problems. This project plays a demonstrative and radiative role in the national mechanical innovation talents cultivation.

1.2 Course Contents and Specific Forms

(a) Construction and practice of the new system of "golden course" group—optimizing and reorganizing the mechanical design courses and teaching process and forming a new golden course group system of "Design Principles and Methods" in the three-in-one "theoretical teaching, practical teaching and independent research", as shown in Fig. 1.

(b) Construction and practice of the new teaching mode of "golden course" group—studying and creating and systematically implementing a new teaching mode of "Design Principles and Methods" golden course group facing the innovative talent cultivation in new engineering of machinery. This mode is supported by "four systems (course system, practical teaching system, independent research system, teaching quality assurance system) and two mechanisms (innovation incentive mechanism, enterprise participation mechanism)", and reflects the three-in-one "theoretical teaching, practical teaching and independent research" in Fig. 2.

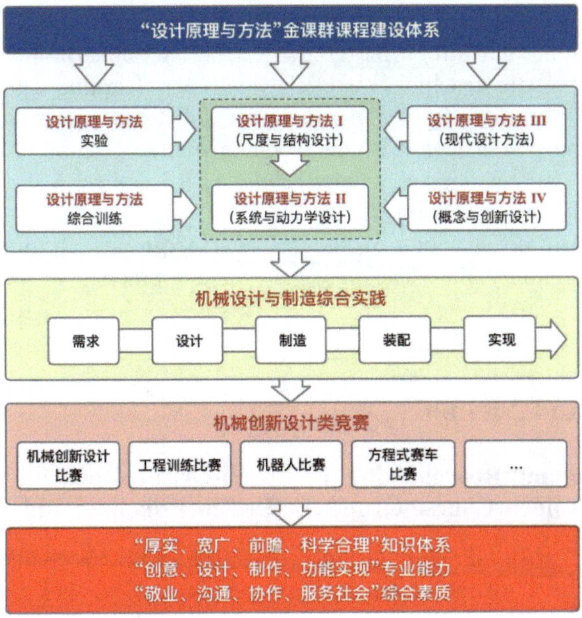

Fig. 1. New golden course group system of “Design Principles and Methods”

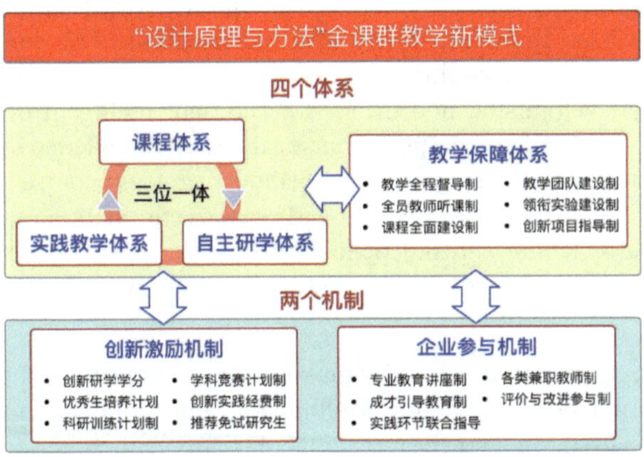

Fig. 2. New teaching mode of “Design Principles and Methods” golden course group

(c) Optimization and reorganization of the knowledge unit of “golden course” group —appropriately refining the traditional content, expanding the new content that reflects the diversity, complexity, comprehensiveness of the engineering design and the networked and intelligent design technology, and forming a modular and menu-based knowledge unit, which is convenient for different teachers to customize the curriculum knowledge reflecting the professional characteristics of the college.

(d) Development and application of teaching resources and teacher-student interactive platform online and offline in "golden course" group—offline paper textbooks, online multimedia teaching materials and open learning and communication websites.
(e) Research and practice on the new teaching mode of online and offline mixing courses of "golden course" group—cloud class, course APP, learning process management and evaluation.
(f) Teaching team construction of "golden course" group—forming a faculty team with high morality that leads the reform and practice of mechanical design courses.

2 Thought and Method

2.1 Optimizing and Reorganizing the Epoch-Oriented and Forward-Looking Golden Course Group of "Design Principles and Methods"

Taking the school positioning and training objectives of the mechanical faculty of our university and other cooperative universities into account, we will carefully select and transform traditional teaching content, which is based on the five-year complete practice of the original "Design Principles and Methods" series of courses to the students at Southeast University, from the perspective of the whole machine and system. We focus on mechanical system design and innovation ability cultivation and integrate the new teaching content into the golden course. The content presents the diversity and complexity of engineering design, the contemporaneity and perspective of the discipline development, and the networked and intelligent design technology. Besides, we will unify and amend the mechanical design knowledge system and build a new system of "Design Principles and Methods" golden course group, which is "precise positioning, clear labor division" and embodying the three-in-one "theoretical teaching, practical teaching and independent research", as shown in Fig. 1. Through systematic practice, we will achieve comprehensive innovation in mechanical talent cultivation.

(a) When combing and optimizing design knowledge units, we will break the original curriculum boundaries and subject affiliation, unify thoughts, integrate the planning and construction, and coordinate the development, which fully demonstrates the integrity of the course group in cultivating students with the ability of innovativeness, practice and solving complex mechanical engineering problems. For example, a reasonable division of different knowledge units of the same object is done and the common objects and knowledge points that should be mastered from the perspective of engineering application are organized, then we can clarify the teaching focus, teaching objectives and integration effects of each course. At last, the scattered knowledge points in the series are connected in series and applied flexibly through engineering cases.
(b) Forming a modular and menu-based knowledge unit, which is convenient for different teachers to customize the curriculum knowledge reflecting the

professional characteristics of the college. When combining the courses in "Design Principles and Methods" course group, we should pay attention to the organic connection, comprehensive application and intelligent development of the knowledge unit, so that the "gold course" group, which shows professional features and realizes the knowledge, ability and quality cultivation to students, will become an important main line in the training plan on the whole. For example, the knowledge unit of "Precision Design" can be associated with the design object and decomposed into "Design Principles and Methods I (Scale and Structure Design)" and "Design Principles and Methods II (System and Dynamics Design)" instead of being set up separately, thus forming a new object-oriented teaching system.

2.2 Constructing the Experimental Practice Teaching System of "Design Principles and Methods" Golden Course Group for Solving Complex Problems

Our cultivation concept is "innovation and creativity" and "engineering ability". Taking the construction of the national experimental teaching demonstration center of "Mechatronics Comprehensive Engineering Training Center" as a platform, we will build a "study and practice integration" normalized practice teaching system through "studying in practice" in course teaching, as shown in Fig. 3, which is based on the "practicing in theoretical learning" comprehensive training and the "study and practice integration" research platform test. This can enable students to develop engineering awareness and learning skills throughout a continuous course. Therefore, we can break the boundaries between disciplines and courses, and create a hierarchical, multi-module, open-ended experimental practical teaching system and a new experimental teaching mode of "online and offline" to meet the requirements of personalized education and research learning.

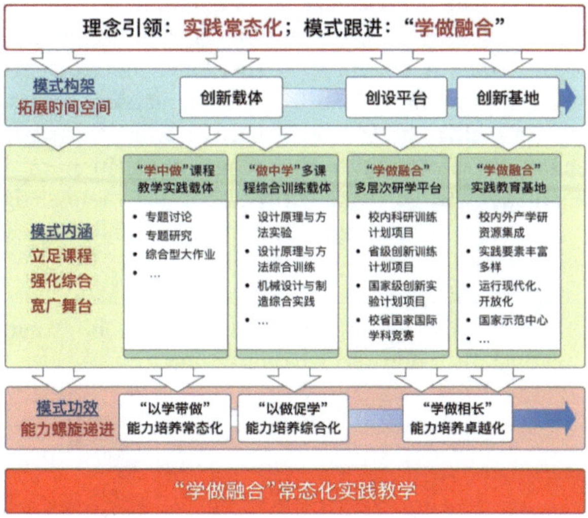

Fig. 3. Normalized practical teaching system of "Design Principles and Methods" golden course group

(a) We take four emphases as a construction idea, including emphasizing comprehensiveness, research and exploration based on basic training, innovation and integration of internal and external curriculum, and then built a "four-level practical teaching system" consisting of "experiment, course design, design and manufacturing comprehensive practice and extracurricular research". Taking the courses as the entry point for the cultivation of students' practical ability, we will construct various forms of "studying in practice" practice carriers, such as "thematic discussion", "special research", "comprehensive large-scale assignment", and so on, to realize the real-time combination of course knowledge learning, application and the normalization of practical training, so that the engineering consciousness and learning ability of students will be cultivated throughout the continuous course teaching. Therefore, students will benefit a lot from such learning results.
(b) Experiments are set up independently and the credits are listed separately. The setting of the experimental project reflects the intersection and integration of the experimental teaching contents of the "golden course" group. The innovative design experiments, independent design experiments and comprehensive application experiments are mainly used, and different difficulty levels are set for each experimental project. Through the establishment of prescribed experimental projects, limited experimental projects, optional experimental projects and self-developed experimental projects combined with the open management of laboratories, the implementation of teaching students in accordance with their aptitude will be done and this can be beneficial to individual development of students.
(c) Domestic exclusive creative project "Integrated Practice of Mechanical Design and Manufacturing"—requiring students to comprehensively apply the knowledge of each course. Students need to customize practical projects and functional objectives under the mechanical engineering background and then realize the principle and scheme based on extensive research, consulting data and full argumentation. Subsequently, they are expected to complete: designing the transmission mechanism and mechanical structure, processing parts and preparing accessories, assembling and debugging physical prototypes, finishing technical data and writing practice reports, etc. This "demand → design → manufacturing → assembly → commissioning → implementation" process experienced by every student can realize the purpose "practice promotes learning" and comprehensively enhance their abilities including "acquisition of knowledge (self-study)", "shared knowledge (teamwork)", "applying knowledge (solving problems)", "summarizing knowledge (innovation)" and "spreading knowledge (communication)".
(d) Making the experimental guidance work of this "golden course" group organically comprehensive. Teachers should participate in the experimental guidance and master many skills while specializing in one.

2.3 Building an Investigative and Personalized Independent Research System of "Golden Course" Group

(a) In accordance with the basic ideas of "refining classroom teaching, strengthening practical teaching and activating extracurricular research", we insist on taking students as the main body, highlighting the early stage of innovation ability training, the gradual progress of innovative activities, the diversity of innovative activities, and the international nature of the innovation activity environment. In addition, we will create and form an independent research system that is nationwide leading and runs through the whole process of teaching/learning, and create a nationwide leading new engineering innovation education environment.
(b) The independent research system consists of special research activities, scientific research practice activities, various discipline competitions and exhibition activities. The theme of the activity not only highlights the discipline characteristics of mechanical engineering, but also focuses on the integration of relevant subject information. The difficulty and evaluation methods of the project emphasize diversity and stratification so as to attract as many low-grade students as possible. In order to improve the interdisciplinary and technical integration of the project, in research activities, students are encouraged to build cross-professional, interdisciplinary, inter-faculty, and inter-aged teams in the research activities. By organizing students to participate in project competition and academic research activities on related topics at home and abroad, students will be led to the international technological and cultural exchange platform.
(c) Giving full play to the platform role of three provincial-level competitions initiated by our project team on behalf of Southeast University and approved by the Jiangsu Provincial Department of Education (Jiangsu Engineering College Advanced Manufacturing Technology Internship Teaching Innovation Production Competition, Jiangsu University Students Mechanics Innovation Design Competition, Jiangsu College Students Engineering Training Comprehensive Ability Competition) and one competition in East China (East China University Students CAD Application Skills Competition) so as to expand the depth and breadth of practical teaching content and sections and implement the education mode of the integration of teaching and competing, making the practice and innovation activities which involve students walk out of the classroom and go to the integration, and walk out of school and serve the society.

2.4 Constructing a Challenging Teaching Quality Assurance System of "Golden Course" Group

We implement curriculum responsibility system of professors, PhD supervisors and teachers, listening system by all faculties, supervision system in teaching process, professor-stationed laboratory and professor-leading experimental construction mechanism. Also, we establish student practice "workshop" and student innovation practice key project system supervised by professors. The quality assurance system of "golden course" group teaching and innovative talent cultivation is formed with the support of high-quality teaching resources and the practical environment, as shown in Fig. 2.

The evaluation results will be used as an important indicator for assessing the performance of team members, and the evaluation and elimination mechanism will be resolutely implemented.

Aiming at the real learning outcome of students, the evaluation mechanism is designed to adapt to the corresponding online and offline teaching platform. We will break the assessment mechanism based on the courses with the theoretical knowledge system and the mastery of practical ability as the evaluation indicators of the progress of each stage and take the ability to solve complex engineering problems and the innovative design thinking as the final evaluation criteria for the cultivation effect of the course group.

Through the construction of a complete teaching quality assurance system, a comprehensive evaluation of the teaching effect of teachers and the learning outcome of students in the teaching process is carried out. We also analyze the impact of the construction quality of each section of the project on the teaching effect and take the completing degree of "high-level, innovation and challenge" as the basic requirement for testing the construction effect of the gold group. Furthermore, we will improve and optimize the project construction in time to form a closed loop of project planning, implementation, evaluation and feedback.

2.5 Constructing a Pioneering and Modern Innovation Incentive Mechanism of "Golden Course" Group

By implementing innovative incentive mechanism such as "Innovative Research Credits", excellent student development program, Student Research Training Program (SRTP), discipline competition program, innovation practice funding support and reward and recommend exemption-free graduate policy, as shown in Fig. 2, we can encourage students' participation in innovation activities and questioning awareness, and cultivate distinctive personality, good engineering quality and innovative will.

2.6 Establishing a Teaching Process of "Golden Course" Group with Enterprise Participation Mechanism

By hiring business and field experts as part-time teachers to participate in the improvement of the "golden course" group system and content, the supervision and evaluation of the teaching sections and effects of "golden course" group, to open modern design techniques and application lectures and jointly guide teaching sections including project research, curriculum design and comprehensive practice, we establish a teaching process of "golden course" group with enterprise participation mechanism.

2.7 Building High-Quality Online and Offline Teaching Resources of "Golden Course" Group

Our teacher team of "golden course" group has a good working foundation in the development and application of teaching resources. We have won 6 provincial-level quality courses (including 1 national-level elaborate course) and edited 9 national-level planning materials, 1 elaborate textbook of the Ministry of Education, 2 elaborate textbooks of Jiangsu Province, 2 national quality courses and 2 resource sharing classes online. In combination with the construction of the "golden course" group, the project focuses on the formulation and preparation of the "13th Five-Year" national-level planning textbook ("online and offline" hybrid) and developing teaching and practical materials that match the update of teaching content with Focus Technology Co., Ltd., Higher Education Press, etc. The materials include application and development teaching cases of pictures, animations, videos and intelligent equipment, curriculum teaching platforms with content customization, teaching process management, tracking and evaluation system (generating the curriculum learning assessment report of students, the curriculum teaching evaluation report of teachers, etc.), experimental equipment, etc. Additionally, we will build the "golden course" group website, making it play a demonstrative and radiative role in colleges.

2.8 Building an Innovative "Online and Offline" Teacher-Student Interactive Platform of "Golden Course" Group

Based on the "Internet+" technology, the "Design Principles and Methods" golden course group APP and cloud platform will be built. Combined with the online teaching materials, textbooks and teaching process management platform of the "golden course" group, the time and space of knowledge learning and the class teaching dimension are expanded. We will explore the new teaching mode of "Internet+" and "golden course", including classroom rollcall, assignment, correction, statistics, study, practice, assessment, evaluation and other aspects of the whole process of quantitative management.

2.9 Building an Experienced Teaching Team of "Design Principles and Methods" Golden Course Group

(a) Building a "golden course" team. The construction of the online and offline parts is in the charge of famous teachers and well-known scholars. Young teachers will be the backbone backup and a high-level teaching echelon will be formed, which are uniformly led and organized by the project leader. The members are mainly from mechanical disciplines, together with professionals in automation, information, electronics, and computers as well. We explore cross-school and inter-agency teamwork mechanisms and organizational models to give full play to the teamwork of talents. Furthermore, we promote the youthful construction of the team through the imparting, helping and leading, and pay attention to the training of young teachers in teaching ability and teaching research, and take advantage of the superiority of young talents in learning information technology. During the construction process, the teaching and research groups and institutions of each course need to ensure the consistency of team building standards and construction progress through online team community, and online teaching and research rooms. In addition, we will make effective use of social resources to strengthen cooperation with social institutions and hire full-time technical personnel responsible for the communication between school and the off-campus technology team, the hardware development of the curriculum resources, and solving technical problems such as classroom design and construction.

(b) Morality building of teachers. We should insist on the concept of "textbook-based" and teaching-educating, lead by example and help each other. We must treat the moral level of teachers as an important indicator to evaluate the work of team members, thereby ideologically eliminating the emergence of "water course".

(c) The "golden course" team building is an important part of the project research. In the process of project practice, we will explore ideas, modes and methods for team building, and then analyze and summarize the practical experience, which can provide references for the construction of golden courses and excellent courses for colleges, even those across the country.

(d) Through the construction of the "golden course" team, we will promote the formation of first-class teaching content, first-class teaching methods and first-class textbooks, and make "golden course" play the guiding and radiative role in colleges.

3 Summary

Design courses are the core curriculum with development priority in the mechanical talent cultivation at home and abroad. The construction of the "golden course" group in mechanical design can bring about the pioneering and modern course content, the advanced and interactive teaching form, and the inquiring and individualized learning process, thereby fully enhance the comprehensive ability and high-level thinking skills of students to solve complex mechanical design problems.

References

1. Peihua, G.: New engineering and new paradigms: concepts, frameworks and implementation paths. Higher Eng. Educ. Res. **2017**(6), 1–13 (2017)
2. Denghua, Z.: The connotation and action of new engineering construction]. Higher Educ. Res. **2017**(3), 1–6 (2017)
3. Jian, L.: Future-oriented construction of new engineering in China. Tsinghua Univ. Educ. Res. **38**(2), 26–35 (2017)
4. Yan, W.: Building China's "Golden Course". Univ. Teach. China **2018**(12), 4–9 (2018)

Exploration and Practice of Project-Based Teaching in the Course Theory of Machine Design and Mechanism

Yuanxiong Cheng(✉), Jiajun Yang, Jianfeng Xu, and Kuanmin Mao

Huazhong University of Science and Technology, Wuhan 430074, China
yuanxcheng@hust.edu.cn

Abstract. In order to qualify Emerging Engineering talents for the ability of the mechanism type selection and design of modern mechanical products, the exploration and practice of project-based teaching is carried out in the course Theory of Machine Design and Mechanism. The project-based teaching encourages students to involve in engineering design and practical projects for sharping students practice ability based on the characteristics of the course. The method integrates the basic knowledge, the basic theory and the methodology of mechanism innovation design. The teaching boundary has been redesigned and the training on innovative thinking and engineering design ability is emphasized. Consequently, teaching outcome is obviously enhanced that students have greatly improve their ability of work independently and problem-solving through the actual design of the project.

Keywords: Theory of Machine Design and Mechanism · Project-based teaching · Emerging Engineering · Innovative thinking · Boundary redesign

1 Introduction

"Emerging Engineering Talents" is a high-quality compound talent with strong engineering practice ability, innovative ability, and adapting to need of the international competitiveness of future emerging industries and new economy. The teaching reform of the course Theory of Machine Design and Mechanism is confronted by question how the course adapts to needs of "Emerging Engineering Talents" which we should seriously consider. Project-based teaching is a training method for practical teaching activities conducted by a relatively complete project. It integrates the basic knowledge, basic method and practical ability on mechanism design and type selection for modern mechanical products under reconstructed teaching boundaries. By training of the project-based teaching process and experience of problem analysis in the actual design of the project, students can improve the ability of independent work and solving practical engineering problems.

D. Wang et al. (Eds.): MeTrApp 2019, MMS 79, pp. 627–635, 2020.
https://doi.org/10.1007/978-981-15-0142-5_61

2 Project-Based Teaching in Course Theory of Machine Design and Mechanism

2.1 Main Features of Project-Based Teaching

The Course Theory of Machine Design and Mechanism is a specialty technical curriculum, which mainly researches common problems existed in design of mechanical products, and conducts study and application of mechanism design and innovation [1]. Project-based teaching combines teaching from lecturers with active practice of project design practice from students. It is student-oriented teaching and learning method, which can fully consider individual differences between students and realize individualized training. Furthermore, learning achievement to design mechanical products with certain functions and practical value can be reached by project-based learning. Through independent exploration and thinking, students will grasp the subject expertise hidden behind the project, and develop their effective abilities of problem-solving, self-learning, and work in teams. A new course curriculum system by project-based teaching is conceived. We redesign the curriculum boundary, and reorganize teaching content by increased or decreased materials according to social needs. In this way, we develop a new learning platform for the individualized training of students, and the in-depth reform of the course teaching is promoted.

To put forward the countermeasures on project-based teaching, we take the project as the main thread, teacher as the guider, and students as the main body. The proposed projects should include several aspects of the knowledge requirements that the students need to master. According to the project characteristics, the students will be divided into groups. The group discussion will be organized and tutoring is available from teacher or tutors. This measure pushes students to participate directly in the whole process of teaching through student-centered learning, and enable students actively to experience, percept and explore the process and knowledge of the mechanical product design [2].

As key point of this teaching reform, a perfect and improved knowledge system needs to be designed for project-based teaching at the top level design of the curriculum system, which targets on transition from knowledge classroom to ability classroom, from closed classroom to opened classroom, from passive classroom to dialogue classroom, from period-mark classroom to question-mark classroom based on characteristics of the curriculum. This is important for deepening the reform of course teaching, improving the quality of course teaching, and cultivating "Emerging Engineering Talents".

2.2 Redesign of Teaching Boundary

Two essential and core issues of course teaching are "teaching what" and "how to teach". Figure 1 shows main characteristics of project-based teaching. According to such characteristics, redesigning teaching boundary is an important part of the implementation of project-based teaching.

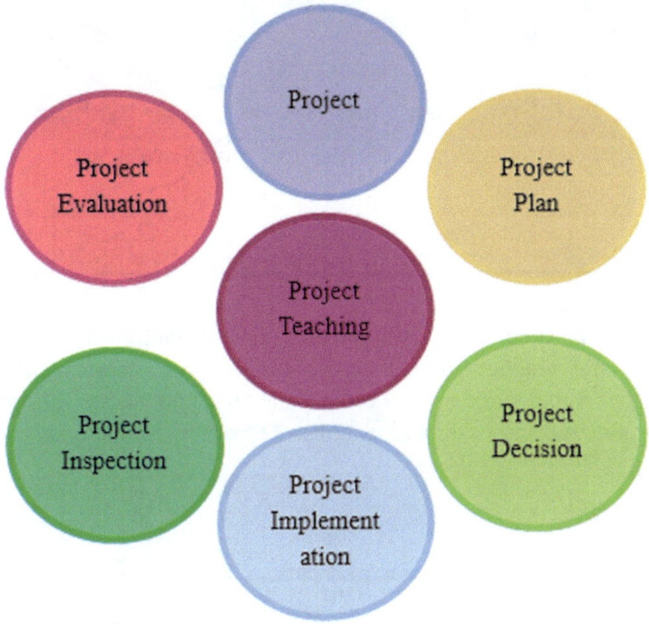

Fig. 1. Characteristics of Project-based Teaching

The project-based teaching of the course will firstly propose suitable projects oriented to products with practical application value. Projects should also adapt to requirements from different teaching aspects, such as the teaching content of the course, teaching procedure of theory and practice, course experiment, curriculum design, and innovation activities. It will depend on re-designing of the teaching boundary, including of the method of course evaluation, scope of basic mechanism and mechanical system motion scheme, mechanism type selection and their synthesis, mechanical innovation design, etc. [3]. The project topic should not be too big, but with certain difficulty and with engineering application value. This will ensure students plan their project independently. It not only promotes students the application of existing knowledge, but also more importantly, it enables students to expand their thinking and solve new problems within a certain extent. Teachers will provide a project assignment paper with functional requirements, clear tasks and time period. Students organize and arrange their own learning and design work within the specified period as required. Finally, students will be demanded clearly and specifically to exhibit their results on projects. In this way, we can obtain direct and effective evidence of students' learning outcomes and examine the effectiveness of the curriculum reform. Teacher's role in the course teaching is shown in Fig. 2.

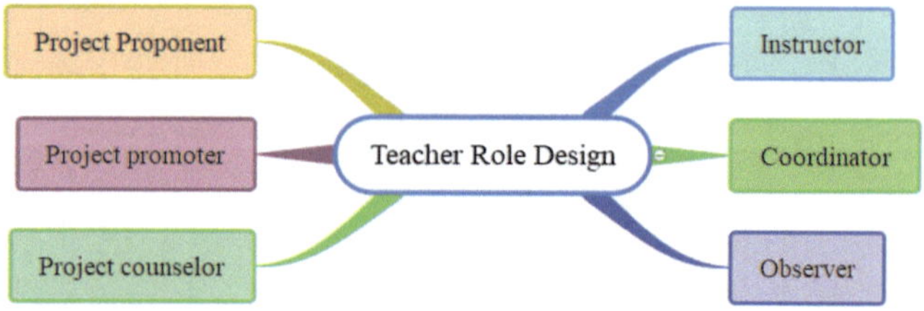

Fig. 2. Teacher's role

Students' role is to complete the project, keep constantly critically thinking about analysis, study and practice, find solutions of problems, learn by themselves, and practice activly. So that students can build their own knowledge framework from the project implementation process, and establish their core competence through project-based teaching. (Shown as Fig. 3).

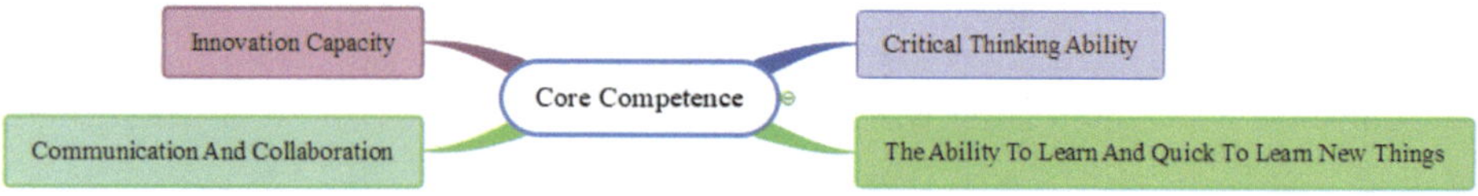

Fig. 3. Students' core competencies

3 Topics Selection of Project-Based Teaching

3.1 Topics Features of Project-Based Teaching

How to combine project-based teaching with the characteristics of course is one of the keys to the success of project-based teaching. It is often more important to ask a question than to solve a problem. It is necessary to select the "question" from the "problem". The good topic is half the battle. How to choose a topic? How to find problems, especially to find high-quality problems? We should observe the phenomena in social life, nature and daily life, study seriously, learn to observe and think, and develop the habit of constantly asking questions. Often people sigh with feeling that I had also seen this phenomenon but I did not think of it any more from the very beginning. The important reason for this situation is that the ability of observation and thinking is poor. One must have a sense of problem and a spirit of question, and form a strong sense of innovation and desire for innovation. Students should be sensitive to things or phenomena and look at people's incomprehensible, commonplace and even inconspicuous things with an innovative vision and perspective. Keep a curiosity at all times, pay close attention to the appearance of any unexpected phenomena, think about what people do not want, see what people do not see. If you can do it, you will gain something by discovering its potential meaning and value. Although there are various problems that need to be solved in our minds, when we really want to determine a topic to explore, we will find that the problems in our minds are too general. This subject is not worth studying. This is why I feel that no subject can be studied. The attributes of each specific innovation object are infinite, and the same is true for the topic.

Within the project-based teaching, teachers need to do a lot of preparatory work. Their main task is to determine the project content, task requirements (problem sets), work plan, and envisage what may happen in the teaching process and the students' ability to withstand the project. When the students are introduced into the project work, teachers will retreat to the position of study guide, and be ready to help students solve difficult problems all the time.

3.2 Topic Selection and Innovative Thinking

The choice of topics is everywhere. Breaking through the mindset and breaking the mystery of creating inventions will surely be helpful. Everything in the world can become the object of people's innovation. Therefore, we all face an infinite number of thinking objects. Whimsy can be integrated with bold imagination, fantasy and association. Many great inventions and creations start from whimsy, and whimsy is the driving force of scientific creation.

The way of thinking about the topic needs multiple perspectives and fantastic ideas. Different perspective will arrival at different conclusion. The whimsy and the divergent thinking will produce various opinions or ideas. Figure 4 shows the characteristics of the linkage mechanism to realize a table the foldable function.

Fig. 4. Folding table

In the teaching practice, almost all the products in the technical field can be selected as the project's topic if you can keep curiosity at all times. The selected topic must be meaningful, recognized by others and innovative. The purpose of the project research is to recognize the laws that the predecessors have not yet recognized or fully understood. This requires that the research topics have a certain degree of originality and novelty. According to the degree of innovation of new products, there probably are new products, replacement products, and improved products. Innovation includes new theory, viewpoint, methods and applications. The selected subject must have a certain theoretical basis and factual basis, and must conform to the basic laws of human cognition and development of things. If the selected topic is in the wrong direction, no matter how big your intelligence, you will still fail. In addition, the topic should be practical. If a product loses its basic function or is not practical, it will have no value. The topic selection should be feasible. When choosing a topic, we must fully consider the subjective and objective conditions of the subject research. Therefore, teachers should consider the actual situation of students for topic selection, such as learned knowledge and accumulated experience, hobbies, time and energy. This will ensure students can achieve success in the project on their own efforts.

4 Project-Based Teaching Practice

4.1 Mechanical Motion Scheme Design and Type Selection

Teaching Procedures: classroom lecture and curriculum design.

Topic: Branch pipe connection with main pipeline under natural gas transmission [3].

Main parameters: main pipe diameter 300 mm, branch pipe diameter 200 mm, wall thickness of steel pipe 8 mm.

Assessment method: defense meeting and grading by fellow students.

Submissions: Design Specification, 3D Product Drawing, Defense PPT.

1. Putting forward the problem from the actual needs of engineering

As the traditional method when branch pipe of natural gas needs to be connected to source pipeline, it is necessary to cut off all the gas sources by valve-off at consumer side for adding new branch pipeline. Since stopping gas supply will greatly affect normal industrial production and residents' lives, how to connect a new branch pipe under gas flow has great research value.

2. Analysis problem by breaking through the mindset

The cutting methods of steel pipe digging holes include plasma arc cutting, laser cutting, water cutting, and mechanical cutting.

According to the conventional thinking, the water cutting is the best solution, because other methods will cause an explosion. However, due to the large amount of water, the gas pipe becomes a water pipe, and this method is not applicable. Breaking through the mindset, fire burning is conditional. If there is no oxygen, it will not burn. Under anaerobic conditions, the above cutting methods are feasible.

3. Proposing multiple scheme design for the same function

After analyzing and discussing the advantages and disadvantages of each scheme, the team selects the plasma arc cutting according to the actual working conditions. Considering the cutting process, determine the functions of the telescopic movement, the rotary motion, removing of cutting waste, and the valve on the branch pipe. Then draft whole process route, and carry out multi-scheme type selection for the same function and conduct product structure design.

Motion scheme of plasma arc-cutting mechanism shown in Fig. 5 (planetary gear + compound spiral + linkage mechanism).

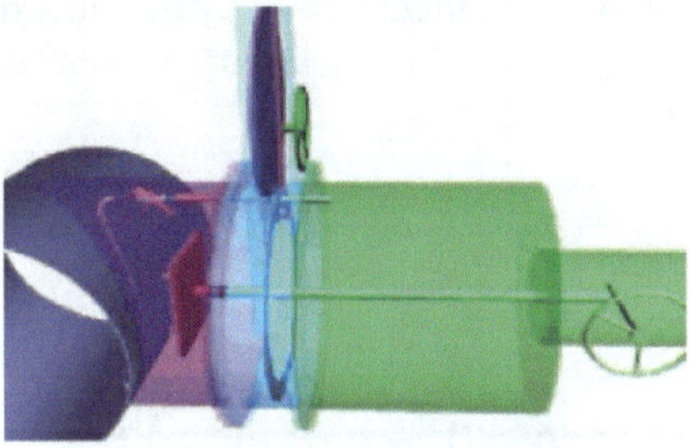

(a) Begin of cutting

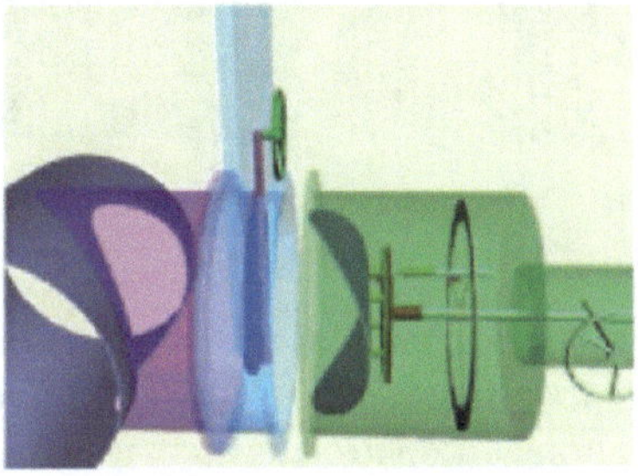

(b) End of cutting

Fig. 5. Motion scheme of plasma arc-cutting for branch pipe connection

4. Effect analysis

Through the project teaching of this topic, the students have been cultivated the innovative thinking ability that is brave to explore, good at expansion, open-minded, and break the routine. The fire burning is conditional, so the cutting method can be optional. After the cutting method is determined, Multi-scheme design with the same function can be conducted. Cutting rotary motion can be double crank, cylindrical gear, planetary gear, etc. The cutting and telescopic action can be pneumatic, rack and pinion, crank slider, double spiral, etc. So that students can understand each specific innovative object, its attributes are infinitely. Just as there are no two identical leaves in the world, this makes each object different from any other object. This improves student capabilities of innovation, flexibility, mechanism selection and innovative design.

4.2 Intelligent Mechanical and Mechanical Design

Teaching procedure: lecture teaching + curriculum design + innovation activities.

Topic: Smart Home - Automatically clean office robots.

Assessment method: defense meeting + grading by fellow students.

Main parameters: The room size is about 80 m^2, the desk height is about 1 m, and the sofa is about 0.6 m high. Other sizes are customizable.

Submissions: Design Specification + 3D Product Drawing + Defense PPT + Model.

This project will develop an intelligent robot by the electromechanical combination design, which can clean the ground dust, sofa and desktop in the environment of the office illustrated in Fig. 6.

Fig. 6. Office environment

The mechanical transmission scheme of robot for cleaning dust on the ground, sofa and desktop hygiene (spiral mechanism + linkage + walking mechanism; image recognition + ranging + automatic control) is shown in Fig. 7.

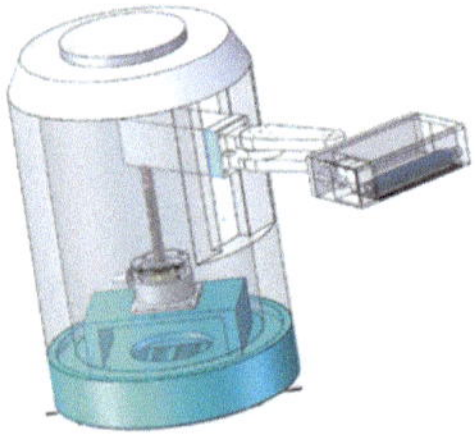

Fig. 7. Cleaning the office hygiene robot

Through the project teaching of the selected topic and redesigning the curriculum boundary, students are encouraged and trained to observe, discover problems, combine electromechanical technology, ingeniously conceive, simplify function properly, solve old problems with new methods, solve new problems with old methods, use the new method solves the new problem. The project help students break the mystery of intelligent machinery and understand characteristics of electromechanical products. In the implementation process of project students have analyzed the limitations of existing sweeping robots, combined walking robots with Cartesian robots, and designed robots to clean office according to mechanism characteristics and environment condition of the office. The project has training comprehensive abilities of students, such as attention to detail, combination with modern information technology, sensing technology, combination with automatic control, to develop intelligent mechanical design capabilities.

5 Conclusion

The project-based teaching is under the guidance of the teacher, and a relatively independent project is handed over to each group of students. The collection of information, the design of mechanism scheme, the implementation of the project and the final evaluation are all the responsibility of the students themselves. Through the project, students understand and grasp the basic requirements of the whole process and each procedure. The project-based teaching is a method that teachers and students work together and achieve progress together. By organically combining theory and practice, the course provides students necessary basic knowledge and methods for mechanical product design. Project-based teaching is also a student-centered teaching method that emphasizes the teaching philosophy and student learning outcomes. The teaching ideas and teaching method of the course have been comprehensively reformed and explored through project-based teaching method, which covers all aspects of the course teaching, such as classroom teaching mode, experimental means, practical procedure, evaluation methods, etc. Project-based teaching includes active practice of curriculum design, updates course boundary, cultivates students' awareness of innovative thinking, stimulates students' desire for knowledge, and promotes student-centered individualized teaching. The project-based case study of intelligent machinery as a teaching carrier improves the ability of students to work independently and solve practical problems in engineering.

References

1. Yang, J.: Theory of Machine Design and Mechanism (Second Edition). Huazhong University of Science and Technology Press, Wuhan (2014)
2. Sun, W.: Exploration of project design based on project teaching method. Time Educ. (Educ. Teach.) **10**, 12–13 (2012)
3. Yang, J.: Mechanical Innovation Design and Practice. Huazhong University of Science and Technology Press, Wuhan (2014)

Project-Based Teaching Skills for University Course of Mechanism and Machine Science

Weizhong Guo(✉)

School of Mechanical Engineering,
Shanghai Jiao Tong University, Shanghai 200240, China
wzguo@sjtu.edu.cn

Abstract. The Mechanism and Machine Science (MMS) course is a fundamental engineering course to enable undergraduates to understand the analysis, design and operation principle of modern machines evolving rapidly nowadays. Modern machines are mechatronic instead of purely mechanical. However, the current course teaching basically focuses on the knowledge of the mechanisms merely instead of the whole machines. A gap exists between the knowledge of the mechanisms and that of real machines for the students that calls a change of the course teaching. To overcome the problems, a Project-based Teaching (PBT) way is investigated with Marathon and Twin-timeline models and Mini research-based training skill for the MMS course aiming to assist students to have both a full image of modern machines and deep understanding of mechanisms.

Keywords: Mechanism and Machine Science · Hierarchical course knowledge · Marathon model · Twin-timeline model · Mini Research-based training skill

The Mechanism and Machine Science (MMS) course is a fundamental engineering course to enable undergraduates to understand the analysis, design and operation principle of machines evolving rapidly [1] nowadays. On one hand, the course is full of concepts and methods somewhat abstract, difficult, and tedious for students to learn. On the other hand, modern machines are mechatronic instead of purely mechanical. Currently, the course teaching basically focuses on the knowledge of the mechanisms merely instead of the whole machines. There is still a gap between the knowledge of the mechanisms and that of real machines for the students.

Obviously, the used MMS teaching style is easily to make students confused and have no complete image of the learnt MMS knowledge in the development of modern machines. Therefore, reforming the teaching style of the MMS course is necessary and a better teaching way is possible to be found. In fact, the course is essentially practice-based since it aims at enabling students to analyze/design/invent machines. Hence, a project-based teaching (PBT) way, i.e. project-based learning (PBL) way can be adopted for the MMS course. In this paper, three skill models are presented to implement the PBT way for the MMS course.

D. Wang et al. (Eds.): MeTrApp 2019, MMS 79, pp. 636–640, 2020.
https://doi.org/10.1007/978-981-15-0142-5_62

1 New Understanding of the MMS Course

1.1 Course Object and Task

The course studies modern machinery/machine. Modern machinery/machine is a high-tech system characterized by mechatronics and intellectualization. In other word, it is an intelligent machine. Therefore, the MMS course should provide students a complete image of modern machines with a focus on the mechanism analysis and synthesis knowledge. In terms of PBL way, a systematic development task can be assigned to students to work out a design/fabrication of some kind of machine, where mechanism design, analysis and motion control are necessary.

1.2 Hierarchical Course Knowledge

To serve the development of modern machines in industries, the content of the curriculum is sorted out comprehensively in terms of unit and system. A three-level knowledge structure can be formed, as shown in Fig. 1, including basic knowledge composed of basic concepts and methods, application knowledge consisting of scheme analysis and synthesis, and machine innovation knowledge consisting of machine analysis and design, with a teaching list of knowledge points and capability points respectively.

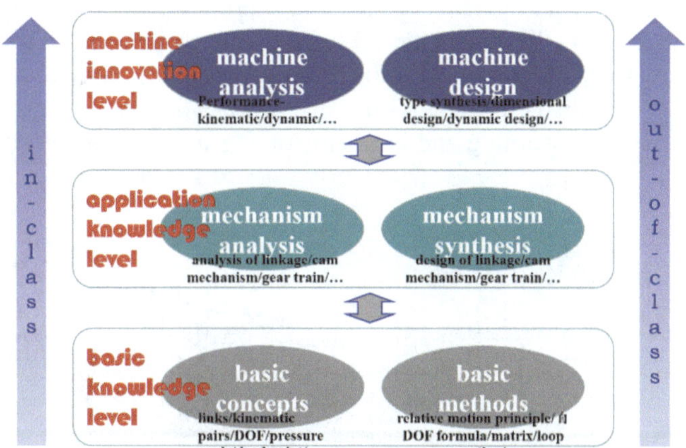

Fig. 1. Hierarchical knowledge and capability points.

By constructing the list of knowledge points and capability points of the curriculum, it is reasonable to divide the teaching/learning task between teachers' in-class teaching and students' self-study outside class, and makes it possible to arrange the course teaching in limited credit hours.

2 Skills for PBT-Based Teaching of the MMS Course

The project-based teaching/learning methodology is suitable for the MMS course's teaching featured with student-centered teaching/learning concept. Two important relationships need to be defined carefully. One is the relationship between teachers and students. The other is the relationship between in-class and out-of-class teaching activities. In this paper, a Marathon model is put forward to define the teacher-student relationship in curriculum teaching. A Twin-timeline model is proposed to define the relationship between in-class and out-of-class coordinating both in-class teaching and project guidance. Besides, a Mini research-based training skill is applied to take advantage of each section of the project guidance.

2.1 The Marathon Model to Define Teacher-Student Relationship

Conventional project-based teaching methods often apply short-term centralized project training. The interaction between teaching and learning is insufficient, superficial and ineffective that the project-based teaching functions inefficiently.

In order to avoid similar problems, the MMS course puts forward a new solution that students enter the project process from the beginning of the semester and PBT activities run through the whole procedure of teaching. Accordingly, a Marathon mode is put forward for the teaching as shown in Fig. 2. That's to say that students are the main body of PBT activities and can be thought as 'athletes', and teachers act as coaches to do leading, accompanying, coaching, promoting and judging and can be regarded as the 'soul' of PBT activities. Teachers can stimulate students' enthusiasm, initiative and creativity by innovating project-setting methods, give full play to the role of students as the main body of teaching activities, and enable the combination of interest-driven research learning and problem-solving training for the students.

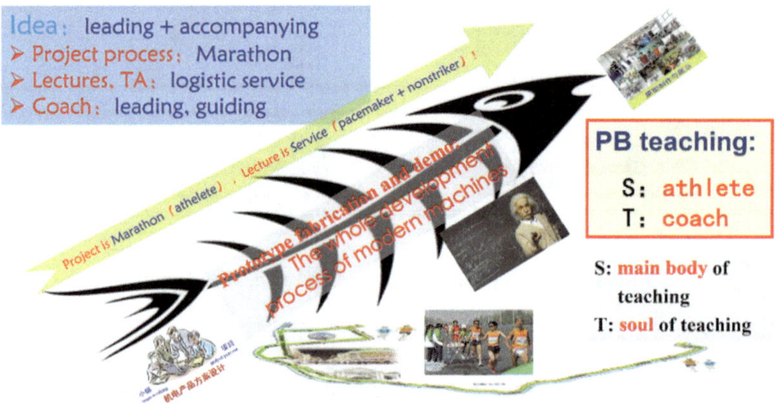

Fig. 2. Marathon model to define teacher-student relationship in PBT.

2.2 The Twin-Timeline Model to Define in-Class and Out-of-Class Relationship

In order to arrange the teaching activities reasonably in limited credit hours, a Twin-timeline model is proposed to coordinate the in-class teaching and out-of-class learning activities (see Fig. 3) in parallel. In-class lecture is a bit ahead of the out-of-class project progress and so lecturers can lead/guide the development of the project. The twin-timeline model effectively realizes the deep interaction and virtuous cycle between lecturers' teaching and students' *learning via doing*.

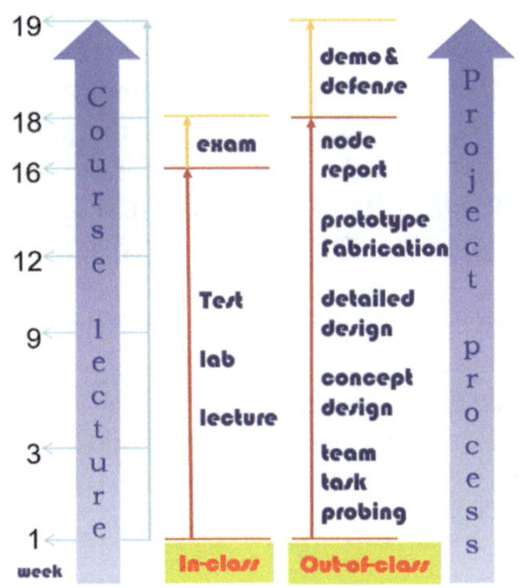

Fig. 3. Twin-timeline model to define in-class and out-of-class relationship in PBT.

The Twin-timeline model not only ensures the completion of teaching tasks within limited in-class hours, but also ensures the full implementation and good results of curriculum projects. In-class teaching provides theoretical tools and methodological support for extracurricular project making while imparting professional knowledge. Extracurricular project training not only fosters students' innovative and practical abilities, but also deepens students' understanding and mastery of knowledge points delivered in in-class teaching. This effectively realizes the deep interaction between teachers' teaching and students' learning by doing.

2.3 Mini Research-Based Training Skill to Guide the Project

The cultivation of students' ability includes the cultivation of innovative consciousness, innovative ability, exploratory ability and research ability. The aim is to build up students' ability of problem's solving. A mini research-based training skill is put forward by flexible problem-setting to cover the whole process from market investigation, literature review, till final scheme design, prototype fabrication and demonstration. The project task is defined by the students through independent market/technical investigation and reviewed by the teacher to judge the correlation to the MMS course. The research thinking is encouraged in each step of the project procedure. Such tactics activates the students' initiative, curiosity and innovation impulse effectively.

3 Closure

Engineering courses are the cornerstone and foundation of industrial development. Adaptation to the times is necessary to take place in terms of technical content and teaching style. As detailed above, a PBT-based teaching practice has been going on earlier from the year of 2001 in my academic life in Shanghai Jiao Tong University, China, and the teaching skills are updating gradually through the years. Good achievements are arrived. Tens of designs are patented and numerous awards are awarded. Especially, the course students have won the personal first place and team second place in the first International Students' Olympiad on Mechanism and Machine Science sponsored by IFToMM (SIOMMS 2011, Russia), both personal and team first places in the second Olympiad (SIOMMS 2013, China), both personal and team second places in the third Olympiad (SIOMMS 2016, Spain), and both personal and team first places in the fourth Olympiad (SIOMMS 2018, Peru).

Acknowledgement. The author thanks the partial financial supports under the projects from Shanghai Jiao Tong University (Grant No. ZXDF602037/002), Ministry of Education, China (Grant No. 201801283003), and Outstanding Online Course Project of Shanghai Universities from Shanghai Municipal Education Commission.

References

1. CecCarelli, M.: Evolution of TMM (Theory of Machines and Mechanisms) to MMS (Machine and Mechanism Science): an illustration survey. In: Proceedings of the 11th World Congress in Mechanism and Machine Science, Tianjin, China (2004)
2. Guo, W.Z., Hongying, Y.U.: Mechanism and Machine Theory. Tsinghua University Press, Beiing (2010). (in Chinese)
3. Zou, H.J., Guo, W.Z.: Mechanism and Machine Theory, 3rd edn. Higher Education Press, Beijing (2016). (in Chinese)
4. Guo, W.Z.: Course teaching is an art and a lore (in Chinese). http://www.sjtupmm.com/article-1883.html. Accessed 13 July 2019
5. Guo, W.Z.: Insight on teaching and reforming of mechanism and machine theory course from a point of view of mechanisms. In: Ruiqin, L., Weizhong, G.U.O. (eds.) Progress on Theoretic and Application Research of Modern Mechanisms, pp. 445–459. Higher Education Press, Beijing (2014) (in Chinese)
6. Guo, W.Z.: On project based teaching of mechanical design course for undergraduates in shanghai jiao tong university. In: Proceedings of the 15th IFToMM World Congress on Mechanism and Machine Science, Krakow, Poland (2019)

An Approach to Foster Student Creative Thinking Ability in the Lectures of Mechanical Principles

Yunpeng Gong(✉), Shujun Li, Zhang Yu, Haiyang Li, Cuiling Li, and Yu Guo

College of Mechanical Engineering and Automation, Northeastern University, Shenyang 110004, China
ypgong@mail.neu.edu.cn

Abstract. The module of Mechanical Principles aims to foster students' abilities both of mechanical design and creative thinking, by providing students with a foundation in the knowledge and principles of mechanical mechanisms. Focusing on the teaching and learning of the design and analysis of linkage mechanisms in the module, this paper proposes a new linkage mechanism with controllable linkage length. The new linkage mechanism is synthesized by adding length-control electromechanical components on the traditional linkage mechanisms with constant linkage length. The motion characteristics of the new linkage mechanism is verified by experimental testing and Matlab simulations. The new linkage mechanism is employ as an experimental setup in the lectures of Mechanical Principles, and the application can foster student creative thinking ability to meet the requirements of the New Engineering education.

Keywords: Mechanical Principle · Linkage mechanism · Variable linkage length · Motion pattern

1 Introduction

Developed engineering education is urgently needed to bring up a large number of creative and outstanding engineering science and technology talents with solid engineering practice foundation and comprehensive interdisciplinary design abilities, so as to provide intellectual and talent supports for China's industrial development and international competition.

As a basic compulsory course in mechanical engineering discipline, the module of Mechanical Principles mainly studies the composition principle, working characteristics, static and dynamic motion performance of mechanical mechanisms. The aim of the module of Mechanical Principles is to synthesize mechanical mechanisms under different design requirements, and foster student abilities of designing mechanical equipment and devices to solve engineering problems.

Linkage mechanisms play a very important role in the lectures of Mechanical Principles. In the teaching process of mechanical principles, it is often required to design linkage mechanisms with constant linkage length, according to some technological

D. Wang et al. (Eds.): MeTrApp 2019, MMS 79, pp. 641–646, 2020.
https://doi.org/10.1007/978-981-15-0142-5_63

parameters. However, a linkage mechanism with constant length linkages has a single motion pattern, which limits student thinking to be a fixed pattern, resulting in the lack of divergent and creative thinking. Therefore, it is a hot research topic on the methods to foster student creative thinking ability in the teaching and learning process of mechanical principles.

In order to foster student creative thinking ability, this paper proposes a new linkage mechanism with controllable linkage length. The new linkage mechanism can obtain different motion patterns by varying the linkage length which is controlled by electromechanical control components. The application of the new linkage mechanism in the lectures of Mechanical Principles, as an experimental setup, is convenient to show different motion patterns of the mechanical mechanism, and also helpful for fostering student creative thinking [1].

2 Design of a Linkage Mechanism with Controllable Linkage Length

2.1 Background

In the teaching and learning process of mechanical principles, the crank linkage mechanism is referred to be one of the basic linkage mechanisms. A crank linkage mechanism is often designed with constant linkage length, which offers a single fixed motion pattern under the constant-velocity rotary actuation exerted on the crank. The motion pattern of the crank, including the acceleration, velocity and displacement, cannot be changed once the linkage length and rotary actuation are fixed. Therefore, the design of a crank linkage mechanism with controllable linkage length is of practical significance for the teaching and learning of mechanical principles.

2.2 Design

Figure 1 shows the design of the linkage mechanism with controllable linkage length, and details the main components of the design.

The crank linkage mechanism with controllable linkage length is mainly composed of frame, crank input motors (stepper motor), variable length components, six-bar mechanism (crank rocker and crank slider mechanism), acceleration sensor and control system.

The crank input motor-2 actuates the six-bar mechanism-4. The output motion of the slider can be adjusted by changing the length of linkage-3. Acceleration of the slider is detected by the acceleration sensor-5, and the speed and displacement of the slider are obtained by the integration calculation based on the acceleration. The mechanical structure, speed, displacement and acceleration of the slider can be programmed and displayed on a computer screen.

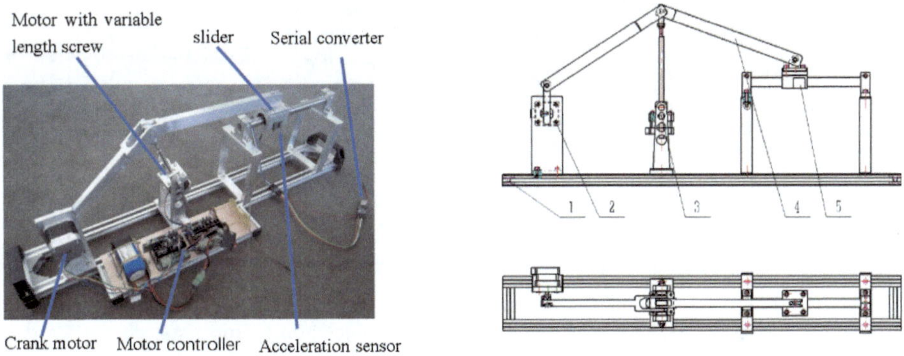

(a) Prototype (b) Schematic diagram

Fig. 1. The crank linkage mechanism with controllable linkage length: 1- frame, 2- crank input motor (stepper motor), 3- variable length component, 4- six-bar linkage mechanism (crank rocker and crank slider mechanism), 5- acceleration sensor

3 Analysis of the Proposed Linkage Mechanism

3.1 Geometric Parameter

The geometric parameters of the linkage mechanism with controllable linkage length can be determined as shown in Fig. 2, under the consideration of geometric dimensions, motion ranges and other design requirements. It can be seen that the linkage length varies from 160 mm to 200 mm.

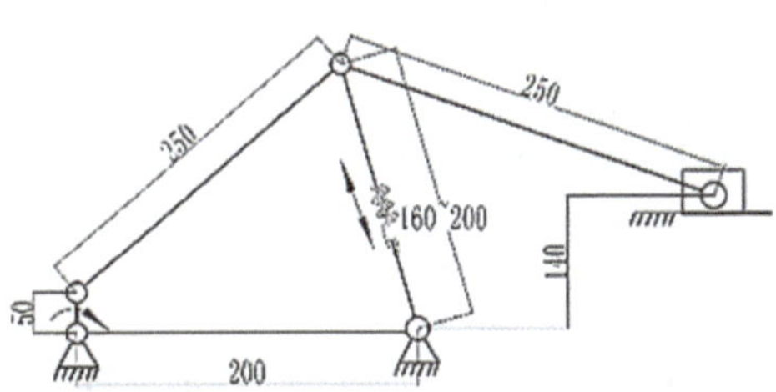

Fig. 2. The model of the linkage mechanism

3.2 Matlab Simulations

The influence of crank speed and rocker length on the output motion can be analyzed employing the developed design and simulation system of linkage mechanism.

- When the crank rotates at a constant speed, the motion of the slider is shown in Fig. 3. Changing the rotating-speed can only change the maximum speed and acceleration.

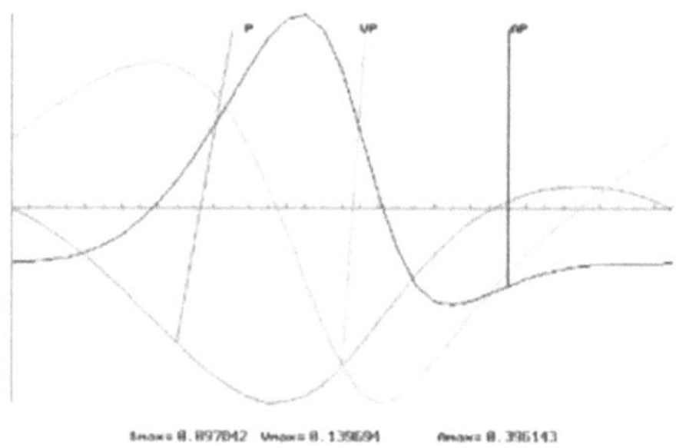

Fig. 3. Slider motion when the crank rotates at a uniform speed

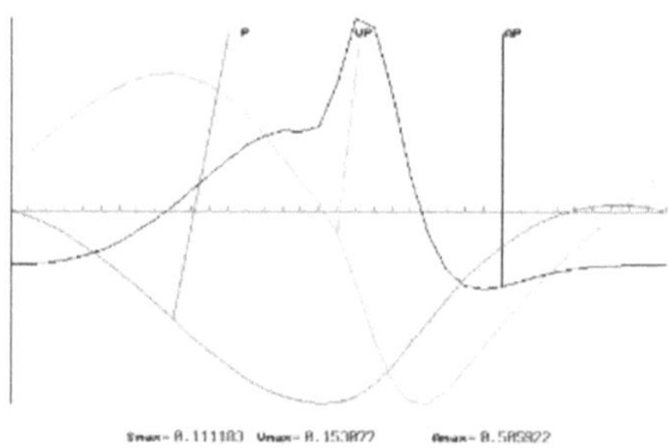

Fig. 4. Slider motion when the crank rotates at a non-uniform speed and the rocking-length varies

- When adjusting the linkage length, the motion of the slider is derived as shown in Fig. 4, both the motion and maximum pressure angle can be changed.
- When the linkage length and the crank rotation speed are controlled, complex motion can be obtained, and the motion performance can be improved as well.

3.3 Parameter Regulation

1. The motion of the linkage mechanism with controllable linkage length is programed using the crank variable speed and the variable linkage length, based on the input requirements of the proposed experiment.
2. Download the program to MCU for offline operation. The rotation speed of the two stepper motors can be controlled by the offline operation system.
3. Calibrate the initial linkage length, and acquire acceleration data of the slider from the acceleration sensor when varying the linkage length and crank rotation speed.
4. Compare the motion parameters. The acceleration parameters of the slider are collected by MPU6050 six-axis accelerometer, and the data collected is processed by Matlab. The experimental testing results of the slider motion are shown by Fig. 5, and the simulation results are detailed in Fig. 6. The results derived from experimental testing and simulations match each other well, which further verifies the design of the linkage mechanism with controllable linkage length.

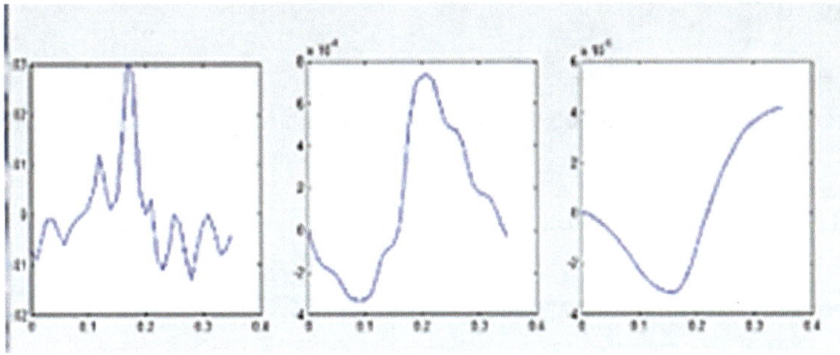

Acceleration Velocity Displacement

Fig. 5. Experimental results of the slider motion

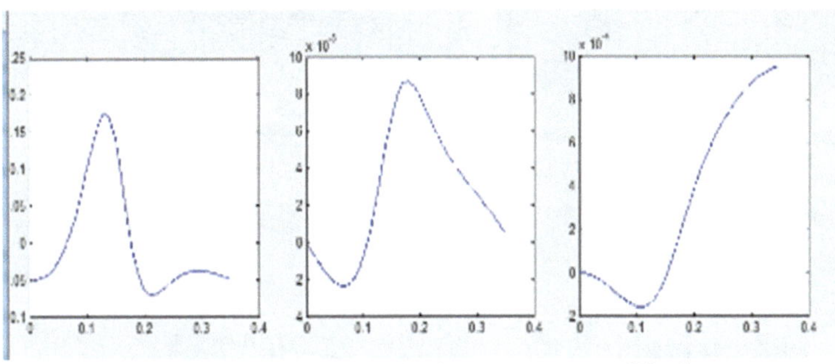

Acceleration Velocity Displacement

Fig. 6. Simulation results of the slider motion

4 Conclusion

1. Due to the limitations of the design of fixed-linkage-length mechanisms in the lecture of the module of Mechanical Principles, a variable-linkage-length mechanism is introduced to design an experimental setup. The new experimental setup can be employed to operate various motion patterns by changing the actuation speed and linkage length. The relationships between the motion patterns and the controlled parameters, the linkage length and the crank rotation speed, can be obtained.
2. The application of the proposed variable-linkage-length mechanism in the lecture of mechanical principles can benefit for figuring out new design schemes, and finally fostering student creative thinking ability to meet the requirements of the New Engineering education.

References

1. Li, S.: Mechanical Principles. Science press, Beijing (2009)
2. Wen, B.: Mechanical Design Handbook. Machinery industry press, Beijing (2015)
3. Xue, D., Chen, Y.: MATLAB Solution of Advanced Applied Mathematics Problems. Tsinghua University Press, Beijing (2009)
4. Wang, S.: Course Design of Mechanical Principle. Science press, Beijing (2006)

Author Index

A
Ao, Hongrui, 376

B
Bai, Le-Le, 123
Bai, Shaoping, 192
Bi, Kedong, 617

C
Cai, Sichen, 533
Cai, Xu, 464
Cai, Yongwu, 70, 99
Can Dede, M. İ., 344
Caro, Stéphane, 332
Chen, Cong, 607
Chen, Jun-Rong, 123
Chen, Li, 419, 440
Chen, Shimin, 62
Chen, Shouqi, 141
Chen, Weixing, 512
Chen, Yuan, 172, 484, 566
Chen, Zihao, 419
Chen, Zilu, 152
Cheng, Yuanxiong, 627
Chu, Jinkui, 12, 113
Chu, Lulu, 524

D
Ding, Jiaxin, 70
Dong, Huimin, 91, 213, 533, 589, 598, 607
Dong, Li, 141

E
Erez, Melih, 42

F
Fang, Hairong, 247

G
Gao, Feng, 464, 512
Gao, Wenlin, 503
Gao, Yuan, 589, 598, 607
Ge, Wenjie, 32, 233, 556
Gong, Yunpeng, 365, 641
Gruescu, Corina-Mihaela, 182
Guang, Chenhan, 152
Guo, Denghui, 484
Guo, Sheng, 290
Guo, Weizhong, 453, 636
Guo, Yu, 641

H
Haiping, Ai, 397
Han, Youcheng, 453
Hang, Lu-Bin, 123
Hao, Guangbo, 365
Hao, Zhixiu, 3, 161
He, Gaosheng, 576
He, Juewei, 3
He, Jun, 464
Hou, Hongjuan, 322
Huang, Liyong, 322
Huang, Xiao-Bo, 123

J
Jiang, Hongyuan, 376
Jiang, Tao, 542
Jing, Xianling, 259

D. Wang et al. (Eds.): MeTrApp 2019, MMS 79, pp. 647–649, 2020.
https://doi.org/10.1007/978-981-15-0142-5

K
Kan, Yajing, 617
Kaneko, Tatsuya, 81
Kiper, Gökhan, 42, 344

L
Lan, Zhi, 3
Li, Chen, 397
Li, Cuiling, 641
Li, Haiyang, 365, 641
Li, Hecheng, 357
Li, Ji, 62
Li, Jiaxun, 322
Li, Lu, 141
Li, Ruiqin, 192
Li, Shujun, 365, 641
Li, Su-Long, 123
Li, Wendong, 223
Li, Xiangyun, 21
Li, Yan, 576
Li, Yang, 113
Liang, Guangshun, 3
Lin, Shi-Sen, 123
Lin, Zirong, 280, 409
Ling, Mingxiang, 385
Liu, Jian, 322
Liu, Ming, 51
Liu, R. C., 474
Liu, Sizhe, 556
Liu, Wenrui, 12
Liu, Xiangyang, 51
Liu, Zhaohui, 430
Liu, Ziwei, 192
Liu, Zi-Yu, 123
Lovasz, Erwin-Christian, 182
Luo, Xueling, 247
Luo, Yi, 524
Luo, Zirong, 542
Lv, Hao, 21

M
Ma, G. Y., 474
Ma, Yong, 376
Makino, Koji, 81
Maniu, Inocenţiu, 182
Mao, Kuanmin, 627
Matsumura, Kenji, 81
Mei, Youen, 301
Meng, Fanyu, 161
Meng, Qiaoling, 365
Mi, Wenbo, 192
Mo, Xiaojuan, 32, 233, 556
Moldovan, Cristian Emil, 182

N
Ni, Zhonghua, 617
Niu, Xiaoru, 542

O
Ouyang, Simin, 99
Özkahya, Merve, 344

P
Paksoy, Erkan, 344
Petuya, Victor, 133

Q
Qian, Feng, 533, 589, 598
Qian, Ruiming, 617

R
Riabtsev, Mykhailo, 133
Riera, Aitor, 133

S
Shan, Xiaqi, 70
Shang, Jianzhong, 542
Shen, Huiping, 280
Shi, Ke, 21
Song, Majun, 290
Sticlaru, Carmen, 182
Su, Xulei, 203
Sun, Jianwei, 12
Sun, Jing, 357
Sun, Zhili, 203

T
Tan, Rulong, 270
Taner, Barış, 344
Terada, Hidetsugu, 81
Tuca, Petrisor Laurentiu, 182

U
Urízar, Mónica, 133
Uzunoğlu, Emre, 344

W
Wang, Delun, 91, 589, 598, 607
Wang, G. M., 474
Wang, Jing, 91
Wang, Kai-long, 494
Wang, Keming, 259
Wang, Ming-Yuan, 123
Wang, Rui, 161
Wang, Wei, 503
Wang, Xianbiao, 376
Wang, Xiangyang, 290
Wang, Xiaodong, 524

Wu, Guanglei, 62, 280, 332, 409, 589, 598
Wu, Kaiyu, 357
Wu, Xiaoyong, 270
Wu, Xin, 223

X
Xiang, Zhaowei, 270
Xiaoyan, Yu, 397
Xie, Jin, 430
Xie, Zhiyu, 99
Xu, Jianfeng, 627
Xu, Yundou, 301
Xu, Zhenghe, 301

Y
Yan, Ran, 270
Yan, Yinpo, 566
Yan, Yutao, 203, 223
Yang, Fan, 301
Yang, Jiajun, 627
Yang, Yang, 152
Yao, Jiantao, 301
Yao, Ligang, 70, 99
Yao, Y. L., 474
Ye, Bin, 512
Yi, Zhongqing, 430
Yin, Guodong, 617
You, Chang-tang, 494
You, Hui-hui, 494
Yu, Fujie, 172
Yu, Hongmin, 503
Yu, Jie, 113
Yu, Jinghu, 312
Yu, Xiaoyan, 419, 440
Yu, Xiao-yan, 494
Yu, Yao, 203
Yu, Zhang, 641
Yue, Yi, 301

Z
Zhang, Chang-ling, 494
Zhang, Chu, 213
Zhang, Dan, 247
Zhang, Fan, 357
Zhang, Haiqiang, 247
Zhang, Hong, 607
Zhang, Jianwei, 192
Zhang, Jianyu, 440
Zhang, Jun, 70, 99
Zhang, Kui, 213
Zhang, Peng-Cheng, 123
Zhang, Sida, 409
Zhang, Yanbin, 259
Zhang, Yaqing, 32, 233
Zhang, Yibo, 223
Zhang, Yonghong, 233, 556
Zhang, Ziang, 233
Zhao, Donglai, 32
Zhao, Fuqun, 247, 290
Zhao, Wenkang, 280
Zhao, Wenxiang, 161
Zhao, Yong, 365
Zhao, Yongqiang, 51
Zhao, Yongsheng, 301
Zheng, Fangyan, 270
Zheng, Lei, 556
Zheng, Minhua, 357
Zhou, Xingyu, 312
Zhou, Yulin, 301